Tracing Tectonic Deformation Using the Sedimentary Record

Special Publication reviewing procedures

The Society makes every effort to ensure that the scientific and production quality of its books matches that of its journals. Since 1997, all book proposals have been refereed by specialist reviewers as well as by the Society's Books Editorial Committee. If the referees identify weaknesses in the proposal, these must be addressed before the proposal is accepted.

Once the book is accepted, the Society has a team of Book Editors (listed above) who ensure that the volume editors follow strict guidelines on refereeing and quality control. We insist that individual papers can only be accepted after satisfactory review by two independent referees. The questions on the review forms are similar to those for *Journal of the Geological Society*. The referees' forms and comments must be available to the Society's Book Editors on request.

Although many of the books result from meetings, the editors are expected to commission papers that were not presented at the meeting to ensure that the book provides a balanced coverage of the subject. Being accepted for presentation at the meeting does not guarantee inclusion in the book.

Geological Society Special Publications are included in the ISI Index of Scientific Book Contents, but they do not have an impact factor, the latter being applicable only to journals.

More information about submitting a proposal and producing a Special Publication can be found on the Society's web site: www.geolsoc.org.uk.

GEOLOGICAL SOCIETY SPECIAL PUBLICATION NO. 208

Tracing Tectonic Deformation Using the Sedimentary Record

EDITED BY

T. McCANN

Geologisches Institut, Bonn, Germany

and

A. SAINTOT

Instituut voor Aardwetenschappen, Vrije Universiteit, The Netherlands

2003
Published by
The Geological Society
London

THE GEOLOGICAL SOCIETY

The Geological Society of London (GSL) was founded in 1807. It is the oldest national geological society in the world and the largest in Europe. It was incorporated under Royal Charter in 1825 and is Registered Charity 210161.

The Society is the UK national learned and professional society for geology with a worldwide Fellowship (FGS) of 9000. The Society has the power to confer Chartered status on suitably qualified Fellows, and about 2000 of the Fellowship carry the title (CGeol). Chartered Geologists may also obtain the equivalent European title, European Geologist (EurGeol). One fifth of the Society's fellowship resides outside the UK. To find out more about the Society, log on to www.geolsoc.org.uk.

The Geological Society Publishing House (Bath, UK) produces the Society's international journals and books, and acts as European distributor for selected publications of the American Association of Petroleum Geologists (AAPG), the American Geological Institute (AGI), the Indonesian Petroleum Association (IPA), the Geological Society of America (GSA), the Society for Sedimentary Geology (SEPM) and the Geologists' Association (GA). Joint marketing agreements ensure that GSL Fellows may purchase these societies' publications at a discount. The Society's online bookshop (accessible from www.geolsoc.org.uk) offers secure book purchasing with your credit or debit card.

To find out about joining the Society and benefiting from substantial discounts on publications of GSL and other societies worldwide, consult www.geolsoc.org.uk, or contact the Fellowship Department at: The Geological Society, Burlington House, Piccadilly, London W1J 0BG. Tel: +44 (0)20 7434 9944; Fax: +44 (0)20 7439 8975; E-mail: *enquiries@geolsoc.org.uk*.

For information about the Society's meetings, consult *Events* on *www.geolsoc.org.uk*. To find out more about the Society's Corporate Affiliates Scheme, write to *enquiries@geolsoc.org.uk*.

Published by The Geological Society from:
The Geological Society Publishing House
Unit 7, Brassmill Enterprise Centre
Brassmill Lane
Bath BA1 3JN, UK

(*Orders:* Tel: +144 (0)1225 445046
Fax: +144 (0)1225 442836)
Online bookshop: http//bookshop.geolsoc.org.uk

British Library Cataloguing Publication Data
A catalogue record for this book is available from the British Library.

ISBN 1-86239-129-7

Typeset by Exe Valley Dataset Ltd, Exeter, UK
Printed by Antony Rowe, Chippenham, UK
Project management by Swales & Willis, Exeter, UK

Distributors

USA
AAPG Bookstore
PO Box 979
Tulsa
OK 74101-0979
USA
Orders: Tel: +1 918 584-2555
Fax: +1 918 560-2652
E-mail: bookstore@aapg.org

India
Affiliated East-West Press PVT Ltd
G-1/16 Ansari Road, Daryaganj,
New Delhi 110 002
India
Orders: Tel: +91 11 327-9113
Fax: +91 11 326-0538
E-mail: affiliat@nda.vsnl.net.in

Japan
Kanda Book Trading Company
Cityhouse Tama 204
Tsurumaki 1-3-10
Tama-shi
Tokyo 206-0034
Japan
Orders: Tel: +81 (0)423 57-7650
Fax: +81 (0)423 57-7651
E-mail: geokanda@ma.kcom.ne.jp

Contents

It is recommended that reference to all or part of this book should be made in one of the following ways:

MCCANN, T. & SAINTOT, A. (eds) 2000. *Tracing Tectonic Deformation Using the Sedimentary Record.* Geological Society, London, Special Publications, **208.**

FERNÁNDEZ-FERNÁNDEZ, E., JABALOY, A. & GONZÁLEZ-LODEIRO, F. 2000. Middle Jurassic to Cretaceous extensional tectonics and sedimentation in the Eastern External Zone of the Betic Cordillera *In*: MCCANN, T. & SAINTOT, A. (eds) *Tracing Tectonic Deformation Using the Sedimentary Record.* Geological Society, London, Special Publications, **208,** 30–53.

Preface

The twin themes of tectonics and sedimentation require the application of many different theoretical, experimental and empirical resources provided by structural geology, sedimentology, geochemistry, geophysics, scale modelling, and field geology. Following this philosophy, we have edited this volume with the intention of providing an integrated approach to the study of linked tectonic–sedimentological systems, rather than to concentrate on individual aspects. This volume was the outcome of an European Union of Geosciences Session entitled 'Tectonics and Sedimentation' held in Strasburg in April 2001. The editors wish to acknowledge the helpful and informed reviews by the following colleagues, without whose interest and support, this volume would not have been possible: C. Betzler, C. Breitkreuz, P. Burgess, E. Burov, O. Clausen, A. Crespo-Blanc, R. Gaupp, J. R. Graham, M. Ford, N. Froitzheim, A. J. Hartley, P. Haughton, C. Krawczyk, P. Krzywiec, O. Lacombe, J. Lamarche, A. Läufer, F. Mouthereau, C. Pascal, J. Platt, W. Ricken, S. Sherlock, S. Sliaupa, R. Stephenson, I. Valladares Gonzalez, M. Wagreich, J. Walsh, J. Wijbrans and N. White.

Tom McCann
Aline Saintot

Tracing tectonic deformation using the sedimentary record: an overview

TOM McCANN[1] & ALINE SAINTOT[2]

[1]*Geologisches Institut, Universität Bonn, Nußallee 8, 53115 Bonn, Germany (e-mail: tmccann@uni-bonn.de)*

[2]*Vrije Universiteit, Instituut voor Aardwetenschappen, Tektoniek afdeling, De Boelelaan 1085, 1081 HV Amsterdam, The Netherlands*

Abstract: Tectonic activity, on a range of scales, is a fundamental control on sedimentary activity. The range of structural deformation within a region extends from the plate tectonic scale, governing, for example, rift initiation, to the basin scale, with the formation of basin-bounding faults. Internal basin configuration is also strongly influenced by tectonic activity. However, the relationship between tectonic activity and sedimentation is a complex one, given the many additional factors which can also influence sedimentary activity, including erosion, sediment transport, source area lithology, groundwater chemistry, range of depositional environments, climate, eustasy, and the relative location of an area and its distality to marine influences. In this paper we provide a selective overview of the issues associated with the interlinked themes of tectonics and sedimentation, examining the main basin types forming in both extensional and compressional plate settings. We then review the various models of sedimentation in the selected basins, both on a local and a basinal scale. Finally, we look to the future – providing a series of possible research areas, almost exclusively multidisciplinary, which would help to improve existing models of interlinked sediment–tectonics systems.

Sedimentary basins, and the depositional successions within them, provide the most tangible and accessible records of the lithospheric, geographical, oceanographic and ecological developments which occur in a specific area over a specific period of time. Tectonic activity, on a range of scales, is a major control on sedimentary activity. In recent years there has been an increase in the number of studies aiming to unravel the links between tectonic events and sedimentary response, both on a basin and intrabasinal scale (e.g. Blair & Bilodeau 1988; Heller *et al.* 1988; Cas & Busby-Spera 1991; Fisher & Smith 1991; MacDonald 1991; Williams & Dobb 1993; Schwans 1995; Cloetingh *et al.* 1997; Gupta 1997).

The range of structural deformation within a region extends from a plate-tectonic scale (e.g. rift initiation to oceanic-ridge formation) – affecting the changing pattern of the oceans and continents, and controlling the size and nature of large source areas, sediment transport pathways and the locations of sediment depocentres – down to the basin scale, where tectonics control the formation of major basin-bounding faults which determine basin form and location. Additionally, tectonic activity also controls the internal basin configuration, for example through the development of smaller intra-basinal faults (both synthetic and antithetic as well as transfer faults) that influence the internal structure of the basin, segmenting it into related but separate depocentres (e.g. Jeanne d'Arc Basin, Tankard *et al.* 1989). These intrabasinal structures also influence the development of topography within a basin by controlling the location of both highs and lows which respectively act as potential sediment sources and sinks, and help to determine channel pathways for sediment (e.g. Alexander & Leeder 1987; Leeder & Jackson 1993; Anders & Schlische 1994; Burbank & Pinter 1999). The broad pattern of faulting within a basin is determined by both the overall geodynamic setting (i.e. divergent, convergent or strike-slip), and by pre-existing crustal weaknesses which can strongly influence fault initiation and location.

Sedimentation results from the interaction of the supply of sediment, its reworking and modification by physical, chemical and biological processes and the availability of accommodation space, i.e. the space available for potential sediment accumulation. Many of these factors have a tectonic component. For example, sediment supply may vary in volume, composition and grain size, as well as in the mechanism and rate of delivery. These variations are largely controlled by the processes noted above. Similarly, accommodation space is controlled by sea-level variation, although relative sea-level changes may have a significant tectonic component.

From: McCANN, T. & SAINTOT, A. (eds) *Tracing Tectonic Deformation Using the Sedimentary Record*. Geological Society, London, Special Publications, **208**, 1–28. 0305-8719/03/$15.00

Tectonic activity, therefore, is a very fundamental control on sedimentation and sedimentary activity. Similarly, the origin of the sedimentary sequences which are deposited within a basin can be related back to the tectonic activity which controlled them. The relationship between tectonic activity and sedimentation, however, is a complex one, given the large range of factors which can influence sedimentation within a basin, including: the rate and magnitude of tectonic activity, the number of faults which are active at any specific time within a basin (including their deformation histories), the rate and magnitude of sediment production (including erosion and sediment transport), the lithological composition of the source area(s), the chemistry of basinal waters, the range of depositional environments, climate, eustasy, and the location of the depositional area and its distance from marine influences (i.e. continentality). Given the inherent variability of all of these factors (together with the fact that many of them are interlinked, e.g. climate and erosion), any basin system is, by default, a complex one. Therefore, modelling of the evolution of the basin infill is difficult, since each individual basin will have its own particular tectonosedimentary signature. Additionally, there is the problem of differentiating between the various factors which influence the composition and distribution of the sedimentary succession within a basin.

Changes in our understanding of the interrelationship of tectonic activity and sedimentation have occurred in several disciplines which play a central role in basin analysis. These include plate-tectonic theory (e.g. Cox & Hart 1986), new geodynamic models, as well as a revolution in our understanding of modern depositional systems, and consequent major advances in the sophistication of actualistic depositional models (e.g. Walker & James 1992; Miall 1997; Reading 1998; Leeder 1999). Petrological models relating sediment composition, especially sand and sandstone, to plate tectonic settings have also been developed (e.g. Dickinson & Suczek 1979; Dickinson 1988; Bahlburg & Floyd 1999), and this work has been extended into the fields of sedimentary geochemistry (e.g. Bhatia 1983; Roser & Korsch 1986; Clift *et al.* 2001) and single grain analysis (e.g. M. Smith & Gehrels 1994; Götze & Zimmerle 2000). New exploration techniques, especially seismic and sequence stratigraphy (e.g. Vail *et al.* 1977; Brown & Fisher 1979; Wilgus *et al.* 1988; Van Wagoner *et al.* 1990; Thorne & Swift 1991; Emery & Myers 1996) have led to a greater understanding of the importance of viewing basins as broad units (in a chronostratigraphic sense) rather than isolated regions. Analysis of the sedimentary succession within a basin, therefore, enables us to determine some of the possible controls on the sedimentary record, and at a range of scales ranging from provenance or the examination of sedimentary structures, up to the recognition and classification of architectural elements and sedimentary sequences, and the reconstruction of depositional environments. Thus, the sedimentary record provides us with a unique opportunity to investigate the tectonic controls which are of significant interest in basin analysis.

Our objective in this paper is to provide a selective review of the linkages between tectonics and sedimentation, and more specifically, studies that have used evidence from the sedimentary record to reconstruct the tectonic history of a region. This overview will initially focus on the main types of tectonic settings and the sediments that are found in conjunction with them. Subsequent sections will examine the various models in use, summarizing with an overview of the current gaps in our knowledge and suggestions for future research areas.

Basin classification

Basin classification schemes vary according to the particular needs of the user. For example, schemes which originate in the field of hydrocarbon exploration (e.g. Kingston *et al.* 1983*a*, *b*) are designed to be used in a predictive manner and tend to be limited to the main basin types (particularly those of interest to the hydrocarbon exploration industry). In contrast, academic classification schemes (e.g. Ingersoll & Busby 1995) tend to be more complex, since they tend towards inclusivity and completeness (Table 1). In this latter scheme, basin types are broadly grouped into those which are formed in divergent plate geodynamic settings (including continental rift basins), those which occur in intraplate settings (including intracratonic basins, oceanic islands and dormant ocean basins), those which form in convergent plate geodynamic settings (including arc-related basins, foreland basins, and trenches), those which are found in transform settings (including transtensional and transpressional tectonics) and a final group which includes basins located in hybrid settings (Ingersoll & Busby 1995). In our overview of basin types, we have chosen to follow the scheme proposed by Ingersoll & Busby (1995) but have simplified it by subdividing the basins into broad geodynamic contexts. Using this approach, it is clear that there are a number of different processes occurring within basins and that these

Table 1. *Basin classification (modified after Dickinson, 1974, 1976a; Ingersoll, 1988b; Ingersoll & Busby, 1995).*

Tectonic setting	Basin type
Divergent	Terrestrial rift valleys
Intraplate	Intracratonic basins Continental platforms Active ocean basins Oceanic islands, aseismic ridges and plateau Dormant ocean basins
Convergent	Trenches Trench-slope basins Fore-arc basins Intra-arc basins Back-arc basins Retro-arc foreland basins Remnant ocean basins Peripheral foreland basins Piggyback basins
Transform	Transtensional basins Transpressional basins Transrotational basins
Hybrid	Intracontinental wrench basins Successor basins

processes are mainly determined by the geodynamic context, but are also influenced by the locations of pre-existing weaknesses and intrabasinal processes (e.g. generation of synthetic and antithetic faults). We have divided our basins into two main groups – those which are formed within broadly extensional tectonic settings (and which would include basins found in convergent plate zones but which exhibit an extensional character, i.e. which involve some component of rifting) and those from compressional settings. Such a subdivision greatly simplifies the characterization of the particular tectonic and sedimentary features of each basin type.

Extensional settings

Introduction

Basins that form within an extensional tectonic setting are characterized by the development of depressions, bounded by normal faults, within which there is a direct relationship between fault activity and sedimentation. In their landmark paper, Leeder & Gawthorpe (1987) provided a clear outline of the influence of movement along an individual fault on the resultant sedimentary unit. The surface length of individual historical normal-fault ruptures range from 10–15 km (Leeder & Gawthorpe 1987), although longer basin-bounding faults (up to 50 km) occur in parts of the East African Rift (Ebinger 1989). In active extensional areas, individual fault displacements are of the order of several metres, although displacement varies from a maximum at the centre of the fault surface to zero at an elliptical tip-line (Fig. 1). Fault displacements vary systematically and there is a clear relationship between the amount of displacement and the size of the individual fault (Walsh & Watterson 1988) (Fig. 2). An exception to this rule would be the so-called 'superfaults', which are characterized by very large displacements occurring during a single slip event (Spray 1997).

Fault activity leads to the superimposition of a tectonically-induced gradient, the magnitude of which is determined by fault displacement, on to a pre-existing topographic one. As noted by

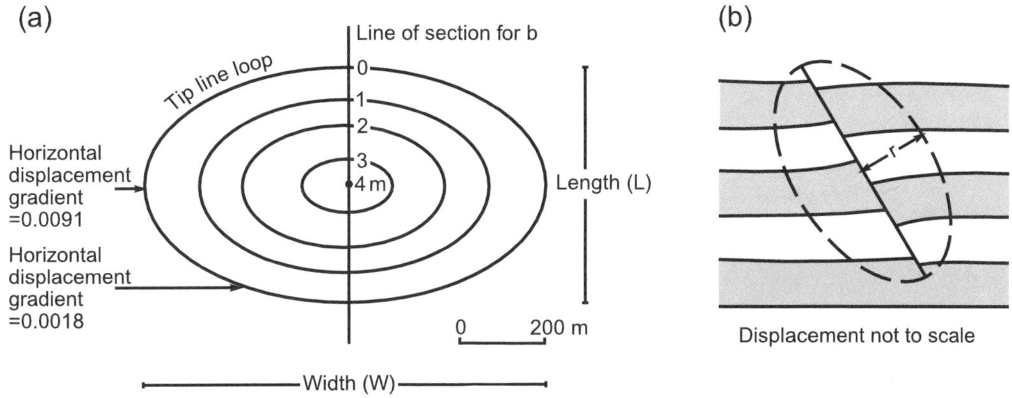

Fig. 1. (**a**) Schematic displacement contour diagram for a simple fault. View is normal to fault surface. (**b**) Cross-section showing perceptible reverse drag associated with simple fault (after Barnett *et al.* 1987).

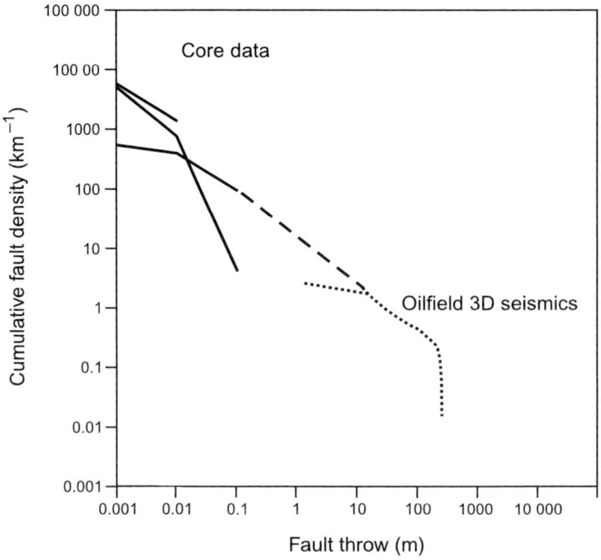

Fig. 2. Comparison of fault displacement measurements on core data and oilfield three-dimensional seismics. Numbers of faults are normalized to cumulative fault density (number of faults per unit length of sample line), and displacements are displayed as fault throw. The core data were corrected to account for the fact that they are from vertical rather than horizontal sample lines. Despite the broad range of fault density, overall the measurements are consistent with a single power-law relationship (dashed line) spanning both core and seismic data, with a slope of *c.*–0.8 (after Walsh *et al.* 1991).

Leeder & Gawthorpe (1987) many geomorphological processes are influenced by gravity, and thus the increase in slope produced as a result of tectonic activity tends to directly influence a variety of sedimentation-related processes (e.g. Alexander & Leeder 1987; Collier *et al.* 1995; Burbank & Pinter 1999) (Fig. 3). The influence of fault deformation on surface processes has recently been confirmed by geodetic measurements which have characterized regional interseismic strain fields in many actively deforming areas (e.g. Norabuena *et al.* 1998). These measurements help to provide a more accurate picture of the tectonic forcing function at regional scales which drive long-term landscape development through the combination of tectonic and topographic gradients.

As noted above, faults increase their length with time, since fault displacement and length are positively related (Walsh & Watterson 1988). Fault segmentation, and the resultant interlinkage between various fault segments, however, complicate this relationship. Recent modelling has shown that fault interaction and linkage can lead to temporal variability within an evolving fault array (Cowie 1998). In addition, the segmented nature of normal fault zones suggests that two structural styles can occur contemporaneously along any one fault segment – the central parts of normal fault segments are characterized by surface fault breaks while growth folding dominates the ends of fault segments where the fault is blind (Gawthorpe *et al.* 1997).

Normal faults control the creation of accommodation space for syntectonic deposition in rift basins (Schlische 1991; Gawthorpe *et al.* 1994). Thus, the displacement history of a series of linked faults would be recorded within the synrift stratigraphy. However, because the spatial extent of the fault interaction is determined by the scale of the fault segments, synrift sequences will vary spatially along fault systems (Dawers & Underhill 2000). Thus, high displacement rates near segment centres may promote rift climax stratal patterns (cf. Prosser 1993) and facies associations, whereas shallow marine conditions may persist at fault tips and in overlap zones between unlinked faults (Dawers & Underhill 2000).

The overall effect of fault displacement on sedimentation and related processes (e.g. erosion, sediment transport) is a cumulative one, and one made more complex by the segmented nature of fault activity within fault zones. Thus, while there will be a close relationship between the history of

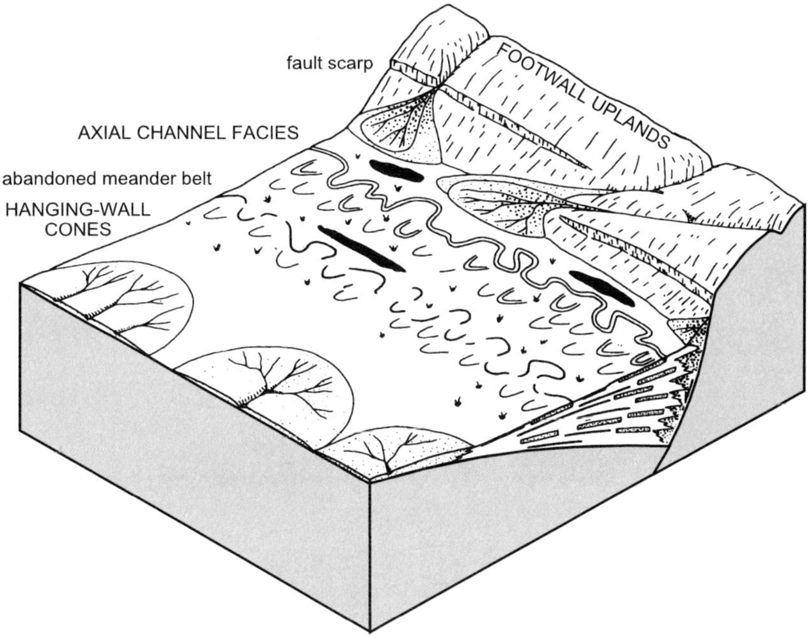

Fig. 3. Block diagram summarizing the major clastic environments present in a continental half graben system with through flowing axial drainage (after Leeder & Gawthorpe, 1987).

fault activity and the lithostratigraphic signal of the basin infill, the precise history is not always easy to determine.

Rift basins

As noted by Ingersoll & Busby (1995), any model of continental rifting must consider the various ways in which the lithosphere behaves, including, for example, rheological differences within the lithosphere, contrasts in the composition and structure between the crust and the mantle, the differences between oceanic and continental crust, pre-existing heterogeneities, and the period of time over which strain operates. Two basic models have been proposed for the development of rift basins – the pure shear model of McKenzie (1978) which involves the development of a symmetrical rift structure flanked by major boundary faults (with associated antithetic and synthetic faults) and that of Wernicke (1981) which results in the development of an asymmetrical basin, associated with a deep (listric) fault along which associated antithetic and synthetic structures develop (Fig. 4). However, these two should only be seen as end members, and the variety of actual rift basin forms is much greater. In addition, modifications of these models exist (e.g. Lister *et al.* 1986) (Fig. 4). Morphologically, rifts may be classified as:

(1) solitary – e.g. Cambrian Tesoffi Rift, Africa;
(2) rift stars – e.g. triple junctions, Nakuru junction, Africa, North Sea area;
(3) rift chains, where several rifts are aligned end to end along linear/arcuate belts of rifting – e.g. East African Rift System, opening of Atlantic Ocean; and
(4) rift clusters, where several subparallel rifts occur in roughly equant areas – e.g. Basin & Range, Aegean (Sengör 1995).

Active extension or stretching of continental lithosphere leads to surface deformation, volcanism and high heat flow due to the effects of normal faulting and the resultant changes in crustal and mantle thickness, structure and state (Ingersoll & Busby 1995). The tectonic environment of stretching is controlled by regional plate motions. Extension may occur in a variety of geodynamic settings, including continental crust adjacent to young oceans, back-arc basins, continental interiors and thickened crustal orogens (Ingersoll & Busby 1995). As defined by Sengör & Burke (1978) rifting may be passive (i.e.

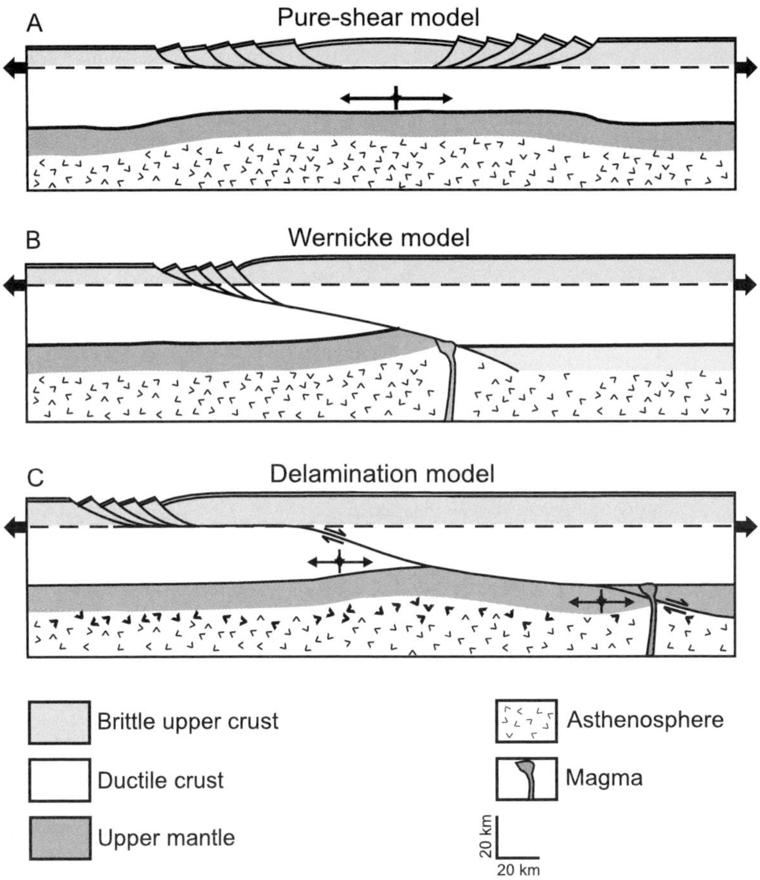

Fig. 4. Three end member models for continental extension (after Lister *et al.* 1986).

closed system, where the input of asthenospheric mass from outside the stretched lithosphere occurs passively as a response to lithospheric thinning) or active (i.e. open system, where rifting is accompanied by the eruption of voluminous volcanics, and the initial rising of the asthenosphere is independent of the magnitude of lithospheric extension) (Fig. 5). However, it is more probable that many rifts evolve under a combination or succession of these two processes (see discussion in Leeder 1995).

The basic structural element of a continental rift is now thought to be a half graben, comprising a single basin-bounding fault. Surface observations in the Basin and Range area have revealed that upper crustal extension is spatially very variable, resulting in local tectonic domains where the upper one-third to one-half of the crust has been removed (Wernicke, 1992). Several structural models have been proposed for half-graben development. These include:

(1) domino faulting, where high-angle normal faults extend deep into the upper crust with nearly constant dip;
(2) listric normal faults, which terminate downwards into subhorizontal detachment faults of regional extent and fault blocks are highly rotated; and
(3) the flexural-rotation (rolling hinge) model, where an initially high-angle normal fault is progressively rotated to lower dips by isostatic uplifting resulting from tectonic denudation (Lucchitta & Suneson 1993) (Fig. 6).

Beneath these areas of extension, however, there is no upwarping of the Moho as would be anticipated if isostatic compensation of the extension occurred within the mantle. Thus, it is possible to find both heterogeneous upper crustal strain and uniform deep crustal structure across extensional domain boundaries resulting from

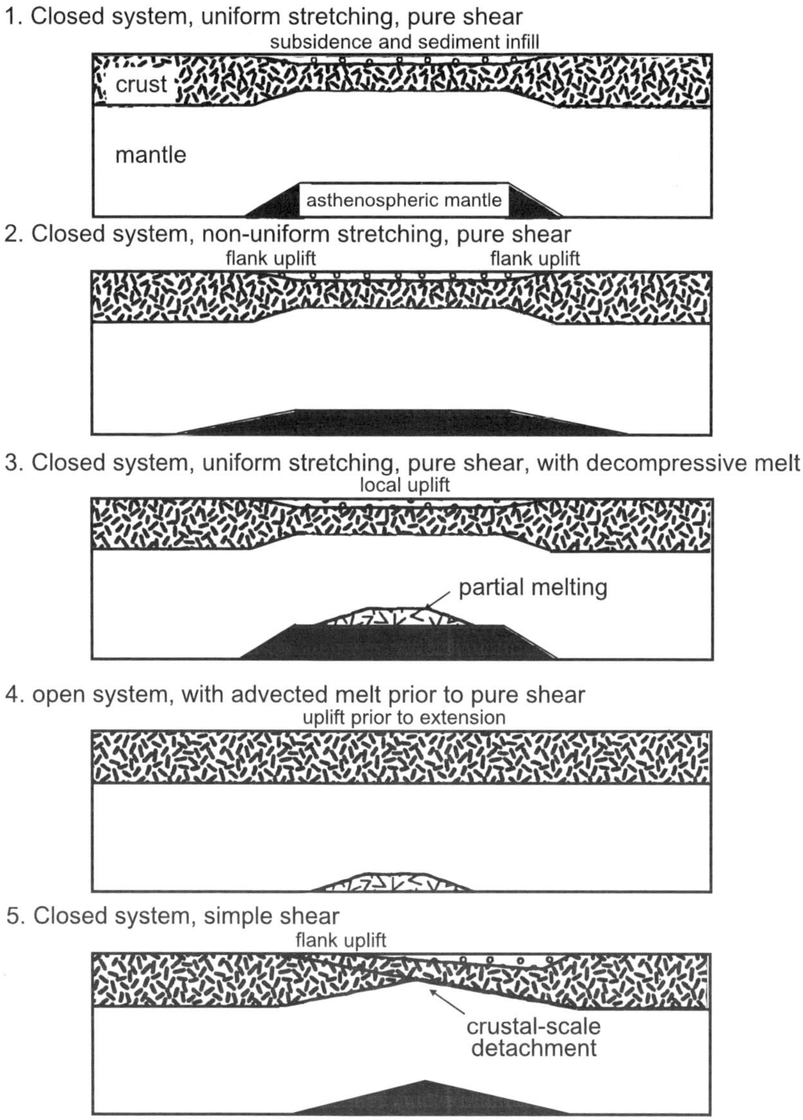

Fig. 5. Schematic diagrams to illustrate possible combinations of pure and simple shear, uniform or non-uniform stretching and magma generation. Local (Airy) isostatic compensation assumed throughout (i.e. lithosphere has small elastic thickness). Surface and upper crustal deformation by faulting not shown (after Leeder, 1995).

the effects of intracrustal isostasy (Ingersoll & Busby 1995). Models of rift basin evolution, incorporating a component of lower crustal flow, proceed from a core-complex mode (involving thick crust – *c*. 50 km, and high heat flow) to a wide-rift mode (weaker crust – *c*. 40 km, and high heat flow) to narrow-rift mode (thin crust – *c*. 30 km, and low heat flow) (Fig. 7).

Arc-related basins

Volcanic arcs are generally arcuate or linear bodies, typically exceeding 1000 km in length and ranging from 50–250 km in width, which parallel subduction-zone trenches (see G. A. Smith & Landis 1995, and references therein for precise terminology of arc complexes). Arc-related basins are found in a convergent geodynamic

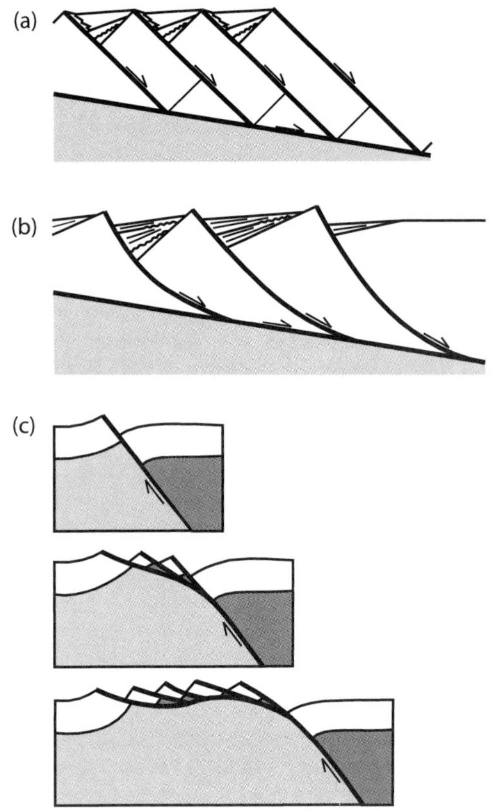

Fig. 6. Styles of upper-plate faulting. (**a**) Domino faulting where initial movement occurs on planar, high-angle, normal faults that subsequently rotate to lower angles with continued extension. The faults do not merge within the detachment zone, and the zone of intersection is brecciated and sheared. (**b**) Listric faulting where movement occurs on curviplanar faults which flatten with depth and merge with the detachment fault. (**c**) Rolling-hinge model where an initially high-angle normal fault is progressively rotated to lower dips by isostatic uplift resulting from tectonic and erosional denudation. New high-angle faults are produced when the original faults are too rotated to accommodate extension (after Lucchitta & Suneson 1993).

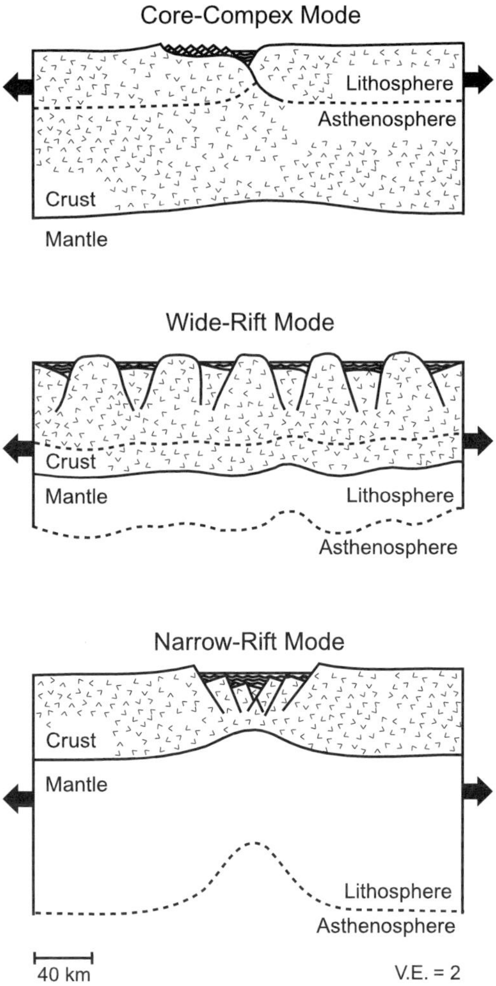

Fig. 7. Cartoon of the lithosphere in three modes of continental extension, emphasizing regions undergoing the greatest amount of continental strain. Lithosphere represents areas with effective viscosities greater than 1021 Pa s. Crustal thicknesses vary from top to bottom, 50 km, 40 km and 30 km respectively. Modified after Buck (1991) and Ingersoll & Busby (1995).

context but are all broadly extensional in terms of their tectonic activity. The three main basin types are related to the location of the volcanic arc, being located on the trench side of the arc (fore-arc), behind the arc (back-arc) or within this structure (intra-arc) (Fig. 8). Fore-arc basins are located between the trench axes, which mark the subduction zone, and the parallel magmatic arc where igneous activity is induced by the inclined descent of oceanic lithosphere (Dickinson 1995). Intra-arc basins are defined as basins located within or including the arc platform, which is the typically positive feature formed by the volcanogenic edifices which cap part or all of the arc massif. This latter feature is the region overlain by crust which has been generated by arc magmatic processes (G. A. Smith & Landis 1995). At the time of their formation intra-arc basins are spatially distinct from both fore-arc and back-arc basins (Fig. 8), but they may be just

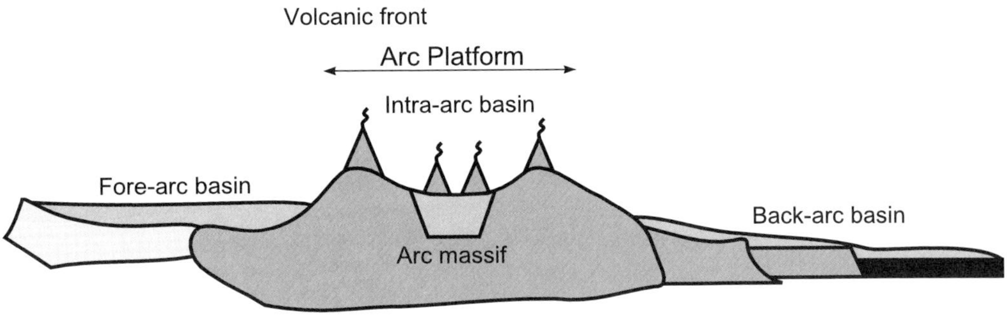

Fig. 8. Diagrammatic cross-section through a convergent plate margin, showing location of arc platform relative to fore-arc and back-arc basins. Areas underlain by arc crust include basement to parts of forearc and backarc basins. Arc volcanoes are typically dispersed over a wide zone perpendicular to plate boundary, but most active volcanoes are aligned along a distinct volcanic front (after G. A. Smith & Landis 1995).

an evolutionary stage for the development of other basin types (e.g. back-arc basins). Back-arc basins occur behind volcanic island arcs and are common along continental margins as well as along convergent plate margins (Marsaglia 1995).

Because of arc migration, however, a single site may change between fore-arc, intra-arc, and back-arc settings, for example, the Luzon Central Valley, which, as a result of changes in subduction polarity and other processes, has successively occupied all three positions over the last 40 Ma (Bachman *et al.* 1983). It may, therefore, be difficult to recognize the precise basin type, i.e. fore-arc, intra-arc or back-arc, based only on their sedimentary and volcanogenic fill (G. A. Smith & Landis 1995). An additional complicating factor is the possibility of intrusion- or collision–accretion-related deformation. Such deformation, and any subsequent uplift and erosion, means that the spatial relationships of volcanogenic materials relative to the arc axis and the distinction – from a geodynamic point of view – of fore-arc, intra-arc and back-arc positions is not always clear (e.g. Lower Palaeozoic Welsh Basin and Lake District). Thus, basins containing volcanogenic material may, more generally, be referred to as arc-related basins.

Volcanic arcs produce large volumes of clastic material that may form much of the arc edifices, in addition to providing a variety of intrusive and extrusive igneous rocks. Intrusive igneous rocks are normally in the form of elongate composite granitoid batholiths. The extrusives include andesitic and dacitic rocks from stratovolcanoes, basalts from intraoceanic arcs and silicic ignimbrites from collapse calderas in continental-margin arcs. Arc settings, therefore, have a significant component of volcanic debris in the resultant sediments.

Fore-arc basins. Forearc basins are extremely variable features, ranging in size from 25 to 125 km wide and between 50 and 500 km long. This variability is a result of the diversity of the controlling factors which govern their genesis (Dickinson 1995). Basins may be simple or compound (i.e. multiple fore-arc basins which lie parallel to one another). These latter features are comprised of strings of interlinked fore-arc depocentres which can extend for 2000–4000 km along modern arc–trench systems. The maximum thickness of sediment fill within a fore-arc basin ranges from 1 to 10 km.

Dickinson & Seely (1979) provided a classification of arc–trench systems, similar to that of Dewey (1980), and outlined plate-tectonic controls governing subduction initiation and forearc development. The factors which control forearc basin geometry include:

(1) initial setting;
(2) sediment thickness on the subducting plate;
(3) the rate of sediment supply to trench;
(4) the rate of sediment supply to the fore-arc area;
(5) the rate and orientation of subduction; and
(6) the time since initiation of subduction.

Subduction environments are extremely variable, although Jarrard (1986*a*) has recognized a number of distinct zones based on a series of factors, including arc curvature, the geometry of the Wadati–Benioff zone, the strain regime of the overriding plate, the convergence rate, 'absolute' motion (relative to hot spots), slab age, arc age and trench depth. His work demonstrated that the strain regime within a subduction environment is probably determined by a combination of convergence rate, slab age and slab dip (Ingersoll & Busby 1995).

Fore-arc basins are bounded by volcano-plutonic assemblages with associated metamorphic rocks on the arc margin, and on the trench margin by uplifted subduction complexes composed of varying proportions of deformed and partly metamorphosed oceanic crust, sea-floor sediments, trench fill, and trench slope deposits (Fig. 4). Within the interior of fore-arc basins, either compressional or extensional deformation may occur during forearc sedimentation leading to the development of syndepositional folding, half-graben sub-basins, etc. Extension can be both normal to the subduction direction, or parallel to it (related to variable rates of lateral slippage along the arc–trench gap; McCaffrey 1992), although it is only likely within the fore-arc region within the first 10–20 Ma of initiation of an intraoceanic subduction zone (Stern & Bloomer 1992). Subduction obliquity can lead to strike slip movement (Jarrard 1986*b*). Deformational contrasts lead to corresponding contrasts in the subsidence history of the basin axis and in the uplift history of the trench–slope break, resulting in complex patterns of sediment distribution in both time and space. This complexity means that no single evolutionary model is applicable to all fore-arc basins (Beaudry & Moore 1985).

Intra-arc basins. Intra-arc basins are thick volcanic–volcaniclastic–sedimentary accumulations that are found along the arc platform, a region formed of overlapping or superposed volcanoes (Fig. 4). G. A. Smith & Landis (1995) suggest that there are two end member types for intra-arc basins, namely:

(1) volcano bounded, which have poorly defined margins, thin sediment infill, and are not associated with arc rifting or the formation of oceanic crust (e.g. Larue *et al.* 1991); and
(2) fault bounded, which are rapidly subsiding, arc parallel or arc-transverse basins caused by tectonically-induced subsidence of segments of the arc platform (e.g. Busby-Spera 1988).

Hybrid basins with characteristics of both types can also be found.

The structural histories of intra-arc basins can vary over time as the arc platforms undergo their complex histories of alternating uplift and subsidence related to angle, obliquity and rate of subduction, which in turn is partly related to the age, thickness and crustal type of the subducting lithosphere (G. A. Smith & Landis 1995). Uplift in the magmatic arc may be associated with crustal thickening and the thermal and physical effects of rising magma. Mechanisms for subsidence, however, are poorly understood, largely hypothetical, and more complex than can be explained by thermal-contraction and flexural-loading models typically applied to other basin types. Six possible mechanisms, acting singly or in combination, may be responsible, including plate boundary forces at the subduction zones, relative plate motions, variations in asthenospheric flow, regional isostasy, magmatic withdrawal and gravitational collapse (G. A. Smith & Landis 1995).

Back-arc basins. A back-arc basin is defined by Ingersoll & Busby (1995) as being either an oceanic basin located behind an intraoceanic magmatic arc, or a continental basin situated behind a continental-margin arc that lacks foreland fold–thrust belts. Back-arc basins initiate by crustal extension, firstly producing rifts and then new ocean crust by sea-floor spreading (Karig 1971; Packham & Falvey 1971). Various active and passive methods have been proposed, but no one theory adequately explains the formation of all back-arc basins, with different interpretations being proposed for different geographic regions (e.g. Carey & Sigurdsson 1984). A number of models for back-arc spreading have been proposed. These include extensive magma intrusion, mantle convection or mantle-wedge flow induced by the subducting slab, and thermal upwelling of a mantle diapir (see Marsaglia 1995 and therein for references). Three other types of back-arc basin are also recognized, including:

(1) non-extensional, which include old ocean basins trapped during plate reorganization which causes a shift of the subduction zone;
(2) back-arc basins which develop on continental crust and are transitional with retro-arc foreland basins; and
(3) so-called 'boundary' basins, which can be produced by extension along plate boundaries with strike-slip components (Marsaglia, 1995).

Intracratonic rift basins

Intracratonic basins are saucer-shaped features which are found within continental interiors away from plate margins, and are floored with continental crust and often underlain by failed or fossil rifts (Klein 1995) (Fig. 9). The development

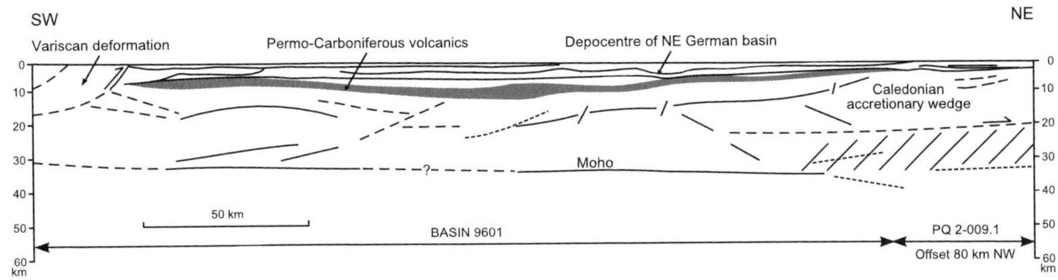

Fig. 9. Interpreted line drawing of part of the BASIN 9601 profile and its offshore extension PQ2.009.1, showing the main tectonic and stratigraphic features. Of particular interest is the saucer-shaped profile of the NE German Basin (after DEKORP-BASIN Research Group, 1999).

of an intracratonic basin involves a combination of basin-forming processes, including continental extension, thermal subsidence over a wide area, and later isostatic readjustments. From studies carried out on a number of intracratonic basins it is clear that their formation followed similar patterns. The processes, in order of occurrence, are:

(1) lithospheric stretching;
(2) mechanical, fault controlled subsidence;
(3) thermal subsidence and contraction; and
(4) merging of slower thermal subsidence with reactivated subsidence due to the isostatically uncompensated excess mass (see Klein 1995 for details).

The precise origin of intracratonic basins, however, is controversial and a variety of different hypotheses have been proposed, including factors which involve an increase in crustal density (due to eclogite phase transformation or thermal modification to the greenschist and amphibolite facies), or magmatic activity (related to igneous intrusions or partial melting and drainage of melt to mid-ocean ridge volcanism) (Klein 1991). Other factors, for example rifting-related hot-spot activity (e.g. Wilson & Lyashkevich 1996), the reactivation of pre-existing structures, far field effects, or changes in intraplate stress, may also occur.

Subsidence analysis studies from North America have shown that the initiation of subsidence of the Illinois, Michigan and Williston basins, and the initiation of subsidence of latest Precambrian and earliest Palaeozoic passive margins were coeval with late Precambrian-age supercontinent break-up (Bond & Kominz 1991). A similar relationship between supercontinental break-up and intracontinental basin formation is also noted from the late Proterozoic of Australia (Lindsay & Korsch 1989) and the Mesozoic and Cenozoic of Europe and India (Klein & Hsui 1987). Additionally, the sedimentary sequences within intracratonic basins have coeval interregional unconformities and similar trends in thickness and volume (Sloss, 1963; Zalan *et al.* 1990; Klein 1995). The supercontinent break-up model, however, is not accepted by everyone. Some authors suggest that the subsidence histories of the basins are independent, and question the existence of anorogenic granites (which would result in crustal discontinuities) beneath the basins (e.g. Bally 1989). However, supercontinent break-up is not an instantaneous process. Instead, it occurs over a long period of geological time, and this can lead to variations within both basin formation times, and subsidence rates and magnitude across the cratonic area (Klein 1995).

Strike-slip basins

The variability and complexity of sedimentary basins associated with strike-slip faults are almost as great as for all other types of basins (Nilsen & Sylvester 1995, 1999*a*, *b*). Christie-Blick & Biddle (1985) provided a comprehensive summary of the structural and stratigraphic development of strike-slip basins, based largely on the work of Crowell (1974*a*, *b*) (Fig. 10). The primary controls on structural patterns within strike-slip basins include:

(1) the degree of convergence and divergence of adjacent blocks;
(2) the magnitude of displacement;
(3) the material properties of deformed rocks; and
(4) the existence of pre-existing structures (Nilsen & Sylvester 1995, 1999*a*, *b*).

The formation of strike-slip basins depends largely on the orientation of the principal direc-

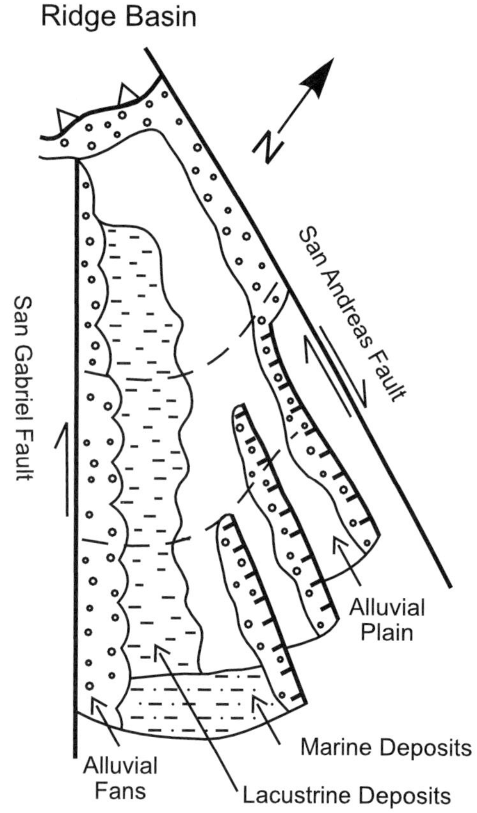

Fig. 10. Map of the Ridge Basin, California, a strike-slip controlled fault-bend basin, showing the asymmetrical basin morphology, the variable depositional facies, the combination of both axial (predominant) and longitudinal fill patterns, and distribution of sedimentary facies relative to the main basin-bounding faults (after Link 1982; Nilsen & McLaughlin 1985).

tion of extension relative to the direction of bulk shear strain, the overstepping arrangement of discontinuous and discrete fault segments, and on the bending geometry of the fault (Nilsen & Sylvester 1995). Transtensional (including pull-apart) basins form near releasing bends (Crowell 1974*b*), while basins associated with crustal rotations about vertical axes, within the rotating blocks (transrotational basins, after Ingersoll, 1988) may experience any combination of extension, compression and strike-slip.

Compressional settings

In the Introduction to extensional settings the clear relationship between extensional fault activity and sedimentation/geomorphic processes was noted. While there are many studies examining the nature of this relationship, there are few which do the same for reverse or thrust faults (see 'Understanding fault activity' below). Instead, displacement on compressional faults tends to be viewed more in terms of the overall geodynamic setting than in terms of its effect on a single fault. However, basins formed in compressional settings will have an abundance of folding and reverse fault activity.

Foreland basin

Foreland basin systems are complex, large-scale features that develop in response to tectonic loading of a foreland plate by the emplacement of large fold–thrust sheets on their margins (Jordan 1981; Allen *et al.* 1986). The increase in thickness as a result of crustal loading leads to a corresponding isostatic adjustment in the crust, resulting in the formation of a down-flexed moat, which is the foreland basin *sensu stricto*. Subsequent erosion transfers mass from the thrust belt to the basin, resulting in uplift of the orogenic belt and increased subsidence in the basin area. Thus sediment-driven load subsidence amplifies and modifies the tectonic-driven subsidence (Jordan 1995). The stratigraphic record of a foreland basin, therefore, reflects the controlling mechanisms on basin formation, namely, regional subsidence related to flexure of the lithospheric plate on which the basin is located, and secondary controls such as local lithology, climate and eustatic sea level (Jordan *et al.* 1988).

Foreland basins may be broadly subdivided into two types:

(1) peripheral or collisional foreland basins, which result from arc–arc, arc–continent, or continent–continent collision; and
(2) retro-arc foreland basins, which form on the continental side of the magmatic arc formed during the subduction of oceanic plates (Dickinson 1976).

Distinguishing between the two types of foreland basin in the ancient record, however, may be difficult, since most orogens undergo several phases of accretion, changes in subduction polarity and changes in the angle of convergence, all of which lead to complications such as strike-slip displacement of the basin and source areas, or even the superposition of basins controlled by different tectonic mechanisms (Miall 1995). Changes in the tectonic style over the course of basin evolution may result in the formation of a hybrid basin that is difficult to classify in terms of its original plate tectonic origin.

Basin-related magmatism

Magmatic activity within basins and the role and extent of mantle involvement in basin formation has been mentioned in the section on rifting (see above). Magmatic activity, both in terms of intrusion (dykes, sills and plutons), extrusion and withdrawal plays an important role in terms of broad basin dynamics and also in terms of the evolution of the basin infill. Evidence of magmatic activity provides important information on the relationship between heat, magma, pressure and the development of stresses (e.g. using volcanic alignments dyke/sill orientations as kinematics/palaeostress indicators) within basins (Sundvoll *et al.* 1992). Periods of active magmatism during basin formation are probably due to the combined effect of tectonic stress and heat flux. Subsequent magmatism can modify the stress distribution in a basin and lead to non-linear transient rheological heterogeneity in the lithosphere, affecting the lithosphere stress transmission on a regional scale (Ingersoll & Busby 1995).

As previously noted, rifting may be 'active' i.e. where the rifting process necessitates the presence of an upwelling convective plume at the base of the lithosphere prior to crustal extension, or 'passive' as a result of lithospheric extension, and without the need for any magmatic upwelling (cf. Sengör & Burke, 1978). Frostick & Steel (1993) have noted that 'active' and 'passive' rifts should be distinguishable on the basis of their sedimentary history. However, many rifts have features diagnostic of both types (Ingersoll & Busby 1995) since volcanism is present in many rifts. Thus, in order to fully understand the evolutionary history of a region it is necessary to understand the precise chronology of the magmatic, topographical, depositional and structural events.

Models of sedimentation

The complexity and variability of tectonic settings gives rise to a corresponding complexity and variability within the basin infill of any given tectonic setting. Prediction of the types of sedimentary sequences which might be produced in each of the various basin types, therefore, is difficult. This predictive problem is further complicated by firstly the similarity of some of the basin types (e.g. intra-arc basin, fore-arc basin), and, by extension, the types of sedimentary sequences which will be produced within them; and, secondly the particular post-depositional history of an individual basin (including deformation, diagenesis, strike-slip movement altering original geographical relationships, etc.).

Models of sedimentation in an extensional setting (basin scale)

Extensional basins are formed under tensional stress regimes and their tectonic evolution can be subdivided into the various extensional phases, namely, pre-rift, synrift, and post-rift. The sequential nature of the tectonic activity leads to the production of a correspondingly characteristic sediment sequence which can be related to the different phases of basin formation. The characteristic structural asymmetry of many rift basins exerts a fundamental control on the distribution of sedimentary environments and lithofacies (e.g. Gibbs 1984; Frostick & Reid 1987; Leeder & Alexander 1987; Leeder & Gawthorpe 1987; Alexander & Leeder 1990; Schlische & Olsen 1990; Lambiase 1991). This is particularly true along the basin margins, where transverse drainage systems evolve on the footwall and hanging-wall uplands, transferring clastic sediments toward the basin centre.

Along the basin-bounding fault, the area of the newly created tectonic uplands is controlled by the length of the tectonic slope produced during extension (Leeder *et al.* 1991). Coarse-grained cones, or aprons, of sediment are located along the length of these boundary faults. Within the resultant sediment sequences, however, there may be evidence of progradational–retrogradational cycles, the nature of which remains controversial. Some workers believe that clastic wedge progradation occurs during times of minimum tectonic activity along the basin margin, and that fine-grained intervals (lacustrine/shallow marine) correspond with times of high rates of basin subsidence (e.g. Leeder & Gawthorpe 1987; Blair & Bilodeau 1988; Heller & Paola 1992). These models assume constant sediment supply, where progradation results from reduction in accommodation during times of decreased subsidence. In contrast, Surlyk (1990) suggested that sedimentary architecture is controlled by episodicity in footwall-generated sediment discharge into depocentres subjected to continuous deepening.

Models of sedimentation in an extensional setting (local scale)

The sediments which occur in fault-bounded half-graben basins have been widely studied in recent years (e.g. Coward *et al.* 1987). These basins develop progressively during extension, significantly controlling the local geomorphology and sediment transfer mechanisms (Leeder & Gawthorpe 1987; Alexander & Leeder 1990). During the development of an extensional basin, distinct evolutionary sequences of basin fills may develop. The predominant symmetry of a half-graben basin is asymmetrical, with a steep footwall slope and a shallower hanging-wall slope (Fig. 3). The pattern of sediment distribution within the basin reflects this basic asymmetry, with the thickest sediment sequence being deposited adjacent to the region of maximum fault throw. Coarser sediments tend to be concentrated along the basin margin (e.g. McCann & Shannon 1993) where the decrease of gradient into the basin from the bounding fault causes rapid deposition and the construction of talus cones, alluvial fans, fan deltas and submarine fans (dependent on the prevailing water depths). In contrast, the hanging-wall source area has a broader, gentler slope and the sediments deposited in this region show a wider distribution. Basin centre environments are strongly controlled by climatic influences, with lake or playa systems forming according to the level and availability of local fresh water relative to evaporation. In arid closed basins, aeolian sand complexes may form. Where extension occurs at, or close to, sea-level then basin flooding may occur, leading to the formation of a marine gulf setting (Leeder & Gawthorpe 1987).

In arc-related environments the sedimentary input is characterized by the presence of volcaniclastic detritus, which in some cases (particularly that of the intra-arc setting) may be dominant. Furthermore, the complexity of arc-related settings makes it difficult to provide a single sequence which can be produced as a response to tectonic variables. In forearc areas the sediment infill comprises mainly interbedded sandstone and shale, with rarer conglomeratic intervals being restricted to proximal sites near the basin margins and along the sediment transport paths (e.g. submarine or fluviodeltaic channels). While clastic sediments usually predominate, carbonate sedimentation (related to water depth and geographical location) may also occur. Within intra-arc basins, the majority of the sediment is volcaniclastic in origin. These sediments are produced independent of weathering processes, and thus the sediment volumes and dispersal distances are larger than those found in other clastic depositional systems. Non-volcaniclastic sediments may be locally significant. Facies associations within the intra-arc setting, however, are not unique to these basins, and thus, the presence of vent-proximal volcanic rocks and related intrusions within the central facies association is critical to the correct identification of intra-arc basin settings (G. A. Smith & Landis 1995). The main sediment types recognized from back-arc basins are those derived from pelagic fallout, airborne ash and submarine gravity flows (Klein 1985). The characteristic lithofacies are variable, reflecting the controls on their distribution. Volcanic components, however, are also present and include lava flows, breccias, pyroclastic rocks and reworked volcaniclastic materials.

The sedimentary infill of a strike-slip basin may be very complex and variable, depending on whether the basins are submarine, lacustrine, subaerial or a combination, either spatially or temporally. Strike-slip basins tend to be asymmetrical, with diverse depositional environments (with characteristically abrupt facies changes), and an axial pattern of basin fill (Nilsen & McLaughlin 1985). Furthermore, the basin fill is derived from multiple basin margin sources that change over time, which may also mean that the basin sediments are petrologically diverse. In addition, basin fill is characterized by abundant synsedimentary slumping and deformation. Distinctive aspects of sedimentary basins associated with strike-slip faults include:

(1) mismatches across basin margins;
(2) longitudinal and lateral basin asymmetry;
(3) episodic rapid subsidence;
(4) abrupt lateral facies changes and local unconformities; and
(5) marked contrasts in stratigraphy, facies geometry, and unconformities among different basins in the same region (Nilsen & Sylvester 1995).

Models of sedimentation in a compressional context (basin scale)

While models for extensional areas are well developed, this is not the case for regions where compressional activity is predominant. The main models that exist for compressional settings are those that describe the evolution of foreland basin successions (e.g. Beaumont 1981; Jordan 1981). The first evidence of an arc–arc or arc–continent collision in the stratigraphic record may be the transfer of sediments, primarily derived from the fold–thrust belt, into a remnant ocean basin from a point of collision along strike. As the foreland basin develops and fills with sediment, the main trend is that of shallowing and coarsening of the sediment (Fig. 11).

DeCelles & Giles (1996) note that a foreland basin system is an elongate region of potential sediment accommodation (Fig. 12). Within a foreland basin system four discrete depozones, comprising wedge top, foredeep, forebulge and backbulge areas, may be recognized. As a result of the continuing evolution of the belt and the

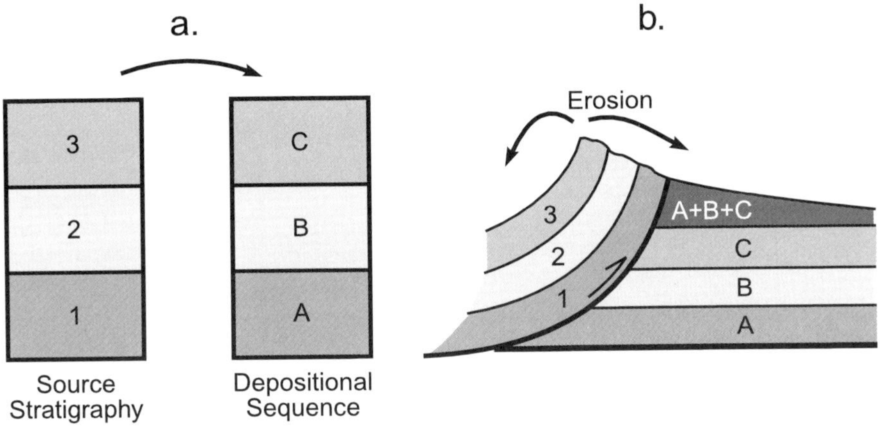

Fig. 11. Composition of basin fill in terms of detrital source petrography: (**a**) classic inverted stratigraphy, (**b**) blended composition (after Steidtmann & Schmitt 1988).

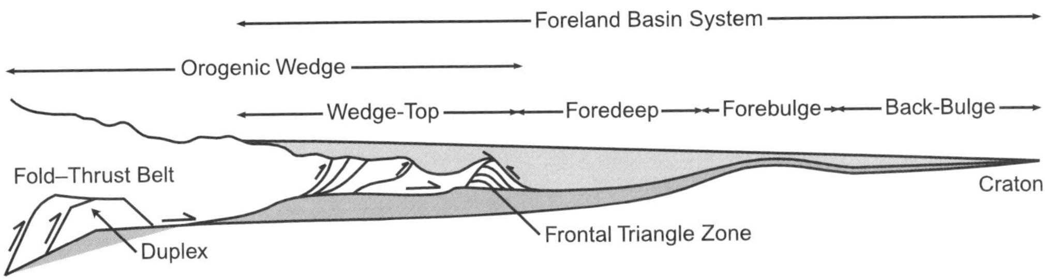

Fig. 12. Schematic cross-section depicting a revised concept of a foreland basin system, with the wedge-top, foredeep, forebulge and back-bulge depozones shown at approximately true scale. The foreland basin system is shown in coarse stipple, and the diagonally-ruled area indicates pre-existing miogeoclinal strata, which are incorporated into the fold–thrust belt towards the left of the diagram. A schematic duplex is depicted in the hinterland part of the orogenic wedge, and a frontal triangle zone and progressive deformation (short fanning lines associated with thrust tips) in the wedge-top depozone are also shown. Note the substantial overlap between the front of the orogenic wedge and the foreland basin system (after DeCelles & Giles 1996).

basin itself, these zones are not fixed in either space or time and the interaction between them can result in an extremely complex sediment distribution pattern within any foreland basin system. Both subsidence and uplift can cause significant local variations in sediment erosion and deposition, while the relative sense of thrust movement can have significant influence on sediment transport pathways. Most suture zones form by the consumption of an ocean between irregular continental margins that do not match in shape when they collide. The suturing process, therefore, is a diachronous one, such that collision is progressive as the uplift and closure of the remnant ocean basin proceeds. Sediment transport is both axial to the fold–thrust belt and normal to it (Jordan 1995). Variations in basin geometry and the composition of the stratigraphic fill may thus be interpreted in terms of the global geodynamic evolution.

Models of sedimentation in a compressional context (local scale)

The evolution of the basin fill in a foreland basin system in terms of sedimentary environment, succession thicknesses and vertical trends, is strongly dependent on the degree of compressional tectonic activity (Muñoz-Jimenez & Casas-Sainz 1997). Generally, foreland basins are initially marine, due to rapid downflexing (Jordan 1981; Flemings & Jordan 1989). At later stages, sedimentation rates exceed subsidence rates, giving rise to continental sedimentation (Allen *et al.* 1986). Variations within the basin fill may be partially related to variations in the flexural response to loading, differences in the type of crust underlying the basin, the particular type of foreland basin that forms (peripheral or retroarc) or the age of the rifted margin underlying the foreland basin (Miall 1995). These variations will affect the different depozones (see previous section) in different ways, leading to a degree of basin segmentation where the subsidence pattern at any single point is distinctive. Depositional sequences, therefore, can show a high degree of lateral variation in their sedimentary architecture making regional correlation difficult. Such problems are only compounded by the structural complexity that can occur in these settings, for example, in structurally segmented foreland basins, such as thrust-top regions where numerous growth anticlines (related to underlying blind thrusts) are present (de Boer *et al.* 1991; Butler & Grasso 1993; Krystinik & Blakeney DeJarnett 1995). This lack of precise correlation can lead to problems in trying to establish the true relationship between the evolution of the fold–thrust belt and the foreland basin.

Sequence stratigraphic models

Analysis of the sedimentary successions within basins tends to focus on the development of sequences separated by major interregional unconformities, and which record an almost complete transgressive–regressive cycle. In 1963 Sloss recognized a series of broad sequences from the cratonic succession of North America. These Sloss sequences, as they came to be known, were subdivided from each other by major tectonic events. According to Klein (1995) the sequences recognized from the North American Craton are comparable to the classic European geological systems and are unique to intracratonic settings in other regions of the world, including the Russian Platform, Brazil and Africa.

Subsequently, the development of the concept of sequence stratigraphy (e.g. Vail *et al.* 1977) concentrated on the subdivision of units into sequences which could be interpreted in terms of particular genetic parameters (i.e. lowstand, highstand etc.). Sequence stratigraphic concepts were initially developed in eustasy-driven passive margin settings. More recently, there have been attempts to extend this work both into continental settings (e.g. Emery & Myers 1996) and, of particular interest for this work, into tectonically active settings. For rift basins, a number of recent papers have explored the extent to which it is possible to recognize rift episodes using characteristic sedimentary sequences (e.g. Prosser 1993; Gawthorpe *et al.* 1994; Howell *et al.* 1996; Ravnas & Steel 1998). The sequence stratigraphic models presented in these examples depart from the traditional passive margin sequence in a number of ways, most notably with regard to the spatial variability of sequence architecture (and, by extension, sedimentary architecture). Within rift basins there can be significant variations in subsidence, sediment supply, and physiography adjacent to extensional rift-basin margins, yet the variability in sequence development in three dimensions has only recently been investigated and is poorly understood (e.g. Dart *et al.* 1994; Gawthorpe *et al.* 1997). Typical two dimensional numerical models of tectonics and sedimentation do not account for along-strike variations in structural style and deposition. However, because many tectonic processes are inherently three dimensional, to be truly predictive and applicable, models are required that attempt to address this three dimensional nature (Hardy & Gawthorpe 1998). Such models allow quantification of the variability of stratigraphy and a better understanding of how different controls interact in three dimensions to generate spatially complex stratigraphy.

An additional point is the relative lack of sequence stratigraphic models for other tectonic settings. Some models have been created for foreland basin successions, especially those formed in broad ramp-like foredeep-forebulge type of settings (e.g. Weimer 1960; Lawton 1986; Miall 1991; Cant & Stockmal 1993; Deramond *et al.* 1993; Lopez-Blanco 1993; Plint *et al.* 1993; Posamentier & Allen 1993; Van Wagoner & Bertram 1995). For other tectonic settings, however, there is a lack of studies (see below).

Source area

Much work has been carried out on the provenance of sedimentary rocks in order to differentiate between the various controlling factors, and to constrain the underlying tectonic controls on sediment production (e.g. Dickinson 1970, 1988; Zuffa 1985; Fontana 1991; Morton *et al.* 1991; Graham *et al.* 1993; Garzanti *et al.* 1996; Bahlburg & Floyd 1999). Sediment supply may be strongly asymmetrical (e.g. half-graben and foreland basin systems), and derived from either a few point sources or where these coalesce to approximate a line source. Models predicting sediment distribution within a particular tectonic setting are by necessity simplified versions of complex realities (e.g. Leeder & Gawthorpe 1987;

Steidtmann & Schmitt 1988) (Fig. 3). More precise evaluation of source-area geology can be determined by the analysis of specific minerals (e.g. Götze & Zimmerle 2000), or textures within lithic fragments (e.g. McPhie *et al.* 1993). In addition, specific heavy minerals (e.g. garnet, zircon), or heavy mineral associations (e.g. rutile–zircon, spinel-zircon) may be used to identify specific source rocks (e.g. Morton 1987; Morton & Hallsworth 1994). Geochemical and isotopic whole rock analysis (e.g. Henry *et al.* 1997) or specific chemical elements (e.g. Roser & Korsch 1986) can also be used to determine facts about the geology of the source region and to provide insight into the relative contributions of individual source rocks. More specifically, sediment infill can be analysed to investigate the geological evolution of the source area, for example, the degree of melting of volcanic source rocks (e.g. Najman & Garzanti 2000), pressure–temperature(–time) conditions of metamorphic source rocks (e.g. von Eynatten *et al.* 1996), or the proportion of juvenile to differentiated crustal materials in the source area using Sr and Nd isotopes (e.g. Najman *et al.* 2000).

Influence of climate on sedimentation

Climate can exercise a very significant control on sedimentation. For example, in foreland basins the climate in which the rising orogen develops is of great importance, both in terms of the tectonic style of the orogen and the architecture of the adjacent foreland basin. Areas of high precipitation (e.g. monsoonal areas) are characterized by rapid erosional unroofing, leading to a corresponding rapid uplift, deep erosion and the development of a foreland basin overfilled with non-marine sediments. In contrast, an arid environment would lead to less erosion, and so erosional unroofing would not compensate uplift, leading to the preservation of the fold–thrust belt and an underfilled foreland basin (Miall 1995). In addition, if a basin is located in a tropical/subtropical area where siliciclastic supply is reduced, then a carbonate template can be superimposed on the distribution of depositional environments within the basin (e.g. Leeder & Gawthorpe 1987; Burchette 1988).

Influence of sea-level change

Changes in relative sea-level influence the proportions of sediment deposited in a particular basin setting. For example in a foreland basin setting, where sea-level rise coincided with flexure-related subsidence, there would be a corresponding increase in the percentages of marine sediments relative to non-marine. Indeed, eustatic sea-level is the prime control on whether a retro-arc foreland basin is marine or non-marine, since the thrust-load driven subsidence is not sufficient to submerge normal thickness continental lithosphere during low eustatic sea level. This is in marked contrast to peripheral foreland basins where subsidence commonly places the surface of the underlying plate beneath sea-level (Jordan 1995). Given the same amount of tectonic subsidence, retro-arc foreland basins may be marine (e.g in the Cretaceous, a period of elevated sea levels), or non-marine (e.g. present day Andes region) (Jordan 1995). Similarly, the location of a rifted basin close to sea level would allow a very different sedimentary succession to evolve (e.g. Leeder & Gawthorpe 1987).

From sediments to tectonics

From the previous sections it is clear that the very complexity of basin models makes it difficult to fully ascertain the predominant controls on a particular setting. However, it is also very clear that the record contained within the sedimentary infill within a basin is of prime importance in being able to evaluate the tectono-sedimentary evolution of a region. Approaches to the analysis of the sedimentary infill are varied (see previous sections), but all share a common goal – to elucidate our understanding of the shared tectonic and sedimentary history of the basin under investigation. The following section will outline some problems associated with trying to trace tectonic evolution using the sedimentary record, as well as the varied techniques which can be applied, as well as introducing the various studies presented in this volume.

Basin type and preservation potential

The preservability of tectonostratigraphic assemblages is an important but seldom-discussed factor in basin analysis and palaeotectonic reconstruction. Some modern basin types are common and volumetrically important, whereas others are rare and volumetrically minor. In addition, even some common modern basin types are rarely found in the geological record because they are prone to uplift and erosion, and/or deformation and destruction (e.g. remnant ocean, back-arc basin). Their rarity in ancient orogenic belts is related to their suceptibility to erosion and deformation. Ingersoll & Busby (1995) have illustrated the typical life span of a selection of sedimentary basins versus their post-sedimentation preservation potential

(Fig. 13). It is clear that those basins which have a relatively high post-sedimentation preservation potential (e.g. intracratonic basins, terrestrial rift valleys) have a better chance for evaluation in terms of their tectonosedimentary characteristics than those basins where the sediment fill has a poor preservation potential (e.g. back-arc, transpressional and inverted basins).

Mapping

Detailed mapping of an area generally involves a combined approach using a variety of mapping techniques (structural, sedimentological, magmatic) in order to provide a broad picture of the geological evolution of a particular region. Such work also includes the mapping of specific features, for example, the architecture of infill v. time, or the spatial distribution of facies v. time, can be invaluable for the elucidation of the tectonic evolution of a region. Fernández-Fernández *et al.* (2003) have used a combination of structural and sedimentological mapping to investigate the Middle Jurassic to Cretaceous history of extension in the Betic Cordillera, Spain. The work of McCann *et al.* (2003) uses a similar approach, again using structural analysis and sedimentological investigation but also incorporating detailed mapping of the magmatic units in order to examine the Mid-Devonian-early Lower Carboniferous succession on the southern margin of the Donbas Basin, Ukraine. This work provides insight into the early phases of basin evolution in this complex region and shows the value of a multidisciplinary approach to such studies. Christophoul *et al.* (2003) mapped a region in the foreland basin of the northern Pyrenees (France) in order to examine the tectono-sedimentary evolution of the thrust belt. Thrust wedge advance and the corresponding loading resulted in basin flexure and sediment infill. Similar work from the Variscan succession of Poland (Lamarche *et al.* 2003) has demonstrated the complexity of orogenic activity in this region. It has also been possible to subdivide the various tectonic episodes into pre-, syn- and post-orogenic phases, thus clarifying the tectonic evolution of this important region.

Studying facies changing time and space

Tracing the changes in sedimentary facies evolution over time and space within an area can provide detailed information about the subtle ways in which tectonics and sedimentation interact in producing complex facies mosaics. Rieke *et al.* (2003) have used this approach to examine the upper Rotliegend succession from northeastern Germany in order to evaluate the importance of tectonic activity in terms of basin evolution. Previous models had suggested that basin evolution was controlled by a series of tectonic events. Rieke *et al.* (2003), however, clearly demonstrate that basin evolution was largely related to thermal subsidence within the region, although facies development was significantly influenced by climate. On a larger scale, Golonka *et al.* (2003) have examined the entire Polish Carpathian region, providing a

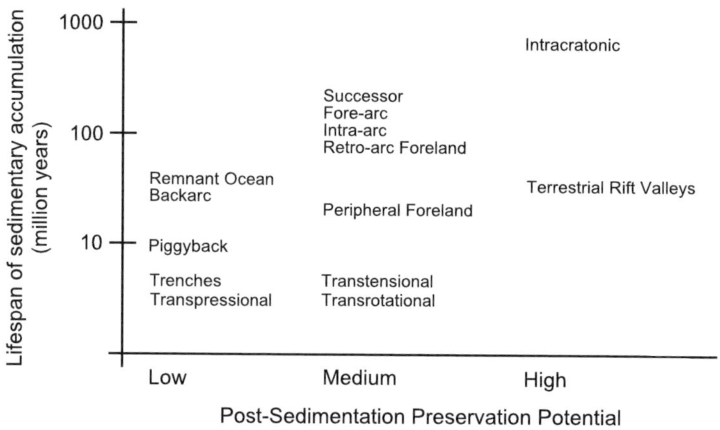

Figure 13. Typical life spans for sedimentary basins versus their post-sedimentation preservation potential. This latter term refers to the average amount of time during which basins will not be uplifted and eroded, or be tectonically destroyed during and subsequent to sedimentation. Sedimentary or volcanic fill may be preserved as accretionary complexes during and after basin destruction (modified after Ingersoll & Busby 1995).

series of maps which outline the changing palaeogeography of this region during latest Triassic–earliest Cretaceous times, a period of pronounced tectonic activity.

Grain-dating – exhumation/erosion, source area

Actualistic petrological models relating sediment composition especially sand and sandstone, to plate tectonic settings have been developed (e.g. Dickinson & Suczek, 1979). Cibin *et al.* (2003) have used petrography to characterize piggy-back basin fill successions and thereby to examine the evolution of thrusting within the northern Apennines, Italy. Augustsson & Bahlburg (2003) use the contrasting geochemical (including Nd and Sm), signatures from the sediment infill within an accretionary wedge sequence to differentiate the signature from the source area and that of the basin itself. Von Eynatten & Wijbrans (2003) have concentrated on a single mineral approach, in this case the Ar/Ar geochronology of detrital white mica, in the evaluation of the exhumation history of the Central Alps.

Sequence analysis

Sequence analysis is an important tool in exploring the broad evolution of a sedimentary basin. It enables different facies to be correlated and the underlying controls to be determined. Lazauskiene *et al.* (2003) have used this approach in the intracratonic Baltic Basin to investigate the Silurian succession, the period of maximum basin subsidence in the region, and relate basin development to tectonic activity along the Caledonian thrust front. On a smaller scale, Derer *et al.* (2003) have used sequence mapping across the Rhine Graben, Germany, to investigate the interrelationship of between fault activity and sequence formation within the region. Of particular interest is the fact that fault activity led to basin compartmentalization, leading to the evolution of different sedimentary successions on either side of the tectonic divide. Wartenberg *et al.* (2003) have used sequence analysis to investigate the evolution of a fore-arc basin succession within the developing collisional zone of western Australia.

Basin modelling

Increasingly, basin modelling is used in order to test certain ideas concerning the evolution of a basin. Carrapa *et al.* (2003) have integrated structural and subsidence analysis data in order to investigate the Oligocene-Miocene history of a basin in northwestern Italy. On a larger scale, Artyushkov & Chekhovich (2003) employed subsidence analysis to investigate the evidence of tectonic subsidence in a region where major eustatic sea-level changes are not recognized. Similarly, Nalpas *et al.* (2003) have analysed the geometries of developing compressional structures, using both mathematical and analogue models, in terms of differing rates of sedimentation. Such work (see below) is of great importance in terms of broadening our knowledge base on compressional tectonic settings.

Problems and future research directions

It is clear from the previous sections that there are a large number of different tectonic settings and that the sediment sequences contained within them can be extremely variable. While there are particular sequences that are characteristic of particular tectonic settings (e.g. the broad marine-to-non-marine succession produced within peripheral foreland basins), it is not always easy to precisely determine from a particular sediment sequence what the dominant tectonic setting was. Some of these have been outlined above (e.g. influence of climate or sea-level), but there are other factors – broadly related to our lack of understanding of the relationship between sedimentation and tectonics – which are more problematic. In an overview of basin modelling problems, Cloetingh *et al.* (1994) noted that although the success of any individual basin model is often gauged by its ability to reproduce the observed sedimentary record, few models deal realistically with sediment transport and preservation. A lack of understanding of these factors can lead to false or oversimplified interpretations. It is, therefore, clear that there is a great need for additional research, preferably multidisciplinary, in these areas in order to improve interlinked sediment–tectonics models.

Understanding fault activity

There is now much better understanding of normal faulting (e.g. Roberts *et al.* 1991) and the scaling relationships that operate (e.g. Walsh *et al.* 1991; Walsh & Watterson 1991, 1992; Dawers *et al.* 1993; Dawers & Anders 1995), which provides some basis for the understanding of how faults nucleate, progagate and link together over time. These faults and their displacements are fundamental building blocks for uplift. However, similarly detailed data on the scaling and linkage of reverse and thrust faults do not exist at present. Studies that fill this gap or define strain partitioning in transtensional and transpressional settings will help with strain quantification in regions of tectonic activity. Such research would greatly aid quantification models of sediment production, for example in basins evolving in compressional settings.

Fault segmentation and the resultant effect on sedimentation patterns is another area which requires investigation. Fault segmentation has been recognized from a variety of settings, including extensional (e.g. Larsen 1988; Peacock & Sanderson 1994; Walsh *et al.* 1999), compressional (e.g. Aydin 1988) and strike-slip settings (e.g. Peacock 1991). However, the precise interaction of the variations in stress generated by either the loss of displacement on individual faults or the transfer of displacement between fault segments, and the effects of these changes in displacement both on sediment basin location and sediment transfer patterns, remain to be studied.

Understanding specific basin types

Some basins are better understood and researched than others. This is particularly true of rift basins of the graben or half-graben type. However, other basin types require much additional research if we are to be able to really understand even the fundamental aspects of basin evolution in such systems. For example, the processes leading to crustal extension and subsidence in strike-slip settings are generally not as well understood as they are in other tectonic settings (Nilsen & Sylvester 1995). Furthermore, the complexity of strike-slip basins can vary according to their scale. Existing thermomechanical models for their formation as well as their structural and stratigraphic evolution are generally poorly developed.

Similarly, existing models for the development of intracratonic basins are largely related to ideas about supercontinent break-up and the resultant changes in heat flow. However, many intracratonic basins do not conform to the predicted subsidence histories. This, coupled with the fact that these basins have not been drilled to basement, leads to much speculation but little clarity.

Within arc-related settings, the situation is even more difficult. G. A. Smith & Landis (1995) note, with some degree of truth, that of all of the basin types considered by most workers involved in basin analysis, intra-arc basins remain the most poorly known. Dickinson (1995), for example, notes that in fore-arc basins little is known about the precise relationship of intrabasinal structures to relevant subduction parameters, such as plate convergence rate, the dip of the subducted slab and the motion of the arc massif relative to the roll-back of the subducted slab into the asthenosphere. All of these factors can influence tectonism within fore-arc basins. In addition, syndepositional deformation within fore-arc basins is varied and not well understood. The deformation may be partly related to the basin fill being underthrust by the subduction complex, or associated with backthrusting, both of which processes result in differential subsidence within the area (Dickinson 1995). In intra-arc settings there is little work done by sedimentologists, since the active processes within these basins are predominantly volcanic. In active arcs, young volcanic rocks may obscure older stratigraphic units and structures. Where more information exists (based on seismic evidence), there is a corresponding lack of information of the nature of the sedimentary and volcanic fills. In ancient sequences, the rocks are highly deformed and/or metamorphosed by later tectonic dismemberment or plutonism (G. A. Smith & Landis 1995).

Marsaglia (1995) notes that more detailed studies of the sedimentary facies architecture of backarc basins is lacking, partly because the depositional environments lack two or three-dimensional exposure upon which models could be constructed. There is also a lack of studies on particular sub-environments, particularly that of the volcanic apron, which, according to Carey & Sigurdsson (1984) could be the most diagnostic feature of back-arc basin sedimentation. In summary, the origin of basins within volcano-plutonic (magmatic) arcs is, in general, poorly understood, largely due to the paucity of studies that integrate volcanology, sedimentology and basin analysis (Ingersoll 1988).

Differential tectonic response

This occurs when parts of the basin are in compression while other parts are in extension. Thus the basin infill provides different tectonic signatures, which need to be compared and contrasted in order to be able to fully ascertain the overall basin history. A corollary of this is the increasingly recognized complexity of normal faults and their movement histories (e.g. Gawthorpe *et al.* 1997). It is extremely probable that such complexity also exists in compressional settings.

Basin compartmentalization

Basin compartmentalization is where a sedimentary basin is sub-divided by structural or other barriers and where the various subbasins may produce different tectonosedimentary signatures. Within trench regions, for example, the subduction of interlinked fore-arc basins can lead to buckling of the basin chain, resulting in segmentation and differential subsidence. This relative isolation of the sub-basins has marked consequences for sedimentation patterns (including facies distribution). Similarly, in back-arc or intra-arc basins sediment transport and deposition patterns may be influenced by the locations of volcanic ridges and variable subsidence of rift blocks.

Problems associated with basin compartmentalization can be even more marked when the pattern is overlain by such secondary factors as sea-level variations. A study from northern Spain revealed the presence of a series of unconformities which had a very segmented nature (resulting from the boundary between zones of uplift and zones of subsidence). This pattern of segmentation was related to structural activity that alternated periods of synrotational forced regression (carving of surface below the prograding shoreface) and post-rotational transgression (accumulation of shale wedges prior to the next increment of tilting) (Dreyer *et al.* 1999). In effect, the segmentation of these unconformities demonstrated that there was insufficient time available for the formation of laterally extensive bounding surfaces in the region.

Phase of basin development

Basin evolution follows a general pattern of tectonic and sedimentary evolution. For example, in rift events we have the production of three clear sequences – the pre-rift, synrift and post-rift successions. Thus, in basin evolutionary models, for each phase of basin evolution (where basins are well understood) a characteristic succession will be produced, and a sediment sequence produced within a syn-rift regime will be very different to one produced in the post-rift thermal subsidence phase.

Sediment budget within a basin

Hovius & Leeder (1998) and Leeder (1999) note that, more than any other issue in basin research, there is a need to explore the consequences of temporal and spatial changes in water and sediment supply and to intersect time series of these variables with other basin-defining variables such as basin subsidence rate, sea- and lake-level change, catchment uplift rate and climate. Sediment budget or mass-balance methods aim at calculating the volumes of eroded sediment (Leeder 1991) and can be used, in conjunction with other information (for example, catchment area size), to calculate the average erosion (Einsele *et al.* 1996) or discharge rates (Kuhlemann *et al.* 2001). While there have been a number of studies done in this area (see Burbank & Pinter 1999 for details), there is still a lack of understanding of the controls on sediment budget within a basin. Burbank & Pinter (1999) also noted that there was a need for better numerical models for erosion, and in particular, models which are supported by real data. Additionally, Schlunegger *et al.* (2001) have noted that when dealing with ancient settings, errors on budget methods can be very high, and that the results may be contrasting.

Sediment transport and post-depositional alteration within the basin also have a significant influence on the evolution of large-scale basin architecture through time, because the basin load modifies basin subsidence, and because post-depositional compaction and diagenesis of sediment affects accommodation space available for additional sediment (Schlager 1993).

Differentiating between tectonic and other controls

This is a very fundamental problem in terms of basin analysis. Sediments are, for the most part, preserved in basins, and the resultant succession records information related both to the depositional mechanisms operating within the basin, and tectonic mechanisms which control basin dynamics and determine the larger scale depostional setting within the basin. The sedimentary record preserved in a basin is thus a product of the interplay of these complex variables. Such factors would include sediment supply, continentality, sea-level variations and climate (e.g. Lindsay & Korsch 1989; Leeder *et al.* 1998; Mack & Leeder 1999). Interpretation of any particular basinal succession, therefore, involves understanding the many different controls on sedimentation. This can be problematic, however, in settings where different controls produce similar effects. In arc-related environments, for example, it can be difficult to distinguish between the interrelationship between tectonic activity and eustatic sea-level change, since tectonic deformation may result in significant changes in relative sea level. In such situations, it is necessary to use as varied an approach to basin analysis as possible in order to rigorously examine the various controls.

The authors would like to thank all of those who submitted their work to this Special Publication. We would also like to thank all of the many scientists who acted as reviewers for the articles within this volume, and through their work have helped to make this volume what it is. We would also like to thank Angharad Hills and Andy Morton from the Geological Society for their help and support in the realization of this project. I. Wolfgramm is thanked for drafting all of the diagrams.

References

ALEXANDER, J. A. & LEEDER, M. R., 1987. Active tectonic control of alluvial architecture. In: ETHRIDGE, F. G., FLORES, R. M. & HARVEY, M. D. (eds) *Recent Developments in Fluvial Sedimentology.* Society of Economic Paleontologists and Mineralogists, Special Publications **39**, 243–252.

ALEXANDER, J. & LEEDER, M. R. 1990. Geomorphology and surface tilting in an active extensional basin, SW Montana, USA. *Journal of the Geological Society, London,* **147,** 461–467.

ALLEN, P. A., HOMEWOOD, P. & WILLIAMS, G. D. 1986. Foreland basins: an introduction. *In:* ALLEN, P. A. & HOMEWOOD, P. (eds) *Foreland Basins.* International Association of Sedimentologists, Special Publication, **8**, 3–12.

ANDERS, M. H. & SCHLISCHE, R. W. 1994. Overlapping faults, intrabasin highs and the growth of normal faults. *Journal of Geology,* **102**, 165–180.

ARTYUSHKOV, E.V. & CHEKHOVICH, P.A. 2003. Silurian sedimentation in East Siberia: evidence for variations in the rate of tectonic subsidence occurring without any significant sea-level changes. *In:* McCann, T. & Saintot, A. (eds) *Tracing Tectonic Deformation Using the Sedimentary Record.* Geological Society, London, Special Publications, **208**, 321–350.

AUGUSTSSON, C. & BAHLBURG, H. 2003. Active or passive contintental margin? Geochemical and Nd isotope constraints of metasediments in the backstop of a pre-Andean accretionary wedge in southernmost Chile (46°30′–48°30′S). *In*: McCann, T. & Saintot, A. (eds) *Tracing Tectonic Deformation Using the Sedimentary Record.* Geological Society, London, Special Publications, **208**, 253–268.

AYDIN, A. 1988. Discontinuities along thrust faults and the cleavage duplexes. *In:* MITRA, G. & WOJTAL, S. F. (eds) *Geometries and Mechanism of Thrusting, with Special Reference to the Appalachians.* Geological Society of America, Special Paper, **222**, 223–232.

BACHMAN, S .B., LEWIS, S. D. & SCHWELLER, W. J. 1983. Evolution of a forearc basin, Luzon Central Valley, Philippines. *AAPG Bulletin*, **67,** 1143–1162.

BAHLBURG, H. & FLOYD, P. A. (eds) 1999. Advanced techniques in provenance analysis of sedimentary rocks. *Sedimentary Geology,* **124**.

BALLY, A. W. 1989. Phanerozoic basins of North America. *In:* BALLY, A. W. & PALMER, A. R. (eds) *The Geology of North America: an Overview.* Geological Society of America, Boulder, Colorado, 397–447.

BARNETT, J. A. M., MORTIMER, J., RIPPON, J. H., WALSH, J. J. & WATTERSON, J. 1987. Displacement geometry in the volume containing a single normal fault. *AAPG Bulletin,* **71,** 925–937.

BEAUDRY, D. & MOORE, G. F. 1985. Seismic stratigraphy and Cenozoic evolution of west Sumatra forearc basin. *AAPG Bulletin,* **69,** 742–759.

BEAUMONT, C. 1981. Foreland basins. *Geophysical Journal of the Royal Astronomical Society,* **65,** 291–329.

BHATIA, M. R. 1983. Plate tectonics and geochemical composition of sandstones. *Journal of Geology,* **91**, 611–627.

BLAIR, T. C. & BILODEAU, W. L. 1988. Development of tectonic cyclothems in rift, pull-apart, and foreland basins: sedimentary response to episodic tectonism. *Geology,* **16,** 517–520.

BOND, G. C. & KOMINZ, M. A. 1991. Disentangling middle Paleozoic sea level and tectonic events in cratonic margins and cratonic basins of North America. *Journal of Geophysical Research,* **94,** 6619–6639.

BROWN, L. F. & FISHER, W. L. 1979. *Seismic Stratigraphic Interpretation and Petroleum Exploration.* American Association of Petroleum Geologists, Continuing Education Series, Course Notes, 16.

BUCK, W. R. 1991. Modes of continental lithospheric extension. *Journal of Geophysical Research,* **96,** 20 161–20 178.

BURBANK, D. W. & PINTER, N. 1999. Landscape evolution: the interactions of tectonics and surface processes. *Basin Research,* **11,** 1–6.

BURCHETTE, T. P. 1988. Tectonic control on carbonate platform facies distribution and sequence development, Gulf of Suez. *Sedimentary Geology,* **59,** 179–204.

BUSBY-SPERA, C. J. 1988. Speculative tectonic model for the early Mesozoic arc of the southwest Cordilleran United States. *Geology,* **16**, 1121–1125.

BUTLER, R. W .H. & GRASSO, M. 1993. Tectonic controls on base-level variations and depositional sequences within thrust-top and foredeep basins: examples from the Neogene thrust belt of central Sicily. *Basin Research,* **5**, 137–151.

CANT, D. J. & STOCKMAL, G. S. 1993. Some controls on sedimentary sequences in foreland basins: examples from the Alberta Basin. *In:* FROSTICK, L. E. & STEEL, R. J. (eds) *Tectonic Controls and Signatures in Sedimentary Successions.* International Association of Sedimentologists, Special Publication **20**, 49–65.

CAREY, S. N. & SIGURDSSON, H. 1984. A model of volcanogenic sedimentation in marginal basins. *In:* KOKELAAR, B. P. & HOWELLS, M. F. (eds) *Marginal Basin Geology: Volcanic and Associated Sedimentary and Tectonic Processes in Modern and Ancient Marginal Basins.* Geological Society of London, Special Publication, **16**, 37–58.

CARRAPA, B., BERTOTTI, G. & Krijgsman, W. 2003. Subsidence, stress regime and rotation(s) of a tectonically active sedimentary basin within the Western Alpine orogen: the Tertiary Piedmont Basin (Alpine domain, Northwest Italy). *In:* McCann, T. & Saintot, A. (eds) *Tracing Tectonic Deformation Using the Sedimentary Record.*

Geological Society, London, Special Publications, **208**, 205–227.

CAS, R. & BUSBY-SPERA, C. J. (eds) 1991. Volcaniclastic sedimentation. *Sedimentary Geology*, **74**.

CHRISTIE-BLICK, N. & BIDDLE, K. T. 1985. Deformation and basin formation along strike-slip faults. *Society of Economic-Paleontologists and Mineralogists, Special Publication*, **37**, 1–34.

CHRISTOPHOUL, F., SOULA, J.-C., BRUSSET, S., ELIBANA, B., RODDAZ, M., BESSIERE, G. & DERAMOND, J. 2003. Time, place and mode of propogation of foreland basin systems as recorded by the sedimentary fill: examples of the late Cretaceous and Eocene retro-foreland basins of the north-eastern Pyrenees. *In*: McCann, T. & Saintot, A. (eds) *Tracing Tectonic Deformation Using the Sedimentary Record.* Geological Society, London, Special Publications, **208**, 229–252.

CIBIN, U., DI GIULIO, A. & MARTELLI, L. 2003. Oligocene-early Miocene tectonic evolution of the Northern Appenines (Northwestern Italy) traced through provenance of piggy-back basin fill successions. *In*: McCann, T. & Saintot, A. (eds) *Tracing Tectonic Deformation Using the Sedimentary Record.* Geological Society, London, Special Publications, **208**, 269–287.

CLIFT, P. D., SHIMIZU, N., LAYNE, G. D. & BLUSZTAJN, J. 2001. Tracing patterns of erosion and drainage in the Paleogene Himalaya through ion probe Pb isotope analysis of detrital K-feldspars in the Indus Molasse, India. *Earth and Planetary Science Letters,* **188**, 475–491.

CLOETINGH, S., SASSI, W. & TASK FORCE TEAM 1994. The origin of sedimentary basins: a status report from the task force of the International Lithosphere Program. *Marine and Petroleum Geology*, **11**, 659–683.

CLOETINGH, S., FERNÀNDEZ, M., MUNOZ, J. A., SASSI, W. & HORVÁTH, F. (eds) 1997. Structural controls on sedimentary basin evolution. *Tectonophysics*, **282**, 1–442.

COLLIER, R. E. L., JACKSON, J. A. & LEEDER, M. R. 1995. Quaternary drainage development, sediment fluxes and extensional tectonics in Greece. *In:* LEWIN, J., MACKLIN, M. G. & WOODWARD, J. C. (eds) *Mediterranean Quaternary River Environments*. Balkema, Rotterdam, 31–44.

COWARD, M. P., DEWEY, J. F. & HANCOCK, P. L. (eds) 1987. *Continental Extensional Tectonics.* Geological Society, London, Special Publication, **28**.

COWIE, P. A. 1998. A healing–reloading feedback control on the growth rate of seismogenic faults. *Journal of Structural Geology*, **20**, 1075–1087.

COX, A. & HART, R. B. 1986. *Plate Tectonics: How it Works.* Blackwell Scientific Publications, Oxford, 392 pp.

CROWELL, J. C. 1974*a*. Origin of late Cenozoic basins in southern California. *Society of Economic Paleontologists and Mineralogists, Special Publication*, **22,** 190–204.

CROWELL, J. C. 1974*b*. Sedimentation along the San Andreas fault, California. *Society of Economic Paleontologists and Mineralogists, Special Publication*, 19, 292–303.

DART, C. J., COLLIER, R. E. L. L., GAWTHORPE, R. L., KELLER, J. V. & NICHOLS, G. 1994. Sequence stratigraphy of (?)Pliocene-Quaternary synrift, Gilbert-type fan deltas, northern Peloponnesos, Greece. *Marine and Petroleum Geology*, **11**, 545–560.

DAWERS, N. H. & ANDERS, M. H. 1995. Displacement–length scaling and fault linkage. *Journal of Structural Geology,* **17**, 607–614.

DAWERS, N. H. & UNDERHILL, J. R. 2000. The role of fault interaction and linkage in controlling synrift stratigraphic sequences: Late Jurassic, Statfjord East Area, Northern North Sea. *AAAPG Bulletin,* **84,** 45–64.

DAWERS, N. H., ANDERS, M. H. & SCHOLZ, C. H. 1993. Growth of normal faults: displacement-length scaling. *Geology*, **21,** 1107–1110.

DE BOER, P. L., PRAGT, J. S. J. & OOST, A. P. 1991. Vertically persistent sedimentary facies boundaries along growth anticlines and climate-controlled sedimentation in the thrust-top South Pyrenean Tremp–Graus Foreland Basin. *Basin Research*, **3,** 63–78.

DECELLES, P. G. & GILES, K. G. 1996. Foreland basin systems. *Basin Research*, **8**, 105–123.

DEKORP–BASIN RESEARCH GROUP 1999. Deep crustal structure of the Northeast German Basin: New DEKORP–BASIN'96 deep-profiling results. *Geology,* **27,** 55–58.

DERAMOND, J., SONQUET, P., FONDECAVE-WALLEZ, M.-J. & SPECHT, M. 1993. Relationships between thrust tectonics and sequence stratigraphy surfaces in foredeeps: model and examples from the Pyrenees (Cretaceous–Eocene, France, Spain). In: WILLIAMS, G. D. & DOBB, A. (eds). *Tectonics and Seismic Sequence Stratigraphy.* Geological Society, London, Special Publication, **71**, 193–219.

DERER, C., KOSINOWSKI, M., LUTERBACHER, H.P., SCHÄFER, A. & SUß, M.P 2003. Sedimentary responses to tectonics in extensional basins: the Pechelbronn formation (Late Eocene to Early Oligocene) in the Northern Upper Rhine Graben, Germany. *In*: McCann, T. & Saintot, A. (eds) *Tracing Tectonic Deformation Using the Sedimentary Record.* Geological Society, London, Special Publications, **208**, 55–69.

DEWEY, J. F. 1980. Episodicity, sequence and style at convergent plate boundaries. *Geological Society of Canada, Special Paper*, **20**, 553–573.

DICKINSON, W. R. 1970. Interpreting detrital modes of graywacke and arkose. *Journal of Sedimentary Petrology,* **40,** 695–707.

DICKINSON, W. R. 1974. Plate tectonics and sedimentation. *In:* DICKINSON, W. R. (ed.) *Tectonics and Sedimentation.* Society of Economic Paleontologists and Mineralogists, Special Publication, **22**, 1–27.

DICKINSON, W. R. 1976. *Plate Tectonic Evolution of Sedimentary Basins.* American Association of Petroleum Geologists, Continuing Education Course Notes Series 1, 62 pp.

DICKINSON, W. R. 1988. Provenance and sediment dispersal in relation to paleotectonics and paleogeography of sedimentary basins. *In:* KLEINSPEHN, K. L. & PAOLA, C. (eds) *New Perspectives in Basin Analysis.* Springer-Verlag, New York, 3–25.

DICKINSON, W. R. 1995. Forearc basins. *In:* BUSBY, C. J. & INGERSOLL, R. V. (eds) *Tectonics of Sedimentary Basins.* Blackwell Science, Oxford, 221–262.

DICKINSON, W. R. & SEELY, D. R. 1979. Structure and stratigraphy of forearc regions. *AAPG Bulletin,* **63,** 2–31.

DICKINSON, W. R. & SUCZEK, C. A. 1979. Plate tectonics and sandstone compositions. *AAPG Bulletin,* **63**, 2164–2182.

DREYER, T., CORREGIDOR, J., ARBUES, P. & PUIGDEFABREGAS, C. 1999. Architecture of the tectonically influenced Sobrarbe deltaic complex in the Ainsa Basin, northern Spain. *Sedimentary Geology*, **127**, 127–169.

EBINGER, C. J. 1989. Geometric and kinematic development of border faults and accommodation zones, Kivu–Rusizi Rift, Africa. *Tectonics,* **8,** 117–133.

EINSELE, G., CHOUGH, S. K. & SHIKI, T. 1996. Depositional events and their records: an introduction. *In:* SHIKI, T., CHOUGH, S. K. & EINSELE, G. (eds) Marine sedimentary events and their records. *Sedimentary Geology*, **104,** 1–9.

EMERY, D. & MYERS, K. J. (eds) 1996. *Sequence Stratigraphy*. Blackwell Science, Oxford.

FERNÁNDEZ-FERNÁNDEZ, E., JABALOY, A. & GONZÁLEZ-LODIERO, F. 2003. Middle Jurassic to Cretaceous extensional tectonics and sedimentation in the Eastern External Zone of the Betic Cordillera. *In*: MCCANN, T. & SAINTOT, A. (eds) *Tracing Tectonic Deformation Using the Sedimentary Record.* Geological Society, London, Special Publications, **208**, 29–53.

FISHER, R. V. & SMITH, G. A. (eds) 1991. *Sedimentation in volcanic settings*. Society of Economic Paleontologists and Mineralogists, Special Publication, **45**, 257.

FLEMINGS, P. B. & JORDAN, T. E. 1989. A synthetic stratigraphic model of foreland basin development. *Journal of Geophysical Research*, **94**, 3851–3866.

FONTANA, D. 1991. Detrital carbonate grains as provenance indicators in the Upper Cretaceous Pietraforte Formation (northern Apennines). *Sedimentology,* **38**, 1085–1095.

FROSTICK, L. E. & REID, I. 1987. Tectonic control of desert sediments in rift basins ancient and modern. Geological Society, London, Special Publication, **35**, 53–68.

FROSTICK, L. E. & STEEL, R. J. 1993. Sedimentation in divergent plate-margin basins. International Association of Sedimentologists, Special Publication, **20**, 111–128.

GARZANTI, E., CRITELLI, S. & INGERSOLL, R. V. 1996. Paleogeographic and paleotectonic evolution of the Himalayan Range as reflected by detrital modes of Tertiary sandstones and modern sands (Indus transect, India and Pakistan). *Geological Society of America Bulletin,* **108**, 631–642.

GAWTHORPE, R. L., FRASER, A. J. & COLLIER, R. E. L. 1994. Sequence stratigraphy in active extensional basins: implications for the interpretation of ancient basin fills. *Marine and Petroleum Geology*, **11**, 642–658.

GAWTHORPE, R. L., SHARP, I., UNDERHILL, J. R. & GUPTA, S. 1997. Linked sequence stratigraphic and structural evolution of propagating normal faults. *Geology*, **25**, 795–798.

GIBBS, A. 1984. Structural evolution of extensional basin margins. *Journal of the Geological Society, London,* **141,** 609–620.

GOLONKA, J., KROBICKI, M., OSZCZYPKO, N., SLACKA, A. & SŁOMKA, T. 2003. Geodynamic evolution and paleogeography of the Polish Carpathians and adjacent areas during Neo-Cimmerian and preceding events (latest Triassic – earliest Cretaceous). *In*: MCCANN, T. & SAINTOT, A. (eds) *Tracing Tectonic Deformation Using the Sedimentary Record.* Geological Society, London, Special Publications, **208,** 137–158.

GÖTZE, J. & ZIMMERLE, W. 2000. *Quartz and Silica as Guide to Provenance in Sediments and Sedimentary Rocks.* Contributions to Sedimentary Geology, **21**, 91.

GRAHAM, S. A., HENDRIX, M. S., WANG, L. B. & CARROLL, A. R. 1993. Collisional successor basins of western China: Impact of tectonic inheritance on sand composition. *Geological Society of America Bulletin,* **105,** 323–344.

GUPTA, S. 1997. Tectonic control on paleovalley incision at the distal margin of the early Tertiary Alpine Foreland Basin, southeast France. *Journal of Sedimentary Research*, **67,** 1030–1043.

HARDY, S. & GAWTHORPE, R. L. 1998. Effects of variations in fault slip rate on sequence stratigraphy in fan deltas: Insights from numerical modelling. *Geology*, **26**, 911.914.

HELLER, P. L. & PAOLA, C. 1992. The large-scale dynamics of grain-size variation in alluvial basins. 2: application to syntectonic conglomerate. *Basin Research*, **4**, 91–102.

HELLER, P. L., ANGEVINE, C. L., WINSLOW, N. S. & PAOLA, C. 1988. Two-phase stratigraphic model of foreland-basin sequences. *Geology,* **16,** 501–504.

HENRY, P., DELOULE, E. & MICHARD, A. 1997. The erosion of the Alps: Nd isotopic and geochemical constraints on the sources of the peri-Alpine molasse sediments. *Earth and Planetary Science Letters,* **146,** 627–644.

HOVIUS, N. & LEEDER, M. 1998. Clastic sediment supply to basins. *Basin Research,* **10,** 1–5.

HOWELL, J. A., FLINT, S. S. & HUNT, C. 1996. Sedimentological aspects of the Humber Group (Upper Jurassic) of the South Central Graben, UK North Sea. *Sedimentology*, **43,** 89–114.

INGERSOLL, R. V. 1988. Tectonics of sedimentary basins. *Geological Society of America, Bulletin,* **100**, 1704–1719.

INGERSOLL, R. V. & BUSBY, C. J. 1995. Tectonics of sedimentary basins. *In:* BUSBY, C. J. & INGERSOLL, R. V. (eds) *Tectonics of Sedimentary Basins.* Blackwell Science, Oxford, 1–51.

JARRARD, R. D. 1986*a*. Relations among subduction parameters. *Reviews of Geophysics*, **24**, 217–284.

JARRARD, R. D. 1986*b*. Terrane motion by strike-slip faulting of forearc slivers. *Geology,* **14**, 780–783.

JORDAN, T. E. 1981. Thrust loads and foreland basin evolution, Cretaceous, western United States. *AAPG Bulletin,* **65**, 2506–2520.

JORDAN, T. E. 1995. Retroarc foreland and related basins. *In:* BUSBY, C. J. & INGERSOLL, R. V. (eds) *Tectonics of Sedimentary Basins*. Blackwell Science, Oxford, 331–362.

JORDAN, T. E., FLEMINGS, P. B. & BEER, J. A. 1988. Dating thrust fault activity by use of foreland-basin strata. *In:* KLEINSPEHN, K. L. & PAOLA, C. (eds) *New Perspectives in Basin Analysis*. Springer, New York, 307–330.

KARIG, D. E. 1971. Origin and development of marginal basins in the western Pacific. *Journal of Geophysical Reseach*, **76,** 2542–2561.

KINGSTON, D. R., DISHROON, C. P. & WILLIAMS, P. A. 1983*a*. Global basin classification system. *AAPG Bulletin*, **67**, 2175–2193.

KINGSTON, D. R., DISHROON, C. P. & WILLIAMS, P. A. 1983*b*. Hydrocarbon plays and global basin classification. *AAPG, Bulletin*, **67,** 2194–2198.

KLEIN, G. D. 1985. The control of depositional depth, tectonic uplift, and volcanism on sedimentation processes in the back-arc basins of the western Pacific Ocean. *Journal of Geology*, **93,** 1–25.

KLEIN, G. D. 1991. Origin and evolution of North American cratonic basins. *South African Journal of Geology*, **94,** 3–18.

KLEIN, G. D. 1995. Intracratonic basins. *In:* BUSBY, C. J. & INGERSOLL, R. V. (eds) Tectonics of Sedimentary Basins. Blackwell Science, Oxford, 459–477.

KLEIN, G. D. & HSUI, A. T. 1987. Origin of intracratonic basins. *Geology*, **15**, 1094–1098.

KOKELAAR, P. 1992. Ordovician marine volcanic and sedimentary record of rifting and volcanotectonism, Snowdon, Wales, United Kingdom. *Geological Society of America Bulletin*, **104,** 1433–1455.

KRYSTINIK, L. F. & BLAKENEY DEJARNETT, B. 1995. Lateral variability of sequence stratigraphic framework in the Campanian and lower Maastrichtian of the Western Interior Seaway. *In:* VAN WAGONER, J. C. & BERTRAM, G. T. (eds) *Sequence Stratigraphy of Foreland Basin Deposits*. American Association of Petroleum Geologists Memoirs **64**, 11–25.

KUHLEMANN, J., FRISCH, W., DUNKL, I. & SZEKELY, B. 2001. Quantifying tectonic versus erosive denudation by the sediment budget: the Miocene core complexes of the Alps. *Tectonophysics*, **330**, 1–23.

LAMARCHE, J., LEWANDOWSKI, M., MANSY, J.-L. & SZULCZEWSKI, M. 2003. Partitioning pre-, syn- and post-Variscan deformation in the Holy Cross Mountains, eastern Variscan foreland. *In:* MCCANN, T. & SAINTOT, A. (eds) *Tracing Tectonic Deformation Using the Sedimentary Record*. Geological Society, London, Special Publications, **208,** 159–184.

LAMBIASE, J. 1991. A model for tectonic control of lacustrine stratigraphic sequences in continental rift basins. *In:* KATZ, B. J. (ed.) *Lacustrine Basin Exploration: Case Studies and Modern Analogues*. APPG Memoir, **50**, 265–276.

LARSEN, P. H. 1988. Relay structures in a Lower Permian basement-involved extension system, East Greenland. *Journal of Structural Geology*, **10,** 3–8.

LARUE, D. K., SMITH, A. L. & SCHELLEKENS, J. H. 1991. Oceanic island arc stratigraphy in the Caribbean region: don't take it for granite. *Sedimentary Geology*, **74**, 289–308.

LAWTON, T. F. 1986. Compositional trends within a clastic wedge adjacent to a fold-thrust belt: Indianola Group, central Utah, USA. *International Association of Sedimentologists, Special Publication*, **8**, 411–423.

LAZAUSKIENE, J., SLIAPUA, S., BRAZAUSKAS, A. & MUSTEIKIS, P. 2003. Sequence stratigraphy of the Baltic Silurian succession: tectonic control on the foreland infill. *In*: McCann, T. & Saintot, A. (eds) *Tracing Tectonic Deformation Using the Sedimentary Record*. Geological Society, London, Special Publications, **208**, 95–115.

LEEDER, M. R. 1991. Denudation, vertical crustal movements and sedimentary basin infill. *Geologische Rundschau*, **80,** 441–458.

LEEDER, M. R. 1995. Continental rifts and proto-oceanic rift troughs. In: BUSBY, C. J. & INGERSOLL, R. V. (eds) *Tectonics of Sedimentary Basins*. Blackwell Science, Oxford, 119–148.

LEEDER, M. R. 1999. *Sedimentology and Sedimentary Basins. From Turbulence to Tectonics*. Blackwell Science, Oxford.

LEEDER, M. R. & ALEXANDER, J. 1987. The origin and tectonic significance of asymmetrical meander belts. *Sedimentology*, **34**, 217–226.

LEEDER, M. R. & GAWTHORPE, R. L. 1987. Sedimentary models for extensional tilt-block/half-graben basins. *In:* COWARD, M. P., DEWEY, J. F. & HANCOCK, P. L. (eds) *Continental Extensional Tectonics* Geological Society, London, Special Publication, **28**, 139–152.

LEEDER, M. R. & JACKSON, J. A. 1993. The interaction between normal faulting and drainage in active extensional basins with examples from the Western United States and Greece. *Basin Research*, **5,** 79–102.

LEEDER, M. R., SEGER, M. J. & STARK, C. P. 1991. Sedimentology and tectonic geomorphology adjacent to active and inactive normal faults in the Megara Basin and Alkyonides Gulf, central Greece. *Journal of the Geological Society, London*, **148,** 331–343.

LEEDER, M. R., HARRIS, T. & KIRKBY, M. J. 1998. Sediment supply and climate change: implications for basin stratigraphy. *Basin Research*, **10,** 7–18.

LINDSAY, J. F. & KORSCH, R. J. 1989. Interplay of tectonics and sea-level changes in basin evolution: an example from the intracratonic Amadeus basin, central Australia. *Basin Research*, **2,** 3–25.

LINK, M. H. 1982. Provenance, paleocurrents, and paleogeography of Ridge Basin, southern California. *In:* CROWELL, J. C. & LINK, M. H. (eds) *Geologic History of Ridge Basin, southern California*. Society of Economic Paleontologists and Mineralogists, Pacific Section, Los Angeles, 265–276.

LISTER, G. S., ETHERIDGE, M. A. & SYMONDS, P. A. 1986. Detachment faulting and the evolution of passive continental margins. *Geology*, **14**, 246–250. (see also Comments & Reply, *Geology*, 14, 890–892)

LOPEZ-BLANCO, M. 1993. Stratigraphy and sedimentary development of the Sant Llorent de Munt fan-delta complex (Eocene, Southern Pyrenean foreland basin, northeast Spain). *In:* FROSTICK, L. E. & STEEL, R. J. (eds) *Tectonic Controls and Signatures in Sedimentary Successions.* International Association of Sedimentologists, Special Publication, **20**, 67–88.

LUCCHITTA, I. & SUNESON, N. H. 1993. Dips and extension. *Geological Society of America Bulletin,* **105,** 1346–1356.

MCCAFFREY, R. 1992. Oblique plate convergence, slip vectors, and forearc deformation. *Journal of Geophysical Research,* **97,** 8905–8915.

MCCANN, T., SAINTOT, A., CHALOT-PRAT, F., KITCHKA, A., FOKIN, P. & ALEKSEEV, A. 2003. Evolution of the southern margin of the Donbas (Ukraine), from Devonian to Early Carboniferous times. *In:* MCCANN, T. & SAINTOT, A. (eds) *Tracing Tectonic Deformation Using the Sedimentary Record.* Geological Society, London, Special Publications, **208,** 117–135.

MCCANN, T. & SHANNON, P. M. 1993. Lower Cretaceous seismic stratigraphy and fault movements in the Celtic Sea Basin, Ireland. *First Break,* **11**, 335–344.

MACDONALD, D. I. M. (ed.) 1991. *Sedimentation, Tectonics and Eustasy. Sea Level Changes at Active Margins.* International Association of Sedimentologists, Special Publication, **12**.

MACK, G. H. & LEEDER, M. R. 1999. Climatic and tectonic controls on alluvial-fan and axial-fluvial sedimentation in the Plio-Pleistocene Palomas half graben, southern Rio Grande Rift. *Journal of Sedimentary Research,* **69,** 635–652.

MCKENZIE, D. P. 1978. Some remarks on the development of sedimentary basins. *Earth and Planetary Science Letters,* **40,** 25–32.

MCPHIE, J., DOYLE, M. & ALLEN, R. 1993. *Volcanic Textures – A Guide to the Interpretation of Textures in Volcanic Rocks.* CODES, University of Tasmania, Australia.

MARSAGLIA, K. M. 1995. Interarc and Backarc Basins. *In:* BUSBY, C. J. & INGERSOLL, R. V. (eds) *Tectonics of Sedimentary Basins.* Blackwell Science, Oxford, 299–330.

MIALL, A. D. 1991. Stratigraphic sequences and their chronostratigraphic correlation. *Journal of Sedimentary Petrology,* **61**, 497–505.

MIALL, A. D. 1995. Collision-related foreland basins. *In:* BUSBY, C. J. & INGERSOLL, R. V. (eds) *Tectonics of Sedimentary Basins.* Blackwell Science, Oxford, 119–148.

MIALL, A. D. 1997. *The Geology of Stratigraphic Sequences.* Springer-Verlag, Berlin.

MORTON, A. C. 1987. Influences of provenance and diagenesis on detrital garnet suites in the Paleocene Forties Sandstone, central North Sea. *Journal of Sedimentary Petrology,* **57,** 1027–1032.

MORTON, A. C. & HALLSWORTH, C. R. 1994. Identifying provenance-specific features of detrital heavy mineral assemblages in sandstones. *Sedimentary Geology,* **90,** 241–256.

MORTON, A. C., TODD, S. P. & HAUGHTON, P. D. W. (eds) 1991. *Developments in Sedimentary Provenance Studies.* Geological Society, London, Special Publication, **57**.

MUÑOZ-JIMENEZ, A. & CASAS-SAINZ, A. M. 1997. The Rioja Trough (N Spain): tectonosedimentary evolution of a symmetric foreland basin. *Basin Research*, **9**, 65–85.

NAJMAN, Y. & GARZANTI, E. 2000. Reconstructing early Himalayan tectonic evolution and paleogeography from Tertiary foreland basin sedimentary rocks, northern India. *Geological Society of America Bulletin,* **112,** 435–449.

NAJMAN, Y. M. R., BICKLE, M. & CHAPMAN, H. 2000. Early Himalayan exhumation: isotopic constraints from the Indian foreland basin. *Terra Nova,* **12**, 28–34.

NALPAS, T., GAPAIS, D., VERGES, J., BARRIER, L., GESTAIN, V., LEROUX, G., ROUBY, D. & KERMAREC, J.-J. 2003. Effects of rate and nature of Synkinematic sedimentation on the growth of compressive structures constrained by analogue models and field examples. *In*: McCann, T. & Saintot, A. (eds) *Tracing Tectonic Deformation Using the Sedimentary Record.* Geological Society, London, Special Publications, **208**, 307–319.

NILSEN, T. H. & MCLAUGHLIN, R. J. 1985. Comparison of tectonic framework and depositional patterns of the Hornelen strike-slip basin of Norway and the Ridge and Little Sulphur Creek strike-slip basins of California. *Society of Economic Paleontologists and Mineralogists, Special Publication* **37**, 79–103.

NILSEN, T. H. & SYLVESTER, A. G. 1995. Strike-slip basins. *In*: BUSBY, C. J. & INGERSOLL, R. V. (eds) *Tectonics of Sedimentary Basins*. Blackwell Science, Oxford, 425–457.

NILSEN, T. H. & SYLVESTER, A. G. 1999a. Strike-slip basins: Part 1. *The Leading Edge*, **18**, 1146, 1148–1159.

NILSEN, T. H. & SYLVESTER, A. G. 1999b. Strike-slip basins: Part 2. *The Leading Edge,* **18**, 1258–1262, 1264–1267.

NORABUENA, E., LEFFLER-GRIFFIN, L. *et al.* 1998. Space geodetic observations of Nazca-South America convergence across the central Andes. *Science*, **279,** 358–362.

PACKHAM, G. H. & FALVEY, D. A. 1971. An hypothesis for the formation of the marginal seas in the western Pacific. *Tectonophysics,* **11**, 79–109.

PEACOCK, D. C. P. 1991. Displacements and segment linkage in strike-slip fault zones. *Journal of Structural Geology,* **13**, 1025–1035.

PEACOCK, D. C. P. & SANDERSON, D. J. 1994. Geometry and development of relay ramps in normal faults systems. *AAPG Bulletin*, **78,** 147–165.

PLINT, G., HART, B. S. & DONALDSON, W. S. 1993. Lithospheric flexure as a control on stratal geometry and facies distribution in Upper Cretaceous rocks of the Alberta foreland basin. *Basin Research,* **5,** 69–77.

POSAMENTIER, H. W. & ALLEN, G. P. 1993. Variability of the sequence stratigraphic model: effects of local basin factors. *Sedimentary Geology,* **86,** 91–109.

PROSSER, S. 1993. Rift-related linked depositional systems and their seismic expression. *In:* WILLIAMS,

G. D. & DOBB, A. (eds) 1993. *Tectonics and Seismic Sequence Stratigraphy*. Geological Society, London, Special Publication, **71**, 35–66.

RAVNAS, R. & STEEL, R. J. 1998. Architecture of marine rift basin successions. *AAPG Bulletin,* **82,** 110–146.

READING, H. G. 1998. *Sedimentary Environments: Processes, Facies and Stratigraphy.* Blackwell Science, Oxford.

RIEKE, H., MCCANN, T., KRAWCYK, C.M. & NEGENDANK, J.F.W. 2003. Evaluation of controlling factors on facies distribution and evolution in an arid continental environment – an example from the Rotliegend of the NE German basin. *In:* MCCANN, T. & SAINTOT, A. (eds) *Tracing Tectonic Deformation Using the Sedimentary Record.* Geological Society, London, Special Publications, **208,** 71–94.

ROBERTS, A. M., YIELDING, G. & FREEMAN, B. (eds) 1991. *The Geometry of Normal Faults*. Geological Society, London, Special Publication, **56**.

ROSER, B. P. & KORSCH, R. J. 1986. Determination of tectonic setting of sandstone-mudstone suites using SiO_2 content and K_2O/Na_2O ratio. *Journal of Geology*, **94**, 635–650.

SCHLAGER, W. 1993. Accommodation and supply – a dual control on stratigraphic sequences. *Sedimentary Geology,* **86**, 111–136.

SCHLISCHE, R.W. 1991. Half-graben basin filling models: new constraints on continental extensional basin development. *Basin Research,* **3**, 123–141.

SCHLISCHE, R. W. & OLSEN, P. E. 1990. Quantitative filling model for continental extensional basins with applications to early Mesozoic rifts of eastern North America. *Journal of Geology*, **98**, 135–155.

SCHLUNEGGER, F., MELZER, J. & TUCKER, G. E. 2001. Climate, exposed source-rock lithologies, crustal uplift and surface erosion: a theoretical analysis calibrated with data from the Alps/North Alpine Foreland Basin system. *International Journal of Earth Sciences*, **90**, 484–499.

SCHWANS, P. 1995. Controls on sequence stacking and fluvial to shallow-marine architecture in a foreland basin. *In:* VAN WAGONER, J. C. & BERTRAM, G. T. (eds) *Sequence Stratigraphy of Foreland Basin Deposits.* AAPG Memoir, **64**, 55–102.

SENGÖR, A. M. C. 1995. Sedimentation and tectonics of fossil rifts. *In:* BUSBY, C. J. & INGERSOLL, R. V. (eds) *Tectonics of Sedimentary Basins.* Blackwell Science, Oxford, 53–117.

SENGÖR, A. M. C. & BURKE, K. 1978. Relative timing of rifting and volcanism on Earth and its tectonic implications. *Geophysical Research Letters*, **5**, 419–421.

SLOSS, L. L. 1963. Sequences in the intracratonic interior of North America. *Geological Society of America Bulletin,* **74**, 93–114.

SMITH, G. A. & LANDIS, C. A. 1995. Intra-arc Basins. *In:* BUSBY, C. J. & INGERSOLL, R. V. (eds) *Tectonics of Sedimentary Basins.* Blackwell Science, Oxford, 263–298.

SMITH, M. & GEHRELS, G. 1994. Detrital zircon geochronology and provenance of the Harmony and Valmy formations, Roberts Mountains Allochthon, Nevada. *Geological Society of America Bulletin,* **106,** 968–979.

SPRAY, J. G. 1997. Superfaults. *Geology*, **25**, 579–582.

STEIDTMANN, J. R. & SCHMITT, J. G. 1988. Provenance and dispersal of tectogenic sediments in thin-skinned thrusted terrains. In: KLEINSPEHN, K. & PAOLA, C. (eds) *New Perspectives in Basin Analysis.* Springer-Verlag, New York, 353–366.

STERN, R. J. & BLOOMER, S. H. 1992. Subduction zone infancy: examples from the Eocene Izu–Bonin–Mariana and Jurassic California arcs. *Geological Society of America Bulletin*, **104**, 1621–1636.

SUNDVOLL, B., LARSEN, B. T. & WANDAAS, B. 1992. Early magmatic phase in the Oslo Rift and its related stress regime. In: ZIEGLER, P. (ed.) *Geodynamics of Rifting. Volume 1 – Case Studies on Rifts, Europe and Asia*. Tectonophysics, **208**, 37–54.

SURLYK, F. 1990. Mid-Mesozoic synrift turbidite systems: controls and predictions. *In:* COLLINSON, J. D. (ed.) *Correlation in Hydrocarbon Exploration.* Graham & Trotman, London, 231–241.

TANKARD, A. J., WELSINK, H. J. & JENKINS, W. A. M. 1989. Structural styles and stratigraphy of the Jeanne d'Arc Basin, Grand Banks of Newfoundland. *In:* TANKARD, A. J. & BALKWILL, H. R. (eds) *Extensional Tectonics and Stratigraphy of the North Atlantic Margins.* AAPG Memoir, **46**, 265–282.

THORNE, J. A., & SWIFT, D. J. P. 1991. Sedimentation on continental margins, VI. A regime model for depositional sequences, their component systems tracts, and bounding surfaces. *In:* SWIFT, D. J. P., OERTEL, G. F., TILLMAN, R. W. & THORNE, J. A. (eds) *Shelf Sand and Sandstone Bodies – Geometry, Facies and Sequence Stratigraphy.* International Association of Sedimentologists, Special Publication, **14**, 189–255.

VAIL, P. R., MITCHUM, R. M., JR, TODD, R. G., WIDMIER, J. M., THOMPSON, S., III, SANGREE, J. B, BUBB, J. N. & HATELID, W. G. 1977. Seismic stratigraphy and global change of sea level. *In:* PAYTON, C. E. (ed.) *Seismic Stratigraphy – Applications to Hydrocarbon Exploration.* AAPG Memoir, **26**, 49–212.

VAN WAGONER, J. C. & BERTRAM, G. T. 1995. *Sequence Stratigraphy of Foreland Basin Deposits.* AAPG Memoir, **64**.

VAN WAGONER, J. C., MITCHUM, R. M., CAMPION, K. M. & RAHMANIAN, V. D. 1990. *Siliciclastic Sequence Stratigraphy in Well Logs, Cores and Outcrops.* American Association of Petroleum Geologists, Methods in Exploration Series, **7**.

VON EYNATTEN, H., GAUPP, R. & WIJBRANS, J. R. 1996. $^{40}Ar/^{39}Ar$ laser-probe dating of detrital white micas from Cretaceous sedimentary rocks of the Eastern Alps: evidence for Variscan high-pressure metamorphism and implications for Alpine Orogeny. *Geology*, **24**, 691–694.

VON EYNATTEN, H. & WIJBRANS, J.R. 2003. Precise tracing of exhumation and provenance using $^{40}Ar/^{39}Ar$ geochronology of detrital white mica: the example of the Central Alps. *In*: McCann, T. & Saintot, A. (eds) *Tracing Tectonic Deformation Using the Sedimentary Record.*

Geological Society, London, Special Publications, **208**, 289–305.

WALKER, R. G. & JAMES, N. (eds) 1992. *Facies Models: Responses to Sea-level Changes*. Geological Association of Canada, Toronto, Canada.

WALSH, J. J. & WATTERSON, J. 1988. Analysis of the relationship between displacements and dimensions of faults. *Journal of Structural Geology*, **10**, 239–247.

WALSH, J. J. & WATTERSON, J. 1991. Geometric and kinematic coherence and scale effects in normal fault systems. *In:* ROBERTS, A. M., YIELDING, G. & FREEMAN, B. (eds) *The Geometry of Normal Faults*, Geological Society, London, Special Publication, **56**, 193–203.

WALSH, J. J. & WATTERSON, J. 1992. Populations of faults and fault displacements and their effects on estimates of fault-related regional extension. *Journal of Structural Geology*, **14**, 701–712.

WALSH, J. J., WATTERSON, J. & YIELDING, G. 1991. The importance of small-scale faulting in regional extension. *Nature*, **351**, 391–393.

WALSH, J. J., WATTERSON, J., BAILEY, W. R. & CHILDS, C. 1999. Fault relays, bends and branch lines. *Journal of Structural Geology*, **21**, 1019–1026.

WARTENBERG, W., KORSCH, R.J. & SCHÄFER, A. 2003. The Tamworth Belt in Southern Queensland, Australia – thrust-characterised geometry concealed by Surat Basin sediments. *In*: McCann, T. & Saintot, A. (eds) *Tracing Tectonic Deformation Using the Sedimentary Record*. Geological Society, London, Special Publications, **208**, 185–203.

WEIMER, R. J. 1960. Upper Cretaceous stratigraphy, Rocky Mountain area. *AAPG Bulletin*, **44**, 1–20.

WERNICKE, B. 1981. Low-angle normal faults in the Basin and Range province: nappe tectonics in an extending orogen. *Nature*, **291**, 645–648.

WERNICKE, B. 1992. Cenozoic extensional tectonics of the US Cordillera. *In:* BURCHFIEL, B. C., LIPMAN, P. W. & ZOBACK, M. L. (eds) *The Cordilleran Orogen: Conterminus US*. Geological Society of America, Boulder, Colorado, 553–581.

WILGUS, C. K., HASTINGS, B. S., KENDALL, C. G. ST C., POSAMENTIER, H. W., ROSS, C. A. & VAN WAGONER, J. C. 1988. *Sea-level Changes: An Integrated Approach*. Society of Economic Paleontologists and Mineralogists, Special Publication, **42**.

WILLIAMS, G. D. & DOBB, A. (eds) 1993. *Tectonics and Seismic Sequence Stratigraphy*. Geological Society, London, Special Publication, **71**, 226.

WILSON, M. & LYASHKEVICH, Z. M. 1996. Magmatism and the geodynamics of rifting of the Pripyat–Dnieper–Donets rift, East European Platform. *Tectonophysics*, **268**, 65–81.

ZALAN, P. V., WOLFF, S. *et al.* 1990. The Parana Basin, Brazil. *In:* LEIGHTON, M. W. & KOLATA, D. R., OLTZ, D. F. & EIDEL, J. J. (eds) *Interior cratonic basins*. AAPG Memoir, **51**, 681–708.

ZUFFA, G. G. (ed.) 1985. *Provenance of Arenites*. D. Riedel Publishing Co., Dordrecht, The Netherlands.

Middle Jurassic to Cretaceous extensional tectonics and sedimentation in the eastern external zone of the Betic Cordillera

E. FERNÁNDEZ-FERNÁNDEZ, A. JABALOY & F. GONZÁLEZ-LODEIRO

Departamento di Geodinamica, Universidad de Granada, Campus Fuentenueva s/n, 18071 Granada, Spain (e-mail: emferfer@inicia.es)

Abstract: In the External Zones of the eastern Betic Cordillera, two sets of Mesozoic high-angle normal faults can be observed, one with ENE–WSW strikes and the other with N–S strikes. Both sets of faults generate half-grabens and grabens, infilled with wedge-shaped and lens-shaped formations deposited during the Late Jurassic to Early Cretaceous. The relationships of these formations indicate progressive tilting of the hanging walls during deposition of the rocks. The largest basins are related to the ENE–WSW faults.

The rocks of Middle Jurassic age, which predate the faulting stage, are shallow-marine oolitic limestones. The Lower Cretaceous Fardes Formation shows evidence of deposits closer to the carbonate compensation depth (CCD). This evidence indicates that normal faulting was related to very significant thinning of the continental crust.

Palaeomagnetic studies in the area demonstrate the existence during the Miocene of clockwise and counterclockwise rotations with vertical axes. Restoring the faults to their original orientation, the present-day ENE–WSW faults and their main basins had original N–S strikes, while the N–S faults originally had WNW–ESE strikes.

This extensional stage occurred at the same time as the rifting of Iberia and North America, the opening of the Gulf of Biscay and the aborted rifting of the Iberian chain.

In the External Zones of the orogens it is possible to determine the geometry and kinematics of the early synsedimentary extensional events if later compressional deformations are taken into account. However, the existence of vertical rotations poses a special problem in the reconstruction of the original orientation of the older structures. In this study we present an example of these early structures, from the External Zones of the Betic Cordilleras, which show deformation by compressional events and vertical axis rotations.

In the western Mediterranean area, there are several Alpine mountain chains, for example, the Betic Cordillera, Rif, Kabylias, Pyrenees, etc. Wide basins with an oceanic crust basement, such as the Provençal Basin, the Algerian Basin and the Tyrrhenian Sea, or a thinned continental crustal basement, such as the Alboran Sea, separate most of these mountain chains from one another. All these mountain chains and basins define a wide band with an approximate east–west trend that accommodates the convergence between Europe and Africa.

The Betic Cordillera has an ENE–WSW trend and is located in the south and southeast of the Iberian Peninsula. Fallot (1948) grouped the rocks of the Betic Cordillera into the External Zones and Internal Zones (Fig. 1). The External Zones are essentially formed of Mesozoic and Cenozoic sedimentary rocks that include several bodies of basic igneous rocks. The Internal Zones comprise several tectonic units, and most of the rocks belonging to these units have undergone Alpine metamorphism. The External Zones comprise sedimentary rocks that were originally deposited close to the Iberian margin, while the Internal Zones rocks were deposited far from the Iberian margin in a location that most authors locate somewhere in the northwestern Mediterranean.

Most of the studies concerning the structural geology of the Betic Cordillera have focused on the structure of the Internal Zone rocks, and only a very few studies have addressed the question of the structure and kinematics of the rocks of the External Zones during their deposition and subsequent orogenesis. However,

From: McCANN, T. & SAINTOT, A. (eds) *Tracing Tectonic Deformation Using the Sedimentary Record*. Geological Society, London, Special Publications, **208**, 29–53. 0305-8719/03/$15.00

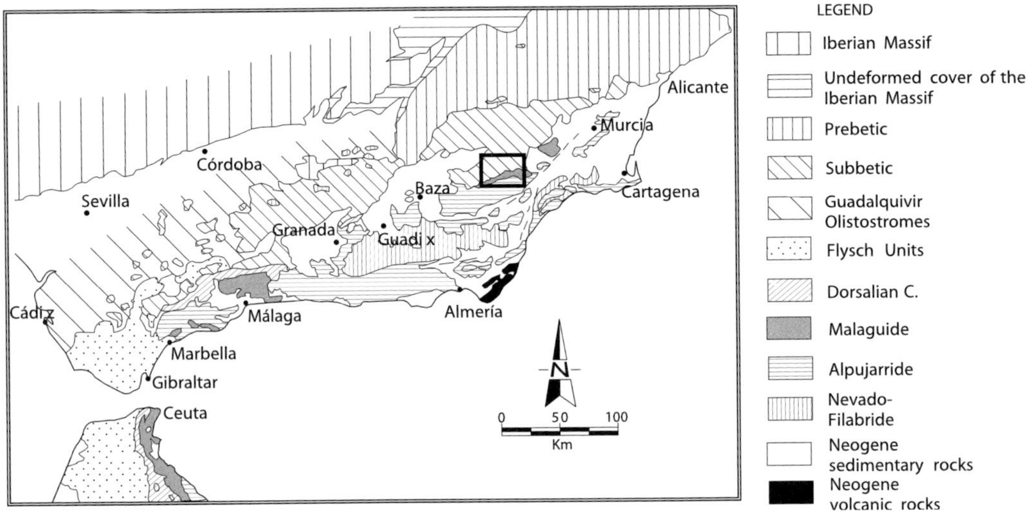

Fig. 1. Geological map of the Betic Cordillera. The small rectangle marks the location of the study area.

extensive stratigraphic and palaeontological studies of the External Zones were undertaken during the twentieth century, and the integration of those data with structural and tectonic information continues to be necessary. Bodies of External Zone rocks do not usually possess tabular geometries and display quite significant thickness variations. The rocks normally show frequent, abrupt lateral changes of facies in small areas. In several regions these features can be observed to be associated with tectonic structures that indicate the relationships between them.

This study analyses the early deformational structures and their relationships with the deposition of the rocks of the External Zones in the eastern sector of the Betic Cordillera during the Mesozoic and Early Cenozoic. The principal aim is to determine the main stages in the tectonic evolution of the Iberian margin during this period and to try to relate them to the relative motion of the Iberian and African plates.

Geological setting

The External Zones crop out north of the Internal Betic Zones in an ENE–WSW-trending belt (Fig. 1). Azema *et al.* (1979) and García-Hernández *et al.* (1980) have divided it into two major palaeogeographical units on the basis on differing Jurassic facies. These units are the Prebetic Zone (Blumenthal 1927) and the Subbetic Zone (Fallot 1945). The latter is separated from the former by the palaeogeographical domain of the Intermediate Units (Foucault 1960, 1962).

The Triassic successions of the Prebetic and Subbetic are very similar and comprise rocks with Germanic facies (i.e. Muschelkalk and Keuper facies), indicating that the depositional conditions were uniform over the ancient basin that developed in the southeastern part of the Iberian Massif (Pérez-López 1991). At the end of the Triassic, a wide, shallow carbonate shelf formed throughout the External Zones (García-Hernández *et al.* 1980). Differentiation of the Prebetic and Subbetic Zones began during the Early Pliensbachian, when the carbonate shelf fragmented due to a period of extension (García-Hernández *et al.* 1980) related to Early Jurassic rifting (García-Hernández *et al.* 1989). From this time on, the Prebetic Zone was characterized by shallow-marine paralic deposits during the Mesozoic, including several coastal and continental episodes. The most important continental episode corresponds with the deposition during the Albian, in nearly the entire Prebetic Zone, of the sandstones of the Utrillas Formation. The Prebetic Zone was separated from the Subbetic Zone by a subsiding trough known as the Intermediate Units (Azema *et al.* 1979; Ruiz-Ortiz 1980), which was infilled by a thick succession of alternating shallow-marine and pelagic facies during the Jurassic, overlain by 1 to 2.5 km of pelagic marls and terrigenous turbidites of Early Cretaceous age.

In contrast, pelagic marine facies are common in the Jurassic rocks of the Subbetic Zone. In the central Betic Cordillera, this Subbetic Zone is subdivided into several palaeogeographical domains (although these are not as evident in other sectors of the cordillera). These domains are the External, Middle and Internal Subbetic (García-Dueñas 1967; Azema *et al.* 1979), named after their relative positions with respect to the emerged continent. The old carbonate shelf that existed prior to the Early Pliensbachian evolved during the rest of the Jurassic into a central subsiding basin (Middle Subbetic) located between two swells (External and Internal Subbetic). In the central part of the Middle Subbetic basin, both volcanic and intrusive basic rocks accumulated during the Jurassic–Early Cretaceous (Morata Céspedes 1993).

All previous works indicate that, during the Cretaceous and Cenozoic, the evolution of the Prebetic continued under similar conditions to those in the Jurassic, while in the Subbetic the differentiation into the External, Middle and Internal Subbetics disappeared, resulting in an essentially flat pelagic basin (Vera 1986). In the Subbetic, the Early Jurassic rifting stage ended during the Middle and Late Jurassic, and younger rocks belong to the post-rift stage (García-Hernández *et al.* 1989).

The External Zones of the cordillera were deformed by compression, mainly from the Early Miocene (Burdigalian) to the Late Miocene (Middle Tortonian) (Lonergan 1991; Kirker & Platt 1999; Galindo-Zaldívar *et al.* 2000; Crespo-Blanc & Campos 2001), producing a fold-and-thrust belt with two deformational fronts. The NNW front faces towards the foreland (Iberian Massif), while the SSE front faces towards the hinterland (Internal Zones). Associated with this compressive deformation, variable rotations with vertical axes took place (i.e. Osete *et al.* 1988, 1989; Villalain *et al.* 1992; Allerton *et al.* 1993; Platzman 1992, 1994).

The study area is located in the eastern Betic Cordillera (Fig. 1) and comprises essentially rocks belonging to the Internal Subbetic. The main structure is the basal thrust of the deformational front facing towards the hinterland, which superposes Subbetic rocks over the Internal Zones. This thrust became inactive in the Late Burdigalian (Lonergan 1993). The internal structure of the Subbetic in the study area corresponds with a basal unit (La Muela Unit), showing a succession from Lower Jurassic to Lower Miocene rocks, and an upper unit (the Maimon Unit), cropping out in a tectonic klippe in the southwestern study area and comprising a succession composed of Lower Jurassic to Palaeogene rocks (Fig. 4). This work will focus on the relationship between sedimentation and tectonics in the La Muela Unit, which was selected due to the quality of its outcrops.

The rocks of the La Muela Unit are folded, and the fold trends have a curved shape in plan view, varying from NNE–SSW trends in the northeastern part of the study region to ENE–WSW trends in the western part. This curved pattern is a result of Neogene compression, as suggested by the palaeomagnetic study of Allerton *et al.* (1993). This study indicates that the greater part of the study area underwent vertical axis rotations.

Out of the different palaeomagnetic declinations determined by Allerton *et al.* (1993), there are six declinations located within the study area. Five of them are within the La Muela Unit, while one is within the Maimon Unit. Out of the five declinations located in the La Muela Unit, three are located in the Sierra del Pericay (Fig. 4), in a region where the main structures have a NNE–SSW trend. They indicate counterclockwise rotations of $-12° \pm 13.8°$, $-8° \pm 7.4°$, $-8° \pm 12.9°$. The other two determinations were obtained on the northern border of the Sierra Larga (Fig. 4), where the structures have an ENE–WSW trend, and indicate, respectively, clockwise rotations of: $64° \pm 9.8°$ and $80° \pm 5.8°$. In addition, Platt (pers. comm.) has obtained several palaeomagnetic declinations from the study area – from the same La Muela Unit. These declinations are distributed throughout the area, and there is one from the Sierra del Pericay that indicates counterclockwise rotations of $-15° \pm 5°$ (Fig. 4). Others located in Sierra Larga indicate a clockwise rotation of $62° \pm 7.3°$, and another in the Gabar shows an average clockwise rotation of $63° \pm 5.5°$. All these palaeomagnetic declinations have been determined in the Middle to Upper Jurassic Upper Ammonitico Rosso Formation, and the age of the rotations is supposed to be Neogene (Allerton *et al.* 1993).

Rock stratigraphy

The stratigraphic nomenclature used in this work derives mainly from the work of Rey (1993), who provided a review of previous stratigraphic studies in the area and a correlation with other areas of the cordillera, producing a clarifying study of the previous local names.

The Subbetic rock succession in the study area (Fig. 2) begins with Sinemurian to Lower Pliensbachian limestones and dolostones of the Gavilán Formation (Van Veen 1966; Rey 1993).

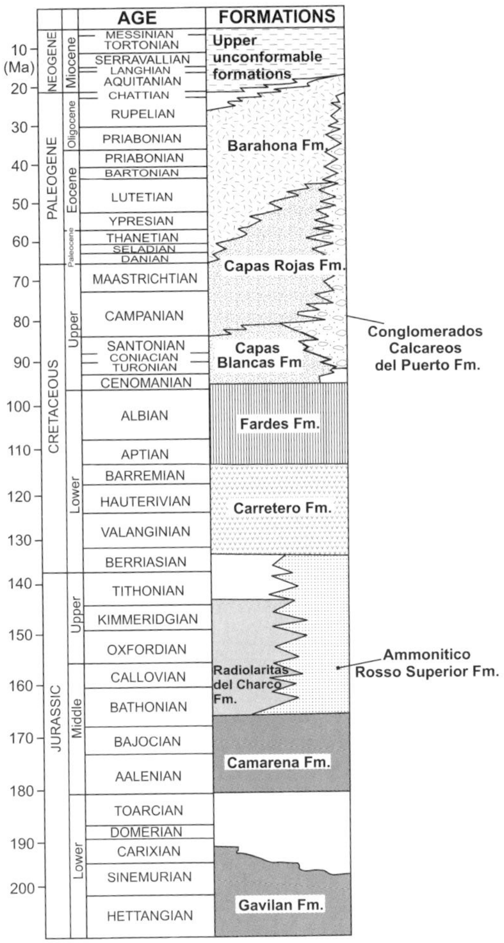

Fig. 2. Stratigraphic column for the study area.

The thickness of this formation varies from 340 to 640 metres and all the rocks show shallow-marine facies. The upper part of this formation is sometimes lacking and there are some neptunian dykes present; these dykes are 1 m wide and are infilled with rocks younger than the Early Jurassic. These circumstances suggest that, at the end of the Early Pliensbachian, the shelf rose and may have been eroded in subaerial conditions (Rey 1993).

Overlying this formation is the white oolitic carbonate of the Camarena Formation (Molina 1987), which is equivalent to the upper part of the Maimón Formation of Geel (1973), interpreted as having been deposited on a shallow-marine carbonate shelf during the Middle Jurassic (Rey 1993). The top of this formation marks the transition from shallow-marine conditions towards more pelagic deposits.

The base of the following formation, the Ammonitico Rosso Superior Formation (Molina 1987; herein Ammonitico Rosso Formation), is another paraconformity. The formation is composed of red nodular limestones (hence the name 'Ammonitico Rosso' facies). These rocks are typical pelagic fossiliferous limestones and contain Callovian to Tithonian fossils (Rey 1993), although Baena *et al.* (1977) propose a Late Jurassic age – from Kimmeridgian to Tithonian. This formation evolves by a lateral change of facies into the Radiolaritas del Charco Formation (Rey 1993). The latter comprises chert-bearing limestones, green limestones, marly limestones and marls rich in radiolaria, with facies that are typical of an open pelagic environment, in stark contrast to the oolitic limestones of the Camarena Formation. Its thickness is extremely variable and can reach a maximum of 100 m. The biostratigraphic data of Rey (1993) indicate an age ranging from Bajocian to Middle Bathonian at the base of the formation, and from Late Callovian to Early Oxfordian in the upper part. However, throughout the study area the base of the Radiolaritas del Charco Formation onlaps both the Ammonitico Rosso and Camarena formations and is erosional in places. These data suggest that the base of the Ammonitico Rosso Formation may be older than the Callovian, and that it is probably Bajocian–Bathonian in age.

Overlying the rocks described above are white marls and marly limestones containing pyrite and chert, known as the Carretero Formation, which can reach 50 m in thickness (Vera *et al.* 1982). The formation is Late Berriasian to Late Barremian in age (Aguado *et al.* 1991) and contains slumps in its upper part. The fauna indicates pelagic deposition in a deep-marine setting (Vera *et al.* 1982), and the presence of slumps indicates that the basin was adjacent to steep unstable slopes.

The next formation, the Fardes Formation (Comas, 1978), comprises dark-green marls, marls, marly clays and clays that include several turbiditic layers. The succession begins with dark-green marls containing radiolaria, and these marls have occasional intercalated layers made up of calcirudites and calcarenites. Above this succession there are marls, marly clays and clays alternating with turbiditic layers composed mainly of oolites from the Camarena Formation. (Fig. 3) Several of these turbiditic layers are extremely thick, such as the Megabed of Rambla Seca (Aguado *et al.* 1991) (Fig. 5). The age of this formation ranges from the Late Barremian to the Early Cenomanian (Aguado *et al.* 1991).

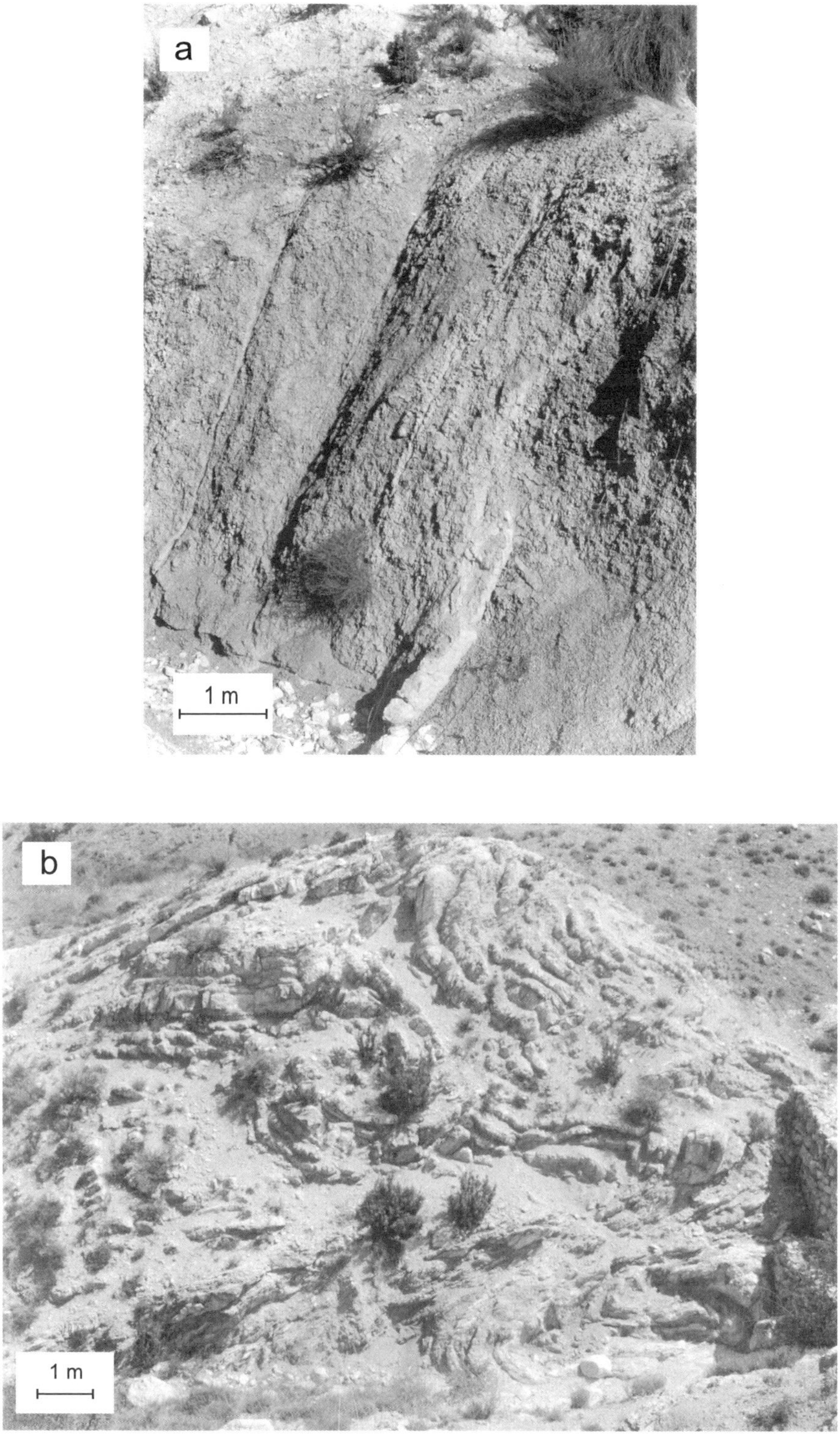

Fig. 3. (**a**) Detail of the expansive clays of the Lower Cretaceous 'Fardes' Formation with a turbiditic layer. (**b**) Slumps in the Upper Cretaceous–Lower Eocene Capas Rojas Formation.

The clay layers have virtually no carbonate and contain smectite, illite, palygorskite and kaolinite (López-Galindo 1986). They alternate with marl and oolitic turbiditic layers where the carbonate is conserved. These features have been interpreted by López-Galindo (1986) and Reicherter (1994) as being produced by the deposition of the formation near the carbonate compensation depth (CCD). During the Albian–Cenomanian, the CCD in the Central North Atlantic was located from 3200 to 3500 m in depth (Van Andel 1975).

The Fardes Formation is overlain by the Conglomerados Calcáreos del Puerto Formation (Rey 1993), which is laterally equivalent to the Capas Blancas Formation. The former is composed of marls and marly limestones that include reworked oolites from the Camarena Formation (Fig. 3), with abundant olistostromes and slumps. Most of the olistostromes were derived from the Gavilán and Camarena Formations. Locally, there are several layers of turbidites. The rocks of the Conglomerados Calcáreos del Puerto Formation were probably deposited close to palaeorelief in which most of the older formations crop out. The lower part of this formation is Late Turonian–Early Coniacian in age (Rey 1993), while the top has a Late Santonian age (Aguado *et al.* 1991) in most of the study area, although in the Arroyo de Taibena Basin it can reach a Palaeogene age.

The Conglomerados Calcáreos del Puerto Formation passes laterally into the Capas Blancas Formation (Martín Algarra 1987). The latter comprises white marly limestones alternating with white marls and is characteristic of a pelagic deep-marine setting. It has been assigned different ages: an Early Cenomanian to Late Santonian age by Rey (1993) and a Cenomanian to Turonian and possibly younger age by Allerton *et al.* (1994) and Reicherter (1994).

The Capas Rojas Formation ('Red Beds' Formation.) (Vera *et al.* 1982) overlies all the Middle Jurassic to Lower Cretaceous forma-

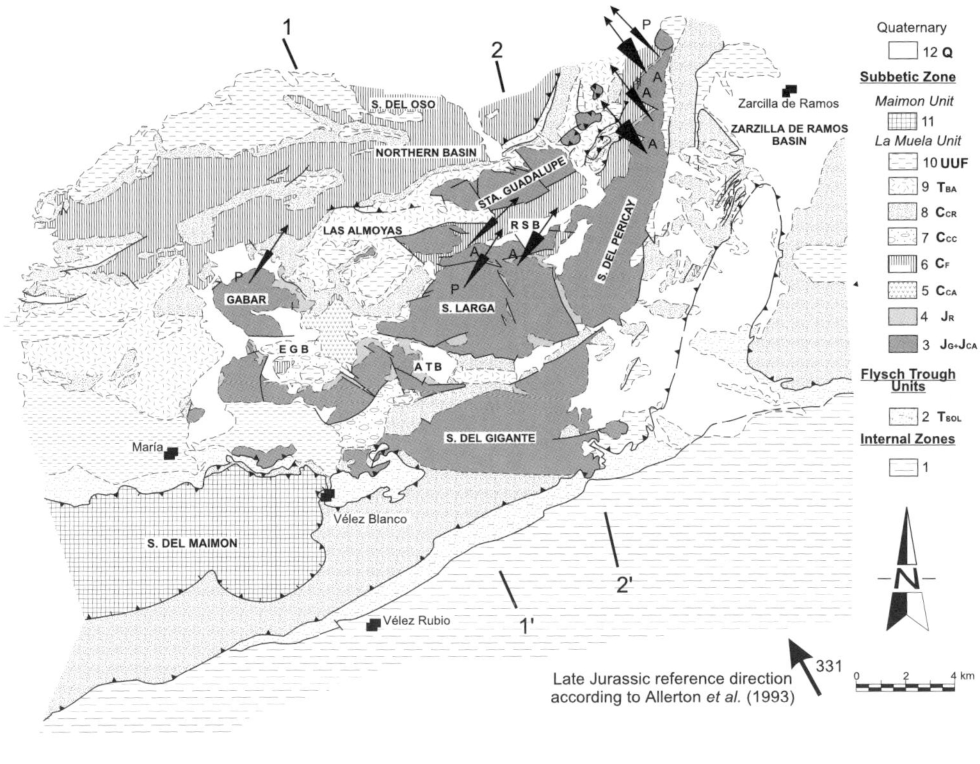

Fig. 4. Simplified geological map of the study area with the palaeomagnetic declinations determined by Allerton *et al.* (1993) [A] and Platt (pers. comm.) [B]. Legend: 1, Internal Zones; 2, Flysch Trough Units; 3, Gavilán and Camarena Formations; 4, Ammonitico Rosso and Radiolaritas del Charco Formations; 5, Carretero Formation; 6, Fardes Formation; 7, Conglomerados Calcáreos del Puerto Formation; 8, Capas Blancas and Capas Rojas Formation; 9, Barahona Formation; 10, Upper Unconformable Formations; 11, Maimon Unit; 12, Quaternary. RSB, Rambla Seca Basin; EGB, Eastern Gabar Basin; ATB, Arroyo de Taibena Basin.

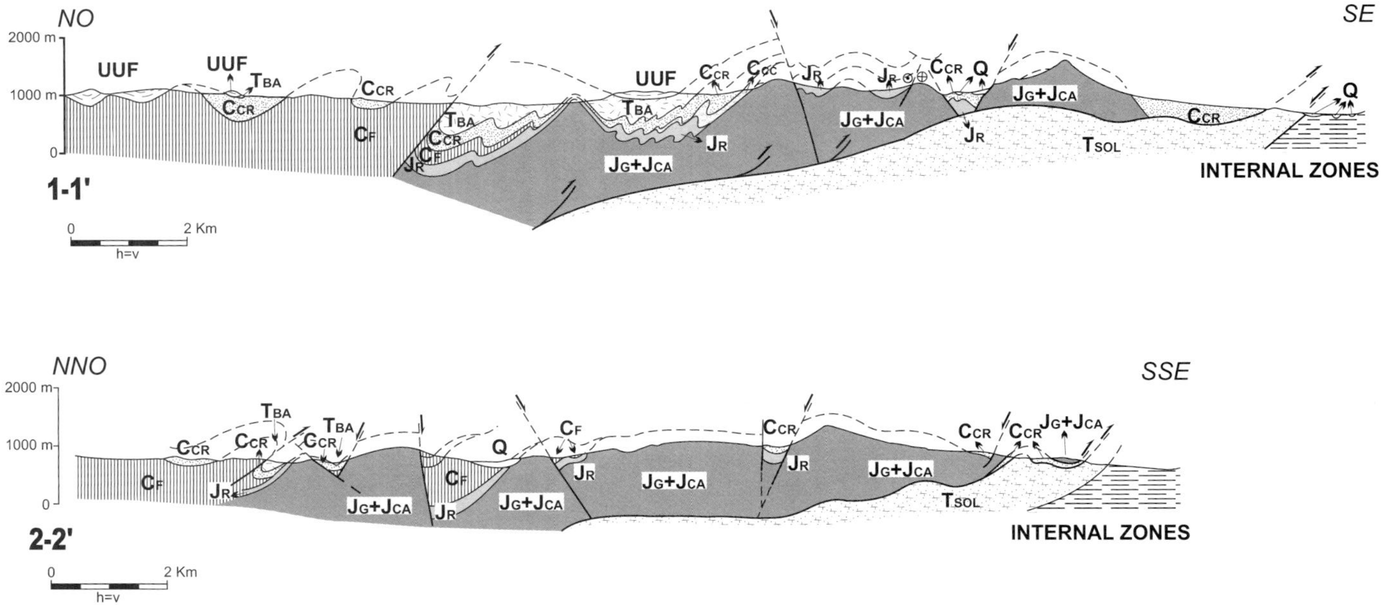

Fig. 5. NE–SW cross-sections of the study area showing the thickness variations associated to normal faults cut by the thrust surface. See the location in Figure 4. The horizontal scale is the same as the vertical one.

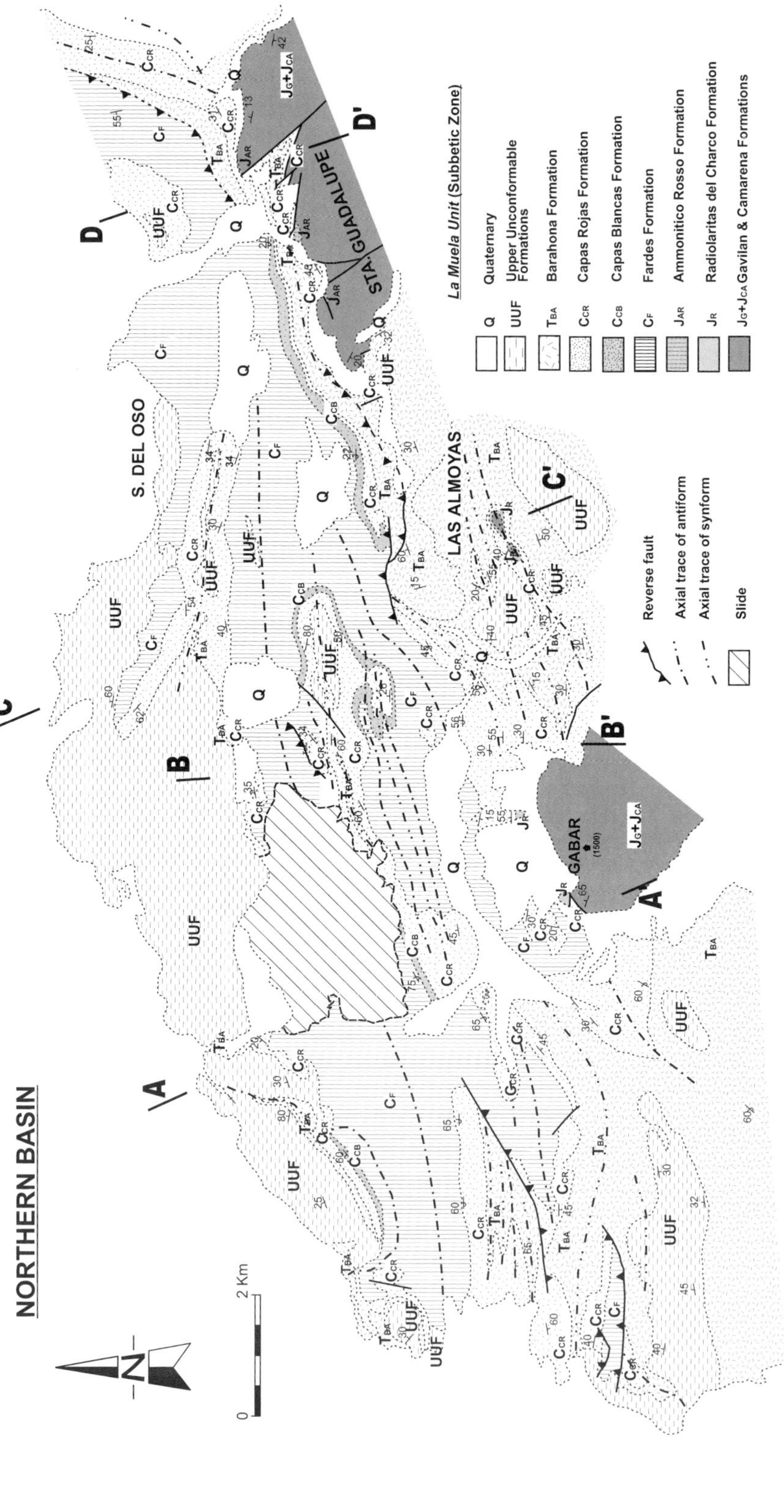

Fig. 6. Geological map of the Northern Basin.

tions (Allerton *et al.* 1994) due to their discontinuous character. This formation consists of red and pink limestones and marly limestones with some turbidite intercalations and slumps (Fig. 3). It was deposited in an environment similar to that of the preceding formations, i.e. deep marine with strong topographic relief and unstable slopes. The base and top of this formation are diachronous. The base has ages ranging from Late Santonian–Campanian to the Early Cenomanian or even Maastrichtian. The top ranges in age from the Maastrichtian (Baena *et al.* 1977; Allerton *et al.* 1994; Reicherter 1994; Martínez-Pérez pers. comm.) to the Early Eocene (Baena *et al.* 1977; Martínez-Pérez pers. comm.).

The Barahona Formation (Wittink 1975) is composed of green marls and marly limestones that alternate with turbiditic layers. The upper part of the succession has thick layers of brown sandstones. The overall thickness of the formation can reach 500 m. It contains fossils typical of shallow-marine environments (Wittink 1975), but the presence of turbiditic layers indicates a deeper marine setting. The base ranges in age from the Maastrichtian to the Early Eocene, while the top of the formation is Aquitanian in age (Martín Pérez pers. comm.).

Tectonics and sedimentation

The synsedimentary structures of the Mesozoic and Cenozoic rocks are normal faults and open joints. Those faults affecting only the Lower and Middle Jurassic rocks have small displacements, but those affecting the Middle Jurassic and Cretaceous rocks usually have important offsets and associated half-grabens.

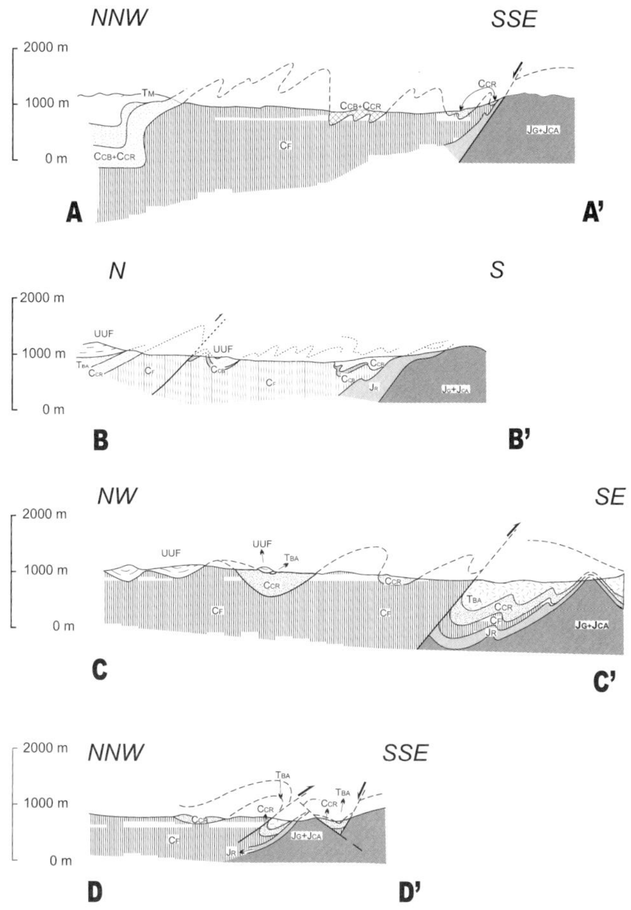

Fig. 7. Cross-sections of the Northern Basin. See location and legend in Figure 6. The horizontal scale is the same as the vertical one.

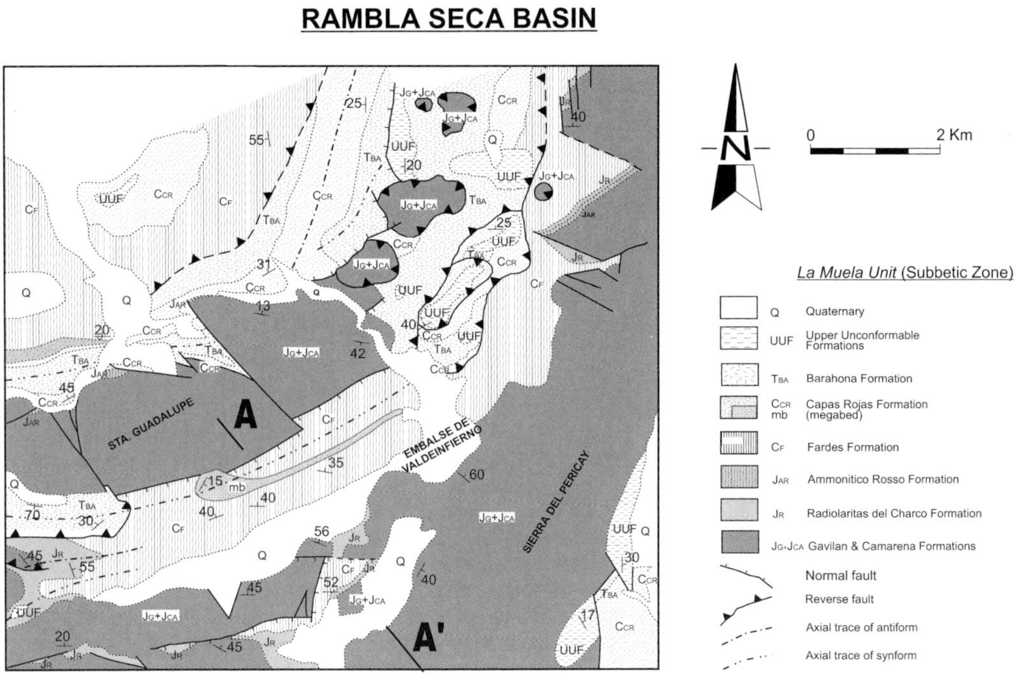

Fig. 8. Geological map of the Rambla Seca Basin. Legend: dashed lines are old normal fault surfaces. Dotted lines are the boundaries of the megabed.

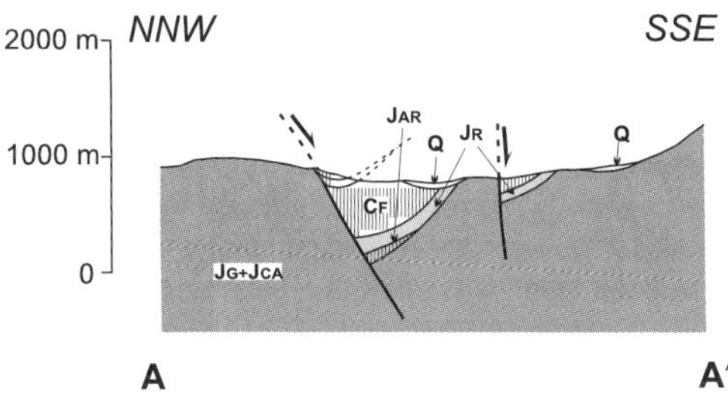

Fig. 9. Cross-section of the Rambla Seca Basin. See location and legend in Figure 8. The horizontal scale is the same as the vertical one.

The Lower to Middle Jurassic formations are cut by normal faults that have caused the tilting of the beds and produced half-grabens and grabens (Figs 4 and 5). The shape of formations younger than the Middle Jurassic can be wedge-shaped, lens-shaped, or tabular. The upper formations are usually more extensive than the lower ones.

There are five main basins: (1) the Northern Basin, located in the region north of the Serrata de Guadalupe and Gabar and at the western end of the study area (Fig. 4). In the central region, there are two basins alienated in a NE–SW trend: (2) The Rambla Seca Basin in the east and (3) the Eastern Gabar Basin in the central and western part (Fig. 4). In the southwest of the study area, there is (4) the Arroyo de Taibena Basin, while (5) the Zarzilla de Ramos Basin (Fig. 4) is located in the eastern part of the area.

The Northern Basin

The Northern Basin is the widest basin in the area and is bounded in the south by two ridges made up of Middle Jurassic limestones of the Camarena Formation: the Serrata de Guadalupe and Gabar (Fig. 6). The northern border of the basin cannot be observed in the study area. The sedimentary succession of this basin is characterized by the absence of the Lower Cretaceous to Palaeogene Carretero and Conglomerados Calcáreos del Puerto Formations, while the Lower Cretaceous to Palaeogene Fardes, Capas Blancas and Capas Rojas Formations are well represented. These three formations show very large variations in thickness (Fig. 7).

The Middle and Upper Jurassic formations (Radiolaritas del Charco and Ammonitico Rosso) can only be observed in the southern border of the area. The Radiolaritas del Charco Formation crops out in the Gabar area, while the Ammonitico Rosso Formation crops out towards the east in the Serrata de Guadalupe area. These formations are not superposed in any part of this basin and they seem to be related by a lateral change of facies. They decrease in thickness towards the south and disappear in the Serrata de Guadalupe region (Figs 6 & 7).

The Fardes Formation has a minimum thickness of 500 m in the north and disappears in the southern border of the basin (Figs 6 & 7). Another characteristic of this formation is the poor development of turbidites. The lack of this Lower Cretaceous Fardes Formation means the Capas Blancas Formation is in contact with the Middle and Upper Jurassic Radiolaritas del Charco Formation in the Gabar (Figs 6 & 7).

The Capas Blancas Formation crops out only in the Northern and Western Basins. In the Northern Basin it has a maximum thickness of 100 m in the central part of the basin, but disappears towards the south and north (Figs 6 & 7). The thinning of this formation towards the south allows the Capas Rojas Formation to overlie the Middle and Upper Jurassic Ammonitico Rosso and Radiolaritas del Charco Formations.

The thickness of the Capas Rojas Formation also increases towards the north (320 m in the northern part of the basin, Sierra del Oso, Figs 6 & 7), whereas it is around 100 m thick on the southern border. Only the lower part of the Palaeogene Barahona Formation is well represented in this basin, with the top only being observed in one outcrop, which does not allow variations in thickness to be determined.

The aforementioned thickness variation can be gradual in the western part of the basin (cross-section A–A′ and B–B′, Fig. 7), but is abrupt in the central and eastern parts (cross-sections C–C′ and D–D′, Fig. 7). In the western part, the Capas Rojas Formation onlaps the Radiolaritas del Charco and the Fardes Formations; the latter formation is wedge shaped and thins southwards until it disappears (cross-section A–A′ and B–B′, Fig. 7). In the central and eastern part of the basin a similar pattern of thinning can be observed near the outcrops of the Middle Jurassic Camarena Formation. The formations thin southwards without disappearing in the central part (cross-sections C–C′, Fig. 7), while the Capas Rojas Formation onlaps all the above formations in the eastern part (cross-sections D–D′, Fig. 7). However, north of this fan-shaped onlap, two reverse faults with a relay pattern can be observed (Figs 6 & 7). The western one has an East–West strike and dips around

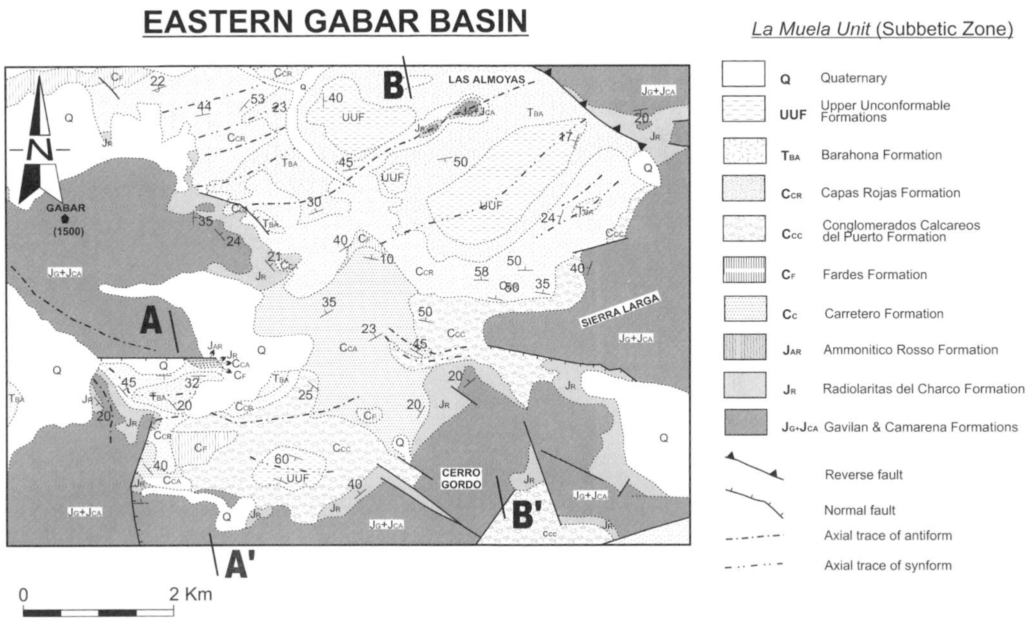

Fig. 10. Geological map of the Eastern Gabar Basin.

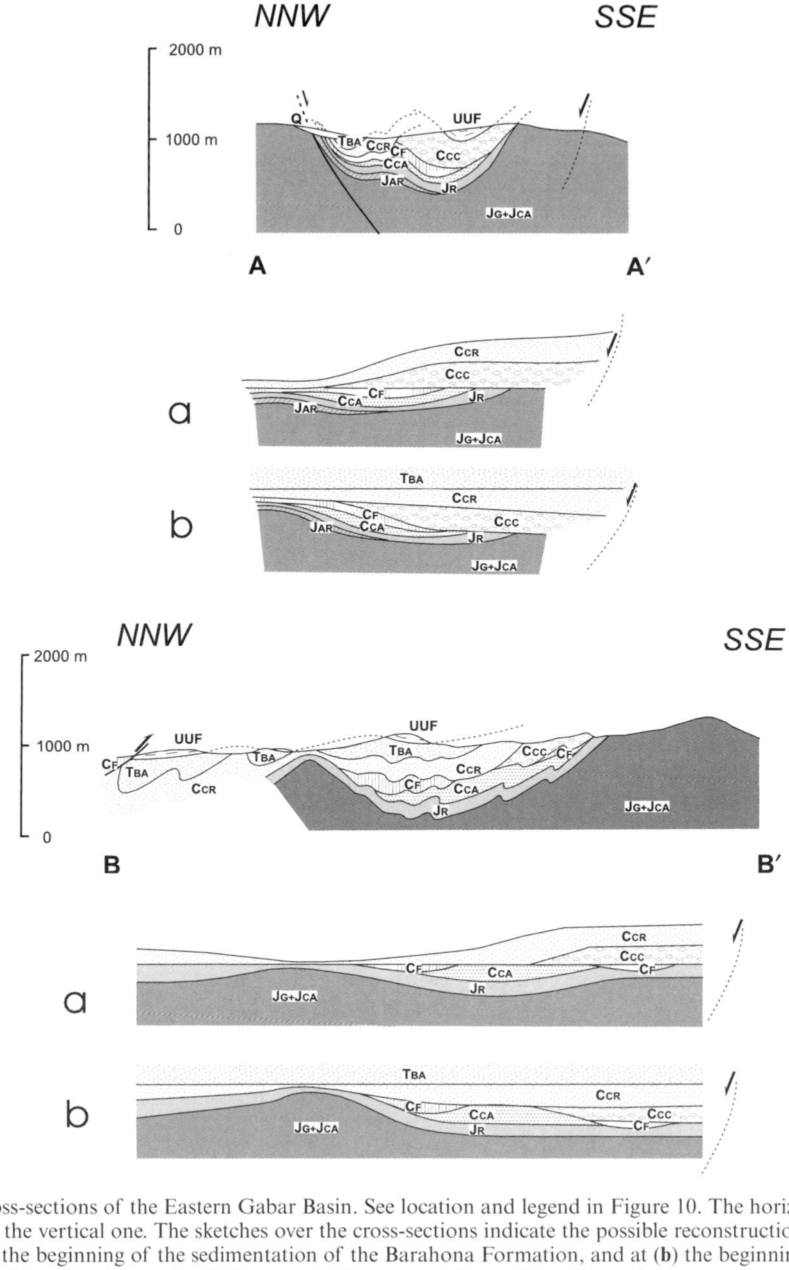

Fig. 11. Cross-sections of the Eastern Gabar Basin. See location and legend in Figure 10. The horizontal scale is the same as the vertical one. The sketches over the cross-sections indicate the possible reconstruction of the basin at (**a**) the beginning of the sedimentation of the Barahona Formation, and at (**b**) the beginning of the deposition of the Conglomerados Calácareos del Puerto and Capas Rojas Formations.

60° towards the north; the eastern one has a very poor outcrop and only the trace on the topography can be determined. In cross-sections C–C′ and D–D′ (Fig. 7), the abrupt thickness variations of the Fardes Formation in both walls of the fault surfaces can be clearly seen.

The relationships between the different formations suggest that the Lower Cretaceous Carretero and Fardes Formations have never overlain the Gabar and the Serrata de Guadalupe. Between these two aforementioned uplands, there is a relative high formed by the limestones of the Camarena Formation, which is capped by the formations of the Cretaceous and Palaeogene, which thin near the high and define an open antiform (Fig. 7).

The Rambla Seca Basin

The Rambla Seca Basin has a curved in plan view that varies from east to west, from a N10°E to a N60°E trend (Fig. 8). This basin is bounded to the west and north by the relief of the Serrata de Guadalupe, and to the east and south by the ridges of the Sierra Larga and the Sierra del Pericay (Fig. 8). The limestones of the Middle Jurassic Camarena Formation form all of these ranges. The basin is an asymmetrical half-graben bounded by a high-angle normal fault to the west and north, while its southern and eastern boundaries comprise unconformities th basin fill overlying the Middle Jurassic Camarena Formation (Fig. 9).

The Rambla Seca Basin is infilled with rocks from the Middle to Upper Jurassic Ammonitico Rosso to the Palaeogene Barahona Formation, reaching a thickness greater than 1200 m (Figs 8 & 9). There are also outcrops of the upper unconformable formations, which are Early Miocene in age. There is no evidence in this basin of deposits from the Lower Cretaceous Carretero Formation or from the Upper Cretaceous to Palaeogene Capas Blancas and Conglomerados Calcáreos del Puerto Formations.

The Middle to Upper Jurassic Ammonitico Rosso Formation crops out in the eastern and southern borders of the basin. The formation is one to two metres thick, is always observed to overlie the Middle Jurassic Camarena Formation, and is capped unconformably by the Middle to Upper Jurassic Radiolaritas del Charco Formation. The Radiolaritas del Charco Formation onlaps the Ammonitico Rosso rocks, and, on the southern and eastern borders of the basin it directly overlies the oolitic limestones of the Camarena Formation (Figs 8 & 9). The Middle to Upper Jurassic Radiolaritas del Charco Formation thins from west to east and from north to south. In the west the formation is around 70 metres thick, while in the east the thickness is reduced to two metres.

Due to the absence of the Carretero Formation in this basin, the Lower Cretaceous Fardes Formation directly overlies the Radiolaritas del Charco Formation. The base of the Fardes Formation can be observed clearly, but the roof is absent due to several reverse faults. The minimum thickness of this formation is 1000 metres. The rocks of the Fardes Formation in this basin include the best examples of turbidites in this region, including the megabed of Rambla Seca, made up of reworked oolites from the Camarena Formation (Aguado *et al.* 1991) (Figs 8 & 9). The bedding in the Fardes Formation defines an open synform with a narrow northern limb and immersion towards the east.

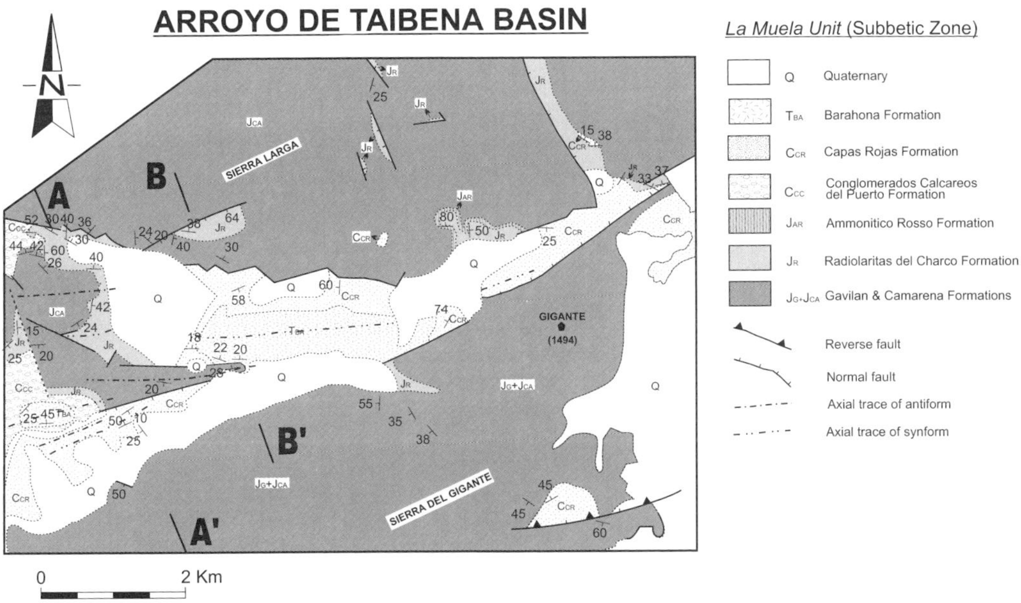

Fig. 12. Geological map of the Arroyo de Taibena Basin.

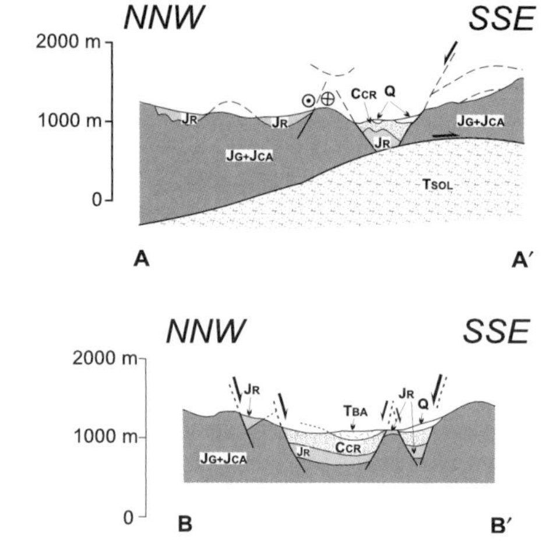

Fig. 13. Cross-sections of the Arroyo de Taibena Basin. See location and legend in Figure 12. The horizontal scale is the same as the vertical one.

The upper part of the succession in this basin, from the Capas Rojas Formation upwards, is detached from the Fardes Formation. Several reverse faults that crop out in the western and northern extremes of the basin are responsible for the detachment. The reverse faults are folded by the open synform and, westwards, cut higher formations in their hanging walls (Figs 8 & 9). The Capas Rojas Formation has a minimum thickness of 200 m and crops out in the hanging wall of these faults, where there is no evidence of the Capas Blancas and the Conglomerados Calcáreos del Puerto rocks.

The Eastern Gabar Basin

The Eastern Gabar Basin has a J shape in plan view. The mountain of Gabar and the antiform of Las Almoyas bound this basin to the north and east (Fig. 10). Towards the south, the basin is bounded by the highs of the Camarena

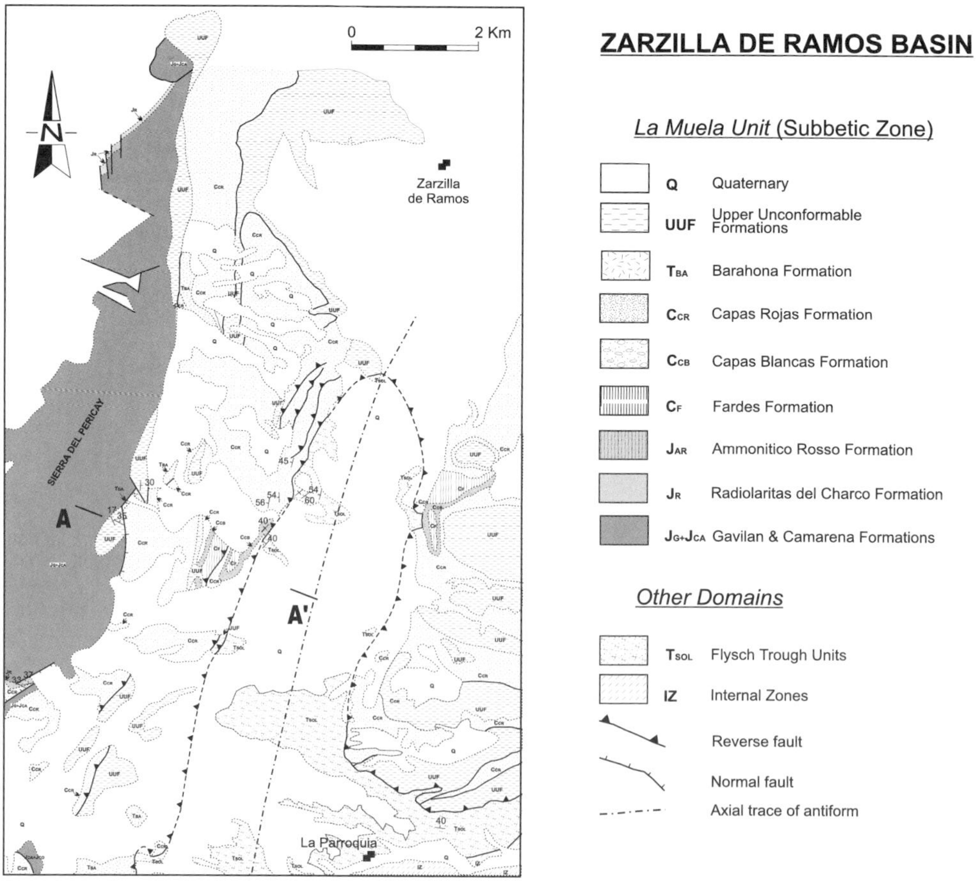

Fig. 14. Geological map of the Zarzilla de Ramos Basin.

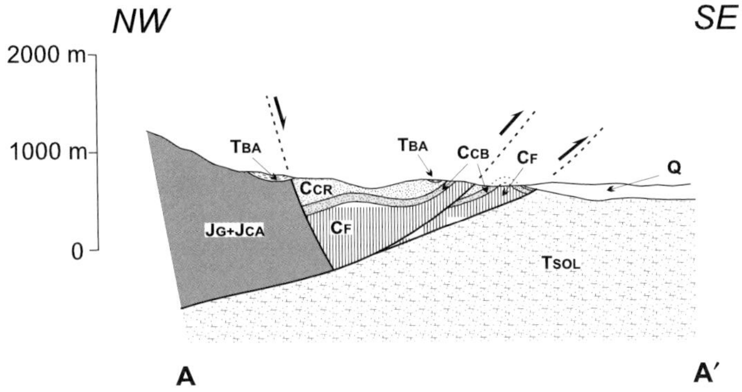

Fig. 15. Cross-section of the Zarzilla de Ramos Basin. See location and legend in Figure 14. The horizontal scale is the same as the vertical one.

Formation, which form the Sierra Larga and Cerro Gordo (Fig. 10). In the southwestern part of this basin, there is an E–W high-angle normal fault dipping towards the south. Towards the east, this normal fault cannot be recognized and only several unconformities can be seen (Fig. 10).

The basin is filled with the entire succession of rocks from the Ammonitico Rosso to the Barahona Formations (Figs 10 & 11). The formations are lens-shaped and show frequent thickness variations and omissions.

The Ammonitico Rosso rocks only crop out south of Gabar, where they are one metre thick (Figs 10 & 11), and they are absent south and eastwards. The Middle–Upper Jurassic Radiolaritas del Charco Formation is one of the few continuous formations in this basin, and can be recognised in all its borders as a rim one to 100 metres thick. The greatest thickness occurs on the southeastern border of the basin.

The Lower Cretaceous Carretero Formation outcrop in the centre of the basin is lens shaped, with a maximum thickness of 200 metres in the centre of the outcrop, and is omitted in the upper formations towards the borders of the basin (Figs 10 & 11). The Lower Cretaceous Fardes Formation can be observed in three small outcrops, indicating that the formation is discontinuous. The formation appears in small lens-shaped bodies with a maximum thickness of 10 metres. The discontinuous character of this outcrop is due to the unconformities that developed at the base of the upper formations (Figs 10 & 11). The Conglomerados Calcáreos del Puerto Formation is disposed unconformably over the above formations in the south of the basin with a total thickness of 600 metres. They are formed by olistostromes, mainly from the Camarena Formation. Towards the north, the facies of this formation changes laterally towards the facies of the Capas Rojas Formation. The Capas Rojas Rojas Formation has a maximum thickness of 250 m in the centre of the basin and thins northwards to 30 m (Figs 10 & 11). As a whole, the Upper Cretaceous to Palaeogene level composed of the Capas Rojas and Conglomerados del Puerto Formations thins from 600 m in the southern border of the basin to 30 m in the northern border. The Barahona Formation is well developed in the basin, with a minimum thickness of 500 m in the main synform (Figs 10 & 11). Its top is eroded by the upper unconformable formations.

The Arroyo de Taibena Basin

The Arroyo de Taibena Basin is a graben bounded by two conjugate normal faults. The northern fault has a mean N100°E strike and dips towards the south. The southern fault has a mean N70°E strike and dips towards the north. The width of the basin increases towards the west. In the western end, a small east–west horst has developed. The fill of the basin defines an open east–west synform (Figs 12 & 13). The basin is filled with Middle–Upper Jurassic to Palaeogene deposits, but there is no evidence of the rocks with Early Cretaceous ages: the Carretero and Fardes Formations. Neither is there any evidence of the Upper Cretaceous Capas Blancas Formation.

The Ammonitico Rosso Formation appears as a very thin succession in the northern border of the basin in the footwall of the northern normal fault. The Radiolaritas del Charco Formation can be observed in the footwalls of the northern and southern faults. Moreover, at

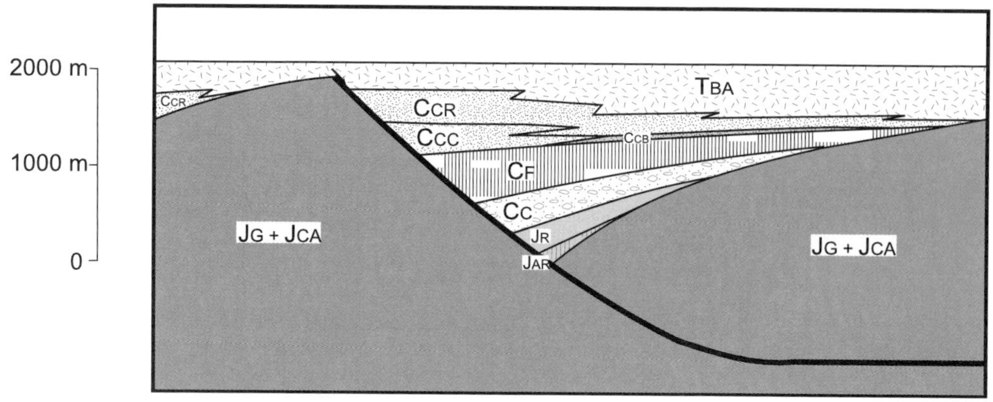

Fig. 16. Tectono-stratigraphic model of the relationships between the Lower Jurassic to Aquitanian rocks.

the western end of the basin the radiolarites can be observed directly overlying the oolitic limestones of the Middle Jurassic Camarena Formation (Fig. 12). The roof of this radiolarite formation cannot be observed in this basin, thus preventing an estimate of the total thickness.

As we pointed out above, there is no evidence of Lower Cretaceous formations and the Upper Cretaceous Capas Rojas Formation covers the Middle–Upper Jurassic rocks. The Capas Rojas Formation has a minimum thickness of 150 metres and is overlain by the Palaeogene Barahona Formation (Fig. 13). At the western end of the basin, the Capas Rojas rocks are laterally in contact with the Conglomerados Calcáreos del Puerto Formation (Fig. 12). This Conglomerados Calcáreos del Puerto Formation has a wedge shape and the Barahona Formation caps the distal part of the wedge. The dating by Aguado & Rey (1996) of these rocks shows that they have the same age as the Capas Rojas and Barahona Formations, suggesting that in this basin the conglomerates are a lateral facies variation of these Upper Cretaceous to Palaeogene formations.

The Zarzilla de Ramos Basin

The Zarzilla de Ramos Basin is located in the eastern part of the study area and is limited to the west by the ridges of the Sierra del Pericay, made up of Middle Jurassic oolitic limestones. The boundary coincides with a normal fault striking N10°E and dipping 60° towards the east (Figs 14 & 15). This fault is the cartographical prolongation of the east–west normal fault that constitutes the northern boundary of the Arroyo de Taibena Basin. The eastern boundary of this basin cannot be observed in the study area (Fig. 4).

In the hanging wall of the normal fault, the succession begins with the Lower Cretaceous Fardes Formation and there is no evidence of older formations. The succession continues upection with the Capas Blancas, Capas Rojas and Barahona Formations. The normal fault is cut by the basal thrust that superposed the Subbetic over the Internal Zones of the cordillera (Figs 14 & 15). The formations filling this basin show no significant thickness variation in the studied area. The Capas Blancas Formation has a thickness of 50 metres, while the Capas Rojas Formation have a thickness of around 300 metres (Figs 14 & 15).

Small secondary basins

In the east–west elongated range of the Sierra Larga (Fig. 12), there are several small normal

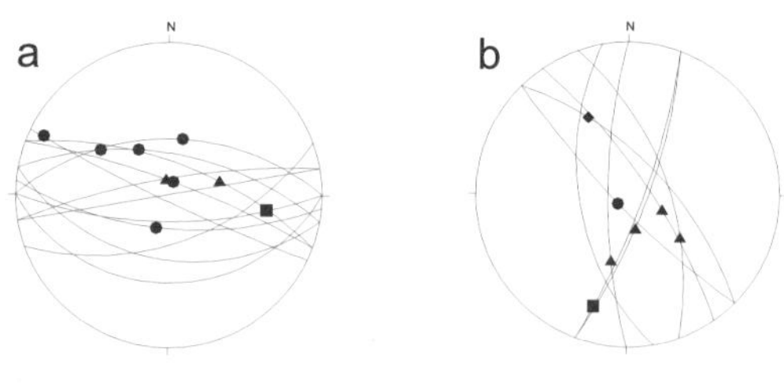

Fig. 17. Diagrams of normal fault orientations. (**a**) Orientation of the east–west set. (**b**) Orientation of the NNW–SSE set. Wulff stereonet, lower hemisphere. Triangle: striae of the normal faults; square: striae of the dextral strike-slip faults, diamonds: striae of the sinistral strike-slip faults; circle: striae with unknown sense of movement.

Fig. 18. Details of two synsedimentary normal faults: (**a**) Detail of a fault scarp in the lower Middle Jurassic limestones fossilized by the 'Conglomerados Calcáreos del Puerto' Formation. (**b**) Details of a normal fault in the 'Capas Rojas' Formation.

faults. These faults can be associated in two sets of conjugate faults. The first one strikes N160°E to N170°E and dips towards the east and west. The other set strikes N80°E to N70°E and dips towards the north and south. In both sets the hanging walls of the faults have associated small, elongated basins subparallel to the trend of the fault. Most of these small basins usually contain rocks from the Middle–Upper Jurassic Radiolaritas del Charco Formation, although two basins contain thin deposits from the Fardes and Capas Rojas Formations.

Syn-sedimentary deformations and structures

Normal faults associated with very important thickness and geometrical variations of the Middle Jurassic to Palaeogene formations can be identified from the above descriptions of the basins. The most obvious are: the normal fault located at the northern boundary of the Rambla Seca Basin (Figs 8 & 9), the fault at the southern border of the Arrroyo de Taibena Basin (Figs 12 & 13), and the fault at the northern border of the Arroyo de Taibena Basin that extends to the western border of the Zarzilla de Ramos Basin (Figs 12 & 13).

The fault located at the northern boundary of the Rambla Seca Basin has a mean N70°E strike and is cut and displaced by NNW–SSE left-handed strike-slip faults (Fig. 8). The fault surface dips towards the south. The oolitic limestones of the Serrata de Guadalupe form the footwall of this fault. North of this range, the onlap of the Capas Rojas Formation over these oolitic limestones can be observed, forming the border of the Northern Basin. However, the Middle–Upper Jurassic Radiolaritas del Charco and the Lower Cretaceous Fardes Formations, which are omitted in the footwall, crop out again in the hanging wall, between the oolitic limestones and the Capas Rojas Formation (Fig. 9). The observed offset of this normal fault can exceed 1500 m. In the hanging wall, these formations have a wedge-shaped geometry and the upper formations extend farther than the lower ones. The megabed of Rambla Seca, dated as Late Aptian by Aguado *et al.* (1991), unconformably caps this normal fault. All these observations allow this normal fault to be interpreted as a synsedimentary fault with a half-graben in the hanging wall. This normal fault can be used to propose a model of the relationship between the formations in the study area and the deformations illustrated in Figure 16.

The northern border of the Arroyo de Taibena Basin is a normal fault with a mean strike of N100°E in its western and central parts and a mean strike of N70°E in its eastern part. This fault continues towards the north into the normal fault that constitutes the western border of the Zarzilla de Ramos Basin (Fig. 4). The former fault is cut by several small faults. It dips towards the south in the Arroyo de Taibena Basin and towards the east in the Zarzilla de Ramos Basin. In the western part of the fault, the footwall is formed by the carbonates of the Lower Jurassic Gavilán Formation overlain by the Upper Cretaceous to Palaeogene Conglomerados Calcáreos del Puerto Formation, while the Middle Jurassic to Lower Cretaceous Formations are omitted. However, in the hanging wall, the Middle and Upper Jurassic Camarena and Radiolaritas del Charco Formations are preserved. The Middle–Upper Jurassic Radiolaritas del Charco Formation has a wedge-shaped geometry and is overlain in an onlap by the Upper Cretaceous to Palaeogene Conglomerados Calcáreos del Puerto Formation. In the eastern part of the fault (Zarzilla de Ramos Basin) is evidence that the fault is not Neogene because its surface is cut by the main thrust superposing the Subbetic over the Internal Zones. In this eastern part, there are also differences in the thickness of the formations in both blocks of the fault. In the footwall, the Capas Rojas Formation has a thickness of 200 metres, while it is around 300 metres in the hanging wall.

The third fault is located at the southern border of the Arroyo de Taibena Basin. This fault is not Neogene because it is cut by the surface of the main thrust of the Subbetic, as can be seen in the tectonic window west of the Sierra del Gigante (Fig. 4). Also in the footwall of the fault (Sierra del Gigante), the Middle–Upper Jurassic Radiolaritas del Charco Formation is omitted in practically the entire block, and the Upper Cretaceous–Palaeogene Capas Rojas Formation directly overlies the Middle Jurassic Camarena Formation. In the hanging wall, however, the Radiolaritas del Charco Formation is preserved below the Capas Rojas Formation.

Although they cannot be directly seen, the existence of other east–west synsedimentary normal faults can be deduced from the thickness variations of the formations, such as the one located inside the Northern Basin. The aforementioned reverse faults in this basin separate two blocks of the Fardes Formation with a very important difference in thickness. This variation in thickness can be explained if the reverse faults developed on an ancient synsedimentary normal fault.

Usually, the basins associated with the

NNW–SSE faults are small and have poor outcrops. Only the largest NNW–SSE fault, which cuts between Sierra Larga and Sierra del Pericay (Figs 4 & 12) shows a basin filled with Middle–Upper Jurassic to Palaeogene formations that suggest its synsedimentary character. In the basin associated with this fault, the Fardes Formation has a wedge-shaped geometry and the Capas Rojas Formation onlaps the Camarena and Radiolaritas del Charco formations (Fig. 12). In this region the cross-cutting relationship between the east–west and the NNW–SSE faults can also be observed: in one location the east–west faults cut the NNW–SSE faults, while in one location a NNW–SSE fault cuts an east–west fault.

The measurements of major and minor faults that can be associated with these synsedimentary deformations are represented in the diagrams of Figure 17. The faults (Fig. 18) can be grouped into two sets of conjugate faults. The most abundant set is characterized by a mean east–west strike, although the individual fault strike varies from N70°E to N115°E. Most of the faults are high-angle faults with dips varying from 60° to 90°, and only one is a low-angle fault. The striations observed in the fault surface are dip-slip, although there are three strike-slip striations that may be the product of the Neogene compressions. The senses of movement, which can be deduced only for three surfaces, are two normal faults and one right-handed strike-slip fault. The other set has a mean NNW–SSE (N165°E) strike, although the strikes of the faults vary between N135°E and N200°E. The faults dip towards the east or west and are all high-angle faults with dips varying from 54° to 80°. Five of these faults have a normal regime while two are left-handed strike-slip faults.

Other synsedimentary deformations are recorded in the Eastern Gabar Basin and are very open folds. The clearest is the antiform that constitutes the northern border of the basin. This antiform, which is clearly modified by the Neogene compressional event, shows a thinning of the Middle–Upper Jurassic to Palaeogene formations in both limbs, suggesting that the fold was growing during the deposition of these formations.

The present-day structure of the Eastern Gabar Basin is an open synform, but the Carretero and Fardes Formations, with their lens-shaped geometry, indicates that the synform was active during the Lower Cretaceous. In fact, if we suppose that the base of the Conglomerados del Puerto Formations is horizontal, the Middle–Upper Jurassic and Lower Cretaceous Formations define a very open synform. Over this synform, the upper formations, the Conglomerados del Puerto and Capas Rojas, have a wedge shape, thinning towards the north (Fig. 11). This variation in thickness can be explained by a normal fault at the south of the basin.

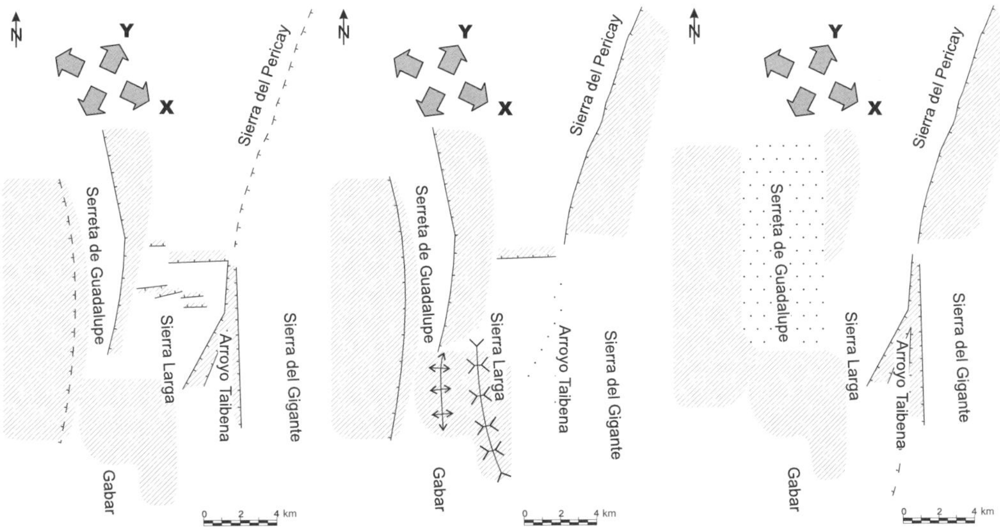

Fig. 19. Palaeogeographical sketches of the evolution of the area from Middle Jurassic to Palaeogene times.

Discussion

The northern basin is characterized by the absence of the Conglomerados Calcáreos del Puerto Formation and also by the development of few turbidite levels in the Fardes Formation. These features suggest the absence of important steep, unstable slopes close to the basin or inside it. The absence of outcrops of the Carretero Formation in this basin can be related to the onlaps that developed in the southern border of the basin. In these onlaps, the Fardes and Capas Rojas Formations onlap the Radiolaritas del Charco Formation; if the Carreretero Formation constitutes part of the same onlap, then it may be located northwards in the basin below the Fardes Formation.

The aforementioned onlap of the formations above the oolitic limestones in the southern border of the basin and the wedge-shaped geometry of the formations (Fig. 7) suggest that the southern border experienced progressive tilting during the Late Jurassic to Cretaceous, contemporaneous with the deposition of the rocks and the activity of the internal fault in the basin.

The geometry of the Rambla Seca Basin clearly corresponds with a half-graben related to the movement of the northern normal fault during the Middle Jurassic to Lower Cretaceous

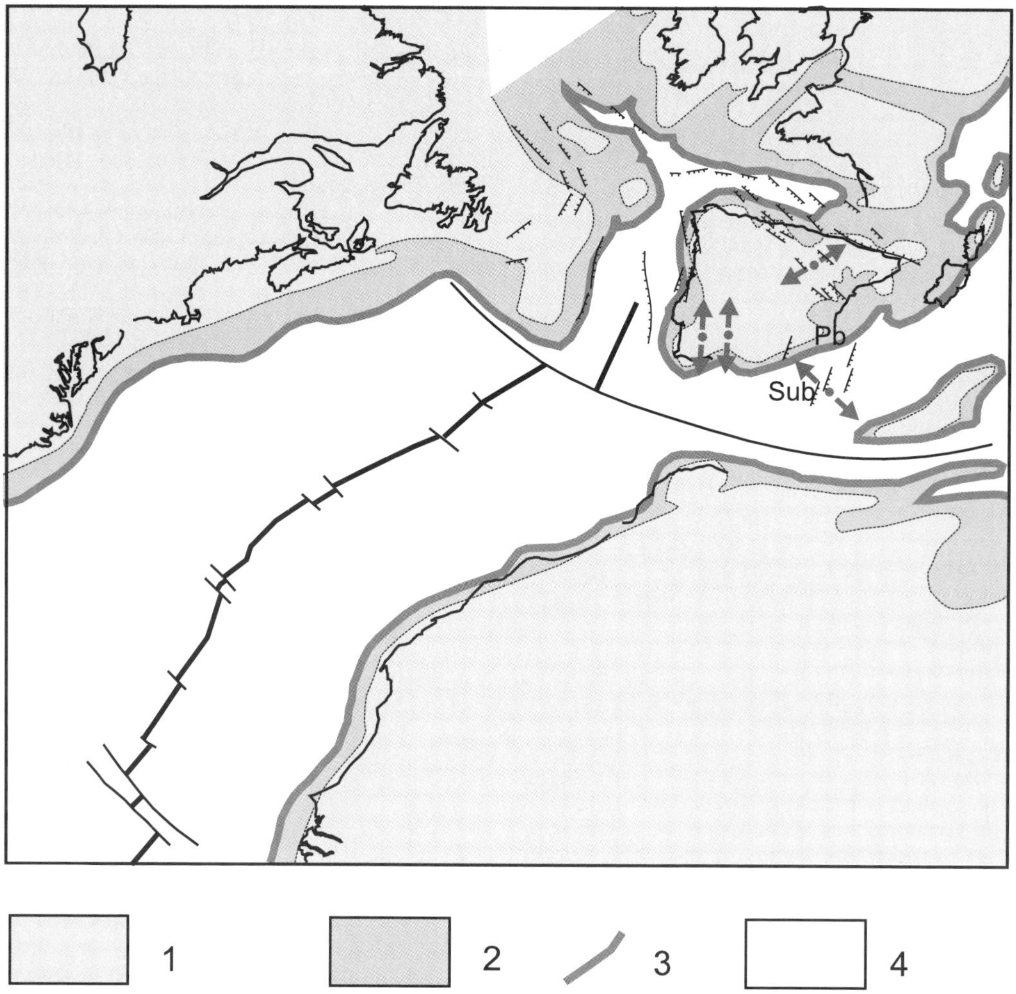

Fig. 20. Plate-tectonic reconstruction during the Early Aptian. Pb, Prebetic Zone; Sub, Subbetic Zone. Modified from Masse *et al.* (1993). Arrows indicate the trend of the extension during the Early Cretaceous rifting. Legend: 1, emerged areas; 2, continental shelf; 3, continental slope; 4, oceanic crust.

times (Fig. 9). The tilting of the hanging wall during the movement of the fault could explain the thickness of the different formations and the onlaps of the successions. The presence of turbidites in the Fardes Formation (including the megabed), formed by reworked oolites from the Camarena Formation, suggest that the fault may have had an unstable steep scarp that was eroded during the Early Cretaceous.

In the Eastern Gabar Basin, the lens shape of the formations suggests that two folds, an antiform and a synform, were active during the Lower Cretaceous. The coexistence of these folds with a generalized extensional regime can be explained by several different hypotheses. The first hypothesis is the existence of blind conjugated normal faults with a graben geometry that produced the very open synform and one or two blind normal faults responsible for the antiform. Another hypothesis is that the folds are fault-bend folds associated with a listric and antilistric staircase geometry of the normal faults in depth. The lack of observations of the deep part of the cross-section does not allow us to decide which of these hypotheses are correct. However, in Figure 16 we have assumed the second hypothesis for the geometry in depth of the faults.

The variations in thickness of the Upper Cretaceous to Palaeogene formations in this Eastern Gabar Basin suggest that the folds became inactive during the Late Cretaceous, and that the deposits filling this basin are controlled by a normal fault that developed south of the basin (Fig. 11). The considerable thickness, around 600 m, of the Conglomerados Calcáreos del Puerto Formation, formed essentially by olistostromes from the Gavilan and Camarena Formations, suggests that this fault may have had a steep unstable slope where the lower part of the succession was exposed.

The Arroyo de Taibena Basin is characterized by the absence of Lower Cretaceous rocks, while the Middle–Upper Jurassic and the Upper Cretaceous to Palaeogene formations are well developed (Fig. 13). This feature can be explained if the two faults that bound the graben had at least two stages of movement – the first one during the Middle to Late Jurassic, when the Radiolaritas del Charco Formation was deposited, completely filling the basin. Later, all the basin and surrounding areas were probably a high during the Early Cretaceous, and later the faults moved again during the Late Cretaceous to Palaeogene.

The Zarzilla de Ramos Basin was active at least during the Early Cretaceous to Palaeogene, and shows no great thickness variations that would suggest the existence of internal synsedimentary faults and folds or tiltings (Fig. 15).

The two systems of synsedimentary normal faults described seem to be contemporaneous due to the cross-cutting relationships and the age of the sedimentary rocks filling the associated basins. The mapped geometry and the orientation of both sets suggest an orthorhombic symmetry (Fig. 6), similar to that predicted by the slip model of Reches (1978) for triaxial deformation. In this model, the main axes of the strain are located in the axis of symmetry of this orthorhombic system; the Z-axis is vertical, the Y-axis is horizontal and trends NW–SE, while the X-axis is horizontal and has a NE–SW trend.

The different palaeomagnetic declinations determined by Allerton *et al.* (1993) and also by Platt (pers. comm.) suggest a correlation between the mean trend of the structures and the vertical axis rotation determined for the area. In the areas where the structures have an ENE–WSW trend, the rotations are always clockwise and have values of $64° \pm 9.8°$, $80° \pm 5.8°$, $62° \pm 7.3°$ and $63° \pm 5.5$ (Allerton *et al.* 1993; Platt pers. comm.). The data clearly show that the present-day ENE–WSW trend is not the original one and, in order to determine the original orientation, a necessary counterclockwise rotation of about 65° must be made. In the areas where the structures have a NNE–SSW trend, the vertical axis rotations are always counterclockwise and their values are: $-12° \pm 13.8°$, $-8° \pm 7.4°$, $-8° \pm 12.9°$ and $-15° \pm 5°$ (Allerton *et al.* 1993; Platt pers. comm.). These values indicate that the NNE-SSW trend must also be corrected in order to determine the original trend. When these vertical axis rotations are restored, the arched pattern structures in plan view disappear and the structures acquire a trend near north–south, which must be the original one. When rotated, the small NNW–SSE faults acquired a new trend of around N80°E. Due to the different ages of activity of the synsedimentary faults, we can propose a model for the evolution of this area during the Middle Jurassic to the Palaeogene (Fig. 19). The evolution shows that the whole area was in extension during the Middle Jurassic to the Early Cretaceous, while only the westernmost part of the area was active from the Late Cretaceous onwards.

The variations of facies from the shallow-marine oolitic limestones with Early–Middle Jurassic ages to the rocks deposited near the CCD level in the Barremian–Cenomanian indicate that the entire region underwent significant tectonic subsidence. This subsidence was greater than 3000 m in the ancient sea bottom of the basins associated with faults with an original north–south orientation. Moreover, the onlaps of the different formations suggest that there may have been important palaeo-relief. All these features indicate that the region underwent significant crustal thinning that may have reached 30% of the thickness of the crust if we suppose an original continental crust with a uniform thickness and a standard density.

Traditionally, only one rifting stage has been recognized in the External Zones of the Betic Cordillera (García-Hernández *et al.* 1989). This rifting was Lower to Middle Jurassic in age, and was associated with the rifting stage that caused the separation of Iberia, then part of Laurasia, from Gondwana. As a result of this rifting, the Tethys extended westwards and the south and southeastern Iberian margins were created (including the Subbetic). After the fragmentation of Pangaea, Gondwana underwent mainly sinistral transtensional displacement with respect to Eurasia during the Late Jurassic–Early Cretaceous (Dewey *et al.* 1989) (Fig. 20).

However, the data presented here and new studies in the Prebetic Zone (Vilas *et al.* 2001) and in the Subbetic (Nieto *et al.* 2001) indicate the existence of another important stage of extension during the Middle Jurassic–Early Cretaceous. In fact, although the above model indicates that the South Iberian margin acted as a passive margin during the Late Jurassic and Cretaceous (Vera 1988), there is strong evidence that the External Zones were undergoing important deformations and vertical movements that reorganised the entire basin.

The data from Vilas *et al.* (2001) show the existence of synsedimentary NE–SW normal faults, active during the Berriasian–Late Albian in the Prebetic, and during the Barremian–Late Albian in the undeformed cover of the foreland of the cordillera. During the Albian, these extensional deformations coincided with the deposition in practically the entire Prebetic of the continental sandstones of the Utrillas Formation, which marks the largest regression in the Prebetic Zone during the Mesozoic and the Palaeogene. This local regression occurred during the sea-level rise that culminated in the highstand of the Turonian transgression, suggesting that the whole Prebetic was rising during the Albian.

In the Intermediate Units, a very important half-graben system associated with a normal listric fault developed during the Early Cretaceous, while the Upper Cretaceous and Palaeogene deposits reflect the post-rift subsidence (Banks & Warburton 1991).

Nieto *et al.* (2001) suggest an interaction between extensional fracturing and saline tectonics in order to explain the coexistence of diapirs, olistostromes and slump in the transition area between the External and the Middle Subbetic, during the Late Jurassic and Earliest Cretaceous. During the same period, in the centre of the Middle Subbetic, an elongated high constituted by basaltic rocks was built (Comas & García-Dueñas 1984; Morata-Céspedes 1993). Lower Cretaceous and Palaeogene rocks onlap this high. Comas & García-Dueñas (1984) show how this relief was cut by normal faults during the Barremian–Aptian in association with olistostromes and slump. In the study area, extensional deformation can also be recognized, associated with subsidence of the area. This extensional stage coincided with the rifting during the Late Jurassic–Early Cretaceous that produced the separation of Iberia from North America and Europe and led to the development of the western and northern margins of Iberia.

During the Late Jurassic–Early Cretaceous, the Iberian chain was affected by an extensional stage with a maximum extension in a NE–SW trend (Giraud & Seguret 1984; Platt 1990; Marqués *et al.* 1996). In the Early Cretaceous, the rifting continued in the western sector of the north–Iberian margin and also started in the Galicia Bank. This rifting process may have extended to the southeastern Iberian margin and affected the External Zones of the Betic Cordillera, and may have been responsible for the aforementioned deformations and vertical movements (Fig. 20). In most of the extensional areas that surrounded Iberia, including the Subbetic domain, the subsidence curves from the Early to Late Cretaceous reflect high subsidence (Reicherter & Pletsch 2000) that may be related to this rifting stage.

In the Late Cretaceous, the extensional stage ended, as reflected by the fact that most of the extensional deformations in the External Zones became inactive and the volcanism ceased. Only in the southern basin of the study area (Arroyo de Taibena Basin) did the normal faults remain active during the Palaeogene. In the study area, compressional deformations clearly developed after the Aquitanian.

Conclusions

We have documented an extensional fracturing stage that produced the subsidence of part of the Subbetic during the Middle Jurassic–Early Cretaceous in the External Zones of the Betic Cordilleras. The fracturing stage generated five small basins associated with half-grabens, grabens and fold structures, where the sedimentary fill shows great thickness variations and onlaps, although it corresponds with rocks with pelagic facies. The synsedimentary normal faults are grouped into two sets of conjugate faults. The most important set has an ENE–WSW trend and is associated with the basins, while the secondary conjugate set of normal faults has a NNW–SSE trend and is associated with small basins. The present-day trends of both the faults and the basins are not the original ones, due to the vertical axis rotations associated with the Neogene compressional deformation experienced by these rocks. When these vertical axis rotations are removed, the original trend of the structures, large normal faults and basins, is nearly north–south, while the secondary normal faults and basins were around N80°E. The main direction of extension of this stage was close to east–west and it influenced the sedimentation in this area.

Fault activity changed the sedimentation conditions from a shallow marine shelf to a basin deeper than the CCD level, probably more than 3000 m in depth, during the Early Cretaceous. The end of the extensional stage caused the sedimentation to become homogeneous, with pelagic marine conditions in the Late Cretaceous, except in the southern basin (Arroyo de Taibena Basin), where several faults remained active during the Palaeogene. This extensional stage can also be recognized in the whole External Zones associated with vertical movements and volcanism that produced the complete reorganization of the basin during the Late Jurassic-Early Cretaceous.

On the margins of Iberia and the Iberian Chain, extensional deformations also occurred during the Late Jurassic–Early Cretaceous, associated with the opening of the North Atlantic. We propose that the extensional stage documented in this work represented the prolongation towards the east of the deformations associated to the opening of the North Atlantic, as occurred in other areas of Iberia.

We want to thank P. Haughton, and J. P. Platt for their help in the review of the manuscript. We also want to thank C. Laurin for the English version of the paper. This work was supported by the 'Grupo de Investigación de la Junta de Andalucía: Geología Estructural y tectónica' and by the CICYT project BET2000-1490-C02-01.

References

AGUADO, R., O'DOGHERTY, L., REY, J. & VERA, J. A. 1991. Turbiditas calcáreas del Cretácico al Norte de Vélez Blanco (Zona Subbética): Bioestratigrafía y génesis. *Revista de la Sociedad Geológica de España*, **4**, 271–304.

ALLERTON, S., LONERGAN, L., PLATT, J. P., PLATZMAN, E. S. & MCCLELLAND, E. 1993. Palaeomagnetic rotations in the eastern Betic Cordillera, Southern Spain. *Earth and Planetary Science Letters*, **119**, 225–241.

AGUADO, R. & REY, J. 1996. Consideraciones sobre la edad del techo de las calizas oolíticas del Jurásico medio de Subbético Interno oriental (Cordillera Bética). *Geogaceta*, **20**, 35–38.

ALLERTON, S., REICHERTER, K. & PLATT, J. P. 1994. A structural and palaeomagnetic study of a section through the eastern Subbetic, southern Spain. *Journal of the Geological Society of London*, **151**, 659–668.

AZEMA, J., FOUCAULT, A., FOURCADE, F., GARCÍA-HERNÁNDEZ, M., GONZÁLEZ-DONOSO, J. M., LINARES, A., LINARES, D., LÓPEZ-GARRIDO, A. C., RIVAS, P. & VERA, J. A. 1979. *Las Microfacies del Jurásico y Cretácico de las Zonas Externas de las Cordilleras Béticas*. Secretariado de Publicaciones Universidad de Granada, Granada, Spain.

BAENA, J., TORRES, J., GEEL, T. & ROEP, T. B. 1977. *Mapa geológico y memoria explicativa, no. 952 (Velez-Blanco)*. Instituto Geológico y Minero de España, Madrid.

BANKS, C. J. & WARBURTON, J. 1991. Mid-crustal detachment in the Betic system of southeast Spain. *Tectonophysics*, **191**, 275–289.

BLUMENTHAL, M. 1927. Versuch einer tektonischen Gliederung der betischen Cordilleren von Central und Südwest Andalusien, *Eclogae Geologicae Helvetiae*, *XX*, 487–592.

COMAS, M. C. 1987. *Sobre la geología de los Montes Orientales: sedimentación y evolución paleogeográfica desde el Jurásico al Mioceno inferior (Zona Subbética, Andalucia.* Doctoral Thesis, Universidad de Bilbao, Spain.

COMAS, M. C. & GARCÍA DUEÑAS, V. 1984. Sobre la evolución fisiográfica del paleomargen mesozoico correspondiente a las zonas externas centrales de las Cordilleras Béticas. *In: El Borde Mediterráneo Español, Evolución del Orógeno bético y Geodinámica de las depresiones neógenas.* Universidad de Granada, Granada, 41–43.

CRESPO-BLANC, A. & CAMPOS, J. 2001. Structure and kinematics of the South Iberian palaeomargin and its relationship with the Flysch Trough units: extensional tectonics within the Gibraltar Arc fold-and-thrust belt (western Betics). *Journal of Structural Geology*, **23**, 1615–1630.

DEWEY, J. F., HELMAN, M. L., TURCO, E., HUTTON, D. H. W. & KNOTT, S. D. 1989. Kinematics of the western Mediterranean. *In:* COWARD, M. P., DIETRICH, D. & PARK, R. G. (eds) *Alpine Tectonics.* Geological Society Special Publication, **45**, 265–283.

FALLOT, P. 1945. Estudios geológicos en la zona subbética entre Alicante y el Río Guadiana Menor. *Memorías del Instituto Lucas Mallada*. Consejo Superior de Investigaciones Científicas, Madrid.

FALLOT, P. 1948. Les Cordillères bétiques. *Estudios geológicos*. **IV**, 259–279.

FOUCAULT, A. 1960. Découverte d'une nouvelle unité tectonique sous le massif subbétique de la Sierra Sagra (Andalousie). *Comptes Rendus de L'Académie des Sciences, Paris*, **250**, 2038.

FOUCAULT, A. 1962. L'Unité du Rio Guardal (Prov. De Grenade, Espagne) et les liaisons entre prébétique et subbétique. *Bulletin de la Societie Géologique de la France*, **4**, 446–452.

GALINDO ZALDÍVAR, J., RUANO, P., JABALOY, A. & LÓPEZ-CHICANO, M. 2000. Kinematics of faults between Subbetic Units during the Miocene (central sector of the Betic Cordillera). *Comptes Rendus de L'Académie des Sciences*, **331**, 811–816.

GARCÍA-DUEÑAS, V. 1967. *La Zona Subbética al Norte de Granada*. Doctoral Thesis, Universidad de Granada, Spain.

GARCÍA-HERNÁNDEZ, M., LÓPEZ GARRIDO, A. C., RIVAS, P., SANZ DE GALDEANO, C. & VERA, J. A. 1980. Mesozoic paleogeographic evolution of the External Zones of the Betic Cordillera. *Geologie en Mijnbouw*, **59**, 155–168.

GARCÍA-HERNÁNDEZ, M., LÓPEZ-GARRIDO, A. C., MARTÍN-ALGARRA, A., MOLINA, J. M., RUIZ-ORTIZ, P. A. & VERA, J. A. 1989. Las discontinuidades mayores del Jurásico de las Zonas Externas de las Cordilleras Béticas: análisis e interpretación de los ciclos sedimentarios. *Cuadernos de Geología Ibérica*, **13**, 35–52.

GEEL, T. 1973. *The geology of the Betic of Malaga, the Subbetic and the zone between these two units in the Velez Rubio area (Southern, Spain)*. GUA Paper of Geology, **5**.

GIRAUD, M. & SEGURET, M. 1984. Releasing solitary overstep model for the Late Jurassic–Early Cretaceous (Wealdien) Soria strike-slip basin (North Spain). *In:* BIDDLE, K. T. & CHRISTIE-BLICK, N. (eds) *Strike-slip deformation, basin formation and sedimentation*. Society of Economic Paleontologists and Mineralogists Special Publication, **37**, 159–175.

KIRKER, A. I. & PLATT, J. P. 1998. Unidirectional slip vectors in the western Betic Cordillera: implications for the formation of the Gibraltar arc. *Journal of the Geological Society of London*, **155**, 193–207.

LONERGAN, L. 1991. *Structural Evolution of the Sierra Espuña, Betic Cordillera, SE Spain*. Doctoral Thesis, Oxford University, UK.

LÓPEZ-GALINDO, A. C. 1986. *Las facies oscuras del Cretácico medio en la Zona Subbética. Mineralogía y sedimentación*. Doctoral Thesis, Universidad de Granada, Spain.

MARQUÉS, L., MAESTRO, A., GIL, A. & CASAS, A. M. 1996. Aportaciones del análisis microestructural a la evolución tectónica del extremo oriental de la Cuenca de Cameros. *Geogaceta*, **20**, 767–769.

MARTÍN ALGARRA, A. 1987. *Evolución geológica alpina del contacto entre las Zonas Internas y las Zonas Externas de las Cordillera Béticas (Sector central y occidental)*. Doctoral Thesis, Universidad de Granada, Spain.

MASSE, J. P., BELLION, Y., BENKHELIL, J., BOULIN, J., CORNEE, J. J., DERCOURT, J., GUIRAUD, R., MASCLE, G., POISSON, A., RICOU, L. E. & SANDULESCU, M. 1993. Lower Aptian palaeoenvironments (114–112 Ma). *In:* DERCOURT, J., RICOU, L. E. & VRIELYNCK, B. (eds) *Atlas Tethys Palaeoenvironmental Maps*. BEICIP–FRANLAB, Rueil-Malmaison.

MOLINA, J. M. 1987. *Análisis de Facies del Mesozoico en el Subbético Externo (Provincias de Cordoba y sur de Jaén)*. Doctoral Thesis, Universidad de Granada, Spain.

MORATA CÉSPEDES, D. 1993. *Petrología y Geoquímica de las Ofitas de las Zonas Externas de las Cordilleras Béticas*. Doctoral Thesis, Universidad de Granada, Spain.

NIETO, L. M., GEA DE, G. A., AGUADO, R., MOLINA, J. M. & RUIZ-ORTIZ, P. A. 2001. Procesos sedimentarios y tectónicos en el tránsito Jurásico/Cretácico: pricisiones bioestratigráficas (Unidad del Ventisquero, Zona Subbética). *Revista de la Sociedad Geológica de España*, **14**, 35–46.

OSETE, M. L., FEEMAN, R. & VEGAS, R. 1988. Preliminary paleomagnetic results from the Subbetic Zone, Betic Cordillera, southern Spain): Kinematic and structural implications. *Physics of the Earth and Planetary Interiors*, **52**, 283–300

OSETE, M.L., FREEMAN, R. & VEGAS, R. 1989. Paleomagnetic evidence for block rotations and distributed deformation of the Iberian–African plate boundary. *In:* KISSEL, C. & LAJ, C. (eds), *Paleomagnetic Rotations and Continental Deformation*, Kluwer Academic Publishers, The Hague, 381–391.

PÉREZ-LÓPEZ, A. D. 1991. *El Trias de Facies Germánica en el Sector Central de la Cordillera Bética*. Doctoral Thesis, Universidad de Granada, Spain.

PLATT, N. 1990. Basin evolution and fault reactivation in the Western Cameros Basin, Northern Spain. *Journal of the Geological Society of London*, **147**, 165–175.

PLATZMAN, E. S. 1992. Paleomagnetic rotations and the kinematics of the Gibraltar Arc. *Geology*, **20**, 311–314.

PLATZMAN, E. S. 1994. East–west thrusting and anomalous magnetic declinations in the Sierra Gorda, Betic Cordillera, southern Spain. *Journal of Structural Geology*, **16**, 11–20.

RECHES, Z. 1978. Analysis of faulting in three-dimensional strain field. *Tectonophysics*, **47**, 109–129.

REICHERTER, K. 1994. The Mesozoic tectono-sedimentary evolution of the central Betic Seaway (External Betic Cordillera, southern Spain). *Tübinger Geowissenschaftliche Arbeiten*, Reihe A, **20**, 265.

REICHERTER, K. & PLETSCH, T. 2000. Evidence for a synchronous circum-Iberian subsidence event and its relation to the African–Iberian plate convergence in the Late Cretaceous. *Terra Nova*, **12**, 141–147.

REY, J. 1993. *Análisis de la Cuenca Subbética Durante el Jurásico y el Cretácico en la Transversal Caravaca Vélez Rubio*. Doctoral Thesis, Universidad de Granada, Spain.

RUIZ-ORTIZ, P. A. 1980. Análisis de facies del Mesozoico de las Unidades Intermedias (entre Castril, prov. de Granada y Jaén). Doctoral Thesis, Granada, Spain.

VAN ANDEL, T. J. 1975. Mesozoic/Cenozoic calcite compensation depth and global distribution of calcareous sediments. *Earth and Planetary Science Letters*, **26**, 187–194.

VAN VEEN, G. W. 1966. Note on a Jurassic–Cretaceous section in the Subbetic SW of Caravaca (prov. Murcia, SE Spain). *Geologie in Mijnbouw*, **45**, 391–397.

VERA, J. A. 1986. Las Zonas Externas de las Cordilleras Béticas. *In: Libro Homenaje J. M. Rios*. Instituto Geológico y Minero, España, **2**, 205–218.

VERA, J. A. 1988. Una modificación al modelo genético para la formación Molicias (Tortoniense sup., Depresión de Guadix, S. España). *Geogaceta*, **5**, 26–29.

VERA, J. A., GARCÍA-HERNÁNDEZ, M., LÓPEZ-GARRIDO, A. C., COMAS, M. C., RUIZ-ORTIZ, P. A. & MARTÍN-ALGARRA, A. 1982. La cordillera Bética, *In:* GARCÍA, A., (ed.) *El Cretácio en España*. Universidad Complutense de Madrid, Madrid, 515–632.

VILAS, L., DABRIO, C. J., PELÁEZ, J. R., & GARCÍA HERNÁNDEZ, M. 2001. Dominios sedimentrarios generados durante el período extensional Cretácico Inferior entre Cazorla y Hellín (Béticas externas). Su implicación en la estructura actual. *Revista de la Sociedad Geológica de España*, **14**, 113–122

VILLALAIN, J. J., OSETE, M. L., VEGAS, P. & GARCÍA-DUEÑAS, V. 1992. Nuevos resultados paleomagneticos en el Subbetico Interno; implicaciones tectonicas. *In: Actas de las Sesiones Cientificas; III congreso Geologico de España*, **1**, 308–312.

WITTINK, R. J. 1975. *A note on an Upper Jurassic to Miocene section in the Subbetic North of Vélez Blanco (Province of Almería, SE Spain)*. GUA Papers Geol., **7**, 89–101.

Sedimentary response to tectonics in extensional basins: the Pechelbronn Beds (Late Eocene to early Oligocene) in the northern Upper Rhine Graben, Germany

C. DERER[1], M. KOSINOWSKI[2], H. P. LUTERBACHER[3], A. SCHÄFER[1] & M. P. SÜß[3]

[1]*Geologisches Institut, Universität Bonn, Nußallee 8, 53115 Bonn, Germany (e-mail: cderer@uni-bonn.de)*
[2]*Niedersächsisches Landesamt für Bodenforschung, Stilleweg 2, 30655 Hannover, Germany*
[3]*Institut für Geowissenschaften, Universität Tübingen, Sigwartstrasse 10, 72076 Tübingen, Germany*

Abstract: The deposition of the late Eocene to early Oligocene Pechelbronn Beds in the northern Upper Rhine Graben was controlled by changes in accommodation space, sediment supply and basin physiography, imposed by the syn-rift tectonic framework. Base-level cycles, defined by variations of the ratio of accommodation space to sediment supply (A/S ratio), allow untangling of the depositional history in this complex structural setting. A transfer zone divided the northern part of the Upper Rhine Graben into a southern and a northern sub-basin and created major depositional gradients. The low A/S ratio in the transfer zone led to sediment bypassing and cannibalisation. Only asymmetric cycles of fluvial and alluvial fan deposits developed, as the sediment was transported to the sub-basins. The higher A/S ratio on the major gradient of the southern sub-basin, which increased from the transfer zone to the south, allowed the formation of symmetric delta/shoreface and lacustrine cycles. At times starvation occurred in the transfer-zone-distal parts of the sub-basin. On subordinate scale, within the southern sub-basin, tilt-blocks bounded by growth faults created halfgrabens with inferior depositional gradients. On the footwall crest, due to low A/S ratio, bypassing and erosion occurred. Here asymmetric cycles of coarse-grained channel fill deposits were preserved. On the hangingwall, close to the normal fault, high A/S conditions were present and symmetric cycles developed. The creation of accommodation space kept pace and even outpaced the footwall-derived sediment supply, which created thick shallow water deposits.

The deposition of sedimentary sequences and the distribution of the depositional environments in active extensional basins are controlled by the interaction of sediment supply and tectonic activity (Leeder & Gawthorpe 1987; Frostick & Steel 1993). The aim of this paper is to bring new insights into the depositional history of the northern Upper Rhine Graben, using a sequence stratigraphic approach. The method combines the accommodation models in extensional basins of Gawthorpe *et al.* (1994) and Howell & Flint (1996) and the principles of genetic stratigraphic base-level cycles (Cross & Lessenger 1998). During the Eocene and Oligocene, the Upper Rhine Graben was a basin characterised by syn-sedimentary tectonics. Thus, we focus on the local tectonic control on sediment dispersal and accumulation.

Regional geology and study area

The Upper Rhine Graben belongs to the European Cenozoic rift system (e.g. Ziegler 1992). It has an almost N-S strike and extends 300 km in length and 40 km in width (Fig. 1). The formation of the Upper Rhine Graben began in the middle to late Eocene and was followed by two main phases of subsidence: late Eocene to early Oligocene and early Miocene. Subsidence was continuous in the northern graben (although at various rates), whereas in the southern part subsidence gave way to inversion from the middle Miocene to the middle Pliocene. Beginning with the middle Pliocene and continuing in the Quaternary, the graben was subjected to sinistral shear (Illies 1978; Teichmüller & Teichmüller 1979; Ziegler 1992).

From: McCann, T. & Saintot, A. (eds) *Tracing Tectonic Deformation Using the Sedimentary Record.* Geological Society, London, Special Publications, **208**, 55–69. 0305-8719/03/$15.00

The lithostratigraphic chart and the gross interpretation of the sedimentary environments of the graben fill are presented in Figure 2. Several transgressions took place during the Tertiary, connecting the Upper Rhine Graben to adjacent marine basins such as the Molasse Basin and the North Sea Basin (e. g. Doebl 1967; Doebl 1970; Pflug 1982; Sissingh 1998; Reichenbacher 2000).

This paper focuses on the Pechelbronn Beds (van Werveke 1904) (latest Eocene to early Oligocene, Fig. 2) of the northern part of the Upper Rhine Graben (Fig. 1). The Pechelbronn Beds represent syn-rift deposits which, in many areas of the northern Upper Rhine Graben, rest directly on the Permian pre-rift sediments. In the study area their thickness varies between zero and 250 metres. Based on litho- and biostratigraphy, as well as on palaeoecology (Schnaebele 1948) the Pechelbronn Beds were subdivided into three units: Lower, Middle and Upper Pechelbronn Beds. Deposition of the Lower Pechelbronn Beds took place under terrestrial conditions: alluvial systems alternated with lacustrine and swamp environments. The main drainage direction of the fluvial systems was probably toward the southwest (Gaupp & Nickel 2001). The deposits vary from high-energy conglomerates and sandstones to organic-rich mudstones. Gaupp & Nickel (2001) also note the presence of a volcanoclastic layer, which was probably derived from the Eocene alkali basaltic volcanism of that area. Towards the top of the Lower Pechelbronn Beds brackish influences become present (Gaupp & Nickel 2001). During the period of deposition of the Middle Pechelbronn Beds, the sea advanced from the South (Doebl 1967) creating brackish/marine environments (Gaupp & Nickel 2001). Based on nannofossils, Martini (1973) attributed the Middle Pechelbronn Beds to the nannoplankton zone NP 22 (earliest Rupelian). At that time offshore mud was deposited in the depocentres, whereas fine-

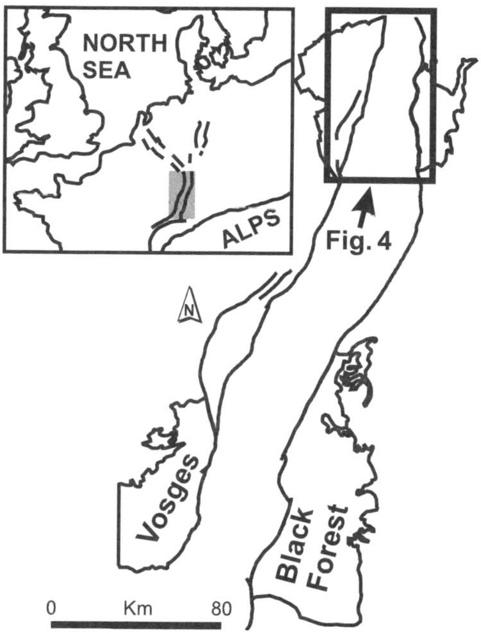

Fig. 1. Location of the Upper Rhine Graben within the European Cenozoic rift system. The study area is the northern Upper Rhine Graben.

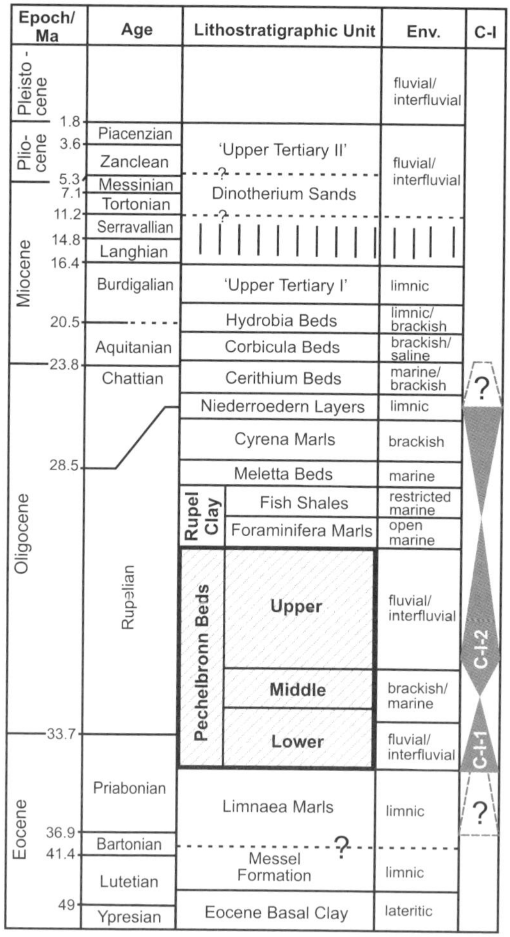

Fig. 2. Lithostratigraphic chart of the northern Upper Rhine Graben. Modified after Hüttner (1991); Sissingh (1998); Martini (2000); chronostratigraphy after Berggren *et al.* (1995). C-I-1 and C-I-2 represent the two large-scale base-level cycles discussed. The studied Pechelbronn Beds are marked by a frame.

grained coastal and deltaic sands occurred in landward positions. The Upper Pechelbronn Beds were deposited in a terrestrial environment (alluvial fans, fluvial/interfluvial, lacustrine) with sediments advancing from the west and interfingering with the remnant brackish/marine settings (lagoons?) of the graben centre and its eastern border (Gaupp & Nickel 2001). The terrestrial deposits consist of conglomerates, lithic sandstones and mudstones. In the brackish/marine setting mudstones and fine-grained quartz sandstones alternate. The Upper Pechelbronn Beds pass gradually upwards into the offshore-marine deposits of the Rupel Clay (late Rupelian, Fig. 2). The Rupel Clay was deposited at a time, when the Upper Rhine Graben was connected with the North Sea Basin (Doebl 1970; Sissingh 1998). During the deposition of the Lower and Middle Pechebronn Beds the palaeoclimate was tropical to subtropical, becoming cooler during the formation of the Upper Pechelbronn Beds (Nickel 1996).

Applied stratigraphic principles

The 'base-level approach' (Wheeler 1964; Cross & Lessenger 1998) takes into account that the deposition of sedimentary sequences is controlled by accommodation space (A) and sediment supply (S). It is the variation of the accommodation space to sediment supply ratio (A/S ratio), equivalent to the upward and downward movement of the base level (*sensu* Wheeler 1964), which leads to the formation of cycles (Fig. 3). A base-level cycle is composed of two hemicycles: during a base-level rise hemicycle (accommodation space to sediment supply ratio increasing) the capacity of the basin to store sediment increases. During a base-level fall hemicycle (accommodation space to sediment supply ratio decreasing) the capacity of storing sediment moves down-gradient, leading to sediment bypassing and erosion. The transition between hemicycles is characterised by turnarounds: fall-to-rise (minimum A/S) and rise-to-fall (maximum A/S).

A distinction may be made between symmetric cycles (where rise and fall hemicycles have comparable thickness) and asymmetric cycles (where sediment thickness of one of the hemicycles dominate). In addition to the symmetry of the cycles, the preservation potential of sedimentary facies and environments changes. At a position on the gradient with a minimum A/S ratio, the preservation potential is low and a hiatus forms as

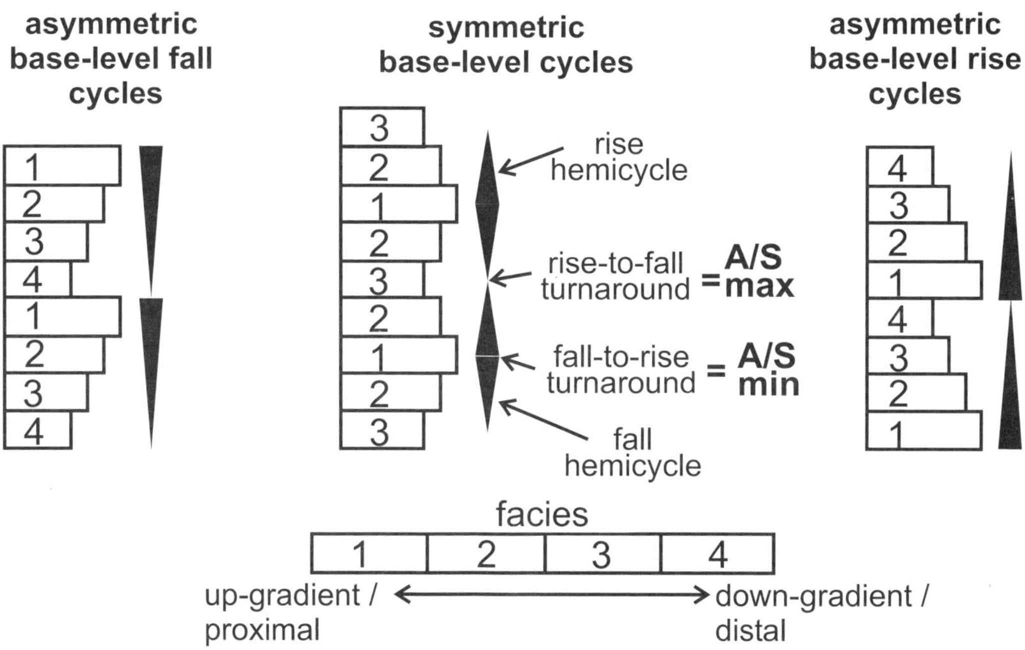

Fig. 3. Definition of the base-level cyclicity and its symmetry. Base-level cycles are defined by the sediment accumulated and preserved during periods of base-level fall and rise. Asymmetric fall and rise cycles contain sediment deposited only during base-level fall or rise respectively. Symmetric cycles contain comparable proportions of sediment deposited during base-level fall and rise. Modified from Cross & Lessenger (1998).

a result of bypassing. At a position with a maximum A/S ratio, a hiatus can occur due to starvation. Thus, the preservation potential and the diversity of depositional elements increase only when the creation of accommodation space and sediment input are in equilibrium.

As shown in the models of Gawthorpe *et al.* (1994) and Howell & Flint (1996), transfer zones, strike and dip variations of accommodation space, different sediment fluxes (axial, footwall- and hangingwall-derived) and a complex basin physiography have to be considered when studying sedimentation in extensional basins. Thus, base-level cycles and turnarounds of different scales can be correlated only if they respond to the same causes influencing accommodation and sediment supply (e.g. if they belong to the same depositional gradient).

It is, therefore, possible to apply the base-level approach only if the tectonic framework of the basin is known.

The present analysis in the northern Upper Rhine Graben is based on several reflection seismic lines and well data (wire-line logs, descriptions of cores and cuttings), which have been made available by German oil companies. The seismic data are mainly used for the interpretation of the tectonic setting, whereas the well information allows base-level cyclicity to be determined.

Tectonic framework

A large-scale transfer zone controls the basin physiography of the northern part of the Upper Rhine Graben, subdividing it into a northern and a southern sub-basin, which have opposing subsidence patterns and tilt directions (Fig. 4). The northern sub-basin is a westward tilted halfgraben, with the main depocentre located along the dominant western border fault. The sub-basin south of the transfer zone is also asymmetric, but tilted toward the eastern border fault, which is dominant in this area. The transfer zone represents a structural high lying between the overlapping ends of the basin-bounding faults defining the northern and southern sub-basins. According to the nomenclature of Morley *et al.* (1990), such a structure may be termed a 'conjugate convergent transfer zone'. The presence of this positive palaeostructural feature (identified in this work as a transfer zone) was previously recognised by Doebl & Olbrecht (1974) and Pflug (1982). This paper will concentrate on the transfer zone and the southern sub-basin (Fig. 5).

In the southern sub-basin, minor faults are oriented sub-parallel to the strike of the graben and dip eastward (Fig. 4). Within the transfer zone these faults interfinger with some westward-dipping minor faults of the northern halfgraben, which are also sub-parallel to the graben margins. These graben-sub-parallel subordinate faults, and faults oblique to the graben margins form a series of tilted fault blocks and horst structures within the transfer zone. Normal faulting influenced deformation within the transfer zone. Oblique-slip and strike-slip movements could not be observed on the available data (2D-reflection seismic), but Gaupp & Nickel (2001) note the presence of approximately north-south-oriented strike-slip faults in the area.

Subordinate growth faults were active during the deposition of the Pechelbronn Beds and a series of tilted fault blocks formed within the sub-basins, and partly in the transfer zone (Fig. 5). These tilt-blocks acted as subordinate half-grabens (Fig. 6). In the hangingwall of the growth faults, the Pechelbronn Beds are wedge-shaped, with the thickness increasing toward the fault plane. The bedding in the lower part of this lithostratigraphic unit is rotated and dips toward the fault. The amount of rotation decreases toward the top of the Pechelbronn Beds and the reflectors of the overlying Rupel Clay usually dip away from the fault. A similar pattern is also observed on segments of the border faults (Fig. 4).

The studied part of the southern sub-basin includes several fault blocks: block B-C, block C-D and block D-E (Fig. 5 & Fig. 6). Block B-C is tilted westward and bounded by one of the main faults delimiting the graben (fault B). The throw of fault B varies considerably along its strike. In the north of the transfer zone, this fault becomes the dominant border fault of the northern sub-basin. In the east, fault block B-C is bordered by fault C, which has a southward increasing throw. Fault block C-D dips toward the south as a result of the presence of the transfer zone, and also toward fault C in the west. The south-directed tilt of this block, parallel to fault C, forms a ramp-type strucure, dipping away from the transfer zone. Block D-E is similarly tilted towards the south and the west (towards fault D). The tilted blocks successively occupy lower structural positions from the western graben margin towards its centre.

The transfer zone represented a structural and topographic high during the deposition of the Pechelbronn Beds. It separated the northern and southern sub-basins and created a major depositional gradient (i.e. the ramp) dipping away from it and into the southern sub-basin. The transfer zone also acted partly as a source area, delivering sediment to the depocentre, axially along the major depositional gradient.

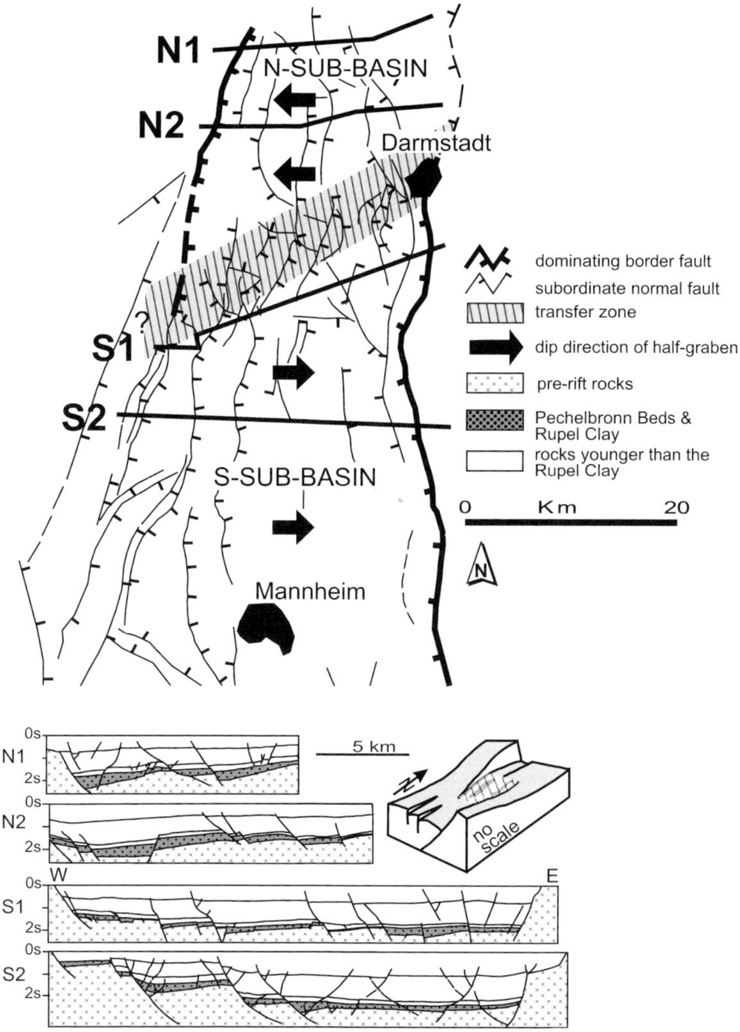

Fig. 4. The northern Upper Rhine Graben. The transfer zone controls the rift geometry, creating two halfgrabens with opposing tilt directions. The cross sections are derived from seismic reflection profiles with two-way travel time in seconds shown on the left. The 3D-block shows a simplified model of the two halfgrabens and the transfer zone. Structure modified from: Andres & Schad (1959); Straub (1962); Stapf (1988); Durst (1991); Plein (1992); Mauthe *et al.* (1993); Jantschik *et al.* (1996). Line S2 on the basis of data from the GeoForschungsZentrum Potsdam with the kind permission of the GeoForschungsZentrum Potsdam.

Tilted fault blocks within the southern sub-basin formed secondary halfgrabens with minor depositional gradients. Sediment input was derived from the footwall or the hangingwall areas. Thus, the syn-sedimentary tectonic framework of the northern Upper Rhine Graben, resulting from the presence of the large-scale transfer zone and from a series of subordinate fault blocks, exercised a significant control on different scales on the development of accommodation space and sediment dispersal.

Cycle hierarchy

In the northern Upper Rhine Graben, a threefold hierarchy of stratigraphic cycles is recognised (C-I, C-II, C-III). Criteria for their recognition include facies changes both within and between individual depositional systems, changes of depositional systems in stratigraphic section, and the areal extent of cycle recognisability.

Small-scale cycles (C-III) are 3 to 15 metres thick and record minor lateral facies shifts within

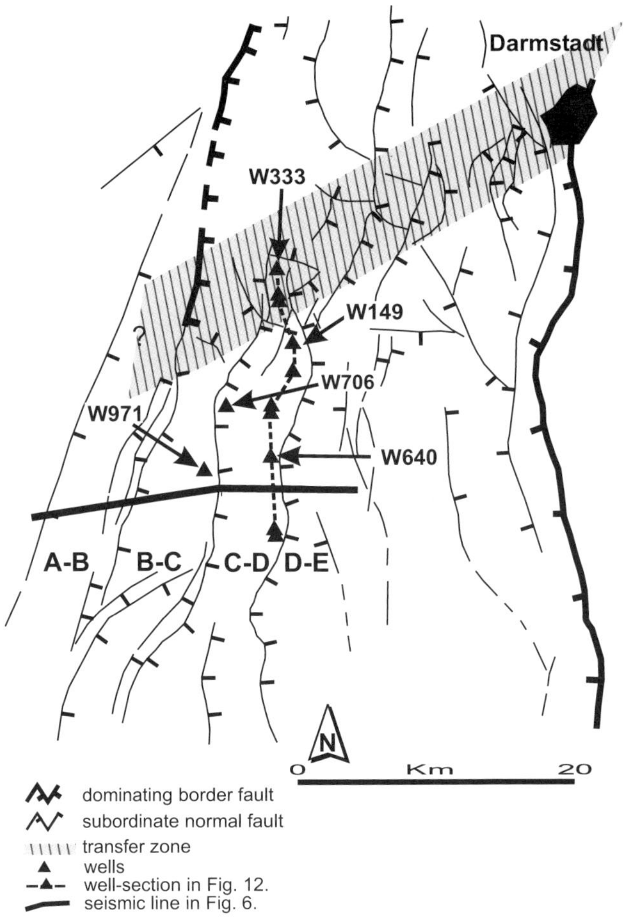

Fig. 5. Structural map of the transfer zone and the southern sub-basin, showing fault blocks (A-B, B-C, C-D and D-E). Locations of the discussed wells, the seismic line in Fig. 6 and the cross-section shown in Figure 12. Structure modified from: Andres & Schad (1959); Straub (1962); Stapf (1988); Durst (1991); Plein (1992); Mauthe *et al.* (1993); Jantschik *et al.* (1996).

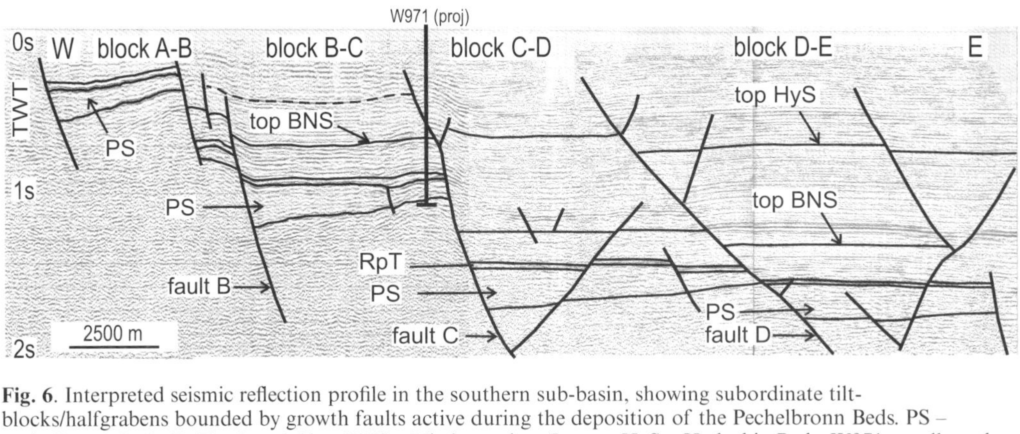

Fig. 6. Interpreted seismic reflection profile in the southern sub-basin, showing subordinate tilt-blocks/halfgrabens bounded by growth faults active during the deposition of the Pechelbronn Beds. PS – Pechelbronn Beds, RpT – Rupel Clay, BNS – Niederroedern Layers, HyS – Hydrobia Beds, W971 – well on the block crest. Two-way travel time in seconds shown on the left (TWT); location in Figure 5.

a depositional system. They can be confidently correlated only locally.

The thickness of the intermediate-scale cycles (C-II) varies between 15 and 50 metres; they can be recognized and correlated over a part of the sub-basin. These cycles reflect an up- or down-gradient shift of depositional systems.

The large-scale cycles (C-I) are between 25 and 200 metres in thickness. They can be traced and correlated basin-wide and are generated by major changes in sedimentation. The large-scale cycles correspond to the lithostratigraphic units, at least those of the Eocene and Oligocene (cycles C-I-1 and C-I-2 in Fig. 2). However, the sequence stratigraphic approach used in this paper leads to a redefinition of boundaries and symmetries of the units within the study area. In the following, the large-scale cycles of the Pechelbronn Beds and the Rupel Clay will be discussed (C-I-1 and C-I-2 in Fig. 2).

Base-level cycles in the Pechelbronn Beds and their spatial variation

In the late Eocene to middle Oligocene deposits of the northern Upper Rhine Graben (Pechelbronn Beds and Rupel Clay) two large-scale cycles are identified (C-I-1 and C-I-2, Fig. 2). The characteristics of these C-I-cycles vary considerably as a function of the local tectonic setting within the basin.

As previously stated, large-scale control on the tectono-sedimentary evolution of the area is exercised by the transfer zone, which influences both sediment dispersal and accumulation by separating two main depocentres (the northern and the southern sub-basins, Fig 4). The transfer zone was characterised by low accommodation space and high sediment input, leading to low preservation conditions, as the sediment was transported into the sub-basins. As a consequence, C-I-1 and C-I-2 were not able to develop as two distinct cycles. Instead, a single, less than 50 metres thick base-level rise cycle formed (Fig. 7). In contrast to the situation in the transfer zone, higher accommodation to sediment supply ratios developed in the two sub-basins (Fig. 8 & 9), which allowed the formation and preservation of both cycles (C-I-1 and C-I-2). These deposits, with a total thickness of more than 200 metres, onlap onto the margins of the transfer zone.

The deposition of the Middle Pechelbronn Beds (containing the first large-scale rise-to-fall turnaround) was dominated by brackish/marine conditions, extending across almost the entire Upper Rhine Graben (Doebl 1967). Offshore mudstones can be identified in most parts of the two sub-basins (where they represent a good

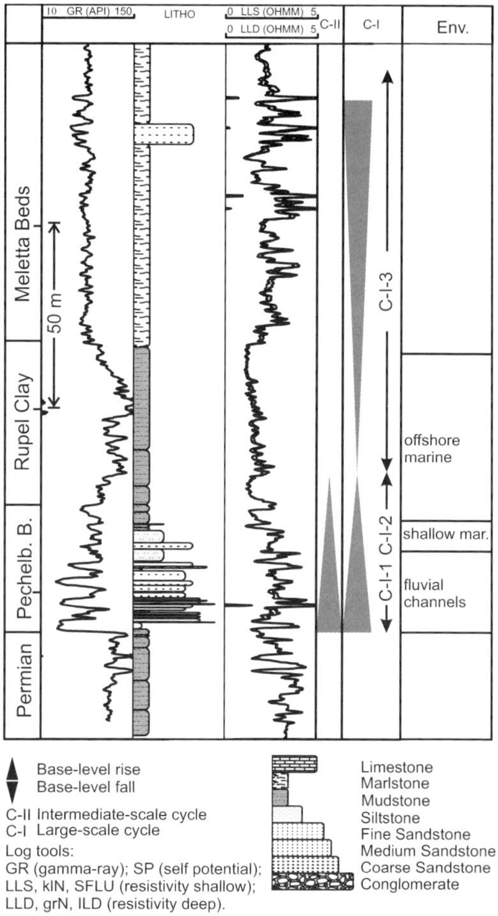

Fig. 7. Well W333 located in the transfer zone. Low accommodation space and high sediment supply. Only one large-scale asymmetric base-level rise cycle is preserved. Location in Fig. 5. Legend for figures 7, 8, 9, 10 and 11.

correlation marker), but not in the transfer zone. Here, either only the equivalent terrestrial deposits were formed, or the sediments were eroded during the following base-level fall. In contrast to the Pechelbronn Beds, the open-marine Rupel Clay shales, containing the second major rise-to-fall turnaround (between C-I-2 and the following C-I-3), extend with a relatively constant thickness and with only one facies type over both sub-basins and the transfer zone.

Within the southern sub-basin, the active tilted fault blocks created an intra-basinal palaeotopography, which controlled deposition (Fig. 6). The fault block crests had low accommodation conditions and moderate sediment input, whereas on the hanging wall, proximal to the growth faults, accommodation space creation was high and partly balanced by footwall-derived clastic material. Thus, the local absence of the offshore Middle Pechelbronn mudstones on parts of the fault block crests can be explained.

Five wells (W333, W149, W640, W971, W706) are chosen for a more detailed illustration of the accommodation space to sediment supply variation in different structural locations within the transfer zone and the southern sub-basin. Their characteristics are summarized in Tables 1 and 2. The locations of the wells are marked in Figure 5.

Accommodation to sediment supply ratio in the transfer zone (W333, Fig. 7)

The transfer zone succession, as represented by well W333, is characterised by high sediment input and low accommodation conditions. Thus, sediment bypassing and cannibalisation occurred. A single, 40 m thick asymmetric base-level rise cycle is preserved and the diversity of depositional environments is low. High-energy, proximal environments (aggrading fluvial channels) formed a thinning-upward sequence. Thin coastal and shallow marine sediments, marking the transition to the offshore Rupel Clay, overlie the fluvial deposits.

Accommodation to sediment supply ratio in the transfer-zone-proximal southern sub-basin (W149, Fig. 8)

Well W149 is positioned on block C-D, in the proximal part of the southern sub-basin, relative to the transfer zone. The moderate accommodation space and the relatively high sediment input via the transfer zone allowed for the development and preservation of both of the large-scale cycles (C-I-1 and C-I-2). In contrast to the transfer zone area, a higher diversity of depositional environments occurred. Due to the high accommodation space, low-energy interfluvial and lacustrine systems could also form at the base of the C-I-1 cycle. In the upper part of cycle C-I-1 and in the lower part of the C-I-2-cycle (C-I-2-fall hemicycle) the brackish/marine environments of the Middle Pechelbronn Beds developed (dashed line in Fig. 8). Here, prograding delta/shoreface systems suggest the proximity of a coastline, and sediment input from the transfer zone onto the ramp-like block C-D. At the fall-to-rise turnaround of the C-I-2 cycle, the top of the shoreface is capped by a subaerial exposure surface. This surface acted as a bypassing area for axial sediment flux from the transfer zone to the south. During the rise of the base-level in the C-I-2-cycle only thin fluvial/interfluvial deposits accumulated, and were subsequently overlain by the shallow and offshore marine sediments of the Rupel Clay.

Accommodation to sediment supply ratio in the transfer-zone-distal southern sub-basin (W640, Fig. 9)

Well W640 is located distally on the depositional gradient of block C-D (i. e. the ramp), created by the transfer zone in the southern sub-basin. As a consequence, the accommodation space increased relative to the proximal conditions, but the amount of clastic sediment reaching this site was subordinate. Thus, the diversity of the depositional environments decreased, being dominated by fine-grained, low-energy deposits. At this distal location offshore Middle Pechelbronn deposits developed (dashed line in Fig. 9) which were time-equivalent to the delta/shoreface systems in well W149 (dashed line in Fig. 8). Due to the higher accommodation space at this location, the exposure at the fall-to-rise turnaround of cycle C-I-2 was not so significant. Thus, thicker fluvial/interfluvial deposits could accumulate during the base-level rise of the C-I-2-cycle.

Accommodation to sediment supply ratio on the fault block crest (W971, Fig. 10)

The position of well W971 is on the crest of the subordinate tilted fault block B-C, in the western

Table 1. *The variation of the accommodation space and sediment supply on the gradient created by the transfer zone*

	Transfer zone	Transfer-zone-proximal southern sub-basin	Transfer-zone-distal southern sub-basin
A/S ratio	Very low	Moderate	High
Accommodation	Low	Moderate	Moderate-high
Sediment supply	High	Moderate-high	Moderate-low
Facies diversity	Low	High	Moderate

Table 2. *The variation of the accommodation space and sediment supply on the footwall crest and on the hangingwall, close to the fault plane*

	Footwall block crest	Hanging wall, fault-proximal
A/S ratio	Moderate-low	High
Accommodation	Moderate-low	High
Sediment supply	High	Moderate
Facies diversity	Low	High

part of the southern sub-basin (Fig. 5, Fig. 6). Here, accommodation space was moderate to low and sediment supply high. During falls in base level this elevated palaeotopographic position was probably exposed and incision, sediment bypassing and sediment amalgamation occurred. Thus, cycles C-I-1 and C-I-2 could not be differentiated. The diversity of environments was low: mainly coarse-grained, high energy sediments of river channels and alluvial fans were deposited and preserved during the rise of the base-level. They were progressively drowned by the marine Rupel Clay transgression.

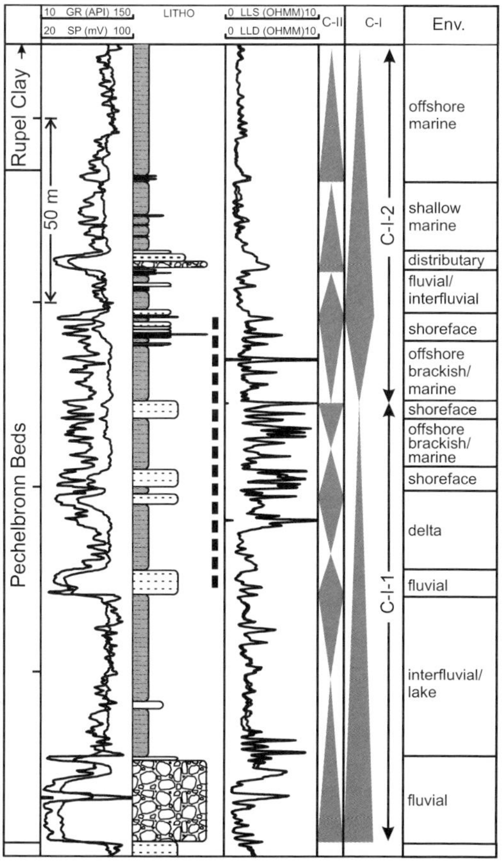

Fig. 8. Well W149 located in the transfer-zone-proximal southern sub-basin. Moderate accommodation space and moderate to high sediment supply. Dashed line marks prograding delta/shoreface systems of the brackish/marine Middle Pechelbronn Beds. Both large-scale cycles (C-I-1, C-I-2) are preserved. Location in Figure 5. Legend in Figure 7.

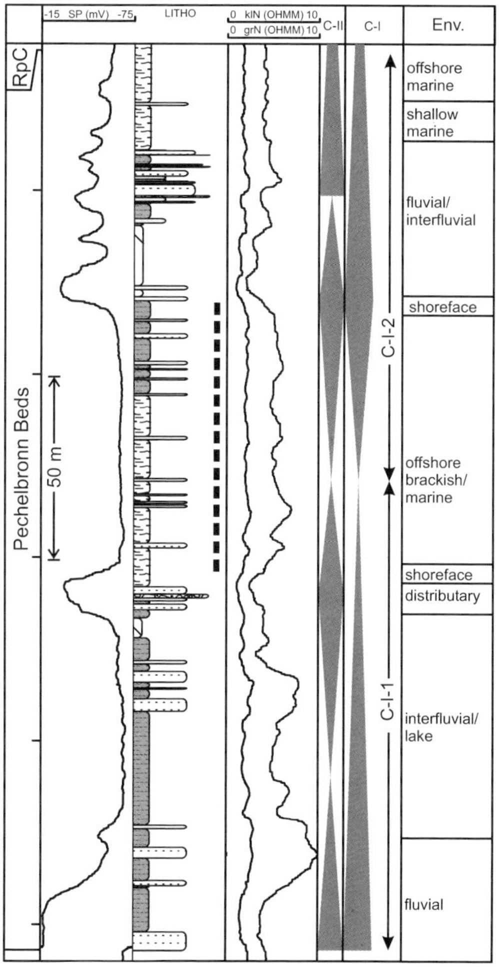

Fig. 9. Well W640 located in the transfer-zone-distal southern sub-basin. High accommodation space and moderate to low sediment supply. Both large-scale cycles (C-I-1, C-I-2) are preserved. The offshore mudstones of the brackish/marine Middle Pechelbronn Beds (dashed line) are the distal equivalents of the delta/shoreface sands of the proximal sub-basin (W149, dashed line in Fig. 8.). Location in Figure 5. Legend in Figure 7.

Accommodation to sediment supply ratio on the hanging wall, proximal to the growth fault (W706, Fig. 11)

Well W706 is located in the hangingwall C-D, close to the active normal fault C, in the western part of the southern sub-basin. The well probably intersects the fault plane, so parts of the Lower Pechelbronn Beds are missing. In this area accommodation space was high due to significant syn-depositional subsidence. Sediment supply was moderate and both of the large-scale cycles were preserved. In the lower part of the C-I-1-cycle fluvial systems occurred, but the creation of accommodation space also allowed the development of thick interfluvial and lacustrine sediments. A transition to brackish/marine environments followed. In the upper part of cycle C-I-1 (dashed line in Fig. 11) high footwall-derived sediment input generated thick shallow water sandstones, which were coeval with the offshore Middle Pechelbronn mudstones in the more central part of the southern sub-basin (well W640 in Fig. 9). The shallow water deposits were topped by a thin succession of offshore sandstones and mudstones. The base-level fall of the C-I-2-cycle led to exposure and during the subsequent rise, fluvial and interfluvial systems aggraded. These were gradually replaced by the marine conditions of the Rupel Clay.

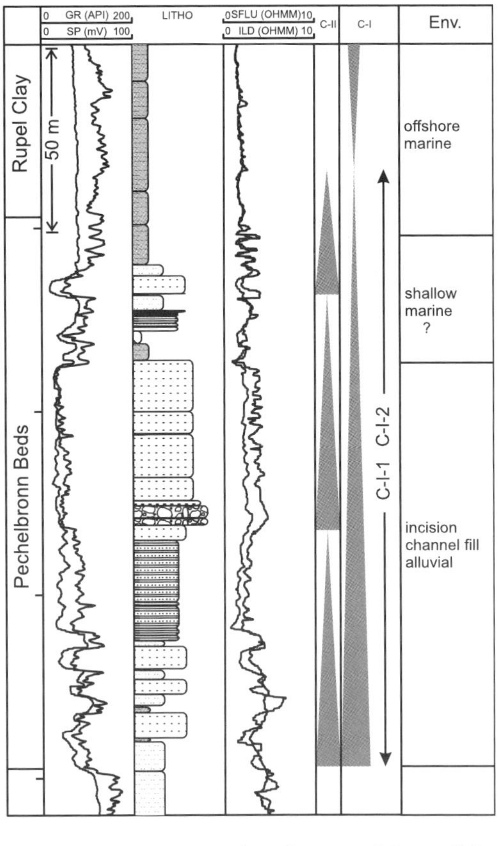

Fig. 10. Well W971 located on the crest of footwall B-C. Moderate to low accommodation space and high sediment supply. During base-level fall exposure and incision occurred. Cycles C-I-1 and C-I-2 cannot be differentiated. Location in Figure 5 and 6. Legend in Figure 7.

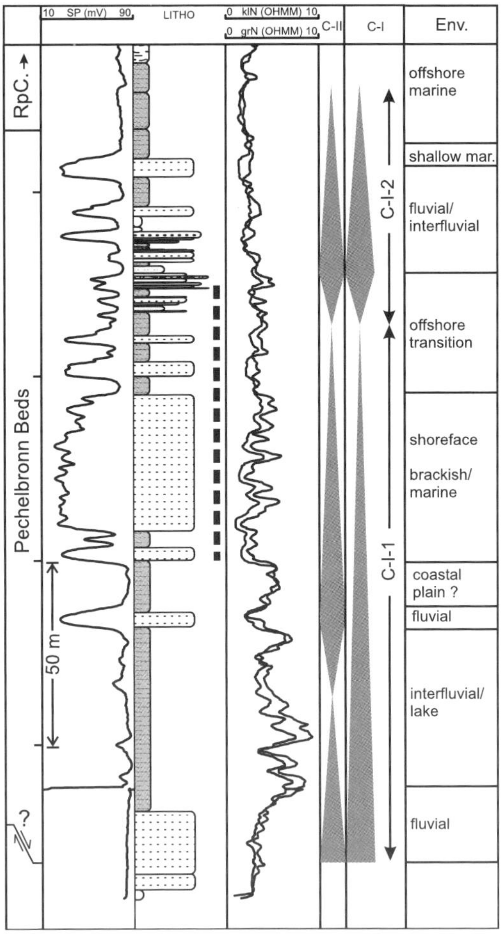

Fig. 11. Well W706 located on the hangingwall C-D, proximal to the active normal fault. High accommodation space due to syn-sedimentary subsidence and moderate sediment input. Both large-scale cycles are preserved (C-I-1 and C-I-2). Thick shoreface deposits (with sediment supplied from the footwall, dashed line) are time equivalent with the Middle Pechelbronn offshore mudstones in well W640 (dashed line in Fig. 9.). Location in Figure 5. Legend in Figure 7.

The locations of wells W333, W149 and W640 belong to the same major depositional gradient that was initiated by the transfer zone. On a proximal-distal profile, these wells show an increase in accommodation to sediment supply ratio. In well W149, accommodation space and sediment supply were closest to equilibrium (accommodation space creation approximately equal to sediment supply). In well W333 (proximal part of the gradient) the clastic input dominated (creation of accommodation space was outpaced by sediment supply). At a more distal location on the gradient (well W640), the creation of accommodation space was higher than sediment supply, leading to periods of sediment starvation.

Wells W971 and W706 belonged to different depositional gradients. However, it is obvious that the accommodation to sediment supply ratio on the crest of the footwall B-C (W971) is lower than that on the immediate hanging wall C-D. Thus, erosion and sediment bypassing was frequent on the block crest, whereas the creation of accommodation space within the downthrown area adjacent to the footwall kept pace or even outpaced the input of clastic material from the footwall.

The subordinate fault blocks within the southern sub-basin have successively lower positions towards the graben centre thus, base-level in halfgraben C-D was higher than on block B-C. So even on the crest of block C-D (well W640) high A/S conditions were possible. The five examples presented above clearly show that due to the tectonic style, the sedimentation pattern significantly changed within a relative small area.

Correlation of base-level cycles

The large-scale base-level cycles of the Pechelbronn Beds (C-I-1 and C-I-2) are correlated along the major depositional gradient of the southern sub-basin, i.e. on the ramp setting of block C-D (Fig. 12, for location see Fig. 5). The cross-section illustrates the stratigraphic variation and the spatial linkage of depositional systems as a function of base-level fluctuations. The cross-section passes from the transfer zone (proximal part, up-gradient) into the southern sub-basin (distal part, down-gradient) thus, the following discussion of the cycles is provided for these two structural elements.

Cycle C-I-1 in the transfer zone

In parts of the transfer zone (well W333 in Fig. 7 and Fig. 12) due to extreme low A/S ratio, only one asymmetric cycle developed. The lack of palaeontologic data and marker horizons makes it difficult to establish with certitude, whether this single base-level rise cycle belongs to C-I-1 or C-I-2. The cycle is composed of aggrading distributary channels, which transported clastic sediments further down-gradient into the sub-basin. Higher up in the stratigraphic section, the terrestrial deposits are followed by thin shallow marine sediments and finally by the offshore facies of the open marine Rupel Clay (maximum A/S).

Down the depositional gradient, at the transition from the transfer zone to the southern sub-basin, where accommodation space was higher, both cycles (C-I-1 and C-I-2) can be differentiated (wells W144, W143). The C-I-1 cycle has a maximum thickness of 25 m and was also formed by aggrading distributary channels. It is topped by thin shallow-water deposits, representing the coastal equivalents of the brackish/marine Middle Pechelbronn Beds, developed in the southern sub-basin.

Cycle C-I-1 in the southern sub-basin

After the formation of the Upper Rhine Graben, sedimentation was predominantly fluvial. The main drainage direction of these fluvial systems was from the transfer zone toward the south (Gaupp & Nickel 2001), as it can also be interpreted from the southward increasing accommodation space. In contrast to the transfer zone, relatively thick deposits of interfluvial and lacustrine sediments were deposited over extended areas of the sub-basin.

During base-level rise, the terrestrial systems passed through shallow-water conditions to the brackish/marine offshore environment of the Middle Pechelbronn Beds (rise-to-fall turnaround). In the transfer-zone-proximal southern sub-basin (dashed line in well W149 of Fig. 8 and wells W149, W138 in Fig. 12), sediment was supplied from the transfer zone by prograding delta/shoreface systems. These conditions were replaced down-gradient, towards the south (dashed line in well W640 of Fig. 9 and wells W899 to W240 in Fig. 12), by coeval offshore sedimentation and starvation, marking the A/S maximum.

In the upper part of the C-I-1-cycle, during the brackish/marine conditions, an axial, north-south sediment flux existed on the ramp-like setting. In the southern sub-basin the thickness of the C-I-1 cycle reaches 150 m.

Cycle C-I-2 in the transfer zone

Sediment was supplied from the western margin of the transfer zone (Gaupp & Nickel 2001). Due

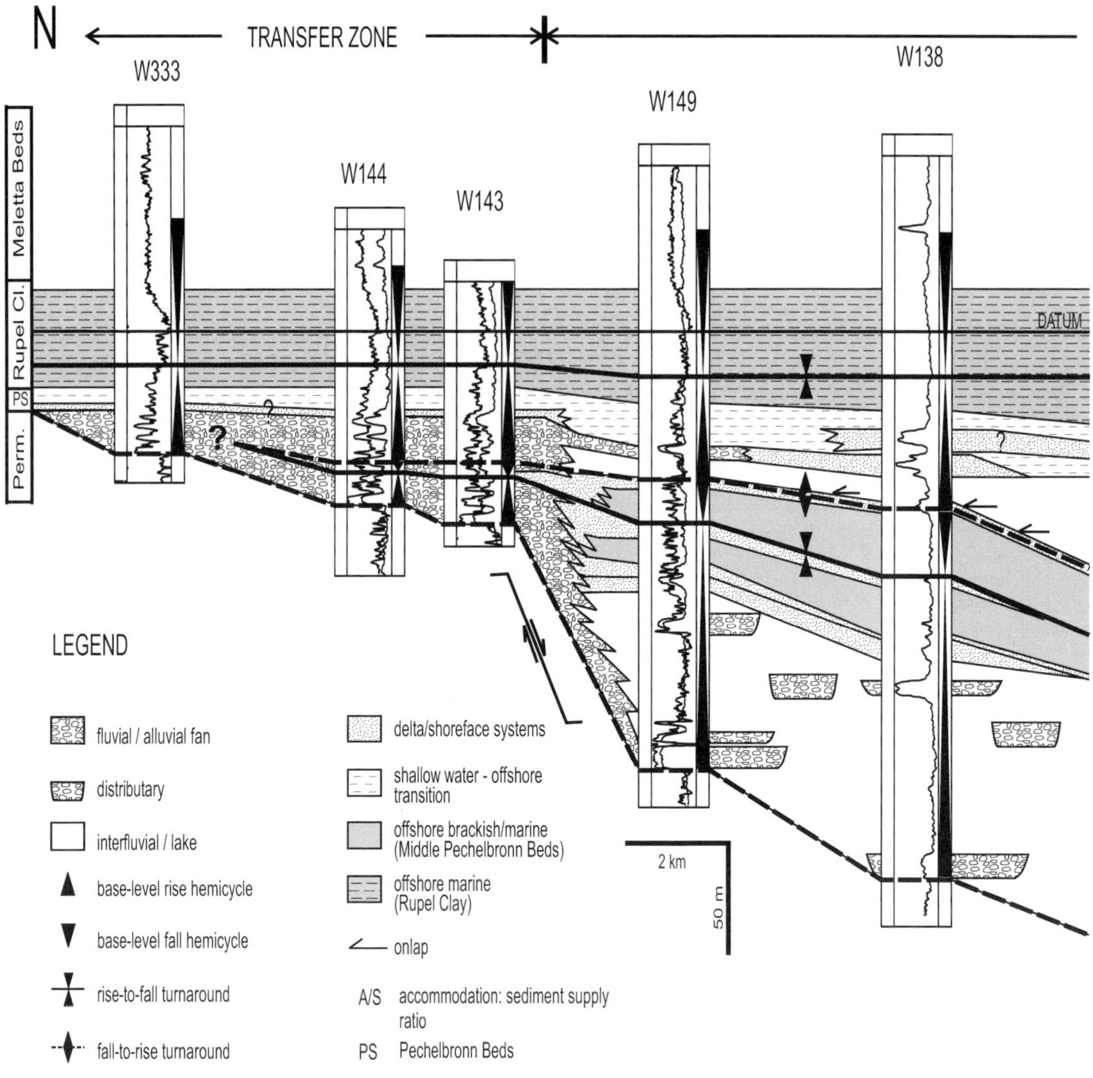

Fig. 12. Cross-section on the ramp setting of fault block C-D, from the transfer zone into the southern sub-basin, showing depositional environments and correlation of the large-scale cycles (C-I-1, C-I-2) in the

to the relative lack of accommodation space most of the clastic material was transported further towards the southern sub-basin. At the southern margin of the transfer zone (W144, W143 in Fig. 12) and mainly during the dominating rise hemicycle, only 50 metres of alluvial fan and distributary channel deposits were preserved. These are overlain by the shallow marine and offshore marine sediments of the Rupel Clay.

Cycle C-I-2 in the southern sub-basin

The rise-asymmetry of the C-I-2-cycle in the transfer zone is gradually replaced in the southern sub-basin by a symmetric pattern. The development of the fall hemicycle was caused by the retreat of the brackish/marine environments towards the south. Sediments were delivered from the western transfer zone into the basin (Gaupp & Nickel 2001), forming prograding

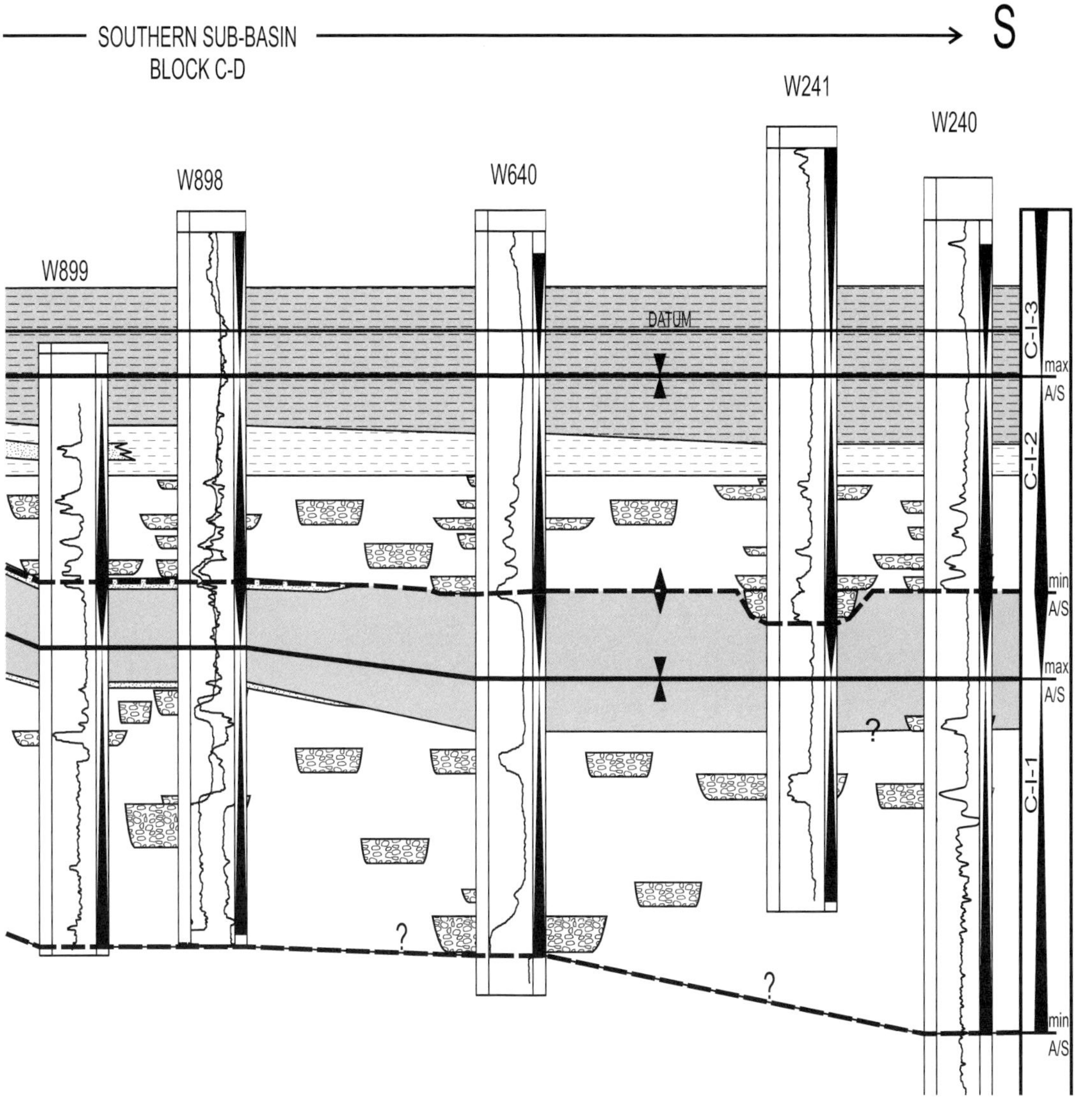

Pechelbronn Beds and the Rupel Clay. Location in Figure 5. The represented logs are gamma ray and self potential. Section datum is a gamma-ray maximum in the offshore shales of the Rupel Clay.

wedges on the ramp. The fall-to-rise turnaround (minimum A/S) was marked by subaerial exposure and sediment bypassing.

The following rise hemicycle within C-I-2 created new accommodation space for aggrading fluvial and interfluvial systems, which onlapped toward the north on the transfer zone. In the south, accommodation space was higher thus, thicker terrestrial deposits than in the vicinity of the transfer zone accumulated. Here the time of exposure and non-deposition was longer. Marginal marine environments, preceding the marine Rupel Clay transgression, gradually replaced the fluvial and lacustrine systems. Due to low sediment input, shallow-water deltas, coastal bars or sandwaves developed only in the vicinity of the transfer zone (wells W138, W899 in Fig. 12). These were subsequently drowned and capped by offshore-marine deposits of the Rupel Clay (rise-to-fall turnaround).

The southward increase of accommodation space to sediment supply ratio on the major depositional gradient created by the transfer zone induced an increase of cycle thickness and symmetry and the decrease of depositional energy downdip.

The drainage of the fluvial systems during the lower C-I-1-cycle was toward the southwest (Gaupp & Nickel 2001). During the period of brackish/marine conditions of the Middle Pechelbronn Beds (upper part of C-I-1-cycle and C-I-2 fall hemicycle), an axial north–south flux prevailed on the ramp of block C-D, as the sediment was delivered through the transfer zone. The footwall-derived sediment (i.e. supplied from the footwall B-C in the west) was confined to the neighbourhood of the fault plane (e.g. well W706, Fig. 11) and did not influence the axial sediment transport from the transfer zone.

The transfer zone also influenced the coastline developed during fluctuations of the base-level.

Conclusions

The Pechelbronn Beds in the northern Upper Rhine Graben display abrupt changes in depositional style, which were controlled by extensional, syn-sedimentary tectonics.

A large-scale conjugate convergent transfer zone divides the study area into two asymmetric halfgrabens of opposite polarity. The transfer zone created two depozones (a northern and a southern sub-basin) and acted partly as a source area. It also created a major depositional gradient dipping from the transfer zone into the southern sub-basin, leading to an axial sediment flux on a ramp-like setting.

The physiography of the southern sub-basin is controlled on a subordinate scale by a series of tilted blocks/halfgrabens, bounded by growth faults. The blocks occupy successively lower structural positions from the western graben margin towards its centre. These blocks induced their own depositional gradients with footwall-derived sediment flux. Thus, the basic structural element, controlling sediment dispersal in the northern Upper Rhine Graben, is the halfgraben, as it was observed for extensional basins by Gibbs (1984).

Accommodation space and sediment supply are controlled on different scales by their relative position within the tectonic framework. The transfer zone had a low accommodation space to sediment supply ratio, in contrast to the southern sub-basin where this was higher. On the ramp-like setting of the southern sub-basin the A/S ratio increased away from the transfer zone. Footwall crests had a lower A/S ratio and delivered sediment to the immediate hangingwall, close to the growth fault. Here accommodation space was created by syn-sedimentary subsidence.

Even though the development of the large-scale cycles may be partially controlled by mechanisms operating outside the studied area (e. g. eustasy, regional transgressions, open communication with neighbouring marine basins), the distribution of depositional environments was controlled by local syn-sedimentary tectonic structures.

The combination of the accommodation models of Gawthorpe *et al.* (1994) and Howell & Flint (1996) and the base-level variation (Cross & Lessenger 1998) used here has proved to be a reliable tool for the interpretation and correlation of strata in this structurally-controlled basin. Before correlation, however, it is important to understand the tectonic framework. The rapid spatial variation of the accommodation space and sediment supply conditions, resulting from tectonic activity, led to the creation of several distinct depositional gradients, which acted independently.

The authors express their thanks to the Deutsche Forschungsgemeinschaft for financing the project SCHA 279/17, which is part of the EUCOR-URGENT (Upper Rhine Graben Evolution and Neotectonics), and to the Wirtschaftsverband Erdöl- und Erdgasgewinnung e.V. for the permission to use the data and to publish the results. We would also like to thank M. Wagreich and W. Ricken for their helpful reviews. M. Böhm is thanked for the drafting support.

References

ANDRES, J. & SCHAD, A. 1959. Seismische Kartierung von Bruchzonen im mittleren und nördlichen Teil des Oberrheintalgrabens und deren Bedeutung für die Ölansammlung. *Erdöl und Kohle*, Hamburg, **5**, 323–334.

BERGGREN, W. A., KENT, D. V., SWISHER, C. C. & AUBRY, M.-P. 1995. A revised Cenozoic geochronology and cronostratigraphy. *In:* BERGGREN, W. A., KENZ, D. V., AUBRY, M.-P. & HERDENBOL, J. (eds) *Geochronology, Time Scales and Stratigraphic Correlation.* Society for Sedimentary Geology, Tulsa, Oklahoma, Special Publications, **54**, 129–212.

CROSS, T. A. & LESSENGER, M. A. 1998. Sediment volume partitioning: rationale for stratigraphic model evaluation and high-resolution stratigraphic correlation. *In*: GRADSTEIN, F. M., SANDVIK, K. O. AND MILTON, N. J. (eds) *Sequence Stratigraphy - Concepts and Applications.* Norwegian Petroleum Society, Amsterdam, Special Publications, **8**, 171–195.

DOEBL, F. 1967. The Tertiary and Pleistocene Sediments of the Northern and Central Part of the Upper Rhinegraben. *Abhandlungen des Geologischen*

Landesamtes in Baden-Württemberg, Freiburg i. Br., **6**, 48–54.

DOEBL, F. 1970. Die tertiären und quartären Sedimente des südlichen Rheingrabens. *In*: ILLIES, J. H. AND MUELLER, S. (eds) *Graben Problems*. Stuttgart, 56–66.

DOEBL, F. & OLBRECHT, W. 1974. An Isobath Map of the Tertiary Base in the Rhinegraben. *In*: ILLIES, J. H. & FUCHS, K. (eds) *Approaches to Taphrogenesis*. Stuttgart, 71–72.

DURST, H. 1991. Aspects of exploration history and structural style in the Rhine graben area. *In*: SPENCER, A. M. (ed) *Generation, accumulation and production of Europe's hydrocharbons*. The European Association of Petroleum Geoscientists, Oxford, Special Publications, **1**, 247–261.

FROSTICK, L. E. & STEEL, R. J. 1993. Sedimentation in divergent plate-margin basins. *In*: FROSTICK, L. E. & STEEL, R. J. (eds) *Tectonic Controls and Signatures in Sedimentary Successions*. International Association of Sedimentologists, Oxford, Special Publications, **20**, 111–128.

GAUPP, R. & NICKEL, B. 2001. Die Pechelbronn-Schichten im Raum Eich-Stockstadt (Nördlicher Oberrheingraben; Blatt 6216 Gernsheim). *Geologisches Jahrbuch Hessen*, Wiesbaden, **128**, 19–27.

GAWTHORPE, R. L., FRASER, A. J. & COLLIER, R. E. L. 1994. Sequence stratigraphy in active extensional basins: implications for the interpretation of ancient basin-fill. *Marine and Petroleum Geology*, Amsterdam, **11/6**, 642–658.

GIBBS, A. D. 1984. Structural evolution of extensional basin margins. *Journal of the Geological Society*, London, **141**, 609–620.

HOWELL, J. A. & FLINT, S. S. 1996. A model for high resolution sequence stratigraphy within extensional basins. *In*: HOWELL, J. A. & AITKEN, J. F. (eds) *High Resolution Sequence Stratigraphy: Innovations and Applications*. Geological Society, London, Special Publications, **104**, 129–137.

Hüttner, R. 1991. Bau und Entwicklung des Oberrheingrabens. Ein Überblick mit historischer Rückschau. *Geologisches Jahrbuch*, Stuttgart, **E48**, 17–42.

ILLIES, J. H. 1978. Two Stages Rheingraben Rifting. *In*: RAMBERG, I. B. & NEUMANN, E.-R. (eds) *Tectonics and Geophysics of Continental Rifts*. Dordrecht, 63–71.

JANTSCHIK, R., STRAUß, C. & WEBER, R. 1996. Sequences-Stratigraphy as a Tool to Improve Reservoir Management of the Eich / Koenigsgarten Oil Field (Upper Rhine Graben, Germany). Society of Petroleum Engineers Inc., European Petroleum Conference, Milan 22–24 October 1996, 71–80.

LEEDER, M. R. & GAWTHORPE, R. L. 1987. Sedimentary models for extensional tilt-block/half-graben basins. *In*: COWARD, M. P., DEWEY, J. F. & HANCOCK, P. L. (eds) *Continental Extensional Tectonics*. Geological Society, London, Special Publications, **28**, 139–152.

MARTINI, E. 1973. Nannoplankton-Massenvorkommen in den Mittleren Pechelbronner Schichten (Unter-Oligozän). *Oberrheinische Geologische Abhandlungen*, Karlsruhe, **22**, 1–12.

MARTINI, E. 2000. Nannoplankton-Gemeinschaften in den Cerithien- und tieferen Inflata-Schichten des Mainzer Beckens und des Oberrheingrabens (Ober-Oligozän/Unter-Miozän). *Jahresberichte und Mitteilungen des Oberrheinischen Geologischen Vereins, N.F.*, Stuttgart, **82**, 251–259.

MAUTHE, G., BRINK, H.-J. & BURRI, P. 1993. Kohlenwasserstoffvorkommen und -potential im deutschen Teil des Oberrheingrabens. *Bulletin der Vereinigung Schweizerischer Petroleum-Geologen und Ingenieure*, Riehen-Basel, **60/137**, 15–29.

MORLEY, C. K., NELSON, R. A., PATTON, T. L. & MUNN, S. G. 1990. Transfer Zones in the East African Rift System and Their Relevance to Hydrocarbon Exploration in Rifts. *American Association of Petroleum Geologists Bulletin*, Tulsa, Oklahoma, **74**, 1234–1253.

NICKEL, B. 1996. Palynofazies und Palynostratigraphie der Pechelbronn Schichten im nördlichen Oberrheintalgraben. *Palaeontographica*, Stuttgart, **B/240**, 1–151.

PFLUG, R. 1982. Bau und Entwicklung des Oberrheingrabens, Darmstadt, 1–145.

PLEIN, E. 1992. Das Erdölfeld Eich-Königsgarten. (Exkursion E am 23. 4. 1992). *Jahresberichte und Mitteilungen des Oberrheinischen Geologischen Vereines N.F.*, Stuttgart **74**, 41–54.

REICHENBACHER, B. 2000. Das brackisch-lakustrine Oligozän und Unter-Miozän im Mainzer Becken und Hanauer Becken: Fischfaunen, Paläoökologie, Biostratigraphie, Paläogeographie. *CourierForschungsinstitut Senkenberg*, Frankfurt a. M., **222**, 1–222.

SCHNAEBELE, R. J. 1948. Monographie Géologique du Champ Pétrolifère de Pechelbronn. *Mémoires du Service de la Carte Géologique d'Alsace et de Lorraine,* Strasbourg, **7**, 1–254.

SISSINGH, W. 1998. Comparative Tertiary stratigraphy of the Rhine Graben, Bresse Graben and Molasse Basin: correlation of Alpine foreland events. *Tectonophysics*, Amsterdam, **300**, 249–284.

STAPF, K. R. G. 1988. Zur Tektonik des westlichen Rheingrabenrandes zwischen Nierstein am Rhein und Wissembourg (Elsaß). *Jahresberichte und Mitteilungen des Oberrheinischen Geologischen Vereines N.F.,* Stuttgart, **70**, 399–410.

STRAUB, E. W. 1962. Die Erdöl- und Erdgaslagerstätten in Hessen und Rheinhessen. *Abhandlungen des Geologischen Landesamtes Baden-Württemberg*, Freiburg i. Br., **4**, 123–136.

TEICHMÜLLER, M. & TEICHMÜLLER, R. 1979. Zur geothermischen Geschichte des Oberrhein-Grabens. Zusammenfassung und Auswertung eines Symposiums. *Fortschritte in der Geologie von Rheinland und Westfalen*, Krefeld, **27**, 109–120.

VAN WERVEKE, L. 1904. Elsass. *In*: ENGLER, K. AND VON HÖFER, H. (eds) *Das Erdöl*. Leipzig, **2**, 209–234.

WHEELER, J. 1964. Baselevel, lithosphere surface, and time-statigraphy. *Geological Society of America Bulletin,* Boulder, Colorado, **75**, 599–610.

ZIEGLER, P. A. 1992. European Cenozoic rift system. *Tectonophysics*, Amsterdam, **208**, 91–111.

Evaluation of controlling factors on facies distribution and evolution in an arid continental environment: an example from the Rotliegend of the NE German Basin

H. RIEKE[1], T. MCCANN[2], C. M. KRAWCZYK[3] & J. F. W. NEGENDANK[3]

[1]*PanTerra Geoconsultants B.V., Veerpolder 5, 2361 KX Warmond, The Netherlands (e-mail: h.rieke@panterra.nl)*
[2]*Geologisches Institut, Rheinische Friedrich-Wilhelms-Universität, Nussallee 8, 53115 Bonn, Germany,*
[3]*GeoForschungsZentrum, Telegrafenberg, 14473 Potsdam, Germany*

Abstract: About 3 km of core material from 14 wells together with additional data from several hundred wells across the NE German Basin (NEGB), have been investigated in order to reconstruct the facies architecture and the evolution of the Upper Rotliegend II. Special attention has also been given to the verification of various controlling factors and their influence on sedimentation in an arid continental environment. The facies architecture within the logged profiles comprises five main environments, namely braided plain, ephemeral stream floodplain, sand flat, mudflat and playa lake. The evolution can be subdivided into four distinct basin-wide correlatable periods – Parchim, Mirow, Dethlingen and Hannover formations – with each of them being characterized by a specific basin geometry and interplay of controlling factors. The deposition of the basal Parchim Formation largely took place within a tectonically created basin, whereas the facies evolution displayed an initial less-arid climatic period and later shift to an arid climate. The succeeding Mirow Formation marks the beginning of thermally induced basin subsidence. However, sedimentation itself clearly reflects a period in which the climate was relatively less arid. The overlying Dethlingen Formation was largely controlled by the increasing thermal subsidence of the basin, leading to broad extension towards the south and east. Internally, the strata can show the effects of climatic variability, depending on their position within the basin. The uppermost Hannover Formation was the product of ongoing basin subsidence, a reduction in sediment supply and an increasingly peneplaned topography. In summary, evolution of the Upper Rotliegend II within the NEGB reveals a variety of factors which have a significant influence on sedimentation, such as climate variations, the creation rate and amount of accommodation space, wind direction, sediment budget and source area lithology. An understanding of how these various factors interlink in controlling basin infill is of great significance in understanding the complex depositional history of arid continental successions.

Exclusively continental strata from recent or ancient formations have been the focus of many researchers with regard to the evolution of sedimentary basins world-wide (e.g. Leeder & Gawthorpe 1987; Blair & Bilodeau 1988; Nemec & Steel 1988; Mountney *et al.* 1999). These studies have tended to focus on the facies architecture, tectonic setting, sequence stratigraphy, evolution of sedimentary infill and provenance analysis. While much effort has been given to the fundamental question of the various factors controlling the sediment supply and evolution of such basins, the question as to whether synsedimentary tectonics or climate change, or a combination of both, is of greater importance to the sediment distribution, thickness and resulting facies architecture, remains to be determined (e.g. Blair 1987; Frostick & Reid 1989; George & Berry 1993; Clemmensen *et al.* 1994; Dorn 1994; Blakey *et al.* 1996; Mack & Leeder 1999). The examination of the Rotliegend sediments within the Northeast German Basin (NEGB) provides an excellent opportunity to contribute to this ongoing debate.

The <2 km thick succession of Rotliegend (i.e. Upper Permian) sedimentary strata within the NE German Basin (NEGB) have been the target of extensive exploration since considerable finds of gas were made in the late 1960s, and these units represent Germany's largest hydrocarbon reservoirs (Müller *et al.* 1993). Until the early 1990s, approximately 1500 deep wells were

From: McCann, T. & Saintot. A. (eds) *Tracing Tectonic Deformation Using the Sedimentary Record.* Geological Society, London, Special Publications, **208**, 71–94. 0305–8719/03/$15.00

drilled within the NEGB and adjacent areas, with several hundred of them penetrating Rotliegend-age formations. Thus, a unique database across the entire NEGB, comprising kilometres of core material (e.g. Hoth *et al.* 1993), geophysical wireline logs and an extensive network of seismic profiles (particularly in NE Germany and the Altmark, see Fig. 3 for location), together with the DEKORP-BASIN 9601 profile (DEKORP-Basin Research Group 1999), has been acquired. However, detailed sedimentological studies of the Rotliegend succession, focusing on facies architecture and facies evolution within the NEGB, were lacking, since previous scientific work in the area focused mainly on the lithostratigraphy (see Plein 1995, and references therein). The stratigraphy of the Permian clastics has been mainly assumed to be controlled by a series of tectonic events, termed 'Altmark events I–IV', which separated the sedimentary evolution of the Upper Rotliegend II into four distinct periods (Hoffmann 1990; Gebhardt *et al.* 1991; Hoffmann *et al.* 1997). However, a detailed analysis of the sedimentary facies has never been undertaken in the region to fully establish, firstly, the precise temporal and regional development and evolution of the various continental depositional environments over time, and secondly, the interrelationship between the development of these sedimentary units and the prevailing tectonic and other controls which were active at that time.

Our understanding of the sedimentology, facies and stratigraphy of the unfossiliferous siliciclastic Rotliegend-age formations within the Southern Permian Basin (Fig. 1) has made fundamental progress since the early 1990s (e.g. George & Berry 1993, 1997; Yang & Nio 1994; Howell & Mountney 1997; Kiersnowski 1997; Gast *et al.* 1998; Glennie 1998; Sweet 1999). Therefore, the aim of this paper is twofold. Firstly, a high resolution facies architecture subdivision will be presented, resulting from detailed facies interpretations. Based on these results, the evolution of the Upper Rotliegend II will be reconstructed as a series of four time slices for the NEGB. This dynamic evolution scheme will thus reveal the complexity of the interacting factors, such as tectonism, climate, sediment supply and basin subsidence, which controlled sedimentation in the area.

The database for this study consists of core material, lithological profiles and geophysical wire-line logs from about two hundred wells across the basin (Fig. 2). In order to reconstruct the facies architecture and evolution, *c.*3.4 km of Rotliegend core material from 14 deep wells across the northern and southeastern margins were logged in detail to determine composition, provenance, facies types, distribution and poss-

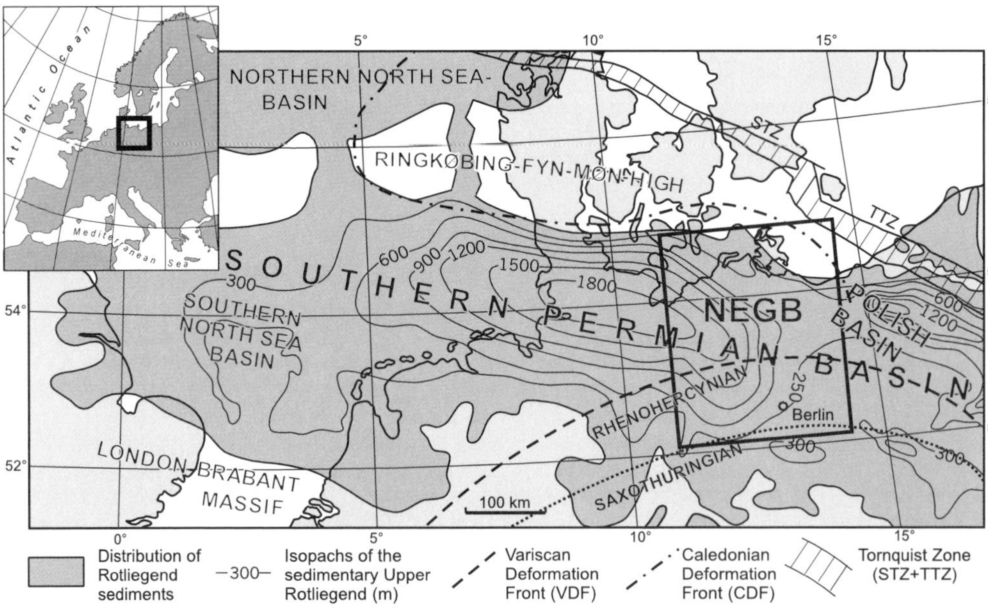

Fig. 1. Location of the NE German Basin (NEGB) within the Southern Permian Basin (after Kiersnowski *et al.* 1995 and Ziegler 1990).

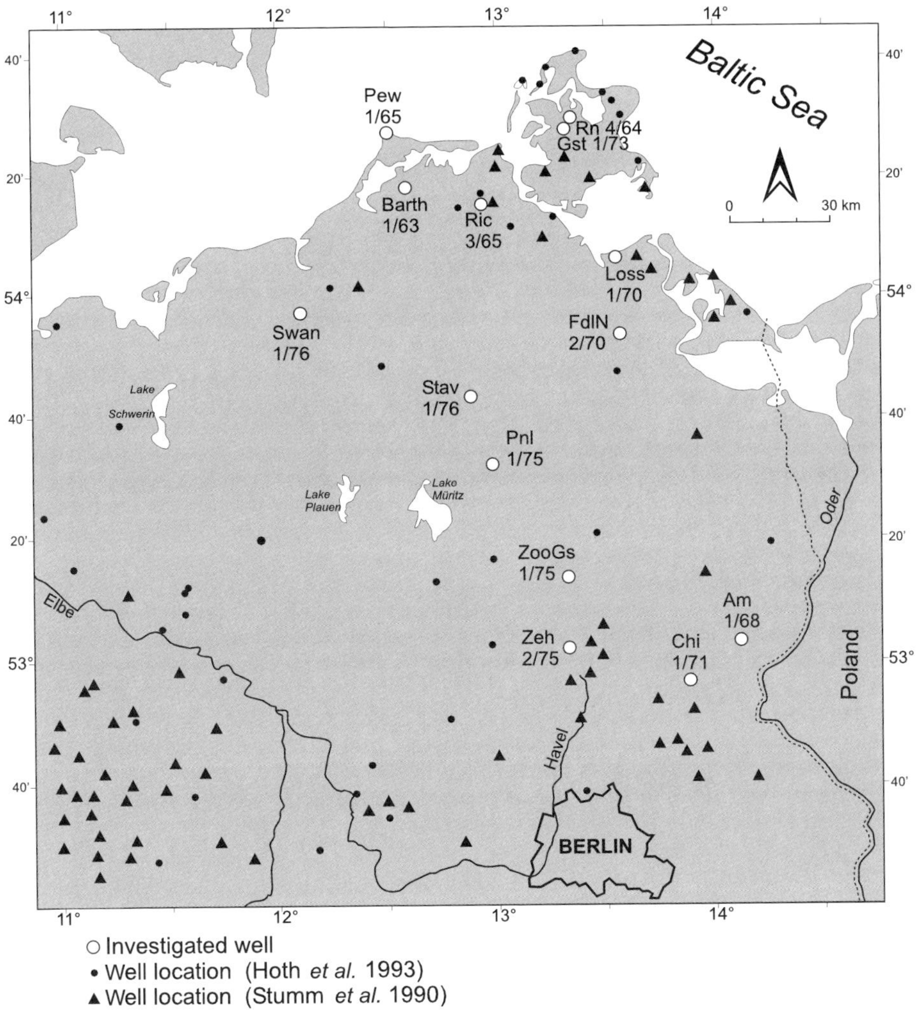

Fig. 2. Geographical map of the investigated area showing the database used in this study. (after Hoth *et al.*, *1993;* Stumm *et al.*, 1990 and this work). Key: Am, Angermünde; Barth, Barth; Chi, Chorin; FdlN, Friedland; Gst, Gingst; Loss, Loissin; Pnl, Penzlin; Pew, Prerow; Ric, Richtenberg; Rn, Rügen; Stav, Stavenhagen; Swan, Schwaan; Zeh, Zehdenick; ZooGs, Zootzen.

ible source areas. Additionally, predominantly unpublished extensive data-sets comprising detailed petrographical investigations and litho-facies maps from former workers provided useful background information.

Regional geology

The NE German Basin is one of a series of interconnected sub-basins, together termed the Southern Permian Basin (SPB), which extended *c.*1500 km from Poland in the east to England in the west (Fig. 1). The German part of the SPB is situated between the Precambrian Baltic Shield to the north and Variscan-influenced areas to the south (Berthelsen 1992; Franke *et al.* 1989; DEKORP-BASIN Research Group 1999). Three major, approximately NW–SE-striking fault zones occur in the area of the NEGB: the Caledonian Deformation Front (CDF), the Tornquist Zone (TZ), comprising the Sorgenfrei–Tornquist Zone (STZ) in the northwest and the Tornquist–Teisseyre Zone (TTZ) in the southeast, and the Variscan Deformation Front (VDF), whose exact position is still under discussion.

The present-day NE German Basin is bounded to the north by the Fyn–Møn–Arkona High which forms the eastward continuation of the Ringkøbing High (Fig. 3). The southern boundary comprises the Variscan fold and thrust belt. The transition to the adjacent Polish Basin terminates the NEGB to the east, whereas the NW German Basin represents the western limit of the study area.

In Permo-Carboniferous times the intracontinental NEGB developed in the area of the post-Variscan foreland depression (Plein 1993). In its initial phase, the collision of the northward-moving African and European plates established

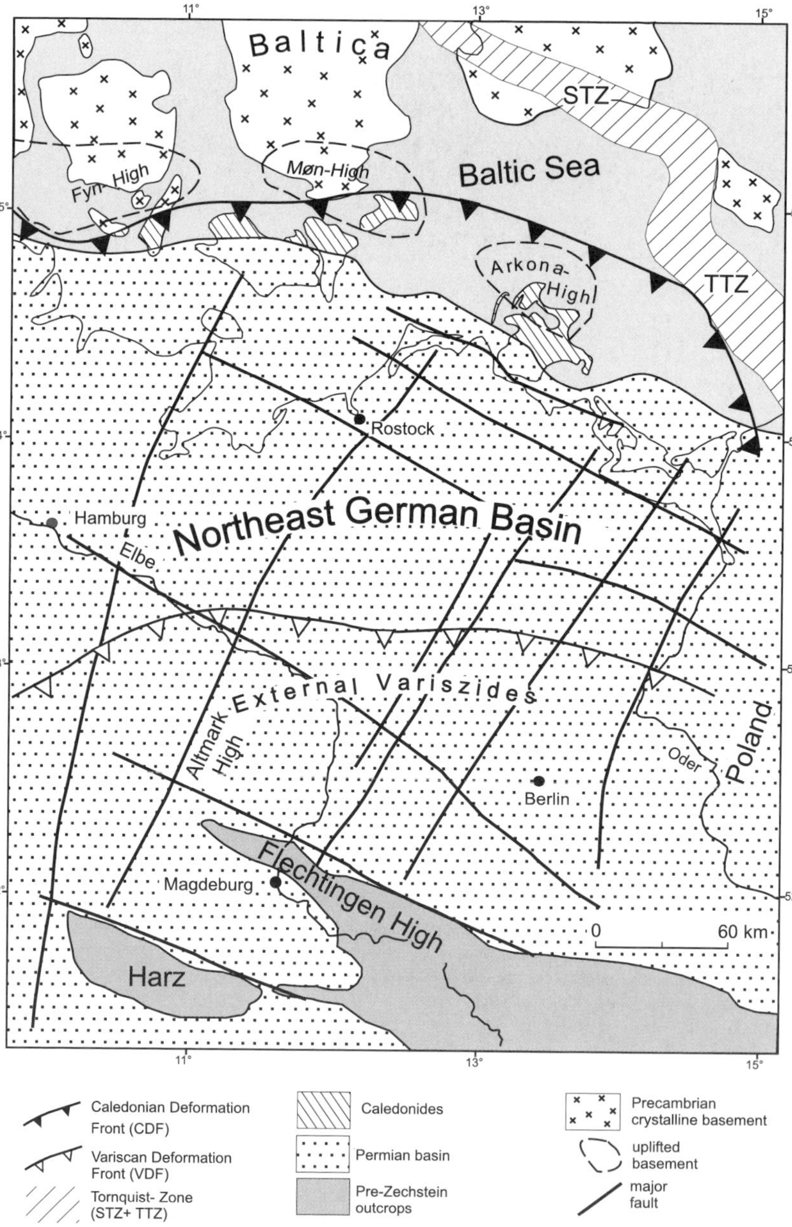

Fig. 3. Geological map of the NE German Basin (NEGB) and the main tectonic elements in the region (modified from Ziegler 1990 and Plein 1995).

a dextral strike-slip regime across the region, with a predominant east–west extensional component (Arthaud & Matte 1977; Gast 1988; Ziegler 1990; Glennie & Underhill 1998). Within the NEGB, wrench-induced normal faulting along a network of NE–SW- and NW–SE-oriented tectonic lineaments took place, which has been termed 'Frankonian movements' (Fig. 3; Franke *et al.* 1989; Bachmann & Hoffmann 1995, 1997). This tectonic activity was partially responsible for the period of extensive magmatism across the NEGB, resulting in an up to 2000-m thick Permo-Carboniferous-age succession of predominantly calc-alkaline SiO_2-rich volcanics (Benek *et al.* 1996). The extrusions of the effusives throughout most of the NEGB are assumed to have taken place within a short time-span between 297–302±3 Ma (i.e. the magmatic flare-up described by Breitkreuz & Kennedy 1999). Large amounts of material derived from the mantle suggest a destabilization of the lithospheric crust (Benek *et al.* 1996).

The cessation of tectonism and magmatic activity coincided with the onset of Rotliegend sedimentation, which was more or less continuous over a period of *c.*40 Ma. Therefore the Rotliegend succession within the NEGB represents the time-span from *c.*297 Ma (i.e. subsequent to volcanic activity) to 258 Ma (i.e. the Rotliegend–Zechstein boundary), and has a cumulative thickness of up to 2 km (McCann 1998). Based mainly on lithostratigraphy, the Rotliegend-age strata can be subdivided into four main subgroups, namely, Altmark, Müritz, (for Rotliegend I) and Havel, Elbe (for Rotliegend II) (Fig. 4; Plein 1995). The latter two can be further

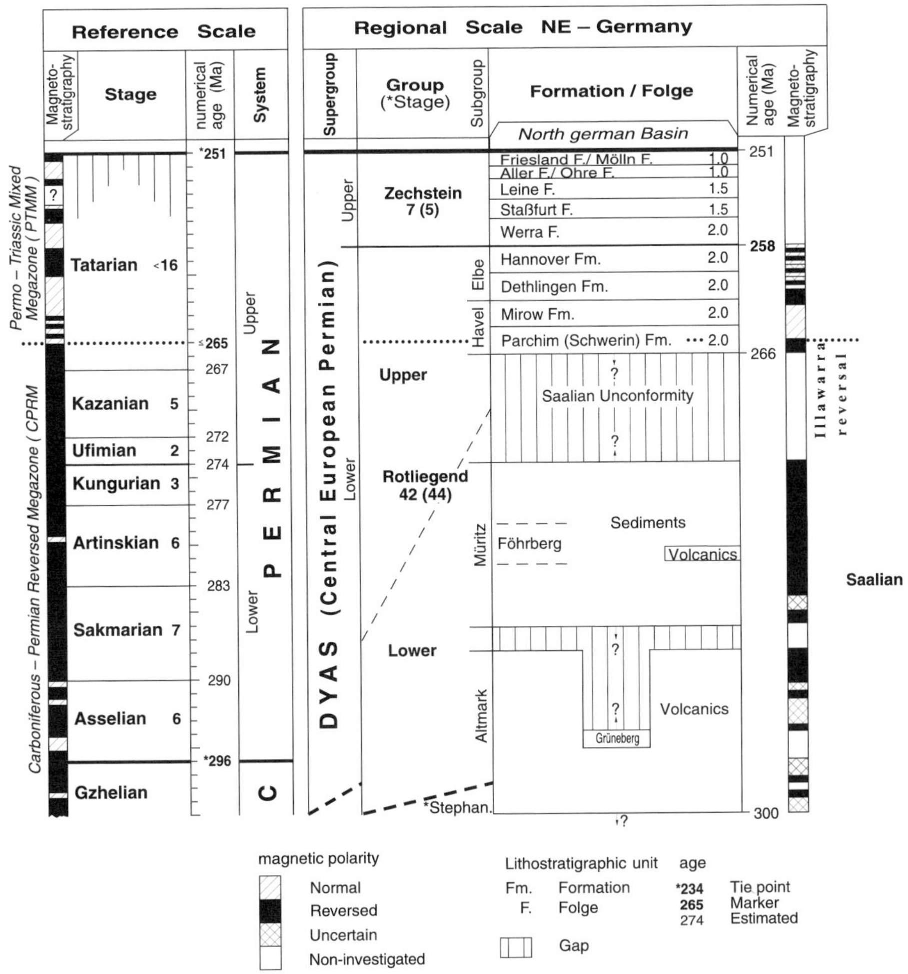

Fig. 4. Regional Permian stratigraphy of the NE German Basin (modified after Menning, in Plein 1995).

divided into the Parchim and Mirow formations and the Dethlingen and Hannover formations, respectively.

Initial deposition of clastics in the NEGB (i.e. Altmark and Müritz subgroups) was confined to a series of restricted and isolated grabens and sub-basins (e.g. Plein 1993; Rieke *et al.* 2001). However, little is known about the spatial distribution of these structures due to their rare occurrence in boreholes and the distinct lack of lateral control. With the exception of rare profiles which contain marine and lacustrine fossil associations, and which have been dated as Permo-Carboniferous and Lower Permian (e.g. Grüneberg Formation; Gaitzsch 1995; Schneider *et al.* 1995) in age, much of the Altmark- and Müritz-age succession is exclusively continental and its subdivision, based purely on lithostratigraphy, is questionable.

The basin-wide correlatable Saalian Unconformity (Fig. 4), representing a time span of uncertain age, separates the overlying Upper Rotliegend II (Havel and Elbe subgroups) from the underlying older Permian to Carboniferous strata (Fig. 4). These so-called Saalian movements reactivated the pre-existing tectonic regime with a predominant east–west extensional component (Bachmann & Hoffmann 1995). A series of roughly NE–SW-trending graben-structures was created and/or rejuvenated across the basin margins during this period (Baltrusch & Klarner 1993; Klarner 1993, this volume). At the Variscan-influenced southern margin of the NEGB, tectonism was accompanied by the extrusion of a tholeiitic to alkaline series of basaltic lavas (Marx *et al.* 1995).

The overlying successions of the Upper Rotliegend II are exclusively continental in character, except for some rare marine ingressions within the Elbe Subgroup (Plein 1995). The general lack of datable fossils precludes more definite biostratigraphic correlation. The magnetostratigraphic Illawarra reversal in the lower part of the Parchim Formation is the only chronostratigraphic time marker. This has a postulated age of *c.*265 Ma (Menning 1995).

In post-Illawarra reversal times, the increasing lithospheric cooling gradually extended to encompass the entire NEGB, with prograding overstepping of the basin margins (Van Wees *et al.* 2000). The ingression of the Zechstein Sea at 258 Ma (magnetostratigraphic date, after Menning, 1995) terminates the Rotliegend evolution within the entire NEGB.

In general, the Rotliegend environment was controlled by the palaeoposition of the Southern Permian Basin at 15°N during Permian times and by its position located in the central parts of the continent Laurasia, several thousands of kilometres distance from the southerly located Tethyan Ocean. The climate was therefore arid to semi-arid, with a predominant palaeo-wind direction from the NE (Northern Hemisphere trade wind; Glennie 1982, 1990).

Facies interpretation

A detailed interpretation and classification of the facies associations within the Rotliegend of the NEGB was possible due to the large amount of available core material. Five main facies, which are interpreted as specific environments following Hardie *et al.* (1978), were determined in the NEGB, namely: braided plain, ephemeral stream floodplain, sand flat, mudflat and playa lake. Specific sedimentological processes operating within the ephemeral stream floodplain environment allowed further subdivision into distinct facies associations, which will be outlined below.

Braided plain environment

The majority of the conglomerates present within the investigated profiles occur at the very base of the Upper Rotliegend II succession and represent *c.*8% of the entire logged clastics. The thickness varies from several metres up to some tens of metres (e.g. 80 m in the Swan 1/76 well).

The conglomerates are predominantly clast-supported, disorganized and massively bedded, with rare weakly developed parallel bedding. Subordinate intercalations of thin beds (<1–2 m) of matrix-supported conglomerates were noted. Open framework layers, some of which show rare poikilitic calcite cementation, also occur. The conglomerate units contain a wide range of clast sizes, ranging from single boulders to mud, although the majority are composed of a fine- to medium-sized pebble fraction. Clast shapes are generally subrounded to subangular.

The conglomerates are interpreted as having been deposited as channel bed-load within a series of fluvial channels within a braided plain environment. The subordinate matrix-supported beds were presumably deposited as the terminal lobes of dense sheet floods.

It is generally difficult to correctly determine the depositional environment of coarse-grained sediments based purely on borehole data, since the lack of boreholes precludes the establishment of the lateral geometry. Thus, it is difficult to distinguish between deposits of an alluvial fan and those of a braided plain. The precise lateral geometry of the sediment bodies – in this situation a decisive criterion – cannot be reconstructed solely from down-hole data. Supportive information, for example from seismic profiles, is also lacking in the NEGB, due to the absence of reflectors within the Rotliegend clastics and due to masking effects of the overburden Zechstein-evaporites (DEKORP-Basin Research Group 1999; Rieke *et al.* 2001). Nevertheless, coarse-grained clastic facies such as the one described are generally either derived from a tectonically controlled local alluvial fan, or they were deposited as part of a broad braided plain environment.

The conglomerates from this study show no typical sedimentological structures, for example, large-scale fining and/or coarsening upward trends, a wide spectrum of grain sizes, development of soil horizons, and changes in depositional style from sheet flow to mass flow or vice versa, which would be typical of an alluvial fan environment. The widespread occurrence of the sediments across the basin, the relatively low thicknesses (*c.*80 m) of the individual units, their predominant stream-flow character and the fact that fault-controlled source areas have not been definitively established, would all strongly suggest that these conglomerates should be referred to the braided plain environment.

Ephemeral stream floodplain environment

Almost all of the logged fine- to coarse-grained sandstones of fluvial origin can be grouped into the ephemeral stream floodplain environment which represents 33% of the examined profiles. The predominant sedimentation processes are channelized and unconfined ephemeral streams and sheet flows involving a mixed load of pebbles, sand and mud. The downstream reduction in the fluvial transport energy and discharge as a result of transmission loss, evaporation and the prograding diversion/bifurcation into a distributary network of fluvial streams are key sedimentological factors for the detailed subdivision of the ephemeral stream floodplain environment. Three distinct facies associations can be distinguished, namely: *proximal fluvial*, *medial fluvial* and *distal fluvial*. The lateral transitions between these facies associations are gradational in character and are determined by their specific lithologies and bedding types.

Proximal fluvial facies association. A typical litholog of the proximal fluvial facies association is shown in Figure 5. The lithologies are dominated by a variety of poorly to moderately sorted coarse-, medium- and fine-grained sandstones showing sediment structures typical of the middle–upper flow regime, such as medium- to large-scale cross-sets (inclinations up to 25°; Fig. 5) or planar lamination. The sandstones are predominantly multistorey/amalgamated in character and have a thickness of up to 30 m, with internally single sets of up to 3 m. Normal grading and subordinately reverse grading occur presently within the sets usually lacking any fine-grained layers. Couplets of very poorly sorted coarse-grained sandstones and sandy mudstones were noted regularly showing intense soft-sediment deformation structures such as convolutions or single water escape structures which almost destroyed the primary fabrics. Rare intercalated muddy fine-grained sandstones (up to 1 m thick) with ripple cross-lamination or flaser structures, are indicative of lower flow regimes.

These complex successions are interpreted to represent a dense network of channelized, high-energy fluviatile streams in the proximal fluvial facies of the ephemeral stream floodplain environment. The predominantly multistorey arranged sandstone-sets bounded by erosive fluvial discordances are typical fabrics for intra-channel facies. The preservation potential for deposits of moderate to low transport energies is generally poor except for the secondary reworked couplets of sand and mud, which are interpreted as the product of overbank-flooding adjacent to the main fluvial channels.

Medial fluvial facies association. The lithology of the medial fluvial facies, as shown in Figure 6, is dominated by fine-grained sandstones with a grain-size spectrum ranging from mud to coarse-grained sand. The sandstones are generally moderately to poorly sorted. The sedimentary structures present – for example, ripple-cross bedding, heterolithic alternations of mud and sand, weakly horizontal bedding/lamination or structureless – are indicative of moderate to low fluvial transport energies. Fining-upward sequences with an average thickness of 2 m are common. They are usually composed of fine-grained sandy layers at the bottom which grade upwards to homogeneous mudstones or claystones at the top. Rip-up clasts within the basal parts of these fining-upward sequences are predominantly composed of mudstones while desiccation cracks are common in the mudstones at sequence tops. Poorly sorted medium- and coarse-grained sandstones are limited to rare single beds (up to 3 m thick). Internally, these sandstones exhibit weakly horizontal bedding to completely disorganized fabrics. Homogeneous, or irregular- to lenticular-bedded sandy mudstones of less than 2.5 m thickness were rarely noted.

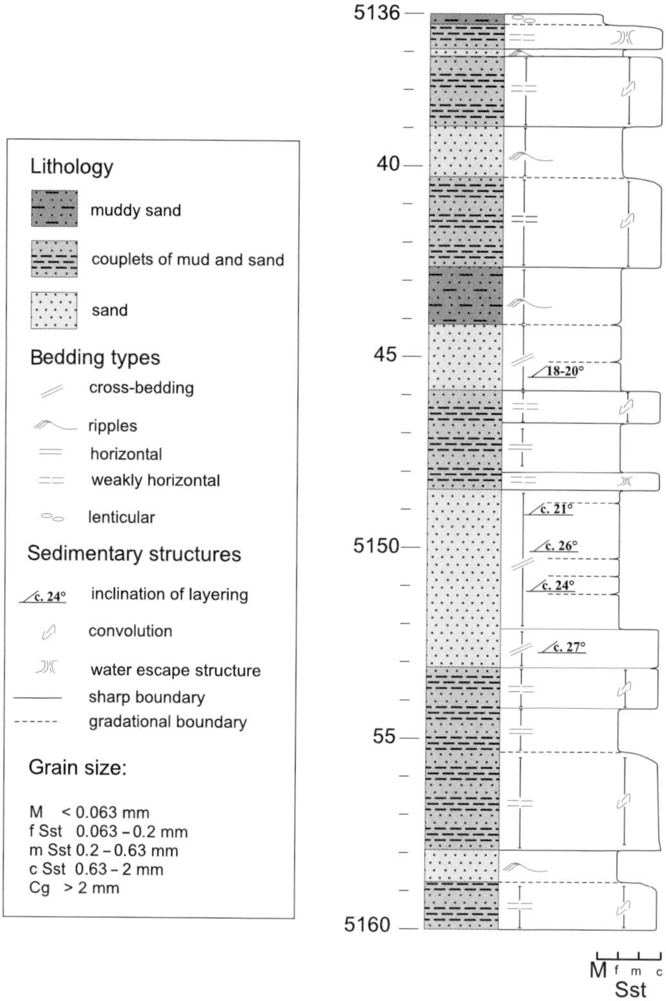

Fig. 5. Litholog from the Penzlin 1/75 well showing the proximal fluvial facies of the ephemeral stream floodplain environment. Scale: cored interval in metres.

The highly variable lithologies observed within these units may be interpreted as having been deposited within the medial fluvial facies association of the ephemeral stream floodplain environment. The predominant accumulation process is periodically unconfined sheet floods which carry mixed sand/mud loads and are indicative of an overall reduction in fluvial transport energy. The commonly occurring graded beds represent these single fluvial waning flow events and are, therefore, a very characteristic feature of this facies. However, the incision of single storey fluvial streams, related to periods of higher transport energy (i.e. flash floods after heavy rainfalls) which is likely to occur within such a environment, is rarely noted. Rip-up clasts and desiccation cracks provide evidence of periods of subaerial exposure and indicate the ephemeral character of these streams.

The intercalation of sandy mudstones marks distinct periods of non-fluvial sedimentation within this environment. The origin of these deposits is discussed in detail in the mudflat environment section (see below).

Distal fluvial facies association. The logged core-section (Fig. 7) shows a very typical example for the distal fluvial facies of the ephemeral stream/floodplain environment. The profiles are usually composed of interdigitated mudstones

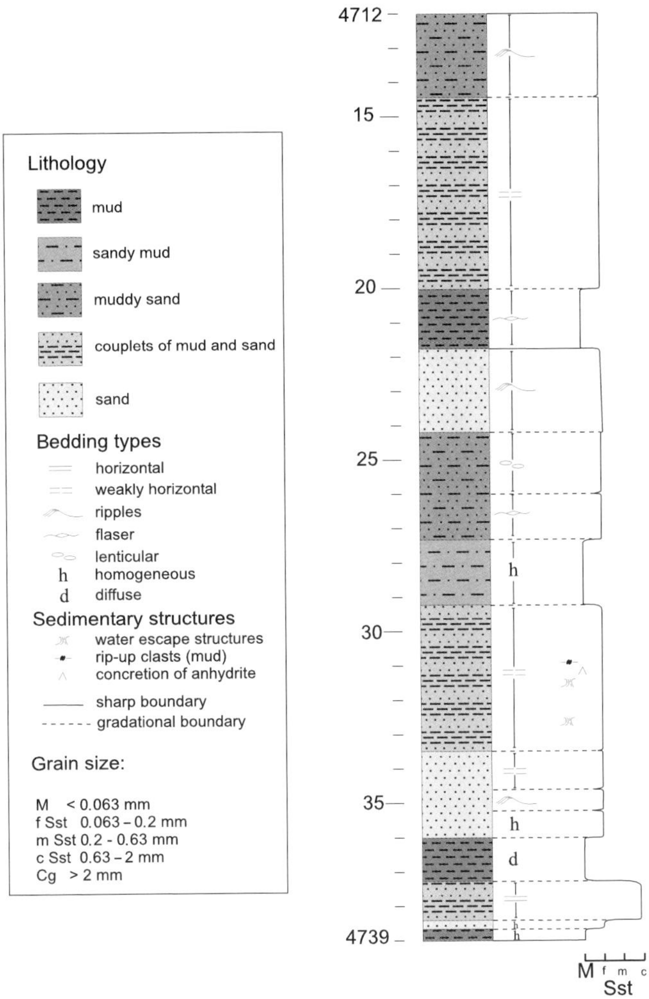

Fig. 6. Logged section from the Stavenhagen 1/76 well illustrating the medial fluvial facies of the ephemeral stream floodplain environment. Scale: cored interval in metres.

with varying contents of sand and single fine-grained sandstone beds with a maximum observed thickness of up to 4 m. Internally, the beds include a variety of sedimentary structures, for example, small-scale ripple cross-bedding, and planar, heterolithic alternations of sand and mud, and wavy bedding. Homogeneous or completely disorganized sandstones (<0.6 m thick), are also observed. Normal grading is common, with sandstones fining upwards into mudstones or claystones. Rip-up clasts, convolute and water-escape structures and desiccation cracks are frequently observed.

These sediments are interpreted to represent a relatively topographically flat depositional environment, which was subjected to episodic incursion by sandy sheet flows or mud flows of the most distal fluvial regime of the ephemeral stream floodplain environment. The surficially unconfined sheet flows rapidly wane due to transmission loss resulting in the deposition of a widespread fine-grained sandstone bed which contains a variety of sedimentary structures indicative of low-energy conditions. Intercalated mudstones represent deposition in ephemeral freshwater lakes. The frequency of the mudstones increases downstream and is related to the decreasing periodicity of the fluvial streams. The distal fluvial facies of the ephemeral stream/floodplain environment passes laterally

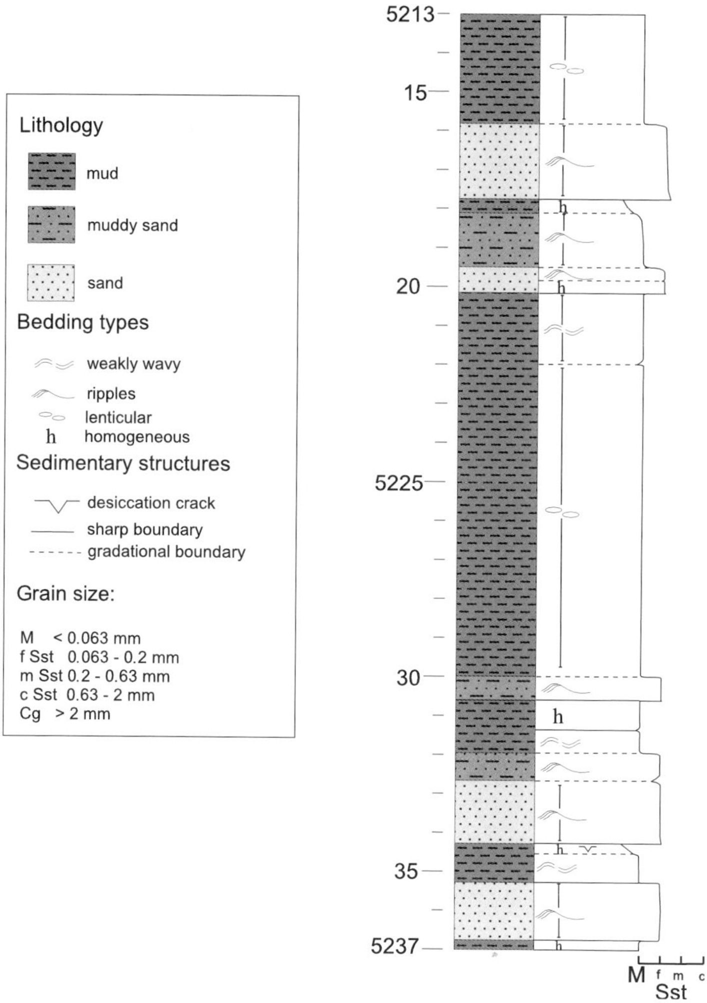

Fig. 7. Litholog from the Schwaan 1/76 well displaying a typical section of the distal fluvial facies of the ephemeral stream floodplain environment. Scale: cored interval in metres.

into the mudflat, sand flat or playa lake environments.

Mudflat environment

The sediments of the mudflat environment are the most frequent within the investigated profiles and represent *c*.45% of the logged cores. The lithologies comprise fine-grained muddy sandstones, sandy mudstones and mudstones. Sorting is generally poor. Sedimentary structures vary markedly from irregular lenticular bedding to units which are very homogeneous and exhibit no internal structures. Boundaries tend to be diffuse rather than sharp. Isolated lenses of either fine-grained sandstones with subrounded to rounded individual grains, or sandy mudstone, are present within the mudstones. These lenses have a mean size of 2 x 1.5 cm. Concretions of anhydrite and halite pseudomorphs were frequently noted. Desiccation cracks are rare. Fine-grained sandstone units, which are up to 2 m thick, occur sporadically within the profiles. Internally, these sandstones range from structureless, to weakly developed ripple cross-bedded or with heterolithic alternations of sand and mud.

The sediments of this environment were deposited on a flat topography where the groundwater level was either close to, or even temporarily at, the surface. The semi-arid climate generated an evaporitic milieu. Thus, the main depositional process is assumed to be the accumulation of wind-blown clastic material (clay to fine-grained sand) on surficially developed efflorescent and precipitated salt crusts (cf. Goodall *et al.*, 2000). The uppermost few centimetres of the sedimentary profile are almost completely cemented by evaporites leading to a muddy to fine-sandy substrate with embedded irregular lenses of fine-grained sand (i.e. popcorn texture). However, the salt crust itself never remains preserved and the primary sediment structures generally have a relatively low preservation potential due to very intense secondary reworking caused by deflation, halo-turbation, an oscillating water table and periodic, surficial flooding following heavy rainfall. The latter is indicated by the regular occurrence of fine-grained sandy layers (sheet flows) and by units of homogeneous to laminated claystones (<2 m thick) of highly variable colours from brown-red to green-grey which were deposited in ephemeral or perennial playa lakes. The frequent occurrence of anhydrite concretions is indicative of the presence of original surface precipitates of gypsum which were altered to anhydrite following burial.

Sand flat environment

Sediments belonging to the sand flat environment are much less abundant than those of the mudflat, comprising approximately 9% of the logged profiles. They consist mainly of fine-grained sandstones with varying contents of mud and subordinately medium- to coarse-grained sandstones. The lithology and sedimentary structures (Fig. 8) indicate that the environment comprised a fine-grained sandy substrate with occasional mud drapes, which produced the resultant irregular wavy to crinkled laminae. Additional structures, for example, small-scale pseudo-cross-lamination, planar adhesion lamination, and vertically climbing adhesion ripples (cf. Goodall *et al.* 2000) were also noted. Intercalations of weakly cemented, poorly sorted, fine- to coarse-grained sandstones, showing no clear structure rarely occur. Individual bed thickness does not exceed 2 m. Anhydrite concretions, reaching a maximum size of *c.*10×20 cm, are rare.

In terms of their genesis, sand flat sediments are similar to those of mud flats. The predominant sedimentation process for the sand flat deposits is the accumulation of wind-blown clastic material on surficially developed salt ridges as described by Fryberger *et al.* (1983, 1984). In addition, small-scale pseudo-cross lamination, planar adhesion-lamination and vertically climbing adhesion ripples provide clear proof of subordinate non-evaporitic deposition.

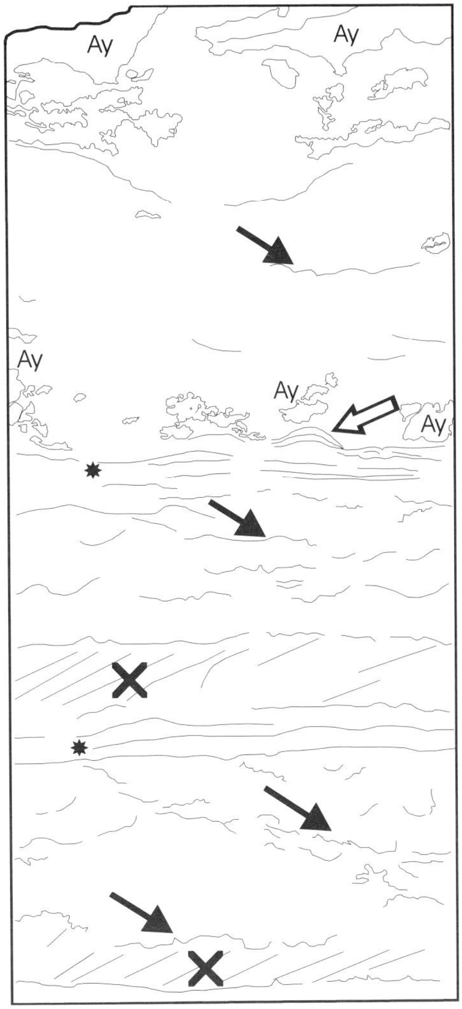

Fig. 8. Draft taken from the Penzlin 1/75 well indicating the characteristic textures of evaporitic and non-evaporitic sediment accumulation. Legend: Ay, concretion of anhydrite; black arrow, irregular to crinkled muddy laminae of former salt ridges; asterisk, planar adhesion lamination; white arrow, vertically climbing adhesion ripples; cross, small-scale pseudo-cross-lamination.

In general, the sand flat environment is characterized by the input of wind-blown sand and, thus, has relatively higher accumulation rates than that of the mudflat environment. This could indicate a more proximal or marginal position of the sand flats relative to the source area. The existence of fragile, and thus rarely preserved, sedimentary structures such as pseudo-cross laminae would suggest a lower level of reworking of the sand flat sediments. Temporarily inundation is marked by the occurrence of poorly sorted fluvial sandstone beds of low-energy sandy sheet floods.

Playa lake environment

The successions of the playa lake environment form *c*.4% of the entire logged strata. Playa lake deposits mainly consist of homogeneous to laminated carbonate-rich claystones of varying colours from green–grey to red–brown. Thinly bedded to even laminated couplets of clay and silt to fine-sand were noted. Interdigitation of homogeneous to laminated claystones and sandy mudstones was also noted. Intercalations of clay laminae and clastic layers composed of fine sand to coarse sand are extremely rare. The claystones contain rare ostracode and conchostracae fossils, whereas *Hydromedusae limnica* (Müller 1978) are abundant in some profiles.

The sediments of this environment were deposited within temporary playa lakes located on the basin floor (cf. Gast 1991; Gaupp *et al.* 2000). Differentiation between ephemeral and perennial water bodies is rarely possible, and the term ‘perennial’ should only be applied to a succession comprising complete sequences of laminated claystones and evaporites (e.g. Gast 1991; Gaupp *et al.* 2000). The lakes within the NEGB were generally closed, with water being supplied from underground springs. Surficial clastic input by fluvial streams can thus be largely excluded owing to the generally arid climate which would have limited the lateral range of these ephemeral streams.

Upper Rotliegend II palaeogeography and basin evolution

Using the detailed facies analysis presented here, it is possible to trace the sedimentary evolution of the Upper Rotliegend II within the NEGB. Four main evolutionary stages (i.e. Parchim, Mirow, Dethlingen and Hannover formations) can be separated throughout the basin, with each of them characterizing a specific basin geometry and facies architecture.

Parchim Formation. Basin morphology at the beginning of Parchim Formation times had been largely established by the regional stress regime which comprised a major E–W-trending extensional component (Fig. 9). The main eastern boundary of the basin is the north–south trending intrabasinal Altmark High, whereas the Wollstein High forms the limit to the SE (and partly east). Towards the north the basin is bounded by the Richtenberg High (Fig. 9). Along the basin margins several NE-trending graben-like structures developed or were rejuvenated by the existing stress regime (Baltrusch & Klarner 1993; Klarner 1993; Helmuth & Süssmuth 1993; Helmut & Schretzenmayr 1995). The main depocentre comprised two distinct sub-basins, each containing up to 600 m of sediments. These sub-basins have been interpreted as transtensional strike-slip basins (e.g. Bachmann & Grosse 1989).

Facies distribution during Parchim Formation times was controlled by the tectonically generated morphology. The initial basin-wide period of sedimentation was indicative of a semi-arid climate and was marked by the extensive progradation of braided plain systems towards the sub-basin centres (Fig. 10a). The simplified palaeogeographical map shows the maximum extent of the braided plain environment. The basin margin graben systems were activated as feeder channels supplying the fluvial streams with clastic material from the surrounding Permo-Carboniferous volcanic hinterlands. Fluvial sedimentation was more active along the southern margin (residual Variscan highland areas), from where most of the sedimentary input was derived. Thus, a few wells in the area penetrated coarse-grained profiles, which show several coarsening and fining-upward cycles and strong variations in the inclination of the beds (up to 20°). These sediments might be interpreted as sheet flood-dominated alluvial fan deposits indicating a proximal deposition close to the source area.

The basin centre shows a prevailing sand flat environment. In the northwesterly sub-basin, homogeneous and laminated claystones of the ephemeral playa lake environment have also been cored.

During the succeeding arid period the sediment supply was almost cut off and the fluvial streams were restricted to thin belts across the basin margins (Fig. 10b). Most parts of the basin were covered by sand flat sediments. In the NW sub-basin, up to 70 m of evaporites (i.e. Schwerin Salina) represent this dry phase. Along the southern margin, the north–south trending Altmark High acted as a morphological barrier to sediment transport. Thus, the NE-directed palaeo-wind accumulated an extensive erg at the eastern flank of the high, comprising dunes, interdune areas and aeolian sheet sands (Kleditzsch & Kurze 1993; Helmuth & Süssmuth 1993).

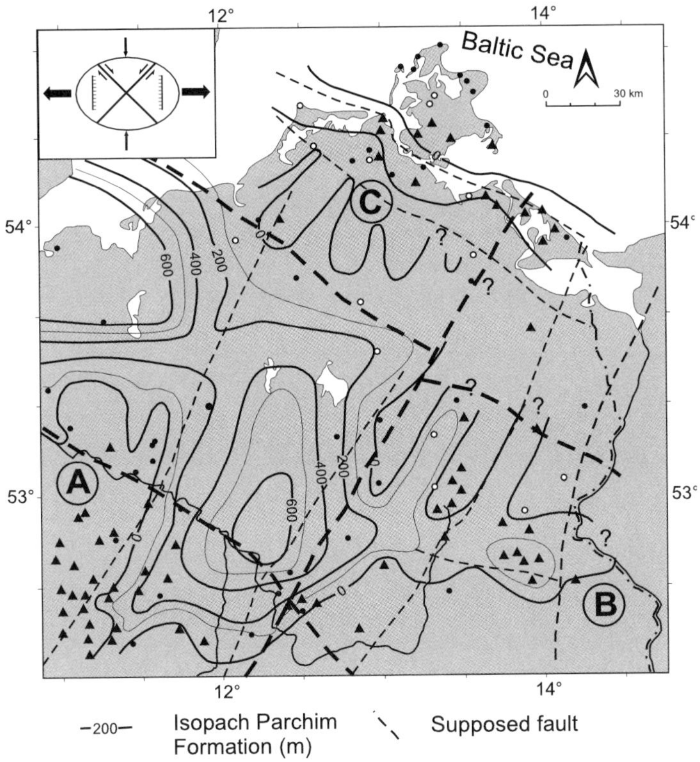

Fig. 9. Isopach map showing the distribution of the Parchim Formation (contours are in metres). Key: A, Altmark High; B, Wollstein High; C, Richtenberg High.

Mirow Formation. By Mirow Formation times, thermally induced basin subsidence had led to the formation of a more uniform depositional area extending both to the east and south (Fig. 11). The basin-bounding highs were progressively eroded over time during this period. The original graben-like features are preserved along the northern and southern margins. The main depocentre is located in the northwest and contains up to 600 m of clastics.

Sediments of the Mirow Formation discordantly overlie those of the Parchim Formation within the central parts of the basin, and Permo-Carboniferous volcanics, or locally older Carboniferous strata, along the basin margins. The lowermost boundary of this formation is a sharply developed one in almost all of the cored profiles, and can thus be easily correlated across the basin. It is characterized by the erosive occurrence of fluvial ephemeral stream floodplain sediments suggestive of a shift to a less-arid climate. The palaeogeographical map (Fig. 12a) indicates the maximum basinward progression of the fluvial belts. The fluvial-dominated strata (up to 200 m thick) at the southern basin margin may be subdivided into proximal, medial and distal fluvial zones. Vertically, the profiles show one or two fining-upwards trends depending on their relative position within the basin. Across the northern basin margin, the older graben structures were reactivated as feeder channels. However, due to the already relatively smooth morphology, the sediments display mainly moderately to low fluvial energies and the sedimentary input was reduced. Cored profiles from the basin centre show intercalations of mudflat sediments and homogeneous claystones of ephemeral playa lakes. The latter claystones contain large quantities of *Hydromedusea*, indicative of freshwater conditions.

Localized small ergs established at the southern margin suggest that the periods of arid climate were relatively short. It is also possible that the laterally restricted occurrence of these erg bodies was related to remnant topography. Coeval reduction of discharge limits the fluvial sedimentation to small belts along the margins and basin centre shows continuous mudflat facies

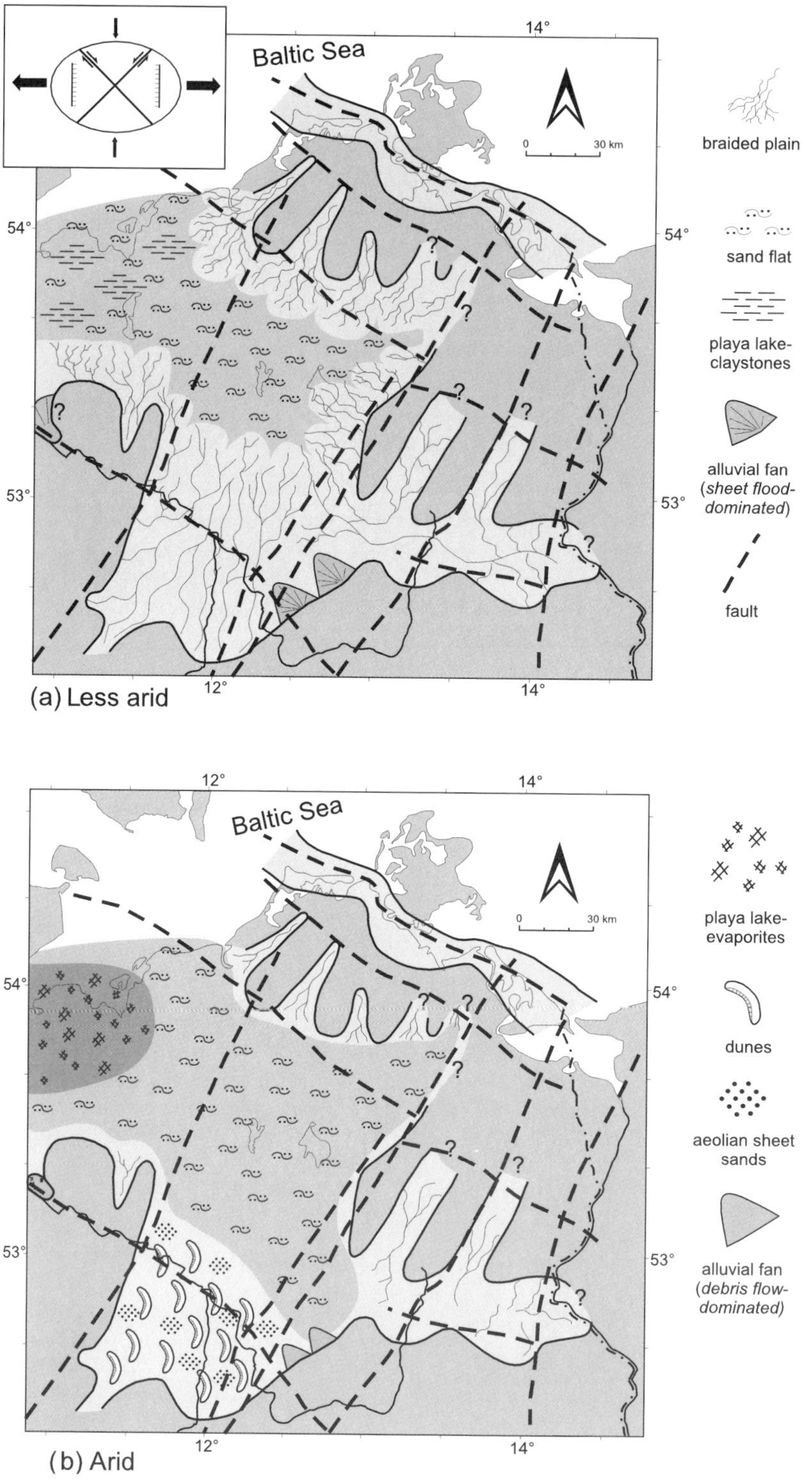

Fig. 10. Palaeogeographical map illustrating the facies architecture during Parchim Formation times for less arid (**a**) and arid (**b**) climate.

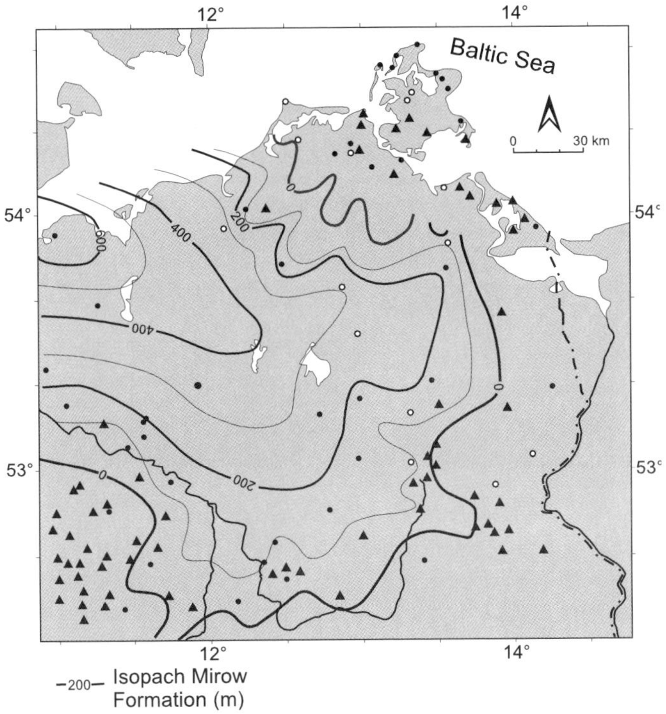

Fig. 11. Isopach map showing the distribution of the Mirow Formation (contours are in metres).

(Fig. 10b). However, these periods are not correlatable, due to a low preservation potential and the unchanged sedimentation.

Dethlingen Formation. Continued thermal subsidence of the NEGB led to further expansion towards the S and E (Fig. 13), essentially eliminating the former Richtenberg and Altmark highs, and leading to the westward connection of the basin to the NW German Basin. The depocentre during Dethlingen Formation times was located in the northwest, where up to 600 m of sediment have been drilled.

As a result of the high apparent rate of accommodation space creation during this period, the facies associations are strongly differentiated. Depending on the specific location within the basin, up to seven possibly climatic-induced cycles can be recognized (Gast & Gebhardt, 1995). Thus, the palaeogeographical maps (Fig. 14) only show the maximum extent of the various environments, and should not be considered as representative of specific spatial or temporal palaeogeographical situations.

Thermal subsidence led to the development of a groundwater-controlled perennial playa lake within the central part of the basin. This lake extended over time to both the east and the south. Grey-green to red-brown laminated claystones represent this environment during the less arid periods (Fig. 14a). Along the southern margin, subsidence and a high sediment budget resulted in extensive progradation of the ephemeral stream floodplain environment. In contrast, the gently dipping northern margin of the NEGB was characterized by a much lower sediment budget, resulting in a weakly developed floodplain environment dominated by moderate to low transport energies. Thus, the base of the Mirow Formation corresponds with a relative retreat of the floodplain environment and a corresponding increase in mudflat sedimentation towards the basin centre.

The precipitation of lacustrine evaporites (up to a few metres thick) within the perennial playa lake environment dominated the basin centre during arid periods (Fig. 14b). Much of the basin was covered by mudflats and only narrow belts of weakly developed floodplains occur along the margins. Subordinate erg deposition, dominated by sand sheets, occurred along the southern margin (Kleditzsch & Kurze 1993). These aeolian

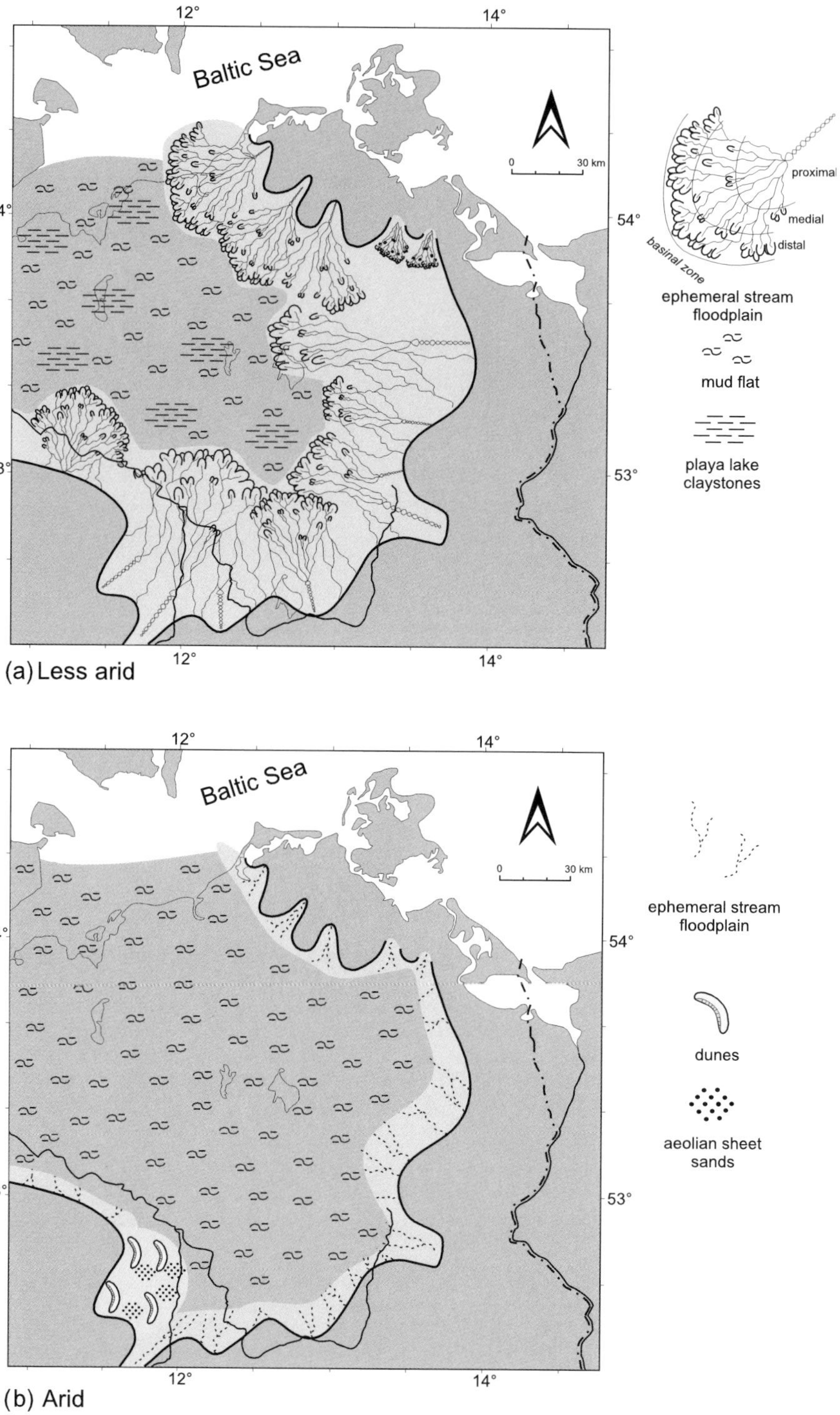

Fig. 12. Palaeogeographical map illustrating the facies architecture during Mirow Formation times for less arid (**a**) and arid (**b**) climate.

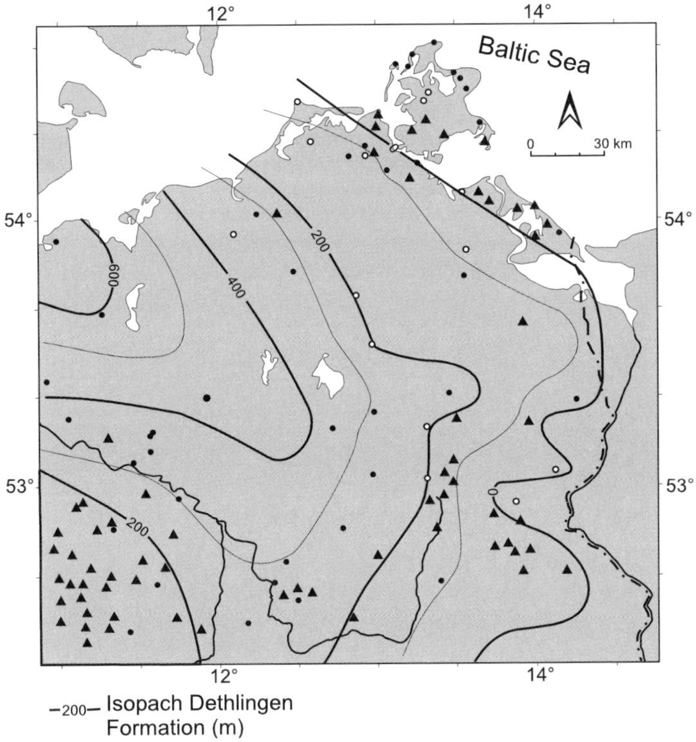

Fig. 13. Isopach map showing the distribution of the Dethlingen Formation (contours are in metres).

facies often occur as metre-scale intercalations within the predominant fluvial facies.

Hannover Formation. Continued basin expansion towards the south and the east characterizes the final phase of Rotliegend deposition (Fig. 15). During Hannover Formation times the NEGB was now a broad uniform feature, extending both to the NW German Basin and to the easterly located Polish Basin. The main depocentre remained in the NW, with a maximum sediment thickness of up to 500 m.

The facies architecture of this final evolutionary stage in the Rotliegend of the NEGB was strongly influenced by the ongoing subsidence, the reduced topography and the consequently reduced clastic sediment supply (Fig. 16). Thus, even during less arid intervals, fluvial deposition was confined to thin belts along the basin margins (Fig. 16a). The perennial playa lake environment prograded towards the east and the south and comprised seven lake-level highstands which developed during this period (Gast & Gebhardt 1995). Cored profiles from the transitional areas consist of interdigitated mudflat sediments and ephemeral playa lake deposits.

During arid phases, fluvial activity ceased and mudflat deposition was extensive over large areas of the basin (Fig. 16b). The presence of more extensive evaporites (up to a few decametres thick) represent the perennial playa lake environment in the most central parts of the basin. The NE-oriented palaeo-wind led to the accumulation of extensive, sand-sheet-dominated aeolian facies along the southern margin. The ingression of the Zechstein Sea terminates the Rotliegend evolution within the entire basin.

Discussion

The evolution of the NE German Basin during Rotliegend II times and its attendant facies architecture have been outlined in detail above. From our understanding of the structural framework within which the NEGB developed, it is clear that there is only real evidence for the events which determined the actual framework of the basin itself. This series of events commenced with the major Permo-Carboniferous volcanic event and continued throughout the periods of deposition of the Altmark and Müritz subgroups. However, the restricted sedimentary

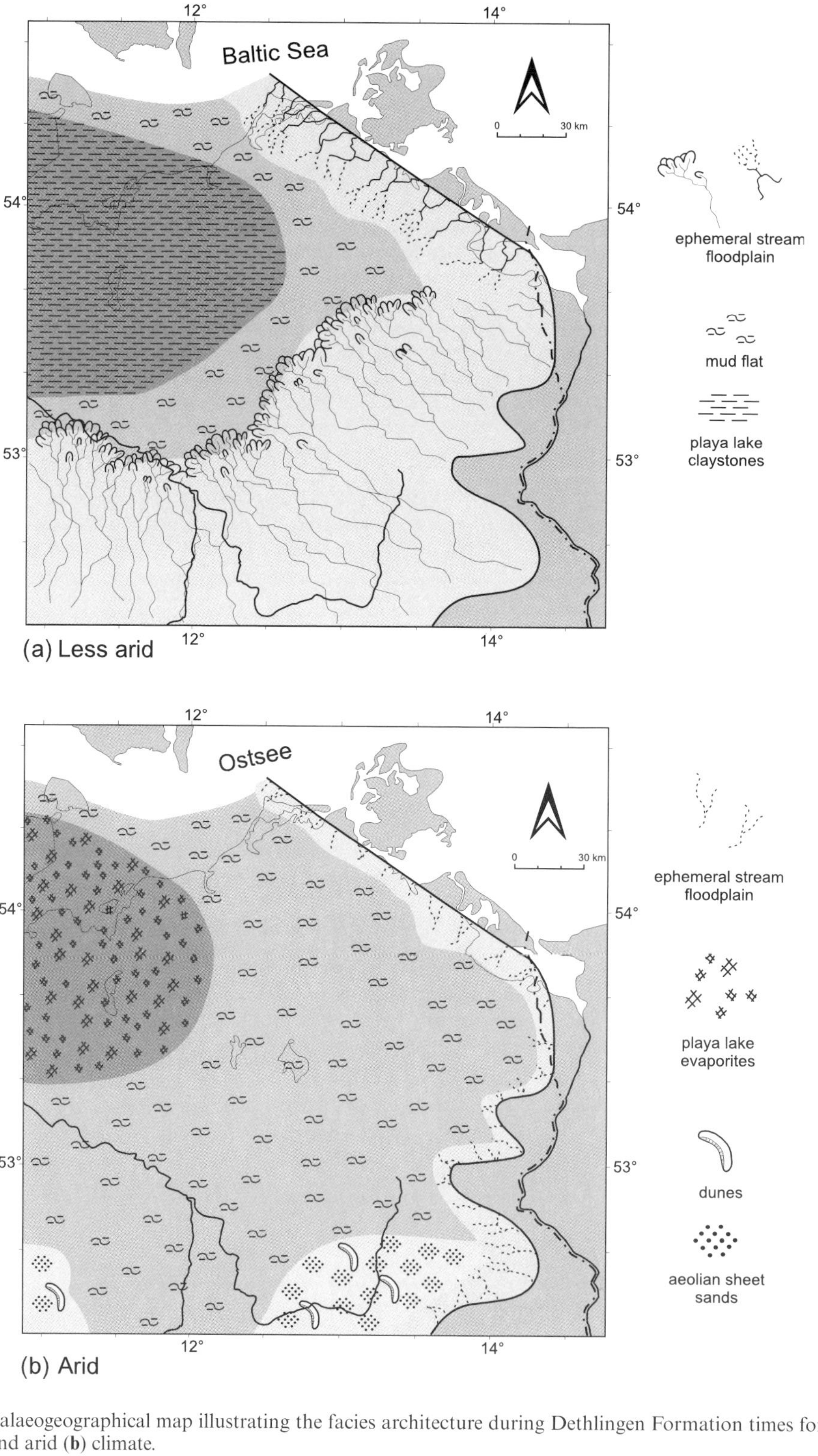

Fig. 14. Palaeogeographical map illustrating the facies architecture during Dethlingen Formation times for less arid (**a**) and arid (**b**) climate.

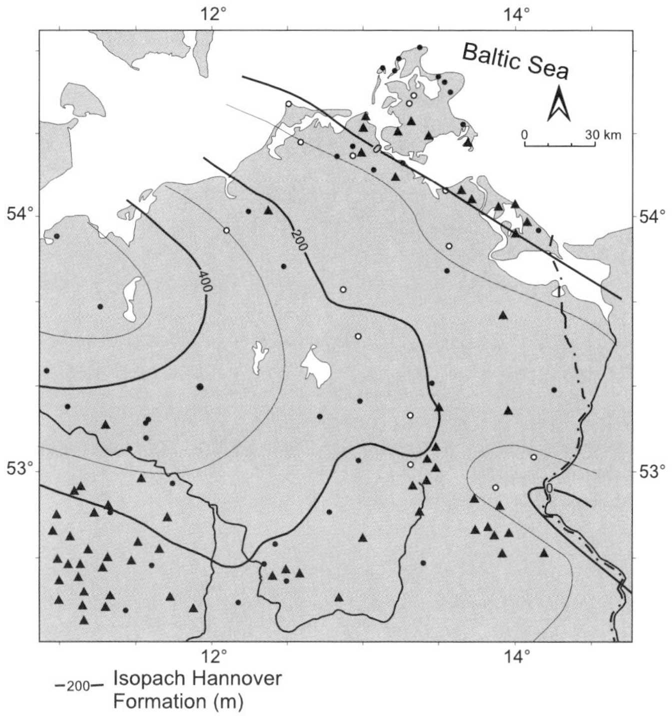

Fig. 15. Isopach map showing the distribution of the Hannover Formation (contours are in metres).

record for this time period, related to the isolation of the individual basins from one another, provides evidence of a disjointed distribution pattern across the region of the NEGB. At the onset of the uppermost Rotliegend (i.e. Rotliegend II), there was a clear change in terms of the basin geometry. As noted above, during the period of deposition of the Parchim Formation, the basin comprised two distinct sub-basins, although it is not clear to what extent these basins were isolated from one another. By Mirow Formation times, however, there was a clearly unified depositional area across the entire NEGB, with highs at the margins, with sediment being sourced from these highs (and adjacent orogenic piles), and transported towards the basin centre. The distribution pattern within the basin itself broadly resembles the models of closed-basin sedimentation as outlined by Leeder & Gawthorpe (1987), albeit with the significant difference that the NEGB was not a half-graben structure. In Dethlingen Formation times, strongly increasing thermal subsidence modified the facies architecture as well as the basin geometry. In the basin centre a large climatically driven playa lake developed and the basin-bounding highs were progressively inundated. In Hannover Formation times the pattern of sedimentation within the NEGB remained broadly similar to that of Dethlingen times, with a number of exceptions. These include the fact that the basin itself continued to grow, and also, that sediment distribution patterns were strongly influenced by the now broadly smooth topography, a resultant decrease in sediment supply and the extensive expansion of the playa lake environment across almost the entire basin.

The evolution of the Rotliegend II within the NEGB has been interpreted as being subdivided by a series of tectonic events – termed Altmark I–IV (see the introductory section of this paper for references). These events supposedly begin at the onset of the Parchim Formation (Altmark I), separate the Parchim and Mirow formations (Altmark II), Mirow and Dethlingen formations (Altmark III) and the Detlingen from the Hannower Formation (Altmark IV). It would, therefore, be expected that there should be evidence of these events both within the structural record, and more importantly in the record of the distribution of sedimentary facies around the basin itself. Periods of tectonic activity, coupled

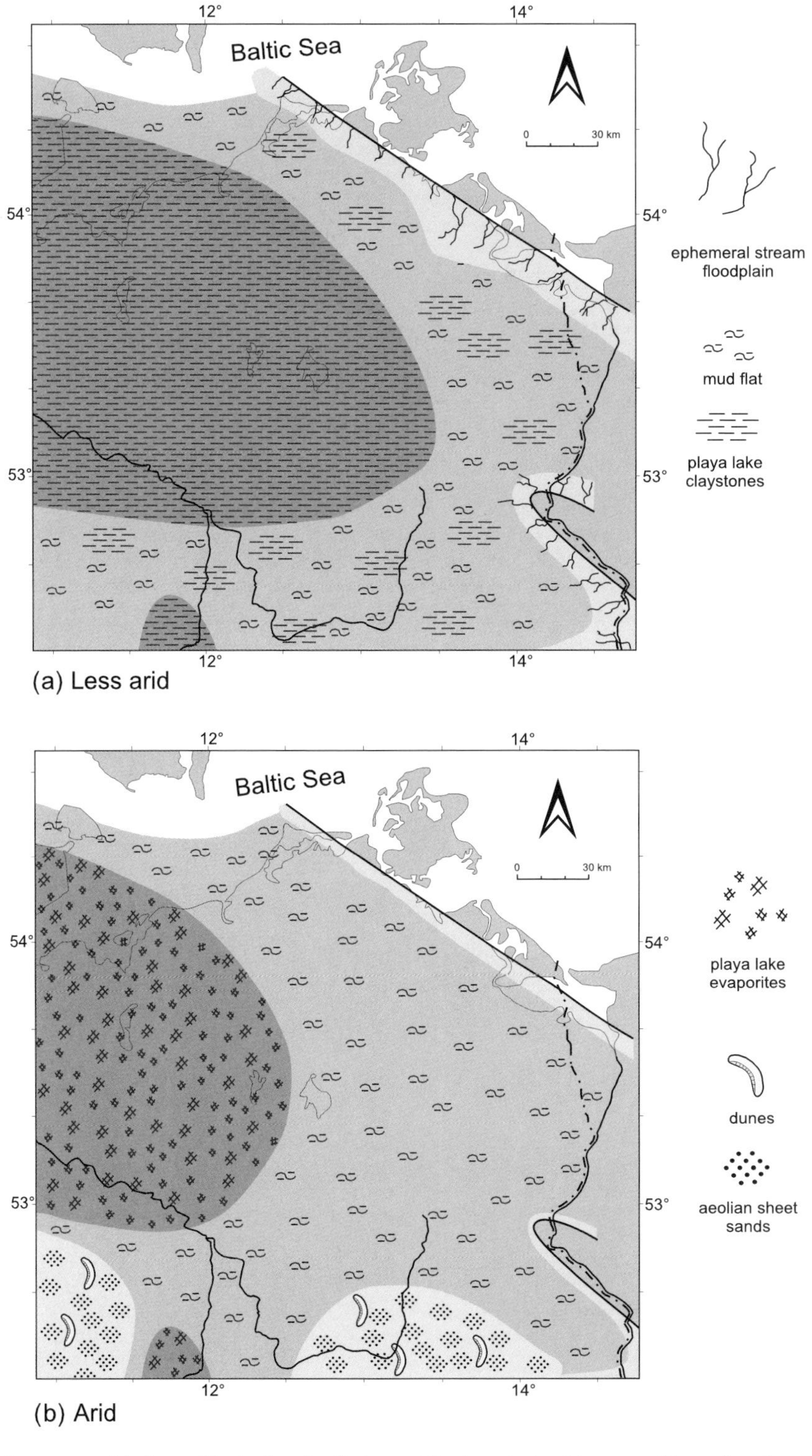

Fig. 16. Palaeogeographical map illustrating the facies architecture during Hannover Formation times for less arid (**a**) and arid (**b**) climate.

with renewed uplift, would be expected to provide new source areas, alter sediment transport directions and broadly affect the distribution of sedimentary facies (and, by extension, depositional environments) within the basin.

When the tectono-sedimentary evolution of the NEGB at this time is examined, a very different pattern is seen. The evolution of the Upper Rotliegend II within the NEGB begins with the initial transtensional-related creation of accommodation space during Parchim Formation times. Despite this, sedimentation was predominantly controlled by climate variations reflecting an initial wet phase and a succeeding dry phase. Tectonic control on sedimentation is restricted to locally developed alluvial fan facies along the southern basin margin.

The subsequent basin-wide change in facies at the beginning of the Mirow Formation is characterized by the extensive progradation of fluvial facies. This is clearly related to a general change in climate to relatively less-arid conditions. Such a conclusion is supported by the fact that fluvial deposition requires both an increase in the amount and periodicity of precipitation. Additionally, no evaporites have been found within the central parts of the basin. The fluvial sandstones at the southern margin contain significant amounts of authigenic anatase, which can be interpreted in terms of a wetter climate during Mirow Formation times (Helmuth & Süssmuth 1993). The playa lake claystones contain abundant fossils of *Hydromedusae limnica,* indicative of freshwater conditions.

The uppermost Dethlingen and Hannover formations were strongly affected by the basin-wide thermally induced subsidence as the controlling factor for the final period of the Rotliegend. The sediment facies were very dependent on climate change. However, climate-induced cyclicities (cf. George & Berry 1993, 1997; Howell & Mountney 1997) can only be found within the most central playa lake successions. At the basin margins, except for localized areas adjacent to the Altmark High, the absence of any evidence of synsedimentary tectonics would preclude the subdivision of the various sedimentary facies, and their differentiation – both of which would be prerequisites for any climatically based correlation. In addition, the predominance of fluvial deposits at the basin margins does not allow any definitive climatic interpretations, and neither does it facilitate further refinement of the existing lithostratigraphy (Gast *et al.* 1998). Furthermore, the broad dish-like morphology of the NEGB (see DEKORP-BASIN Research Group, 1999) and the relative lack of any significant internal morphology would presumably have led to intensified secondary reworking of the sediments within the basin, thus destroying the majority of the climate-change marker horizons.

The NEGB shows no evidence of significant synsedimentary tectonism, as described for the NW German Basin, the Dutch Basin (e.g. Verdier 1996) and the English Basin (e.g. George & Berry 1997). This absence presumably reflects the superior trend to increasing tectonic activities towards the west which can be explained with the beginning of the breakup of Pangea in the more western parts of the Southern Permian Basin in Upper Permian times.

The NEGB was part of the more extensive Southern Permian Basin, and directly bordered the NW German Basin to the west and the Polish Trough to the east. However, the sedimentary record of the NEGB as outlined in this study corresponds with the broad sedimentary framework of the Southern Permian Basin which developed across Northern and Central Europe. The Rotliegend of the NEGB reflects the transition from the fluvial-dominated formations, where evaporites were distinctly absent, in the easterly Polish Basin (e.g. Kiersnowski 1997; Karnkowski 1999) to the westerly NW German Basin, where fluvial facies were absent (e.g. Gast 1988; Gralla 1988). This trend, from a fluvial-dominated eastern region (i.e. wet) to an evaporite-dominated (i.e. dry) regime in the west is a reflection of climatic variations within the Southern Permian Basin.

In summary, the Rotliegend evolution within the NEGB reveals a variety of factors which have a significant influence on sedimentation, such as climate variability, the creation rate and amount of accommodation space, wind direction, sediment budget and source area lithology. Temporally changing interactions generate a depositional evolution separated into four distinct stages which are correlatable almost throughout the entire basin. The knowledge of these combinations should be strongly considered in order to understand the complex depositional history of arid continental successions.

We would to thank J. Kopp (GLA Brandenburg) and W. Von Bülow, A. Janke and J. Nielson (LUNG Mecklenburg-Vorpommern) for providing access to core material, and for their help and assistance. We would also like to thank J. Graham and R. Gaupp for their helpful reviews.

References

ARTHAUD, F. & MATTE, P. 1977. Late Paleozoic strike-slip faulting in southern Europe and northern Africa: result of a right-lateral shear zone between the Appalachians and the Urals. *Geological Society of America Bulletin*, **88**, 1305–1320.

BACHMANN, G. H. & GROSSE, S. 1989. Struktur und Entstehung des Norddeutschen Beckens – geologische und geophysikalische – Interpretation einer verbesserten Bouguer–Schwerekarte. *Niedersächsische Akademie der Geowissenschaften Veröffentlichungen*, **2**, 23–47.

BACHMANN, G. H. & HOFFMANN, N. 1995. Paläotektonik – Bildung und Entwicklung des Norddeutschen Rotliegend-Beckens. *In*: PLEIN, E. (ed.) *Stratigraphie von Deutschland I Norddeutsches Rotliegend-Becken Rotliegend Monographien Teil II*. Senckenbergische Naturforschende Gesellschaft, Frankfurt a. M., 156–169.

BALTRUSCH, S. & KLARNER, S. 1993. Rotliegend-Gräben in NE-Brandenburg. *Zeitschrift der Deutschen Geologischen Gesellschaft*, **144**, 173–186.

BENEK, R., KRAMER, W., MCCANN, T., SCHECK, M., NEGENDANK, J., KORICH, D., HUEBSCHER, H.-D. & BAYER, U. 1996. Permo-Carboniferous magmatism of the Northeast German Basin. *Tectonophysics*, **266**, 379–404.

BERTHELSEN, A. 1992. From Precambrian to Variscan Europe. *In*: BLUNDELL, D., FREEMAN, R. & MUELLER, ST. (eds) *A Continent Revealed – The European Geotraverse*. Cambridge University Press, Cambridge, 153–164.

BLAIR, T. C. 1987. Tectonic and hydrologic controls on cyclic alluvial fan, fluvial, and lacustrine rift-basin sedimentation, Jurassic–Lowermost Cretaceous Todos Santos Formation, Chiapas, Mexico. *Journal of Sedimentary Petrology*, **57(5)**, 845–862.

BLAIR, T. C. & BILODEAU, W. L. 1988. Development of tectonic cyclothems in rift, pull-apart, and foreland basins: sedimentary response to episodic tectonism. *Geology*, **16**, 517–520

BLAKEY, R. C., HAVHOLM, K. G. & JONES, L. S. 1996. Stratigraphic analysis of eolian interactions with marine and fluvial deposits, Middle Jurassic Page Sandstone and Carmel Formation, Colorado Plateau, USA. *Journal of Sedimentary Research*, **66(2)**, 324 342.

BREITKREUZ, C. & KENNEDY, A. 1999. Magmatic flare-up at the Carboniferous/Permian Boundary in the NE German Basin revealed by SHRIMP zircon ages. *Tectonophysics*, **301**, 307–326.

CLEMMENSEN, L. B., OXNEVAD, I. E. I. & DE BOER, P. L. 1994. Climatic controls on ancient desert sedimentation: some late Palaeozoic and Mesozoic examples from NW Europe and the Western Interior of the USA. *In*: DE BOER, P. L. & SMITH, D. G. (eds) *Orbital Forcing and Cyclic Sequences*. Blackwell Science, Oxford, 439–457.

DEKORP-BASIN RESEARCH GROUP 1999. Deep crustal structure of the Northeast German basin: New DEKORP-Basin'96 deep profiling results. *Geology*, **27(1)**, 55–58.

DORN, R. I. 1994. The role of climatic change in alluvial fan development. *In*: ABRAHAMS, A. D. & PARSONS, A. J. (eds) *Geomorphology of Desert Environments*. Chapman & Hall, London, 593–615.

FRANKE, D., HOFFMANN, N. & KAMPS, J. 1989. Alter und struktureller Bau des Grundgebirges im Nordteil der DDR. *Zeitschrift für Angewandte Geologie*, **35(10–11)**, 289–296.

FROSTICK, L. E. & REID, I. 1989. Climatic versus tectonic controls of fan sequences: lessons from the Dead Sea, Israel. *Journal of the Geological Society of London*, **146**, 527–538.

FRYBERGER, S. G., AL-SARI, A. & CLISHAM, T. 1983. Eolian dune, interdune, sand sheet, and siliciclastic sabkha sediments of an offshore prograding sand sea, Dhabran, Saudi Arabia. *AAPG Bulletin*, **67**, 280–312.

FRYBERGER, S. G., AL-SARI, A. M., CLISHAM, T., RIZVI, S. A. R. & AL-HINAI, K. G. 1984. Wind sedimentation in the Jafurah sand sea, Saudi Arabia. *Sedimentology*, **31**, 413–431.

GAITZSCH, B. 1995. Grüneberg-Formation. *In*: PLEIN, E. (ed.) *Stratigraphie von Deutschland I Norddeutsches Rotliegendbecken Rotliegend Monographien Teil II*. Senckenbergische Naturforschende Gesellschaft, Frankfurt a. M., 102–106.

GAST, R. 1988. Rifting im Rotliegenden Niedersachsens. *Die Geowissenschaften*, 4, 115–122.

GAST, R. 1991. The perennial Rotliegend saline lake in NW Germany. *Geologisches Jahrbuch*, **A 119**, 25–59.

GAST, R. & GEBHARDT, U. 1995. Elbe-Subgruppe. *In*: PLEIN, E. (ed.) *Stratigraphie von Deutschland I Norddeutsches Rotliegendbecken Rotliegend Monographien Teil II*. Senckenbergische Naturforschende Gesellschaft, Frankfurt a. M., 121–135.

GAST, R., PASTERNAK, M., PISKE, J. & RASCH, H.-J. 1998. Das Rotliegend im nordostdeutschen Raum: Regionale Übersicht, Stratigraphie, Fazies und Diagenese. *In*: KARNIN, W.-D., MERKEL, D., PISKE, J. & SCHRETZENMayer, S. (eds) *Geowissenschaftliche Ergebnisse der Kohlenwasserstoff-Exploration im Land Brandenburg und im Thüringer Becken in den Jahren 1991–1996 (Zechstein und Rotliegend)*. Geologisches Jahrbuch, **A 149**, 59–79.

GAUPP, R., GAST, R. & FORSTER, C. 2000. Late Permian playa lake deposits of the Southern Permian Basin (Central Europe). *In*: GIERLOWSKI-KORDESCH, E. H. & KELTS, K. R. (eds) *Lake Basins Through Space and Time*. AAPG Studies in Geology, **46**, 75–86.

GEBHARDT, U. 1994. Zur Genese der Rotliegend-Salinare in der Norddeutschen Senke (Oberrotliegend II, Perm). *Freiberger Forschungshefte*, **C 452**, 3–22.

GEBHARDT, U., SCHNEIDER, J. & HOFFMANN, N. 1991. Modelle zur Stratigraphie und Beckenentwicklung im Rotliegenden der Norddeutschen Senke. *Geologisches Jahrbuch*, **A 127**, 405–427.

GEORGE, G. T. & BERRY, J. K. 1993. A new lithostratigraphy and depositional model for the Upper Rotliegend of the UK sector of the southern North Sea. *In*: NORTH, C. P. & PROSSER, D. J. (eds) *Characterization Of Fluvial And Aeolian Reservoirs*. Geological Society, London, Special Publications, **73**, 291–319.

GEORGE, G. T. & BERRY, J. K. 1997. Permian (Upper Rotliegend) synsedimentary tectonics, basin development and palaeogeography of the Southern North Sea. *In*: ZIEGLER, K., TURNER, P. & DAINES,

S. R. (eds) *Petroleum Geology of the Southern North Sea; Future Potential*. Geological Society, London, Special Publications, **123**, 31–61.

GLENNIE, K. W. 1982. Early Permian (Rotliegendes) palaeowinds of the North Sea. Sedimentary Geology, **34**, 245–265.

GLENNIE, K. W. 1990. Lower Permian – Rotliegend. *In*: Glennie, K. W. (ed.) *Introduction to the Petroleum Geology of the North Sea*. Blackwell Scientific Publications, London, 120–152.

GLENNIE, K. W. 1998. Lower Permian – Rotliegend. *In*: Glennie, K. W. (ed.) *Petroleum Geology of the North Sea – Basic Concepts and Recent Advances*. Blackwell Science, London, 137–173.

GLENNIE, K. W. & UNDERHILL, J. R. 1998. Origin, development and evolution of structural styles. *In*: GLENNIE, K. W. (ed.) *Petroleum Geology of the North Sea – Basic Concepts and Recent Advances*. Blackwell Science, London, 42–84.

GOODALL, T., NORTH, C. P. & GLENNIE, K. W. 2000. Surface and subsurface sedimentary structures produced by salt crusts. *Sedimentology*, **47**, 99–118.

GRALLA, P. 1988. Das Oberrotliegende in NW-Deutschland – Lithostratigraphie und Faziesanalyse. *Geologisches Jahrbuch*, **A 106**, p. 59 S.

HARDIE, A. L., SMOOT, J. P. & EUGSTER, H. P. 1978. Saline lakes and their deposits: a sedimentological approach. *In*: MATTER, A. & TUCker, M. E. (eds) *Modern and Ancient Lake Sediments*. Special Publications, International Association of Sedimentologists, London, 7–41.

HELMUTH, H. J. & SÜSSMUTH, S. 1993. Die lithostratigraphische Gliederung des jüngeren Oberrotliegenden (Oberrotliegendes II) in Nordostdeutschland. *In*: MÜLLER, E. P. & PORTH, H. (eds) *Perm im Ostteil der Norddeutschen Senke*. Geologisches Jahrbuch, **A 131**, 31–55.

HELMUTH, H. J. & SCHRETZENMAYR, S. 1995. Zur raum-zeitlichen Genese der Gräben. *In*: PLEIN, E. (ed.) *Stratigraphie von Deutschland I Norddeutsches Rotliegendbecken Rotliegend Monographien Teil II*. Senckenbergische Naturforschende Gesellschaft, Frankfurt a. M., 169–174.

HOFFMANN, N. 1990. Zur paläodynamischen Entwicklung des Präzechsteins in der Nordostdeutschen Senke. *In*: NIEDERSÄCHSISCHE AKADEMIE DER GEOWISSENSCHAFTEN (ed.) Geologie und Kohlenwasserstoff-Erkundung im Präzechstein der DDR – Nordostdeutsche Senke. *Niedersächsische Akademie der Geowissenschaften Veröffentlichungen*, Hannover, 5–18.

HOFFMANN, N., POKORSKI, J., LINDERT, W. & BACHMANN, G. 1997. Rotliegend stratigraphy, palaeogeography and facies in the eastern part of the Central European Basin. *In*: PODEMSKI, M., DYBOVA, J. S., JAWOROWSKI, K., JURECZKA, J. & WAGNER, R. (eds). *Proceedings of the XIII International Congress on the Carboniferous and Permian: Prace Panstwowego Instytutu Geologicznego (1988)*, Warsaw, Poland, Wydawnictwa Geologiczne, pp. 75–86.

HOTH, K., RUSBUELDT, J., ZAGORA, K., BEER, H. & HARTMANN, O. 1993. Die tiefen Bohrungen im Zentralabschnitt der Mitteleuropäischen Senke – Dokumentation für den Zeitabschnitt 1962–1990. *Gesellschaft für Geowissenschaften e.V.*, Berlin, 145 pp.

HOWELL, J. & MOUNTNEY, N. 1997. Climatic cyclicity and accommodation space in arid to semi-arid depositional systems: an example from the Rotliegend Group of the UK southern North Sea. *In*: TURNER, K. & DAINES, S. R. (eds) *Petroleum Geology of the Southern North Sea: Future Potential*. Geological Society, London, Special Publications, **123**, 63–86.

KARNKOWSKI, P. H. 1999. *Origin and Evolution of the Polish Rotliegend Basin*. Polish Geological Institute, Special Papers, **3**, 93 pp.

KIERSNOWSKI, H. 1997. Depositional development of the Polish Upper Rotliegend Basin and evolution of its sediment source areas. *Geological Quarterly*, **41(4)**, 433–456.

KIERSNOWSKI, H., PAUL, J., PERYT, T. M. & SMITH, D. B. 1995. Facies, paleogeography and sedimentary history of the Southern Permian Basin in Europe. *In*: SCHOLLE, P. A., PERYT, T. M. & ULMER, S. D. S. (eds) *The Permian of Northern Pangaea; Vol. 2, Sedimentary Basins and Economic Resources*. Springer-Verlag, Berlin, 119–136.

KLARNER, S. 1993. Analyse des strukturellen Aufbaus (Grabenproblematik) des Rotliegenden im Raum Zehdenick-Gransee. *In*: MÜLLER, E. P. & PORTH, H. (eds) *Perm im Ostteil der Norddeutschen Senke*. Geologisches Jahrbuch, **A 131**, 239–251.

KLEDITZSCH, O. & KURZE, M. 1993. Ergebnisse petrographischer Untersuchungen an Sandsteinen des tieferen Oberrotliegenden im Raum Altmark/Westmecklenburg. *In*: Müller, E. P. & Porth, H. (eds) *Perm im Ostteil der Norddeutschen Senke*. Geologisches Jahrbuch, **A 131**, 141–178.

LEEDER, M. R. & GAWTHORPE, R. L. 1987. Sedimentary models for the extensional tilt-block/half-graben basins. *In*: COWARD, M. P., DEWEY, J. F. & HANCOCK, *P. L.* (eds) *Continental Extensional Tectonics*. Geological Society, London, Special Publications, **28**, 139–152.

MCCANN, T 1998. Sandstone composition and provenance of the Rotliegend of the NE German Basin. *Sedimentary Geology*, **116**, 177–198.

MACK, G. H. & LEEDER, M. R. 1999. Climatic and tectonic controls on alluvial-fan and axial-fluvial sedimentation in the Plio-Pleistocene Palomas half graben, Southern Rio Grande Rift. *Journal of Sedimentary Research*, **69(3)**, 635–652.

MARX, J., HUEBSCHER, H.-D., HOTH, K., KORICH, D. & KRAMER, W. 1995. Vulkanostratigraphie und Geochemie der Eruptivkomplexe. In: PLEIN, E. (ed.) *Stratigraphie von Deutschland I Norddeutsches Rotliegendbecken Rotliegend Monographien Teil II*. Senckenbergische Naturforschende Gesellschaft, Frankfurt a. M., 54–83.

MENNING, M. 1995. Magnetostratigraphie. *In*: PLEIN, E. (eds) *Norddeutsches Rotliegendbecken – Rotliegend-Monographie Teil II*. Senckenbergische Naturforschende Gesellschaft, Frankfurt a. M., 40–45.

MOUNTNEY, N., HOWELL, J., FLINT, S. & JERRAM, D. 1999. Climate, sediment supply and tectonics as controls on the deposition and preservation of the aeolian–fluvial Etjo Sandstone Formation, Namibia. *Journal of the Geological Society of London*, **156**, 771–777.

MÜLLER, A. H. 1978. Über Hydromedusen (Coelenterata) und medusoide Problematica aus dem Rotliegenden von Mitteleuropa. *Freiberger Forschungsheft*, **C 342**, 29–44.

MÜLLER, E. P., DUBSLAFF, H., EISERBECK, W. & SALLUM, R. 1993. Zur Entwicklung der Erdöl- und Erdgasexploration zwischen Ostsee und Thüringer Wald. *In*: MÜLLER, E. P. & PORTH, H. (eds) *Perm im Ostteil der Norddeutschen Senke*. Geologisches Jahrbuch, **A 131**, 5–30.

NEMEC, W. & STEEL, R. 1988. What is a fan delta and how do we recognize it? *In*: NEMEC, W. & STEEL, R. (eds) *Fan Deltas: Sedimentology and Tectonic Settings*. Blackie, London, 3–13.

PLEIN, E. 1993. Bemerkungen zum Ablauf der paläogeographischen Entwicklung im Stefan und Rotliegend des Norddeutschen Beckens. *Geologisches Jahrbuch*, **A 131**, 99–116.

PLEIN, E. 1995. *Norddeutsches Rotliegendbecken Rotliegend-Monographie Teil II*. Senckenbergische Naturforschende Gesellschaft, Frankfurt a. M., 193 pp.

RIEKE, H., KOSSOW, D., MCCANN, T. & KRAWCZYK, C. 2001. Tectono-sedimentary evolution of the northernmost margin of the NE German Basin between uppermost Carboniferous and Late Permian (Rotliegend). *Geological Journal*, **36(1)**, 19–38.

SCHNEIDER, J., GEBHARDT, U., GAITZSCH, B. & DÖRING, H. 1995. Fossilführung und Biostratigraphie. *In*: PLEIN, E. (ed.) *Stratigraphie von Deutschland I Norddeutsches Rotliegendbecken Rotliegend Monographien Teil II*. Senckenbergische Naturforschende Gesellschaft, Frankfurt a. M., 25–35.

STUMM, M., LINDERT, W. & LANGE, G. 1990. *Forschung Erdgas Beckenzentrum Rotliegendes IV – Lithofazieskarte Parchim-Schichten (1:500.000)*. Zentrales Geologisches Institut – Kombinat Erdöl-Erdgas, Berlin, Unpublished.

SWEET, M. L. 1999. Interaction between aeolian, fluvial and playa environments in the Permian Upper Rotliegend Group, UK southern North Sea. *Sedimentology*, **46**, 171–187.

WEES, J.-D., VAN, STEPHENSON, R. A. *ET AL.* 2000. On the origin of the Southern Permian Basin, Central Europe. *Marine and Petroleum Geology*, **17**, 43–59.

VERDIER, J. P. 1996. The Rotliegend sedimentation history of the southern North Sea and adjacent countries. *In*: RONDEEL, H. E., BATJES, D. A. J. & NIEUWENHUIJS, W. H. (eds) *Geology of Gas and Oil under the Netherlands*. Kluwer Academic Publishers, The Hague, 45–56.

YANG, C. S. & NIO, S. D. 1994. Application of high-resolution sequence stratigraphy to the Upper Rotliegend in the Netherland Offshore. *In*: WEIMER, P. & POSAMENTIER, H. W. (eds) *Siliciclastic Sequence Stratigraphy: Recent Developments and Applications*. AAPG Memoirs, **58**, 285–316.

ZIEGLER, P. 1990. *Geological Atlas of Western and Central Europe*. Shell International Petroleum Maatschappij BV, Amsterdam.

Sequence stratigraphy of the Baltic Silurian succession: tectonic control on the foreland infill

J. LAZAUSKIENE[1], S. SLIAUPA[2], A. BRAZAUSKAS[3] & P. MUSTEIKIS[3]

[1]*Geological Survey of Lithuania, S. Konarskio 35, LT-2600 Vilnius, Lithuania; Vilnius University, Ciurlionio 21/27 LT-2009 Vilnius; Lithuania (e-mail: jurga.lazauskiene@lgt.lt)*

[2]*Institute of Geology, Sevcenkos 13, LT-2600, Vilnius, Lithuania*

[3]*Vilnius University, Ciurlionio 21/27 LT-2009 Vilnius, Lithuania*

Abstract: The Baltic basin overlaps the SW margin of the East European Craton (EEC). During the Silurian its subsidence was governed by the flexural bending of the EEC margin due to the collision of Eastern Avalonia and Baltica. Two mechanisms – orogenesis and dynamic loading – were responsible for the flexural subsidence of the basin. Lithofacies, sequence- and cyclo-stratigraphic analysis were applied to reveal the tectonosedimentary evolution of the Silurian Baltic Basin, focusing on the imprint of geodynamic processes in adjacent orogens on the sedimentary architecture. Adopting a sequence stratigraphic approach, 10 depositional sequences superimposed by the lower rank cycles were identified in the Silurian Baltic Basin. The Llandovery sequences correspond with the initial stage of flexuring. The low terrigenous influx to the basin is explained by the lack of relief in the fold belt and its location at a distance from the orogenic front. The Wenlock–Lower Ludlow sequences reflect the accelerating flexuring. An increase in orogenic-sourced terrigenous material indicates the advancement of the Caledonian orogen. The Late Ludlow–Pridoli sequences comprise the final stages of flexuring and basin infilling. Two major provenances – cratonic and orogenic – competed to supply terrigenous sediment to the basin during Silurian times.

The Baltic sedimentary basin originated on the southwestern margin of the East European Craton (EEC; Fig. 1) as part of a marginal sedimentary basin system during the Early Palaeozoic (Sliaupa & Lazauskiene 1997). Initially it was established as a passive continental margin basin during latest Vendian–Early Cambrian times in response to the breaking apart of the Rodinia megacontinent. The Cambrian–Middle Ordovician passive-margin setting changed to that of a convergent continental margin during the Late Ordovician–Silurian (Sliaupa *et al.* 1997; Poprawa *et al.* 1999). Convergence reached its peak in the Late Silurian, and resulted in the overthrusting of Eastern Avalonian accretionary wedges on to the southwestern margin of the Baltica plate. As implied by three-dimensional geodynamic modelling, the Baltic foreland basin originated as a result of the complex interaction of two controlling mechanisms, namely orogenesis and dynamic loading (Lazauskiene *et al.* 2002).

The tectonosedimentary evolution of the basin during the Silurian is recorded in a more than 3-km thick sedimentary succession (Tomczykowa, 1988; Lapinskas 2000). The stratigraphic section is almost complete, with no hiatuses recognized in the central part of the basin (Paskevicius 1997). The thickness of the Silurian succession increases towards the southwest. The reconstructed palaeothicknesses exceeded 5 km adjacent to the Teisseyre–Tornquist Zone (TTZ), while the succession thins to *c.*50–100 m towards the eastern basin margin (Vejbaek *et al.* 1994; Milaczewski & Modlinski 1998; Lapinskas 2000; Fig. 2). Graptolite shales dominate the sedimentary succession in the western and central parts of the basin, grading into marlstones in the transitional zone and to the carbonate platform in the shallow eastern margin of the Baltic Basin (Fig. 3).

Both geodynamic constraints and lithofacies analysis suggest that there was a close relationship between the basin fill processes in the Silurian foreland basin and tectonic activity in the Caledonian orogenic belt. Despite this, the lithology of the Silurian succession is rather different from that of the typical foreland basin 'flysch' and 'molasse' evolutionary stages. Peripheral foreland basins are commonly infilled by turbiditic deep marine 'flysch' sediments,

From: McCann, T. & Saintot, A. (eds) *Tracing Tectonic Deformation Using the Sedimentary Record.* Geological Society, London, Special Publications, **208**, 95–115. 0305-8719/03/$15.00

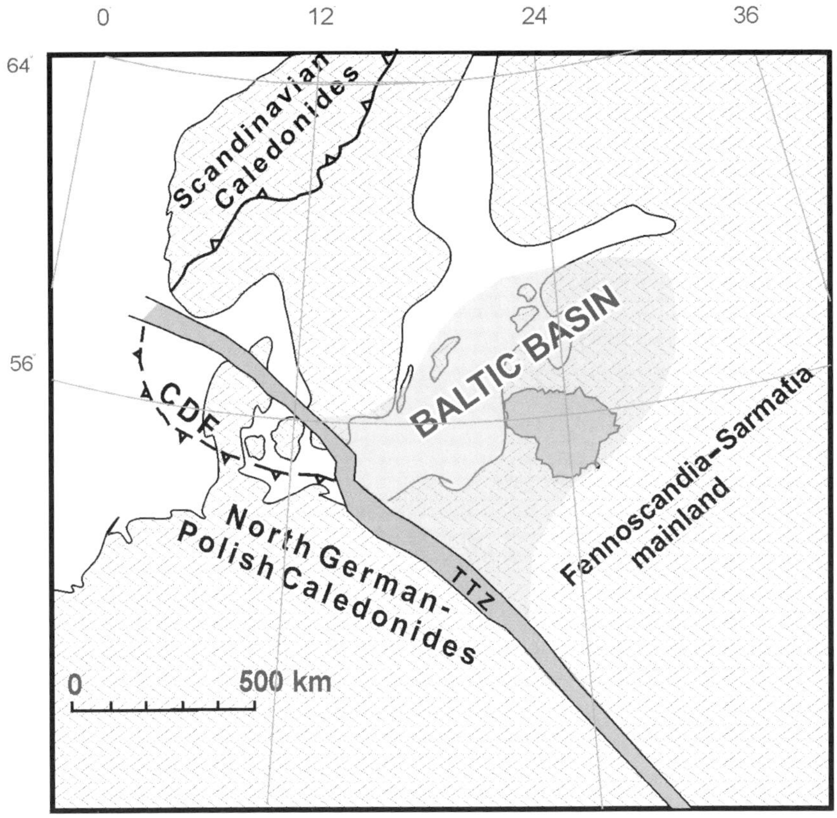

Fig. 1. Location of the Silurian Baltic basin along the western margin of the East European Craton. CDF, Caledonian Deformation Front; TTZ, Teisseyre–Tornquist Zone.

grading upwards into shallow-marine and continental 'molasse' deposits (Mitchell & Reading 1969; Allen *et al.* 1986; Jordan & Flemings 1989; Zoetemeijer 1993; Proust *et al.* 1998). Unlike in typical foreland basins, fine-grained terrigenous material was infilling the Baltic Basin during the Silurian. The deposition of fine-grained siltstones, resulting from the activity of distal turbidites (Jaworowski 1971, 2000), was recorded only in the westernmost part of the basin along the TTZ. Geochemical studies of the Silurian shales have suggested that the source orogen was a recycled one (Sliaupa, 1999a). Furthermore, the lithological composition of the sediments, as supported by three-dimensional flexural modelling, suggests that relief was low in the source area (Lazauskiene *et al.* 2002).

Based on previous work, it was clear that more detailed studies were needed in order to determine how the observed patterns of sedimentation in the Silurian Baltic Basin could be related to the tectonic mechanisms inferred from geodynamic analysis.

The stratigraphy and lithofacies of the Silurian succession of the Baltic Basin are well studied (Lapinskas 1987, 1996, 2000; Tomczykowa 1988; Nestor & Einasto 1997; Paskevicius 1997; Jaworowski 2000). In addition, the major aspects of the geodynamic evolution of the basin have been analysed recently (Sliaupa *et al.* 1997; Stephenson *et al.* 1997; Maletz *et al.* 1997; Poprawa *et al.* 1999; Beier *et al.* 1999; Lazauskiene *et al.* 2002). The present study, therefore, aims to examine the tectono-sedimentary development of the Baltic Basin in a foreland setting, based on the analysis of the sequential architecture and cyclicity of the Silurian succession.

Biostratigraphic, well-log and core data of reference wells from Lithuania, representing the main lithofacies zones of the basin, were studied to construct the sequence stratigraphic framework of the Silurian succession and specify the

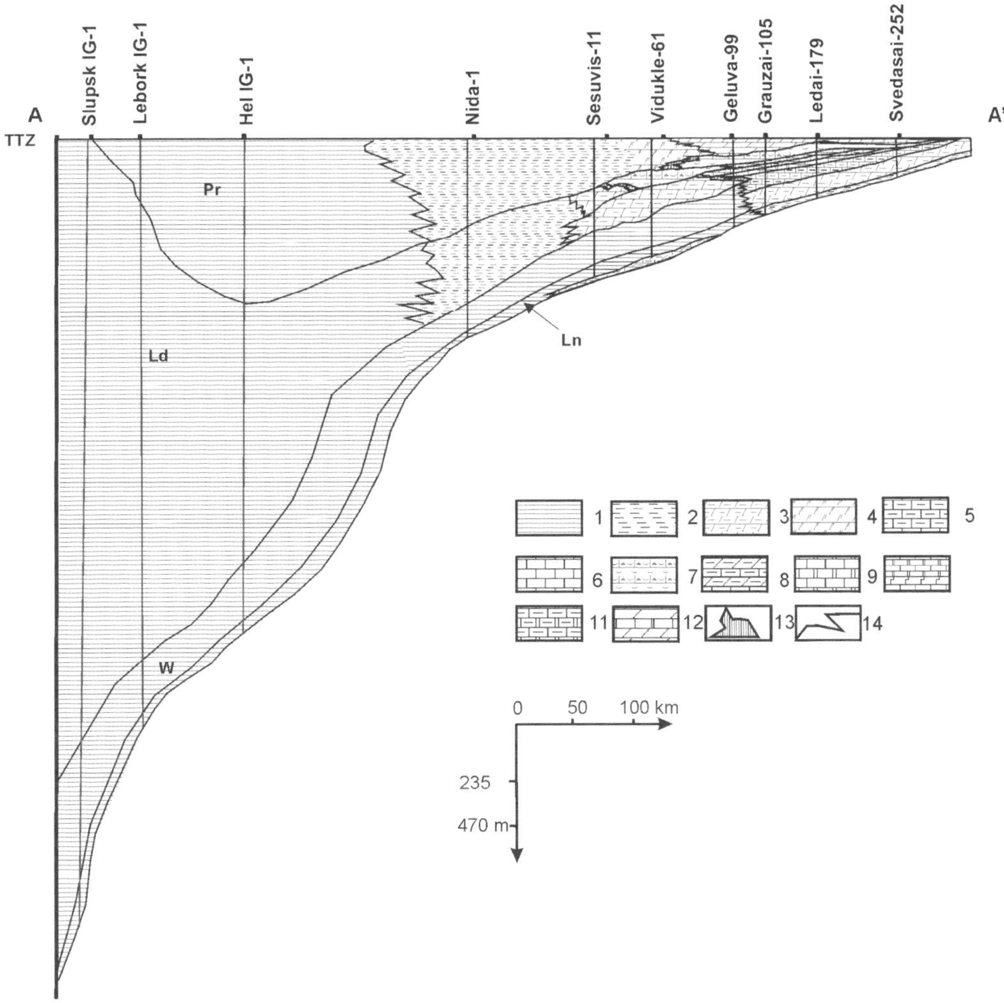

Fig. 2. Generalized lithofacies cross-section along the profile *A–B–C* (location of line is shown in Fig. 3); TTZ, Teisseyre–Tornquist Zone. 1, shales; 2, carbonaceous claystones; 3, clayey marlstones; 4, carbonaceous marlstones; 5, clayey limestones; 6, limestones; 7, organogenic-detrital limestones; 8, clayey limestones and marlstones; 9, dolomites; 10, dolomites and gypsum; 11, clayey dolomites; 12, dolomites interbedded with marlstones; 13, reefs; 14, boundaries of Silurian stages.

trends of tectono-sedimentary evolution in the Baltic foreland basin (Fig. 3).

Geological setting

Stratigraphy

A detailed stratigraphic scheme of the Silurian of the Baltic basin has been built up over decades of extensive studies (e.g. Tomczykowa & Tomczyk 1979; Nestor 1994; Musteikis 1993; Paskevicius *et al.* 1994; Lapinskas 2000). The Silurian succession comprises Llandovery, Wenlock, Ludlow and Pridoli series subdivided into stages, regional stages, groups, formations and beds (Fig. 4).

The thickness of the Llandovery strata varies from tens of metres, up to 160 m in South Estonia (Paskevicius 1997) and exceeding 300 m adjacent to the Caledonian Deformation Front (CDF; McCann 1996*a*). Early Llandovery basal shallow-water organogenic–detrital carbonate sediments, overlying a Ordovician passive-margin carbonate platform, accumulated in the eastern marginal zone, and pass into greenish and dark claystones in the southwest (Figs 4 & 5). The upper parts of the Rhudanian, Aeronian and

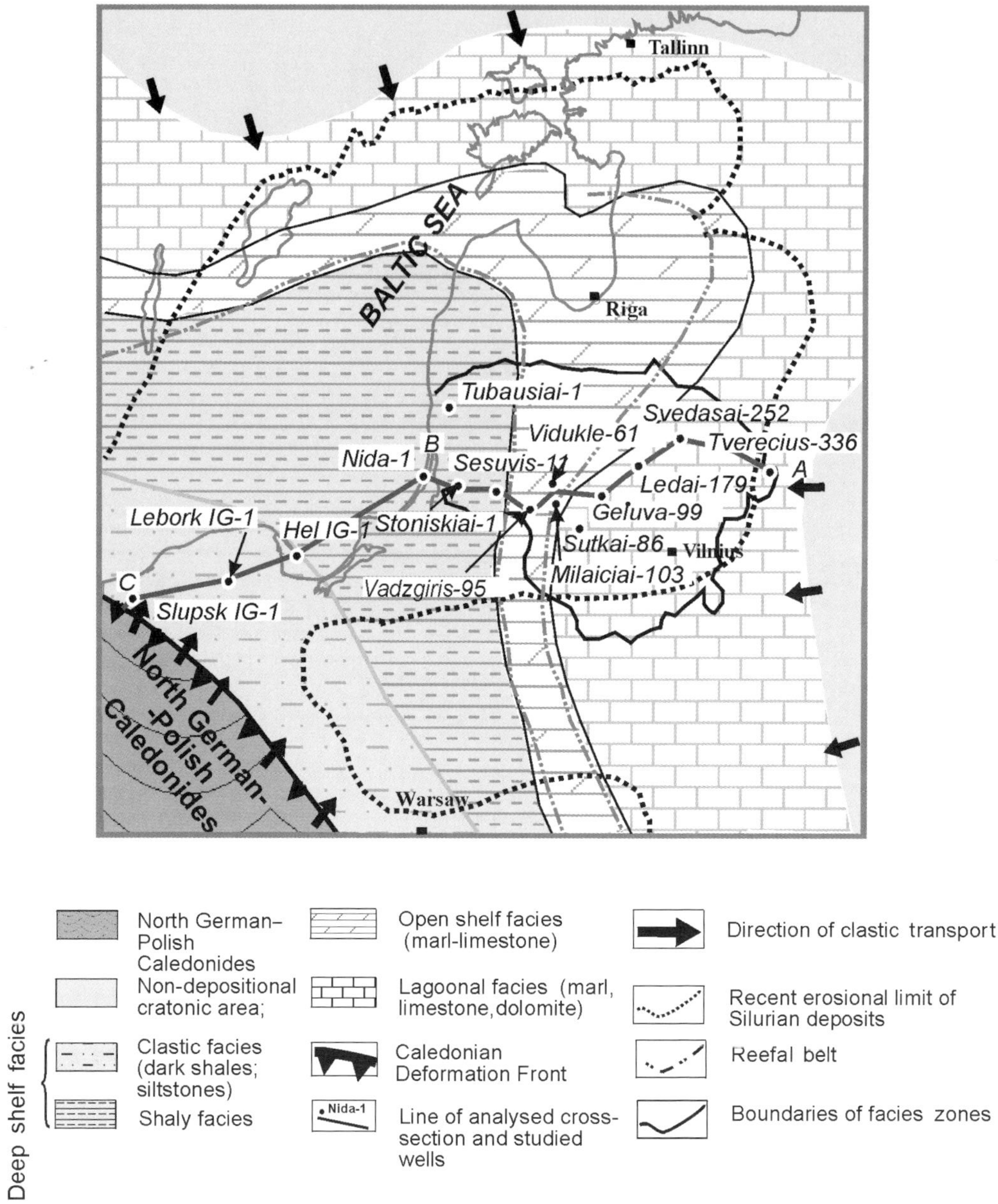

Fig. 3. Silurian lithofacies of the Baltic foreland basin (after Kaljo *et al.* 1991; Valiukevicius & Katinas 1995; Lapinskas 1996; Paskevicius 1997) and location of wells. *A–B–C*, line of analysed cross-section.

Telychian successions are composed of greenish-grey calcareous claystones interbedded with clayey marlstones. The strata of the former two stages are eroded in the eastern part of the basin, while the Telychian sediments are represented by a rhythmic alternation of greenish-grey and red claystones (Fig. 5).

The Wenlock succession ranges from 40 m, to 600 m in thickness close to the Teisseyre–Tornquist Zone (Fig. 2). Dark graptolite shales dominate the westernmost part of the Silurian Baltic basin. Clayey sediments give way to greenish-grey marlstones and fine-grained limestones, while detrital limestones and dolomites

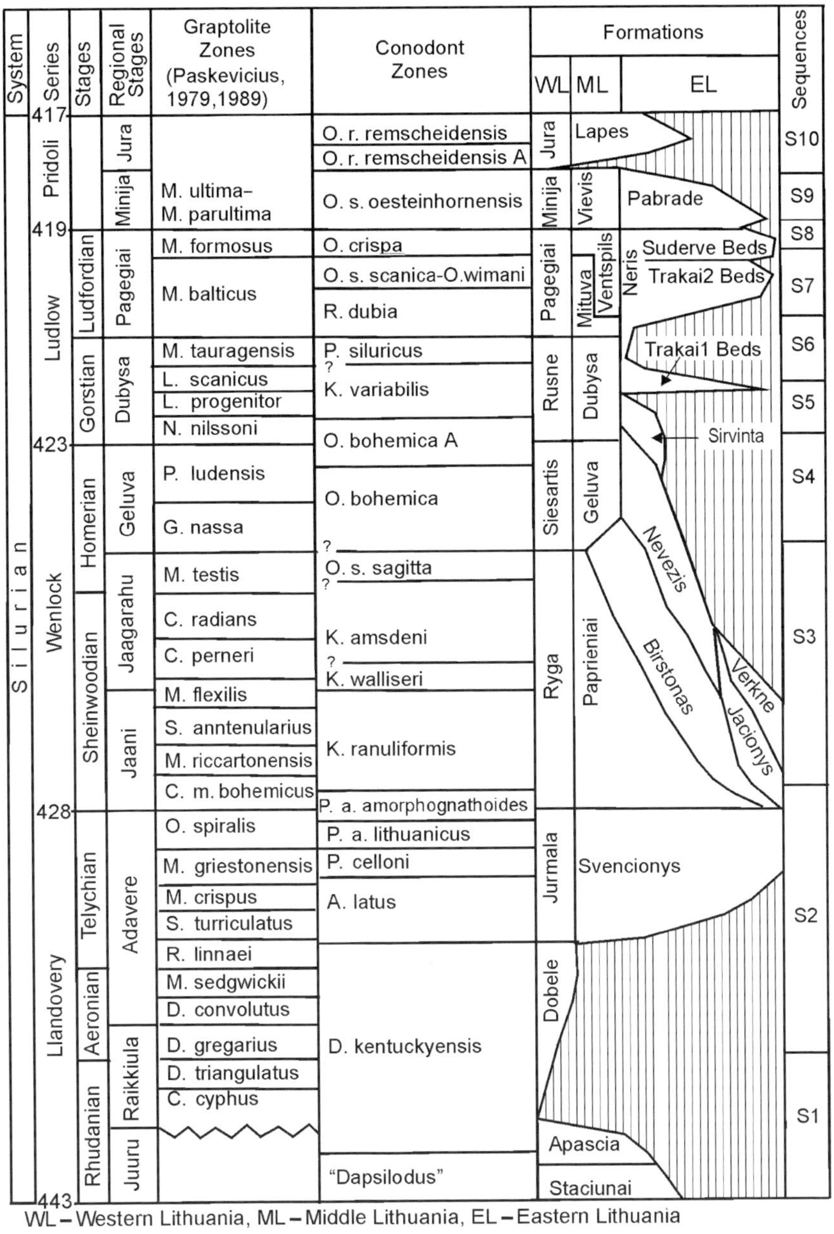

Fig. 4. The stratigraphic scheme of the Silurian deposits (after Paskevicius 1997; Lapinskas 2000). Time-scale adopted from Gradstein & Ogg (1996).

represent the nearshore environments of the easternmost periphery of the basin (Fig. 5).

The thickness of the overlying Ludlow sediments ranges from 50 m in the east to 2400 m in the west. During the Early Ludlow the regressive shallow-water lithofacies – represented by marlstones, limestones, dolomitic marlstones and dolomites – extended over the eastern shelf and passed into dark graptolite shales in the west. During Late Ludlow times grey calcareous–clayey sediments, dolomitic marlstones and dolomites were deposited, while dark graptolite shales were restricted to the southwesternmost part of the basin (Fig. 5).

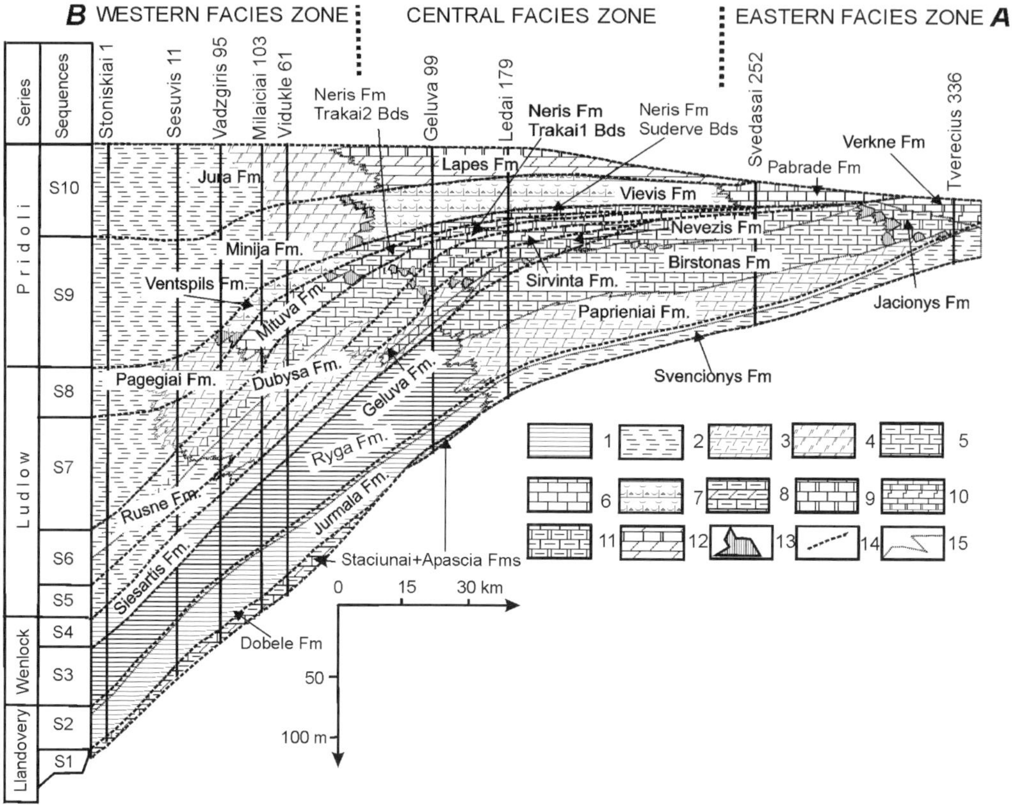

Fig. 5. Lithofacies cross-section throughout the central and eastern parts of the Silurian Baltic Basin (line of cross-section *A–B* is shown in Fig. 3), distribution of depositional sequences and main tectono-sedimentary stages of foreland basin development. 1, shales; 2, carbonaceous claystones; 3, clayey marlstones; 4, carbonaceous marlstones; 5, clayey limestones; 6, limestones; 7, organogenic-detrital limestones; 8, clayey limestones and marlstones; 9, dolomites; 10, dolomites and gypsum; 11, clayey dolomites; 12, dolomites interbedded with marlstones; 13, reefs; 14, boundaries of sequences; 15, boundaries of formations.

A similar trend of sedimentation is documented for Pridoli times. The thickness of Pridoli sediments does not exceed 700 m in the west, although the original thickness was much greater (Tomczykowa & Tomczyk 1979; Paskevicius 1997). The nearshore carbonaceous sediments pass westwards into a deeper water clayey succession. The latter part of the Pridoli shows a shift of the clayey and shallow-water carbonaceous facies boundary to the eastern nearshore zone (Figs 3 & 5).

Geodynamic evolution of the Silurian Baltic basin

The Early Palaeozoic tectonic evolution of the Baltic Basin was intimately related to tectonic activity along of the SW and NW margins of the Baltica plate. The Baltic Basin developed as a passive-margin basin during the latest Precambrian–Middle Ordovician (Poprawa *et al.* 1997). The change from a passive to convergent margin setting is recorded, since Late Ordovician times, by the gradually increasing subsidence which took place during the Late Ordovician–Early Silurian. Accelerating subsidence – a feature which is characteristic for basins developed under a compressional regime (Kominz & Bond 1986; Angevine *et al.* 1990; King 1994) – culminated during Pridoli times (Poprawa *et al.* 1999; Fig. 6). The tectonic subsidence rates increased towards the southwest, attaining nearly 500 m/Ma adjacent to the CDF, while a low rate of subsidence is recorded in the eastern part of the basin. The acceleration of subsidence during the Silurian was related to the overthrusting of accretionary wedges of the North German–Polish (NGP) Caledonides on

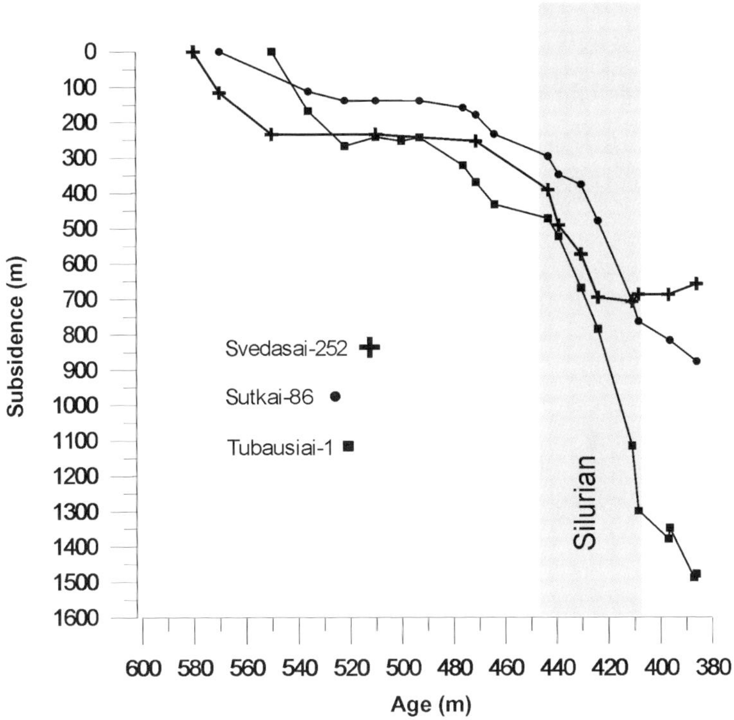

Fig. 6. Tectonic subsidence curves modelled for the selected wells of central and eastern parts of the Baltic foreland basin (location of wells is shown in Fig. 3).

to the western margin of Baltica. The influence of the Scandinavian Caledonides along the NW margin of the East European Craton may have resulted in additional subsidence. The results of one-dimensional subsidence analysis and three-dimensional flexural modelling have allowed the Silurian Baltic Basin to be interpreted as a foreland basin whose development was related to collisional processes along the SW rim of the Baltica plate (Poprawa *et al.* 1999; Lazauskiene *et al.* 2002).

Three-dimensional geodynamic modelling implies that the subsidence of the Silurian Baltic foreland basin was invoked by two mechanisms driving the flexural bending of the lithosphere (Lazauskiene *et al.* 2002). The loading of the Caledonian Orogenic belt induced moderate wavelength subsidence of the westernmost part of the basin. Here, the influence of the orogenic load was restricted to only a 250 km wide zone along the Caledonian Deformation Front. The additional subsidence mechanism contributed to the long-wavelength (*c.*650 km) subsidence of the foreland basin. This mechanism was related to the 'dynamic' load, probably caused by viscous mantle corner flow above a subducting plate. The isostatic uplift of marginal parts of the Baltic basin took place during the Late Silurian and Early Devonian and was related to the destruction of the Caledonian orogen and cessation of the subduction-induced dynamic load (Sliaupa & Poprawa 2000).

Data and methods

The present study is based on an integration of the available drill core, well log and conodont biostratigraphy data, and using standard sequence- and cycle-stratigraphic techniques and concepts as developed by Van Wagoner *et al.* (1988), Posamentier & James (1993) and Posamentier & George (1993). These methods were used to identify the sequence boundaries, sequence packages and system tracts. For comparison with the units distinguished, the eustatic sea-level curve by Ross & Ross (1996), corrected after the regional sea-level data (Johnson *et al.* 1991; Loydell 1998; Nestor & Einasto 1997), was adopted. In order to unravel the evolution of the depositional systems in the Silurian Baltic Basin, the analysis of ten reference wells from the Lithuanian part of the basin along an east–west trending cross-section, which represents the main structural-facies zones of the Silurian Baltic basin, was undertaken (Fig. 3). Detailed stratigraphic and lithofacies correlation of wells was carried out in order to analyse the distribution of the various lithofacies, together with spatial and lateral variations in sediment thicknesses, and changes in sequence architecture throughout the eastern half of the Silurian Baltic Basin.

More than 4000 m of drill cores of the representative wells were analysed in order to recognize the cyclical variations in lithology and, most importantly, to identify any evidence of possible erosional boundaries within the sedimentary succession, such as drowning and pyrite-mineralized surfaces, hardgrounds, etc., since compelling evidence for subaerial exposure is rarely available in the core.

The boundaries of the depositional sequences in the Silurian succession were identified using the complex integration of biostratigraphic constraints, gamma ray and resistivity logs and lithofacies data. A total of 3100 conodont faunal samples were collected and analysed in order to reconstruct the sequence-stratigraphic framework of the Silurian succession. Eleven conodont biostratigraphic zones, also supported by the lithological and log data, were interpreted as the correlative bounding surfaces of the major sequences along the margins of the basin. Respectively, the system tracts within the depositional sequences were identified based on characteristic patterns of gamma ray, chronostratigraphic position and bounding surface attributes.

The cyclicity of the Silurian succession was analysed as a record of depositional processes in the basin, using the methodology described in detail by Karagodin & Armentrout (1996). Based on variations of the lithological composition of the sediments, four major types of cycles, i.e. transgressive, regressive, re-transgressive and trans-regressive, can be distinguished in the sedimentary successions. Following such an approach, within the depositional sequences we identified a number of lower rank sedimentary cycles of different type, based on the sedimentological interpretation of gamma ray and resistivity logs and combined with well core analysis (Fig. 7).

Sequence- and cyclo-stratigraphy of the Silurian succession

Sequence definition and description

Ten depositional sequences of different durations were identified in the Silurian succession of the Baltic Basin. These correspond with the Early Llandovery, Middle–Late Llandovery, Early Wenlock, Late Wenlock, early Early Ludlow, late Early Ludlow, early Late Ludlow, late Late Ludlow, Early Pridoli and Late Pridoli successions. These sequences will be described in detail below.

The base of the S1 Sequence (Early Llandovery) formed in response to the major sea level fall which occurred at the Ordovician–Silurian boundary. The top of the sequence is marked by an erosional surface which formed at the end of Early Llandovery times. Initial deposition of this mixed terrigenous–calcareous succession was coincident with the onset of the slow Rudhanian sea-level rise (Loydell 1998). The sequence comprises micritic nodular limestones with associated lenses of marlstones and clayey limestones which were deposited in an open-shelf environment on the margin of the basin. In the southwestern part of the basin, the same sequence comprises dark-coloured claystones, which are cyclically interbedded with grey- and red mudstones and marlstones (Figs 5 & 7).

The S2 Sequence (Middle–Late Llandovery) is composed of thin laminated black and dark-grey-coloured shales which contain a suite of graptolites. The dark shales are cyclically interbedded with grey and red claystones and marlstones, and the entire succession is interpreted as having been deposited below wave base under dysaerobic to anaerobic conditions. This succession represents the maximum extent of the Silurian Baltic Basin during Late Llandovery times, and was presumably related to the continuing, global eustatic sea-level rise which continued throughout the Llandovery (Johnson *et al.* 1991). The upper part of the S2 Sequence is characterized by black and dark-grey graptolitic shales in the central and western parts of the basin, while greenish-grey calcareous claystones and marlstones predominate along the eastern margin of the basin (Figs 4 & 5).

The deposition of the S3 Sequence (Early Wenlock) was coincident with the generally regressive sea-level trend. This pattern, however, was interrupted by some minor transgressive episodes (Johnson *et al.* 1991). The sediments of the S3 Sequence are similar to those of the underlying S2 Sequence. Dark-coloured graptolitic shales were deposited in the western part of the basin. To the east, these graptolitic shales give way to grey marlstones (Paprieniai Formation) and fine-grained nodular limestones interbedded with greenish-grey marlstones (Birstonas Formation). The deposition of detrital limestones (Jacionys and Nevezis formations) occurred in the lagoonal and tidal-range environments. Red

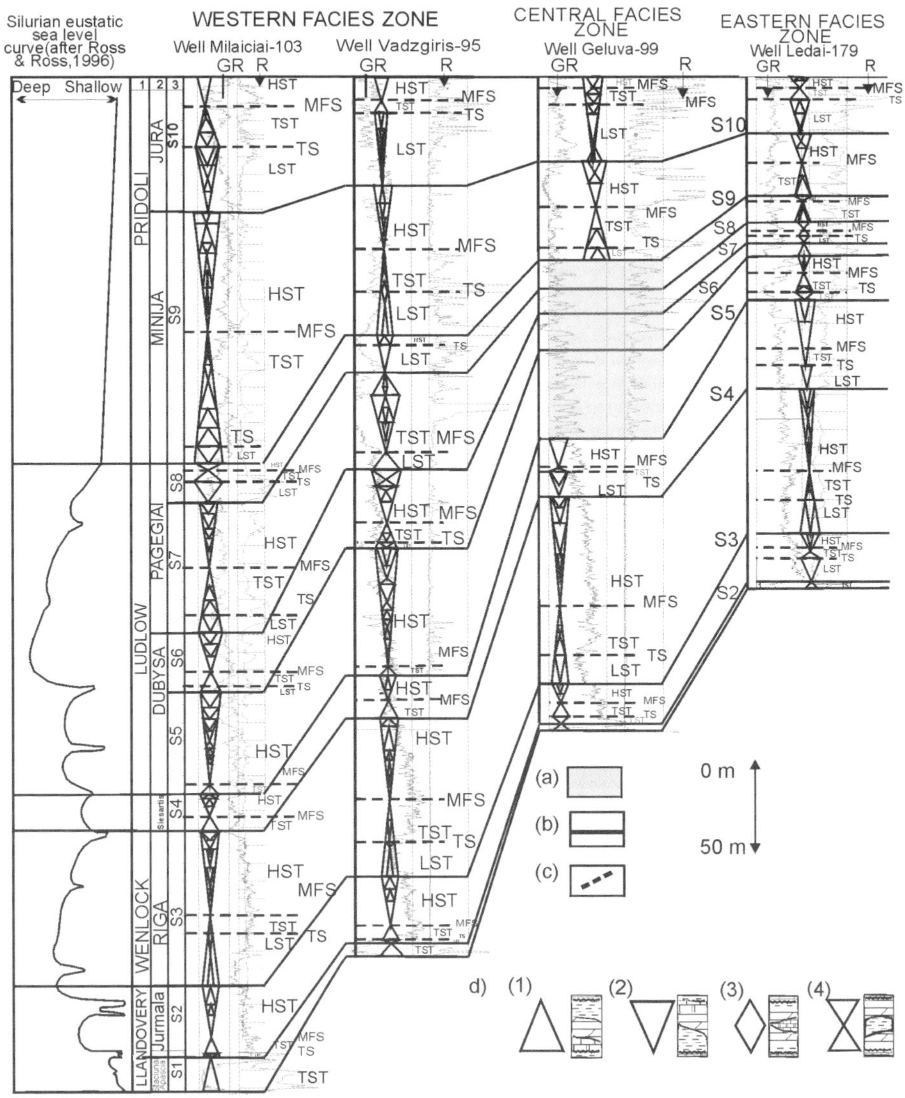

Fig. 7. Sequence- and cyclo-stratigraphy of the Silurian succession: 1, system; 2, formation; 3, depositional sequences; (a) blank, reef growth zone; (b) correlative boundaries of sequences and major supercycles; (c) boundaries of system tracts; (d) different types of cycles: (1) transgressive; (2) regressive; (3) regressive–transgressive; (4) transgressive–regressive (modified from Karagodin & Armentrout 1996). Abbreviations: HS, highstand system tract; LST, lowstand system tract; TST, transgressive sytem tract; SB, sequence boundary; TS, transgressive surface; MS, maximal flooding surface; GR, gamma ray log; R, resistivity log.

dolomitic marlstones and light-grey laminated dolomites with gypsum (Verkne Formation) were dominant along the easternmost basin margin. The building outwards of the carbonate sediments during the Silurian resulted in the westward progradation of the carbonate platform succession. This can be clearly seen in the changes in sequence geometry during the basin development. The upward-shallowing portion of limestones and dolomites in the S3 Sequence was confined only to the eastern part of the Silurian Baltic Basin (Fig. 5).

The Sequences S4 (Late Wenlock) and S5 (early Early Ludlow) are rather similar in terms of their lithologies and lithofacies. The S4 Sequence is represented by greenish-grey and red dolomitic marlstones which contain nodules of very fine-grained limestones and interbeds of organogenic–detrital limestones in the eastern and central facies zones. To the southwest, these limestones are replaced by dark-grey massive graptolite shales and clayey marlstones with rare interlayers of clayey limestones (Siesartis Formation; Fig. 5).

Greenish-grey and red massive dolomitic marlstones with rare interbeds of dolomite and gypsum concretions, pass westwards into dark grey- and black, bedded claystones, which show a weak internal parallel lamination. These claystones alternate with very fine-grained clayey limestones (Rusne Formation) and are interpreted as belonging to the S5 Sequence (early Early Ludlow). The global Wenlock–Early Ludlow eustatic sea-level fall strongly influenced the pattern of sedimentation in the Baltic Basin area, as can be seen in the prominent upwards increase in the numbers of microcrystalline limestone interbeds within the shaley package of the S5 Sequence. These limestones are interpreted as representing part of this regressive trend and are indicative of an overall shallowing of the basin.

The S6 Sequence (late Early Ludlow) is represented by the deposition of the regressive shallow-water limestones and dolomitic marlstones of the Dubysa Formation and dolomites from the lower part of the Trakai Beds (Neris Formation) in the eastern part of the basin. These calcareous sediments pass westwards into dark grey-coloured claystones. The cyclical alternation of lagoonal winnowed skeletal grainstones to micro-laminated brownish-grey clayey dolomites reflects the maximum period of basin shallowing. Some minor transgressive phases, as recorded by the deposition of dark grey shales with lenses of limestones and horizontally laminated dolomitic marlstones (lowermost part of the S6 Sequence) stand out from the overall regressive trend.

A rapid shift of shallow-water carbonate lithofacies to the west, recorded in the changed architecture of sequences S4–S6, indicates continuing progradation of the carbonate platform (Fig. 5).

Light greenish-grey microcrystalline, organogenic–detrital and nodular limestones with interbeds of greenish-grey marlstones and grey dolomites (Mituva Formation) comprise the S7 Sequence (early Late Ludlow). These sediments were deposited along the margin of the basin. In contrast, the southwesternmost part of the basin was dominated by black and grey, thinly laminated graptolite shales which contain concretions of clayey limestones.

The S8 Sequence (late Late Ludlow) is represented by greenish-grey calcareous marlstones interbedded with light organogenic–detrital limestones (Pagegiai and Ventspils Formations) along the western facies zones, and dominant grey organogenic–detrital limestones and dolomites (the uppermost part of the Trakai and Suderve beds), deposited in the eastern and central parts of the basin. The maximum extent of carbonate facies progradation to the west is clearly reflected in the changes of the geometry of S7–S8 sequences and also represented by the migration of the Sutkai Reef Belt (Lapinskas 2000) during Ludlow times (Fig. 5).

The depositional architecture of the S9 Sequence (Early Pridoli) represents the general shift of the lithofacies zones to the southwest which occurred at this time. The grey dolomites and dolomitic marlstones (Pabrade Formation), which represent a tidal flat environment, as well as dark-grey fragmental limestones (Vievis Formation) deposited in a lagoonal setting, gave way to the more open shelf biomicritic marlstones (Minija Formation) and, to the west, to grey deeper-water shales. The coastward shift of the Sutkai Reef Belt to the north-east occurred during Early Pridoli times (Fig. 5). The geometry of the depositional S9 Sequence clearly indicates the eastward-directed retrogradation of the carbonate platform during the Early Pridoli, that followed the preceding Wenlock-Ludlow progradation (Figs 5 & 8).

Light grey-coloured limestones with marlstone interbeds (Lapes Formation) comprise the S10 Sequence (Late Pridoli) in the eastern part of the study area. These pass westward into the greenish-grey marlstones and calcareous claystones of the Jura Formation. The upper part of the S10 Sequence in the eastern part of the basin is comprised of shallow-water grey dolomites. These pass into light-grey limestones and greenish-grey marlstones and clays (the upper part of Jura Formation) to the west (Figs 3 & 5). The changes in the sedimentary architecture of the Baltic Basin during the Late Pridoli, are reflected in the geometry of the S10 Sequence, and represent a narrowing of the basin. A coincident increase in influx of terrigeneous sediment in the central part of the basin suppressed carbonate accumulation on the open shelf in that region.

Systems tracts within the Silurian Baltic Basin

The sedimentary packages, related to the particular phases of marine deepening or shallowing, have been interpreted as lowstand, transgressive

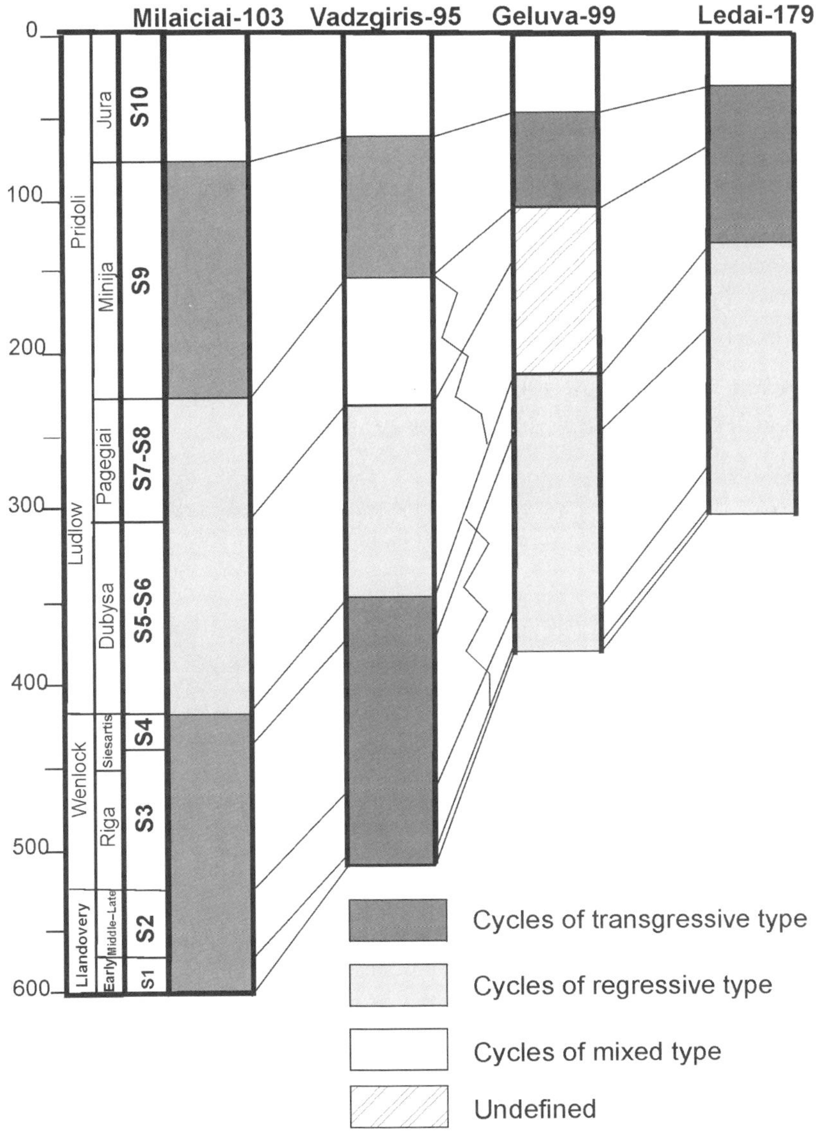

Fig. 8. Distribution of the transgressive and regressive cycles.

and highstand systems tracts within the depositional sequences of the Silurian Baltic Basin (Fig. 7).

In the Silurian succession, lowstand system tract (LST) intervals are represented by shallower water sediments which show distinctive shallowing-upwards trends both in terms of the lithologies present and also in that they exhibit a gradually decreasing gamma ray signal (Fig. 7). The LST portion of the S1–S4 sequences is dominated by shale-rich sediments in the western part of the basin, whereas along the eastern margin of the basin the same LST interval is marked by mostly carbonate deposition. Here the lithologies are mainly organogenic–detrital limestones, micrite and microcrystalline dolomites. No distinct lateral lithofacies variations are recorded in the LST succession of the S5–S10 sequences, composed mostly of clayey microcrystalline limestones and dolomitic clayey marlstones, which indicates an increase in the amounts of carbonate upwards in the Silurian section (Fig. 7). The abundant pyrite-mineralized and non-pyritized hardground surfaces, drowned surfaces, subaerial exposures and, most importantly, the stacking pattern of depositional units, have enabled us to identify these portions of the sequences as lowstand system tracts, which bear many similarities to those described in the literature (Jacquin *et al.* 1991; Posamentier & James 1993).

Deepening trends within the basin are recorded in transgressive system tracts (TST), which are indicated by a rise in the shale content and an associated abrupt increase in the gamma-ray log pattern. TST of the Lower Silurian sequences (S2–S4) in the western part of the basin tend to be dominated by finer-grained deeper water sediments (black shales, claystones and clayey dolomitic marlstones). By Late Silurian times (S5–S10 sequences), however, the lithological composition has changed from a prevailing terrigeneous to a mixed terrigenous–carbonate one, and TST sediments are mostly composed of interbedded microcrystalline organogenic limestones and clayey dolomitic marlstones. This could be related to the progradation of the carbonate facies, as indicated by the geometry of the S4–S6 sequences. TST intervals in the western part of the basin differ from the ones of the LST in their carbonate content, which is less than 15% in the TST intervals and *c.* 40% in the LST portion. A general trend of deepening is also observed in the marginal part of the Silurian Baltic basin. The transition of clayey and organogenic–detrital limestones to dolomitic marlstones (transition from clayey–carbonaceous dolomites to organogenic–detrital limestones within the S7 Sequence) up the section, describes the TST intervals on the eastern periphery of the basin (Fig. 7).

The highstand systems tracts (HST) of the Silurian Baltic Basin are represented mostly by shallow-water carbonate deposition, and show a general trend of shallowing as recorded in lithological changes and the gradually decreasing gamma-ray trace (Fig. 7). This general shallowing pattern is observed in HST sequences across the basin, although in detail the HST may be represented by differing lithologies, depending on the precise position within the basin. For example, in the western part of the basin there is a progressive upwards increase in the carbonate content. Here the HST interval of the lower S1–S4 sequences comprises only clayey dolomitic marlstones, while organogenic–detrital and fragmental, sometimes oolitic, limestones dominate the uppermost S5–S10 sequences (Figs 5 & 7). Along the eastern margin, the HST portions are represented by predominantly carbonaceous sediments (dark-coloured dolomitic marlstones, dolomitic limestones and clayey microcrystalline dolomites).

Remnants of thin crinoidal faunal detritus, pyrite/mica-mineralized surfaces, a number of burrowed hardgrounds, the stacking of the system tracts relative to each other, and the up to 40–50% increased carbonate content compared with equivalent LST sediments, made it possible to distinguish the above-mentioned portions of sequences as highstand system tracts, which are basically similar to the described examples (Jacquin *et al.* 1991; Ehrenberg *et al.* 2000).

Sequence architecture and the cyclicity of the Silurian Baltic Basin

The defined depositional sequences were analysed in terms of cyclicity of sedimentation. Based on drill-core analysis and well-log correlation, a number of superimposed lower order cycles have been distinguished within the sequences (Fig. 7). The cycles are of different lithological compositions and reflect the development of particular lithofacies zones of the Silurian Baltic Basin. The type, symmetry, geometry and distribution of the cycles record the changes in the tectono-sedimentary evolution of the foreland basin.

The Silurian succession starts from the lowermost Llandovery S1 Sequence, which shows marked asymmetry in terms of its cyclicity. In the studied area S1 comprises only a TST succession (Fig. 7). It is bounded, both above and below, by marked stratigraphic gaps which represent periods of erosion. The succession of the Middle–Late Llandovery S2 Sequence reflects the asymmetrical pattern represented by thinned LST and TST intervals in the western area of the basin. It passes into the prominent HST portion that has a number of superimposed lower order cycles of regressive type. In the central part of the basin the HST portion is reduced, therefore the S2 Sequence is symmetrical in character and comprises a stack of incomplete transgressive and complete transgressive–regressive cycles. In the marginal area it is described by thick LST on which are superimposed asymmetrical cycles of regressive type.

The Wenlock succession is characterized by a series of symmetrical transgressive–regressive cycles (sequences S3 and S4). No any distinct lateral variations were observed in the general trend of cyclicity and sequence geometry, while lower rank cycles show considerable differences between western and eastern parts of the basin. In the western half, superimposed series of transgressive cycles characterize the S3 Sequence, whereas the regressive-type cycles dominate in the central and eastern periphery of the basin (Fig. 7). The S4 Sequence is rather symmetrical in the western part of the basin. LST sediments are missing in this zone. Further to the east a change in the pattern of sedimentation is revealed by the asymmetrical cycles of regressive type. Thus, a stack of lower rank cycles of regressive type comprises the S4 section in the eastern nearshore part of the foreland, while the transgressive cycles dominate in the west (Fig. 7).

In the Ludlow succession, the dominant pattern of cycles is an asymmetrical one. The cyclicity within the Ludlow succession in the central area (Well Geluva-99) is complicated by reefal build-ups (Fig. 7). In the western part of the basin, reduced TST and expanded HST portions of the lowermost Ludlow S5 sequence show an asymmetrical pattern. The lower rank cycles of regressive type dominate the sequence in this zone, while the S5 section on the eastern periphery shows asymmetrical transgressive cyclicity.

The S6 Sequence is described by a different pattern in the cycles in distinct parts of the basin. In the western half of the basin this succession shows a regressive trend, and superimposed on this are a series of regressive cycles (Fig. 7). In the eastern marginal zone of the basin it is comprised of a set of transgressive cycles.

Lateral changes in the pattern of cyclicity are observed also within the S7 Sequence. A symmetrical transgressive–regressive succession is recorded in the westernmost well Milaiciai-103, whereas the HST interval is increasingly reduced to the east (Fig. 7). The superimposed cycles are of the predominantly regressive type in the western half of the basin. A pattern of transgressive cyclicity is identified in the marginal zone.

The regressive trend and asymmetry are characteristic of the S8 Sequence. Nevertheless, the eastern portion of the S8 Sequence is more symmetrical. The general pattern of superimposed cycles is rather similar to that of S7 (Fig. 7).

Two phases of sedimentation could be recognized within the Pridoli succession. A general trend of basin infilling during the Early Pridoli (S9 Sequence) is represented by extended TST and HST. A stack of dominantly transgressive lower-order cycles is observed both in the western and eastern parts of the foreland basin (Fig. 7).

The Late Pridoli S10 Sequence shows considerable changes in the architecture and cyclicity as compared with the underlying Lower Pridoli section. An asymmetrical regressive trend characterizes the S10 section in the studied area (Fig. 7). Expanded LST and reduced TST and HST intervals characterize the uppermost Silurian portion. A series of regressive and transgressive cycles are superimposed on these. The HST sediments are missing in the centre and the eastern marginal part of the Silurian Baltic Basin (Fig. 7).

The analysis of cyclicity within the Llandovery succession indicates a progressive marine transgression interrupted by a short-term regression at the Lower–Middle Llandovery boundary. The Lower Llandovery transgressive cycle most probably represents the transgressive part of a complete cycle, while the upper part is eroded. Based on the high percentage of mudstones present within the Llandovery-age cycles, they were presumably deposited within a deep-marine setting.

A change in sedimentation, represented by increased rates of deposition, completeness and symmetry of cycles, is recognized within the Wenlock succession. This could be explained in terms of stability of deposition within the deep-marine environments across the basin. This more symmetrical pattern might also be interpreted as recording the unidirectional nature of sediment flux into the basin. The observed differences in terms of the basin infill between the western and the central–eastern parts of the study area are related to the increasing differentiation of the flexure-influenced subsidence rates, and a gradual modification of the carbonate ramp to a carbonate platform in the east.

The Ludlow depositional sequences represent a series of tectono-stratigraphic units that can be linked to the intensive phases of the flexural subsidence of the foreland basin. In the western part of the Silurian Baltic Basin the relative increase in mudstone content and the dominant asymmetry implies that there was a change in the pattern of sedimentation (Fig. 7). Correlative conodont faunal extinction patterns, identified as Linde and Lau events (Jeppsson, 1998), during the Early and Middle Ludlow, also suggest considerable changes in the basin environment. The change in infilling of the basin was most probably related to pulses of increased sediment input derived from the Caledonian uplands in the southwest. Such a scenario would envisage increasing tectonic activity along the southwestern margin of Baltica. The asymmetrical nature of the transgressive cycles predominates in the eastern part of the Silurian Baltic Basin. They may be interpreted as the preserved half-cycles of complete sedimentary cycles – where the upper regressive parts may have been removed by subsequent transgressions. The pattern of asymmetry and reduced thicknesses, supported by lithological composition, indicates the increased energy levels within this shallow-water environment in the nearshore area (Fig. 7).

The completeness and symmetry of the Early Pridoli cycle (S9 Sequence) indicates compensated sedimentation, i.e. when increased rates of deposition exceeded the still-accelerating flexural subsidence. The irregularity in the distribution of the cycles within the Late Pridoli S10 Sequence implies pulses of sediments input, presumably related to tectonic activity along the south-western margin of Baltica. Due to the regression at the end of the Silurian, the uppermost parts of the Late Pridoli cycles are missing in sections from the eastern margin of the Baltic Basin.

Comparison of the cyclicity of the Silurian Baltic Basin with the global eustatic sea-level fluctuations (Ross & Ross 1996) suggests that four of the depositional sequences (S1, S3, S4 and S6 respectively) resulted from global eustatic sea-level fluctuations. These fluctuations correspond with third-order cycles of 2-4 Ma duration, as defined by Ross & Ross (1996; Fig. 7). The maximal widening of the Baltic Basin at the Llandovery–Wenlock boundary (S3 Sequence) correlates with a contemporaneous global sea-level rise, which culminated at the very beginning of Wenlock, within the limits of the *C. centrifugus–M. riccartonensis* graptolite zone (Ross & Ross 1996). The uppermost Wenlockian transgressive–regressive cycle also shows a rather good correlation with the eustatic sea-level fluctuations of the Ross & Ross (1996) curve. The second eustatic Silurian sea-level maximum at the beginning of Late Ludlow times is reflected in the geometry of the S6 Sequence. Thin lowstand and transgressive system tracts with only subtle changes in lithologies indicate a rapid rise in sea level, whereas the maximal flooding surface (MFS) corresponds with the peak of eustatic sea-level rise (Fig. 7).

While S1, S3, S4, S6 sequences show evidence of a relationship between their architecture and global eustatic sea-level fluctuations, others (S2, S5, S7, S8, S9 and S10) do not. For this latter group, the tectonic activity provides a better explanation for the origin of the units than the eustatic sea-level changes. There is also some other evidence for the significance of tectonic controls on evolution of the basin. The main stage of structuring of the Baltic Basin took place during Late Silurian–earliest Devonian times. Reactivation of the major fault zones, also involving the crystalline basement, is recorded in Late Ludlow to Pridoli times in a compressional regime. This fault reactivation was related to the orogenic activity and long-distance stress transmission (NW–SE) from the Scandinavian Caledonides (Sliaupa 1999b; Sliaupa *et al.* 2000). The sediments of the S7 Sequence in the central part of the basin provide a record of this series of related tectonic events. The lithostratigraphic section of Well Vadzgiris-95 contains a succession that comprises non-fossiliferous carbonaceous marlstones. The deposition of these specific greenish carbonaceous marlstones, containing no remnants of benthic fauna, is restricted to only a narrow N–E-directed zone. Interestingly, adjacent wells (Milaiciai-103 for example) show a similar lithological succession, but with an abundant benthic fauna (Fig. 7). This infers considerably different rates of sedimentation, possibly related to the local block movements along the reactivated faults.

Lower-rank cycles were defined in the Silurian succession, and provide important information on the basin evolution and its tectonic control. The regressive cycles dominate the Llandovery–lowermost Ludlow succession of the western part of the studied area, while they pass into package of the trangressive-type cycles in the marginal part of the foreland basin (Fig. 8). This pattern was inverted during Ludlow times, with transgressive cycles documented in the west and regressive ones being found in the east. A hinge zone is identified in central Lithuania (Fig. 8). This was interpreted as result of instability of the foreland–forebulge–backbulge triad, in turn related to the evolution of the orogenic system in the west. The regressive package of Llandovery–lowermost Ludlow rocks was probably deposited on a concealed forebulge, while the eastern transgressive set of cycles indicate a back-bulge environment. The origin of the concealed forebulge is explained in terms of the superposition of a short-wave orogenic load on a long-wave dynamic flexuring of the lithosphere. This position also explains the only moderate influence of eustatic fluctuation on the pattern of the sedimentation in the nearshore zone of the Silurian Baltic Basin. With progressing advancement of the North German–Polish Caledonides and orogenic loading, the concealed forebulge was shifted to the east during Ludlow times, which caused marked shallowing on the periphery of the foreland, and also a more pronounced tectonic signal in the sedimentary pattern. The Pridoli cycles are rather uniform throughout the study area (Figs 7 & 8), with the lower part of the Pridoli being dominated by transgressive cycles. This might be interpreted as being indicative of the further migration of the furebulge to the east due to progressive overthrusting along the western margin of the Baltica plate. On the other hand, the rates of sedimentation were very high and the pattern of the cycles might be also related to particular features of the basin infilling. The upper part of the Pridoli does not show any distinct features of cyclicity (Fig. 8).

Discussion and conclusions

Discussion

Foreland basin systems are complex, large-scale features that respond to interacting complexes of variables (Flemings & Jordan 1989; Jordan & Flemings 1991). Foreland basins develop in response to orogenic loading of a foreland plate (Jordan 1981; Allen *et al.* 1986). The geometry of foreland basins in general depends on the distribution of the loading and the strength of the underlying lithosphere. The stratigraphic record of a foreland basin, therefore, reflects the controlling mechanisms on basin formation, namely, regional subsidence related to flexure of the lithospheric plate on which the basin is located, and secondary controls such as local lithology, climate and eustatic sea-level (Jordan *et al.* 1988). By definition, foreland basins are associated with fold–thrust belts and are thus affected by the 'topographic loads' of these adjacent thrust belts, which typically produce flexural responses over lateral distances of several hundred kilometres in the foreland plate (DeCelles & Giles 1996). The evolution of the basin fill in terms of sedimentary environment, succession thicknesses and vertical trends, is strongly dependent on the degree of compressional tectonic activity (Munoz-Jimenez & Casas-Sainz 1997). Generally, foreland basins are initially marine, due to rapid downflexing (Jordan 1981). At later stages the sedimentation rates exceed the subsidence rates, giving rise to continental sedimentation (Allen *et al.* 1986).

The Silurian sedimentary basin is one of the most representative in the sedimentary cover of the Baltic Basin. Current quantitative one-dimensional subsidence (McCann *et al.* 1997; Poprawa *et al.* 1997; Poprawa *et al.* 1999) and three-dimensional geodynamic modelling (Lazauskiene *et al.* 1998, 2002) allowed the Silurian Baltic Basin to be interpreted as a foreland basin of the Caledonian Orogen. The sedimentary and geodynamic evolution of the southwesternmost part of the foreland was analysed in detail (Maletz *et al.* 1997; Beier *et al.* 1999;) on the basis of lithological studies, seismic and palaeomagnetic data (Katzung *et al.* 1993; Hoffman & Franke 1997; Krawczyk & DEKORP-BASIN Research Group, 1997). Two-dimensional sedimentary modelling (Lazauskiene 2000; Lazauskiene *et al.* 2000, 2001) quantitatively analysed infilling trends in the Silurian Baltic foreland basin, implying their close relationships to collisional processes in the adjacent orogen as well as the changeable influx of clastics to the Baltic Basin.

Extensive lithofacies, biostratigraphic, sedimentological studies (Lapinskas 1987, 2000; Dadlez 1978; Tomczykowa 1988; Nestor & Einasto 1997; Paskevicius 1997; Jaworowski 2000) were integrated in a number of lithofacies–sedimentary models of the Silurian sedimentary basin. Based on collected data, the major tectonic–sedimentary stages of basin development during the Silurian were distinguished (Lapinskas 1987, 2000; Nestor & Einasto 1997; Paskevicius 1997). A possible provenance for the terrigenous material supplied to the Baltic Basin Silurian has been considered for a long time. Two different sources of terrigenous material were suggested by P. Lapinskas (Lapinskas 1987). Later these assumptions were accentuated in all of the lithofacies models, pointing to the leading role of Scandinavian Caledonides (Nestor & Einasto 1977; Lapinskas 2000). Beier *et al.* (1999) interpreted the infilling of the western part of the foreland basin during the Llandovery in terms of a dominant terrigenous influx from the Avalonian Orogen. They treated Baltica as a secondary source (Beier *et al.* 1999). The results of geochemical REE studies of the region (Sliaupa, 1999a) implied that one source predominated during the Silurian, which was interpreted either as a prevailing cratonic influx, or as sediments of recycled Caledonian Orogen-type.

The sequence- and cyclo-stratigraphic approach represents a powerful tool for systematic study of the internal architecture of sedimentary basins and the controlling role of different factors, particularly tectonic ones, on the formation and geometry of the depositional sequences. Despite very extensive lithofacies and sedimentological studies in the Baltic region mentioned above, the pattern of sedimentation in the Silurian Baltic Basin has not been analysed in the context of regional and global geodynamic processes.

A series of related Silurian depositional sequences were observed within the Baltic Basin of NE Europe. The various tectono-sedimentary stages within this succession could be correlated with the evolution of the thrust system located to the south of the region, which led to the formation of a foreland basin within the Baltic area. The topography and distance to the orogenic deformation front of the fold belt strongly influenced the lithology and the amount of sediment influx (Maletz *et al.* 1997; Sinclair 1997). The actual differentiation of the topography in the Caledonian orogenic belt is still under consideration. However, detail lithofacies studies

(Jaworowski 1971, 2000; Lapinskas 1978, 2000) and current geodynamic constraints (Lazauskiene *et al.* 2002) are in favour of a low (1-2 km high) topographic elevation of the North German–Polish Caledonides.

The development of the Silurian foreland basin within the Baltic region began with an initial influx of fine-grained terrigenous sediments. These were presumably derived from the flat Fennoscandia–Sarmatia Platform to the east. The initial Llandovery sequences (i.e. S1 and S2) and partly 53 represent this early stage of flexuring, which was related to the transformation of the Baltic Basin area from a passive margin to a more convergent setting. The sedimentary architecture of these two sequences represents an undercompensated stage of foreland basin infilling (Fig. 9). The presence of relatively thin sequences along the periphery of the Silurian Baltic basin is indicative of the reduced amount of siliciclastic input into the basin at this time. The low thickness of the Llandovery mudstone, sandstone and siltstone succession, recorded in well G-14 (McCann 1996a, 1996b; Maletz 1997), also implied that subsidence in the basin exceeded the rates of sedimentation. Increasing subsidence resulted in a deepening of the basin in this area, with concomitant basin widening. Similar trends were recorded in the foreland basin further to the southwest – in the German and Danish area (Vejbaek 1994; Beier *et al.* 1999). The earliest Silurian sequences in Bornholnm area (mudstones and siltstones with carbonate interbeds) were interpreted as a deep water phase of foreland basin infilling (Beier *et al.* 1999). As orogenic activity continued, this led to an increase in subsidence, particularly in the western part of the foreland basin.

The overlying Wenlock and Early Ludlow sequences (S3 to S6) were deposited during a period of intense flexuring. The resultant sedimentary geometry of these particular sequences reflects a gradual increase in sediment supply and infilling of the foreland basin. Considerable thicknesses (exceeding 300-600 m) of shallow-water Wenlockian clastics in the German–Danish sector (Vejbeak *et al.* 1994; Maletz 1997; Beier *et al.* 1999) indicate the same trend of fast infilling of the foreland. A thick Lower Silurian succession of laminated siltstones and mudstones in western Pomerania (Dadlez 1978) reflects a high rate of sedimentation in the collisional tectonic setting (Poprawa *et al.* 1999).

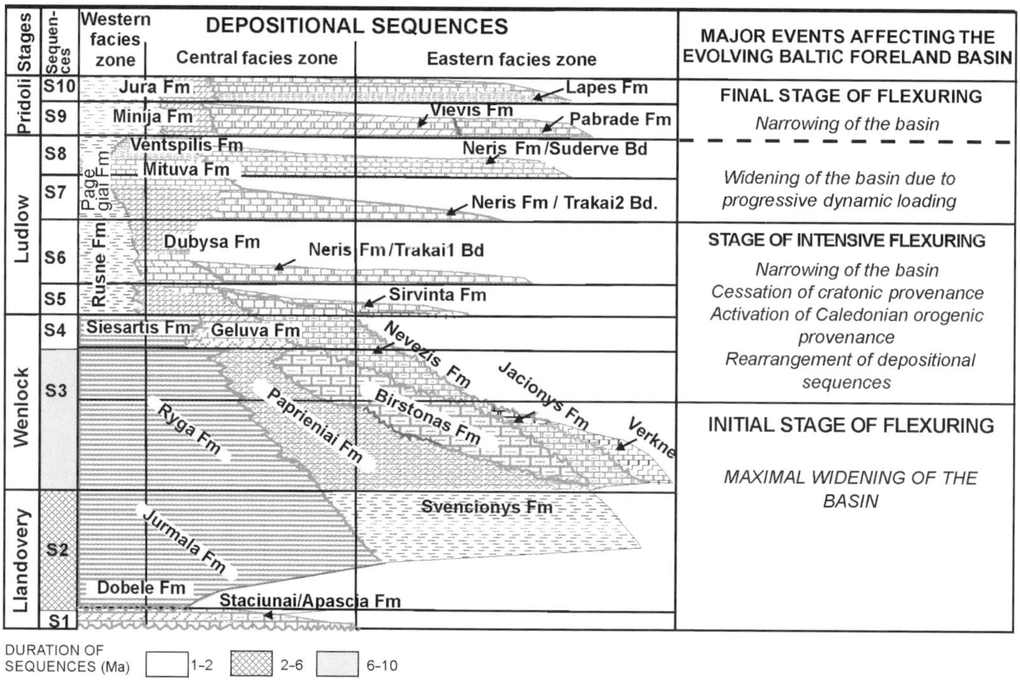

Fig. 9. Generalized chart of the depositional sequences and the main tectonic events of the Silurian Baltic foreland basin (legend is shown in Fig. 3).

Thus, during Wenlock–Early Ludlow times the rates of sedimentation in the central and eastern parts almost equalled those of the western part of the Baltic Basin, where flexure-related subsidence was progressing. The considerable increase in flexural subsidence was associated with narrowing of the basin. The gradual shallowing of the depositional environments in the eastern peripheral parts of the foreland was related to the progressive formation of a shallow isolated carbonate ramp that later evolved into a carbonate platform (Figs 5 & 7). The coeval spread of the clayey facies (S3) to the northeast occurred parallel to the general trend of narrowing of the basin. As suggested by geological data (Lapinskas 2000) and supported by two-dimensional basin infill modelling (Lazauskiene *et al.* 2000), the influx of terrigenous material from a cratonic source began to decline from Late Wenlock times onwards. The reduced amount of clastic input and the accompanying significant aggradation of shallow-marine carbonates along the basin margins led to progradation of the carbonate platform westwards (Fig. 7). This progradation and the coeval restriction in basin area may have been related to the gradual sea-level fall in Wenlockian times, which was coincident with the continuing overthrusting of the North German–Polish Caledonides. The maximal global eustatic sea-level rise at the boundary of the Early Wenlock (Ross & Ross 1996) was followed by a continuous sea-level fall. Global trends of sea level lowering (Johnson *et al.* 1991) in the Baltic Basin were most pronounced during the second half of Wenlock – the minimum sea-level was compatible with the lower boundary of the *lundgreni* graptolite zone (Loydell 1998).

The Late Ludlow–Pridoli sequences (S7 to S10) record the final stages of foreland-basin evolution within the Baltic Basin. At this time subsidence patterns across the Silurian Baltic Basin were highly variable. Flexural subsidence reached values of 200–500 m/Ma close to the Teisseyre–Tornquist Zone (Poprawa *et al.* 1999). At the same time stabilization of subsidence was observed in the central part of the basin. A considerable decrease in subsidence (up to 20 m/Ma) and even minor uplift, was recorded on the easternmost periphery (Sliaupa *et al.* 1997; Poprawa *et al.* 1999). This variability has been interpreted in terms of increased flexural bending of the Baltica plate margin (Poprawa *et al.* 1999; Lazauskiene *et al.* 2002). A coeval increase in the amount of Caledonian-sourced sediments entering the basin, however, was sufficient to outpace the rate of accommodation space provided by flexural subsidence, resulting in a general shallowing of depositional environments across the entire Baltic Basin. Even though Silurian deposits of Ludlow–Pridoli age are mostly eroded further to the southwest of the region (Giese *et al.* 1994; McCann 1996*b*), the shallow-water to terrestrial sedimentation in the Scania area during Ludlow–Pridoli times (Vejbaek *et al.* 1994), and a succession of undeformed Pridoli shales and siltstones preserved in Western Pomerania (Milaczewski & Modlinski 1998), imply a similar degrees of shallowing throughout the basin.

The depositional architecture of the Late Ludlow S7 and S8 sequences is indicative of the increasing size/magnitude of the basin during this time. This period of relative basin growth may be correlated with the progression of subduction-related flexure along the southwestern margin of the Baltica plate. Subsequent narrowing of the basin during Pridoli times can not be explained in terms of eustatic sea level fall alone, and it is suggested that the mechanism of dynamic downflexing, which died out over time, was also involved.

Both tectonic and eustatic mechanisms controlled the formation and location of the various Silurian depositional sequences in the Baltic Basin. Four of the sequences (S1, S3, S4, S6) may be interpreted as having formed in response to global eustatic sea-level fluctuations. The similar duration of these particular sequences and the corresponding cycles (2–3 Ma) would suggest that they can be correlated with the third-order Silurian sea-level fluctuations as first proposed by Ross & Ross (1996). The remaining six sequences, however, do not show any clear relationship with the prevailing sea-level curve fluctuations. Instead, it is herein suggested that these sequences reflect a dominant tectonic signal. There is a strong relationship between sedimentation within the Baltic Basin at this time and the ongoing evolution of the adjacent Caledonian Orogen. Silurian deposition took place in a context of collisional processes between the Eastern Avalonia microcontinental fragment and Baltica. This collision, and the fact that the rate and magnitude of collision varied along the orogenic front, resulted in the progressive differentiation of the subsidence rates throughout the basin, starting from the Early Llandovery.

Terrigenous sediments within the basin were derived from two main areas, namely, the Caledonian Orogen and the Fennoscandian–Sarmatian Platform (Lapinskas 1996, 2000). The dominant provenance, therefore, changed during the development of the foreland in the Silurian, as strongly supported by two-dimensional quantitative basin infill models (Lazauskiene *et al.* 2000). Quantitative forward two-dimensional sedimentary simulation was carried out by the modelling software SEDPAK 2.1 (Strobel *et al.* 1989), with the aim of determining quantitatively the basin-infilling trends of the Baltic Basin in a foreland setting (Lazauskiene *et al.* 2000, 2001). The westernmost orogenic source did not provide a significant amount of sediments during the Llandovery, due to the low morphological relief along the North German–Polish Caledonides and the northerly continental margin. The symmetry of the transgressive–regressive Late Llandovery–Wenlock supercycles suggests that sediment influx occurred over a relatively long time period and was predominantly unidirectional. This would suggest that the adjacent cratonic source controlled the sedimentary pattern in the eastern and central parts of the Baltic Basin. Such a conclusion is supported by the results of quantitative two-dimensional sedimentary modelling (Lazauskiene *et al.* 2000, 2001).

The general trend of tectonic and basin-infill evolution allows the Llandovery–Early Wenlock period to be characterized in terms of the initial stages of an evolving foreland system (Fig. 9). Over the course of collisional processes, the orogenic prism of the NGP Caledonides was thrusted further to the east, on to the margin of Baltica. This overthrusting led to the intensification of the flexuring of the foreland basin. As a result of orogen advancement, fine-grained terrigenous material was increasingly transported into the Baltic Basin from the orogenic wedge since Wenlock (i.e. not after Wenlock, but starting from Wenlock). Due to the evolving orogenic loading, a gradual deepening of the western part of the Silurian basin towards the Teisseyre-Tornquist Zone has been recorded. This began in the latter half of the Wenlock. This deepening was also associated with a decrease in basin size.

During Ludlow times there was a steady increase in subsidence in the western parts of the Baltic Basin. The large thicknesses, lithological composition and the shallow depositional environments of the sediments (Musteikis & Paskevicius 1999; Lapinskas 2000) implied an increase in the amount of terrigenous material, derived presumably from the Caledonian Orogen. This increase would suggest that there may have been some coeval tectonic reactivation in the North German–Polish Caledonides Belt. Considerably increased subsidence, the pattern of basin infilling and the geometry of the depositional sequences characterize the other – final – stage of foreland basin development in the Baltic Basin. The widening of the basin during the later part of the Ludlow can only be explained by combined interaction of the orogenic and dynamic loading. The mechanism of the dynamic loading related to viscous mantle corner flow above a subducting plate (Catuneanu *et al.* 1997) was suggested as one of the forces driving the long-wavelength subsidence of the Silurian foreland basin (Lazauskiene *et al.* 2002). Over the course of collisional processes between the Baltica and Eastern Avalonia plates, increasing dynamic flexure caused the widening of the foreland basin.

The differentiation of the subsidence rates throughout the Silurian Baltic Basin determined the evolution of the Pridoli sequences. The westernmost half of the foreland was characterized by still-accelerating flexural subsidence, whereas decreased rates of flexural subsidence induced rapid carbonate accumulation in the nearshore zone during the Pridoli. This effect was multiplied by a considerably increased influx of orogenic-sourced terrigenous material diluting carbonate sedimentation in open-shelf environments. This forced the carbonaceous facies zone, marked by the Sutkai Reef Belt, to regress back to the periphery of the basin during the first half of the Pridoli. Ongoing collisional processes resulted in the narrowing of the Silurian basin in the Late Pridoli. Quantitative two-dimensional basin infill models suggested that the orogenic source controlled the depositional architecture of the western part of the basin. Whereas the eastern peripheral part of the Baltic foreland basin can be inferred as a subject of the complex interaction of two different sources of sediments, i.e. the cratonic source in the east and the orogenic source in the west during this time (Lazauskiene *et al.* 2000).

The foreland stage of the evolution of the Baltic Basin changed at the Lower–Middle Devonian boundary, which can be explained in terms of isostatic unflexing due to topographic and dynamic unloading.

Conclusions

The sedimentation pattern of the Silurian Baltic Basin was intimately related to the evolution of the orogenic systems in the west that controlled subsidence and sedimentation rates in the foreland. In summary, the foreland basin comprises a mixed clastic and carbonate succession with siliciclastic sedimentation dominating in the westernmost part of the basin adjacent to the Caledonian Deformation Front, and mixed carbonates and clastics in the central and marginal zone.

The sedimentary architecture of the western part of the foreland basin during the Silurian times may be interpreted in terms of the continuing evolution of the Caledonian orogenic belt to the south of the region. It was dominated by shaley sedimentation, whereas carbonates accumulated in the east, showing the transformation of a carbonate ramp to a carbonate platform during the Silurian. The subsidence rate gradual increased from the Early Silurian times onwards, reaching its peak during the Pridoli. This is explained in terms of progressing overthrusting of Eastern Avalonia on to the margin of Baltica. Furthermore, it was associated with progressing dynamic loading that caused broad downflexing of the lithosphere in the 600 km wide Silurian Baltic foreland basin. The initial stage of basin evolution was characterized by a low influx of siliciclastics, which was not enough to compensate for the increasing subsidence. This indicates that the source of the North German–Polish Caledonides was far away and low-lying. The lithofacies distribution also suggests that main influx of clastic material came from the eastern craton. An increasing influx of the terrigenous material from the west resulted in compensated deposition during Wenlock–Early Ludlow times, and overfilling during the latter part of the Silurian. Still, the dominance of clastic material in the foreland implies a low topography for the North German Polish Caledonides, whereas the sandy lithofacies were confined to only the westernmost proximal part of the foreland basin.

Sequence stratigraphic analysis (based on conodont biostratigraphy, well-core and log data), defines 10 different depositional sequences in the eastern half of the Silurian Baltic foreland basin. Four of these sequences (i.e. S1, S3, S4, S6), defined in the Llandovery–Lower Ludlow succession, may be correlated with the global eustatic sea-level curve of Ross & Ross (1996), and it is highly probable that sea-level variations were the predominant controls on the formation of these particular sequences. However, a number of the other sequences (i.e. S2, S5, S7, S8, S9 and S10), do not conform to the sea-level curve pattern, and it is herein suggested that the predominant control on the form and evolution of these sequences was tectonic activity along the evolving Caledonian orogenic front located to the south of the region. There are clear differences between the eustatic- and tectonic-controlled depositional sequences on the periphery of the foreland basin. The eustatic ones, dominating the first part of the Silurian, show only insignificant hiatuses and shifts in the coastline. Moreover, the general sedimentary trend within these sequences indicates a regression of the basin that is related to the outpacing of the carbonate sedimentation and progradation of carbonate platform on the periphery of the Silurian Baltic Basin. By contrast, the tectonically driven changes in the pattern of the basin infilling are more abrupt and more pronounced in the eastern periphery of the foreland basin.

The authors wish to acknowledge the contribution provided by Geological Survey of Lithuania and Lithuanian National Science programme 'Lithosphere' to the realization of this work. Our special thanks go to T. McCann, for improving the organization and the English of the manuscript.

References

ALLEN, P. A., HOMEWOOD, P. & WILLIAMS, G. D. 1986. Foreland basins: an introduction. *In:* ALLEN P. A. & HOMEWOOD P. (eds) *Foreland Basins*. Special Publications of the International Association of Sedimentology, **8**, 3–14.

ANGEVINE, CH. L., HELLER, P. L. & PAOLA, CH. 1990. *Quantitative Sedimentary Basin Modelling.* AAPG Continuing Education Course Note, **32**.

BEIER, H., MALETZ, J., KATZUNG, G. & BOHNKE, A. 1999. The southern Baltic Foreland Basin: interpretation of the Avalonia terrane-accretion events and the Early Palaeozoic lithostratigraphy on the SW-margin of the East European Platform. *Between Eurobridge and TESZ. Eurobridge Workshop*, Suwalki, Poland, 5–8.

CATUNEANU, O., BEAUMONT, C. & WASCHBUSCH, P. 1997. Intraplay of static loads and subduction dynamics in foreland basins: reciprocal stratigraphies and the 'missing' peripheral bulge. *Geology*, **25**, 1087–1090.

DADLEZ, R. 1978. Sub-Permian Rock Complexes in the Koszalin-Chojnice Zone. *Kwartalnik Geologiczny*, **22(2)**, 269–299 [in Polish with English summary].

DECELLES, P. G. & GILES, K. G. 1996. Foreland basin systems. *Basin Research*, **8**, 105–123.

EHRENBERG, S. N., PICKARD, N. A. H., SVANA, T. A., NILSSON I. & DAVYDOV, V. I. 2000. Sequence stratigraphy of the inner Finnmark carbonate platform (Upper Carboniferous–Permian), Barents Sea – correlation between well 7128/6-1 and the shallow IKU cores. *Norsk Geologisk Tidsskrift*, **80**, 129–162.

FLEMINGS, P. B. & JORDAN, T. E. 1989. A synthetic stratigraphic model of foreland basin development. *Journal of Geophysical Research*, **94**, 3851–3866.

GIESE, U., KATZUNG, G. & WALTER, R. 1994. Detrital composition of Ordovician sandstones from the Rugen boreholes: implications for the evolution of the Tornquist Ocean. *Geologische Rundschau*, **83**, 293–308.

GRADSTEIN, F. M. & OGG, J. 1996. A Phan erozoic time scale. *Episodes*, **19**.

HOFFMAN, N. & FRANKE, D. 1997. The Avalonia–Baltica Suture in NE Germany – new constraints and alternative interpretations. *Zeitschrift der Deutsche Geologische Gesellschaft*, **25(1–2)**, 3–14.

JACQUIN, T., ARNOUD-VANNEAU, A., ARNOUD, H., RAVENNE, C. & VAIL, P. R. 1991. System tracts and depositional sequences in a carbonate setting: a study of continous outcrops from platform to basin at the scale of seismic line. *Marine and Petroleum Geology*, **8**, 122–138.

JAWOROWSKI, K. 1971. Sedimentary structures of the Upper Silurian siltstones in the Polish Lowlands. *Acta Geologica Polonica*, **21**, 519–571.

JAWOROWSKI, K. 2000. Facies analysis of the Silurian shale–siltstone succession in Pomerania (northern Poland). *Geological Quarterly*, **44**, 297–316

JEPPSSON, L. 1998. Silurian oceanic events: summary of general characteristics. *In:* LANDING, E. & JOHNSON, M. E. (eds) *Silurian cycles. Linkages of Dynamic Stratigraphy with Atmospheric, Oceanic and Tectonic Changes.* James Hall Centennial Volume. *New York State Museum Bulletin,* **491**, 239–255.

JOHNSON, M. E., KALJO, D. & RONG, J.-Y. 1991. Silurian eustasy. *Special Papers in Paleontology*, **44**, 145–163.

JORDAN, T. E. 1981. Thrust loads and foreland basin evolution, Cretaceous, western United States. *AAPG Bulletin*, **65**, 2506–2520.

JORDAN, T. E. & FLEMINGS, P. B. 1989. From geodynamic models to basin fill – a stratigraphic perspective. *In:* CROSS, T. A. (ed.) *Quantitative Dynamics Stratigraphy*, Prentice Hall, 149–163.

JORDAN, T. E. & FLEMINGS, P. B. 1991. Large-scale stratigraphic architecture, eustatic variation, and unsteady tectonism: a theoretical evaluation. *Journal of Geophysical Research*, **96**, 6681–6699.

JORDAN, T. E., FLEMINGS, P. B. & BEER, J. A. 1988. Dating thrust fault activity by use of foreland-basin strata. *In*: KLEINSPEHN, K. L. & PAOLA, C. (eds) *New Perspectives in Basin Analysis*. Springer, New York, 307–330.

KALJO, D., NESTOR, H. & EINASTO, R. 1991. Aspects of Silurian carbonate platform sedimentation. *Special Papers in Palaeontology*, **44**, 205–224.

KARAGODIN, J. & ARMENTROUT, D. 1996. [Analysis of the basic concepts and terms of lithmonology and sequence stratigraphy.] *Geology and Geophysics*, **37(7)**, 3–11 [in Russian].

KATZUNG, G., GIESE, U., WALTER, R. & VON WINTERFELD, C. 1993. The Rugen Caledonides, northeast Germany. *Geological Magazine*, **130**, 725–730.

KING, L. M. 1994. Subsidence analysis of Eastern Avalonia sequences: implications for Iapetus closure. *Journal of the Geological Society*, **151**, 647–657.

KOMINZ, M. A. & BOND, G. C. 1986. Geophysical modelling of the thermal history of foreland basins. *Nature*, **320**, 252–256.

KRAWCZYK, C. M. & DEKORP-BASIN RESEARCH GROUP 1997. The northern rim of the Central European Basin System: first results of the offshore–onshore survey, Basin '96. *Terra Nostra*, **97(11)**, 64–66.

LAPINSKAS, P. 1987. [Silurian Formations of the Baltic Syneclise]. *In:* GORETSKYJ, R. G. & SUVEIZDIS, P. (eds) *Tectonics, Facies And Formations Of Western East European Platform*. Minsk, Nauka i Technika, 103–116 [in Russian].

LAPINSKAS, P. 1996. [Paleogeographical peculiarities of the Silurian in the Southern Peri-Baltic basin]. *In:* SAKALAUSKAS, K. *et al.* (eds) *Oil in Lithuania*, Institute of Geology, Vilnius, 27–34 [in Lithuanian].

LAPINSKAS, P. 2000. *Structure and Petroliferosity of the Silurian in Lithuania.* Institute of Geology, Vilnius.

LAZAUSKIENE, J. 2000. *Quantitative modelling of the Baltic Silurian Sedimentary Basin*. Abstract of Doctoral Dissertation. Institute of Geology, Vilnius.

LAZAUSKIENE, J., SLIAUPA, S. & STEPHENSON, R. A. 1998. 3D Flexural Model of the Silurian Foreland Basin in the Baltic region. *In:* Suveizdis, P. & Zdanaviciute, O. (eds) *Perspectives of Petroleum Exploration in the Baltic region*. Institute of Geology, Vilnius, 101–103.

LAZAUSKIENE, J., STEPHENSON, R. A., SLIAUPA, S. & VAN WEES, J.-D. 2002. 3D flexural model of the Silurian Baltic Basin. *Tectonophysics*, **346**, 115–135.

LAZAUSKIENE, J., SLIAUPA, S., BRAZAUSKAS A. & MUSTEIKIS, P. 2000. Sequence stratigraphy of the Baltic Silurian succession: tectonic controls on the foreland basin infill evolution. *Geophysical Journal Kiev*, **22(4)**, 109.

LAZAUSKIENE, J., SLIAUPA, S., MUSTEIKIS, P. & BRAZAUSKAS, A. 2001. Sequence Stratigraphy and Tectonic Control of the Baltic Silurian Foreland Basin. *11 EUG Meeting*, Book of Abstracts. Strasburg.

LOYDELL, D. K. 1998. Early Silurian sea-level changes. *Geological Magazine*, **135(4)**, 447–471.

MCCANN, T. 1996a. Silurian facies in northeast Germany. *Zeitschrift der Deutsche Geologische Gesellschaft*, **147**, 209–219.

MCCANN, T. 1996b. Pre-Permian of the north-east German Basin. *Geological Journal*, **31**, 159–177.

MCCANN, T., SLIAUPA, S., POPRAWA, P., NARKIEWICZ, M., MATYJA, H., LIPIEC, M. & LAZAUSKIENE, J. 1997. The Devonian-Carboniferous tectonic evolution of selected intracratonic and passive margin basins from Central and Eastern Europe: constraints from subsidence analysis. *Terra Nostra*, **97/11**, 85–89.

MALETZ, J. 1997. Ordovician and Silurian strata of the G-14 well (Baltic sea): Graptolite faunas and biostratigraphy. *Zeitschrift der Deutsche Geologische Gesellschaft*, **25(1/2)**, 29–39.

MALETZ, J., BEIER, H., KATZUNG, G. & NIEDZWIEDZ, A. 1997. A Lower Palaeozoic (Ordovician–Silurian) foreland basin at the south-western rim of Baltica. *Terra Nostra*, **97(11)**, 81–84.

MILACZEWSKI, L. & MODLINSKI, Z. 1998. The older Palaeozoic in the Western Part of Polish Baltic Aquatorium and adjacent land. *In*: Suveizdis, P. & Zdanaviciute O. (eds) *Perspectives of Petroleum Exploration in the Baltic Region*, Institute of Geology, Vilnius, 17–19.

MITCHELL, A. H. G. & READING, H. G. 1969. Continental margins, geosynclines and oceanfloor spreading. *Journal of Geology*, **77**, 629–646.

MUNOZ-JIMENEZ, A. & CASAS-SAINZ, A. M. 1997. The Rioja Trough (N Spain): tectonosedimentary evolution of a symmetric foreland basin. *Basin Research*, **9**, 65–85.

MUSTEIKIS, P. 1993. Detailed stratigraphy and significance of Silurian Brachiopods in it. *Geologija*, **14**, Vilnius, 131–140.

MUSTEIKIS, P. & PASKEVICIUS, J. J. 1999. Brachiopod communities of the Lithuanian Silurian. *In*: BOUCOT, A. J. & LAWSON, J. D. (eds) *Paleocommunities: a case study from the Silurian and Lower Devonian*. Cambridge University Press, 305–326.

NESTOR, H. 1994 Comments on modernized Silurian Correlation Chart of Estonia and Latvia. *Geologija*, **17**, Vilnius, 88–94.

NESTOR, H. & EINASTO, R. 1997. Ordovician and Silurian carbonate sedimentation basin. *In*: RAUKAS, A. & TEEDUMAE, A. (eds) *Geology and Mineral Resources of Estonia*, Estonian Academy Publishers, Tallinn, 192–205.

PASKEVICIUS, J, LAPINSKAS, P., BRAZAUSKAS, A., MUSTEIKIS, P. & JACYNA, J. 1994. Stratigraphic revision of the regional stages of the Upper Silurian Baltic Basin. *Geologija*, **17**, Vilnius, 64–87.

PASKEVICIUS, J. 1997. *The Geology of the Baltic States*. Geological Survey of Lithuania, Vilnius, 387 pp.

POPRAWA, P., NARKIEWIC, M., SLIAUPA, S., STEPHENSON, R. & LAZAUSKIENE, J. 1997. Caledonian accretion along TESZ. *Terra Nostra*, **11**, 110–117.

POPRAWA, P., SLIAUPA, S., STEPHENSON, R. A. & LAZAUSKIENE, J. 1999. Late Vendian–Early Palaeozoic tectonic evolution of the Baltic Basin: regional implications from subsidence analysis. *Tectonophysics*, **314**, 219–239.

POSAMENTIER, H. W. & GEORGE, P. A. 1993. Variability of the sequence stratigraphy models: effects of local basin factors. *Sedimentary Geology*, **86**, 91–109.

POSAMENTIER, H. W. & JAMES, D. P. 1993. An overiew of sequence-stratigraphic concepts: uses and abuses. *Special Publications of the International Association of Sedimentology*, **18**, 3–18.

PROUST, J. N., CHUASHOV, E., VENNIN, E. & BOISSEAU, T. 1998. Carbonate platform drowning in a foreland setting: the mid-Carboniferous platform in Western Urals (Russia). *Journal of Sedimentary Research*, **68(6)**, 1175–1188.

ROSS, C. A. & ROSS, J. R. P. 1996. Silurian sea-level fluctuations. *In*: WITZZKE, B. J., LUDVIGSON, G. A. & DAY, L. (eds) *Palaeozoic Sequence stratigraphy. Views from the North American Craton: Boulder, Colorado*. Geological Society of America, Special Papers, **306**, 187–192.

SINCLAIR, H. D. 1997. Tectonostartigraphic model for underfilled peripheral foreland basins: An Alpine perspective, *GSA Bulletin*, **109(3)**, 324–346.

SLIAUPA, S. 1999a. REE based provenance study of Silurian sediments in the Baltic basin. *Romanian Journal of Tectonics and Regional Geology*, **77**, 74–75.

SLIAUPA, S. 1999b. Far-field stress transmission indications in Early Palaeozoic structural evolution of the Baltic basin. *Romanian Journal of Tectonics and Regional Geology*, **77(1)**, p. 59

SLIAUPA, S. & LAZAUSKIENE, J. 1997. 1D subsidence analysis along Lithuanian part of the Eurobridge Profile (Vendian–Cretaceous). *Eurobridge Workshop*, Lithuanian Geological Survey, Vilnius, 85–88.

SLIAUPA, S. & POPRAWA, P. 2000. Post-Silurian unflexuring of the Western part of the Baltic sedimentary basin. *Geophysical Journal Kiev*, **22/4**, 129.

SLIAUPA, S., POPRAWA, P., LAZAUSKIENE, J. & STEPHENSON, R. A. 1997. The Palaeozoic subsidence history of the Baltic Syneclise in Poland and Lithuania. *Geophysical Journal Kiev*, **19(1)**, 137–139.

SLIAUPA, S., POPRAWA, P. & JACYNA, J. 2000. Structural analysis of seismic data in the Baltic basin: evidences for Silurian-Early Devonian intra-plate compression in the foreland of Caledonian orogen. *Journal of the Czech Geological Society*, **45/3-4**, 260–261.

STEPHENSON, R. A., SLIAUPA, S., SKRIDLAITE, G. & BEEKMAN, F. 1997. Lithosphere rheology along EUROBRIDGE: what are the geodynamic implications. *EUROBRIDGE Meeting*, Book of Abstracts, Vilnius, p. 78.

STROBEL, J., CANNON, R., KENDALL, CH.G.ST.C., BISWAS, G. & BEZDEK, J. 1989. Interactive simulation (SEDPAK) of clastic and carbonate sediments in shelf to basin settings. *Computers and Geosciences*, **15(8)**, 1279–1290.

TOMCZYKOWA, E. 1988. Silurian and Lower Devonian biostratigraphy and palaeoecology in Poland. *Biuletyn Instytutu Geologicznego*, **359**, 21–41.

TOMCZYKOWA, E. & TOMCZYK, H. 1979. Stratigraphy of the Polish Silurian and Lower Devonian and development of the Proto-Tethys. *Acta Palaeontologica Polonica*, **24(1)**, 165–181.

VALIUKEVICIUS, J. & KATINAS, V. 1995. Atlas of the Lithological–Paleogeographical Maps of Baltic Phanerozoic. Scale 1:2 500 000. Institute of Geology, Vilnius.

VAN WAGONER, J. C., POSAMENTIER, H. W., MITCHEM, R. M. & VAIL, P. R. 1988. An overview of the fundamentals of sequence stratigraphy and key definitions. *In:* WILGUS, C. K., HASTINGS, B., KENDALL, C. G. ST. C., POSAMENTIER, H. W., ROSS, C. A. & VAN WAGONER, J. C. (eds) *Sea-level Changes: an Integrated Approach*. SEPM Special Publication, **42**, 39–45.

VEJBAEK, O. V., STOUGE, S. & POULSEN, K. D. 1994. Palaeozoic Tectonic and Sedimentary Evolution and Hydrocarbon Prospectivity in the Bornholm Area. *Danmarks Geologiske Undersogelse*, **34** Copenhagen, 1–23.

ZOETEMEIJER, R. 1993. *Tectonic Modelling of Foreland basins*. Academic Proefschrift Vrije Universiteit, Amsterdam.

Evolution of the southern margin of the Donbas (Ukraine) from Devonian to Early Carboniferous times

T. MCCANN[1], A. SAINTOT[2], F. CHALOT-PRAT[3], A. KITCHKA[4], P. FOKIN[5], A. ALEKSEEV[5] & EUROPROBE-INTAS RESEARCH TEAM

[1]*Geologisches Institut, Universität Bonn, Nußallee 8, 53115 Bonn, Germany (e-mail: tmccann@uni-bonn.de)*

[2]*Vrije Universiteit, Instituut voor Aardwetenschappen, Tektoniek afdeling, De Boelelaan 1085, 1081 HV Amsterdam, The Netherlands*

[3]*Centre de Recherches Petrographiques et Géochimiques, Université Henri Poincare, Boîte Postale 20, 15 rue Notre Dame des Pauvres, 54501 Vandoeuvre-les-Nancy Cédex, France*

[4]*CASRE, Suite 49, 55b Gonchar Street, Kiev 01601, Ukraine*

[5]*Geological Faculty, Moscow State University, Leninskie Gory, 119899 Moscow, Russia*

Abstract: A Devonian–Early Carboniferous succession comprising thick clastic and carbonate sediments with interbedded volcanics was examined along the southern margin of the Donbas fold belt, Ukraine. Following initial rifting and subsidence, a continental (fluvial, lacustrine) succession was established. This first phase of synrift activity (Eifelian) was accompanied by the extrusion of basalts. In Late Givetian–early Famennian times, half-graben development was pronounced and a series of E–W-trending half-grabens were formed, along with coeval volcanic activity. Subsequent basin subsidence led to the establishment of a broad marine carbonate platform across the region (Late Famennian–Tournaisian-early Viséan). Renewed uplift led to partial exposure and karstification of the platform. This was accompanied by trachyte intrusion and extrusion. The overlying chert-rich unit was probably deposited under lacustrine conditions, although a shelf environment has also been suggested. Renewed tectonic activity along the main basin-bounding fault resulted in the synsedimentary deformation of this unit.

The NW–SE-trending Pripyat–Dniepr–Donets Basin strikes in a southeasterly direction, and over a distance of *c.*2000 km, from Belarus through Ukraine, where it connects with the deformed southern margin of the East European Platform (Karpinsky Swell) in southern Russia (Fig. 1). The basin is bounded by two Precambrian basement massifs – the Priazov Massif (eastern part of the Ukrainian Shield) to the north, and the Voronezh Massif to the south, together forming the Ukrainian craton (Fig. 1). Together, these massifs define a broad domal structure which is transected by the rift (Wilson & Lyashkevich 1996). The basin system can be subdivided from northwest to southeast into the relatively shallow Pripyat Trough, the much deeper Dniepr–Donets Basin (DDB) and the inverted Donbas fold belt (DF). Sediment thicknesses increase from *c.*2 km in the Pripyat Trough to *c.*22 km in the DF (Chekunov *et al.* 1993; Stovba *et al.* 1996). The Pripyat Trough and the DDB are important hydrocarbon provinces, while the DF contains large coalfields. The Donbas fold belt–Karpinsky Swell region forms the inverted part of the intracratonic Dniepr–Donets rift basin, and is therefore, of key importance both for constraining the anomalous subsidence history (since no subsidence model has yet explained the 22-km thick sediment succession) and formation mechanism of the linked Dniepr–Donets–Donbas–Karpinsky–Peri-Caspian sedimentary basin system and also for solving the puzzle of Late Palaeozoic palaeogeography along the southern part of the East European Craton. The Donbas fold belt (DF) is a 150-km wide region

From: McCann, T. & Saintot, A. (eds) *Tracing Tectonic Deformation Using the Sedimentary Record.* Geological Society, London, Special Publications, **208**, 117–135. 0305–8719/03/$15.00

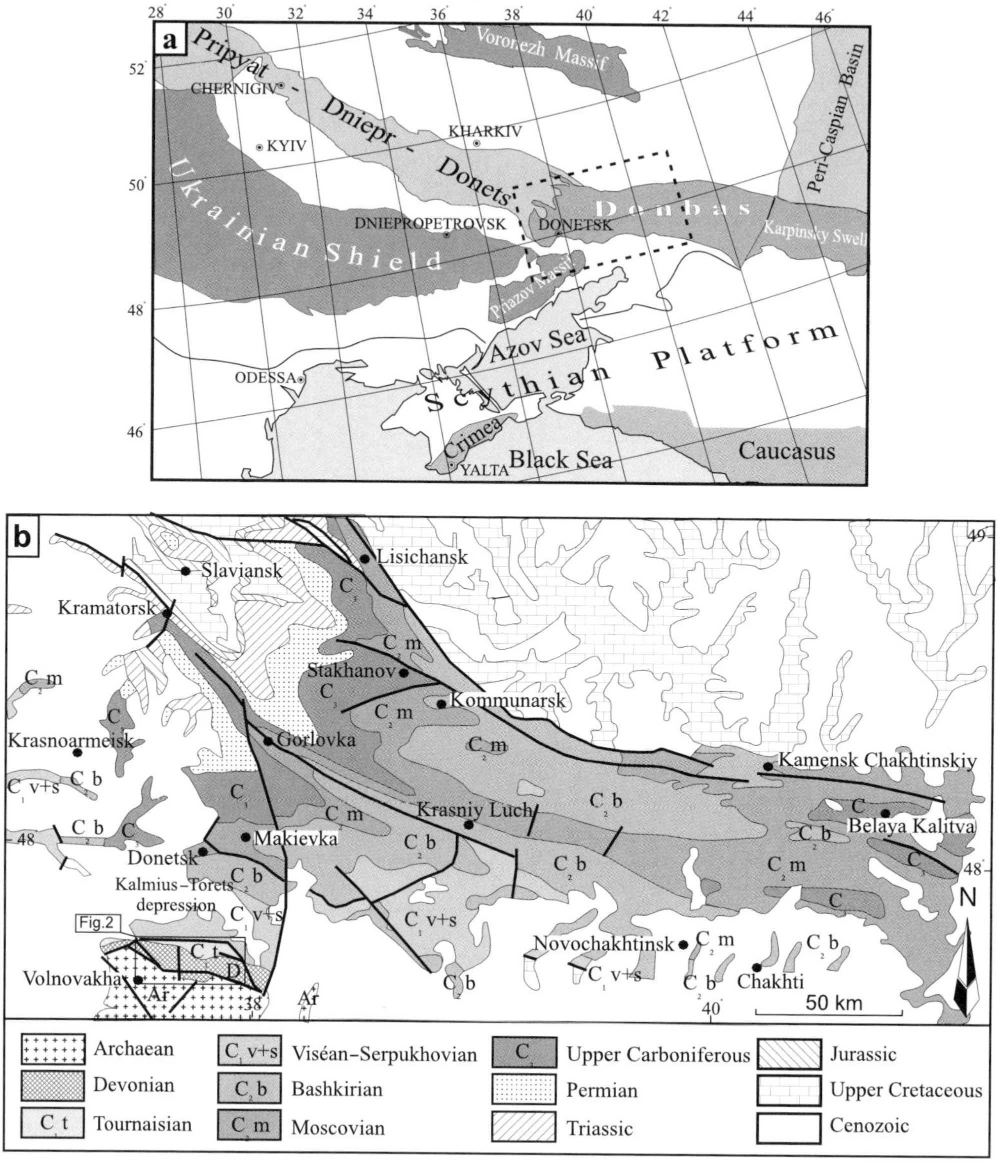

Fig. 1. (a) Location map of the Pripyat–Dniepr–Donetsk–Donbas basin system. The various structural units which compose the southern edge of the East European Platform are indicated by a selection of grey shadings. **(b)** Detailed stratigraphic map of the Donbas fold belt (after *Geological map of the USSR and adjoining water-covered areas 1983*). The location is indicated by the dashed outline in Figure 1(a).

which extends 500 km from the eastern Ukraine to southwest Russia (Fig. 1). The DF has undergone a complex tectonic evolution. Seismic evidence reveals that the Dniepr–Donets segment of the rift is characterized by linear bounding faults which penetrate much of the crust, and by significant crustal thinning (ß of approx 1.3; Kusznir *et al.* 1996) and abundant synrift volcanic activity.

The precise area of study comprises *c.* 500 km^2 and is situated about 30 km to the south of the city of Donetsk, and located between the Donbas fold belt and the crystalline Priazov Massif (Figs 1 & 2). The geology of the region is

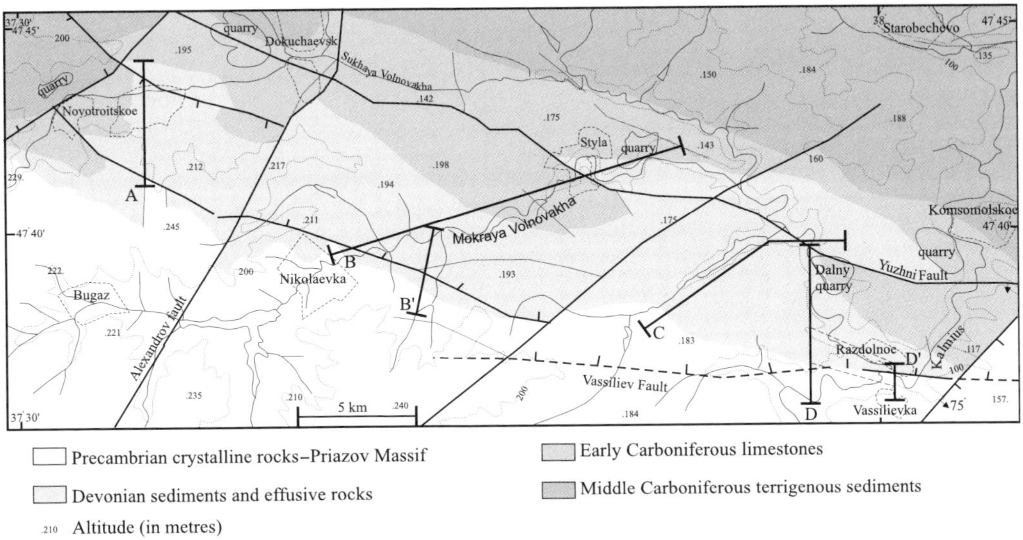

Fig. 2. Detailed map of the southern margin of the Donbas fold belt, showing the locations of the various cross-sections (after the *Geological map of the Ukrainian Donbas fold belt 1:200 000, 1995*).

characterized by a series of half grabens which developed between the Vassiliev Fault, located within the Priazov Massif, and the Yuzhni Fault, marking the southern boundary of the Styla Horst, in response to rifting phases that occurred in Devonian and Early Carboniferous times (Figs 1 & 2). The relationship between sedimentation and volcanism in the southern Donbas region, and at this particular stage in the geological evolution of the entire region, has received little attention in Western literature. Throughout much of the Donbas and Donets region, these strata are overlain by a thick cover of Upper Palaeozoic, Mesozoic and Cenozoic sediments, and thus the precise relationships between the different units are difficult to decipher. In the study area, however, exposure, while not ideal, facilitates a more detailed study of these key exposures for unravelling the early tectonic and sedimentary evolution of the Donbas region. These new data outline the sedimentological, volcanological and tectonic development of this key region from Middle Devonian to Early Carboniferous times.

Geological background

The broad geological evolution of the region is closely linked to the development of the Pripyat–Dniepr–Donets system. Prior to the onset of crustal extension, the Priazov Massif and Voronezh Massif were covered by marine Middle Devonian sediments (Alekseev *et al.* 1996; Nikishin *et al.* 1996). During the Late Devonian, rifting, which was associated with domal basement uplift and magmatism, was widespread from the Pripyat–Dniepr–Donets system in the south to the Eastern Barents Sea in the north (Wilson & Lyashkevich 1996). The evolution of this rift system is thought to be contemporaneous with the development of a major back-arc rift system in Western and Central Europe (Ziegler 1990).

Rift development in the DDB region commenced with a synrift phase in the Devonian and a second rift event in Early Carboniferous times, followed by extensional rejuvenation during the Late Carboniferous and a post-rift phase between the Permian and Quaternary, with thrust and fold development in the Late Cretaceous (Stephenson *et al.* 1993; Stovba & Stephenson 1999). The main extensional phase of the Dniepr–Donets segment occurred between the Late Frasnian (370 Ma) and the end of the Devonian (363 Ma) (Kusznir *et al.* 1996) which is when maximum rates of downwarping occurred. Rifting was accompanied by intense, probably plume-related, volcanic activity (Wilson & Lyashkevich 1996).

Sedimentology of the Southern Donbas region

Middle Devonian (Eifelian–Givetian)

The oldest exposed Devonian sediments in the region (near Styla Quarry and Nikolaevka village) rest unconformably on a Precambrian alkaline porphyritic and/or graphic granite. These sediments are of Eifelian–Givetian age (Aisenverg *et al.* 1975) (Quartz-rich sandstones overlying granite at the base of the Styla section B in Fig. 3). The top of the granite is weathered down to 10–15 m, and is overlain by a feldspar-rich, coarse-grained sandstone. The rocks are clastic, ranging from coarse gravels to micro-conglomerates to fine-grained sandstones and siltstones. In some areas the sandstones are more quartz rich, and may even resemble a quartz-pebble conglomerate. Indeed, in one particular location the rocks were quartz dominated, with mainly quartz clasts set in a quartz cement. Rare carbonates have also been noted (Figs 3 & 4).

Beds are normally graded, and up to 70 cm thick, with sharp upper and lower boundaries,

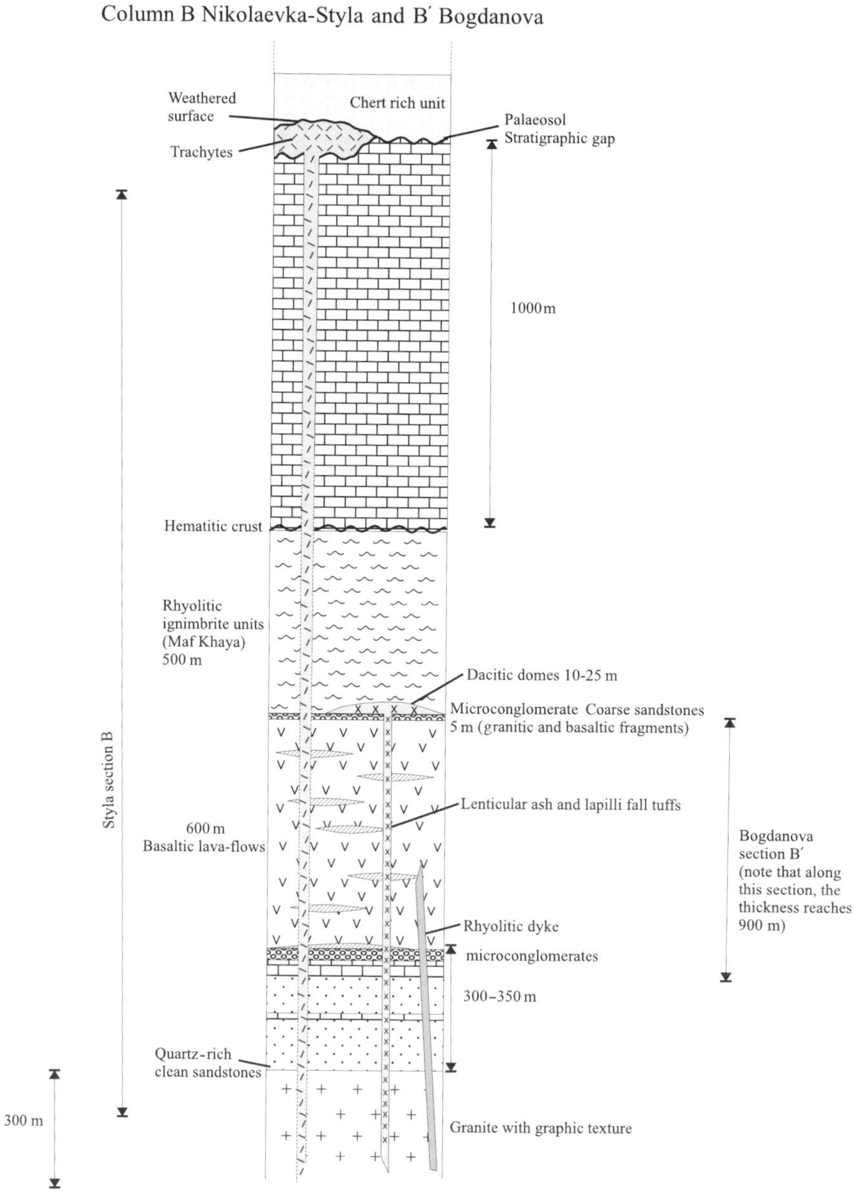

Fig. 3. Schematic stratigraphic column from the Nikolaevka–Styla (column B) and Bogdanova (column B′) areas.

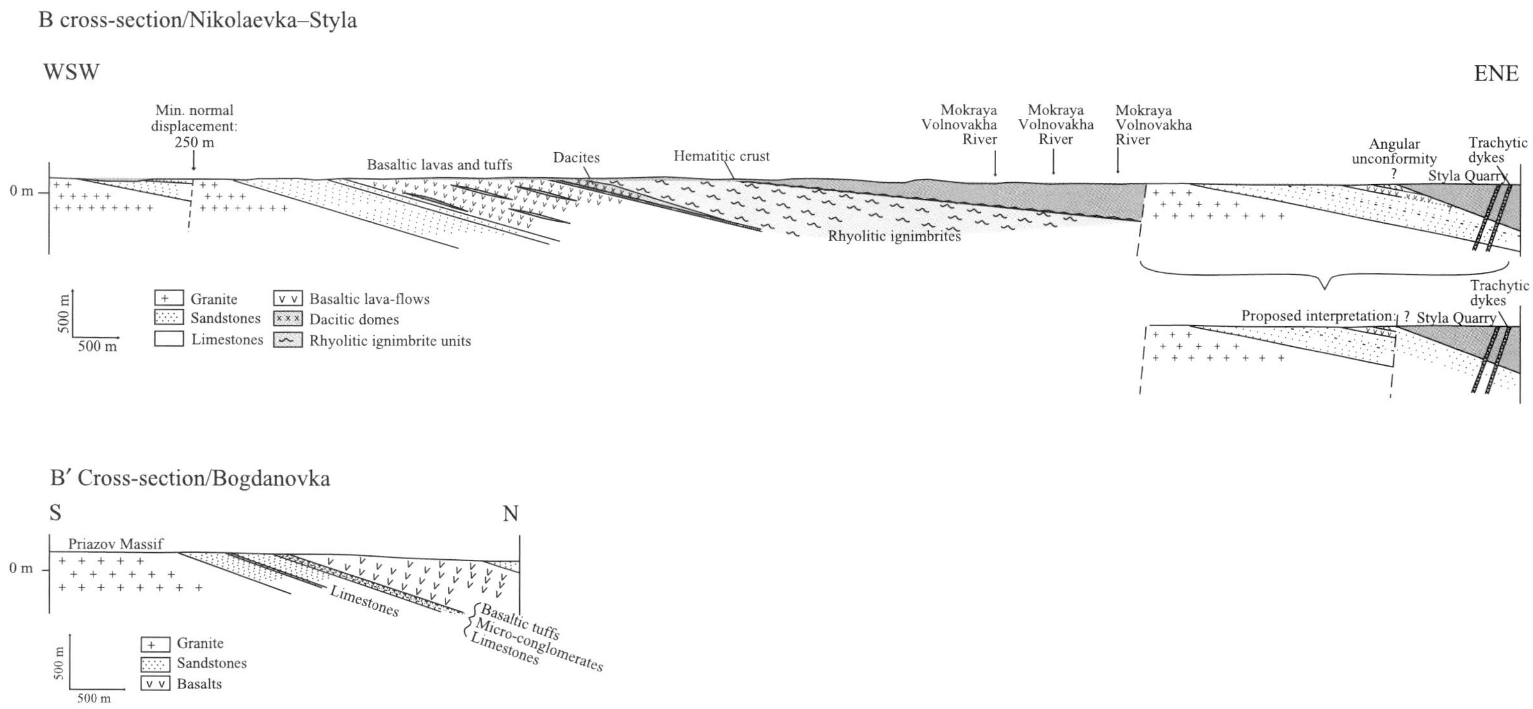

Fig. 4. Geological cross-sections from the Nikolaevka–Styla area (cross-section B) and the Bogdonovka area (cross-section B′). See Figure 2 for a precise location.

and local evidence of scouring and channelling (see below). Internally, the beds are planar and trough cross-bedded (transport direction to the north). There is evidence of local thinning and fining upwards.

Elsewhere within the study area other Middle Devonian outcrops were examined. The sediments here strongly resemble those of the Styla region, comprising mainly coarse-grained sandstones, interbedded with quartz-pebble conglomerates where the individual clasts are up to 3–4 cm in diameter. Bed thickness ranges from 0.5 to 1.0 m and internally they are cross-bedded (080°), becoming increasingly more quartzite-like up section. The beds both thin and fine upwards. Interbedded mudstones are also observed,as are sandstones with some possible tuffaceous interbeds.

As noted above, there is much evidence of channelling within the sandstones of the Middle Devonian, with individual channels cutting down into the underlying sediments. Some of these channels are associated with finer-grained (mud–fine-grained sand) overbank and inter-channel deposits. Bed lenticularity may be pronounced. The channel-fill sandstones (typical orientation – 048°) are up to 3.0 m thick, normally graded and coarse grained. Individual clasts within the coarse channel fill may be up to 15.0 cm in diameter. In some locations, the sediments are intercalated with volcanic units – even showing signs of channels cutting into the underlying volcanics.

Petrographically, the samples are coarse- to medium-grained sandstones comprising monocrystalline quartz (Qm) with subsidiary polycrystalline quartz (Qp), plagioclase and a series of accessory minerals. Some samples, in addition to being rich in Qm, also have a quartz cement (sedimentary quartzites). Volcanic fragments are rare within these sandstones.

The predominance of parallel lamination and trough cross-stratification, together with the presence of channel and overbank deposits, suggest that these sediments were deposited in a continental environment, most probably fluvial or delta plain. There was, however, no evidence of any soil development.

?late Givetian/Frasnian–?early Famennian

The uppermost Devonian sediments are predominantly clastic and reddish, or sometimes light grey, in colour and include microconglomerates and coarse- to medium-grained sandstones (see also Aisenverg *et al.* 1975) (Fig. 5). Pebbles of quartz and effusive rocks are contained within the microconglomerates. Individual sandstone beds are up to 100 cm thick, have sharp boundaries, are normally graded (sometimes from a microconglomerate base) and show both trough cross-stratification and parallel lamination with some possible bed amalgamation. Some sections show evidence of coarsening upward cycles. In some areas the succession is siltstone rich, with thinner cross-bedded sandstones up to 50 cm thick. Some of these cross-bedded sandstones show evidence of channelling. The siltstones are laminated and sometimes contain coalified and silicified plant remains such as *Lepidodendron* fragments. The upper part of the succession has a hematitic crust, several metres in thickness (Fig. 5)

Petrographically, the microconglomerates contain a variety of lithic fragments, including: polycrystalline quartz (Qp) in an altered matrix; sediment lithic fragments (mainly siltstone and mudstone); volcanic fragments (mainly basic, but also acidic and ?pyroclastics), some of which are altered. Grains are contained within a matrix, which may also show evidence of alteration. Associated sandstones may be rich in monocrystalline quartz (Qm), subordinate Qp, with plagioclase also present (plagioclase percentages can be up to 15–20% in some samples). These sandstones also contain rare altered volcanic fragments. In some areas the sandstones contain so much Qm that they are almost quartzite-like.

The sediments of the Upper Devonian were probably deposited in a similar setting to those of the Middle Devonian. The evidence of trough cross-bedding, channel and interchannel deposits and the rare plant remains suggest deposition in a fluvial or delta-plain setting.

?Uppermost Famennian–Tournaisian

At the transition from the ?Uppermost Famennian to the Carboniferous there is a basal horizon (conglomerates, sandstones, limestones) 1.5–3.0 m thick. This horizon is overlain by limestones and micritic shales up to 6.0 m thick and is then replaced by massively bedded, planar limestones of the Tournaisian (upper part of Styla Section B, see Fig. 3). The limestones show marked bed continuity, with individual beds being up to 1.90 m thick.

The overlying Tournaisian succession comprises flat bedded, occasionally knobbly and lenticular, laterally continuous micritic carbonates. Bed thickness ranges from *c.*0.6 to 2.0 m, with the strata arranged in thinning-upward sequences, although there does not appear to be any formal pattern to these sequences. Rare, disarticulated bioclastic material (coral, shelly fragments) is present. The limestones appear to

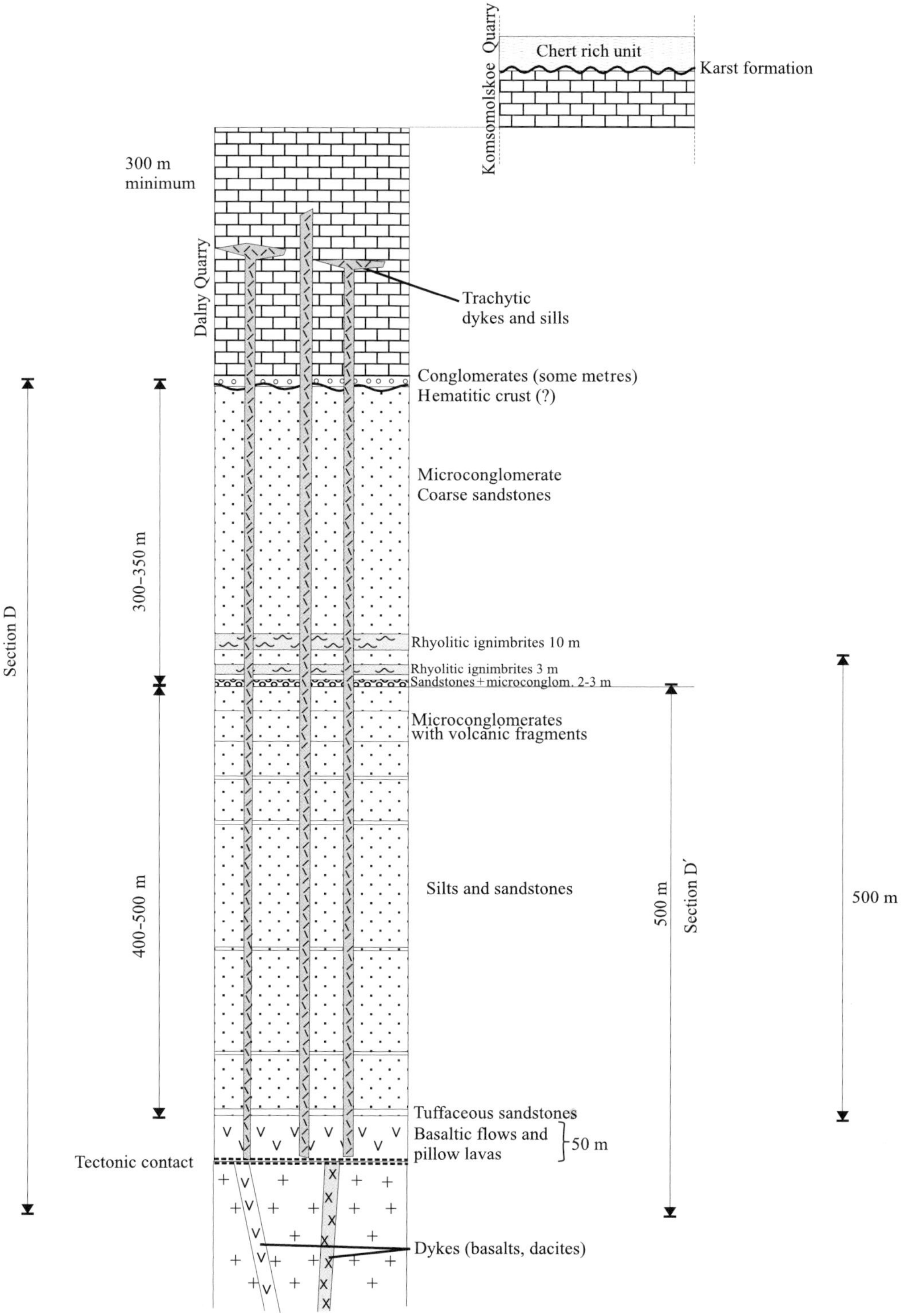

Fig. 5. Schematic stratigraphic column from the Razdolnoe–Dalniy/Vassilievka–Razdolnoe area (column D–D′).

be normally graded, although there is little variation in grain size. Some beds show planar lamination, but there is no evidence of any wave-related structures. Stylolitic layers are often present. Rare, thin mudstones may be intercalated. In some areas, the succession passes upwards into mudstones and fine–medium-grained siltstone.

Petrographically, the limestones of the Tournaisian are mainly pelsparites, with rarer biosparites which include possible ostracods and rare foraminifera. In some areas bioclastic material is more common, and here ooliths and crinoidal material may also be present. Ooliths are generally rare, although they appear to be more common at the base of the succession. The most noticeable feature of these limestones is the lack of any terrigeneous input. The dominance of peloids would suggest a shallow-marine low-energy environment. The presence of shell hash rather than intact bioclastic material suggests periods of increased energy (storms) on the shelf area. In addition, the great lateral continuity of the limestone units (>1 km), together with the lack of any obvious relief would suggest that these carbonate sequences accumulated in a low-energy shelf–platform environment. Carbonate platforms range in size from a few to hundreds of kilometres across, and the larger platforms may have built sediment piles hundreds of metres thick. The platform margin was probably rimmed, since there is some localized evidence of small reefal mounds. This may have restricted circulation with the open ocean, thus sheltering the main platform area.

Viséan

The Viséan succession can be subdivided into two units – early and late Early Viséan in age (see also Poltaev *et al.* 1988) (upper part of Styla section B of Fig. 3). The lowermost Viséan units comprise limestones and are very similar to the underlying Tournaisian strata. Beds are flat-lying and laterally continuous, although bed lenticularity is also noted. Units show evidence of upward-thinning with no sign of any internal structure within the individual beds. There is some rare evidence of parallel lamination. Furthermore, the occurrence of discrete thin, discontinuous layers of bioclastic material within some beds would suggest that bed amalgamation occurred. Beds generally range up to 80.0–150.0 cm in thickness, although in some areas bed thickness is much less (10.0–80.0 cm). Some of the thicker bed bases show evidence of sediment loading. Stylolites are also present.

Bioclastic material (e.g. solitary corals) tends to be much rarer than in the Tournaisian succession. However, in some areas the limestones have a much richer fauna, mainly comprising brachiopods, corals (both colonial and single tetracorals). Some facies successions were also observed. These commence with limestone at the base. These limestones are almost pure micrite, or rarely, sparse biomicrites. The bioclastic material mainly comprises isolated microfossils. These then pass upwards into limestones which are more fossiliferous, contain patchy sparry calcite and have an almost hash-like appearance. Some of these samples contain mixtures of peloids and fossils in a sparry calcite (pel–bio-sparite). In the upper part of the section the limestones are once again micrite rich. The entire unit probably reflects the development of a sequence of varying energy levels.

Petrographically, the limestones are mainly micrites to sparse biomicrites, or occasionally pel–bio-sparites to biosparites. The bioclastic fraction comprises microfossils, with other bioclastic material being in the form of a shell hash. The presence of rare monocrystalline quartz is indicative of proximity to a terrigenous source. Some micritic envelope structures are also present.

The depositional environment for this lowermost part of the Viséan is similar to that of the underlying Tournaisian. The decrease in bioclastic material and the relatively high percentage of micritic material may be indicative of deposition within a quieter-water environment, probably in a sheltered area of the platform (inner platform), although the area was also subjected to rare terrigenous input. The uppermost parts of this succession record evidence of subaerial karstification, suggesting emergence and weathering (Fig. 3).

In some places these Viséan karstified limestones are overlain by a faulted, folded and slumped siliciclastic succession which is late Early Viséan in age (uppermost part of column B in Fig. 3). The sediments here comprise thin shales and mudstones which are interbedded with silts and fine-grained sandstones. Some samples are micritic with rare microfossils. These latter have been infilled with chalcedony, which also replaces some isolated and fragmentary shell fragments. However, the majority of the sediments are siliceous, where the individual beds comprise a dark fine-grained silica-rich groundmass with some microfossils which have been chertified. In parts the sample resembles a very fine-grained polycrystalline quartz (i.e. chert). Cavities formed within the karstified limestone were infilled with deposits of the overlying chert-rich unit (see below).

This chert layer is microcrystalline in character and comprises mainly sponge spicules. There are also some concentric fragments that may be ooids. Other samples comprise pure chert without any internal structure or fossils. In some locations, these chert-rich units overlie the weathered tops of some trachytic domes, although other trachytic domes – also with weathered tops – are overlain elsewhere within the study area by carbonate deposits (Fig. 6). In these latter areas the chert units are missing, while the limestones are mainly biosparite (microspar) with some biomicrite. The limestones contain up to 15% quartz grains. The precise depositional environment for these chert-rich units is unclear. The lack of any carbonate material would suggest possible deposition below the carbonate compensation depth. Such a rapid change in depth (i.e. from subaerial karstification to deep marine), however, is difficult to explain. A possible water depth of *c.* 500 m has been suggested (A. Alekseev, pers. comm.) which corresponds with the calculated subsidence rate (also ± 500 m) for the region in Early Viséan times (Van Wees *et al.* 1996). Another possible explanation, however, is that the cherts were deposited within a lacustrine environment. Lacustrine silica-rich deposits, where the silica is derived from diatoms and sponges, have been reported from a number of areas (e.g. Owen & Crossley 1992). Lakes fed by silica-rich groundwaters are especially favourable sites for the

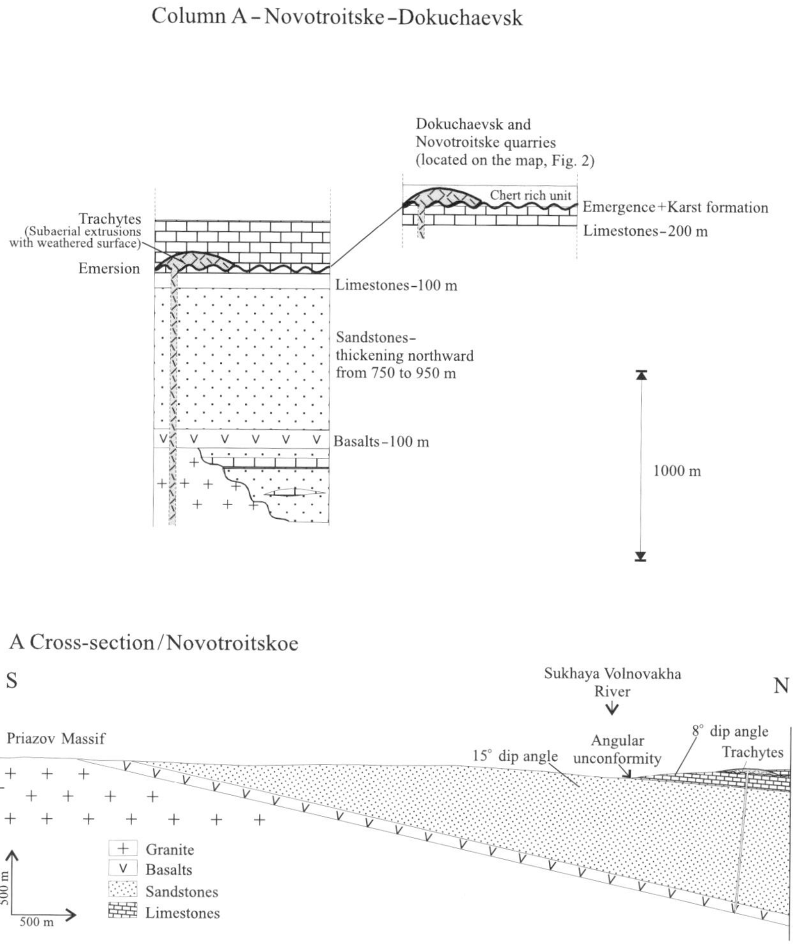

Fig. 6. Stratigraphic column (column A) from the Novotroitske–Dokuchaevsk area, and cross-section from the Novotroitskoe area (cross-section A). See Figure 2 for a precise location.

accumulation of such sediments (Talbot & Allen 1996). Of particular interest in the present situation is the observation that inlakes associated with faults and active hydrothermal systems, alkaline hot springs can be an additional source of siliceous deposits (e.g. Lake Bogoria, Kenya; Renaut & Owen 1988). The lack of regional exposure of these chert-rich units makes it difficult to determine the precise nature of the depositional environment, although a lacustrine setting would be favoured by the predominance of sponge spicules, in association with possible ooids. The extensive volcanic activity, with associated hydrothermal activity, would also have provided suitable conditions for the development of lacustrine silica-rich units.

Volcanism during the initial stages of evolution of the Donbas Basin

Along the southern part of the Donbas fold belt, volcanic rocks are interstratified with Middle Devonian–Lower Viséan clastic and carbonate sedimentary rocks (Fig. 4). The volcanics directly overlie the Proterozoic basement of the Priazov Massif in the eastern part of the studied region. Two eruptive cycles of varying importance are identified: the older one, which is ?late Middle Devonian–Upper Devonian in age, is rather diverse in terms of composition and is well developed at a regional scale; the younger, which is intra-Lower Viséan in age, is less extensive, more homogeneous in composition, and probably much shorter in duration. The volcano-sedimentary series strikes 70° to 100° and dips at 15–20° (up to 30°) degrees to the north. The thickness of the older volcanic sequence is around 1000–1100 m (only 800 m; after Pucharovsky, 1947), while it is up to 30 m for the younger sequence. Four main volcanic lithologies have been mapped. These are described in more detail below.

Basaltic (s.l.) lava-flows, pillow-lavas and pyroclastic fallout

Massive or sometimes vesiculated lava-flows (metres to tens of metres in thickness), interstratified with thin and lenticular flow-breccias, comprise the basaltic pile which has a total thickness of up to 600 m (Figs 3, 4, 7 & 8). Lenticular, thin (1–5 m) and rather poorly-sorted lapilli- and ash-fall tuffs precede, and are interstratified with, this series of lava flows. The scarcity of pyroclastic deposits suggests fissural eruptions occurring over a relatively short period of time. In addition, all of these volcanic products are interstratified and overlain by thin and lenticular tuffaceous microconglomerates and arkoses (some metres in thickness), which were deposited in a fluvial environment (see previous section). Pyroclastic deposits related to fallout in a subaqueous environment were not observed. All of these features suggest continental subaerial volcanic activity. An interesting lateral variation exists in the eastern part of this basin margin (SE of Razdolnoe; Fig. 5) where pillow lavas form a chaotic edifice with a flat shape, *c.*10 m thick at the base of the basaltic sequence, and directly overlying the granitic basement. The absence of any sedimentary deposits between the basement and the basaltic pillows suggests subaqueous emplacement on a rather rapidly subsiding basin floor at the time of eruption. All of these basaltic units are probably ?Lateate Givetian–early Frasnian in age (Fig. 5). Volcanic activity was thus coeval, being subaerial in the west and subaqueous in the east.

Petrographically, these basalts (picrobasalts to basalts, basanites and trachy-basalts (cf. LeBas *et al.* 1986; Chalot-Prat, unpublished data) are microlitic, mostly porphyritic, and frequently microvesiculated. The microlitic texture is often fluidal, suggesting low magma viscosity during emplacement. Vesiculation occurred during and/or after flow. In the pillow basalts, quenched textures on plagioclase microlites are frequent. Phenocrysts (1 to 8 mm in diameter; 1 to 20% in volume) are predominantly Ti-augite (frequent sector-zoning and concentric zoning), and/or pseudomorphs of olivine, and/or plagioclase, and magnetite. The main microlites present include plagioclase, augite and magnetite.

Dacitic (s.l.) dykes and extrusions

These are found either as dykes (metres in thickness) within the Priazov Massif and the basaltic units, or as extrusive domes intercalated between the basaltic flows and the overlying rhyolitic pyroclastic fallout and flow deposits (Figs 5, 7 & 8). Near the centre of the Styla Block, the rhyolitic deposits are absent and there the dacites, and their caps of flow breccias, are directly overlain by intra-Lower Viséan trachytic lava-flows. The dacitic domes are rather flat (500–750 m in diameter and 10–25 m in height) with well-developed columnar jointing perpendicular to the convex upper shape. At the central dome axis, the planar jointing is concordant with the stratification of sedimentary deposits below and above the domes. These edifices result from the localized emplacement of small volumes of highly viscous magma.

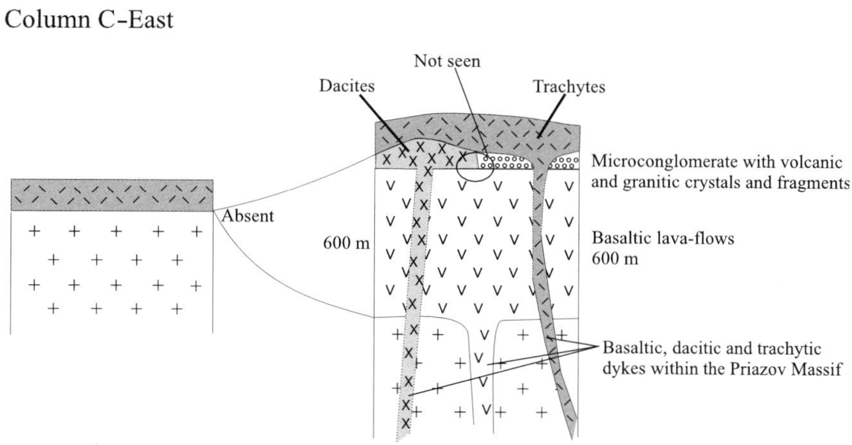

Fig. 7. Stratigraphic column from the eastern part of the study area (column C–east).

In terms of their petrography, these light-grey dacites (cf. LeBas *et al.* 1986; Chalot-Prat unpublished data) are always porphyritic, with a texture which normally varies from microlitic to microgranular towards the core of the extrusion. Phenocrysts (1 to 5 mm in diameter; 10 to 30% in volume) are mainly green–brown amphibole (frequent concentric zoning often around a resorbed blue–green core) associated with Fe-augite (or hedenbergite?), plagioclase, titanite, apatite, magnetite and rare biotites. Microlites are mainly plagioclase (and alkali feldspars?) associated with amphibole and augite. Amphibole and clinopyroxene frequently display resorption features (sinuous contours and resorbed cores), attesting to magma mixing, probably within the same reservoir. Some rounded or shredded metamorphic quartz xenocrysts are also observed.

Based on their geometric relationships with the surrounding rocks, together with their internal structure and texture, these domes are interpreted as extrusive. They are probably Upper Devonian in age, although they had previously been mapped as Permian- or Triassic-age intrusive bodies.

Rhyolitic pyroclastic flow, surge and fallout deposits

These pyroclastic deposits cover a limited area (Fig. 3) in the western part of the study area. Elsewhere they are absent (Fig. 6) or rare (Fig. 5). In Figure 3, it can be seen that these units form an uninterrupted volcanic sequence of about 500 m in thickness above the basaltic flows or the dacitic domes and below the grey Upper Devonian–early Lower Viséan limestones. These green, pink, white, or even black, rocks form lenticular stratified deposits resembling fine-grained shales. They are often massive, but may also have planar jointing. Some regularly stratified layers are well sorted and show direct or inverse graded-bedding; some display internal oblique stratification. No sedimentary deposits are intercalated within these pyroclastic deposits, attesting to the rapidity of deposition of these units. Previous studies (e.g. Makoukina 1961; Rotai 1934; Puscharovsky 1947) were confined to the easternmost part of the study area, and did not attempt to integrate their findings into a broader regional interpretation.

Petrographically, these rhyolites (ultrapotassic rhyolites – cf. LeBas *et al.* 1986: Chalot-Prat unpublished data) are medium to fine grained with an unwelded pyroclastic (lapilli and/or ash size) texture. Varying percentages of phenoclasts are present, including alkali feldspars and/or quartz, well-preserved pumice fragments and shards, and small microlitic lava xenoliths, all of which are in a fine ash matrix. Depending on both the internal structure of the individual layers and the degree of sorting of the pyroclasts, these lithologies may be interpreted as subaerial ignimbritic flow units (*sensu* Freundt & Rosi 2001) including deposits of both pyroclastic flows (massive without any clast sorting at thin section scale), or surges (oblique stratification), and also fallout (well stratified with frequent sorting of clasts according to their composition and size: crystal tuffs, pumice tuffs, etc.). Furthermore, these ash and lapilli fallout layers frequently include well-preserved Upper Famennian plant remnants (Rotai 1944; Puscharovsky 1947) without any sedimentary clasts. The plants were incorporated with the pyroclasts, following explosive eruption. The rare lithic clasts comprise microlitic lava and/or rhyolitic ignimbrites from previous deposits.

It should also be noted that the rhyolitic ignimbrites directly overlie the previously described dacitic domes. The well-preserved texture of pumice fragments and shards of these pyroclastics precludes the possibility of any contact metamorphism by a subsurface intrusion, which is an additional argument in favour of extrusive, rather than intrusive, emplacement of the dacitic domes.

Trachytic sills, dykes and extrusions

These occur as dykes (metres to tens of metres in thickness), or sills, cross-cutting the basaltic flows and the Priazov basement to the south, and as vertical dykes and sills, 50-200 cm in thickness, within the ?upper Famennian–lowermost Viséan limestones. They also form domes or restricted lava flows occurring above the previously described basaltic flows and/or the dacitic domes. These extrusions are directly overlain by the upper Lower Viséan fine-grained silicic or calcareous sediments (Fig. 6). Domes and lava-flows are rather thin (up to 30 m) and typically display columnar jointing. The uppermost *c.* 1 m of the domes is weathered. The overlying sediments display a texture and composition which shows no evidence of any metamorphic alteration (neither baking, nor the growth of contact metamorphic minerals). This would suggest subaerial, rather than intrusive emplacement of these trachytic rocks, as had been previously suggested (Makoukina 1961), followed by their *in situ* alteration (this alteration occurred in a tropical or equatorial climate setting and was synchronous with the karstification of the early Lower Viséan limestones), and subsequent rapid subsidence and marine flooding of the basement. The numerous dykes cross-cutting the basaltic pile, and sometimes the Styla Block (Fig. 8) and the absence of any related pyroclastic deposits, suggest fissural eruptions and a rather short volcanic episode.

In terms of their petrography, these trachytes (cf. LeBas *et al.* 1986; Chalot-Prat unpublished data) are vesiculated, finely microlitic and porphyritic. Phenocrysts (up to 6 mm in diameter; up to 20% in volume) and microlites (0.5 mm in diameter; up to 40% in volume) are mainly plagioclase (with frequent concentric zoning), alkali feldspar, green amphibole, titanite, apatite and magnetite. Phenocrysts are often partially resorbed, which suggests some disequilibrium related to magma mixing within the reservoir.

Structural setting

The region of study on the margins of the southern Donbas is bounded by the WNW–ESE-trending Vassilievka Fault to the south and the WNW–ESE-trending Yuzhni Fault to the north (Fig. 2). The latter fault dips approximately 70° to the south. Structurally, the area adjoins the southeastern flank of the Kalmius–Torets Depression and consists of the oldest exposed sedimentary sequence within the Donbas

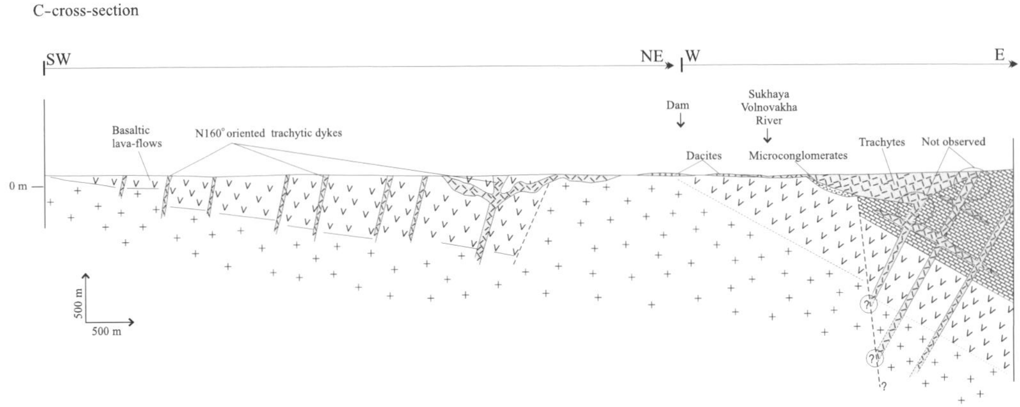

Fig. 8. Geological cross-section from the eastern part of the study area (cross-section C). See Figure 2 for a precise location.

Basin (Fig. 1). The structural setting of the southern Donbas region has been investigated in detail over three field seasons. This work has led to the compilation of a regional map and a series of cross-sections which illustrate the structural setting. The sediments, which are predominantly NNE-dipping (average dip 10–15°), directly overlie a Precambrian basement (granites, granodiorites, migmatites and amphibolites; see Makoukina 1961). A series of cross-sections clearly shows the half-graben structures which developed across the area in response to Devonian and subsequent Viséan rifting (Figs 4, 6, 8 & 9). According to our fieldwork, the typical rift-induced pattern of half-grabens was produced by the northern fault system at a regional scale. Detailed palaeostress analysis (using micro-fault inversion) confirms the normal nature of the faulting. This micro-tectonic study has also recognized a NNE–SSW tensional trend under which the WNW–ESE-trending normal marginal faults have moved (Saintot *et al.* 1999) and no evidence of oblique movements was recorded at the times of rifting phases (i.e. Late Devonian and Early Viséan).

In the western part of the study area there is clear evidence of several normal movements along the approximately E–W-trending faults of the northern part of the region (along which the half-graben developed, Figs 4 & 6). In Figure 6 a clear angular unconformity exists between the sandstones which dip 15°N and the overlying limestones which dip 8°N. This unconformity is even more pronounced on a second cross-section

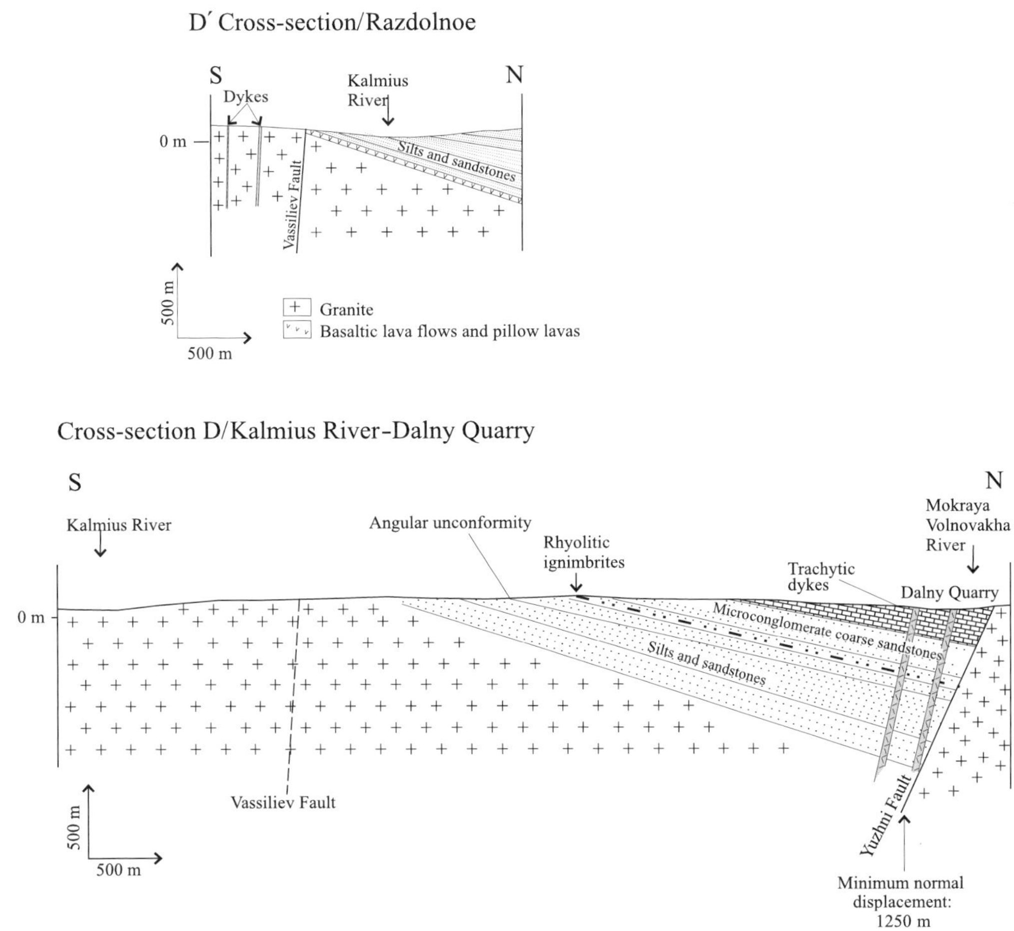

Fig. 9. Geological cross-sections from the Kalmius River–Dalniy Quarry area (cross-section D) and the Razdolnoe area (cross-section D′). See Figure 2 for a precise location.

(Fig. 4) where the sandstones dip 20-25°N and the overlying limestones dip 10-15°N. The development of this unconformity is consistent with published seismic interpretations (Stovba *et al.* 1995, 1996; Stovba & Stephenson 1999). The presence of the unconformity would suggest that tilting to the north occurred both during the Late Famennian (i.e. between the deposition of the youngest late Famennian sandstones and the overlying late Famennian limestones) and after carbonate sedimentation along the northern WNW–ESE fault zone (Fig. 2). In addition, along the northernmost part of cross-section B (Fig. 4), the dip of the limestones (20–25°N) is more than that of the underlying strata (10°N), suggesting the presence of a fault along which another half-graben developed. This interpretation would be consistent with the general structural pattern which developed in the region (see Fig. 4).

In the easternmost part of the region (Fig. 9), a weak angular unconformity exists within the Middle–Late Devonian sediments. There is, however, no angular unconformity at the base of the overlying limestone. In constrast to the western area (see Figs 4 & 6), tilting occurred earlier during Devonian times (possibly in the Givetian to Frasnian) rather than during the Late Famennian.

The cross-sections show that there is abundant evidence of tectonic activity throughout the entire region. This was probably related to normal reactivation of the basin-bounding faults (e.g. the Yuzhni Fault Zone). This phase of tectonic activity would correspond with the Early Viséan phase of rift reactivation which was postulated on the basis of seismic evidence (cf. Stovba and Stephenson 1999). If this is the case, then we can assume that this phase of tectonic tilting produced an efficient slope on which the upper Lower Viséan chert units could have slumped. Moreover, synsedimentary normal faults have been observed within these upper Lower Viséan chert units, suggesting evidence of contemporaneous tectonic activity.

The chronology between the trachytic volcanism (described above from dykes and domes) and the tilting of units, i.e. normal movement along the marginal fault zone, is constrained assuming that the dykes were vertical when they developed (see cross-sections with tilted dykes). Crustal fracturing facilitated magma ascent and eruption in a subaerial environment which was in the process of karstification. Thus, subaerial trachytic volcanism was synchronous with karstification. Tilting related to normal movement along the basin-bounding fault, i.e. the Yuzhni Fault Zone which forms the northern boundary of the half-graben in Figure 9, occurred after uplift (and emergence) and was synchronous with the slumping of the chert-rich unit.

Discussion

Along the southern margin of the Donbas fold belt, the analysis of the interlinked tectonic, sedimentary and magmatic history has revealed the evolution for this area from Middle Devonian through to Early Viséan times (Fig. 10). In some ways this evolution mirrors that of the Dniepr–Donets region to the northwest, but in others it is clearly different. A significant difference is that, within the study area, a synrift event – without any pre-rift event as in the Pripyat–Dniepr–Donets system – occurred at the onset of the Devonian. This initial rifting event was active and not passive inasmuch as it began with a period of uplift and associated volcanism, both being related to mantle-plume activity (Wilson & Lyashkevich, 1996), rather than simple passive extension of the lithosphere. This initial active synrift event is evidenced by numerical models of the region (Van Wees *et al.* 1996), but has not been previously demonstrated by field studies.

Thus, the region was subjected to a series of tectonic events, beginning in Middle Devonian times, which led to the formation of a series of interrelated half-grabens. The development of integrated tectono-sedimentary successions within these half-grabens (*sensu* Leeder & Gawthorpe 1987), however, was complicated by the series of related magmatic events. Within these half-grabens, the volcanic succession comprises *c.* two-thirds of all recorded strata within the basins. This fact must be kept in mind when attempting to interpret the tectonic and sedimentological evolution of the region.

Analysis of the sediments which occur in fault-bounded half-graben basins has been the subject of much research (e.g. Coward *et al.* 1987). In particular, Leeder & Gawthorpe (1987) used the occurrences of discrete sediment packages and predictable sequences to propose tectono-sedimentary models. The model which most closely resembles the present area is a continental basin with axial drainage which evolved over time into a carbonate-dominated shelf succession.

In the study area, initial basin infill occurred following the deposition of an Eifelian–Givetian clastic, mainly continental sequence. Subsequent fault activity led to the initiation of a series of half grabens. The Middle Devonian fluviatile deposits contain mainly material derived from the underlying basement, providing evidence of its exposure and active erosion at this time. The

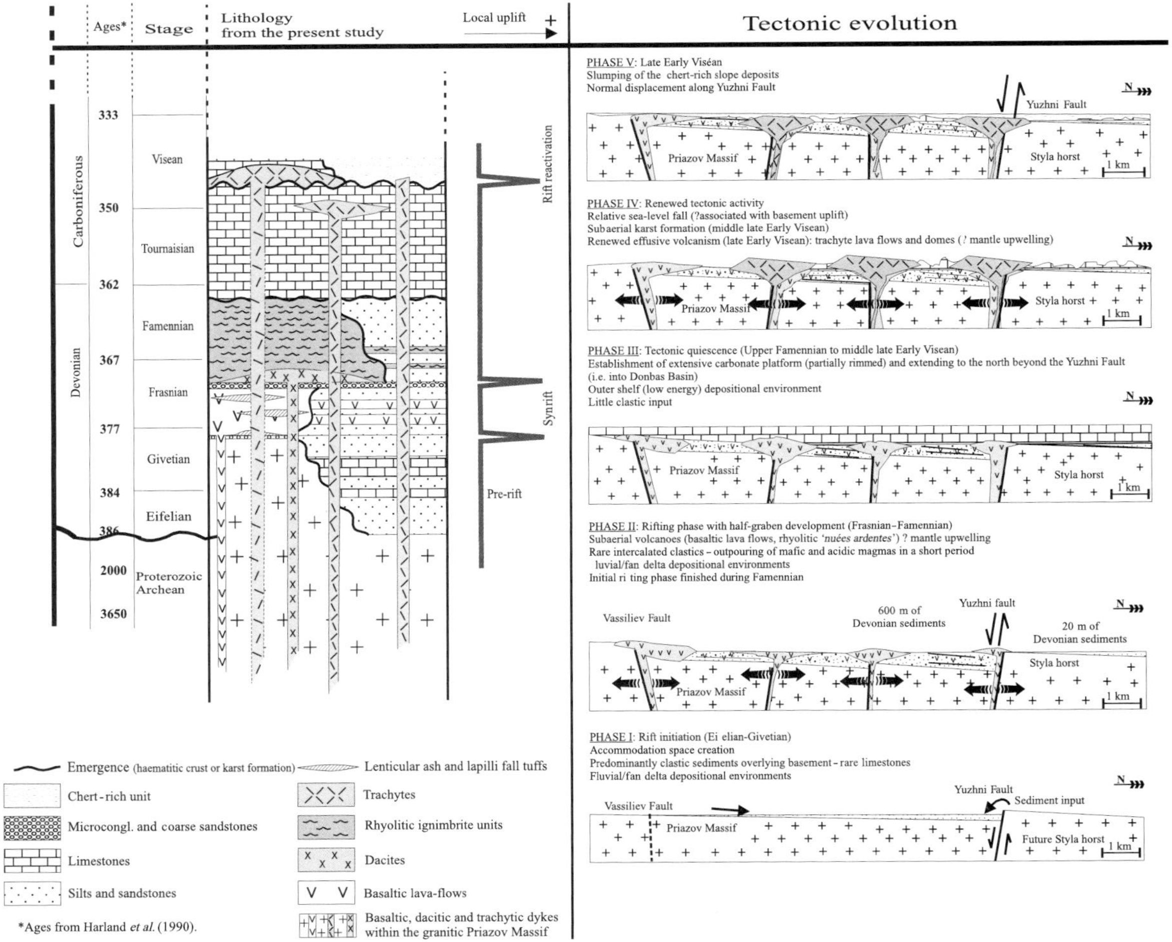

Fig. 10. Schematic stratigraphic section from the southern margin of the Donbas (left-hand side), together with a series of schematic cross-sections illustrating the tectono-magmatic–sedimentary evolution of the region from Eifelian to late Early Viséan times.

northerly Yuzhni Fault was the main graben-bounding fault, with activity along the Vassilievka Fault being more restricted. This latter fault was one of a series of parallel faults which developed along the southern margin of the basin, and within the Priazov Massif. These faults provided a conduit for ascending magmas in the region, and also account for the numerous dykes of doleritic basalts, dacites and trachytes within the Proterozoic basement. (Indeed, removal of a large part of this half-graben system is also a possiblity that must be considered.) During this initial phase of rifting, volcanic activity was not completely absent, as evidenced by rare volcanic lithic fragments, and even a small channel cut into an underlying basaltic unit. This would support the suggestion that crustal-scale fracturing was present right from the initial stages of rifting. Normal faulting was restricted to localized movement along the Yuzhni Fault, leading to the development of a weak intra-Devonian angular unconformity.

The first tectonic pulse was accompanied by subsidence and the establishment of a fluvial depositional system across the area with possible local lacustrine carbonates. Basaltic eruptions were significant, and the outpouring of such large volume of basaltic magma at this early stage of basin formation in the southern Donbas region suggests that the first feeder dykes extended as far as the mantle and were probably related to major crustal fractures. The lack of pyroclastic edifices suggests that the basaltic volcanism was fissural. Furthermore, the absence of intercalated clastic deposits within the entire basaltic unit would suggest a rapid magmatic outpouring, possibly coeval with a significant but short-lived decrease in the subsidence rate during eruptive phases.

Following cessation of the volcanic activity, fluvial depositional environments were re-established across the region. A subsequent phase of volcanic activity involved the emplacement of dacitic dykes and domes sporadically across the region followed by outpourings of rhyolitic nuées ardentes. This second volcanic period was synchronous with the deposition of the fluvial succession further to the east (Fig. 5), with tectonic instability extending until the end of this phase of volcanism. There is a clear angular unconformity between the mainly Devonian synrift succession and the overlying ?Upper Famennian-lowermost Viséan carbonate platform sediments which suggests that the onset of subsidence across the region was not uniform, and involved a degree of tilting of the strata.

Volcanic activity in Devonian times was mainly subaerial, as evidenced both from the sediments and also from features within the volcanics themselves. There is also evidence of some restricted and localised subaqueous eruption (e.g. lacustrine ?pillow lavas). Eruptions were fissural and the duration of each volcanic episode was always quite short (*c.* 1Ma). This would suggest not only the short-lived opening of the magma conduits but also that magma generation was accompanied by brittle crustal deformation whereby the resulting faults facilitated the passage of magmas to the surface.

A subsequent pulse of tectonic activity was recorded for the late Early Viséan, leading to the uplift and exposure of the previously deposited carbonates and localized karst formation. This marked the end of the Early Carboniferous period of basin subsidence. Renewed tectonic activity was accompanied by the sub-aerial extrusion of a trachytic magma. Associated trachytic dykes cross-cut the early Early Viséan carbonates. This phase of trachytic volcanism closely followed, or was perhaps even coeval with, the period of limestone karstification interpreted here in terms of basement uplift, rather than sea-level fall. It should also be noted that the Upper Famennian to Lower Viséan limestones are missing in the south-central part of the Styla Block. The block presumably remained continental and tectonically stable throughout this period. The precise sequence of events which follows is unclear – largely as a result of the restricted exposure of the slumped chert deposits. If these deposits are, indeed, marine in origin (which has been suggested by Russian colleagues), then the phase of renewed subsidence, and possible sea-level rise (Ross & Ross 1988) led to the deposition of this chert-rich siliciclastic unit which overlies the platform carbonates. This period of basin subsidence may also have exhibited a degree of tectonic instability, leading to the formation of major slide deposits produced as a result of destabilization of the sediments and their downslope movement as a result of gravity sliding. However, while other mechanisms for the destabilization of this sediment body must also be considered, the broad regional extent of the deformation would suggest that the underlying cause was one which could cause instability on a broad scale. Given that trachytic volcanism – and uplift – occurred before normal faulting (i.e. tilting of all units), the Viséan phase of rifting is in good agreement with an active rifting model. A second interpretation, however, must also be provided. This is where the chert-rich deposits are interpreted as having been deposited within a lacustrine environment. This would, therefore, mean that the period of emergence, signalled by the karstification of the limestones, continued with the establishment of a lacustrine environment within the region. As already noted, the areal extent of these lakes is impossible to determine due to the restricted outcrop. However, the fine-grained nature of the sediment, together with the general geological setting would support such an interpretation. Silica-rich deposits have been reported from a number of lakes (Talbot & Allen 1996). The source of silica in these environments is both from silica-rich organisms within the lake, but also in association with hydrothermal activity, or as a product of alkaline conditions (e.g. Eugster 1980).

Throughout the synrift period, the observed changes in magma composition and magma volume from one volcanic episode to another (i.e. basalt–dacite–rhyolite–trachyte) indicates that partial melting and magma mixing involved successively different levels of the mantle and crustal lithosphere. In addition, both partial melting and/or rock-source volume decreased with time. For both the rhyolitic and dacitic events, crustal fracturing was short lived and less deep than during the period of basalt eruption (i.e. down to middle/lower crustal levels only). The cause of this shift in the source of partial melting within the lithosphere, together with the rather transient fracturing of the continental crust, could be related to lithospheric thinning at this particular stage of Donbas Basin formation. Such a conclusion would be supported by the work of Van Wees *et al.* (1996), who noted that Late Devonian–Early Carboniferous stretching resulted in a lithosphere thickness of *c.* 50 km. It should also be noted that the eruption of trachytes, which marks the last phase of volcanic activity in the study area, occurred in Early Viséan times, when rift reactivation took place across the entire Dniepr–Donets–Donbas system (Stovba & Stephenson 1999).

The relationship between volcanism and sedimentation is interesting since each eruptive cycle is preceded by a period of clastic deposition, the sediments of which record relief formation and erosion. However, the majority of the lithic fragments within the fluviatile successions were predominantly derived from the underlying crystalline basement rather than from the volcanic successions. This would suggest that the drainage basins supplying these rivers were concentrated within the shield area. Thus, sediment was mainly derived from these crystalline uplands, with limited percentages being derived from the coeval volcanics within the half grabens themselves. An additional point, however, is that the predominance of large rivers would suggest that transport distances were also greater – another factor leading to a decrease in volcanic material.

Conclusions

A series of tectonic pulses has been recorded in the Palaeozoic evolution of the southern Donbas region. Alekseev *et al.* (1996) noted that prior to the onset of crustal extension in the Dniepr–Donets area, the Ukrainian Shield and Voronezh High were covered by marine Middle Devonian sediments (not recorded in the present study area). The onset of crustal extension was contemporaneous with magmatism (Wilson & Lyashkevich 1996) and uplift which was highest to the south of the present southern margin of the Donbas Basin and on the Styla Block. Initial sedimentation involved the establishment of a localized fluvial and lacustrine depositional environment within the first half-grabens which developed between the main basin-bounding faults (Eifelian–Early Givetian). The first phase of synrift activity (during Eifelian times) was probably associated with some initial basaltic extrusion. The Late Givetian–Early Famennian period was one of pronounced tectonic, volcanic and sedimentary activity. Tectonic activity involved both the upper crust – with the formation of a series of half-grabens and the resultant subdivision of the region into several blocks extending in an E–W direction, and the lower crust – with the formation of crustal-scale fractures which enabled the mantle and crustal magmas to ascend to the surface. The resultant volcanic pile covered the central part of the study area, while fluvial deposition was restricted to the margins. Each volcanic episode was preceded by a phase of renewed uplift, which may have been localized (i.e. doming resulting from magmatic upwelling). The end of this major volcanic period marked the onset of a phase of basin subsidence, and the establishment of carbonate deposition across the region. Subsidence variations led to the formation of a slight angular unconformity between the clastic Middle–Upper Devonian succession and the overlying carbonate platform sediments.

A subsequent uplift phase led to the exposure of the entire platform and resultant karstification. This period of karst formation was coeval with the subaerial extrusion of trachytes (also subjected to extreme chemical weathering). It should be emphasized that the numerous trachytic dykes not only occur within the underlying limestones, but also within the basaltic pile and the Priazov Massif – up to 50 km to the south of the present-day margin of the Donbas. The widespread nature of this event attests to a fissural genesis, and also for significant fracturing of the upper crust (trachytes are most likely derived from lower crustal melts: Chalot-Prat, unpublished data) during intra-Early Viséan times. In addition, this spatially extended partial melting of the crust may be linked to basalt intrusion, which supports the idea that emersion was related to thermal uplift of the lithosphere rather than a relative or eustatic sea-level fall.

During the late Early Viséan, two possibilities are suggested for the formation of the siliceous unit. The first involves the development of a lacustrine environment while the second involves a marine transgression which flooded much of the karstified surface and the subsequent deposition of the unit. This phase would correspond with the continuing synrift activity in the region (which began with the trachytic volcanism). Deposits from the overlying carbonates, which are Late Viséan in age, may correspond with the true beginning of the post-rift phase, with related subsidence, of basin development. This siliclastic unit exhibits a variety of wet-sediment deformation features, including slumping, folding and faulting, which can be interpreted as being related to the instability of the basement as a result of renewed activity along the Yuzhni Fault.

This work was partially funded by INTAS-97-0743. We would like to thank all of our INTAS colleagues who participated in various stages of the fieldwork, and who shared their knowledge of the area both in the field and at meetings and conferences. We would also like to thank our two reviewers – C. Breitkreuz and N. Froitzheim – for their many helpful comments on improving the manuscript. This is NSG Publication Number 20020601.

References

AISENVERG, D. E., LAGUTINA, V. V., LEVENSTEIN, M. L. & POPOV, V. S. 1975. *Field Excursion Guidebook for the Donets Basin.* C. R. 8th Congress on International Strat. Geol. Carboniferous, Moscow, 1975, 360 pp.

ALEKSEEV, A. S., KONONOVA, L. I. & NIKISHIN, A. M. 1996. The Devonian and Carboniferous of the Moscow Syneclise (Russian Platform): stratigraphy and sea-level changes. *Tectonophysics,* **268**, 149–168.

CHEKUNOV, A. V., KALUZHNAYA, L. T. & RYABCHUN, L. I. 1993. The Dnieper–Donets Paleorift, Ukraine: deep structures and hydrocarbon accumulations. *Journal of Petroleum Geology,* **16**, 183–196.

COWARD, M. P., DEWEY, J. F. & HANCOCK, P. L. (eds) 1987. *Continental Extensional Tectonics.* Geological Society Special Publication, **28**, 139–152.

EUGSTER, H. P. 1980. Lake Magadi, Kenya, and its precursors. *In:* NISSENBAUM, A. (ed.) *Hypersaline Brines and Evaporitic Environments.* Developments in Sedimentology, **28**. Elsevier, Amsterdam, 195–232.

FREUND, A. & ROSI, M. (eds) 2001. *From Magma to Tephra: Modelling Physical Processes of Explosive Volcanic Eruptions.* Elsevier, Amsterdam, 318 pp.

Geological Map of the USSR and Adjoining Water-covered Areas, Scale 1:2 500 000, Minesterstvo Geologii SSSR, BSEGEI, 1983.

Geological Map of the Ukrainian Donbas Foldbelt, Scale: 1:200 000, Artemovsk Geological Survey, Ukraine, 1995.

HARLAND, W. B, ARMSTRONG, R. L., COX, A. V., CRAIG, L. E., SMITH, A. G. & SMITH, D. G. 1990. *A Geologic Time Scale 1989.* Cambridge University Press, Cambridge, 263 pp.

KUSZNIR, N. J., STOVBA, S. M., STEPHENSON, R. A. & POPLAVSKII, K. N. 1996. The formation of the northwestern Dnieper–Donets Basin: 2D forward and reverse synrift and post-rift modelling. *Tectonophysics*, **268**, 237–255.

LEBAS, M. J., LE MAITRE, R. W., STREICKEISEN, A. & ZANETTIN, B. 1986. A chemical classification of volcanic rocks based on the total alkali–silica diagram. *Journal of Petrology,* **27**, 745–750.

LEEDER, M. R. & GAWTHORPE, R. L. 1987. Sedimentary models for extensional tiltblock/half graben basins. In: COWARD, M. P., DEWEY, J. F. & HANCOCK, P. L. (eds) *Continental Extensional Tectonics.* Geological Society Special Publication **28**, 139–152.

MAKOUKINA, G. O. 1961. Petrography of the dyke effusive complex of the SW Donbas. *In:* TKACHUK, L. G. (ed.) *IGS Proceedings*, Petrography, Mineralogy and Geochemistry, Issue **15**, 172 pp. (in Ukranian).

NIKISHIN, A. M., ZIEGLER, P. A., STEPHENSON, R. A., CLOETINGH, S. A. P. L., FURNE, A. V., FOKIN, P. A., ERSHOV, A. V., BOLOTOV, S. N., KOROTAEV, M. V., ALEKSEEV, A. S., GORBACHEV, V. I., SHIPILOV, E. V., LANKREIJER, A., BEMBINOVA, E. YU. & SHALIMOV, I. V. 1996. Late Precambrian to Triassic history of the East European Craton: dynamics of sedimentary basin evolution. *In.* STEPHENSON, R. A., WILSON, M., DE BOORDER, H. & STAROSTENKO, V. I. (eds) *EUROPROBE: Intraplate tectonics and basin dynamics of the eastern European Platform. Tectonophysics*, **268**, 23–63.

OWEN, R. B. & CROSSLEY, R. 1992. Spatial and temporal distribution of diatoms in sediments of Lake Malawi, central Africa, and ecological implications. *Journal of Palaeolimnology*, **7**, 55–71.

POLETAEV, V. I. *ET AL.* 1988. The guidebook to excursion to the Lower/Middle Carboniferous boundary deposits of Donbas. *In:* AISENVERG, D. YE. (ed.) *IGS Publication*, 92 p. (in Russian and English).

PUSCHAROVSKY, Y. M. 1947. Devonian sediments of southern outskirts of the Donets Basin. *In:* SHATSKY, N. S. (ed.) *Materials on Geology of the Donets Sediments of the Southern Outskirts of the Donbas.* MOIP Press, New Series, **9(13)**, 5–21 (in Russian).

RENAUT, R. W. & OWEN, R. B. 1988. Opaline cherts

associated with sublacustrine hydrothermal springs at Lake Bogoria, Kenya Rift. *Geology*, **16**, 699–702.
ROSS, C. A. & ROSS, J. R. P. 1988. Late Paleozoic transgressive–regressive deposition. In: WILGUS, C. K., HASTINGS, B. S., KENDALL, C. G. ST. C., POSAMENTIER, H. W., ROSS, C. A. & VAN WAGONER, J. C. (eds), *Sea-level Changes: an Integrated Approach.* Society of Economic Paleontologists and Mineralogists, Special Publication **42**, 227–247.
ROTAI, A. P. 1944. *Geological Works in the Donets Basin during the Summer of 1928–1929*. Proceedings VGRO, 356 (in Russian).
SAINTOT, A., PRIVALOV, V., ZHIKALIAK, M., BREM, A. & EUROPROBE/INTAS 1999. Some kinematic indicators for the tectonic evolution of the Donbass fold-and-thrust belt (Ukrainian part). Joint meeting of EUROPROBE TESZ/PANCARDI/GEORIFT, Tulcea (Romania). *Romanian Journal of Tectonics and Regional Geology*, **77**, p. 83.
STEPHENSON, R. A. & EUROPROBE INTRAPLATE TECTONICS AND BASIN DYNAMICS DNIEPER–DONETS AND POLISH TROUGH WORKING GROUPS 1993. Continental rift development in Precambrian and Phanerozoic Europe: EUROPROBE and the Dnieper–Donets Rift and Polish Trough basins. *Sedimentary Geology*, **86**, 159–175.
STOVBA, S. M. & STEPHENSON, R. A. 1999. The Donbas fold belt: its relationships with the uninverted Donets segment of the Dniepr–Donets Basin, Ukraine. *Tectonophysics*, **313**, 59–83.
STOVBA, S. M., STEPHENSON, R. A. & DVORIANIN, E. 1995. Dniepr–Donets Basin, Ukraine: main observations from regional seismic profiles. *Comptes Rendus des de l'Académie des Sciences, Paris*, **321**, Série IIa, 1103–1110.
STOVBA, S., STEPHENSON, R. A. & KIVSHIK, M. 1996. Structural features of the Dniepr–Donets Basin, Ukraine from regional seismic reflection profiles. *Tectonophysics*, **268**, 127–147.
TALBOT, M. R. & ALLEN, P. A. 1996. Lakes. *In:* READING, H. G. (ed.) *Sedimentary Environments: Processses, Facies and Stratigraphy.* Blackwell Science, 83–124.
VAN WEES, J.-D., STEPHENSON, R. A., STOVBA, S. M. & SHIMANOVSKY, V. 1996. Tectonic variation in the Dniepr–Donets Basin from automated modelling of backstripped subsidence curves. *Tectonophysics*, **268**, 257–280.
WILSON, M. & LYASHKEVICH, Z. M. 1996. Magmatism and the geodynamics of rifting of the Pripyat–Dnieper–Donets rift, East European Platform. *Tectonophysics*, **268**, 65–81.
ZIEGLER, P. A. 1990. Collision-related intra-plate deformations in Western and Central Europe. *Journal of Geodynamics*, **11**, 357–388.

Geodynamic evolution and palaeogeography of the Polish Carpathians and adjacent areas during Neo-Cimmerian and preceding events (latest Triassic–earliest cretaceous)

J. GOLONKA[1], M. KROBICKI[2], N. OSZCZYPKO[1], A. ŚLĄCZKA[1] & T. SŁOMKA[2]

[1]*Jagiellonian University, Oleandry Sr. 2a, 30-063 Kraków, Poland (e-mail: golonka@geos.ing.uj.edu.pl)*
[2]*University of Mining and Metallurgy, Mickiewicza 30, 30-059 Kraków, Poland*

Abstract: The aim of this paper is to place the geodynamic and palaeogeographical evolution and position of the major crustal elements of the Polish Carpathians within a global framework. Neo-Cimmerian movements and their synsedimentary consequences are the main objects of our elaboration in relation to sedimentary record. Five time-interval maps are presented, which depict the plate-tectonic configuration, palaeogeography and lithofacies for the circum-Carpathian region and adjacent areas from the Late Triassic through to the Early Cretaceous.

Almost simultaneous tectonic events proceeding within different types of Carpathian sedimentary basins (Pieniny Klippen Belt and Outer Carpathian Silesian Basins) indicate the very important role of the Neo-Cimmerian movements (mainly of the Osterwald Phase) in the geodynamic history of the northernmost margin of the Tethyan Ocean. The global plate reorganization is related to this Tethyan Neo-Cimmerian tectonic activity.

We here present five time interval maps which depict the plate-tectonic configuration, palaeogeography and lithofacies for the circum-Carpathian region (Fig. 1) and adjacent areas from the Late Triassic through Late Jurassic to Early Cretaceous.

The aim of this paper is to place the geodynamic and palaeogeographic evolution and position of the major crustal elements of the Polish Carpathians within a global framework. Neo-Cimmerian (Jurassic/Cretaceous) (Fig. 2) orogenic movements and their synsedimentary consequences are the main subjects considered in relation to the sedimentary record. Therefore, in our model we have restricted the number of plates and terranes, trying to utilize the existing information with some degree of certainty. Using computer technology we have applied kinematic principles to reconstructing the relationships between the tectonic components of the circum-Carpathian area.

The Neo-Cimmerian movements (Fig. 2) are well recorded in the Polish Carpathians, while the older Cimmerian events (Eo- and Meso-Cimmerian) are better known from the regions outside Poland. Two regions of the Polish Carpathians: the Pieniny Klippen Belt and Silesian Unit (Outer Flysch Carpathians), are discussed in this paper, based on the older sedimentological results and the authors' own palaeoecological and sedimentological studies. The record of these two regions is well preserved, and therefore could constitute a good example of the relationships between plate-tectonic movements and sedimentation in these basins.

Outline of the geological setting

The Polish Carpathians form a part of a major arc of mountains, which stretch more than 1300 km from the Vienna Basin to the Iron Gate on the Danube (Fig. 1). To the west the Carpathians are linked to the Eastern Alps, and eastwards to the Balkan chain. Traditionally, the Western Carpathians are subdivided into an older range known as the Inner Carpathians and a younger one, known as the Outer Carpathians. At the boundary of these two ranges the Pieniny Klippen Belt (PKB) is situated – in the suture zone between the European and Alcapa plates.

The Outer (Flysch) Carpathians are composed of flysch sequences ranging in age from Jurassic to Early Miocene (Ślączka 1996). These deposits were folded and overthrust during Miocene times (Alpine Orogeny), forming north-verging nappes detached from their original

From: McCann, T. & Saintot, A. (eds) *Tracing Tectonic Deformation Using the Sedimentary Record.* Geological Society, London, Special Publications, **208**, 138–158. 0305–8719/03/$15.00

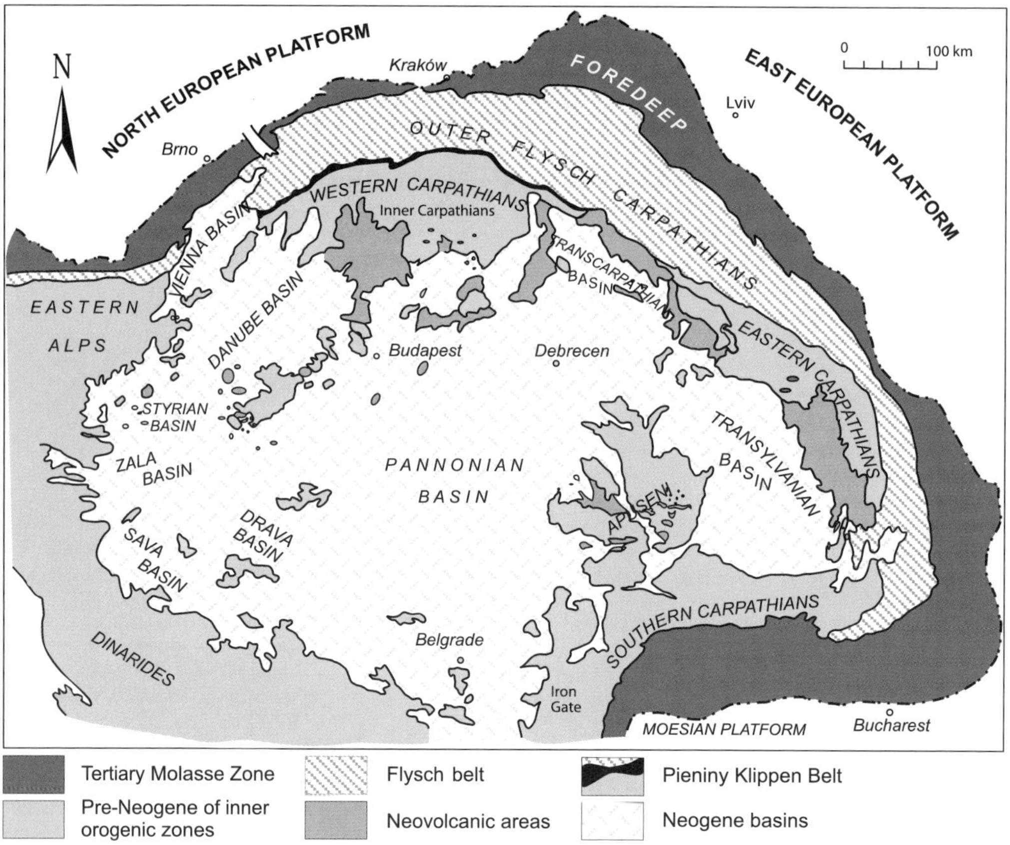

Fig. 1. Tectonic sketch map of the Alpine–Carpathian–Pannonian–Dinaride basin system (after Kováč *et al.*, 1998; simplified).

basement (Ślączka 1996). From south to north, the nappes are: the Magura, Dukla–Fore-Magura, Silesian, Sub-Silesian and Skole Units. Along the outer margin of the Carpathians a narrow belt of folded Miocene deposits was developed. The Pieniny Klippen Belt is composed of several successions of mainly deep- and shallow-water limestones, covering a time-span from the Early Jurassic to the Palaeogene. This strongly tectonized structure is a terrane about 600 km long and 1-20 km wide, which stretches from Vienna in the west to Romania to the east (Fig. 1). The PKB is separated from the Magura Unit by the Miocene subvertical strike-slip fault (Birkenmajer 1986).

Mapping methodology

The presented maps were primarily generated as Intergraph design files and PostScript-Corel Draws files using computer software and databases. The plate-tectonic model used is based on PLATES and PALEOMAP software (see Golonka *et al.* 1994, 2000; Scotese & Lanford 1995; Golonka & Gahagan 1997). A tectonic reconstruction program, used to create palaeocontinental base maps, was utilized to take tectonic features in the form of digitized data files, and assembles them according to user-specified rotation criteria.

The rigid, outer part of the Earth is broken up into many pieces that are referred to as lithospheric plates. These plates, comprising both the continental landmasses and oceanic basins, are in motion relative to each other and to the Earth itself. Assuming the earth is a sphere, the motion of a plate across the Earth's surface can be described as motion about the axis of a pole of rotation that passes through the centre of the Earth. The intersection of the pole's axis with the

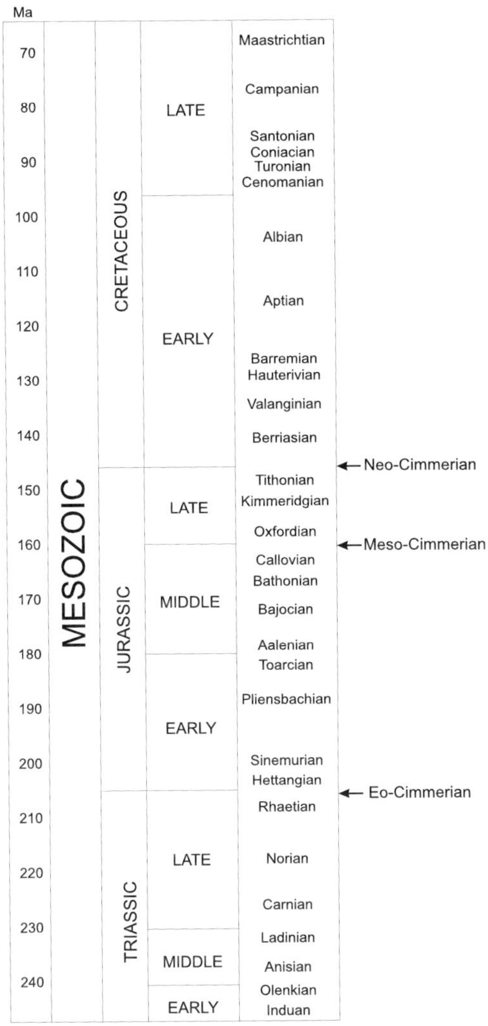

Fig. 2. Mesozoic stratigraphic chart (after Golonka, 2000).

Earth's surface is referred to by its latitude/longitude coordinates. The distance that the plate travels about the pole is an angular distance, and is recorded in degrees. A stage pole of rotation describes the distance that a plate moved from one time to the next (i.e. from 20 Ma to 10 Ma). A finite pole of rotation describes the total distance that a plate moved from some time in the past, to the present day (i.e. from 20 Ma to 0 Ma). A rotation file contains a list of poles of rotation for various plates. The rotation files used by the PLATES software contain finite poles of rotation. Thus, for each plate, there are several finite poles of rotation for different times in the past. Plate models that use rotation files describing the motion of plates relative to other plates are called relative framework models. Among the data that show the relative motions between plates are fracture zones. Fracture zones are essentially flow-lines between plates. For example, in the South Atlantic, fracture zones show the motion of South America relative to (or away from) Africa. The finite pole of rotation describing this motion is a relative pole. South America is referred to as the moving plate and Africa as the fixed plate.

The rotation file contains a list of finite rotations between pairs of tectonic elements, at different episodes of time, with brief bibliographic notes or general comments for each individual rotation. The following is a sample segment taken from a rotation file, describing the motion of Central Carpathian, Eastern Alpine and Tisa plates in relation to the Eastern European plate (Baltica), with column labels added for clarity.

A	B	C	D	E	F	G
335	0.0	0.0	0.0	0.0	302	!Eastern (Calcareous) Alps–Baltica Jan
335	14.0	0.0	0.0	0.0	302	!Eastern (Calcareous) Alps–Baltica Jan
335	36.0	33.4	51.7	4.90	302	
335	65.0	33.4	51.7	4.90	302	
335	145.0	33.36	51.73	12.59	302	
335	180.0	0.0	0.0	0.0	302	!
335	300.0	0.0	0.0	0.0	302	!
335	400.0	0.0	0.0	0.0	302	!
336	0.0	0.0	0.0	0.0	302	!Central Carpathian–Baltica Jan
336	14.0	0.0	0.0	0.0	302	! Central Carpathian–Baltica Jan
336	36.0	33.4	51.7	4.90	302	
336	65.0	33.4	51.7	4.90	302	
336	145.0	33.36	51.73	12.59	302	
336	180.0	0.0	0.0	0.0	302	! Central Carpathian–Baltica Jan
336	300.0	0.0	0.0	0.0	302	! Central Carpathian–Baltica Jan
336	400.0	0.0	0.0	0.0	302	! Central Carpathian–Baltica Jan
999	180.0	3.68	40.03	5.61	302	!
337	0.0	0.0	0.0	0.0	302	!Tisa–Eurasia Jan
337	14.0	0.0	0.0	0.0	302	!Tisa–Eurasia Jan
337	36.0	3.4	51.7	4.90	302	
337	65.0	33.4	51.7	4.90	302	

337	145.0	33.36	51.73	12.59	302	
337	190.0	33.36	51.73	12.59	302	
337	210.0	33.88	38.32	14.87	302	
337	260.0	0.0	0.0	0.0	302	!Tisa–Eurasia Jan
337	300.0	0.0	0.0	0.0	302	!Tisa–Eurasia Jan
337	400.0	0.0	0.0	0.0	302	!Tisa–Eurasia Jan

A – Tectonic plate to be rotated
B – Time in millions of years of rotational stage
C – Latitude of finite pole (Latitudes are positive to the north)
D – Longitude of finite pole (Longitudes are positive to the east)
E – Angle of opening
F – Reference plate relative to which rotation is made
G – General comments and bibliographic information

When the PALEOMAP or PLATES program reads the rotation file, it first searches for all the rotations corresponding to the required time of reconstruction. If there is no finite rotation between a pair of tectonic elements, for the required time, then the program will interpolate the missing rotation from the two bracketing rotations (i.e. corresponding with the times immediately younger and immediately older than the required time). Since each finite rotation is defined relative to different reference plates (tectonic elements), the program restores each individual finite rotation between pairs of plates to a unique reference frame. This reference frame is the Earth's spin axis. This requires the building of a consistent rotation file, i.e. for each pair of tectonic elements, the reference tectonic element can be linked directly or indirectly to the spin axis, as in the following example:

	000 (spin axis)	
	†	
	†	
	701 (Africa)	
101 (North America)	201 (South America)	301 (Europe)
*	*	*
102 (Greenland)	204 (Mexico)	302 (Baltica)
*	*	*
809 (Spitzbergen)	205 (Yucatan)	336 (Central Carpathian)

*finite rotations describing relative motions.
†finite rotations describing absolute motions (i.e. relative to a hot-spot frame or to the geomagnetic axis).

In the above example each individual rotation can be calculated relative to Africa, and then added to the absolute motions of Africa relative to the spin axis. Hot-spot volcanics serve as reference points for the calculation of palaeo-longitudes (Golonka & Bocharova 2000). Palaeomagnetic data have been used to define palaeolatitudinal position of continents and the rotation of plates (see, for example, Besse & Courtillot 1991; Van der Voo 1993; Krs *et al.* 1996). An attempt has been also made to utilize the palaeomagnetic data from minor plates and allochthonous terranes (see for example, Channell *et al.* 1992, 1996; Patrascu *et al.* 1992, 1993; Pechersky & Safronov 1993; Beck & Schermer 1994; Mauritsch *et al.* 1995, 1996; Feinberg *et al.* 1996; Kondopolou *et al.* 1996; Krs *et al.* 1996; Márton & Martin 1996; Haubold *et al.* 1999; Márton *et al.* 1999, 2000; Grabowski 2000; Muttoni *et al.* 2000*a*, *b*). However, the nature of rotation indicated by palaeomagnetism measured in sedimentary rocks in allochthonous terranes remains somewhat uncertain. It could be caused by the rotation of crustal (basement) elements, the rotation of blocks separated by dextral faults (e.g. Márton *et al.* 2000), or the rotation of thrust sheets (e.g. Muttoni *et al.* 2000*b*). Measurements in flysch deposits could also indicate the arrangement of magnetized grains (domains) by turbiditic currents. For example, the magnetic declination of the Podhale Flysch in Poland perhaps records the sedimentological arrangement of grains (see the maps of sedimentological transport in the Carpathian Flysch, e.g. Książkiewicz 1962) rotated by the Inner Carpathian Plate to their present position. The crustal rotation in a range of 20–30° agrees with the direction of the Penninic domain or the relation between the Inner Carpathians and the Pieniny Klippen Belt (see discussion in Golonka & Krobicki 2001; also Aubrecht & Túnyi 2001; Stampfli 2001).

All data files are in the PLATES/PALEOMAP format. The data header consists of 12 fields on two consecutive lines. The first line contains archive information describing the reference source of the data, the position of the string in the file, and a literal description of the data itself. The second line provides key information about the data: the plate identification, when the data appears and disappears in the temporal framework of the plate model, and the data type. A more complete description of the fields is given below:

Line 1:

a. b. c.

1003 183 Central Carpathian Polygon

a. Reference source of the data
b. String number within file
c. Literal description of data

Line 2:

a. b. c. d.

336 545 –999.0 PB

a. Plate number. This number identifies the plate, and hence the rotation parameters that should be applied to the data. 336 is the number of Central Carpathian plate. The basic input was the PALEOMAP model (see Scotese & Lanford, 1995), which describes the relative motions between approximately 300 global plates and terranes. Sixty plates were added in the Mediterranean–Tethys area (Golonka & Gahagan, 1997).
b. Time of data appearance. This is the time in millions of years that the data string will appear on the reconstruction. A default time of 999.9 selects the oldest data available in the database. The 545 in the example above indicates the beginning of the Phanerozoic.
c. Time of data disappearance. This is the time in millions of years that the data string will disappear from the maps. For example, Palaeozoic sutures are removed from Mesozoic and Cenozoic maps. A default time of –999.9 indicates that these data are to remain on the map indefinitely.
d. Alphabetic data (data type). This two-letter alphabetic code enables you to easily distinguish coastlines, ridge, axes, sutures, grid marks, etc. For example PB – means plate boundary

The following is an example of data file:

1003	184	Eastern (Calcareous)	Alps Polygon
335	545.0	–999.0	PB
47.5855	10.0722	3	
47.6840	10.5796	2	
47.7576	11.0412	2	
47.8510	11.5646	2	
47.8985	11.8155	2	
47.8884	12.1815	2	
47.8939	12.6976	2	
47.8974	13.0576	2	
47.8981	13.4480	2	
47.9104	13.8223	2	
47.9330	14.2731	2	
48.0561	15.2024	2	
48.1227	15.7219	2	
48.3240	16.8351	2	
48.3240	16.8351	2	
46.8695	15.7575	2	
46.8695	15.7575	2	
46.4950	14.2522	2	
46.4950	14.2522	2	
46.3191	10.7859	2	
46.1805	8.9617	2	
46.1805	8.9617	2	
47.5855	10.0722	2	
99.0000	99.0000	3	
1004	184	Kirsehir	Polygon
525	545.0	–999.0	PB
40.9105	35.4701	3	
40.8873	35.4851	2	
40.6953	35.2911	2	
40.5271	35.0997	2	
40.3931	34.9309	2	
40.2462	34.7262	2	
40.0754	34.4885	2	
39.7933	30.3930	2	
39.7933	30.3930	2	
38.6435	31.0275	2	
38.5419	31.1157	2	
38.4908	31.1380	2	
38.3287	31.2458	2	
38.0687	31.4561	2	
37.4946	31.9699	2	
36.9450	32.6148	2	
36.5184	33.1786	2	
36.5184	33.1786	2	
36.6632	33.8052	2	
37.1134	34.4950	2	
37.7486	35.2466	2	
38.0945	35.8055	2	
38.0945	35.8055	2	
38.4129	36.4499	2	
38.6393	36.9981	2	
38.6393	36.9981	2	
39.0978	38.0251	2	
39.0978	38.0251	2	
39.4258	39.1362	2	
39.4258	39.1362	2	
39.8461	39.7436	2	
39.8461	39.7436	2	
40.4273	38.0026	2	
40.4273	38.0026	2	
40.7596	36.4697	2	
40.9105	35.4701	2	
99.0000	99.0000	3	

The generalized facies and palaeoenvironment database information was added to base maps. For example the reef data (see Kiessling *et al.* 1999) were rotated together with plates, polygon maps. The calculated palaeolatitudes and palaeo-longitudes were used to generate computer maps in the Microstation design format using the equal-area Molweide projection. Global and regional palaeogeographical papers (e.g. Książkiewicz 1962; Ślączka, 1976; Ronov *et al.* 1984, 1989; Ziegler 1988; Zonenshain *et al.* 1990; Stampfli *et al.* 1991; Dercourt *et al.* 1993, 2000;

Kováč *et al.* 1993, 1998; Şengör & Natalin 1996; Robertson 1998; Plašienka 1999; Golonka *et al.* 1999, 2000; Ślączka *et al.* 1999) as well as the unpublished maps and databases from the PALEOMAP group (University of Texas at Arlington), PLATES group (University of Texas at Austin), University of Chicago, Institute of Tectonics of Lithospheric Plates in Moscow, Robertson Research in Llandudno, Wales, UK, and the Cambridge Arctic Shelf Programme were used to obtain palaeolithological and palaeoenvironmental data.

Description of the maps and a geodynamic interpretation

This section will describe the general setting of the Carpathians plate-tectonic relationships between geodynamic events and sedimentary conditions, according to their palaeogeographical context.

Triassic (see Figs 3 & 4)

Late Palaeozoic collisional events provided a background for the Mesozoic–Cenozoic geodynamic evolution of the circum-Carpathian area. The older, Cadomian and Caledonian basement elements experienced Hercynian tectonothermal overprint (Rakús *et al.* 1988; Dallmeyer *et al.* 1996). During Carboniferous–Permian times the major crustal elements of the region were formed. These crustal elements belonged to the supercontinent Pangaea, which reached maturity in the Early Permian and moved steadily northwards as consequence of post-Variscan movements (Golonka & Ford 2000). The global Pangaean model agrees very well with the position of

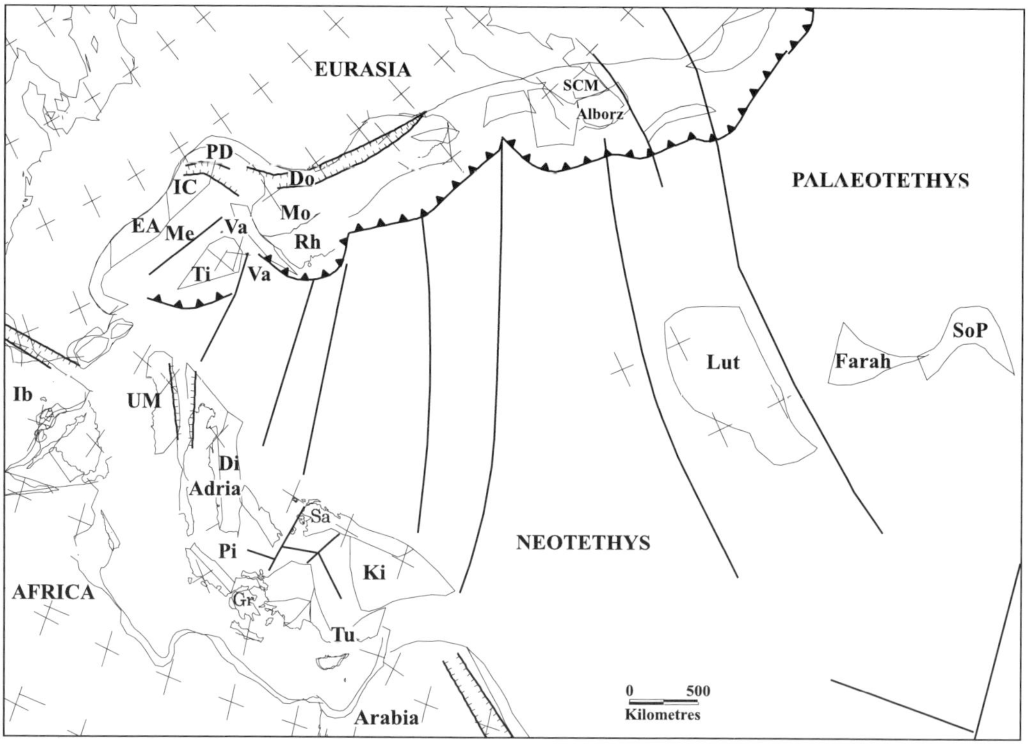

Fig. 3. General palaeogeography of the Western and Central Tethys during the Late Triassic; position of the plates at 225 Ma. Abbreviations of ocean and plate names are as follows: Di, Dinarides; Do, Dobrogea; EA, Eastern Alps; Gr, Greece (Ionian etc.); Ib, Iberia; IC, Inner Carpathians; Ki, Kirsehir plate; Me, Meliata–Halstatt Ocean; Mo, Moesia Plate; PD, Polish–Danish aulacogen; Pi, Pindos Ocean; Rh, Rhodopes; Sa, Sakariya Plate; SCM, South Caspian Microcontinent; SoP, South Pamir Plate; Ti, Tisa Plate; Tu, Taurus Plate; UM, Umbria–Marche; Va, Vardar Ocean.

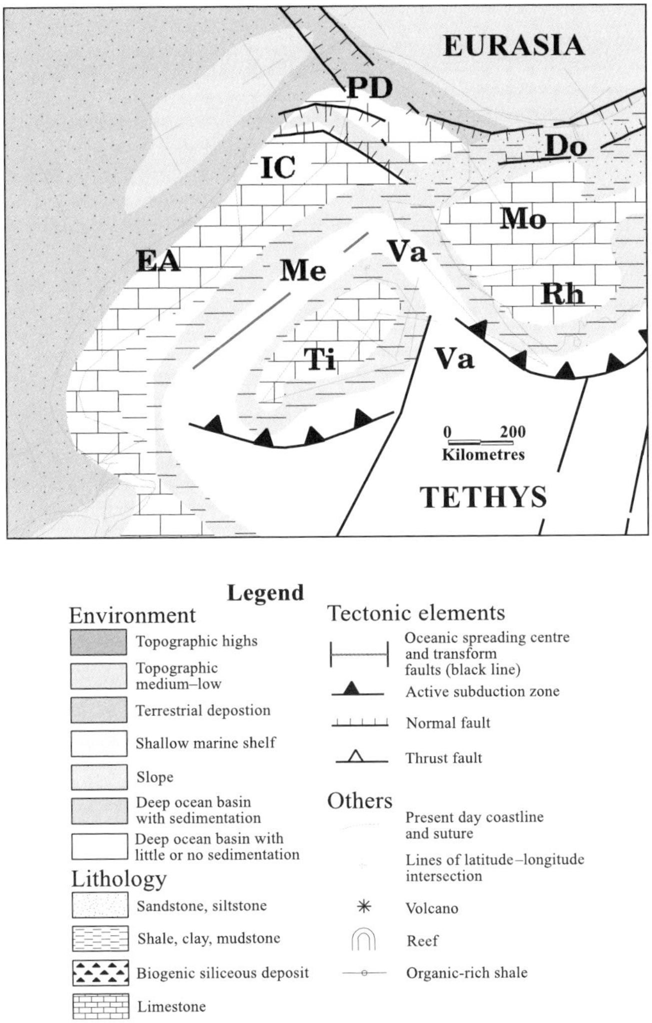

Fig. 4. Palaeogeography of the circum-Carpathian area during the Late Triassic; plate position at 225 Ma. Abbreviations of ocean and plate names are as follows: Do, Dobrogea; EA, Eastern Alps; IC, Inner Carpathians; Me, Meliata–Halstatt Ocean; Mo, Moesia Plate; PD, Polish–Danish aulacogen; Rh, Rhodopes; Ti, Tisa Plate; Va, Vardar Ocean.

Eurasia (Bohemian Massif) (adjacent to the Carpathian plates) as determined from palaeomagnetic studies (Krs *et al.* 1996) i.e. located near the Equator (Van der Voo 1993; Golonka *et al.*, 1994). Moesia, the Rhodopes, the Eastern Alps, the Inner Carpathians, Tisa (Figs 3 & 4 – Mo, Rh, EA, IC, Ti – respectively) and adjacent terranes, were sutured to the Laurasian arm of Pangaea, while Adria (Apulia) and adjacent terranes were situated near the Gondwanan (African) arm (Wortmann *et al.* 2001). The Palaeotethys Ocean divided these two arms of Pangaea (Golonka *et al.* 2000). The subduction zone along the Palaeotethys was formed during the Variscan Orogeny and was an active force causing many Mesozoic geodynamic events. At the same time the passive margin has been formed along the Gondwanan arm of Pangaea. The Neotethys Ocean (Fig. 3) originated during the Permian as aneffect of the Carboniferous–earliest Permian rifting of the Cimmerian Plates (see Dercourt *et al.* 1993; Golonka *et al.* 1994; Şengör & Natalin 1996). This ocean had Arabia, Greater India and

Australia on one side, and the so-called Cimmerian continent – Lut–Farah–South Pamir–Qiantang–Southeast Asia (Fig. 3 – SoP) on the other. The spreading was driven by trench-pulling forces related to the north-dipping subduction, as well as ridge-pushing forces related to mantle upwelling, expressed by hot-spot activity (Golonka & Bocharova 2000). The continued northwards drift of the Cimmerian continent and the opening of the Neotethys Ocean corresponded with the closing and progressive consumption of Palaeotethyan oceanic crust. Rifting and the oceanic type of basin opening could have also occurred in the Mediterranean, recorded by the deep-water sediments of Sicily (Catalano *et al.* 1991; Kozur 1991), Lago Negro (Marsella *et al.* 1993) and Crete (Kozur & Krahl 1987).

The oceanic system was established in Southern and Central Europe during Permian–Triassic times. A narrow branch of the Neotethys separated the Apulia (Adria)–Taurus Platform from the African continent. The Apulia Platform was connected with the European marginal platforms. Its northernmost part was possibly separated from the Umbria–Marche region (Fig. 3 – UM) by a rift. The incipient Pindos Ocean (Fig. 3 – Pi) separated the Pelagonian, Sakariya and Kirsehir blocks (Fig. 3 – Sa, Ki) from the Ionian–Taurus Platform (Fig. 3 – Gr, Tu) (Robertson *et al.* 1991, 1996; Stampfli *et al.* 1991). The Vardar–Transylvanian Ocean (Fig. 3 – Va) separated the Tisa (Bihor–Apuseni) block (Fig. 3 – Ti) from the Moesian–Eastern European Platform (Săndulescu *et al.* 1981; Săndulescu 1988; Săndulescu & Visarion 2000). There is a possibility that an embayment of the Vardar–Transylvanian oceanic zone existed between the Inner Carpathians (Fig. 3 – IC), and the European Platform (Golonka *et al.* 2000). Exotic material from the Triassic pelagic spotty limestones which occur as pebbles within Cretaceous–Palaeogene gravelstones in the Pieniny Klippen Belt (from the enigmatic so-called Exotic Andrusov Ridge – Birkenmajer 1988; Birkenmajer *et al.* 1990) and Magura Unit (Soták 1986) could have originated in this embayment.

The position of the embayment and its relation to the other parts of the Tethys, the Vardar Ocean, the Meliata–Halstatt Ocean, the Dobrogea Rift and the Polish–Danish Aulacogen (Figs 3 & 4 – Va, Me, Do, PD – respectively) remain fairly speculative. According to Rakús *et al.* (1988), two oceanic units were located south of the Inner Carpathian Plate. One was open during Triassic times, but closed during the Late Triassic as a result of the Early Cimmerian collision. Another, represented by sequences at the classic profile of Meliata in Southern Slovakia, opened during the Early–Middle Jurassic as a back-arc basin, and then closed during Late Jurassic times. The position of the Meliata Ocean, the time of closure and the role of the Tisa Unit in Mesozoic collisional events is still the subject of lively discussion (see Kozur 1991; Stampfli 1996, 2001; Channell & Kozur 1997; Plašienka 1999; Golonka *et al.* 2000; Wortmann *et al.* 2001). In our opinion, the Meliata–Halstatt Ocean (Kozur 1991; Kiessling *et al.* 1999; Golonka *et al.* 2000) separated the Tisa (Bihor–Apuseni) Block and the Eurasian margin. The Northern Calcareous Alps (Eastern Alps – Fig. 3 – EA) and Inner Carpathians formed the marginal platform of Europe (Plašienka & Kováč, 1999).

The Late Palaeozoic and Triassic rifting and sea-floor spreading resulted in several separated carbonate platforms. Shallow-water limestones, marls and dolomites with numerous reefs (Kiessling *et al.*, 1999) prevailed in the platform areas underlain by Late Permian and Early Triassic clastic deposits. Following the Permian–Triassic mass extinction, faunal communities recovered during the Middle Anisian (Marcoux & Baud 1996; Kiessling *et al.* 1999.) Rifting and block-fragmentation in the Tethys played an important part in this revival. Dolomitization was fairly common on the carbonate platforms. The sedimentary sequences rest on metamorphic and granitic rocks of Late Palaeozoic age.

Early Jurassic (Fig. 5)

Around the Triassic–Jurassic boundary in the central Tethys area, several blocks of Cimmerian provenance (Şengör 1984; Şengör & Natalin 1996) collided with the Eurasian margin in the Eo-Cimmerian Orogeny. Alborz, and the South Caspian Microcontinent (Fig. 3 – SCM) collided with Eurasia at an earlier time (Carnian), while the Lut, Farah and South Pamir blocks (Fig. 3 – SoP) collided during a later phase (Kazmin 1990, 1997; Zonenshain *et al.* 1990). The collision of the microplates with the southern margin of Eastern Europe and Central Asia resulted in compressional events, which were recorded in a major deformation of Permian–Triassic deposits, and the general uplift of the Fore-Caucasus and Middle Asia regions (Golonka 2000). After the collision of the Cimmerian and Chinese plates, a new northwards dipping subduction zone developed along the northern margin of the Neotethys, south of the accreted continent. Extensive volcanism occurred along this zone.

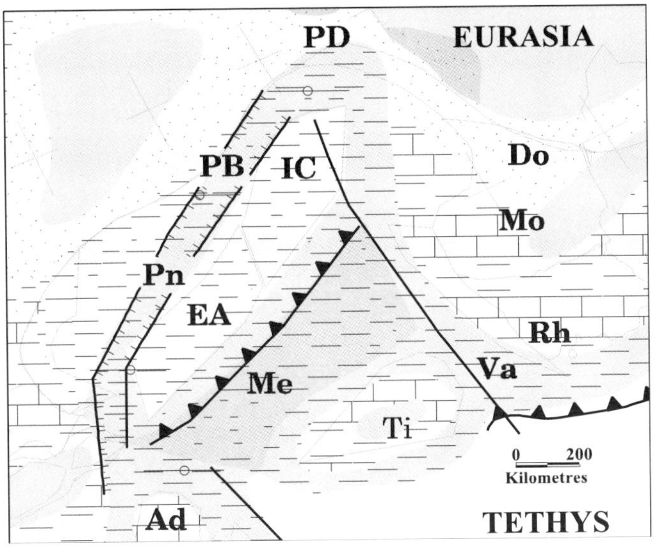

Fig. 5. Palaeogeography of the circum-Carpathian area during the Early Jurassic; plate positions at 195 Ma. Abbreviations of ocean and plate names are as follows: Ad, Adria (Apulia); Do, Dobrogea; EA, Eastern Alps; IC, Inner Carpathians; Me, Meliata–Halstatt Ocean; Mo, Moesia Plate; PB, Pieniny Klippen Belt Basin; PD, Polish–Danish aulacogen; Pn, Penninic Ocean; Rh, Rhodopes; Ti, Tisa Plate; Va, Vardar Ocean. For an explanation of the lithological symbols, see Fig. 4.

In the circum-Carpathian area the Eo-Cimmerian Orogeny was marked by deformations within the Moesian Platform (Fig. 5 – Mo) (Tari *et al.* 1997). The cause of this deformation remains speculative. It may have been a collision of the Serbo-Macedonian Block with Moesia–Rhodopes (Fig. 5 – Mo, Rh), as suggested by Golonka *et al.* (2000), or a collision within the Rhodopean margin. In the Northern Carpathian area the Eo-Cimmerian orogeny is marked by an uplift of the Inner Carpathian Plate (Fig. 5 – IC). In the Late Triassic shallowing of the basin occurred, with the deposition of neritic and lagoonal sediments of the so-called Carpathian Keuper. Sedimentary hiatuses were also common in this area, for example within the Czerwone Wierchy facies (Tatra Mts). Upper Triassic and Lower Jurassic deposits are absent, and Middle Jurassic–Bajocian or Bathonian deposits transgressed on to Middle Triassic limestones. In the Javorinska Široka facies Upper Dogger beds transgressed on to the Keuper or the Middle Triassic.

The subduction zone may have jumped from the southern Tisa margin to the Eastern Alps–Carpathian southern margin during Eo-Cimmerian events. Reorganization of plate systems following the Eo-Cimmerian Orogeny resulted in rifting and incipient spreading in the Alpine–Carpathian area. The Pieniny Klippen Belt Basin (Fig. 5 – PB) originated at that time (Fig. 5). A restricted environment prevailed in this newly formed basin. Black *Bositra* (*'Posidonia'*) shales, spotty marls (Fleckenmergel) and black flysch known from the Pieniny Klippen Belt Basin indicate oxygen-depleted conditions (Birkenmajer 1986; Tyszka 1994, 2001). Drowning of the carbonate platform and anoxia caused a dramatic decline in the number of reefs during Early Jurassic times (Kiessling *et al.* 1999).

Middle Jurassic (Fig. 6)

The Middle Jurassic saw the final closure of Paleotethys and of the opening of the central Atlantic–Penninic oceanic system (Golonka 2000). Extension of the Tethys to the northwest into the proto-Mediterranean produced a connection with central Atlantic, which was in the advanced drifting stage (Withjack *et al.* 1998). The Ligurian Ocean (Fig. 6 – Li) or Southern Penninic Ocean (Fig. 6 – Pn) (Dercourt *et al.* 1993; Channell *et al.* 1996), as well as the Pieniny Klippen Belt Basin (Fig. 6 – PB) (Golonka *et al.* 2000) formed the Jurassic–Cretaceous oceanic system. They were opening during the Early–Middle Jurassic (Fig. 5) following Early Jurassic rifting (Fig. 5). The material analysed (exotic pebbles), which

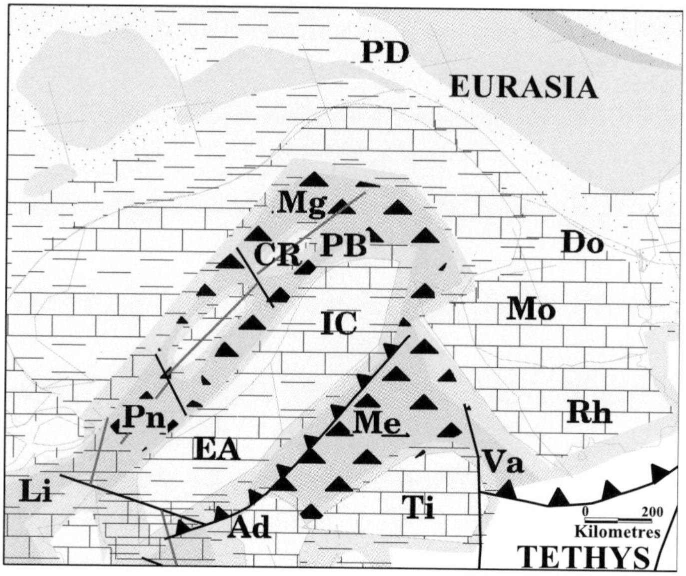

Fig. 6. Palaeogeography of the circum-Carpathian area during the Middle Jurassic; plate position at 166 Ma. Abbreviations of ocean and plate names are as follows: Ad, Adria (Apulia); CR, Czorsztyn Ridge; Do, Dobrogea; EA, Eastern Alps; IC, Inner Carpathians; Li, Ligurian Ocean; Me, Meliata–Halstatt Ocean; Mg, Magura Basin; Mo, Moesia Plate; PB, Pieniny Klippen Belt Basin; Pn, Penninic Ocean; PD, Polish–Danish aulacogen; Rh, Rhodopes; Ti, Tisa Plate; Va, Vardar Ocean. For an explanation of the lithological symbols, see Fig. 4.

occurred both in the Southern Apennines, Western Alps and the Pieniny Klippen Belt, was derived from presumed obducted oceanic crust (Marschalko 1986; Ricou 1996). Its existence suggests a Middle–Late Jurassic age for the oldest Ligurian oceanic crust. According to Winkler & Ślączka (1994), the Pieniny data fit well with the supposed opening of the Southern Penninic (Ligurian) Ocean. The newly opened basin may have overprinted the Triassic embayment of the Transylvanian–Vardar Ocean mentioned above (Fig. 6 – Va). Stampfli (2001) recently postulated single Penninic Ocean separating Apulia (Adria) (Fig. 6 – Ad) and the Eastern Alps blocks (Fig. 6 – EA) from Eurasia. We proposed a similar model for the Pieniny Klippen Belt Basin in the Carpathians. The orientation of the Pieniny–Magura Basin (Fig. 6 – PB, Mg) was NE–SW (see discussion in Golonka & Krobicki 2001; Aubrecht & Túnyi 2001). This oceanic basin was divided into the northwestern and southeastern basins by the mid-oceanic Czorsztyn Ridge. The deepest part of the southeastern basin is documented by extremely deep-water Jurassic–Early Cretaceous deposits (radiolarites and pelagic limestones) of the Złatna Unit (Sikora 1971; Golonka & Sikora 1981) later also described as the Ultrapieninic unit or Vahicum (e.g. Birkenmajer 1986; Plašienka 1999). The transitional slope sequences between deepest basinal units and ridge units (Czorsztyn Ridge – Fig. 6 – CR) are known as Pieniny, Branisko (=Kysuca), Niedzica and Czertezik successions. The shallowest ridge sequences are known as the Czorsztyn Succession. Dark Early Jurassic deposits in this succession are followed by Middle–latest Jurassic crinoidal and nodular limestones and Cretaceous variegated marls (*couches rouge* facies). The deepest part of the northwestern basin is represented by extremely deep-water reduced Late Jurassic–Early Cretaceous deposits (radiolarites and pelagic, strongly condensed Maiolica-type limestones) of the Magura (or Grajcarek or Hulina) unit (Golonka & Sikora 1981; Birkenmajer 1986; Golonka *et al.* 2000). Presumed transitional slope sequences are known from some outcrops located north of the Czorsztyn Ridge (such as Zawiasy and Stare Bystre in Poland) (Golonka & Sikora 1981) where synsedimentary breccias have been found recently within cherty limestones of the Maiolica facies. Ridge sequences as well as transitional slope sequences are also called 'Oravicum' (e.g. Plašienka 1999).

Late Jurassic (Figs 7 & 8)

During the Late Jurassic (Fig. 7) the southern part of the North European Platform, north of the Pieniny/Magura realm (Figs 7 & 8 – PB/Mg) started to be rifted and small basins (e.g. Silesian Basin in the Western Carpathians – Figs 7 & 8 – Si), with black, mainly redeposited marls (of Kimmeridgian–Tithonian age) were created (Pescatore & Ślączka 1984).

The Western Carpathian Silesian Basin probably extended into the Eastern Carpathian Sinaia or 'black flysch' Basin (Fig. 7 – Sn) (Săndulescu 1981; Săndulescu 1988). The black sediments mark the beginning of an euxinic cycle of the Outer Carpathian Basin, which lasted until the Albian. The rapid supply of shallow-water clastic material to the basin could be an effect of the strong tectono-eustatic sea-level fluctuations known from that time. The marls pass gradually upwards into calcareous turbidites (Cieszyn Limestones, see Słomka 1986) which created several submarine fans. The occurrence of deep-water microfauna indicates that subsidence of the basin must have been quite rapid.

The origin of the Outer Carpathian Basin is a part of a global progressive break-up of Pangaea. It was related to the rejuvenation of the Polish–Danish Aulacogen (Figs 7 & 8 – PD) and the origin of the North Sea rifts (Fig. 8) (Golonka 2000; Golonka *et al.* 2000). The Tethys was connected with the Silesian Basin through the Vardar Ocean (Figs 7 & 8 – Si, Va). The junction of the Tethyan, Atlantic–Ligurian–Penninic–Pieniny Klippen Belt–Magura Ocean may have been located in the Eastern Slovakian–Ukrainian Carpathians, and is represented by the Inacovce–Kricevo Zone (Fig. 10 – In) (Soták *et al.* 2000). Thus the Late Jurassic break-up system extended from the Gulf of Mexico through Central Europe to the North Sea and the proto-North Atlantic (Golonka 2000). Simultaneously, the oceanic plate of the Meliata–Halstatt Ocean (Fig. 7 – Me) was subducted northwards beneath the Inner Carpathian and Eastern Alpine plates (Dallmeyer *et al.* 1996).

In the Late Jurassic the Pieniny Klippen Belt–Magura Ocean was in advanced drifting stage with the mid-oceanic Czorsztyn Ridge separating two fully oceanic basins – Pieniny and Magura (Figs 7 & 8 – PB, Mg). The Czorsztyn Ridge (Fig. 7 – CR) could be traced from the vicinity of Vienna through Western Slovakia, Poland and Eastern Slovakia to the Transcarpathian Ukraine (Birkenmajer 1986). It was

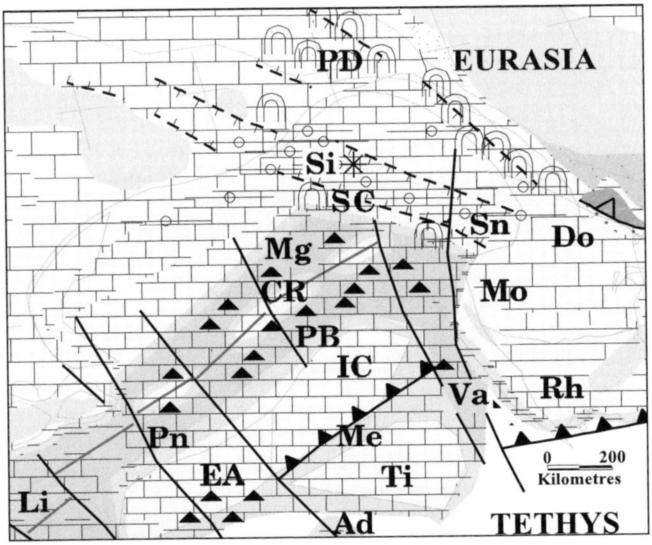

Fig. 7. Palaeogeography of the circum-Carpathian area during the Late Jurassic; plate position at 152 Ma. Abbreviations of ocean and plate names are as follows: Ad, Adria (Apulia); CR, Czorsztyn Ridge; Do, Dobrogea; EA, Eastern Alps; IC, Inner Carpathians; Me, Meliata–Halstatt Ocean; Mg, Magura Basin; Mo, Moesia Plate; PB, Pieniny Klippen Belt Basin; PD, Polish–Danish aulacogen; Pn, Penninic Ocean; Rh, Rhodopes; SC, Silesian Ridge (Cordillera); Si, Silesian Basin; Sn, Sinaia Basin; Ti, Tisa Plate; Va, Vardar Ocean. For an explanation of the lithological symbols, see Figure 4.

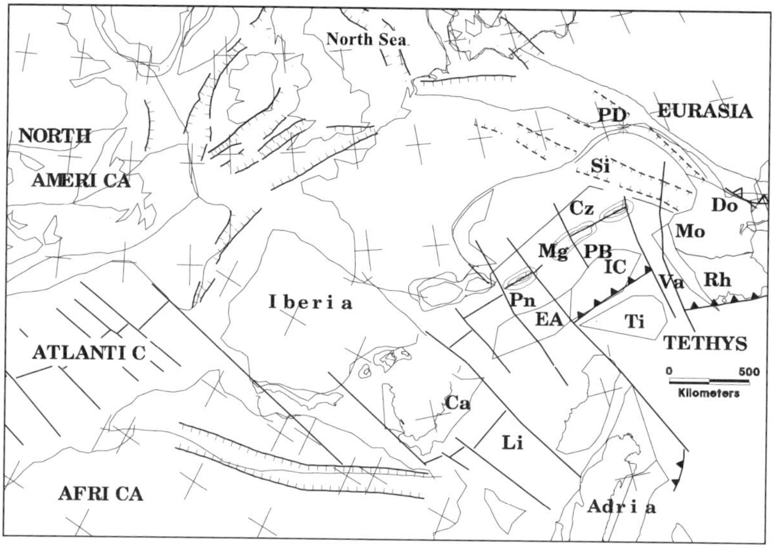

Fig. 8. General Palaeogeography of the Western and Central Tethys during the Late Jurassic; plate position at 152 Ma. Abbreviations of ocean and plate names are as follows: Ca, Calabria; Do, Dobrogea; EA, Eastern Alps; IC, Inner Carpathians; Li, Ligurian Ocean; Mg, Magura Basin; Mo, Moesia plate; PB, Pieniny Klippen Belt Basin; PD, Polish–Danish aulacogen; Pn, Penninic Ocean; Rh, Rhodopes; Si, Silesian Basin; Ti, Tisa Plate; Va, Vardar Ocean.

an elongated structure covered mainly with relatively shallow-water carbonate deposits. The palaeogeographical extent of the Magura Basin remains somewhat enigmatic and speculative. In the Eastern Carpathians Bombiţă *et al.* (1992) found possible Liassic–Early Dogger age sediments, including andesite tuffites. The latter could be an expression of an early stage of hot-spot volcanism. According to Romanian geologists (Săndulescu *et al.* 1981; Bombiţă *et al.* 1992) the Marmarosh Massif (Fig. 10 – Mr) is situated north of the Magura Basin (see also Żytko 1999). The ophiolite blocks in the Marmarosh area (Lashkevitch *et al.* 1995) indicate the existence of Mesozoic oceanic crust in this area.

A rifted fragment of the European Platform separated the Silesian Basin and the Pieniny Klippen Belt–Magura Ocean. This fragment is known as the Silesian Ridge (Cordillera) (Fig. 7 – SC). In the Polish Outer Carpathians it is known only from exotic rocks (Książkiewicz, 1962; Burtan *et al.* 1984) representing Cadomian–Hercynian crystalline basement and Late Palaeozoic, Mesozoic and Palaeogene sedimentary rocks – mainly carbonates.

The European platform adjacent to Silesian Basin in southern Poland and Ukraine was covered by a marine Upper Jurassic–Lower Cretaceous carbonate series exceeding 1000 m in thickness, partially covered by Tertiary molasse series or by thrust units of the Outer Carpathians. A maximum thickness of 1500 m of Jurassic sediments was encountered beneath the Carpathian Thrust south of Rzeszów (Maksym *et al.* 2001; Zdanowski *et al.* 2001). Present-day isopachytes reflect the original sedimentation, with a depocentre along the rift axis, and later erosional sediment removal. From a lithostratigraphic point of view, six facies units can be recognized in these Oxfordian to Valanginian carbonates, namely a calcareous planktonic facies, a calcareous sponge facies, an algal–oolitic facies, a calcareous–marly coquina facies, and a dolomitic–calcareous and algal series (Golonka 1978; Zdanowski *et al.* 2001). The barrier reefs and associated fore-reef and back-reef facies occur within an algal–oolitic series. The facies, depocentres and rift patterns in the European Platform trend NW–SE. The Silesian Basin probably represents probably the same trend. By connecting westernmost occurrence of Silesian Unit in Czech Republic with the Sinaia area in Romania in the global Jurassic palaeogeographical setting, we obtained a clear WNW–ESE or NW–SE direction for the Silesian Basin and Silesian Ridge. This orientation is perpendicular to the direction of the Penninic–Pieniny Klippen Belt–Magura Ocean.

Latest Late Jurassic – earliest Early Cretaceous (Fig. 9)

Major plate reorganization occurred during Tithonian times. The Jurassic Pangaean break-up system in Central Europe and North Sea area was abandoned (Golonka 2000). The North Sea turned from an active rift into an aulacogen. A new Atlantic spreading was initiated in the area between the Newfoundland shelf and Iberia (Fig. 8) (Ziegler 1988; Driscoll *et al.* 1995). The Polish–Danish rift was also converted into an aulacogen (Żytko 1984, 1985) with marginal marine, sometimes evaporitic sediments.

Subduction of the oceanic crust of the Meliata–Halstatt Ocean was completed by the end of the Jurassic. Terrane collision of Tisa (Fig. 9 – Ti) and adjacent blocks with the Inner Carpathians (Fig. 9 – IC) took place during this time (Dallmeyer *et al.* 1996; Froitzheim *et al.* 1996; Plašienka 1999). The closure of the Meliata–Halstatt Ocean corresponds well with the cessation of spreading in the Ligurian–Penninic–Pieniny–Magura Ocean.

The process of closure of the Pieniny Klippen Belt Basin is a matter for discussion, as summarized by Golonka & Krobicki (2001), according to earlier elaborations (e.g. Birkenmajer 1986, 1988; Plašienka 1999; Golonka *et al.* 2000). The models under discussion are based on the different subduction directions under microplate terranes: the northern margin of the Inner Carpathians, the southern margin of the Eurasian Plate and both margins of the Pieniny–Magura Basin. The present authors prefer the first model, connected with the southerly movement of the Czorsztyn Ridge (Fig. 9 – CR) and its adjacent areas. The latest Jurassic blueschists found as pebbles (exotics) in the Albian flysch in the Pieniny–Magura Basin (Fig. 9 – PB–Mg) indicate the existence of a subduction zone beneath the northern margin of the Inner Carpathian Plate (Fayrad 1997). Detailed explanations of this problem have been given by Golonka & Krobicki (2001).

The Silesian Basin (Fig. 9 – Si) was extended the Southern Carpathian Severin zone (Fig. 9–Sv) (Săndulescu 1988). The Jurassic separation of the Bucovino–Getic microplate (Fig. 9 – BG) from the European Plate may be related to this extension (Fig. 9). There is a possibility that an oceanic plate of the Vardar–Transylvanian Ocean (Fig. 9 – Va) was subducted beneath the Bucovino–Getic Plate. The eastward subduction of Getic terranes may be connected with northwards subduction under the above-mentioned

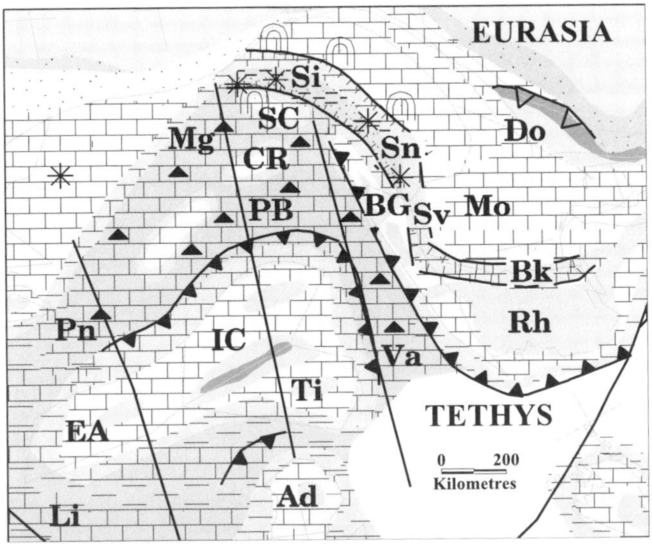

Fig. 9. Palaeogeography of the circum-Carpathian area during latest Late Jurassic–earliest Early Cretaceous times; plate position at 140 Ma. Abbreviations of ocean and plate names are as follows: Ad, Adria (Apulia); BG, Bucovino–Getic Plate; Bk, Balkan Rift; CR, Czorsztyn Ridge; Do, Dobrogea; EA, Eastern Alps; IC, Inner Carpathians; Mg, Magura Basin; Mo, Moesia Plate; PB, Pieniny Klippen Belt Basin; Pn, Penninic Ocean; Rh, Rhodopes; SC, Silesian Ridge (Cordillera); Si, Silesian Basin; Sn,Sinaia Basin; Sv, Severin Basin; Ti, Tisa Plate; Va, Vardar Ocean. For an explanation of the lithological symbols, see Fig. 4.

Rhodopes plate (Fig. 9 – Rh). A back-arc origin is suggested for the Balkan rift between Moesia and Rhodopes (Fig. 9 – Bk, Mo, Rh) (Tchoumatchenko & Sapunov 1994) of the northwards moving subduction of the Tethyan oceanic plate under the Rhodopes.

Subsidence in the Silesian Basin was accompanied by the extrusion of basic lavas (teschenites) which were probably connected with the development of initial rifting in this basin (Narębski 1990). Simultaneously, shallow-water carbonate sedimentation, with coral reefs (the so-called Štramberk Limestones), took place on the Eurasian platform. These limestones represent various types of carbonates formed on platforms, developed along the northern shore of the Tethys or around the intraoceanic Silesian Ridge (Cordillera) (Fig. 9 – SC), separating the Silesian Basin from the Pieniny–Magura realm. The remnants of these carbonate platforms with reefs were the result of the fragmentation of the European Platform in this area.

Tethyan plate reorganization resulted in extensive movement along gravitational faults. Several tectonic horsts and grabens were formed during the Neo-Cimmerian Orogeny, rejuvenating some older, Eo- and Meso-Cimmerian faults. The initial stages of subduction of the oceanic crust of the Pieniny Klippen Belt Ocean, as postulated by Birkenmajer (1986, 1988), may have been related to these movements.

The sedimentary record of Early Cretaceous Neo-Cimmerian movements

Two regions of the Carpathians: Pieniny Klippen Belt and Silesian Unit of the Outer Carpathians (Fig. 1) are mainly discussed in this paper, based on older sedimentological data (e.g. Birkenmajer 1958, 1975, 1977, 1986) and the present author's own palaeoecological (Krobicki 1994, 1996) and sedimentological studies (Słomka 1986). The relations between the Pieniny Klippen Belt and the Silesian Unit are also connected with the Jurassic–Cretaceous stage of development of the Magura Unit. Therefore in the following section we will also briefly discuss palaeodynamic environment of the Magura Basin.

The Pieniny Klippen Belt

The present day Pieniny Klippen Belt (Fig. 1) is a result of the amalgamation of the various palaeogeographical units described above. It has developed as a narrow zone and separates two major structural parts of the Carpathian range: the Inner and Outer Carpathians. It corresponds structurally with one of the main planes of crustal discontinuity within the Carpathians, along which the Moho's surface suddenly dips northwards (Ney 1976; Sikora 1976; Birkenmajer 1986). According to Birkenmajer (1985, 1986) the paragenesis of the Pieniny Klippen Belt is linked to the presence of a destructive plate margin in the northern part of the Tethyan Ocean, where subduction and consumption of lithospheric plates continued, with some breaks, from the Jurassic–Cretaceous boundary to the Miocene. The Pieniny Klippen Belt Basin (Figs 5–9 – PB) and Czorsztyn Ridge (CR) sediments constitute the major part of the Pieniny Klippen Belt tectonic unit. The fragments of the Inner Carpathian Plate (IC) and Magura Basin were also incorporated into the Pieniny Klippen Belt. The remaining part of the Magura Basin was incorporated into the Outer Carpathians (Birkenmajer 1986).

In a palinspastic reconstruction it can be seen that the Pieniny–Magura Ocean is well marked by longitudinal facies zones which correspond with ridges and troughs in the sea-floor. During the Jurassic and Cretaceous the submarine Czorsztyn Ridge was an elongated structure, nearly 500 kilometres long and some tens of kilometres wide. At the boundary of the Jurassic and Cretaceous the Czorsztyn Ridge was bordered to both the north and the south by basins in which the deep-water deposition of cherty limestones of the Maiolica facies was taking place. As suggested by Birkenmajer (1986, 1988), extensive gravitational faulting of this area occurred especially during the Meso-Cimmerian (Middle–Late Jurassic boundary) and the Neo-Cimmerian (Late Jurassic–Early Cretaceous) movements. The effects of the Neo-Cimmerian movements are particularly well seen in both the Tithonian and Berriasian sediments, where several tectonic horsts and grabens were formed – partly based on older, Eo- and Meso-Cimmerian faults. They are documented by facies diversification, condensed beds and hardgrounds with ferromanganese-rich crusts, sedimentary–stratigraphic hiatuses, neptunian dykes, and also by synsedimentary breccias (Birkenmajer 1958, 1975, 1986; Aubrecht *et al.* 1997). The best example of this type of breccia occurs within the wide-spread, exclusively carbonate sedimentation of the Berriasian *Calpionellopsis* zone (D) (Wierzbowski & Remane 1992). This synsedimentary tectonic activity is reflected by light-coloured pelagic limestones containing angular fragments of micritic limestones of older, underlying beds, interpreted as synsedimentary scarp breccia (Birkenmajer 1958, 1975, 1986; Golonka & Krobicki 2001) (Fig. 10b). Sedimentation of this breccia coincides

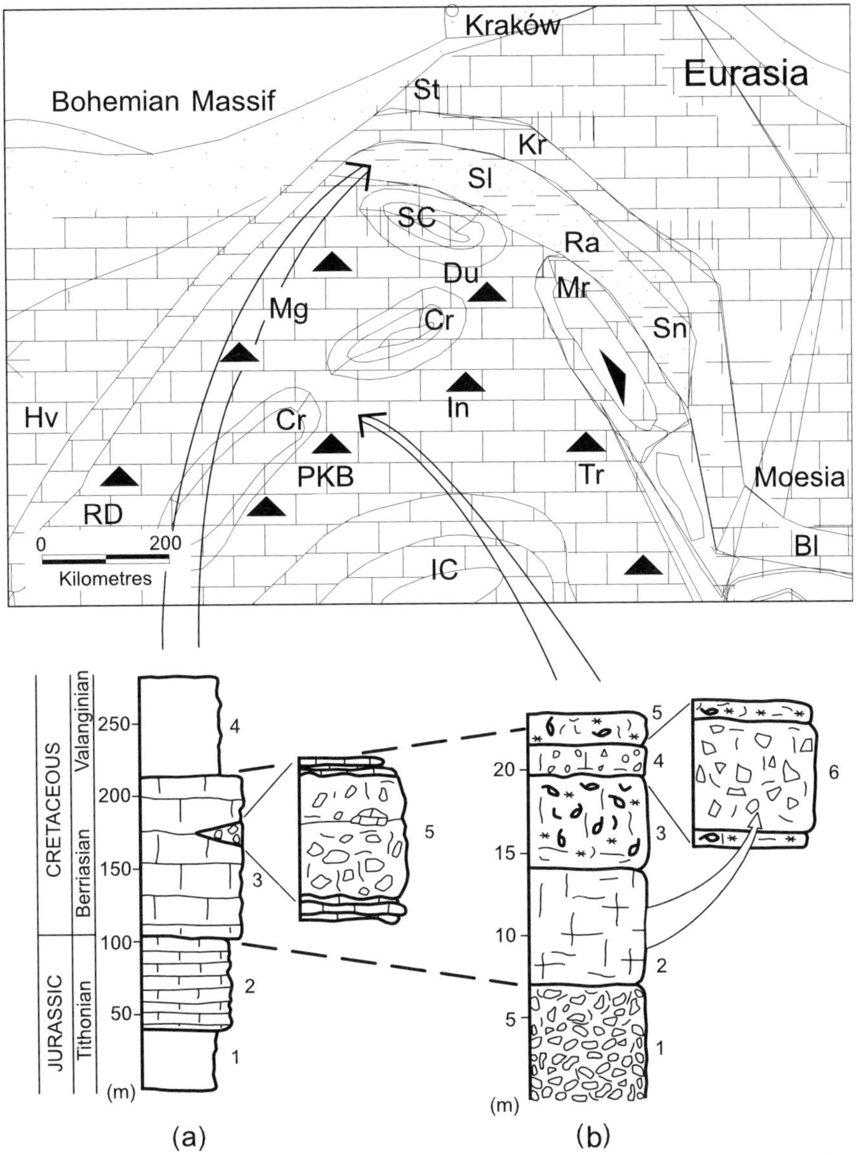

Fig. 10. Sketch of the Berriasian palaeogeography of the Carpathian basins and lithostratigraphic profiles of the Cieszyn beds (**a**) (after Słomka, 1986): 1, Lower Cieszyn Shales; 2, Lower Cieszyn Limestones; 3, Upper Cieszyn Limestones; 4, Upper Cieszyn Shales; 5, debris-flow deposits; and Tithonian–Berriasian units of the Pieniny Klippen Belt (**b**) (after Sobótka Klippe at Czorsztyn – lithostratigraphic units based on Birkenmajer, 1977, stratigraphy after Wierzbowski & Remane, 1992): 1, Czorsztyn Limestone Formation (red nodular limestone – Ammonitico Rosso type); 2, Sobótka Limestone Member of the Dursztyn Limestone Formation (micritic – *Calpionella* limestone); 3–5, Łysa Limestone Formation: 3, Harbatowa Limestone Member (crinoid-brachiopod limestone); 4, Walentowa Breccia Member (limestone breccia); 5, Kosarzyska Limestone Member (crinoid–brachiopod limestone); 6, detailed view of the Walentowa Breccia Member with redeposited clasts of the Sobótka Limestone Member. Abbreviations of oceans and plates names: Bl, Balkans; Cr, Czorsztyn Ridge; Du, Dukla Basin; Hv, Helvetic domain; IC, Inner Carpathians; In, Inacovce-Kricevo; Kr, Kruhel; Mg, Magura Basin; Mr, Marmarosh; PKB, Pieniny Klippen Belt Basin; Ra, Rachov Basin; RD, Rhenodanubian Basin; SC, Silesian Ridge (Cordillera); Sl, Silesian Basin; Sn, Sinaia Basin; St, Štramberk; Tr, Transilvanian. For an explanation of the lithological symbols, see Figure 4.

very well with the moment when the shallowing effect within the Czorsztyn area was strongest (Krobicki 1994, 1996). The Neo-Cimmerian uplift of the sea-bottom, reflected in the shallowing-upwards record of sedimentation connected with this movement, should be estimated as about 100–200 m, as suggested by both the sedimentological and the palaeo-ecological data (cf. Golonka & Krobicki 2001). The initial stages of subduction of oceanic crust under the southern, active margin are probably related to these movements.

The orientation of the Czorsztyn Ridge remains somewhat speculative, but one of the most probable positions was recently suggested to be a SW–NE direction (see above and discussion of Golonka & Krobicki 2001). This was recently supported by palaeomagnetic measurement of the directions of Jurassic neptunian dykes, suggesting a synsedimentary extensional regime on the Czorsztyn Ridge (Aubrecht & Túnyi 2001). Additionally, this direction agrees both with the palaeotectonic reconstruction, presented for example by Plašienka (2000, fig. 3), and the spreading opening trend of Central Atlantic–Ligurian system with the same NE–SW direction as a result of Pangaean break-up (e.g. Dercourt *et al.* 1993; Golonka *et al.* 1994, 1996; Ricou 1996; Kiessling *et al.* 1999; Golonka & Krobicki 2001).

Preliminary palaeomagnetic results from the Ukrainian part of the Pieniny Klippen Belt (Lewandowski *et al.* 2000) support a NE–SW direction. A palaeopole for the Middle Jurassic Czorsztyn-type deposits from the Ukrainian Carpathians matches that for the Eurasian craton in this area. The rocks investigated represent the easternmost end of the Czorsztyn Ridge. A NW–SE direction for the Czorsztyn Ridge and its westernmost termination somewhere within present-day Austria, as assumed by Michalík (1994), would produce quite different results. The easternmost end of the ridge would be somewhere around 30° N, a long distance south of the Eurasian plate. Therefore a SW–NE position shows a much better correspondence with the palaeomagnetic data. Of course, more palaeomagnetic research along the Czorsztyn Ridge will help to better constrain the ridge position.

The Outer Carpathians – Silesian Unit

Within the Outer Carpathian basins the most unusual aspect is the presence of a set of E–W trending troughs, separated by submarine and emergent ridges (cordillera). The Silesian Basin (Fig. 10 – Si) is one of the oldest of these Carpathian basins. It was bounded to the north by a submerged sub-Silesian ridge containing the Baška and Inwałd cordillera, whereas to the south it was bounded by the Silesian Cordillera (Ridge) (Figs 8–10 – Sc).

The Cieszyn Beds (?Kimmeridgian–Hauterivian) form the oldest stratigraphic unit of the Silesian Nappe in the Outer Carpathians. They consist mainly of detrital and pelitic limestones, calcareous sandstones, marls and marly shales. The attain a maximum thickness of more than 800 m. The clastic material for Cieszyn Beds was generally derived from the northern margin of the Silesian Basin (e.g. Kruhel, Štramberk – Fig. 10 – Kr, St) (Książkiewicz 1960; Peszat 1967; Malik 1986). However, part of the clastic source area was situated on the islands at the southern margin of this basin and related to the northern margins of the Silesian Ridge (Cordillera) (Książkiewicz 1962; Ślączka 1976; Eliaš & Eliašova 1984; Słomka 1986; Matyszkiewicz & Słomka 1994).

The Cieszyn Beds cropping out in the Żywiec region comprise several bodies of debris-flow deposits (Fig. 10 – A). They include numerous fragments and pebbles of detrital and pelitic limestones of the Cieszyn Beds, organodetrital limestones, marly shales, the Carboniferous coals and metamorphic rocks (granitic gneisses, gneisses and crystalline schists). The pebbles are randomly arranged in a mass of structureless, hard, marly silt. Both clays and embedded lumps of limestone have local bends and folds closing generally towards the north, which would suggest that the sliding mass was derived from the south (Dżułyński *et al.* 1959). These debris-flow deposits document the evolution of the Silesian Ridge during the initial development of the active cordillera. The Jurassic carbonate platform was developed on the submarine ridge. The basement of the carbonate platform consisted of Palaeozoic sedimentary and metamorphic rocks. The turbiditic Cieszyn Beds were deposited on the slope and bottom of the Silesian Basin (Słomka 1986). During the Early Cretaceous tectonic activity, part of the basin was uplifted together with the Silesian Ridge (Cordillera), and the Cieszyn Beds in this area were redeposited by debris flows (Słomka, 2001). The appearance of mass-movement debris-flow deposits containing fragments of the older Cieszyn Beds and exotics from the basement rocks testify to the higher rate of uplift connected with Neo-Cimmerian activity (the Osterwald Phase). The Early Cretaceous development of the Silesian Basin, perhaps from a rifting into a spreading phase, as suggested by the presence of teschenitic magmatism

(Narąbski 1990; Lucińska-Anczkiewicz *et al.* 2000) was probably another effect of this Osterwald Phase.

The Magura Basin

During the Jurassic and Early Cretaceous the Magura Basin was situated between the Czorsztyn Ridge and the Silesian Cordillera (Figs 8–10) (Golonka *et al.* 2000). The extremely condensed Jurassic–Early Cretaceous sedimentary successions of this basin (Golonka & Sikora 1981) are known today only from scattered localities within the Pieniny Klippen Belt (so-called Grajcarek Unit or Hulina Succession) and from the Late–Cretaceous–Palaeogene Magura Unit. In the central and western part of the Magura Unit, Early Cretaceous sequences are represented by pelitic deposits and fine-grained turbiditic limestones of Barremian–Albian age known from Moravia (Švábenická *et al.* 1997). The existing scattered data do not provide evidence for Neo-Cimmerian movements. The Magura Succession from the Pieniny Klippen Belt, represented by deep-water cherty limestones (Maiolica or Biancone), of the *Nannoconus* facies (Golonka & Sikora 1981) also do not indicate Neocomian tectonic movement during the latest Jurassic–earliest Cretaceous. However, Cimmerian tectonic movements may have been recorded at the easternmost end of the Magura Basin, in the Eastern Carpathians. This problem requires further investigation.

Conclusions

Almost simultaneous tectonic events occurring within different types of sedimentary basin (Pieniny Klippen Belt and Outer Carpathians–Silesian) indicate a very important role for Neo-Cimmerian movements (mainly of the Osterwald Phase) in the geodynamic history of the northernmost margin of the Tethyan Ocean, commonly connected with rejuvenation of older tectonic phases (Eo- and/or Meso-Cimmerian) (Fig. 10). The evolution of several, mainly longitudinal, Carpathian basins is proposed to be connected with subduction processes and the development of the active margins. The presumed initial stages of ocean crust subduction within the Pieniny Klippen Belt Basin (Birkenmajer 1986), riftogenesis (Silesian Basin – see Narębski 1990; Vašíček *et al.* 1994), volcanism (Rakús *et al.* 1988) and even the prevailing palaeoceanographic conditions (Birkenmajer 1986; Golonka & Krobicki 2001) are most probably connected with the Neo-Cimmerian tectonic event. Alternatively, the formation of these allodapic beds can also be interpreted as having been caused by eustatic events (lithohorizon Be-7), and corresponds very well with the Berriasian section of the Nozdrovice Breccia within the Inner Carpathians (Reháková & Michalík 1992; Michalík *et al.* 1995, 1996), which was developed as scarp breccias along active submarine fault slopes (Michalík & Reháková 1995). On the other hand, the eustatic changes may have been connected with the global plate reorganization which took place during Tithonian–Berriasian times (Golonka 2000). This reorganization is also related to Tethyan Neo-Cimmerian tectonic activity.

This work was partly supported by a grant from the Polish Committee for Scientific Research (KBN Grant No. 6 P04D 040 19). It is also a contribution to IGCP 453. We are grateful to the referees for their reviews.

References

AUBRECHT, R. & TÚNYI, I. 2001. Original orientation of neptunian dykes in the Pieniny Klippen Belt (Western Carpathians): the first results. *Contributions to Geophysics and Geodesy*, **31**, 557–578.

AUBRECHT, R., MIŠÍK, M. & SÝKORA, M. 1997. Jurassic synrift sedimentation on the Czorsztyn Swell of the Pieniny Klippen Belt in Western Slovakia. *In:* PLAŠIENKA, D., HÓK, J., VOZÁR, J. & ELLECKO, M. (eds) *Alpine Evolution of the Western Carpathians and Related Areas*, Dionýz Stur, Bratislava, Slovakia, 53–64.

BECK, M. E. & SCHERMER, E. R. 1994. Aegean palaeomagnetic inclination anomalies – is there a tectonic explanation? *Tectonophysics*, **231**, 281–292.

BESSE, J. & COURTILLOT, V. 1991. Revised and synthetic apparent polar wander paths of the African, Eurasian, North American and Indian plates, and true polar wander since 200 Ma. *Journal of Geophysical Research*, **96**, 4029–4050.

BIRKENMAJER, K. 1958. Submarine erosional breaks and Late Jurassic synorogenic movements in the Pieniny Klippen Belt geosyncline. *Bulletin l'Académie Polonaise de Sciences*, **8**, 551–558.

BIRKENMAJER, K. 1975. Tectonic control of sedimentation at the Jurassic–Cretaceous boundary in the Pieniny Klippen Belt, Carpathians. *Mémoir du Bureau de Recherche de Géologie et Minéralogie*, **86**, 294–299.

BIRKENMAJER, K. 1977. Jurassic and Cretaceous lithostratigraphic units of the Pieniny Klippen Belt, Carpathians, Poland. *Studia Geologica Polonica*, **45**, 1–158.

BIRKENMAJER, K. 1985. *In:* BIRKENMAJER, K. (ed.) *Carpatho–Balkan Geological Association, XIII Congress (Kraków, 1985). Guide to Excursion 2: Main Geotraverse of the Polish Carpathians (Kraków–Zakopane)*, **5–13**, 31–36.

BIRKENMAJER, K. 1986. Stages of structural evolution of the Pieniny Klippen Belt, Carpathians. *Studia Geologica Polonica*, **88**, 7–32.

BIRKENMAJER, K. 1988. Exotic Andrusov Ridge: its role in plate-tectonic evolution of the West Carpathian Foldbelt. *Studia Geologica Polonica*, **91**, 7–37.

BIRKENMAJER, K., KOZUR, H. & MOCK, R. 1990. Exotic Triassic pelagic limestone pebbles from the Pieniny Klippen Belt of Poland: a further evidence for Early Mesozoic rifting in West Carpathians. *Annales Societatis Geologorum Poloniae*, **60**, 3–44.

BOMBIȚĂ, G., ANTONESCU, E., MALATA, E., MÜLLER, C. & NEAGU, T. 1992. Pieniny-type Mesozoic formation from Maramures, Romania (second part). *Acta Geologica Hungarica*, **35**, 117–144.

BURTAN, J., CHOWANIEC, J. & GOLONKA, J. 1984. Preliminary results of studies on exotic carbonate rocks in the western part of the Polish Flysch Carpathians [in Polish, English summary]. *Biuletyn Instytutu Geologicznego*, **346**, 147–159.

CATALANO, R., DI STEFANO, P. & KOZUR, H. 1991. Permian circumpacific deep-water faunas from the western Tethys (Sicily, Italy): new evidence for the position of the Permian Tethys. *Palaeogeography, Palaeoclimatology, Palaeoecology*, **87**, 75–108.

CHANNELL, J. E. T. & KOZUR, H. W. 1997. How many oceans? Meliata, Vardar, and Pindos oceans in Mesozoic Alpine paleogeography. *Geology*, **25**, 183–186.

CHANNELL, J. E. T., BRANBER, R., SPIELER, A. & ŞTONER, J. S. 1992. Paleogeomagnetism and paleogeography of the Northern Calcareous Alps (Austria). *Tectonics*, **11**, 792–810.

CHANNELL, J. E. T., TUYSUZ, O., BEKTAS, O. & ŞENGÖR, A. M. C. 1996. Jurassic–Cretaceous paleomagnetism and paleogeography of the Pontides (Turkey). *Tectonics*, **15**, 201–212.

DALLMEYER, R. D., NEUBAUER, F., HANDLER, R., FRITZ, H., MUELLER, W., PANA, D. & PUTIS, M. 1996. Tectonothermal evolution of the internal Alps and Carpathians; evidence from $^{40}Ar/^{39}Ar$ mineral and whole-rock data. *In:* SCHMID, S. M., FREY, M., FROITZHEIM, N., HEILBRONNER, R. & STUENITZ, H. (eds) Alpine Geology; Proceedings of the Second Workshop. 2nd Workshop on Alpine Geology. *Eclogae Geologicae Helvetiae*, **89**, 203–227.

DERCOURT, J., RICOU, L. E. & VRIELYNCK, B. (eds) 1993. *Atlas Tethys Paleoenvironmental Maps*. Gauthier-Villars', Paris.

DERCOURT, J., GAETANI, M., VRIELYNCK, B., BARRIER, E., BIJU-DUVAL, B., BRUNET, M.-F., CADET, J. P., CRASQUIN, S. & SANDULESCU, M. (eds) 2000. *Atlas Peri–Tethys Paleogeographical maps. CCGM/CGMW, Paris, 24 Maps and Explanatory Note*, **I-XX**, 1–269.

DRISCOLL, N. W., HOGG, J. R., CHRISTIE-BLICK, N. & KARNER, G. D. 1995. Extensional tectonics in the Jeanne d'Arc basin, offshore Newfoundland: implications for the timing of break-up between Grand Banks and Iberia. *In:* SCRUTTON, R. A., STOKER, M. S., SHIMMIELD, G. B. & TUDHOPE, A. W. (eds) *The Tectonics, Sedimentation and Palaeoceanography of the North Atlantic Region*. Geological Society, London, Special Publication, **90**, 145–163.

DŻUŁYŃSKI, S. KSIĄŻKIEWICZ, M. & KUEANEN, P. H. 1959. Turbidites in flysch of the Polish Carpathians. *Geological Society of America Bulletin*, **70**, 1089–118.

ELIAŠ, M. & ELIAŠOVA, M. 1984. Facies and palaeogeography of the Jurassic in the western part of the Outer Flysch Carpathians in Czechoslovakia. *Sbornik Geologickich Ved*, **39**, 105–170.

FAYRAD, S. W. 1997. Petrological model for blueschist facies metamorphism in the Pieniny Klippen Belt. *In:* GRECULA, P., HOVORKA, D. & PUTIŠ, M. (eds) *Geological Evolution of the Western Carpathians.* Mineralia Slovaca – Monograph, Bratislava, 155–162.

FEINBERG, H., EDEL, B., KONDOPOULOU, D. & MICHARD, A. 1996. Implications of ophiolite palaeomagnetism for the interpretation of the geodynamics of Northern Greece. *In:* MORRIS, A. & TARLING, D. H. (eds) *Paleomagnetism and Tectonics of the Mediterranean Region*. Geological Society, London, Special Publication, **105**, 289–298.

FROITZHEIM, N., SCHMID, S. M. & FREY, M. 1996. Mesozoic paleogeography and the timing of eclogite facies metamorphism in the Alps: a working hypothesis. *Eclogae Geologicae Helvetiae*, **89**, 81–110.

GOLONKA, J. 1978. Upper Jurassic microfacies in the Carpathians foreland (in Polish: English abstract). *Biuletyn Instytutu Geologicznego*, **310**, 1–38.

GOLONKA, J. 2000. Cambrian–Neogene plate tectonic maps. *Wydawnictwa Uniwersytetu Jagiellońskiego, Kraków*, 1–125.

GOLONKA, J. & BOCHAROVA, N. Y. 2000. Hot spot activity: the break-up of Pangea. *Palaeogeography, Palaeoclimatology, Palaeoecology*, **161**, 49–69.

GOLONKA, J. & FORD, D. 2000. Pangean (Late Carboniferous–Middle Jurassic) paleoenvironment and lithofacies. *Palaeogeography, Palaeoclimatology, Palaeoecology*, **161**, 1–34.

GOLONKA, J. & GAHAGAN, L. 1997. Tectonic model of the Mediterranean terranes. *Bulletin AAPG*, **8**, 1386.

GOLONKA, J. & KROBICKI, M. 2001. Upwelling regime in the Carpathian Tethys: a Jurassic–Cretaceous palaeogeographic and palaeoclimatic perspective. *Geological Quarterly [Warsaw]*, **45**, 15–32.

GOLONKA, J. & SIKORA, W. 1981. Microfacies of the Jurassic and Lower Cretaceous sedimentarily thinned deposits of the Pieniny Klippen Belt in Poland (in Polish: English abstract). *Biuletyn Instytutu Geologicznego*, **31**, 7–37.

GOLONKA, J., ROSS, M. I. & SCOTESE, C. R. 1994. Phanerozoic paleogeographic and paleoclimatic modeling maps. *In:* EMBRY, A. F., BEAUCHAMP, B. & GLASS, D. J. (eds) *Pangea: Global Environment and Resources*. Canadian Society of Petroleum Geologists Memoir, **17**, 1–47.

GOLONKA, J., EDRICH, M., FORD, D. W., PAUKEN, R. B., BOCHAROVA, N. Y. & SCOTESE, C. R. 1996. Jurassic paleogeographic maps of the world. *In:* MORALES, M. (ed.) *The continental Jurassic*. Museum of Northern Arizona Bulletin, **60**, 1–5.

GOLONKA, J., OSZCZYPKO, N, & ŚLĄCZKA, A. 1999. Geodynamic evolution of the Carpathian Foredeep basin – a global perspective. *Biuletyn Państwowego Instytutu Geologicznego*, **387**, 100–101.

GOLONKA, J., OSZCZYPKO, N. & ŚLĄCZKA, A. 2000. Late Carboniferous – Neogene geodynamic evolution and paleogeography of the circum-Carpathian region and adjacent areas. *Annales Societatis Geologorum Poloniae*, **70**, 107-136.

GRABOWSKI, J. 2000. Palaeo- and rock magnetism of Mesozoic carbonate rocks in the Sub-Tatric series (Central West Carpathians) palaeotectonic implications. *Polish Geological Institute, Special Papers*, **5**, 1–88.

HAUBOLD, H., SCHOLGER, R., FRISCH, W., SUMMESBERGER, H. & MAURITSCH, H. J. 1999. Reconstruction of the geodynamic evolution of the Northern Calcareous Alps by means of paleomagnetism. *Physics and Chemistry of the Earth, Part A – Solid Earth and Geodesy*, **24**, 697–703.

KAZMIN, V. G. 1990. Early Mesozoic reconstruction of the Black Sea–Caucasus region. Evolution of the northern margin of the Tethys. *Mémoire Société Géologique de France, Paris, Nouvelle Série*, **54**, 147–58.

KAZMIN, V. G. 1997. Mesozoic to Cenozoic history of the back-arc basins in the Black Sea–Caucasus region. *Unpublished Report: CASP Arctic and Russian Studies III Report* **656**, Cambridge Arctic Shelf Program, Cambridge, UK, 70 pp.

KIESSLING, W., FLÜGEL, E. & GOLONKA, J. 1999. Paleoreef maps: a comprehensive database of Phanerozoic reefs with graphic presentations. *Bulletin AAPG*, **83**, 1552–1587.

KONDOPOLOU, D., ATZEMOGLOU, A. & PAVLIDES, S. 1996. Palaeomagnetism as a tool for testing geodynamic models in the North Aegean: convergences, controversies and a further hypothesis. *In:* MORRIS, A. & TARLING, D. H. (eds) *Palaeomagnetism and Tectonics of the Mediterranean Region*. Geological Society, London, Special Publications, **105**, 277–288.

KOVÁČ, M., NAGYMAROSY, A., SOTÁK, J. & ŠUTOVSKÁ, K. 1993. Late Tertiary paleogeographic evolution of the Western Carpathians. *Tectonophysics*, **226**, 401–415.

KOVÁČ, M., NAGYMAROSY, A., OSZCZYPKO, N., ŚLĄCZKA, A., CSONTOS, L., MARUNTEANU, M., MATENCO, L. & MÁRTON, M. 1998. Palinspastic reconstruction of the Carpathian–Pannonian region during the Miocene. *In:* RAKÚS, M. (ed.) *Geodynamic Development of the Western Carpathians*. Geological Survey of Slovak Republic, Bratislava, 189–217.

KOZUR, H. 1991. The evolution of the Meliata–Halstatt ocean and its significance for the early evolution of the Eastern Alps and Western Carpathians. *Palaeogeography, Palaeoclimatology, Palaeoecology*, **87**, 109–130.

KOZUR, H. & KRAHL, W. 1987. Erster Nachweis von Radiolarien in tethyalen Perm Europas. *Neues Jahrbuch für Geologie und Paleontologie Abhandlungen*, **174**, 357–372.

KROBICKI, M. 1994. Stratigraphic significance and palaeoecology of the Tithonian–Berriasian brachiopods in the Pieniny Klippen Belt, Carpathians, Poland. *Studia Geologica Polonica*, **106**, 89–156.

KROBICKI, M. 1996. Neo-Cimmerian uplift of intraoceanic Czorsztyn pelagic swell (Pieniny Klippen Belt, Polish Carpathians) indicated by the change of brachiopod assemblages. *In:* RICCARDI, A. C. (ed.) *Advances in Jurassic Research*. GeoResearch Forum, **1–2**, 255–264.

KRS, M., KRSOVA, M. & PRUNER, P. 1996. Palaeomagnetism and palaeogeography of the Western Carpathians from the Permian to the Neogene. *In:* MORRIS, A. & TARLING, D. H. (eds) *Palaeomagnetism and Tectonics of the Mediterranean region*. Geological Society, London, Special Publications, **105**, 175–184.

KSIĄŻKIEWICZ, M. 1960. Outline of the paleogeography in the Polish Carpathians. *Prace Instytutu Geologicznego*, **30**, 209–231.

KSIĄŻKIEWICZ, M. (ed.) 1962. *Geological Atlas of Poland; Stratigraphic and Facial problems*. Instytut Geologiczny, Wydawnictwa Geologiczne, Warsaw, Poland.

LASHKEVITSCH, Z. M., MEDVEDEV, A. P. & KRUPSKIY, Y. Z. 1995. *Tectonomagmatic evolution of Carpathians* (in Russian). Naukova Dumka, Kiev.

LEWANDOWSKI, M., KROBICKI, M., MATYJA, B. A. & WIERZBOWSKI, A. 2000. Rekonesans paleomagnetyczny w jurajskie skały pienińskiego pasa skałkowego Ukrainy Zakarpackiej (in Polish only). *Polska Grupa Robocza Systemu Jurajskiego, I Spotkanie, Wiktorowo 28–29.09.2000*; Abstracts, Polskie Towarzystwo Geologiczne, Kraków, 16.

LUCIŃSKA-ANCZKIEWICZ, A. ŚLĄCZKA, A., VILLA, I. M. & ANCZKIEWICZ, R. 2000. $^{39}Ar/^{40}Ar$ dating of the teschenite association rocks from the Polish Outer Carpathians. *Mineralogical Society of Poland – Special Papers*, **17**, 241–242.

MAKSYM, A., BASZKIEWICZ, A., GREGOSIEWICZ, Z., LISZKA, B. & ZDANOWSKI, P. 2001. Facies and reservoir potential of uppermost Jurassic and Lower Cretaceous deposits in the Brzezówka–Zagorzyce region versus geological structure of the Southern Carpathian Foredeep (English summary). *Przegląd Geologiczny*, **49**, 401–407.

MALIK, K. 1986. Turbidite facies and fan-facies associations in the Cieszyn Limestones Upper Tithonian–Berriasian, Northwestern Carpathians, Poland. *In:* TEISSEYRE, A. K. (ed.) *7th European Regional Meeting Excursion Guidebook. Kraków, Poland*. Ossolineum, Wrocław, 53–66.

MARCOUX, J. & BAUD, A. 1996. Late Permian to Late Triassic Tethyan Paleoenvironments. Three snapshots: Late Murgabian, Late Anisian, Late Norian. *In:* NAIRN, A. E. M., RICOU, L.-E. VRIELYNCK, B. & DERCOURT, J. (eds) *The Oceans Basins and Margin, Vol. 8, The Tethys Ocean*. Plenum Press, New York and London, 153–190.

MARSCHALKO, R. 1986. Evolution and geotectonic significance of the Klippen Belt Cretaceous flysch in the Carpathian megastructure. *VEDA, Bratislava*, 1–137.

MARSELLA, E., KOZUR, H. & D'ARGENIO, B. 1993. Monte Facito Formation (Scythian–Middle

Carnian). A deposit of the ancestral Lagonegro Basin in Southern Apennines. *Bolletino de Servisio Geologico Italia*, **119**, 225–248.

MÁRTON, E. & MARTIN, P. 1996. Large scale rotations in North Hungary during the Neogene as indicated by palaeomagnetic data. *In:* MORRIS, A. & TARLING, D. H. (eds) *Paleomagnetism and Tectonics of the Mediterranean Region*. Geological Society, London, Special Publication, **105**, 153–173.

MÁRTON, E., MASTELLA, L. & TOKARSKI, A. K. 1999. Large counterclockwise rotation of the Inner West Carpathian Paleogene Flysch – evidence from paleomagnetic investigations of the Podhale Flysch (Poland). *Physics and Chemistry of the Earth, Part A – Solid Earth and Geodesy*, **24**, 645–649.

MÁRTON, E., KUHLEMANN, J., FRISCH, W. & DUNKL, I. 2000. Miocene rotations in the Eastern Alps – palaeomagnetic results from intramontane basin sediments. *Tectonophysics*, **323**, 163–182.

MATYSZKIEWICZ, J. & SŁOMKA, T. 1994. Organodetrital conglomerates with ooids in the Cieszyn Limestone (Tithonian–Berriasian) of the Polish Flysch Carpathians and their palaeogeographic significance. *Annales Societatis Geologorum Poloniae*, **63**, 211–248.

MAURITSCH, H. J., SCHOLGER, R., BUSHATI, S. L. & RAMIZ, H. 1995. Palaeomagnetic results from southern Albania and their significance for the geodynamic evolution of the Dinarides, Albanides and Hellenides. *Tectonophysics*, **242**, 5–18.

MAURITSCH, H. J., SCHOLGER, R., BUSHATI, S. L. & XHOMO, A. 1996. Palaeomagnetic investigations in Northern Albania and their significance for the geodynamic evolution of the Adriatic Aegean realm. *In:* MORRIS, A. & TARLING, D. H. (eds) *Paleomagnetism and Tectonics of the Mediterranean Region*. Geological Society, London, Special Publication, **105**, 265–275.

MICHALÍK, J. 1994. Notes on the paleogeography and paleotectonics of the Western Carpathian area during the Mesozoic. *Mitteilungen Österreichische Geologische Gesellschaft*, **86**, 101–110.

MICHALÍK, J. & REHÁKOVÁ, D. 1995. Sedimentary records of Early Cretaceous tectonic activity in the Alpine–Carpathian region. *Slovak Geological Magazine*, **2**, 159–164.

MICHALÍK, J., REHÁKOVÁ, D. & VAŠÍCEK, Z. 1995. Early Cretaceous sedimentary changes in West-Carpathian area. *Geologica Carpathica*, **46**, 285–296.

MICHALÍK, J., REHÁKOVÁ, D. & JABLONSKÝ, J. 1996. Geodynamic setting of fluxoturbidites in West Carpathian Upper Jurassic and Lower Cretaceous sedimentary basins. *Slovak Geological Magazine*, **3–4**, 325–329.

MUTTONI, G., GAETANI, M., *ET AL.* 2000*a*. Middle Triassic paleomagnetic data from northern Bulgaria: constraints on Tethyan magnetostratigraphy and paleogeography. *Palaeogeography, Palaeoclimatology, Palaeoecology*, **160**, 223–237.

MUTTONI, G., LANCI, L., ARGNANI, A., HIRT, A. M., CIBIN, U., ABRAHAMSEN, W. & LOWRIE, W. 2000*b*. Paleomagnetic evidence for a Neogene two-phase counterclockwise tectonic rotation in the Northern Apennines (Italy). *Tectonophysics*, **326**, 241–253.

NARĘBSKI, W. 1990. Early rifts in the evolution of western part of the Carpathians; geochemical evidence from limburgite and teschenite rock series. *Geologicky Zbornik – Geologica Carpathica*, **41**, 521–528.

NEY, R. 1976. The Carpathians and plate tectonics. *Przegląd Geologiczny*, **6**, 309–316.

PATRASCU, S., BLEAHU, M., PANAIOTU, C. & PANAIOTU, C. E. 1992. The paleomagnetism of the Upper Cretaceous magmatic rocks in the Banat Area of South Carpathians – tectonic implications. *Tectonophysics*, **213**, 341–352.

PATRASCU, S., SECLAMAN, M. & PANAIOTU, C. 1993. Tectonic implications of paleomagnetism in Upper Cretaceous deposits in the Hateg and Rusca Montana Basins (South-Carpathians, Romania). *Cretaceous Research*, **14**, 255–264.

PECHERSKY, D. M. & SAFRONOV, V. A. 1993. Palinspastic reconstruction of the Mountaneous Crimea position in the Middle Jurassic–Early Cretaceous from paleomagnetic data. *Geotectonika*, **1**, 96–105.

PESCATORE, T. & ŚLĄCZKA, A. 1984. Evolution models of two flysch basins: the Northern Carpathians and the Southern Appenines. *Tectonophysics*, **106**, 49–70.

PESZAT, C. 1967. The lithological development and conditions of sedimentation of the Cieszyn Limestones [in Polish: English abstract]. *Prace Geologiczne Oddział PAN Kraków*, **44**, 1–111.

PLAŠIENKA, D. 1999. *Tectonochronology and Paleotectonic Model of the Jurassic–Cretaceous Evolution of the Central Western Carpathians*. Veda, Bratislava, 1–127.

PLAŠIENKA, D. 2000. Paleotectonic controls and tentative palinspastic restoration of the Carpathian realm during the Mesozoic. *Slovak Geological Magazine*, **6**, 200–204.

PLAŠIENKA, A. & KOVÁČ, M. 1999. How to loop Carpathians – an attempt to reconstruct Meso-Cenozoic palinspastic history of the Carpathian orocline. *Geologica Carpathica*, **50**, 163–165.

RAKÚS, M., MIŠÍK, M., MICHALÍK, J., MOCK, R., DURKOVIC, T., KORÁB, T., MARSCHALKO, R., MELLO, J., POLÁK, M. & JABLONSKÝ, J. 1988. Paleogeographic development of the West Carpathians: Anisian to Oligocene. *Mémoires de la Société Géologique de France, Nouvelle Série*, **154**, 39–62.

REHÁKOVÁ, D. & MICHALÍK, J. 1992. Correlation of Jurassic/Cretaceous boundary beds in West Carpathian profiles. *Földtani Közlöny*, **122**, 51–66.

RICOU, L.-E. 1996. The plate tectonic history of the past Tethys Ocean. *In:* NAIRN, A. E. M., RICOU, L.-E., VRIELYNCK, B. & DERCOURT, J. (eds) *The Oceans Basins and Margin, Vol. 8, The Tethys Ocean*. Plenum Press, New York and London, 3–70.

ROBERTSON, A. H. F. 1998. Mesozoic–Tertiary Tectonic evolution of the easternmost Mediterranean area: integration of the marine and land evidence. *In:* ROBERTSON, A. H. F., RICHTER, C. & CAMERLENGHI, C. A. (eds) *Proceedings of the Ocean Drilling Program, Scientific Results*, **60**, 723–782.

ROBERTSON, A. H. F., CLIFT, P. D., DEGNAN, P. & JONES, G. 1991. Paleogeographic and paleotectonic evolution of Eastern Mediterranean Neotethys. *Palaeogeography, Palaeoclimatology, Palaeoecology*, **87**, 289–344.

ROBERTSON, A. H. F., DIXON, J. E., BROWN, S., COLLINS, A., MORRIS, A., PICKETT, E. A., SHARP, I. & USTAOMER, T. 1996. Alternative tectonic models for the Late Palaeozoic–Early Tertiary development of Tethys in the Eastern Mediterranean region. *In:* MORRIS, A. & TARLING, D. H. (eds) *Palaeomagnetism and Tectonics of the Mediterranean region*. Geological Society Special Publication, **105**, 239–263.

RONOV, A., KHAIN, V. E. & BALUKHOVSKI, A. N. 1989. *Atlas of Lithological Paleogeographical Maps of the World: Mesozoic and Cenozoic of the Continents*. USSR Academy of Sciences, Leningrad, 1–70.

RONOV, A., KHAIN, V. E. & SESLAVINSKI, K. 1984. *Atlas of Lithological Paleogeographical Maps of the World: Late Precambrian and Paleozoic of the Continents*. USSR Academy of Sciences, Leningrad, 1–79.

SĂNDULESCU, M. 1988. Cenozoic tectonic history of the Carpathians. *In:* ROYDEN, L. & HORVÁTH, F. (eds) *The Pannonian Basin: a study in Basin Evolution*. Memoir AAPG, **45**, 17–25.

SĂNDULESCU, M. & VISARION, M. 2000. Crustal structure and evolution of the Carpathian–Western Black Sea area. *Romanian Geophysics*, **7**, 228–231.

SĂNDULESCU, M., KRÄUTNER, H., BALINTONI, I., RUSSO-SĂNDULESCU, D. & MICU, M. 1981. The structure of the East Carpathian (Moldavia–Muramures Area). *Carpathian–Balkan Geological Association, XII Congress Bucharest – Romania 1981, Guide to Excursion B1*, 1–92.

SCOTESE, C. R. & LANFORD, R. P. 1995. Pangea and the paleogeography of the Permian. *In:* SCHOLLE, P. A., PERYT, T. M. & ULMER-SCHOLLE, D. S. (eds) *The Permian of Northern Pangea, Vol. 1: Paleogeography, Paleoclimate, Stratigraphy*. Springer Verlag, Berlin, Heidelberg and New York, 3–19.

ŞENGÖR, A. M. C. 1984. The Cimmeride orogenic system and the tectonics of Eurasia. *Geological Society of America Special Paper*, **195**, 1–82.

ŞENGÖR, A. M. C. & NATALIN, B. A. 1996. Paleotectonics of Asia: fragment of a synthesis. *In:* YIN AN & HARRISON, T. M. (eds) *The Tectonic Evolution of Asia*. Cambridge, Cambridge University Press, 486–640.

SIKORA, W. 1971. Outline of the tectonogenesis of the Pieniny Klippen Zone in Poland in the light of the new geological data (in Russian). *Annales Societatis Geologorum Poloniae*, **61**, 221–238.

SIKORA, W. J. 1976. On lineaments found in the Carpathians. *Annales Societatis Geologorum Poloniae*, **46,** 3–37.

ŚLĄCZKA, A. (ed.) 1976. Atlas of paleotransport of detrital sediments in the Carpathian–Balkan mountain system. *Instytut Geologiczny, Wydawnictwa Geologiczne*, Warsaw, Poland.

ŚLĄCZKA, A. 1996. Oil and gas in the northern Carpathians. *In:* WESSELY, G. & LIEBL, W. (eds) *Oil and Gas in Alpidic Thrustbelts and Basins of Central and Eastern Europe*. Special Publications of the European Association of Geoscientists and Engineers, **5**, Geological Society, London, 187–195.

ŚLĄCZKA, A., OSZCZYPKO, N., MALATA, E. & CIESZKOWSKI, M. 1999. An early history of the Outer Carpathian basin. *Geologica Carpathica*, **50**, 170–172.

SŁOMKA, T. 1986. Statistical approach to study of flysch sedimentation – Kimmeridgian–Hauterivian Cieszyn Beds, Polish Outer Carpathians. *Annales Societatis Geologorum Poloniae*, **56**, 227–336.

SŁOMKA, T. 2001. Early Cretaceous debris flow deposits in the Cieszyn Beds of the Żywiec region (Carpathians, Poland) (In Polish, English summary). *Kwartalnik AGH – Geologia*, **27**, 89–110.

SOTÁK, J. 1986. Stratigraphy and typlogy of the Upper Triassic development in outer units in the West Carpathians (reconstructions from redeposites localised in the Silesian Cordillera area). *Knihovnicka Zemny Plyn a Nafta*, **31**, 1–53.

SOTÁK, J. BIROŃ, A., PROKEŠOVA, R. & SPIŠIAK, J. 2000. Detachment control of core complex exhumation and back-arc extension in the East Slovakian Basin. *Slovak Geological Magazine*, **6**, 130–132.

STAMPFLI, G. M. 1996. The Intra-Alpine terrain; a Paleotethyan remnant in the Alpine Variscides. *In:* SCHMID, S. M., FREY, M., FROITZHEIM, N., HEILBRONNER, R. & STUENITZ, H. (eds) Alpine Geology; Proceedings of the Second Workshop. *Eclogae Geologicae Helvetiae*, **89**, 13–42.

STAMPFLI, G. M. (ed.) 2001. *Geology of the western Swiss alps – a guide book*. Mémoires de Géologie (Lausanne), **361**.

STAMPFLI, G., MARCOUX, J. & BAUD, A. 1991. Tethyan margins in space and time. *In:* CHANNELL, J. E. T., WINTERER, E. L. & JANSA, L. F. (eds) *Palaeogeography, Palaeoclimatology, Palaeoecology*, **87**, 373–409.

ŠVÁBENICKÁ, L., BUBÍK, M., KREJCI, O. & STRANÍK, Z. 1997. Stratigraphy of Cretaceous sediments of the Magura group of Nappes in Moravia (Czech Republic). *Geologica Carpathica*, **48**, 179–191.

TARI, G., DICEA, O., FAULKERSON, J., GEORGIEV, G., POPOV, S., STEFANESCU, M. & WEIR, G. 1997. Cimmerian and Alpine stratigraphy and structural evolution of the Moesian Platform (Romania/Bulgaria). *In:* ROBINSON, A. G. (ed.) *Regional and Petroleum Geology of the Black Sea and Surrounding Regions*. AAPG Memoir, **68**, 63–90.

TCHOUMATCHENKO, P. & SAPUNOV, I. 1994. Intraplate tectonics in the Bulgarian part of the Moesian Platform during the Jurassic. *Geologica Balcanica*, **24**, 3–12.

TYSZKA, J. 1994. Response of Middle Jurassic benthic foraminiferal morphogroups to disoxic/oxic conditions in the Pieniny Klippen Basin, Polish Carpathians. *Palaeogeography, Palaeoclimatology, Palaeoecology*, **110**, 55–81.

TYSZKA, J. 2001. Microfossil assemblages as bathymetric indicators of the Toarcian/Aalenian 'Fleckenmergel'-facies in the Carpathian Pieniny Klippen Belt. *Geologica Carpathica*, **52**, 147–158.

VAN DER VOO, R. 1993. *Paleomagnetism of the Atlantic, Tethys and Iapetus*. Cambridge University Press, Cambridge.

VAŠIČEK, Z., MICHALÍK, J. & REHÁKOVÁ, D. 1994. Early Cretaceous stratigraphy, paleogeography and life in Western Carpathians. *Beringeria*, **10**, 1–170.

WIERZBOWSKI, A. & REMANE, J. 1992. The ammonite and calpionellid stratigraphy of the Berriasian and lowermost Valanginian in the Pieniny Klippen Belt (Carpathians, Poland). *Eclogae Geologicae Helvetiae*, **85**, 871–891.

WINKLER, W. & ŚLĄCZKA, A. 1994. A Late Cretaceous to Paleogene geodynamic model for the Western Carpathians in Poland. *Geologica Carpathica*, **45**, 71–82.

WITHJACK, M. O., SCHLISCHE, R. W. & OLSEN, P. O. 1998. Diachronous rifting, drifting, and inversion on the passive margin of central Eastern North America: an analog for other passive margins. *Bulletin AAPG*, **82**, 817–835.

WORTMANN, U. G., WEISSERT, H., FUNK, H. & HAUCK, J. 2001. Alpine plate kinematics revisited: The Adria problem. *Tectonics*, **20**, 134–147.

ZDANOWSKI, P., BASZKIEWICZ, A. & GREGOSIEWICZ, Z. 2001. Facies analysis of the uppermost Jurassic and the Lower Cretaceous deposits in the Zagorzyca region (southern Poland) (in Polish, English summary). *Przegląd Geologiczny*, **49**, 161–178.

ZIEGLER, P. A. 1988. Evolution of the Arctic-North Atlantic and the Western Tethys. *Memoir AAPG*, **43**, 1–198.

ZONENSHAIN, L. P., KUZMIN, M. L. & NATAPOV, L. N. 1990. Geology of the USSR: a plate-tectonic synthesis. *In:* PAGE, B. M. (ed.) *Geodynamics Series*, Vol. **21**, Washington, D. C., American Geophysical Union, 1–242.

ŻYTKO, K. 1984. The Atlantic, the Indian Ocean and main linear fracture zones of the post-Variscan Europe. *Annales Societatis Geologorum Poloniae*, **52**, 3–38.

ŻYTKO, K. 1985. Some problems of a geodynamic model of the Northern Carpathians. *Kwartalnik Geologiczny*, **29**, 85–108.

ŻYTKO, K. 1999. Correlation of the main structural units of the Western and Eastern Carpathians, [in Polish: English summary]. *Prace Państwowego Instytutu Geologicznego*, **168**, 135–164.

Partitioning pre-, syn- and post-Variscan deformation in the Holy Cross Mountains, eastern Variscan foreland

J. LAMARCHE[1], M. LEWANDOWSKI[2], J.-L. MANSY[3] & M. SZULCZEWSKI[4]

[1]*GeoForschungsZentrum Potsdam, PB 4.3, Telegrafenberg C427, 14473 Potsdam, Germany*
[2]*Institute of Geophysics, Polish Academy of Sciences, Ks. Janusza 64, 01-452 Warsaw, Poland*
[3]*Sédimentologie et Géodynamique, USTL; SN5; 59655 Villeneuve d'Ascq Cedex; France*
[4]*Institute of Geology, Warsaw University, Al. Żwirki i Wigury 93, PL 02-089 Warsaw, Poland*

Abstract: In this study we demonstrate how a combined structural, sedimentological and palaeomagnetic approach provides a new perspective on the tectonic evolution of the Holy Cross Mountains. In the field, we performed a structural and sedimentological analysis of Palaeozoic rocks. Our analysis was complemented by a palaeomagnetic study and by the restoration of balanced cross sections in Palaeozoic and Mesozoic rocks. Different steps of deformation were restored for a *c*.350 Ma period. (1) The extensional tectonics of the Devonian basin was unravelled: the resulting normal fault system constituted the fundamental structural control for the later Variscan tectonic inversion and Alpine deformations. (2) The style of Variscan folding is characterized and quantified by way of a cross section across the Holy Cross Mountains. (3) The role of the reactivation of Variscan faults during the Permo-Triassic initiation of the Polish Basin was examined. (4) The localized Alpine compressive deformation was quantified and shown to contribute only to a minor degree to the present-day state of deformation in the Holy Cross Mountains. The Holy Cross Fault zone is the product of the interplay of changing transtensional and transpressional settings during the Variscan diastrophic cycle, with the final effect of the Variscan evolution being the flower-like structure of the Holy Cross Fault zone.

The Palaeozoic massif of the Holy Cross Mountains represents the easternmost exposure of the Variscan domain in Western and Central Europe (Fig. 1), located in a complex area at the eastern termination of the Variscan domain and the margin of the East European Craton (Berthelsen 1992; Pożaryski *et al.* 1992; Pożaryski & Karnkowski 1992; Pharaoh & Bayer 1999; Dadlez 2001). A long history of geological studies has not led to a common opinion on the tectonic evolution of the Holy Cross Mountains area. Particularly heated discussions, althouth far from reaching a consensus, were focused on the impact of Late Carboniferous versus Early Devonian tectonic movements on the present-day structural framework of the mountains (e.g. Stupnicka 1992; Dadlez *et al.* 1994; Mizerski 1995; Znosko 1996, 2001).

Since it is deeply buried beneath Permian–Mesozoic rocks, the structure of the transition zone between the Variscan and Alpine orogenic domains and the stable palaeo-continent (known as Baltica) in Poland is poorly known, although intensive studies over more than ten years have generated numerous controversial publications (see Pożaryski 1975; Blundell *et al.* 1992; Berthelsen 1993; Pharaoh *et al.* 1997; Thybo *et al.* 1999; Franke & Zelaźniewicz 2000). The current view on the structure of the Palaeozoic basement along Baltica is the *terrane hypothesis* (Pozaryski *et al.* 1992; Franke 1995). In southern Poland, at least three crustal blocks are recognized: the Łysogóry, Małopolska and Upper Silesian blocks, which were amalgamated during the Late Palaeozoic. The controversial origin of the crustal blocks (Baltica- versus Gondwana-derived), the precise timing of their docking along the margin of Baltica (Lewandowski 1993; Nawrocki 1995; Bełka 2000) and the palaeogeography between con-

From: McCann, T. & Saintot, A. (eds) *Tracing Tectonic Deformation Using the Sedimentary Record.* Geological Society, London, Special Publications, **208**, 159–184. 0305-8719/03/$15.00

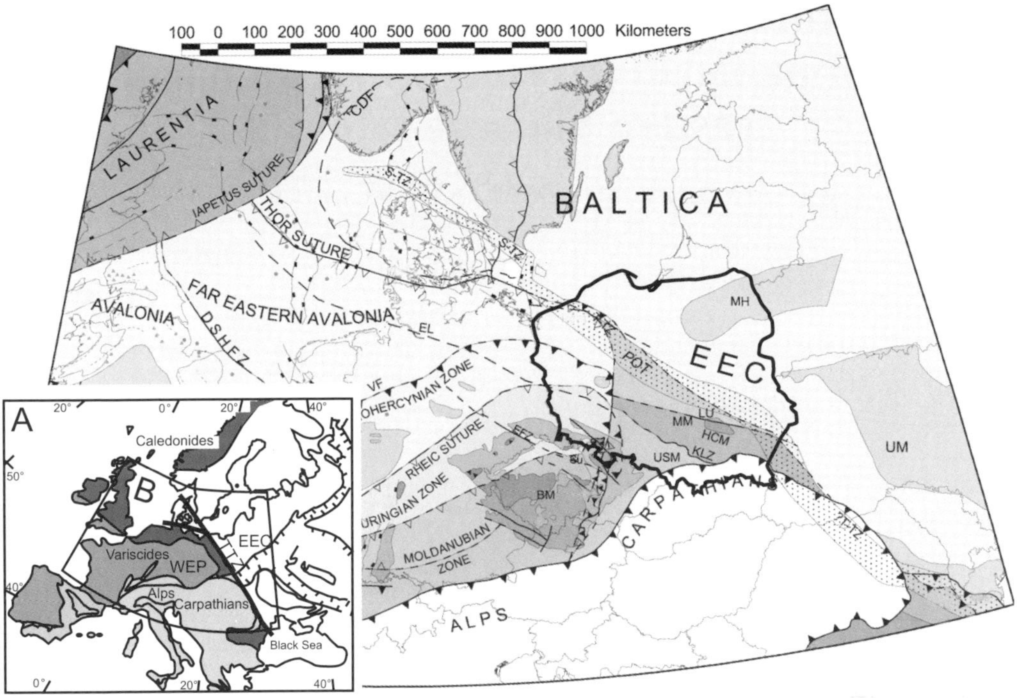

Fig. 1. (**a**) Tectonic overview map of Europe (modified after Gee, D. G. & Zeyen, H. J., 1996), WEP, West European Platform; EEC, East European Craton; (**b**) Palaeozoic basement of Northern and Central Europe (according to the PACE Working Group). Poland is indicated by its border (bold black line). BM, Bohemian Massif; CDF, Caledonian Deformation Front; EFZ, Elbe Fault Zone; EL, Elbe Line; HCM, Holy Cross Mountains; LU, Łysogóry Unit; MH, Mazurska High; MM, Małopolska Massif; POT, Polish Trough; STZ, Sorgenfrei–Tornquist Zone; Su, Sudetes; TTZ, Teisseyre–Tornquist Zone; UM, Ukrainian Massif; VF, Variscan Front.

tinents and micro-plates before the Late Carboniferous (wide oceans versus narrow basins) is still a matter for debate (see Bełka *et al.* 2000; McKerrow *et al.* 2000; Tait *et al.* 2000).

The Łysogóry and Małopolska blocks, separated by a fracture zone identified as the Holy Cross Fault zone, crop out in the Holy Cross Mountains. Thus this region is of major importance for understanding Variscan geology. After Early Palaeozoic extension and before Variscan shortening, a relatively homogeneous Devonian carbonate platform developed on both crustal units. However, the Devonian tectonic conditions of the area are not clear, and whether the Devonian basin sealed the terranes or developed on mobile, still drifting terranes is unclear (see Lewandowski 1993; Bełka *et al.* 2000). In addition, limited data are available because of subsequent erosion, Permo-Mesozoic extension and Alpine basin inversion which caused further superimposed deformation. Hence, one goal of this study is to decipher the changing tectonic context from Early Palaeozoic drifting of exotic crustal blocks to Late Palaeozoic amalgamation and consolidation of the Central European basement. As an area of repeated deformation during different epochs in the Phanerozoic, the Holy Cross Mountains represent a good target for studying the role of structural inheritance deriving from the Variscan tectonic inversion of the basin and from Mesozoic and Cenozoic deformations. In addition, the depositional architecture of the Devonian–Carboniferous stratigraphic succession can be used as a good indicator of pre-Variscan tectonic settings. In order to unravel the polyphase tectonic evolution, we combined a structural analysis of the Palaeozoic rocks with a sedimentological characterization. Both approaches were applied on meso- to basin-scale features. Discrimination between Variscan and Alpine deformations was made from field analysis, combined with balancing of geological cross-sections in the

Palaeozoic massif and Permo-Mesozoic cover. Good exposures of Late Palaeozoic rocks enabled the reconstruction of a palaeogeographical profile parallel to a structural cross-section. The effect of sedimentary patterns and early faults on later deformations was highlighted by comparison of the palaeogeographical and structural profiles. We also used palaeomagnetic data to address issues relating to both the tectonic rotations and the age of folding, and we developed a coherent picture of the post-Silurian tectonic evolution of the Holy Cross Mountains.

Outline of the geology

The Holy Cross Mountains are commonly divided into two units or regions: the Łysogóry Unit to the north and the Kielce Unit to the south (Czarnocki 1957), also called the *North Holy Cross Mts* and *South Holy Cross Mts*, respectively (Fig. 2). According to Pożaryski (1977), the Kielce Unit structurally belongs to the Małopolska Massif, which extends southwards to the boundary with the Upper Silesian Massif (Buła *et al.* 1997). The WNW–ESE-oriented Holy Cross Fault separates the Łysogóry Unit and the Kielce Unit. The Holy Cross Fault, inferred from geological mapping but not accessible in the field for direct observation, is interpreted as a shallow, south-verging thrust fault (Stupnicka 1988, 1992), a steep normal fault (Mizerski 1979; Znosko 1983) or a system of several faults (Znosko 1996). In the present study, the Holy Cross Fault is included in a fault array, the so called 'Holy Cross Fault zone'.

The Devonian–Lower Carboniferous stratigraphic sequence constitutes a distinct high-level depositional cycle in the Holy Cross Mountains. In the Łysogóry Unit, the stratigraphic succes-

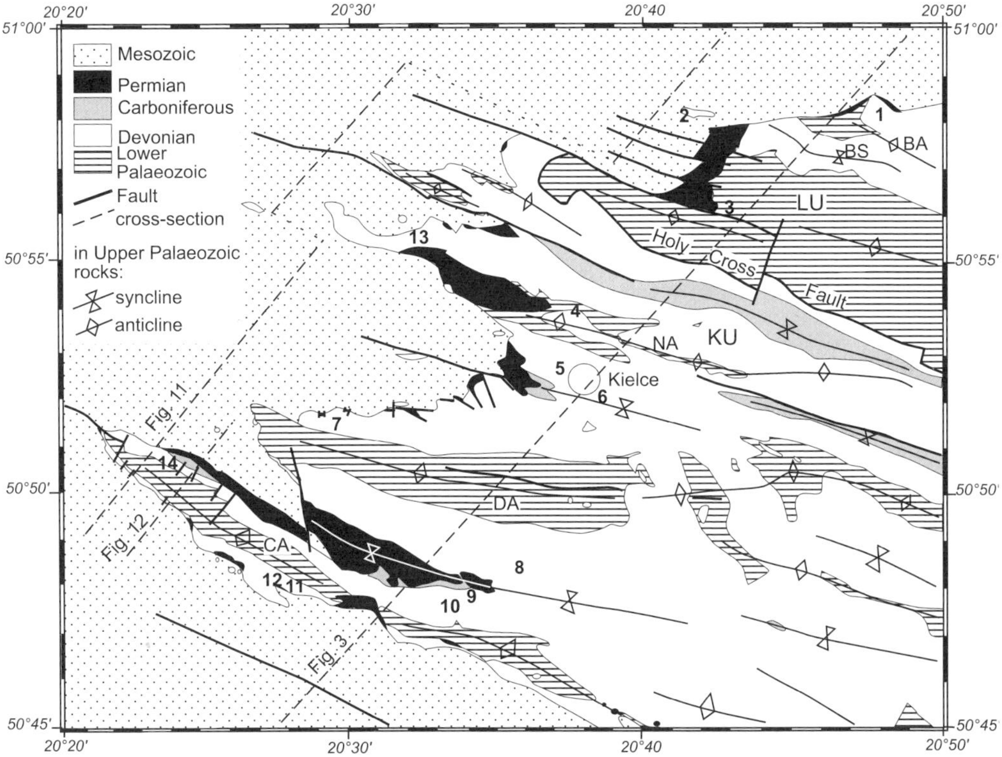

Fig. 2. Simplified geological map of the Western Holy Cross Mountains after Czarnocki (1938) with the location of outcrops described in the text and the location of the cross-sections displayed in Figures 3, 11 and 12. 1, Bukowa; 2, Zachełmie; 3, Wiśniówka; 4, Gruchawka; 5, Czarnów; 6, Wietrznia; 7, Jaworznia; 8, Trzuskawica; 9, Kowala; 10, Radkowice; 11, Chęciny Castle; 12, Rzepka; 13, Kostomłoty; 14, Ostrówka; KU, Kielce Unit; LU, Łysogóry Unit; BA, Bronkowice Anticline; BS, Bodzentyn Syncline; NA, Niewachów Anticline; DA, Deminy Anticline; CA, Chęciny Anticline.

sion is continuous from the Silurian to the Devonian. The first Devonian sediments have been interpreted to either rest conformably on the Silurian rocks (Stupnicka 1992 and previous authors), or with an angular unconformity of 15° to 20° (Dadlez *et al.* 1994; Znosko 1996). Mariańczyk (1973) described Lower Devonian conglomerates sealing Caledonian folds in the Silurian rocks. However, the documentation provided by Mariańczyk (1973) was poor and the unconformity has not been identified elsewhere in the Łysogóry Unit (cf. Jurewicz and Mizerski 1991).

In the Kielce Unit, the Devonian sedimentation started in Siegenian–Emsian times after a sedimentary gap (Tarnowska 1981), with the Devonian rocks unconformably covering the Lower Palaeozoic units. Although the degree of Early Palaeozoic deformation is still a matter of dispute, a pre-Emsian erosion of the uplifted Lower Palaeozoic rocks is obvious (Bednarczyk *et al.* 1970; Kowalczewski 1971; Tarnowska 1981; Głazek *et al.* 1981; Malec 1993; Szulczewski 1995). Palaeomagnetic studies (Schätz *et al.* 2002) suggest that the Lower Palaeozoic formations of the Kielce Unit had already reached their present-day position in the Late Ordovician and have remained stable with respect to Baltica ever since. However, other palaeomagnetic data sets (Lewandowski 1993; Szaniawski 1997; Grabowski & Nawrocki 1996, 2001) may point to vertical-axis rotations of the Devonian formations during Variscan deformation. Taking both sets of data at face value, they imply a tectonic contact between the Emsian rocks and the underlying formations, at least in some places in the Kielce Unit. It should be stressed that this hypothesis does not preclude a primary erosional character for the contact, as indicated by Głazek *et al.* (1981) and Malec (1993). Palaeomagnetic data collected by Nawrocki (2000) from allegedly Silurian diabases of the Bardo syncline do not support vertical-axis block rotations of the Kielce Unit during Variscan movements, but this result has been drawn from rocks of probable post-Silurian age (Migaszewski 2002) and of uncertain structural position.

In Emsian times, detrital sediments were homogeneously deposited in both units, followed by Eifelian shallow-marine carbonates. Starting from Givetian times, the sedimentary facies were spatially diversified due to synsedimentary tectonic activity (Szulczewski 1995). In the Holy Cross Mountains area, the stratigraphic record below the Variscan unconformity is preserved only up to the Upper Viséan. Younger rocks were eroded but can be found in cores farther to the north of the Holy Cross Mountains under the Mesozoic cover (Żakowa and Migaszewski 1995). In both units, the sedimentary record indicates the deepening upward of the basin from the Devonian to the Viséan, while during the Late Viséan this tendency was reversed and a regression is recorded (Szulczewski 1995).

During the Late Carboniferous, the Variscan shortening entailed the tectonic inversion of the Devonian basin, leading to folding of Lower and Upper Palaeozoic rocks. The older Carboniferous sedimentary rocks involved in the Variscan folding are of Viséan age (e.g. Ostrówka quarry) and the folds are unconformably covered by Upper Permian and Lower Triassic rocks. Therefore the deformation is attributed to the Variscan orogeny, an age which has been further constrained by palaeomagnetic methods (Lewandowski, 1981). The NNE–SSW-oriented shortening (Lamarche *et al.* 1999) led to WNW–ESE-trending folds (e.g. Stupnicka 1992; Mizerski, 1995) and to the reactivation of pre-existing faults, resulting in the main present-day structural trends in the Holy Cross Mountains.

Following the Variscan folding event, the Holy Cross Mountains region was exhumed and partially eroded. During Permo-Mesozoic times, the Holy Cross Mountains were located in the southeastern part of the Mid-Polish Trough (Kutek and Głazek 1972; Dadlez *et al.* 1995), which was tectonically inverted at the Cretaceous/Tertiary transition (Jaroszewski 1972; Kutek and Głazek 1972; Dadlez *et al.* 1995; Lamarche *et al.* 1998, Kutek 2001). The Permian–Mesozoic extension, as well as the later tectonic inversion of the area, involved additional brittle and ductile deformation of the Palaeozoic rocks of the Holy Cross Mountains (Lamarche *et al.* 1998).

Methodology

Structural cross-section

A structural cross-section striking NE–SW was constructed through the Holy Cross Mountains, sub-perpendicular to the main Variscan trend (Fig. 2). To elaborate the cross-section, we collected different types of data:

(1) structural data from the field measured mainly in quarries and in a smaller number of natural outcrops;
(2) local structural data described in the literature;
(3) geological maps of the Holy Cross Mountains in order to interpolate the structures between the quarries; and
(4) sedimentological and stratigraphic data of the Upper Palaeozoic rocks derived from the literature and from our own fieldwork.

Twenty-seven quarries provide the field observations over the western Holy Cross Mountains, among which 14 are described in this paper (see Appendix 1). Excavated rocks range in age from the Cambrian to the Carboniferous. In the quarries, we measured bedding planes, fold axes, fault planes, kinematic indicators and thicknesses of the formations. We noted the style, the vergence and the amplitude of the folding. In addition, we analysed the synsedimentary deformations and the relative chronology of the tectonic markers. In places where the quarries are located exactly along the cross-section, structural data were directly drawn on the profile, whereas, in the case of quarries located some kilometres away from the line, the structural data were projected on to the profile in a direction parallel to the axis of the main Variscan trend. In areas where Variscan and Alpine structures interfered, only the quarries close to the cross-section were taken into account. Although quarries located too far from the profile were not projected on to the cross-section, studying them provided useful qualitative and kinematic constraints.

The field observations were complemented by tectonic data from the literature describing outcrops which presently are not accessible. We took into account the age of the rocks and the bedding characteristics, as well as the style and vergence of the folds and thrusts. In particular, we used the following publications: Tomczyk & Turnau-Morawska (1964), Bednarczyk *et al.* (1970), Znosko (1974, 1996, 2001), Kowalczewski (1976), Kowalczewski and Studencki (1983), Kłossowski (1985), Kowalczewski *et al.* (1986), Stupnicka (1988; 1992), Jurewicz & Mizerski, (1991), Malec (1993), Dadlez *et al.* (1994), Mizerski (1995), Orłowski and Mizerski (1995, 1996), Kowalczewski & Dadlez (1996), Kowalczewski *et al.* (1998) Bednarczyk & Stupnicka (2000).

In the areas where data were discontinuous along the cross-section, the dip and the age of the rocks were interpolated and drawn on the base of the structural interpretation of the available geological maps of the Holy Cross Mountains at various scales (Czarnocki, 1938; geological maps by Polish Geological Institute, scale 1:50 000).

The accuracy of the cross-section is given by the topographic level displayed on the profile (continuous black line, B in Fig. 3), by the number and location of quarries directly used to constrain the structures (10 out of 27 quarries analysed in the framework of this study), by the large amount of data from old to most recent literature, and by the quality of the geological maps used to complete the gaps between the observation points.

Lithostratigraphic cross-section

On the basis of a lithostratigraphic cross-section, the deep structure of the Upper Palaeozoic rocks below the present topographic level and the structure of now eroded rocks above the present topographic level can be reconstructed. Quantitative constraints for litho- and biostratigraphic data of the Upper Palaeozoic sedimentary rocks were taken from field observations, not only in scarce quarries but all along the profile and over the Holy Cross Mountains, as well as from the following literature: Freyer & Żakowa (1967), Filonowicz (1968), Kazmierczak (1971), Szulczewski (1971, 1978), Szulczewski & Żakowa (1976), Głazek *et al.* (1981), Racki (1981, 1993), Żakowa (1981), Narkiewicz & Olkowicz-Paprocka (1983), Belka & Skompki (1988), Narkiewicz *et al.* (1990), Romanek & Rup (1990), Orłowski, (1992), Racki & Bultynck (1993), Matyja & Narkiewicz (1995), Szulczewski *et al.* (1996).

Six segments (Fig. 3, A to F) were constructed from north to south along the cross-section, each of them having constant facies and thickness:

- segment A corresponds with the thickest lithostratigraphic column representative of the whole Łysogóry Unit. It is characterized by a thick Lower Devonian clastic succession, shallow-marine platform carbonates restricted to the Eifelian, and carbonates and shales of the Middle and Upper Devonian.
- segment B starts from the Holy Cross Fault zone and extends southwards to the area of the Wietrznia quarry. It corresponds with the Kielce–Łagów (Central) Synclinorium of the Holy Cross Mountains. It is characterized by relatively thick Lower Devonian, and shallow-marine platform carbonates of mainly Middle Devonian age, as well as by basinal Upper Devonian and Lower Carboniferous rocks.
- segment C is restricted to the area of the Wietrznia quarry, located at the flank of the Dyminy Anticline. This segment is marked by a reduced Lower Devonian thickness, by attenuated, marginal platform carbonates in the Middle Devonian, by a condensed sequence restricted to the upper part of the Frasnian and the lower part of the Famennian, and by the expanded, basinal Famennian and Lower Carboniferous.
- segment D extends from the Dyminy Anticline to the centre of the Gałęzice Syncline. It is characterized by a thin Lower Devonian package, by thick carbonate platform sediments encompassing the Middle Devonian and the Frasnian, by the condensed and incomplete Famennian and Tournaisian, and

by a thick interval of the uppermost Lower Carboniferous.
- segment E extends from the centre of the Gałęzice Syncline to the northern limb of the Chęciny Anticline. The main difference to segment D is the expanded thickness of the Frasnian and the Famennian, developed in basinal, carbonate and shaly facies.
- segment F extends to the south of segment E. It is similar to segment E, but with a slightly thicker Eifelian layer.

A comparison of the lithological columns of segments A to F indicates strong and sudden facies and thickness variations, suggesting synsedimentary tectonic activity during the deposition of the Devonian and a difference in basin evolution for the Łysogóry Unit and the Kielce Unit.

Palaeomagnetic analysis

Palaeomagnetic studies enable the detection of crustal block movements with respect to a reference continent (Baltica in our study) if coeval palaeo-poles for both units can be determined from a characteristic remanent magnetization (NRM). Additionally, a secondary palaeomagnetic component, if superimposed on tectonically deformed rocks, may serve as a folding age indicator by comparison of the calculated palaeo-pole with a time-calibrated apparent polar wander path (for more about palaeomagnetic applications in tectonics, see, for example Butler 1992).

Key geological points

Key outcrops were used for constructing the cross-section. The detailed description of structures and kinematic indicators as well as synsedimentary tectonic features in quarries along and across the cross-section can be found in the Appendix. In this chapter, only the main structures are summarized in Table 1 for quarries along the cross-section from north to south. The structures of each quarry are schematically drawn at their respective locations in Fig. 3, and localized in Fig. 2 by their numbers (in brackets).

Interpretation of the geological cross-section

Main structural features

The synthetic structural cross-section through the Holy Cross Mountains is shown in Fig. 3. The main feature of the Variscan deformation is an alternation of synclines and anticlines, involving Devonian sedimentary rocks and showing a wavelength of five to ten kilometres. In most cases, the anticlinal folds display a flat northern and a steeper southern limb resulting in a slight vergence to the south. The Variscan folding is disharmonic, implying a decoupling level under the main competent unit, which can occur at the boundary between the Upper and Lower Palaeozoic complexes. Such a major decoupling can be inferred from the palaeomagnetic data. Indeed, as the Mid–Late Ordovician formations yielded an Ordovician palaeo-pole close to the Ordovician palaeo-pole for Baltica (Schätz *et al.* 2002), while at the same time the overlying Devonian rocks were rotated clockwise during Variscan deformation (Lewandowski 1993; Grabowski & Nawrocki 1996; Szaniawski 1997), the assumption of an intervening detachment plane is inevitable.

Although being regular in general, in detail the wavelength of folds is perturbed, firstly, in the area of the Holy Cross Fault zone, where the anticline is of larger amplitude with a faulted southern limb, as well as, secondly, in the southernmost part of the section, where a northwards vergence along major reverse faults is observed. In this area, south of the Chęciny Anticline the folds and faults affecting the Mesozoic layers indicate that Alpine deformation contributed significantly to the present-day state of deformation. A certain asymmetry in the structural pattern between the Kielce and the Łysogóry Units is observed. In particular, many faults occur in the Kielce Unit on the flanks of the main folds. One reason could be the differences in basin development throughout the whole of the Palaeozoic and particularly in the Devonian. This is evidenced by the number of subordinate palaeogeographical units, which developed due to a higher tectonic mobility of the Kielce Unit (units B–F; Fig. 3), contrasting with rather stable basin development in the Łysogóry Unit (unit A; Fig. 3). If the main faults are compared with the extent of the paleogeographic segments A to F, a good spatial correlation can be recognized. In consequence, it is proposed that the pattern of Devonian synsedimentary faults was reactivated during Variscan shortening that influenced the size and the shape of the folds. In particular, the Holy Cross Fault zone may have been a major boundary during the development of the Devonian basin.

Although detailed analysis of the Early Palaeozoic deformations was not a primary goal of this study, we have depicted structural data for the Lower Palaeozoic (Figs 3 & 11) collected from available literature (Czarnocki 1938; Tomczyk & Turnau-Morawska 1964; Bednarczyk

Table 1. *Main structures in quarries used to constrain the cross-section of Figure 3.*

Number in Fig. 2, 3	Name of quarry	Figure	Age of rocks	Average structure		Additional structures
				Bedding	Dip	
	Bronkowice Anticline		Devonian Silurian	Large scale anticline S-verging detached folds (Znosko, 1996)		Vertical limb
1	Bukowa	Fig. 4	Emsian	N110°	50°N to 10°N	noraml synsedimentary faults
2	Zachełmie		Middle Devonian	N100°	40–45°N	Variscan unconformity
3	Wiśniówka		Upper Cambrian to Tremadocian	N110°	70°N	Tectonic slices
4	Gruchawka		Lower Devonian Upper Silurian		35–40N° 55°N	10° angular unconformity
5	Czarnów	Fig. 6	Frasnian, Famennian	Asymmetrical fold		
6	Wietrznia	Fig. 9	Late Givetian to Famennian	N090°	50-50°N	WNW–ESE normal synsedimentary faults
7	Jaworznia		Frasnian	gentle folds		Variscan unconformity
8	Trzuskawica		Frasnian, Famennian	gentle folds		Normal faults
9	Kowala	Fig. 7	Frasnian, Famennian	N070°	40°N to 80°N	Slumps
10	Radkowice		Eifelian	N095°	80°N	
11	Chęciny		Givetian, Frasnian	N110°	70°N to 80°N	
12	Rzepka		Givetian, Frasnian	N110°	25°N	
13	Kostomłoty	Fig. 5	Frasnian	E-W folds		
14	Ostrówka	Fig. 8	Frasnian to Upper Visean	N100°	15°N to 70°N	Normal synsedimentary faults

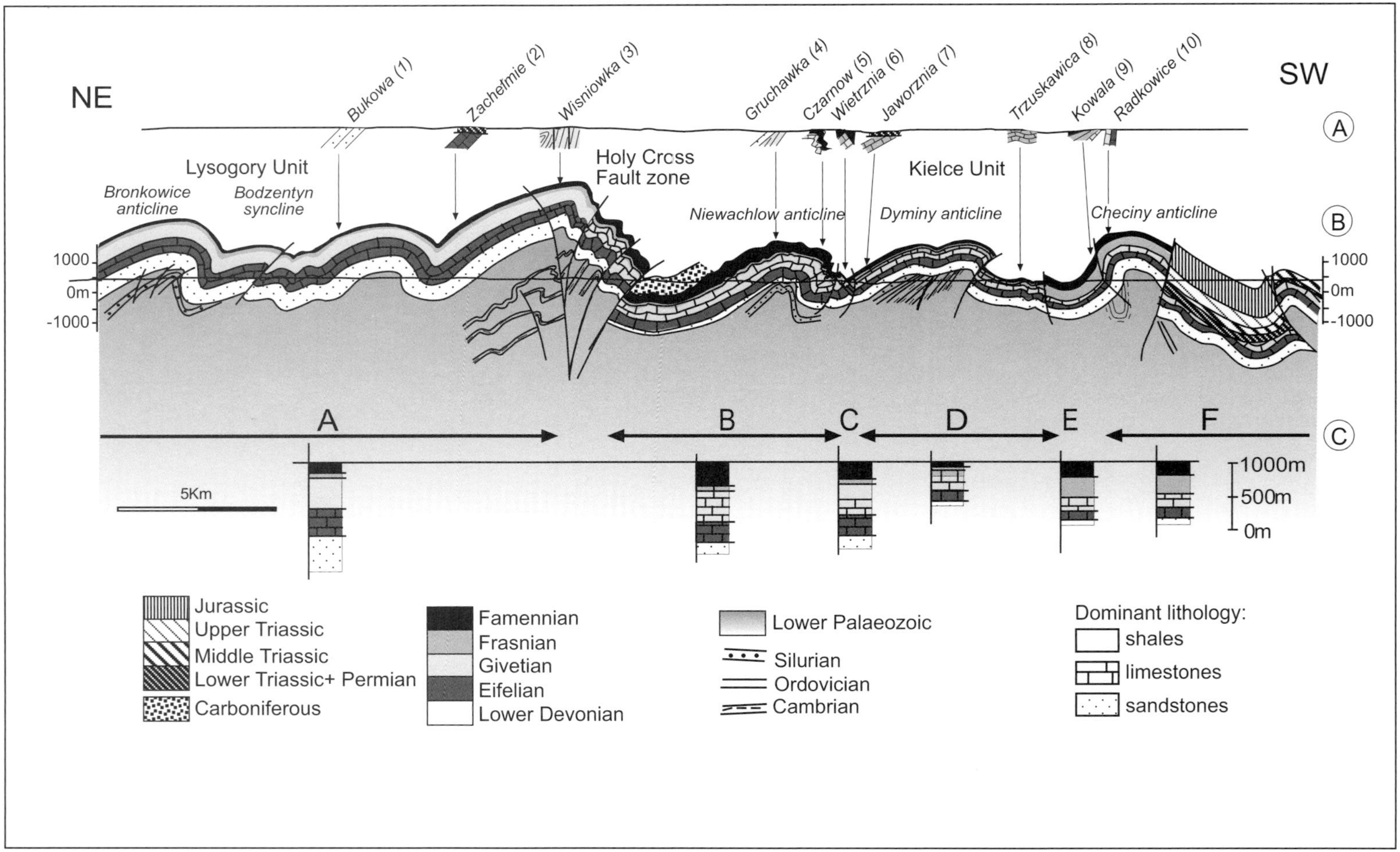

Fig. 3. NE–SW-oriented cross-section of the Western Holy Cross Mountains. (**a**) Schematic view of the outcrops located along the cross-section (vertical arrows) and described in the text. (**b**) Interpretative geological cross-section of the Holy Cross Mountains. (**c**) Lithological columns of segments A to F and their extension along the geological cross-section B (horizontal black arrows).

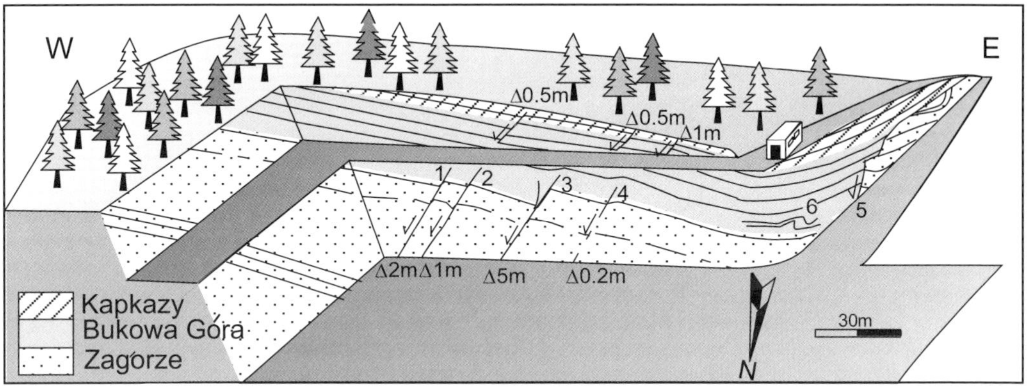

Fig. 4. Sketch of the northwestern part of the Bukowa quarry (Łysogóry Unit, latitude 50°58′; longitude 20°48′, 1 in Fig. 2) exploiting Emsian Zagórze, Bukowa and Kapkazy formations. The southern wall of the quarry displays normal synsedimentary faults (1 to 4). On the entrance path, a normal synsedimentary fault (5) and associated slump structures (6) can be seen (see explanations in the text).

et al. 1970; Kłossowski 1985; Jurewicz & Mizerski 1991; Malec 1993; Kowalczewski & Dadlez 1996; Orłowski & Mizerski 1996; Kowalczewski *et al.* 1998; Znosko 2001) and our own measurements. It is obvious from the cross-section (Fig. 3) that the general pattern of the deformations which occurred within the Lower Palaeozoic mimics the Variscan geometry. For instance, tight anticlines in the Silurian rocks can be observed beneath the Bronkowice and Niewachlów anticlines. Similarly, a tight syncline developed in Ordovician rocks (Znosko 2001) in the Brzeziny Syncline (the southernmost part of the cross-section, see Fig. 3). Although Znosko (2001) considers syn-Caledonian movements to be responsible for folding of the Ordovician in Brzeziny, we are treating this conclusion with caution when it is applied to the similarity of structural trends between deformed Ordovician and Devonian rocks in the Brzeziny area. For instance, closer examination of the map by Czarnocki (1938) makes it possible to conclude that fold axis undulations, inferred by Znosko (2001) for folded Ordovician in the Brzeziny Syncline, are also observed for the Devonian in the same area. Thus, in our opinion, the folds in the Lower Palaeozoic rocks of these two areas can be at least partly attributed to the Variscan deformation. Observed unconformities could be considered an effect of disharmony due to differences in competency between the Lower and Upper Palaeozoic complexes. At least the present-day structural pattern does not require pre-Devonian folding, but only pre-Devonian uplift and erosion. In this context, therefore, it is worthwhile to further study the nature of the well-documented erosional unconformity between the Lower and Upper Palaeozoic rocks (Bednarczyk *et al.* 1970; Kowalczewski 1971; Tarnowska 1981; Głazek *et al.* 1981; Malec 1993; Szulczewski 1995).

Variscan polyphase deformation

The style of large-scale deformation indicates slight southwards vergence of the main Variscan folding. However, we have a more complicated sequence of deformation from our observations in the quarries. For instance, in the Czarnów quarry, early north-verging ramps predate the main south-verging folding (Fig. 6). Similarly, small-scale symmetrical folding in the Kostomłoty quarry predates the general tilting of 20° to the south (Fig. 5), which may result from the formation of the main Niewachlów anticline (Fig. 3). Therefore, although the main Variscan signature is the slightly south-verging folding, a polyphase deformation can be deduced, especially marked by an early phase of shortening, locally being north-verging. As a consequence and in spite of the main south-verging, the geometry of the folds in the Upper Palaeozoic rocks does not necessarily indicate a south-vergent large-scale thrust, as postulated by Stupnicka (1992). In contrast to Stupnicka (1992), our structural interpretations are derived from the presence of disharmonic folding and steep faults. The NNE–SSW-oriented shortening of a strike-slip pre-faulted domain can also explain such a peculiar polyphase and polyvergence structural setting.

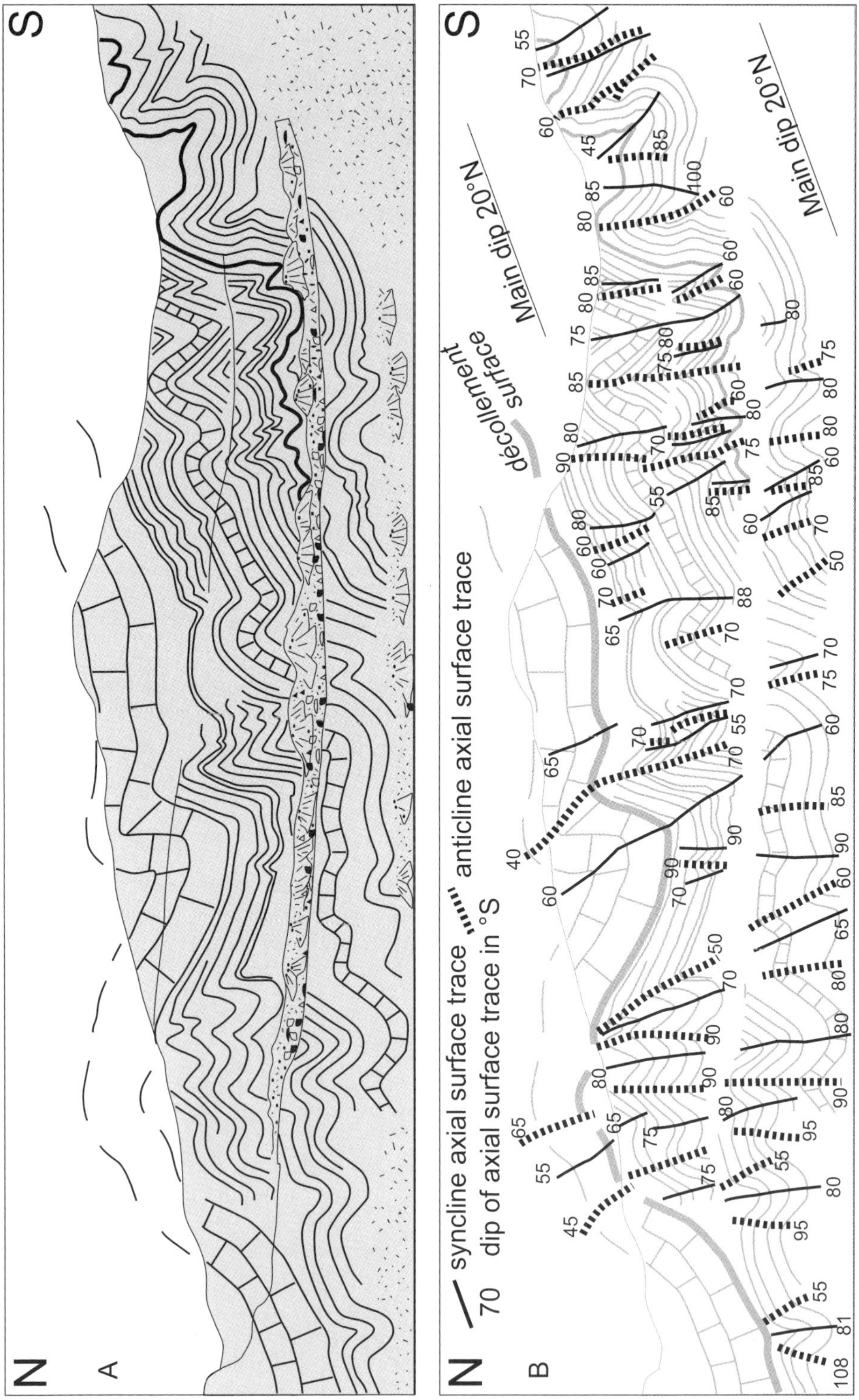

Fig. 5. Sketch of the eastern wall of Kostomłoty quarry (Lat. 50°33′ Long. 20°21′, 13 in Fig. 2) displaying Givetian–Frasnian rocks. The east–west-trending folds are Variscan in age. The mean dip of the folds, axial surface trace is 70°S and the average bedding dip 20°N.

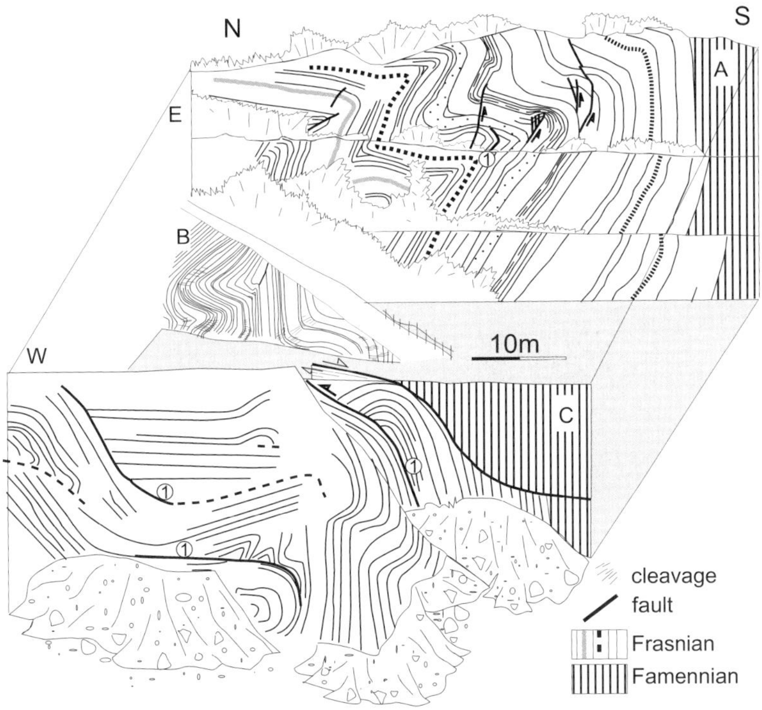

Fig. 6. Three north–south-oriented cross-sections through the Czarnów quarry, displaying Frasnian and Famennian rocks (Kielce Unit, Lat. 50°30°; Long. 20.20°, 5 in Fig. 2). North-verging ramps (1) precede the major Variscan folding with east–west-oriented axes associated with a syn-fold cleavage. See the text for further discussion.

The Holy Cross Fault zone

In the contact zone between Łysogóry and Kielce units, the main structure is a large-scale, slightly asymmetrical anticline, south-vergent, with the northern limb affected by an array of faults arranged in a flower-like pattern (Fig. 3). The flower-like pattern suggests a strike-slip origin for the contact zone, which finds support in palaeomagnetic data (Lewandowski 1993). During the Devonian, the boundary between both units might have already been an active fault zone that generally controlled the facies distribution (cf. Szulczewski 1995). Hence, when the Variscan shortening affected the area, the tectonic inversion occurred favourably in the pre-deformed zone and a larger amount of deformation occurred around the Holy Cross Fault zone.

Deformation of the post-Variscan cover above the Holy Cross Fault indicates slight sinistral brittle reactivation by Alpine tectonics (Jaroszewski 1972; Lamarche 1999; Lamarche *et al.* 1999). However, as shown by the relationships of the Permo-Triassic cover to the underlying Palaeozoic rocks, most of the deformation is Variscan in age. In this scenario, although the Alpine strike-slip reactivation of the Holy Cross Fault is evident (Lamarche *et al.* 1999), the relative left lateral displacement between the Kielce Unit and the Łysogóry Unit is below palaeomagnetic resolution.

The strike-slip component

The small asymmetry of the folds, the relatively steep faults and the flower-like structure around the Holy Cross Fault zone suggest a Variscan strike-slip component between the Kielce Unit and the Łysogóry Unit. A N–S- to NNE–SSW-direction of Variscan shortening was deduced from the direction of fold axes and associated faults in the Holy Cross Mountains (Lamarche *et al.* 1999, 2002) being consistent with a relative dextral displacement of both units along the margin of Baltica. However, the magnitude and the time-span of such a strike-slip movement are disputed (Lewandowski 1993, 1994, 1995, 2000; Dadlez *et al.* 1994; Nawrocki 1995, 2000). In particular, the Emsian sandstones of the Kielce Unit yielded dual-polarity, although poorly clustered palaeomagnetic directions deviated towards the NW, from coeval palaeo-poles for Baltica which could be interpreted as an effect of the post-Emsian strike-slip movements (Lewandowski 1993). On the other hand, the Mid–Late Ordovician limestones from Mójcza village (Kielce Unit) yielded dual-polarity remanent magnetization that plots at the Ordovician segment of the apparent polar wander path (Schätz *et al.* 2002), although underlying Arenigian sandstones show characteristic NRM component rotated westerly by almost 60° (Lewandowski 1987), in line with the north-westerly twisted structural trend in the area. From the field, we gathered indicators for sedimentation in an active tectonic context in the Łysogóry Unit for the Emsian and in the Kielce Unit for the Frasnian–Famennian (see the examples of synsedimentary extension, in the Appendix and Figs 7–9). In addition, changes in the lithostratigraphic columns A to F (Fig. 3) reveal the spatial and temporal variation of uplifted/downwarped blocks implying long-term tectonic activity during the Devonian. As a consequence, though the magnitude of relative movements cannot be unequivocally resolved, it may be concluded that the strike-slip displacement between both units took place during the whole of the Devonian. Taking into account that most kinematic markers indicate normal faulting during the Devonian, the effective tectonic context was rather transtensional.

The continuous transtension may have entailed synsedimentary tectonics – as recorded in the Devonian sedimentary rocks of both units, but especially along the Holy Cross Fault zone and within the Kielce Unit, where the lithostratigraphic columns are highly differentiated. If

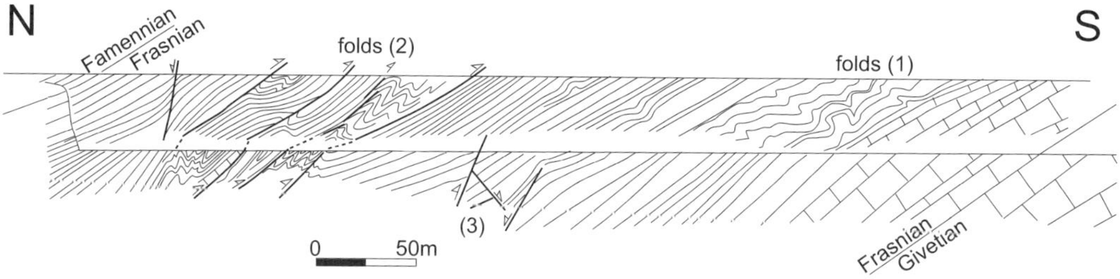

Fig. 7. Schematic cross-section of the eastern walls of the Kowala quarry showing Givetian to Famennian rocks (Kielce Unit, lat. 50°48′; long. 20°30′, 9 in Fig. 2). Synsedimentary structures as slumps (1) and synsedimentary faults (3)are observed, as well as folds (2) related to Variscan reverse faults.

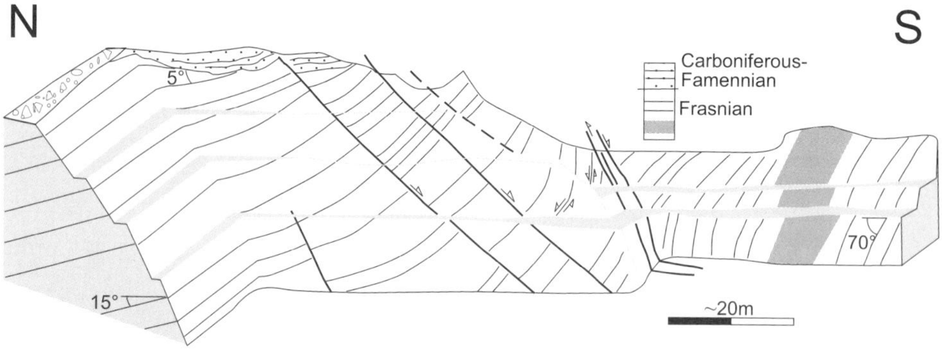

Fig. 8. View of the Ostrówka quarry (Kielce Unit, Lat. 50°30′; Long. 20°12′, 14 in Fig. 2) displaying Frasnian to Carboniferous rocks. The Frasnian rocks, dipping 15°N to 70°N are affected by normal synsedimentary faults, as well as by probable Variscan reverse faults. Tectonic activity during the Devonian is confirmed by an angular unconformity of 5° between Frasnian and Famennian.

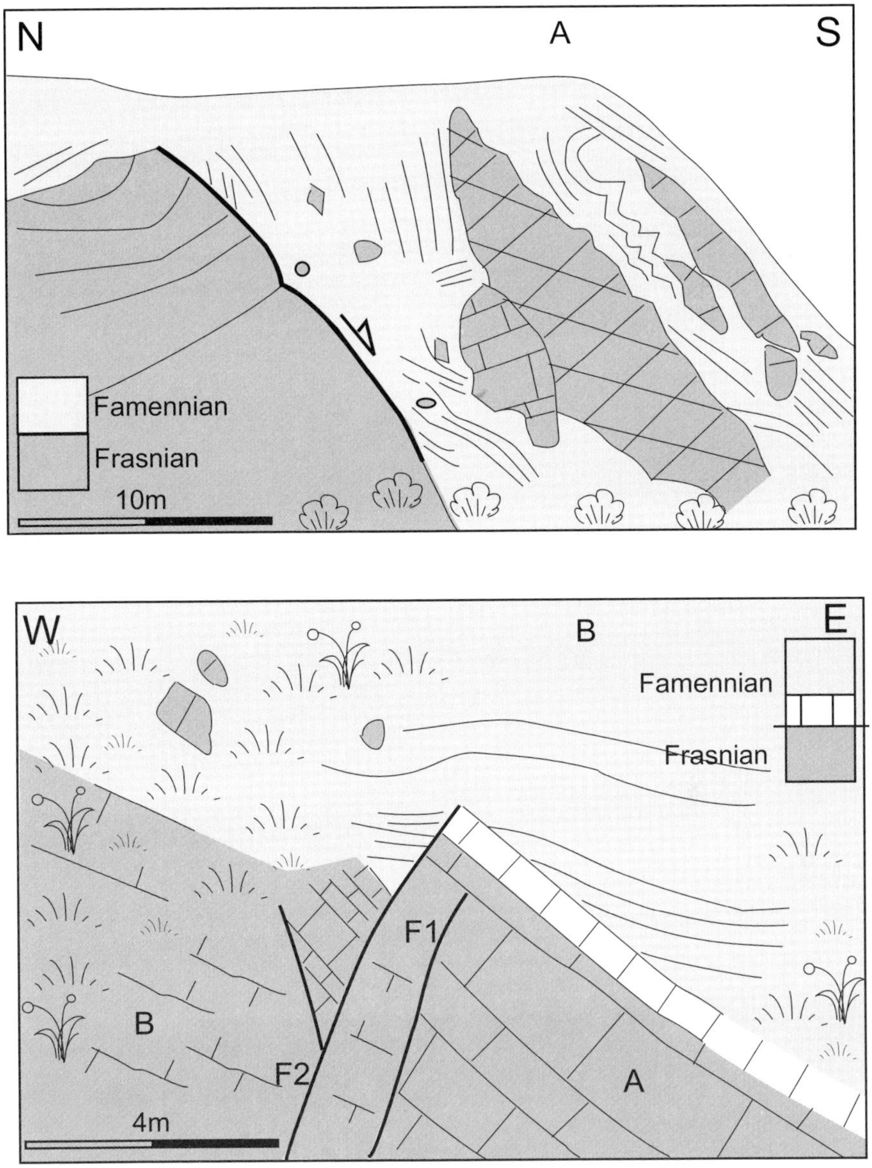

Fig. 9. Normal synsedimentary faults in the Wietrznia quarry (Lat. 50°3°; Long. 20°23°, 6 in Fig. 2). A, N105°-oriented fault affecting the Frasnian. The fault-related trough is filled with a breccia composed of Frasnian and Famennian fragments. B, N160–170°-oriented faults showing syn-Frasnian activity.

true, the strike-slip displacement between Łysogóry Unit and Kielce Unit may have ended with the final north–south to NNE–SSW Variscan shortening during the Late Carboniferous. Some evidence of a transtensional regime in the foreland of Baltica may be found also farther to the southeast of the Holy Cross Mountains (see Narkiewicz *et al.* 1998).

Differentiating Devonian, Variscan and Alpine deformations

As already mentioned, some Palaeozoic rocks of the Holy Cross Mountains were deformed not only due to the Variscan deformation, but also due to pre-Variscan transtension as well as post-Variscan and Alpine tectonics. As a consequence, the cross-section proposed in Figure 3 represents the present day state of Palaeozoic rocks, which is the superposition of pre-, syn- and post-Variscan deformations. Hence, the goal of the following structural analysis will be to estimate the percentage of Alpine versus Variscan deformations, and to restore the pre-folding state of the Late Palaeozoic rocks. Restoration of the different steps of deformation by balancing the cross-section is expected to unravel the tectonic conditions of the Devonian basin as well as the process of Variscan tectonic inversion and structural inheritance during Alpine tectonics.

Balancing Variscan and Alpine deformations

In order to balance the Variscan and Alpine deformations in the Holy Cross Mountains, we have to take into consideration the Lower Devonian to Famennian layers that can be traced from NE to SW (A in Fig. 10). Nevertheless, we have to keep in mind that the Famennian rocks are covered by Carboniferous and younger sedimentary rocks, and that, at least in the southwestern part of the section, the shape of the folds is significantly affected by Alpine deformation.

The deformation of the Palaeozoic rocks can be subdivided into folds and faults. Following our interpretative cross-section, the faults cut the limbs of the folds (A in Fig. 10). Thus, in a first step, we restored the displacement along the faults. As the Famennian is the youngest layer that can be traced over the entire cross-section, we restored the faults so that the displacement of the topmost Famennian becomes zero (B in Fig. 10). At that stage of the restoration, we obtained a continuous wavelength of folds, and two types of faults:

(1) Faults which have no pre-Famennian displacement (dashed faults, a, in Fig. 10).
(2) Faults which displace pre-Famennian layers (faults, b, in Fig. 10).

The first ones (1) correspond with faults developed not earlier than during Variscan deformation, whereas the second ones (2) can be interpreted as pre-existing faults which have been reactivated during Variscan deformation.

In a second step, in order to restore the supposed pre-Variscan folding position of the rocks, we balanced the folds until the top of the Famennian became horizontal, keeping, firstly, the length of layers, and secondly, the angular relationships between faults and layers, unchanged (C in Fig. 10). As a result, we obtained the supposed cross-section during pre-Variscan and post-Famennian times, which is marked by four main palaeogeographical domains (A, B, D and F, Fig. 10) separated by three fault zones: the Holy Cross Fault zone, and fault zones to the north and south of the Dyminy High (Fig. 10). A comparison of the palaeogeographical domains

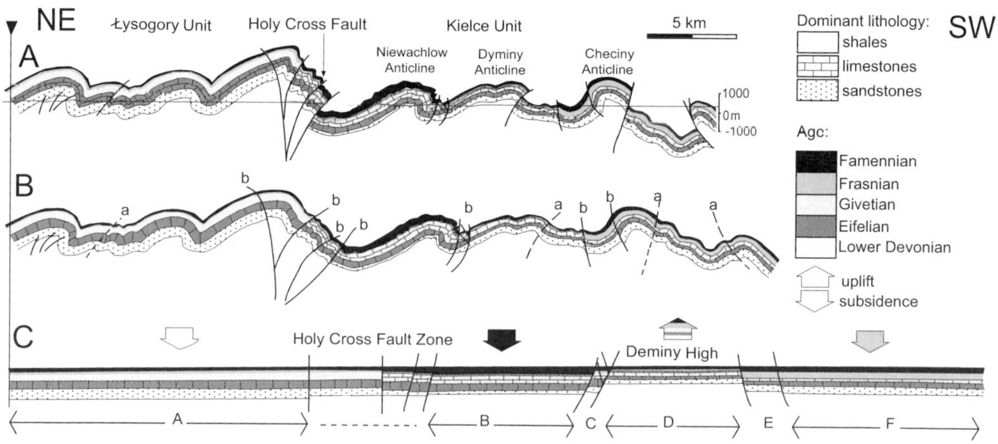

Fig. 10. Restoration of the pre-Variscan folding structure in the western Holy Cross Mountains. (**a**), Geological cross-section of Devonian rocks through the Holy Cross Mountains. (**b**), Restoration of the post-Famennian displacement along the faults. a, post-Famennian faults; b, reactivated pre-Famennian faults. (**c**), Balanced cross-section of Devonian rocks through the Holy Cross Mountains. and main periods of uplift/subsidence (arrows) in the segments A, B, D and F.

allows estimation of the tectonic conditions leading to the development of the Devonian basin in the Holy Cross Mountains (C in Fig. 10). We deduced that:

- During the Lower Devonian, the thicker deposits in the Łysogóry Unit and in the segment F reveal a more subsiding area of sedimentation compared with the centre of the basin.
- During the development of the Eifelian carbonate platform, the tectonic conditions were homogeneous, with the exception of segment D in which the reduced thickness indicates a relatively high position.
- The greater thickness and the shaly facies of the Givetian in the Łysogóry Unit shows a comparably higher subsidence rate of this unit during the Givetian.
- During the Frasnian the strongest subsidence occurred in segment F as shown by the thick shaly Frasnian, while in segments B and D the Frasnian was still partly marly and thin.
- The pattern of the Famennian shale thickness reveals a greater subsidence in segment B than elsewhere.
- From the Lower Devonian to the Famennian, thin and carbonate-rich sediments characterize segment D. This condensed succession indicates a permanent high position of the so-called Dyminy High, bounded by synsedimentary faults.

From this analysis it appears clear that, although the Dyminy High was a constant structure, the area of maximum subsidence was variable in time and space. This observation is typical for strike-slip dominated tectonic regimes, which is in agreement with the hypothesis of a strike-slip context during the Devonian, as inferred from the tectonic analysis and palaeomagnetic data. Our analysis also leads to the conclusion that the different basin evolution for the Łysogóry Unit and Kielce Unit and the tectonically active regime resulted in the main faulted zone between both units as well as within the Kielce Unit. Palaeomagnetic data (Lewandowski, 1981; Szaniawski, 1997) point to a diachroneity in the folding process, with Devonian formations located closer to Holy Cross Fault zone (Kostomłoty) being deformed earlier than those situated some 20 km to the south (Radkowice).

The Devonian tectonic pattern and the Variscan structures correlate well (cf. A and C; Fig. 10). The Holy Cross Fault zone is located at the boundary between the Łysogóry Unit and the Kielce Unit, which was a tectonically active zone during Devonian. Segment B correlates with the Niewachlów Anticline. A correlation was also noticed between segment D and the Dyminy Anticline, as well as between segment E and the Chęciny Anticline. Therefore, it can be concluded that the pre-existing Devonian synsedimentary faults were tectonically inverted due to Late Carboniferous shortening. The pre-existing main fault zones determined the location of most of the major Variscan folds.

Concerning the rate of Variscan shortening, we kept in mind that the balancing of the cross-section is a two-dimensional restoration in which lateral displacements are not constrained. We calculated a rate of apparent shortening of 15%, thereby taking the cumulative effect of both the Variscan and Alpine deformations into account. It is therefore necessary to differentiate between Alpine and Variscan contributions.

Alpine deformations

The tectonic input of Alpine age (Maastrichtian–Palaeogene tectonic inversion, see the section on the 'Outline of Geology' at the beginning of this paper) is evident in the southwest of the cross-section, where the Mesozoic rocks are folded and faulted (see the Chęciny Anticline). However, north of this area, the Permian–Mesozoic cover has been eroded, which complicates the balancing of the Alpine deformation along the entire section, although palaeomagnetic results from Kostomłoty (Lewandowski 1981) clearly indicated that Variscan folds in the centre of the Palaeozoic core of the Holy Cross Mountains were not remodelled during the Alpine cycle. As a consequence, we used a cross-section of the Permo-Triassic cover located some kilometres from the Holy Cross Mountains to the west, in order to portray the post-Variscan extension (Fig. 11) as well as a reduced cross-section across the Chęciny Anticline located to the west of the main cross-section to restore the Alpine folding (Fig. 12).

The Permo-Triassic cross-section consists of three segments which we juxtaposed along a single line to form an artificial continuous profile. Then, we superimposed this profile of the Permo-Triassic cover to the cross-section of Palaeozoic rocks (box in Fig. 11). Both the juxtaposition and the superposition on the Palaeozoic cross-section were projected in a parallel direction to the axis of the Variscan trend. In order to make the small Permo-Triassic structures observable, the vertical scale of the Permo-Triassic profile is exaggerated. Thus, we obtained a profile which is not a real cross-section but an artificial perspective profile in

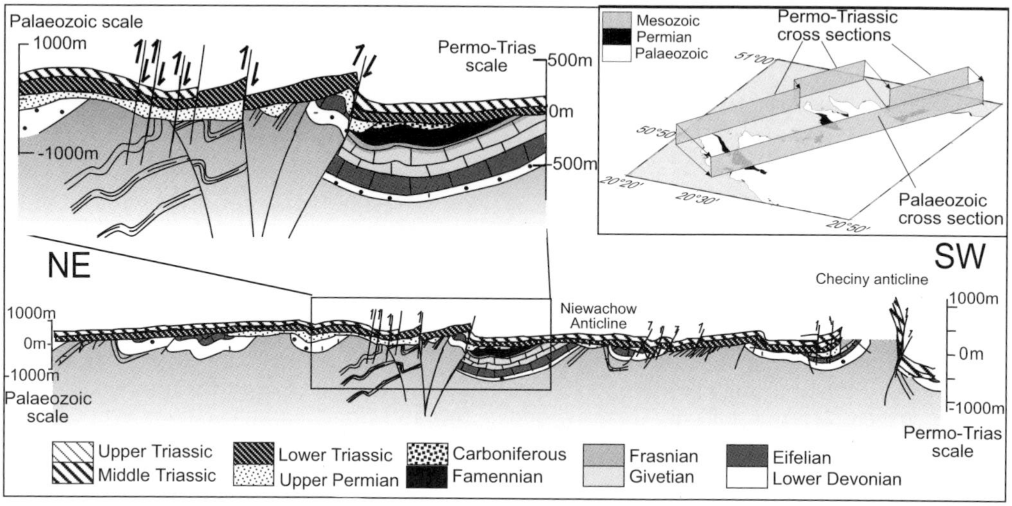

Fig. 11. Superposition of the Permo-Triassic cross-section on the Palaeozoic cross-section, and focus on the Holy Cross Fault zone.

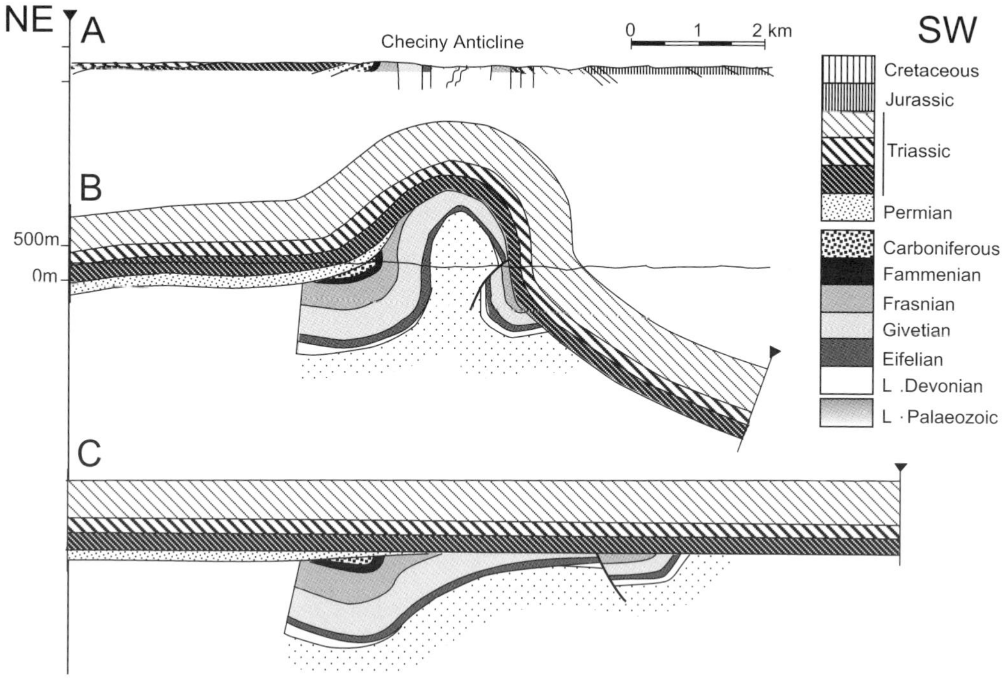

Fig. 12. Restoration of the pre-Alpine folding structure of the Chęciny anticline. A, Structure of the rocks at the topographic surface. B, Interpretative cross-section of the present-day structure. C, Restoration of the horizontal bedding of Mesozoic rocks before the Alpine (post-Cretaceous) folding, and restoration of the former Variscan shape of the fold affecting the Upper Palaeozoic rocks.

which it is possible to compare the location and nature of structures in both profiles.

It can be seen from the Permo-Triassic cross-section that the Upper Permian sediments are discontinuous, being either lens shaped or fault bounded (Fig. 11). Locally, the Permian lenses are located in the axis of a Variscan anticline (Niewachlów Anticline, for instance; Fig. 11). The continuous Triassic cover, which blankets the Permian structures, is affected by gentle folding as well as by reverse faulting. Both types of deformations are compressive and Maastrichtian–Palaeogene in age (Lamarche *et al.* 1998). The detailed view of Figure 11 shows that a reverse fault also corresponds with the boundary of Permian sediments, which leads to the interpretation that the Permian normal fault was reactivated as a reverse fault during the Maastrichtian–Palaeogene shortening. It is interesting to see that the faults bounding or deforming the Permo-Triassic sediments are mostly located in the prolongation of the Variscan faults (see the enlargement on Fig. 11).

Therefore, we deduced from these observations that the former Variscan faults that delineated the extent of Permian sediments were first reactivated as normal faults during Permian times. Subsequently, these faults have been reversely reactivated a second time most probably during the Maastrichtian–Palaeogene shortening. This shortening was also responsible for the folding of the Mesozoic rocks as well as of their Palaeozoic basement at the southwestern margin of the Holy Cross Mountains.

The Maastrichtian–Palaeogene shortening is depicted by the section across the Chęciny anticline shown in Fig. 12B. In Figure 12C, the Mesozoic layers are restored to their pre-folding position, using the same principles as for the Palaeozoic cross-section. After restoration, we calculated an apparent Alpine shortening rate of 17%, which is greater than the Variscan shortening. Because the cross-section of the Palaeozoic rocks (Fig. 3) displays the accumulation of Variscan and Alpine shortening, the calculated Maastrichtian–Palaeogene (or Alpine) shortening rate should logically be the smallest. However, we observe the opposite.

Several hypotheses may explain this apparently controversial result. Firstly, our results could indicate a significant effect of the post-Variscan extension which occurred between both phases of shortening. Indeed, it can be envisaged that the Permo-Triassic extension led to noticeable deformation of the Palaeozoic rocks, reducing the apparent Variscan shortening rate compared with the Maastrichtian–Palaeogene one. However, the cross-sections of Permo-Triassic rocks (Figs. 11) indicate that the Permo-Triassic extension is only brittle and not significant enough to 'un-fold' the Palaeozoic rocks. Moreover, the nature of the Maastrichtian–Palaeogene deformation is heterogeneous within the Holy Cross Mountains. Previous structural analyses of the Permian–Mesozoic cover surrounding the Holy Cross Mountains have shown that the Maastrichtian–Palaeogene folding was mainly restricted to the SW border of the mountains and was genetically linked to the reactivation of deep Palaeozoic and/or Permian–Mesozoic faults (Lamarche 1999). However, palaeomagnetic data from Radkowice quarry (10 in Figs 2 & 3) indicate only a weak Alpine reactivation of the northern limb of the Chęciny Anticline in this area (see Appendix). Furthermore, the comparison of the deformation in Permo-Triassic and Palaeozoic rocks in Fig. 11 also clearly shows that the Permian–Mesozoic rocks are much less folded than the Palaeozoic rocks, except from the Chęciny Anticline where the shortening rate was calculated. In addition, at a larger scale, the palaeo-stress pattern induced by the Maastrichtian–Palaeogene comression is known to be highly heterogeneous (Lamarche *et al.* 2002). Therefore, it can be assumed that the Maastrichtian–Palaeogene rate of shortening was 17% in some parts along the Chęciny Anticline, while it was much less elsewhere in the Holy Cross Mountains. This constitutes evidence for localized Alpine deformation due to reactivation of basement structures. Nevertheless, a third hypothesis can be invoked. Indeed, both Variscan and Maastrichtian–Palaeogene shortening occurred in a transpressive tectonic context. As the structural balancing does not take the lateral displacement into account, the controversial difference between Alpine and Variscan shortening rates may, at least partially, be an artefact induced by lateral escape and simple shearing along strike during the deformation.

Conclusions

In this paper, we have shown that a combined structural and sedimentological approach allows the differentiation of individual steps of the polyphase deformation that the Holy Cross Mountains have undergone since Devonian times. Key outcrop analysis highlights synsedimentary extension during Devonian times, Variscan shortening, Permo-Triassic extension and Maastrichtian–Palaeogene compression. The detailed study of a geological cross-section across the Holy Cross Mountains led to further structural analyses, consisting of a step-by-step restoration. Thus, the present-day state of deformation of the Palaeozoic rocks has been partitioned into pre-Variscan, Variscan, Permo-Triassic and Alpine deformations.

In terms of the geodynamic evolution of the area, the following conclusions have been drawn. The Holy Cross Mountains evolved from Late Palaeozoic to Tertiary times in an alternately transtensional (Devonian times) and transpressional (Late Carboniferous and Alpine times) strike-slip tectonic context, as interpreted from palaeomagnetic and structural data. During Devonian and Lower Carboniferous times, a sedimentary basin developed on the Kielce Unit and the Łysogóry Unit in the context of tectonic instability. This resulted in differentiated sub-basins, interpreted to be due to a long-lasting strike-slip (very likely dextral as shown by the palaeomagnetic study) tectonic regime along the margin of Baltica. Four main sedimentary domains (from north to south): the Łysogóry Unit (segment A), the Niewachlów area (segment B), the Dyminy High (segment D) and the Chęciny area (segment F), are separated by three major fault zones: the Holy Cross Fault zone, and the northern and the southern borders of the Dyminy High. Synsedimentary tectonics are noticeably recorded in the sediments and are also visible on the synthetic lithostratigraphic section across the Holy Cross Mountains. During the Devonian, the Dyminy High was a continuously uplifted area as compared with the neighbouring sub-basins. In addition, the activity of the Holy Cross Fault zone led to the different evolution of the Łysogóry Unit and Kielce Unit. Increasing tectonic activity during Frasnian and Famennian times may be regarded as an early manifestation of the late Variscan tectonic inversion of the Devonian basin. Folding of the Devonian–Carboniferous formations during the final suturing of both units took place during the Late Carboniferous. The Variscan climax induced the reactivation of the Devonian normal faults.

The Variscan folding is characterized by an alternation of slightly south-verging folds, favoured by a decoupling level at the base of Upper Palaeozoic rocks. The folding process was polyphase, marked by early small-scale folds and north-verging ramps, involved in the later large-scale major folding. Inferred previously from palaeomagnetic data, the strike-slip component of the deformation is confirmed by the flower-like structure of the Holy Cross Fault zone, as well as the small fold asymmetry. Following our interpretations, the Holy Cross Fault zone appears as a major strike-slip fault array developed under transtensional conditions within the Devonian basin between the Łysogóry Unit and the Kielce Unit. It was reactivated as a reverse fault during the Variscan tectonic inversion, giving rise to the southern faulted limb of a major anticline.

After erosion of the Variscan relief, Late Permian–Early Triassic extension due to a period of rifting along the Mid-Polish Basin led to the normal reactivation of the Variscan reverse faults. They controlled the scarce deposition of Upper Permian sediments in fault-bounded half-grabens and in palaeogeographically controlled lenses. Following a long period of basin subsidence during the Mesozoic, the Maastrichtian–Palaeogene tectonic inversion of the Mid-Polish Basin significantly overprinted some parts of the Holy Cross Mountains. The compression entailed a reverse reactivation of most of the faults, as well as localized folding along the Chęciny range, which may be superimposed on the pre-existing deformation of Palaeozoic rocks. Although real, the Late Permian extensional reactivation and Maastrichtian–Palaeogene inversion had only minor impact on the rate of deformation in the Holy Cross Fault area. We calculated a cumulative Variscan and Maastrichtian–Palaeogene (Alpine) apparent rate of shortening of 15%. This value is difficult to estimate due to the concentration of Alpine deformations mainly in the Chęciny area.

In summary, in this paper we have successfully distinguished, characterized and quantified the successive steps of major deformation affecting the Holy Cross Mountains over *c*.350 Ma, resulting from varying stress patterns and basin evolution. Notably, we have demonstrated that the Variscan, Permo-Triassic and Maastrichtian–Palaeogene deformations inherited the primary tectonic foundations of the Devonian basins.

The authors wish to express their gratitude to E. Stupnicka, J. Świdrowska, M. Hakenberg and J. Wieczorek for their contribution and discussion in the field. The authors gratefully acknowledge constructive reviews by J. Walsh and A. Läufer, as well as R. Di Primio and Volker Otto, who improved the English of the manuscript. This work was started within the framework of and financially supported by the Peri-Tethys Programme and the French Foreign Affairs Ministry, and was consolidated with a POLONIUM project. The authors are indebted to the EUROPROBE programme (European Science Foundation), which facilitated the necessary international cooperation.

Appendix

Structural indicators

Bronkowice anticline. The Devonian rocks of the Bronkowice Anticline (Fig. 3) form a large-scale anticline (8 to 10 km wide) with gently south-verging asymmetry and a near-vertical southern limb. The Silurian rocks in the core of the fold display a different style of deformation. On the basis of existing geological maps, Znosko (1996) envisaged a set of south-verging detached folds that developed in Upper Silurian rocks. The rocks are occasionally accessible in local road cuttings and stream valleys. Due to alternating competent and incompetent lithologies, disharmonic folding is expected within the Bronkowice Anticline.

Bukowa quarry (1). The Bukowa quarry is located at the northern slope of the Bukowa Mountain (1 in Fig. 2). It is cut into Emsian siliciclastics, shales and sandstones, exploiting the Zagórze, Bukowa Góra and Kapkazy formations, respectively in their stratigraphic order (Fig. 4). This sequence is included in the southern limb of the Bodzentyn Syncline, the broadest syncline in the Łysogóry Unit. The bedding strikes N110° in average and the dip varies from 50°N in the south to 10°N in the north approaching the Bodzentyn syncline. Synsedimentary extensional features were observed in the Bukowa quarry, and are described in detail in the following section. On the basis of palaeomagnetic analyses, Lewandowski (2000) reported a *c*.30° clockwise rotation of the Upper Silurian rocks with respect to Baltica.

Zachełmie quarry (2). The Zachełmie quarry is cut into the Chełm Mountain and situated between the villages of Zagnańsk and Zachełmie, where the Devonian units are elevated compared with the low-lying Permian-Mesozoic cover of the Holy Cross Mountains (2 in Fig. 2). In the quarry, Middle Devonian dolomites truncated by the Variscan unconformity and overlain by a fluvial succession of the Lower Triassic Buntsandstein (Szulczewski, in Lipiec *et al.*, 1995) are exposed. The Devonian succession can be assigned to the Wojciechowice Formation, regarded as being Eifelian in age. It belongs to the northern limb of the Łysogóry anticline and to the southern limb of the Bodzentyn Syncline. The rocks dip monoclinally with 40–45°N and trend N100°. Tilting of the Devonian rocks is deduced to be of Variscan age, contemporaneous with the large-scale folding. As the Lower Triassic is lying horizontally, no Alpine folding occurred in this area.

Wiśniówka quarry (3). Three quarries (Wiśniówka Duża, Wiśniówka Mała and Podwiśniówka) are located close to each other at Wiśniówka Mountain in the Łysogóry Unit, about one kilometre north of the Holy Cross Fault zone (3 in Fig. 2). They are dug into the Upper Cambrian to Tremadocian quartzites, clays and sandstones (Kowalczewski *et al.* 1986). Although complicated in detail, the most remarkable structural features are the nearly vertical position of the bedding, dipping 70°N and trending N110°, as well as the alternation of several slices of Tremadocian and Cambrian rocks tectonically juxtaposed from north to south. The major large-scale faults are subvertical, trending N110° parallel to the bedding and to the main fold axes of the Holy Cross Mountains. No clear kinematic indicators are observable on these faults. Based on these structural observations, we deduced that the area has been intensively folded and, cut by major N110°-trending vertical faults.

The age, style and origin of the folds in the Wiśniówka quarries remain a matter of debate. The lack of younger Palaeozoic rocks covering the folds hampers the relative dating of the folding. The question as to whether the folds are syndepositional, Early Devonian or Late Carboniferous in age is still unclear (see Mizerski 1979; Kowalczewski *et al.* 1986; and Kowalczewski and Dadlez 1996, for discussion). The rocks show no cleavage, which is surprising considering the age of the rocks and the proximity to the Holy Cross Fault zone, which was a major tectonic zone during the Late Palaeozoic. The top of the sequence, cropping out near the northern entrance to the Wiśniówka Wielka quarry, is folded. Palaeomagnetic data collected from this fold structure show that the rocks of Wiśniówka were in a horizontal position during Silurian times, pointing to the absence of a Late Cambrian (so-called Sandomirian) tectonic event in the Łysogóry Unit, otherwise proven to be present in the Kielce Unit (Lewandowski 1993).

Kostomłoty quarry (13). The active Krzemucha (Kostomłoty *sensu stricto*) and Laskowa quarries are located in the Kielce Unit, west from the cross-section (13 in Fig. 2). However, observations made in the Kostomłoty quarry are of interest for depicting the structural style of the Variscan folding, which can be reproduced along the cross-section at the northern limb of the Niewachów anticline (Fig. 3). The exposed rocks are Givetian and Frasnian in age. Thick and competent dolomites constitute the Lower Givetian. The Givetian/Frasnian transition is developed as alternating marls and shales, which are overlain by the Frasnian limestones (Kostomłoty Beds). In the Laskowa quarry, the Variscan unconformity, similar to those of the Zachełmie and Jaworznia quarries, is visible and covered by the Buntsandstein (Szulczewski, in

Lipec *et al.* 1995). In the Kostomłoty quarry, the Givetian to Frasnian succession generally dips to the north with a mean angle of 20° (Fig. 5). In detail, the beds display intensive folding as well as a local syn-fold cleavage (Lamarche *et al.* 1999), folding being Variscan in age (Lewandowski 1981; Grabowski & Nawrocki 1996). The contrasting lithology yields strong disharmonic folding with metric to decametric wavelengths and decoupling levels at the top and at the base of the alternating incompetent Givetian and Frasnian marly–clayey schists (Fig. 5). Fold axes trend N090° to N110°, suggesting a N–S to NNE–SSW direction of Variscan shortening, further constrained by a syn-fold cleavage (Lamarche *et al.* 1999). Although the axial surface of the folds dips 70°S on average, the folds are rather symmetrical (B in Fig. 5). As a consequence, we consider the dip of the fold planes to result from large-scale tilting of the already folded rocks of 20° to the north, as also indicated by the mean dip of the bedding, rather than being due to a vergence of the Variscan folds to the north.

Gruchawka (4). Engineering work for the power station in Kielce (4 in Fig. 2) exposed a suite of the Upper Silurian–Lower Devonian clastics. According to Malec (1993, 2001) the Late Caledonian unconformity is situated within the succession between the Miedziana Góra Conglomerate and the conglomerate from Gruchawka. Malec (2001) recently ascribed the Miedziana Góra Conglomerate to the Ludlow stage and regards it as a molasse deposit of the submarine delta fan, which is however, older than the folding responsible for the sub-Devonian angular unconformity in the Kielce region. This unconformity is demonstrated by the truncation of the depicted fan and caps it unconformably with a veneer of another conglomerate (conglomerates from Gruchawka), which is ascribed to the Prag stage. According to this interpretation of the stratigraphic relationships within the succession, the Lower Devonian sandstones and siltstones dip 35–40°N and cover the Silurian, dipping 55°N with an angular unconformity of about 10°. The interpretation of stratigraphic relationships between the two conglomerates as a manifestation of the Caledonian unconformity, however, remains controversial (Szulczewski 1994; Kowalczewski *et al.* 1998).

Czarnów (Śluchowice) quarry (5). The quarry is located in the Kielce Unit at the outskirts of the city of Kielce, about 8 km south of the Holy Cross Fault zone (5 in Fig. 2). Exposed rocks include the Frasnian thin-bedded limestones (mostly Kostomłoty Beds) to the north and the Famennian marly–shaly deposits to the south (Fig. 6). An intense folding affects the Frasnian limestones. We can distinguish a major large-scale fold as well as small-scale folds. The large-scale fold is asymmetrical, marked by a subhorizontal flat limb to the north and a subvertical limb (locally dipping 70°N) to the south, indicating south-verging major folding. Small-scale folds with metre-scale wavelengths are visible in the limbs and hinges of the major fold. Their geometry varies from the east to the west, from *en chevron*-type folds (A in Fig. 6) to fault-bend and fault-propagation folds (B–C in Fig. 6) (Lamarche 1999). *En chevron* folds display a south-directed vergence and syn-folding cleavage in agreement with the large-scale folds. In contrast, the fault-bend and fault-propagation folds are associated with thrust faults, which are affected by the main folding (1 in Fig. 6). In addition, the cleavage locally cuts through a north-verging fault-bend fold (Lamarche *et al.* 1999). Hence, these faults occurred before the main folding and cleavage. Moreover, a north-directed vergence for these thrusts is deduced after restoration of the large-scale folding. As a consequence, an early stage of north-verging shortening is postulated that took place before the major south-verging Variscan folding.

The large-scale fold observed in the Czarnów quarry is structurally incorporated into the folded vertical limb of the major Niewachów anticline. Measured fold axes and bedding planes indicate an east–west-trend for the fold axes, which points to a north–south direction of Variscan shortening (Lamarche *et al.* 1999).

Wietrznia (6). Located on the northern limb of the Dymimy Anticline, the Late Givetian, Frasnian and Famennian rocks exposed in the abandoned Wietrznia quarries (6 in Fig. 2) strike N090°–095° on average and dip 50°–55°N. Large WNW–ESE-oriented and south-dipping normal faults, paragenetically associated with gravity slides, are attributed to synsedimentary tectonics, being inverted later during Variscan shortening. As they are an expression of synsedimentary tectonics, these structures are described further on in the following section.

Jaworznia (7). In the abandoned Jaworznia quarry, located some kilometres west of the cross-section (7 in Fig. 2), the Frasnian limestones of the Kowala Formation (Upper Sitkówka Beds) are unconformably covered by the horizontally oriented clastic Buntsandstein deposits (Lower Triassic). As described by Głazek and

Romanek (1978), the Buntsandstein clastics here cover the gentle folds developed in the Devonian limestones and penetrate them along the unconformity surface strongly diversified by numerous fissures and a graben. According to these authors, block faulting persisted in this area from Permian times onward and was synsedimentary with regard to the Early Triassic deposition. The stratigraphic relationships in the quarry show that the Frasnian limestones were folded during the Variscan Orogeny, remained uplifted in the Late Permian and were subsequently subjected to NNE–SSW extension and subsidence early in the Triassic.

Trzuskawica quarry (8). Long-wavelengths undulations with a rounded blunt hinge and very small folding angle ϕ (definition after Twiss and Moores 1992) affect the Frasnian carbonates of the Trzuskawica quarry (Kielce Unit). This weak deformation indicates that the quarry is located in a flat monoclinal domain, which separates the two steep limbs of the Dyminy Anticline to the north and the Chęciny anticline to the south (8 in Fig. 2). However, the rocks are affected by NNW–SSE-trending normal faults cutting through the entire quarry, although their throw is only a few decimetres. Similar faults with reverse throw also affect the Frasnian rocks. However, because no sequential relationships are visible between normal and reverse faults, direct relative dating is not possible. Nevertheless, on some of the normal fault planes two cross-cutting generations of slickensides indicate that a strike-slip regime preceded the extension. The absence of sediments blanketing the Frasnian rocks does not allow discrimination between a Variscan and Alpine age for the faults.

Kowala (9). The Kowala quarry (9 in Fig. 2) is cut into Frasnian and the Famennian carbonate rocks (Fig. 7). They form generally monoclinal structures, with the rocks trending N070° and dipping 40° to 80°N. The lower portion of the Frasnian (Kowala Formation, Upper Sitkówka Beds) to the south of the quarry is composed of thick, competent and unfolded beds. The Upper Frasnian and the Famennian to the north are represented by thinner beds with shaly alternations, which are folded and faulted. Two types of folds are observed. The first type (1 in Fig. 7), affects a thick succession (*c*.30 m) of thin-bedded rocks. The folds are chaotic, discontinuous and heterogeneous. Sandwiched between two thick, apparently undeformed, limestone beds, the folds are not truncated and die out laterally into undeformed layers. In addition, the folds are not associated with reverse faulting or cleavage. Thus, they are interpreted as slump structures. The second type of folds (2 in Fig. 7), affecting thin-bedded Frasnian rocks, is related to reverse faults which truncate the beds. In this case, folds and reverse faults display a consistent south-directed vergence. Thus, these second-type folds are interpreted to be related to tectonic events. As the quarry is located near the northern limb of the Chęciny Anticline, the development of the second type of folds and reverse faults can be correlated with Variscan shortening. The folding and reverse faulting, as well as the former slump structures, are held partially responsible for the thickening of the Upper Devonian succession in this area.

Radkowice (10). In the Radkowice quarry Eifelian dolostones and limestones strike N095°, dip 80°N and constitute the northern vertical limb of the Chęciny Anticline (10 in Fig. 3). The rocks are affected by numerous north-dipping, steep, mostly reverse faults. As younger sediments do not seal them, their relative age is not well constrained. However, a Variscan age of folding is confirmed by palaeomagnetic data (Szaniawski 1997) with only a small Alpine rejuvenation.

Chęciny–Rzepka (11–12). The outcrops of Chęciny and Rzepka are located in the southern limb of the Chęciny anticline, to the west of the cross-section line (11 and 12 in Fig. 2). The structures deduced from both outcrops are not directly projected on the cross-section but indicate the shape of the southern limb of the Chęciny anticline. An outcrop of the Givetian to Frasnian carbonate rocks (*Stringocephalus* Beds to Chęciny Beds of the Kowala Formation) is found within the foundations of Chęciny Castle. Thin-bedded limestones prevail in the sucession. Beds dip monoclinally 70°S to 80°S and trend N110° to N120°. About one kilometre west of the castle, the old quarry of Rzepka exposes Middle Devonian dolomites dipping 30° to the NNE. Combining the structural data from both outcrops, we deduced the presence of an asymmetrical large-scale syncline with an angular hinge of 90° composed of a subvertical northern limb (Chęciny) and a subhorizontal southern limb (Rzepka).

Ostrówka (14). Although located to the west of the cross-section line, we also considered the structure of the Ostrówka quarry which is located at the northwestern prolongation of the northern limb of the Chęciny Anticline (14 in Fig. 2) because it gives a good example of the pre-Variscan and Variscan tectonic styles. Details of the Frasnian to the Upper Viséan basin evolution can be found in Szulczewski *et al.* (1996). The layers trend N100° on average, with a dip varying from 70°N in the south (near the fold hinge) to 15°N in the north, showing the curvature of the fold limb (Fig. 8). The rocks are affected by faulting, some of which is reverse due to Variscan shortening, whereas some is normal (Fig. 8). The latter faults are described in the following paragraph.

Examples of Devonian synsedimentary extension

Bukowa quarry (1). The Bukowa Góra quarry exposes three Upper Emsian formations (from older to younger): the sandy–shaly Zagórze formation, the shaly Bukowa Góra Formation and the sandstones of the Kapkazy Formation (Fig. 4). On the northern wall of the quarry, four normal faults (1 to 4, Fig. 4) striking N020° and dipping 50° to 80°W affect the Zagórze Formation with offsets of 0.2 m to 5 m. Three of the faults propagate through the Bukowa Góra and Kapkazy formations, but their downthrow offsets decrease upwards in the Bukowa Góra Formation (Fig. 4). The normal displacement in the Kapkazy Formation does not exceed 1 m per fault. In the entrance path of the quarry, a normal fault (5 in Fig. 4) striking *c.*N160° and dipping 50°W, affects the Zagórze Formation with an estimated offset of more than 5 m. The fault borders a half-graben filled by the Bukowa Góra Formation. At the bottom of the half-graben, a few clayey layers of the Bukowa Góra Formation are folded (6 in Fig. 4) and decoupled on top of the competent Zagórze Formation. The fold is asymmetrical, verging towards the fault labelled 5 in Fig. 4. The fold and the normal fault are covered by the upper beds of the Bukowa Formation which are flat-lying and appear to be undeformed. The overlying beds of the Kapkazy Formation blanket the half-graben. As a result, the Kapkazy Formation lies directly on the Zagórze Formation, to the north of the fault labelled 5 in Figure 4. We interpret the fold as having resulted from gravitational sliding of soft sediments (slump) on a hard slope being tilted when the half-graben was formed. As a result, all the faults and the associated fold can be attributed to synsedimentary extension during the Late Emsian in the Łysogóry Unit. The trend of the faults indicates a nearly east–west direction of the extension.

Wietrznia (6). In the Wietrznia quarry, the Frasnian limestones and the Famennian carbonates and marly shales are exposed. The Upper Devonian is deeply penetrated by the karstified Variscan unconformity surface, which is covered by the Upper Permian conglomerates and Lower Triassic red sandstones of the Buntsandstein (Szulczewski, in Lipiec *et al.* 1995). Facies and stratigraphy of the prevailing part of the Frasnian are significantly influenced by the Late Devonian syndepositional block faulting (Szulczewski 1989; Szulczewski, in Lipiec *et al.* 1995). It resulted in a remarkable differentiation in thickness, stratigraphic completeness and facies development.

Two types of faults have been observed (Fig. 9): (1) A large N105°-trending and 70°S-dipping normal fault affects the Frasnian and Famennian and is associated with sedimentary breccias comprising a shaly Famennian matrix and olistoliths of Frasnian limestones (A in Fig. 9). (2) More discrete faults are N160°–170°-oriented and nearly vertical (B in Fig. 9). These faults separate two limestone blocks (F1 and F2, B in Fig. 9) described by Szulczewski (1989 in Lipiec *et al.* 1995). In the deeper block, the Frasnian succession affected by faulting reaches about 50 m in thickness, while in the most elevated one it is attenuated to 50 cm and markedly discontinuous. The condensed and stratigraphically discontinuous deposition of cephalopod facies rocks developed on the palaeotopographic high around the Frasnian–Famennian boundary. The block-faulted palaeotopography became finally compensated. The deepening of the basin, indicated by a change in facies from platform to pelagic carbonates, was associated with tectonic activity triggering the reactivation of fault F2 and was marked by a small angular unconformity of the Famennian shales lying on the Frasnian limestones. The extension persisted during the beginning of the shale deposition as revealed by their flexure above the fault.

Large N105°-striking normal faults dipping to the south also affect the limestones (A in Fig. 9). The southern sag block is composed of sedimentary megabreccias comprising angular blocks of limestones and shales exclusively of Frasnian and Famennian age (*fitted breccia* of Szulczewski, in Lipiec *et al.* 1995). Although chaotic, a relative layering of the shales and olistoliths parallel to the fault is visible in the breccias (A in Fig. 9). Laterally to the west of the quarry, the sedimentary breccias are more chaotic and rest on top of the Frasnian limestones, without tectonic contact but most probably filling a palaeo-trough. The location and internal structure of the breccias suggest that the trough's formation was caused by gravitational mass movement along a slope of tectonic origin. The occurrence of angular blocks of Frasnian limestones as well as the preservation of internal layering of the Famennian shales imply the presence of a proximal Frasnian uplifted and dismembered relief in the hinterland. Szulczewski (in Lipiec *et al.* 1995) regarded the *fitted* megabreccia in this quarry as younger than the Permian *rock-matrix breccia*, and as possibly coeval with the Buntsandstein deposition. However, a system of horst and grabens formed by the Famennian north–south directed extensional event may have been the origin of the structures observed in the Wietrznia quarry.

To summarize, the extensional regime occurred continuously during the Frasnian and Famennian at the northern margin of the Dyminy anticline with an initial phase of ENE–WSW-oriented extension during the Frasnian–Early Famennian, changing to north–south in the Late Famennian. This tectonic regime resulted in a horst-and-graben system, of which the uplifted areas were dismembered and provided the material for the tectonic breccias during the Famennian deepening of the basin. The Variscan shortening obscured a genuine Late Devonian tectonic pattern, as could still be seen in the late 1960s in the eastern part of the quarry, where the Frasnian limestones are locally thrust over the breccias – the wall being destroyed by mining work (Lewandowski 1971).

Ostrówka (14). Szulczewski *et al.* (1996) described facies and thickness variations across the Ostrówka quarry in areas bounded by faults suggesting the occurrence of Late Devonian synsedimentary tectonics. In addition, the Famennian and Tournaisian succession (post-platform deep water deposits) in the Ostrówka quarry is condensed and rests unconformably with an angle of up to 15° on the Frasnian limestones (shallow-water carbonates) (Fig. 8). According to Szulczewski *et al.* (1996), the angular unconformity is probably related to the tilting of tectonic blocks. On top of the quarry, normal faults bounding a half graben were measured. After restoration of the bedding before folding, the faults appear to be conjugate, NW–SE-oriented with nearly dip-slip normal striations resulting from a NE–SW-oriented extension (Lamarche 1999), which we interpret to be of synsedimentary origin with regard to Late Devonian deposition prior to Variscan shortening.

Kowala quarry (9). In the Kowala quarry, the Frasnian limestones display folds with slump characteristics (folds 1 in Fig. 7). They affect a succession of incompetent layers and are bracketed at the top and bottom by undeformed thick limestone layers. In addition, traces of two normal faults showing synsedimentary characteristics were observed in the Frasnian limestones (3 in Fig. 7). One is truncated and thus post-dated by a reverse fault probably associated with Variscan deformation. Szulczewski (1968) described slump structures and turbidites occurring continuously during the Frasnian and Famennian in the Holy Cross Mountains. The material for these sediments was derived from Frasnian reef complexes and its deposition was probably caused by synsedimentary tectonics.

References

BEDNARCZYK, W. S. & STUPNICKA, E. 2000. Stratigraphy and new data on tectonics of the Ordovician strata in the section at Międzygórz quarry (Eastern Holy Cross Mountains, Poland). *Annales Geologorum Societatis Poloniae*, **70**, 283–287.

BEDNARCZYK, W. S., CHLEBOWSKI, R. & KOWALCZEWSKI, Z. 1970. Budowa geologiczna północnego skrzydla antykliny dyminskiej w Górach Świętokrzyskich. *Biuletyn Geologiczny*, **12**, 197–223.

BEŁKA Z., AHRENDT, H., FRANKE, W., WEMMER, K. 2000. The Baltica–Gondwana suture in central Europe: evidence from K–Ar ages of detrital muscovites and biogeographical data. *In:* FRANKE, W., HAAK, V., ONKEN, O. & TANNER, D. (eds) *Orogenic Processes: Quantification and Modelling in the Variscan belt*. Geological Society, London, Special Publication, **179**, 87–102.

BEŁKA Z. & SKOMPKI, S. 1988. Mechanizm sedymentacji i pozycja facjalna wapienia węglowego w południowo-zachodniej części Gór Świętokrzyskich. *Przegland Geologiczny*, **36(8)**, 442–448.

BERTHELSEN, A., 1992. Mobile Europe. *In*: BLUNDEL, D., FREEMAN, R. & MULLER, S. (eds) *Continent Revealed: the European Geotraverse.* Cambridge University Press, Cambridge, 11–32.

BERTHELSEN, A. 1993. Where different geological philosophies meet: the Trans-European Suture Zone. *Publications of the Institute of Geophysics, Polish Academy of Sciences*, **A20(255)**, 19–31.

BLUNDELL , D., FREEMAN, R. & MUELLER, S. (eds) 1992. *A Continent Revealed: the European Geotraverse*. Cambridge University Press, Cambridge.

BUŁA, Z., JACHOWICZ, M., & ŻABA, J. 1997. Principal characteristics of the Upper Silesian Block and Malopolska Block border zone (southern Poland). *Geological Magazine*, **134(5)**, 669–677.

BUTLER, R. F., 1992. *Paleomagnetism: Magnetic Domains to Geologic Terranes*. Blackwell Scientific Publications, Oxford, 319 pp.

CZARNOCKI, J. 1938. *Carte Géologigue Générale de la Pologne 1:100 000, Feuille 4: Kielce* (in Polish and French). Państwowy Instytut Geologiczny, Warsaw.

CZARNOCKI, J. 1957. Geologia regionu Łysogórskiego. *Prace Inst. Geol.*, **18**, 97pp.

DADLEZ, R. 2001. Holy Cross Mountains area – Crustal structure, geophysical data and general geology. *Geological Quarterly*, **45(2)**: 99–106.

DADLEZ, R., KOWALCZEWSKI, Z. & ZNOSKO, J. 1994. Some key problems of pre-Permian tectonics of Poland. *Geological Quarterly*, **38(2)**, 169–189.

DADLEZ, R., NARKIEWICZ, M., STEPHENSON, R. A., VISSER, M. T. M. AND VAN VEES, J.-D. 1995. Tectonic evolution of the Mid-Polish Trough: modelling implications and significance for central European geology. *Tectonophysics*, **252**, 179–195.

FILONOWICZ, P. 1968. *Objaśnienia do Szczegółowej Mapy Geologicznej Polski, Arkusz Morawica.* Wydawnictwa Geologiczny, Warsaw.

FRANKE, D. 1995. Caledonian terranes along the southwestern border of the East European Platform – evidence, speculation and open questions. *Studia Geophysica et Geodetica*, **39**, 241–256.

FRANKE, W. & ŻELAŹNIEWICZ, A. 2000. The eastern termination of the Variscides: terrane correlation and kinematic evolution. *In:* FRANKE, W., HAAK, V., ONKEN, O. & TANNER, D. (eds) *Orogenic Processes: Quantification and Modelling in the Variscan Belt.* Geological Society, London, Special Publication, **179**, 63–86.

FREYER, G. & ŻAKOWA, H. 1967. Famennian conodonts from borehole Bolechowice 1 – in the Holy Cross Mts (Fameńskie konodonty z wiercenia Bolechowice 1 – Góry Świętokrzyskie). *Acta Geologica Polonica*, **17(1)**, 105–135.

GEE, D. G. & ZEYEN, H. J. (eds), 1996. *EUROPROBE 1996 – Lithosphere Dynamics: Origin and Evolution of Continents.* Uppsala University.

GŁAZEK, J. & ROMANEK, A. 1978. Jaworznia. *In:* PIĄTKOWSKI , T. S. & WAGNER, R. *Symposium on Central European Permian. Guide of Excursions, pt. 2. Zechstein of the Holy Cross Mts.*: 41–49, Warsaw.

GŁAZEK, J., KARWOWSKI, L., RACKI, G. & WRZOŁEK, T. 1981. The early Devonian continental/marine succession at Checiny in the Holy Cross Mountains, and its paleogeographic and tectonic significance. *Acta Geologica Polonica*, **34**, 233–254.

GRABOWSKI, J. & NAWROCKI, J. 1996. Multiple remagnetizations in the Devonian carbonates in the northwestern part of the Kielce region (Holy Cross Mts., southern part). *Geological Quarterly*, **40**, 47–64.

GRABOWSKI, J. & NAWROCKI, J. 2001. Palaeomagnetism of some Devonian carbonates from the Holy Cross Mts. (Central Poland): large pre-Permian rotations or strain modified palaeomagnetic directions? *Geological Quarterly*, **45(2)**, 165–178.

JAROSZEWSKI, M., 1972. Mesoscopic structural criteria of tectonics of non-orogenic areas: an example from north-eastern Mesozoic margin of the Świętokrzyskie Mountains. *Studia Geologica Polonica*, **38**, 9–215.

JUREWICZ, E. & MIZERSKI, W. 1991. Nowe dane o budowie geologicznej antykliny Bronkowic. *Bulletin of Geology*, **32**, 121–144 (with English abstract).

KAZMIERCZAK, J. 1971. Morphogenesis and systematics of the Devonian Stromatoporoidea from the Holy Cross Mountains, Poland. Morfogeneza i systematyka dewonskich Stromatoporoidea z Gór Świętokrzyskich. *Palaeontologica Polonica*, **26**, p. 150.

KŁOSSOWSKI, J. 1985. Sedymentacja środkowego dewonu w regionie łysogórskim (profil Świętomarz – Śniadka). *Przegląd Geologyczny*, **33(5)**, 264–267.

KOWALCZEWSKI, Z. 1971. Podstawowe problemy geologiczne dewonu dolnego Gór Świętokrzyskich. *Kwartalnik Geologiczny*, **15(2)**, 263–283.

KOWALCZEWSKI, Z. 1976. Cefaj etapoj en la tektonika evolvo de Sanktoruca montaro. *Geology International*, **3**, 61–68.

KOWALCZEWSKI, Z. & DADLEZ, R. 1996. Tectonics of the Cambrian in the Wiśniówka area (Holy Cross Mts., Central Poland). *Geological Quarterly*, **40**, 23–46.

KOWALCZEWSKI, Z. & STUDENCKI, M. 1983. Budowa geologiczna góry Krzemianki koło Kielc. *Kwartalnik Geologiczny*, **27(4)**, 695–708.

KOWALCZEWSKI, Z., KULETA, M., LISIK, R. & MOCZYDLOWSKA, M. 1986. Nowe dane o skalach kambru i dolnego ordowiku z okolic Wisniowki w Gorach Świętokrzyskich. *Kwartalnik Geologiczny*, **30(2)**, 201–228.

KOWALCZEWSKI, Z., JAWOROWSKI, K. & KULETA, M. 1998. Klonów Beds (uppermost Silurian–?lowermost Devonian) and the problem of Caledonian deformations in the Holy Cross Mts. *Geological Quarterly*, **42**, 341–378.

KUTEK, J. 2001. The Polish Permo-Mesozoic Rift Basin. *In*: ZIEGLER, P.A., CAVAZZA, W., ROBERTSON, A. H. F. & CRASQUIN-SOLEAU, S. (eds), Peri-Tethys Mem. 6, *Peri-Tethyan Rift/Wrench Basins and Passive Margins.* Mémoire Muséum National Histoire Naturelle, Paris, **186**, 213–236.

KUTEK, J. & GŁAZEK, J. 1972. The Holy Cross area, Central Poland, in the Alpine cycle. *Acta Geologica Polonica*, **22**, 603–653.

LAMARCHE, J. 1999. Evolution tectonique et rôle de la Zone Teisseyre–Tornquist. Importance des structures héritées dans la formation des bassins et la déformation de l'Europe Centrale. Thèse de Doctorat, *Mémoire Sciences Terre Université Pierre et Marie Curie*, Paris, 226 pp.

LAMARCHE, J., MANSY, J.-L., BERGERAT, F., AVERBUCH, O., HAKENBERG, M., LEWANDOWSKI, M., STUPNICKA, E., ŚWIDROWSKA, J., WAJSPRYCH, J. & WIECZOREK, J. 1999. Variscan tectonics in the Holy Cross Mountains (Poland) and role of the structural inheritance during the Alpine tectonics. *Tectonophysics*, **313**, 171–186.

LAMARCHE, J., ŚWIDROWSKA, J., BERGERAT, F., HAKENBERG, M., MANSY, J. L., WIECZOREK, J., STUPNICKA, E. & DUMONT, T. 1998. Development and deformation of a Mesozoic basin adjacent to the Teisseyre–Tornquist Zone: the Holy Cross Mountains (Poland). *In*: CRASQUIN-SOLEAU, S. & BARRIER, E. (eds) *Epicratonic basin of Peri-Tethyan Platform.* Mémoire Museum National Histoire Naturelle, Paris, 75–92.

LAMARCHE, J., BERGERAT, F., LEWANDOWSKI, M., MANSY, J.-L., ŚWIDROWSKA, J., WIECZOREK, J. 2002. Variscan to Alpine heterogeneous paleo-stress field above a major Palaeozoic suture in Carpathian foreland (southeastern Poland). *Tectonophysics* **357**, 55–80.

LEWANDOWSKI, M. 1971. *Stratygrafia górnego dewonu okolic Zagórza i Wietrzni na podstawie konodontów.*

M.Sc. Thesis, Arch. Geol. Faculty, Warsaw University (in Polish only).

LEWANDOWSKI, M. 1981. Post-folding characteristic magnetization of the Upper Devonian Kostomloty Beds in the Holy Cross Mts. *Acta Geologica Polonica*, **31(3–4)**, 265–272.

LEWANDOWSKI, M. 1987. Results of the preliminary paleomagnetic investigations of some Lower Paleozoic rocks from the Holy Cross Mts. (Poland). *Kwartalnik Geologiczny*, **31(4)**, 543–556.

Lewandowski, M. 1993. Paleomagnetism of the Paleozoic rocks of the Holy Cross Mts (Central Poland) and the origin of the Variscan Orogen. *Publications, Institute Geophysics*, **A-23**. Polish Academy of Sciences, Warsaw, 84 pp.

LEWANDOWSKI, M. 1994. Paleomagnetic constraints for Variscan mobilism of the Upper Silesian and Malopolska Massifs, southern Poland. *Geological Quarterly*, **38(2)**, 211–230.

LEWANDOWSKI, M. 1995. Paleomagnetic constraints for Variscan mobilism of the Upper Silesian and Małopolska Massifs, southern Poland – reply. *Geological Quarterly*, **39**, 283–293.

LEWANDOWSKI, M. 2000. Block rotations in the Palaeozoic fold belts of Poland: next palaeomagnetic evidences from the Holy Cross Mts and the Kaczawa Mts. In: *Joint Meeting of EUROPROBE (TESZ) and PACE Projects*, Zakopane, Abstract Volume, 61–62.

LIPIEC, M., MALEC, J., MATYJA, H., MIGASZEWSKI, Z., PASZKOWSKI, M., PROTAS, A., SKOMPSKI, S., SZULCZEWSKI, M., ZBROJA, S., ŻAKOWA, H. & ŻELICHOWSKI, A. 1995. Guide to Excursion A2, Development of the Variscan basin and epi-Variscan cover at the margin of the East European Platform (Pomerania, Holy Cross Mts., Kraków Upland). *XII International Congress Carboniferous–Permian, August 28 – September 2, 1995, Kraków, Poland*, 25–26. Polish Geological Institute, Warsaw.

MCKERROW, W. S., MAC NIOCAILL, C., AHLBERG, P. E., CLAYTON, G., CLEAL, C. J. & EAGAR, R. M. C. 2000. The Late Palaeozoic relations between Gondwana and Laurussia. *In:* FRANKE, W., HAAK, V., ONKEN, O. & TANNER, D. (eds) *Orogenic Processes: Quantification and Modelling in the Variscan Belt*. Geological Society of London, Special Publication, **179**, 9–20.

MALEC, J. 1993. Upper Silurian and Lower Devonian in the western Holy Cross Mts. *Geological Quarterly*, **37(4)**, 501–536.

MALEC, J. 2001. Sedimentology of deposits from around the Late Caledonian unconformity in the western Holy Cross Mts. *Geological Quarterly*, **45 (4)**: 397–415.

MARIAŃCZYK, E., 1973. New data on geology of the Bronkowice region (in Polish with English summary). *Przegląd Geologiczny*, **21**, 158–160.

MATYJA, H. & NARKIEWICZ, M. 1995. Conodont stratigraphy of the Upper Devonian in the Janczyce I borehole section, eastern Holy Cross Mts. *Geological Quarterly*, **39(2)**, 177–205.

MIGASZEWSKI, Z. M. 2002. Datowanie diabazów i lamprofirów Świętokrzyskich metodą K–Ar i Ar–AR. *Przegląd Geologiczny*, 50, 227– 229 (in Polish with English abstract)

MIGASZEWSKI, Z.M., HALAS, S. AND DURAKIEWICZ, T., 1996. The age of the calcite mineralization in the Holy Cross Mts. based on lithologic-petrographic and isotopic evidence (in Polish with English summary). *Przeglad Geologiczny*, 44, 275–280.

MIZERSKI, W. 1979. Tectonics of the Lysogóry unit in the Holy Cross Mts. *Acta Geologica Polonica*, **29**, 1–38.

MIZERSKI, W. 1995. Geotectonic evolution of the Holy Cross Mts in central Europe. *Biuletyn Państwowego Instytutu Geologiczny*, **372**, 5–47.

NARKIEWICZ, M. & OLKOWICZ-PAPROCKA, I. 1983. Stratygrafia dewońskich utworów węglanowych wschodniej części Gór Świętokrzyskich. *Kwartalnik Geoogiczny*, **34(3)**, 225–255.

NARKIEWICZ, M., RACKI, G. & WRZOŁEK, T. 1990. Litostratygrafia dewońskiej serii stromatoporowo–koralowcowej w Górach Świętokrzyskich. *Kwartalnik Geologiczny*, **34(3)**, 433–456.

NARKIEWICZ, M., POPRAWA, P., LIPIEC, M., MATYJA, H., & MIŁACZEWSKI, L. 1998. Pozycja paleogeograficzna i tektoniczna a rozwój subsydencji dewońsko-karbońskiej obszaru pomorskiego i radomsko–lubelskiego. *Prace Państwowego Instytutu Geologiczny*, **CLXV**, 31–46.

NAWROCKI, J. 1995. Paleomagnetic constraints for Variscan mobilism of the Upper Silesian and Malopolska Massifs, southern Poland – comment. *Geological Quarterly*, **38**, 271–282.

NAWROCKI, J. 2000. Late Silurian paleomagnetic pole from the Holy Cross Mountains: constraints for the post Caledonian tectonic activity of the Trans European Suture Zone. *Earth and Planetary Science* Letters, **179**, 325–334.

ORŁOWSKI, S. 1992. Cambrian stratigraphy and stage subdivision in the Holy Cross Mountains, Poland. *Geological Magazine London*, **129**, 471–474.

ORŁOWSKI, S. & MIZERSKI, W. 1995. New data on geology of the Middle Cambrian rocks in the Klimontow Anticlinorium (Holy Cross Mts.). *Geological Quarterly*, **39(3)**, 293–306.

ORŁOWSKI, S. & MIZERSKI, W., 1996. The Cambrian rocks and their tectonic evolution in the Deminy Anticline of the Holy Cross Mts. *Geological Quarterly*, **40(3)**, 353–366.

PHARAOH, T. & BAYER, U. 1999. Deep structure and basin evolution along the Trans European Suture Zone: introduction. *Terra Nostra*, **11**, p. 300.

PHARAOH, T. C., ENGLAND, R. W., VERNIERS, J. & ŻELAŻNIEWICZ, A. 1997. Introduction: geological and geophysical studies in the Trans-European Suture Zone. *Geological Magazine [London]*, **134(5)**, 585–590.

POŻARYSKI, W. 1975. Interpretacja geologiczna wynikow glebokich sondowan sejsmicznych na VII profilu miedzynarodowym. *Przegląd Geologiczny*, **23**, 163–171.

POŻARYSKI, W. 1977. The Variscan epoch in the Epi-Gothian Platform and its border zone. *In*: W. POŻARYSKI (ed.), *Tectonics of Poland*, Wydawnictwa Geologiczne, Warsaw, 207–252.

POŻARYSKI, W. & KARNKOWSKI, P. 1992. Tectonic map of Poland during the Variscan time. 1:1 000 000, *Państwowy Instytut Geologiczny* (ed.), Wydawnictwa Geologiczne, Warsaw.

POŻARYSKI, W., GROCHOLSKI, A., TOMCZYK, H., KARNKOWSKI, P. & MORYC, W. 1992. Przekrój geologiczny przez Polskę południowo-wschodnią. *Przegląd Geologiczny*, **41**, 687–695 [in Polish with English abstract].

RACKI, G. 1981. Stratygrafia i tektonika utworów dewonu środkowego i górnego w kamieniołomach Jażwica. *In: Przewodnik LIII Zjazdu Polskiego Towarzystwa Geologicznego*. Kielce, 6–8 września 1981. ŻAKOWA, H. (ed.), *Wydawnictwa Geologiczny*, 171–179.

RACKI, G. 1993. Evolution of the bank to reef complex in the Devonian of the Holy Cross Mountains. *Acta Palaeontologica Polonica*, **37(2–4)**, 87–182.

RACKI, G. & BULTYNCK, P. 1993. Conodont biostratigraphy of the Middle to Upper Devonian boundary beds in the Kielce area of the Holy Cross Mts. *Acta Geologica Polonica*, **43(1–2)**; 1–25.

ROMANEK, A. & RUP, M. 1990. Podział litostrtygraficzny dewonu w profilu otworu wiertniczego Kowala 1 (in Polish with English summary). *Kwartalnik Geologiczny*, **34**, 221–242.

SCHÄTZ, M., ZWING, A., BELKA, Z. SOFFEL, H. C. & BACHTADSE, V. 2002. Palaeomagnetism of Ordovician carbonate rocks from Malopolska, SE Poland – no evidence for major rotations. *EGS Meeting, Nice*, abstract EGS02-A-03278.

STUPNICKA, E. 1988. Charakter i geneza dyslokacji Świętokrzyskiej. *Przegląd Geologiczny*, **1**, 40–46.

STUPNICKA, E. 1992. The significance of the Variscan Orogeny in the Świętokrzyskie Mountains (Mid Polish Uplands). *Geologische Rundschau*, **81**, 561–570.

SZANIAWSKI, R. 1997. *Paleomagnetyzm a diageneza dolomitów dewońskich z regionu kieleckiego Gór Świętokrzyskich*. M.Sc. Thesis, Arch. Geol. Fac., Warsaw University, 87 pp.

SZULCZEWSKI, M. 1968. Slump structures and turbidites in Upper Devonian limestones of the Holy Cross Mts. *Acta Geologica Polonica*, **18(2)**, 303–330.

SZULCZEWSKI, M. 1971. Upper Devonian conodonts, stratigraphy and facial development in the Holy Cross Mts. *Acta Geologica Polonica*, **21**, 1–135.

SZULCZEWSKI, M. 1978. The nature of unconformities in the Upper Devonian–Lower Carboniferous condensed sequence in the Holy Cross Mts. *Acta Geologica Polonica*, **28**, 283–297.

SZULCZEWSKI, M. 1989. Światowe i regionalne zdarzenia w zapisie stratygraficznym pogranicza franu z famenem Gór Świętokrzyskich (Stratigraphic record of global and regional events across the Frasnian–Famennian boundary in the Holy Cross Mountains). *Przegląd Geologiczny [Warsaw]*, **11**, 551–557.

SZULCZEWSKI, M. 1994. The Holy Cross Mountains during the Devonian and Carboniferous. Excursion Guidebook, the Holy Cross Mountains. *Europrobe, Trans-European Suture Zone Workshop*. Kielce, Poland, September 24th–October 1st, 1994, 19–33.

SZULCZEWSKI, M. 1995. Depositional evolution of the Holy Cross Mts. (Poland) in the Devonian and Carboniferous – a review. *Geological Quarterly*, **39**, 471–488.

SZULCZEWSKI, M. & ŻAKOWA, H. 1976. Nowe dane o famenie synkliny gałęzickiej. *Biuletyn Instytutu Geologicznego*, **296(12)**, 51–73.

SZULCZEWSKI, M., BELKA, Z. & SKOMPSKI, S. 1996. The drowning of the carbonate platform: an example from the Devonian–Carboniferous of the southwestern Holy Cross Mountains, Poland. *Sedimentary Geology*, **106**, 21–49.

TAIT, J., SCHÄTZ, M., BACHTADSE, V. & SOFFEL, H. 2000. Palaeomagnetism and Palaeozoic palaeogeography of Gondwana and European terranes. *In:* FRANKE, W., HAAK, V., ONKEN, O. & TANNER, D. (eds) Orogenic processes: quantification and modelling in the Variscan belt. *Geological Society of London, Special Publication*, **179**, 21–34.

TARNOWSKA, M. 1981. Dewon dolny w centralnej części Gór Świętokrzyskich. Przewodnik Zjazdu: *polskiego towarzystwa geologicznego*, **53**, 57–68.

THYBO, H., PHARAOH, T. & GUTERCH, A. (eds) 1999. Geophysical investigations of the Trans-European Suture Zone. *Tectonophysics*, **314(1–3)**, 1–5.

TOMCZYK, H. & TURNAU-MORAWSKA, M. 1964. Stratygrafia i petrografia ordowiku Brzezin w Górach Świętokrzyskich. *Acta Geologica Polonica*, **14(4)**; 501–546 (in Polish with English abstract).

TWISS, R. J. AND MOORES, E. M. 1992. *Structural geology*, New York, 532 pp.

ZNOSKO, J. 1974. Outline of the tectonics of Poland and the problems of the Vistulicum and Variscicum against tectonics of Europe. *In*: CHABELSKA, E. (ed.), *Tectonic Research in Poland*. Wydawnictwa Geologiczne, Warsaw, 7–38.

ZNOSKO, J. 1983. Tektonika srodkowopołudniowej Polski pozakarpackiej (in Polish with English summary). *Kwartalnyk Geologiczny*, **27**, 73–78.

ZNOSKO, J. 1996. Tectonic style of the Early Palaeozoic sequences in the Holy Cross Mountains. *Geological Quarterly*, **40(1)**, 1–22.

ZNOSKO, J. 2001. New data on Caledonian, alpine-style folding in the Holy Cross Mts., Poland. *Geological Quarterly*, **45(2)**, 155–163.

ŻAKOWA, H. 1981. Rozwój i stratygrafia karbonu Gór Świętokrzyskich. Wykształcenie i stratygrafia karbonu w synklinie gałęzickiej. Przewodnik Zjazdu: polskiego towarzystwa geologicznego, **53**, **89–100**, 197–209.

ŻAKOWA, H. & MIGASZEWSKI, Z. 1995. The Carboniferous in the Holy Cross Mts. *In*: MALEC, J., MATYJA, H., MIGASZEWSKI, Z., PASZOWSKI, M., PROTAS, A., SKOMPSKI, S., SZULCZEWSKI, M., ZBROJA, S., ŻAKOWA, H. & ŻELICHOWSKI, A. (eds) Development of the Variscan basin and epi-Variscan cover at the margin of the East European Platform (Pomerania, Holy Cross Mts., Kraków Upland). *XIII International Congr. Carbonif.–Permian, (XIII ICC-P), August 28 – September 2, Krakov, Poland.*

The Tamworth Belt in Southern Queensland, Australia: thrust-characterized geometry concealed by Surat Basin sediments

W. WARTENBERG[1], R. J. KORSCH[2] & A. SCHÄFER[1]

[1]*Geologisches Institut, Rheinische Friedrich-Wilhelms-Universität Bonn, Nußallee 8, 53115 Bonn, Germany*
(e-mail: wwberg@uni-bonn.de; schaefer@uni-bonn.de)
[2]*Australian Geodynamics Cooperative Research Centre, Geoscience Australia, GPO Box 378, Canberra, ACT 2601, Australia*
(e-mail: Russell.Korsch@ga.gov.au)

Abstract: The subsurface geometry and tectonic development of the Devonian–Carboniferous Tamworth Belt, a fore-arc basin in the New England Orogen, Eastern Australia, has been examined using seismic reflection, aeromagnetic and gravity data. The Tamworth Belt is bounded to the west by the Moonie Fault, a thrust fault, which exhibits a fault-bend-fold geometry. Two major westward-dipping faults form the main eastern boundary, with a series of eastward-dipping backthrusts located farther to the west. The eastern margin also coincides with a gravity and magnetic ridge, similar to the gravity and magnetic pattern of the serpentinites and iron-enriched rocks that are exposed along the Peel Fault to the south. In the investigated area, the Tamworth Belt is over 50 km wide and has been shortened by at least 10 km across strike. The sedimentary succession is at least 12 km thick and is moderately folded. Within the succession, six seismic sequences were identified, each of which is separated by a major sequence boundary.

The New England Orogen (NEO) is the easternmost tectonic unit of the Australian continent, extending over approximately 1300 kilometres from Bowen (22°S) to Newcastle (32°S) (Fig. 1). It consists of magmatic arc, fore-arc and accretionary wedge rocks produced during Late Devonian to Cretaceous plate convergence at the interface of eastern Gondwana and the Panthalassan Ocean (Leitch 1975; C. G. Murray *et al.* 1987; Korsch *et al.* 1990). The New England Orogen may be subdivided into a southern and a northern unit, due to the presence of the Mesozoic Surat and Clarence–Moreton basins that conceal much of the central part of the orogen (Scheibner *et al.* 1996; C. G. Murray *et al.* 1997). Thus, it is difficult to correlate the southern and the northern fore-arc basin successions (Tamworth and Yarrol belts, respectively, in Fig. 1). The accretionary wedge assemblages occur in outcrop across the eastern NEO (e.g. Tablelands Complex assemblage, Woolomin, Beenleigh, D'Aguilar, Wandilla and Shoalwater terranes in Fig. 1), while the magmatic arc component is only exposed in the northern NEO (Connors and Auburn arcs in Fig. 1). In order to provide a better understanding of the geological setting of the Tamworth Belt, the relevant components of the New England Orogen and adjacent basins (e.g. back-arc Bowen Basin), together with some background information on geometry and tectonic deformation, are introduced in the following subsections.

Magmatic arc

There are apparently no arc-related rocks south of the Mesozoic cover. Two models have been proposed to describe the original tectonic setting along the eastern margin of Gondwana for the Silurian to Early–Middle Devonian. Aitchison & Flood (1995) interpret the absence of the arc as being related to a change in subduction direction over time. They envisage a western continental margin and an eastern island arc with the intervening oceanic crust being subducted to the east. After subduction ceased, there was a polarity change, with oceanic crust subducting to the west from Late Devonian to Late Carboniferous. Korsch *et al.* (1997), however, have suggested that the subduction system was westward-dipping and that the arc could now be buried beneath the younger Permian–Triassic sedimentary rocks of the Gunnedah Basin (see Fig. 1 for location).

From: McCann, T. & Saintot, A. (eds) *Tracing Tectonic Deformation Using the Sedimentary Record.* Geological Society, London, Special Publications, **208**, 185–203. 0305-8719/03/$15.00

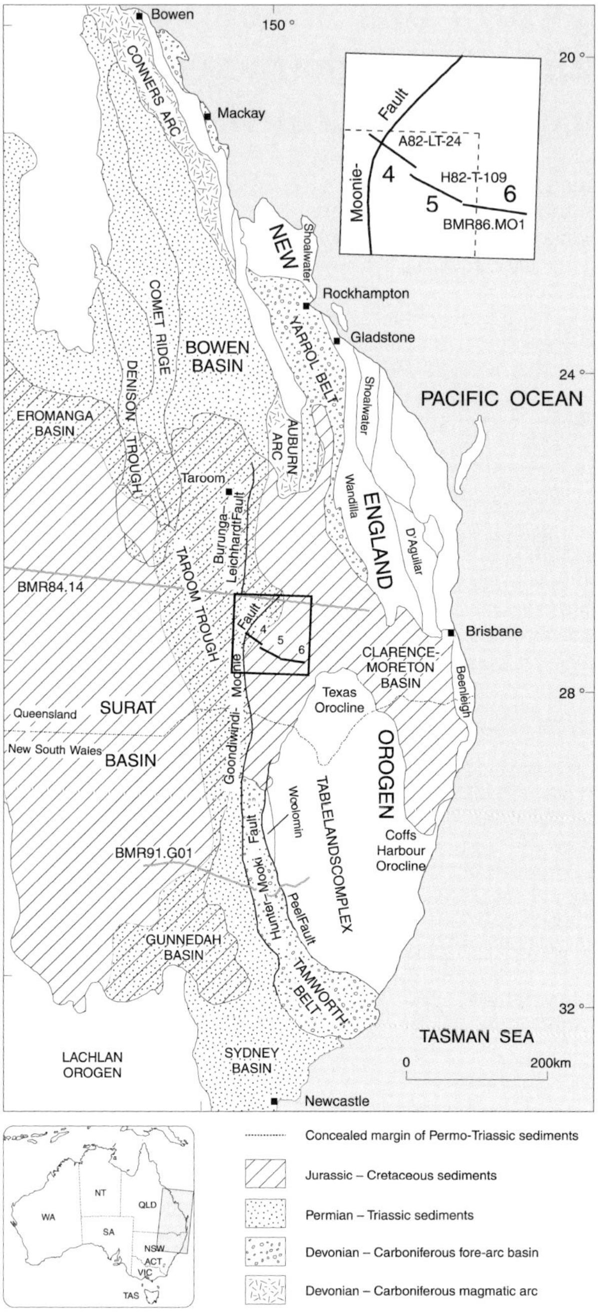

Fig. 1. Map of Eastern Australia, showing the relationship of the New England Orogen to the adjacent sedimentary basins. The Tamworth and Yarrol belts represent the Devonian–Carboniferous fore-arc basin that extends in the subsurface to the east of the Goondiwindi–Moonie and Burunga–Leichhardt faults. The box at the northern end of the Goondiwindi–Moonie Fault identifies the study area and is shown at the top right in further detail where the smaller box marked with a dashed line identifies the seismic study area. The lines numbered 4, 5 and 6 refer to the locations of the seismic profiles and aeromagnetic traverses shown in Figures 4, 5 & 6 respectively. The east–west-running grey heavy solid lines refer to the deep seismic profiles BMR84.14 and BMR91.G01 respectively.

Fore-arc basin

The Late Devonian–Late Carboniferous Tamworth Belt is dominated by volcaniclastic sedimentary rocks deposited predominantly in a shallow-marine shelf environment (McKelvey & McPhie 1995; Scheibner 1998). Volcanic material is more abundant in the western part of the region, closer to the active magmatic arc (McPhie 1987). The geology of the Yarrol Belt has been summarized by Day *et al.* (1983) and recently revised by the Queensland Department of Mines and Energy (C. G. Murray *et al.* 1997).

Back-arc basin

The back-arc Bowen Basin, located to the west of the Tamworth Belt, consists of terrestrial to shallow marine successions. It was initiated by Early Permian back-arc extension and, from the middle Permian, subsidence was driven by foreland loading (Korsch & Totterdell 1996). The Bowen Basin is part of a near-longitudinal Permian–Triassic basin system that includes, from south to north, the Sydney, Gunnedah and Bowen basins (see Fig. 1 for location).

Fault geometry

Petroleum industry seismic lines across the eastern part of the Bowen Basin system show that the approximate eastern limit of this basin system is defined by a major system of discrete Middle Triassic thrust faults and displacement transfer zones (Korsch & Totterdell 1995a). These faults are, from south to north, the Hunter, Mooki, Goondiwindi, Moonie, Leichhardt and Burunga faults (see Fig. 1 for location). Movement on all of these faults is west-directed, apart from the northernmost Burunga Fault, which is east-directed. The eastern limit of the exposed part of the Tamworth Belt is defined by the Peel Fault, to the east of which occur accretionary wedge rocks of the Tablelands Complex (Fig. 1). In outcrop, in northern New South Wales, the Peel Fault dips steeply to the east, and is interpreted in the deep seismic line BMR91.G01 to be a splay off a major westward-dipping structure that cuts to the base of the crust (Korsch *et al.* 1993, 1997).

Deformation

Deformation in the Tamworth and Yarrol belts was mainly controlled by west to west-northwest-directed thrusting, forming part of a foreland thrust belt (Woodward 1995; Holcombe *et al.* 1997). In the Bowen and Surat basins, Korsch & Totterdell (1995b) and Korsch *et al.* (1998) have defined several deformational events, most of which were contractional in nature. At least two of the deformational events seen in the subsurface Tamworth Belt can be related to events in the Bowen and Surat basins. These are the Middle–Late Triassic Goondiwindi Event and the early Late Cretaceous Moonie Event.

In the study area, the Devonian–Carboniferous Tamworth Belt, therefore, is a fore-arc basin which is conformably overlain and concealed by the Early Permian–Middle Triassic back-arc Bowen Basin and, subsequently, unconformably by the Early Jurassic–Early Cretaceous Surat Basin (Fielding *et* al. 1990; O'Brien *et al.* 1990; Korsch *et al.* 1992*a*, 1992*b*) (see Fig. 1). Much of the hydrocarbon-driven research in this area has been concentrated in the Surat Basin. However, the precise geometry and evolutionary history of the underlying Tamworth Belt has been relatively ignored. In this paper, using a database which is mainly restricted to industry seismic profiles (originally shot in the Surat Basin), additional geophysical data (gravity and aeromagnetics), some isolated wells, and two deep seismic profiles (BMR84.14 and BMR91.G01), we will focus on outlining the basin architecture and the tectonic evolution of the Tamworth Belt fore-arc basin (here only present in the subsurface) and its relationship to the overlying Bowen Basin (in the southern part of the orogenic belt and close to the Texas Orocline) (Wartenberg *et al.* 1999) (Fig. 1).

Tectonic evolution of the New England Orogen

As noted above, west-directed subduction-related processes in the Devonian and Carboniferous produced three parallel belts at the eastern Gondwana margin in Australia (Kirkegaard 1974; C. G. Murray *et al.* 1987; Korsch *et al.* 1993; Scheibner and Basden 1996). In the Late Carboniferous, there was a change from a compressional to an extensional tectonic regime, possibly related to a change in the dynamics of the subduction system (Korsch *et al.* 1997; Murray *et al.* 1987). Subsequently, in the Early Permian (*c*.290–280 Ma), the initiation of the Bowen–Gunnedah–Sydney basin system took place to the west of the magmatic arc in a back-arc tectonic setting (Korsch *et al.* 1993; Tadros 1993). Also in the Early Permian, but probably later than the mechanical extension phase, oroclinal bending of the fore-arc basin and accretionary wedge (*c*.285–265 Ma) occurred to produce the Texas and Coffs Harbour oroclines (Korsch & Harrington 1987; C. G. Murray *et al.* 1987) (see Fig. 1 for location).

The late Early to early Late Permian (*c.*265–262 Ma) saw another change in the dynamics of the subduction system, with the back arc changing from an extensional to a contractional regime (Korsch & Totterdell 1995c). This led to the formation of a retro-foreland fold–thrust belt (cf. Catuneanu *et al.* 1997) that was better developed in the Queensland sector of the New England Orogen than in New South Wales (Korsch *et al.* 1997). Moreover, the contractional regime resulted in the development of a significant retro-foreland basin phase in the Bowen–Gunnedah–Sydney basin system that extended until the Middle Triassic (Korsch & Totterdell 1995b) (see Fig. 2 for details of the stratigraphy).

From the Early Jurassic to Early Cretaceous, eastern Australia was still part of the eastern convergent margin of Gondwana, but almost all of it was situated in a back-arc setting, with only minor remnants of the magmatic arc being preserved near the Queensland coast (Korsch & Totterdell 1996). The Surat and Eromanga basins (see Fig. 1 for location) developed at this time within a back-arc setting, inboard of the continental margin volcanic arc. These basins contain a thick, sometimes volcaniclastic, succession which blanketed much of the Devonian–Carboniferous subduction-related units, including the fore-arc basin, part of which, in southern Queensland and northern New South Wales, is still covered by up to nearly 2000 m of sediments (see Fig. 1 for location, and Figs 4, 5 & 6 for the overlying Surat sediments).

Geophysical data

The data used for this study consists of a series of seismic reflection profiles, petroleum exploration well information, gravity and aeromagnetic data.

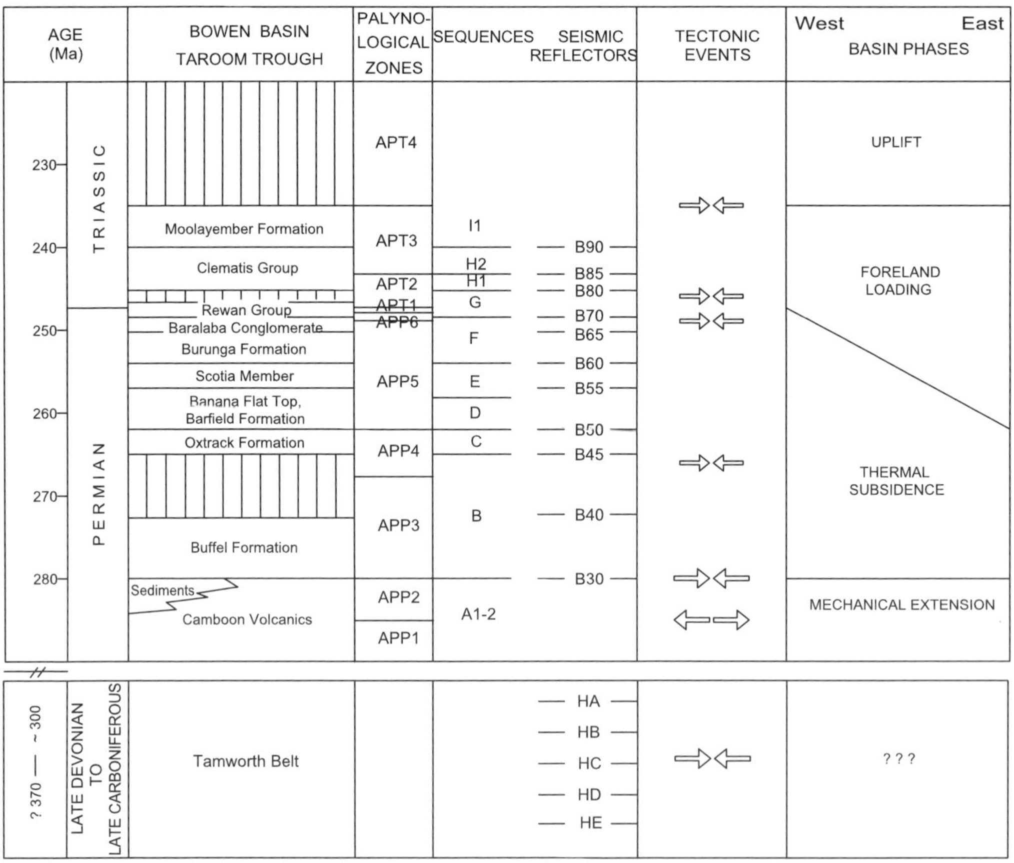

Fig. 2. Chart showing the stratigraphy, seismic reflectors and basin phases of the Bowen Basin and Tamworth Belt successions. For the location of the Taroom Trough, see Figure 1.

These data were used to provide a more complete view of the fore-arc basin in the subsurface in order to define the basin boundaries.

Seismic reflection data

Data used for this paper consists of a series of 109 good quality seismic reflection profiles (Fig. 3) and eight exploration wells (five of which penetrate the Tamworth Belt) in southernmost Queensland, Australia. These profiles were originally acquired by the petroleum industry as part of their exploration of the overlying Surat Basin. The majority of the seismic lines are unmigrated (usually final stacks) and were displayed with a datum of 244 m. Mapping of

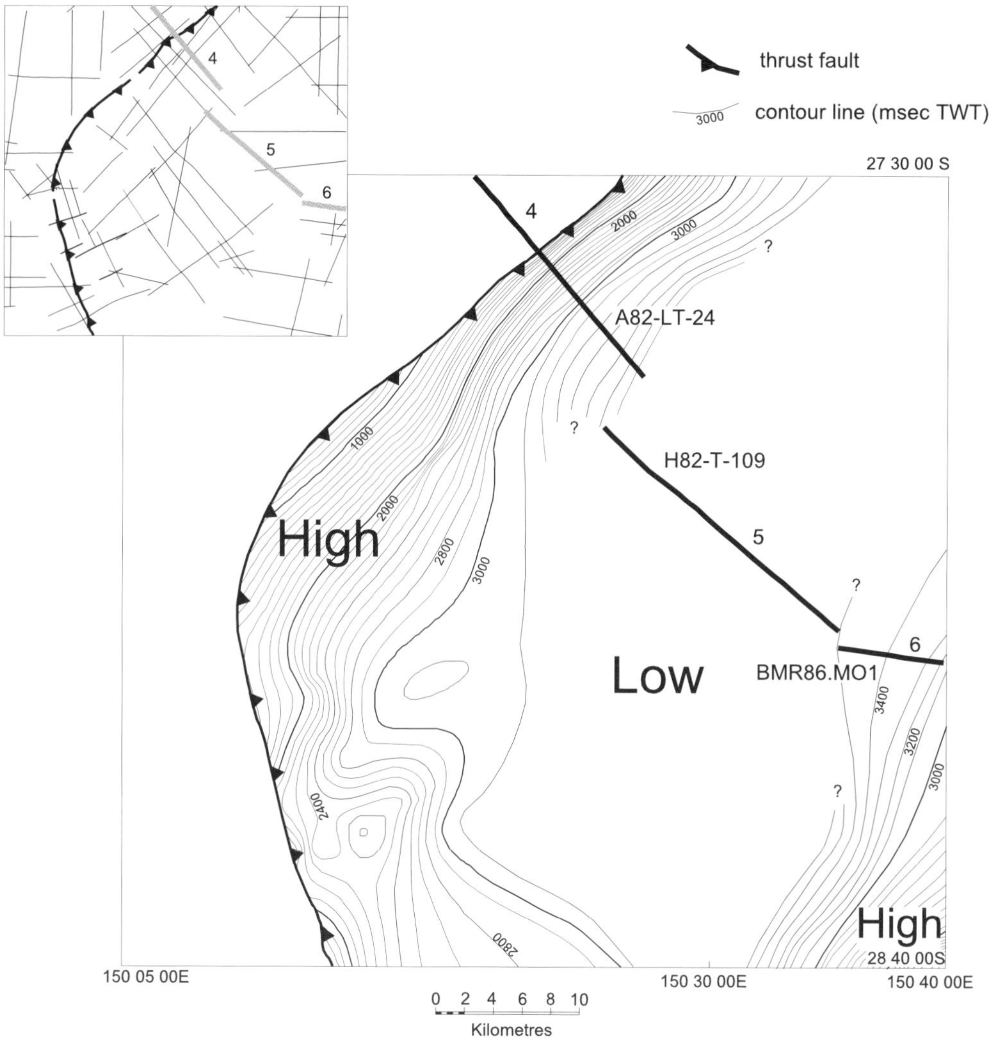

Fig. 3. Schematic structure contour map of the Tamworth Belt situated between the Moonie Fault to the west and the Peel Fault extension to the east, based on the interpretation of the seismic profile network in the investigated area (depths in milliseconds two-way-travel time). The western basin boundary is the Moonie Fault. Sediments were also deposited to the west of this structure but are now mainly deeper than the seismic resolution (see Fig. 4, where the uppermost reflectors HA and HB can be traced west of the fault). Within the Tamworth Belt itself the basement is not always clearly imaged. Thus, the resulting diagram illustrates the broad, rather than the detailed basin trend (see text for details). The black heavy solid lines represent the three representative profiles (Figs 4, 5 & 6). In the top left corner, the seismic profile network is illustrated in light black solid lines.

the interpreted profiles was carried out using PETROSYS software. Within the Tamworth Belt succession, five reflectors (HA, HB, HC, HD, HE) were mapped in the area, although HE is restricted to the eastern part of the region (Fig. 6). The Tamworth Belt reflectors span a Late Devonian to Late Carboniferous age (Fig. 2), although the precise ages of the individual reflectors, however, could not be determined, since it was not possible to correlate them with any of the wells. Therefore, in order to ensure that the contour maps of the region were representative, the original database was restricted to the area containing the highest concentration of seismic profiles (Fig. 3, compare with the inset in Fig. 1).

Two additional reflectors (S10, B30) are also included on the diagrams (Figs 2, 4 & 5). The S10 represents the base of the Surat Basin, while the base of the Bowen Basin is defined by the B30 sequence boundary (terminology of Totterdell *et al.*, 1992) (see Fig. 2).

In this paper, we focus on three representative seismic sections in order to discuss the sequence stratigraphic interpretation and the subsurface geometry of the Tamworth Belt (see Fig. 1 for location). Two industry lines A82-LT-24 (Fig. 4) and H82-T-109 (Fig. 5) were acquired in 1982 by Alliance Minerals and Hartogen Exploration, respectively. The former was recorded down to four seconds and the latter to three seconds two-way travel time (TWT). The third line, BMR86.M01 (Millmerran profile, Fig. 6), was originally acquired by the Bureau of Mineral Resources (now Geoscience Australia). The Millmerran profile was the only seismic line

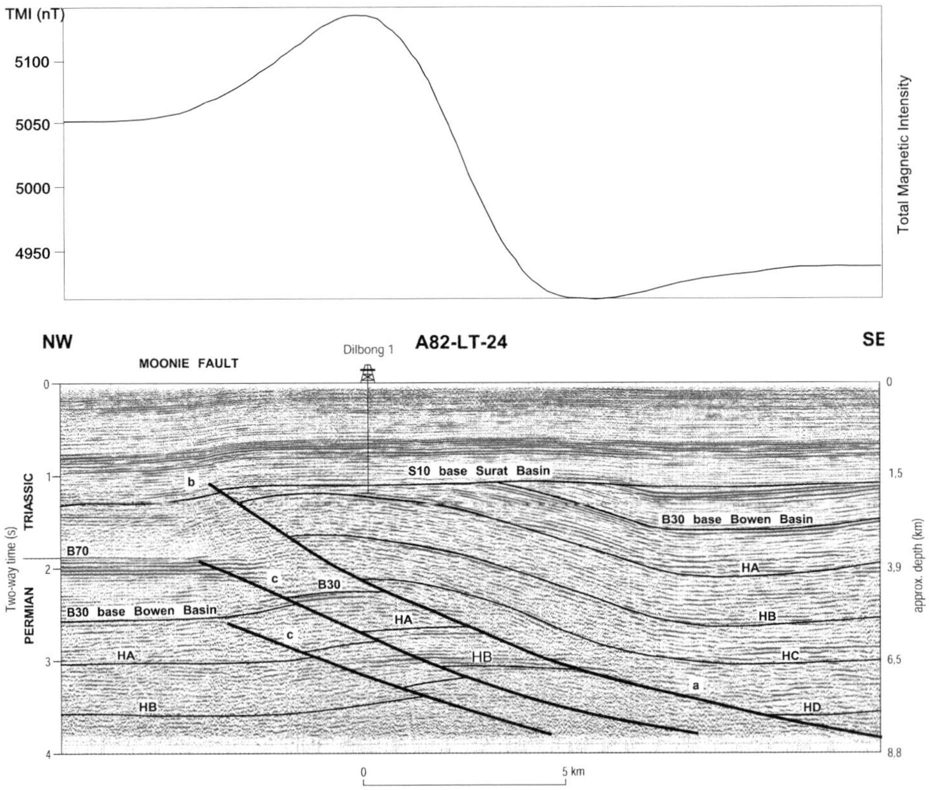

Fig. 4. Interpretation of the unmigrated (final stack) seismic reflection profile A82-LT-24 located at the eastern margin of the Bowen Basin. Heavy solid lines represent the Moonie Fault (a – showing a fault–bend fold geometry) and two synthetic thrust faults with minor displacement (c). The Late Cretaceous reactivation of the Moonie Fault is highlighted as b. The light solid lines represent the interpreted seismic reflectors HA to HD of the Tamworth Belt succession. B30 and S10 represent the sequence boundaries at the base of the Bowen and Surat basins, respectively, and B70 represents the approximate position of the Permian–Triassic boundary. The aeromagnetic traverse is displayed above the seismic section.

within the study area that was recorded to six seconds TWT. The original quality of this profile, however, was poor, and therefore in 1998 Geoscience Australia undertook additional processing of the profile in order to be better able to integrate the data into the existing database. The reprocessed Millmerran profile was subsequently redisplayed at the same vertical scale as the two industrial lines. Together, these three lines form a composite cross-section oriented approximately NW–SE across the Tamworth Belt (Figs 1 & 3). Unfortunately, these sections do not tie to one another, but the basin geometry has allowed correlation between the profiles (gaps between the lines are approximately 1100 m and 3900 m). The only other cross-sections across the Tamworth Belt are the two deep seismic profiles (BMR84.14 to the north and BMR91.G01 to the south, both recorded down to 20 seconds TWT) both of which are located some distance from the seismic study area (Fig. 1).

Interpretation of the seismic reflection data

In the study area, a series of major NW–W-directed thrusts may be recognized on the basis of seismic data (Figs 4, 5 & 6). The most significant of these structures, the Moonie Fault (Fig. 4), is part of a larger Middle Triassic fault system (see above for details). The complexity of this thrust event is indicated by the presence of synthetic thrusts (Fig. 4) and west-directed backthrusts (Fig. 6). Two major westward-dipping faults (Fig. 6) form the eastern boundary of the Tamworth Belt. The western Tamworth Belt boundary could not be defined by the present database. Moreover, the base of the subsurface Tamworth Belt succession could not be defined due to the shallow character of the seismic profiles (in the northeastern part of the study area the recording time of the seismic lines is 3 s TWT) (Fig. 3). However, the hanging-wall geometry is characterized by the Moonie Fault to the west and the westward-dipping fault to the

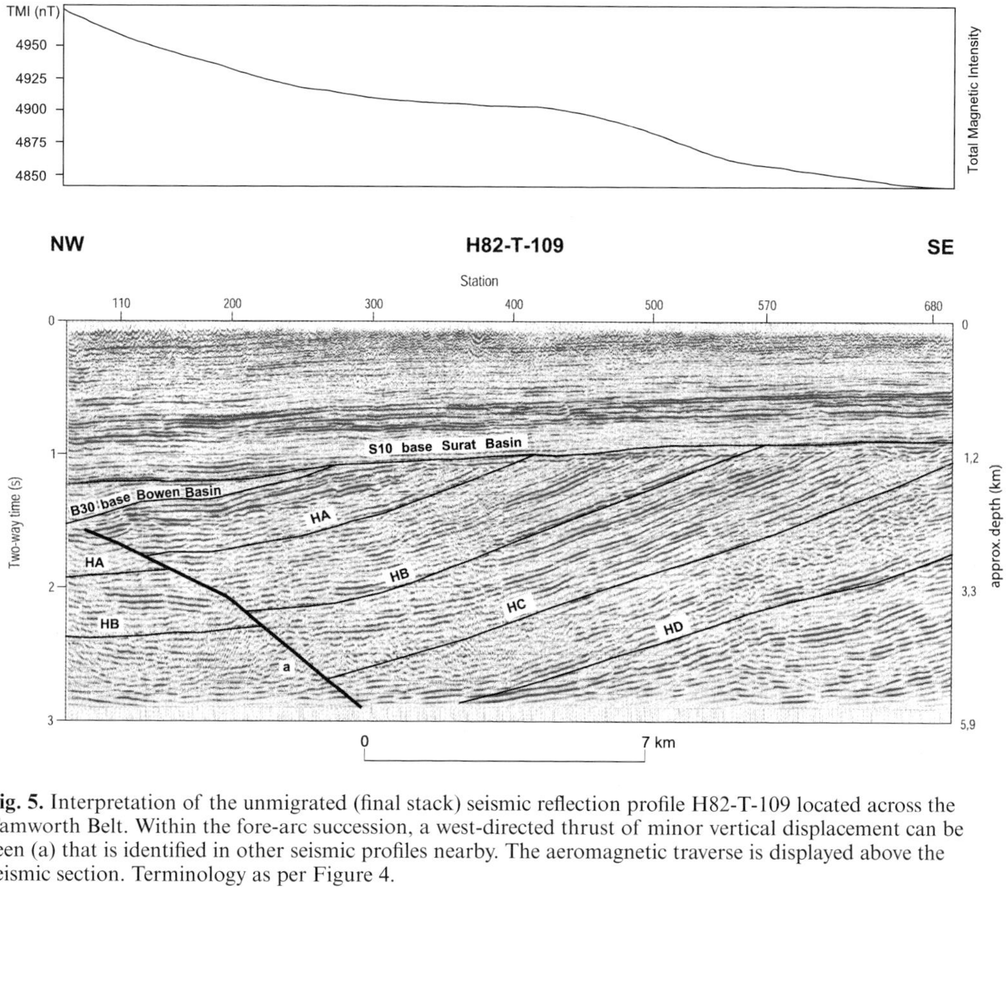

Fig. 5. Interpretation of the unmigrated (final stack) seismic reflection profile H82-T-109 located across the Tamworth Belt. Within the fore-arc succession, a west-directed thrust of minor vertical displacement can be seen (a) that is identified in other seismic profiles nearby. The aeromagnetic traverse is displayed above the seismic section. Terminology as per Figure 4.

east (c_1 in Fig. 6) and thus images the structure of the subsurface Tamworth Belt succession (Fig. 3).

The Moonie Fault which, towards its northern tip, curves to the NE, has a classic fault-bend fold geometry (Fig. 4) (for definition see Suppe, 1983) and appears to transport part of the Bowen Basin as a piggy-back structure in the hanging wall of the thrust (Fig. 4). On the profile A82-LT-24 (Fig. 4), the fault has a vertical displacement of over 3 km. Geometrical restoration of the Moonie Fault, prior to the phase of thrust movement, suggests that the Tamworth Belt succession possibly extends to the west beneath the sedimentary rocks of the Bowen Basin within the Taroom Trough (on seismic line A82-LT-82, reflectors HA and HB were identified west of the thrust fault, Fig. 4). However, the precise western extent of the Tamworth Belt could not be determined. The seismic pattern of the Tamworth Belt succession southeast of the Moonie Fault is clearly of a different nature to that of the succession on the northwestern side of the fault. Within the hanging wall, the Tamworth Belt succession may be characterized by strong, more continuous reflectors, whereas in the footwall the reflectors are weaker and markedly less continuous (Fig. 4).

Tracing particular horizons across the thrust fault reveals that there is a distinct offset. For example, the B30 reflector is positioned at *c.*1650 ms TWT within the hanging-wall succession, whereas on the footwall side it is at a recording depth of 2600 ms TWT. The offset, therefore, between the two reflectors is *c.*950 ms TWT (Fig. 4). Additionally, both reflectors may be traced almost as far as the fault plane, revealing the thrusted nature of reflector repetition. Such a pattern suggests, perhaps, the

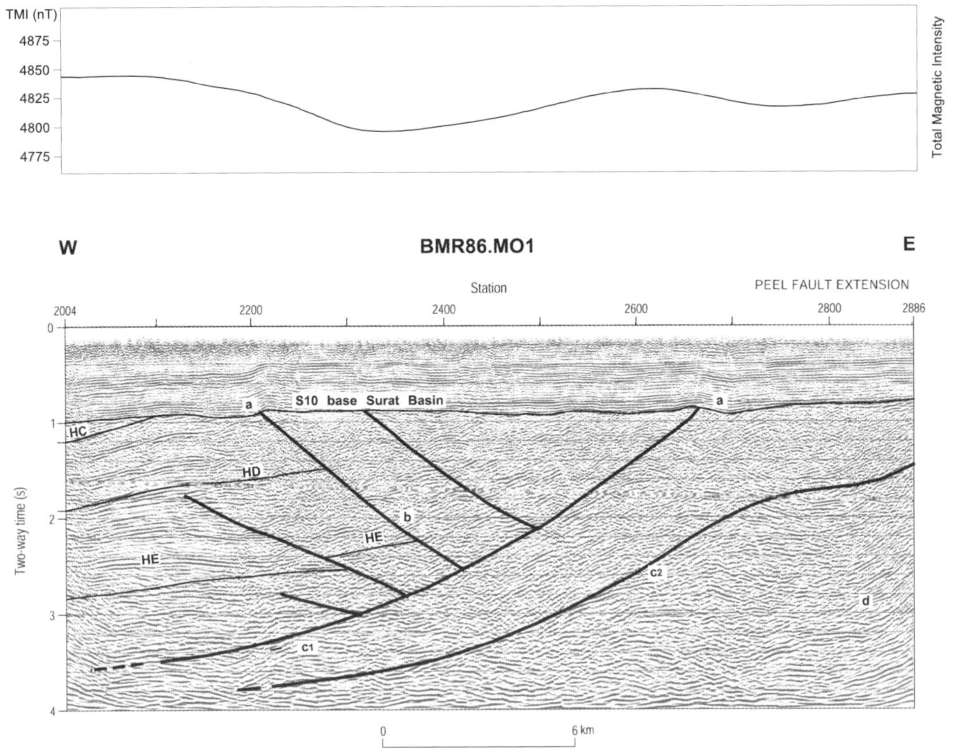

Fig. 6. Interpretation of the migrated seismic reflection profile BMR86.M01 (Millmerran-profile) located at the eastern end of the Tamworth Belt. The ratio of vertical to horizontal scale approximately equals 1. Heavy solid lines represent the two interpreted major westward-dipping structures (c_1, c_2) and some west-directed backthrusts (b). The fault at the right hand-margin of the profile (c2) can be traced on the aeromagnetic anomaly maps to be the northern subsurface extension of the Peel Fault, thus defining the eastern limit of the Tamworth Belt. The easternmost, more disrupted seismic reflection pattern (d) mimics the nature of the geology in the accretionary wedge. The Late Cretaceous Moonie Event is highlighted as a, folding and uplifting the Surat Basin succession. The aeromagnetic traverse is displayed above the seismic section. Terminology as per Fig. 4.

presence of a fault ramp structure farther to the east of the seismic profile A82-L-24 (Fig. 4).

Within the Tamworth Belt, six seismic sequences are recognized on the basis of reflector truncation geometries. Internally, however, the seismic sequences appear to be very similar in terms of individual reflector properties (e.g. amplitude, continuity, etc.), although there are variations both in terms of these particular properties and also the relationship with either the underlying or overlying sequence boundaries. The within-sequence variations may be interpreted in terms of different facies zones within the fore-arc basin succession. The stratigraphic variation of these zones is, therefore, indicative of the changing pattern of facies distribution over time.

Aeromagnetic and gravity data – methodology

Gravity and aeromagnetic data from eastern Australia are held by Geoscience Australia. The digital information has been modified using ER Mapper software and is based on a geodetic Map Projection with a Geodetic Datum of AGD 66 (Australian Grid Data). Total magnetic intensity (TMI), as well as gravity datasets (Bouguer gravity) were produced in pseudocolour mode and are here displayed in shades of grey. The original datasets for both aeromagnetic and gravity information were initially generated to provide an overview of these parameters for the entire Australian continent. As part of this study, however, both the aeromagnetic and gravity data were reprocessed in order to provide more detailed information on the geometry of the area under investigation.

Aeromagnetic data – specific methodology

The aeromagnetic images (gradient-enhanced residuals of total magnetic intensity) are based on the 1996 Geoscience Australia 1:5 000 000 scale Magnetic Anomaly Map of Australia (Tarlowski *et al.* 1996) and on the 1976 Geoscience Australia 1:2 500 000 scale Magnetic Anomaly Map of Australia (BMR 1976). The surveys within the Tamworth Belt are particularly dense, with an average line spacing of about 400 m, with some surveys having a line spacing of 200 m. As part of this study, the generated images were modified by using a Gaussian equalize. This was based on a contrast maximize using a 3×3 north–south Kernel Sunangle Filter. A 3×3 west–east Kernel Sunangle Filter was used in the algorithm's Intensity Layer. An average 5×5 Kernel Lowpass Filter was added to the Pseudo Layer. Additional techniques (e.g. smoothing and draping the shades of grey) were used in order to enhance the anomaly contrast in such a way as to make it possible to view it as a grey-scale image (Figs 7 & 8). The narrow line spacing of maximum 400 m allowed us to construct several magnetic traverses along key seismic lines in order to complete the structural interpretation of the subsurface Tamworth Belt succession across strike (Figs 4, 5, 6 & 9).

Gravity data – specific methodology

The gravity anomaly map of the New England Orogen region (Fig. 10) is based both on onshore residual Bouguer gravity anomalies (A. S. Murray *et al.* 1997) and on offshore free air gravity anomalies (Sandwell & Smith 1997). The transform limits displaying the working area range from -579.05 μm s^{-2} to 488.96 μm s^{-2}. The image contrast was enhanced by Sun shading with an azimuth of 315° and an elevation of 75° and can, therefore, also be expressed as grey-scale anomaly maps (Fig. 10).

Aeromagnetic data – presentation

Aeromagnetic images of the New England Orogen region image the Bowen Basin to have a higher total magnetic intensity. The Tamworth Belt to the east is generally identified through lower magnetic anomaly values (Figs 4, 7 & 8). Several aeromagnetic traverses were constructed in ER Mapper along the key seismic lines (Figs 4, 5 & 6). The aeromagnetic response of the western Tamworth Belt is characterised by a moderate dip in the magnetic field from west to east (Figs 4, 5 & 8). A NE–SW-trending ridge (*c.*5010–5060 nT) – which appears to change to a north–south-oriented feature further south – identifies the easternmost limit of the Tamworth Belt. Off this aeromagnetic ridge, a second, smaller and more disrupted ridge can be seen forging immediately to the west, also trending NE–SW (Figs 7 & 8).

Aeromagnetic data–- interpretation

By interpreting TMI images based on a line spacing of maximum 400 m, we can expect to obtain valuable information on any major geological feature near the surface. As discussed above, the shallow pattern of the seismic reflection profiles illustrates the flat-lying character of the Mesozoic platform cover (Figs 4, 5 & 6). Some distinct markers are seen on the aeromagnetic images (see previous subsection entitled Aeromagnetic data – presentation) although the

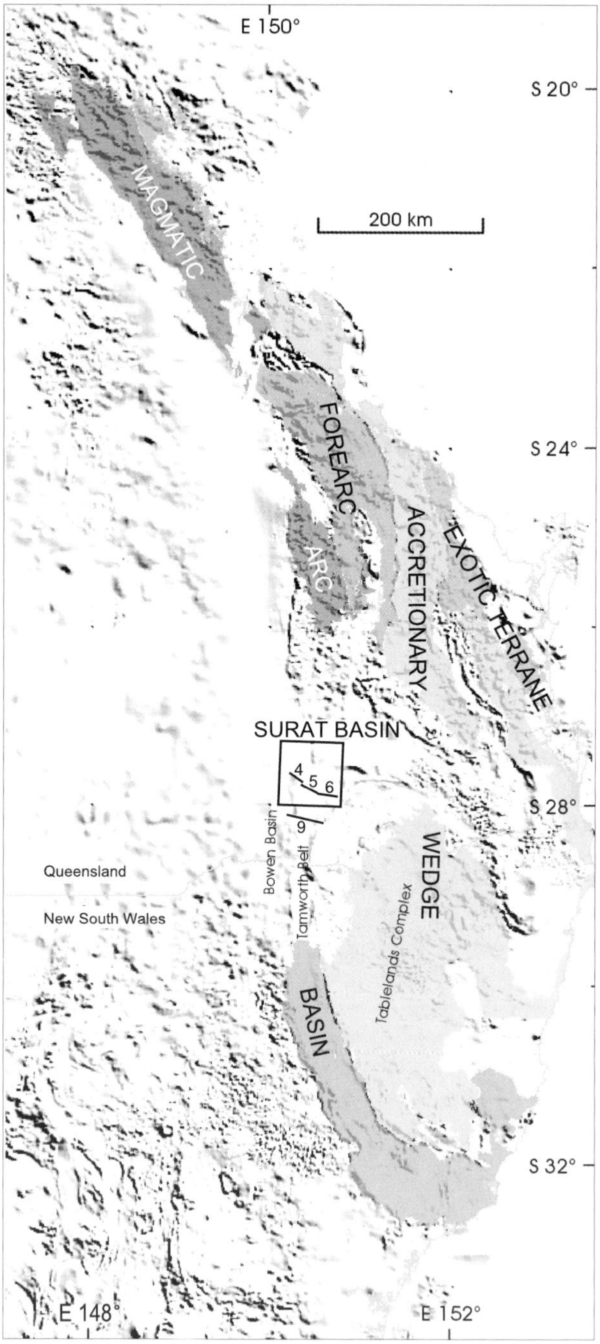

Fig. 7. Aeromagnetic anomaly image of the study area using total magnetic intensity (nT). The map is based on aeromagnetic data collected by Geoscience Australia and the Geological Surveys of New South Wales and Queensland. It is overlain by a simplified terrane map of the New England Orogen, identifying magmatic arc, fore-arc and accretionary complex rocks using different shades of grey. The box frames the study area, showing the positions of the three magnetic profiles as seen in the Figs 4, 5 & 6 respectively (cf. Fig. 1). The profile numbered 9, located directly south of the study area, refers to the position of the aeromagnetic traverse shown in Figure 9.

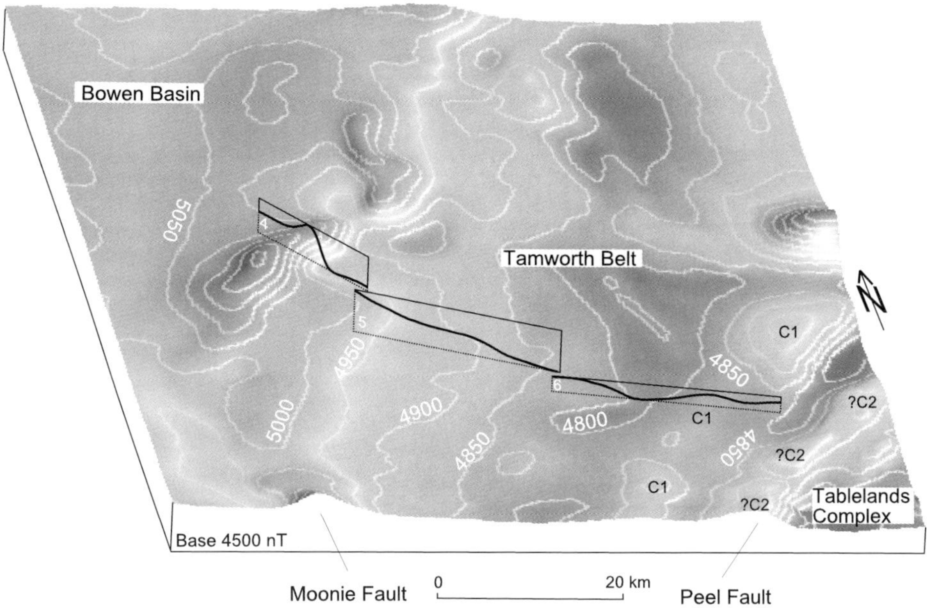

Fig. 8. Three dimensional aeromagnetic image of the study area, displayed in contours with a line spacing of 50 nT. The base of the block diagram is the 4500 nT contour line. The black solid lines numbered 4, 5 and 6 refer to the locations of the seismic profiles and aeromagnetic traverses shown in Figs 4, 5 & 6, respectively. The aeromagnetic high marked as c_1 may be correlated with the westward-dipping fault c_1 on the corresponding seismic profile of Fig. 6. The aeromagnetic ridge marked as ?c_2 may be correlated with the Peel Fault extension (cf. c_2 in Fig. 6).

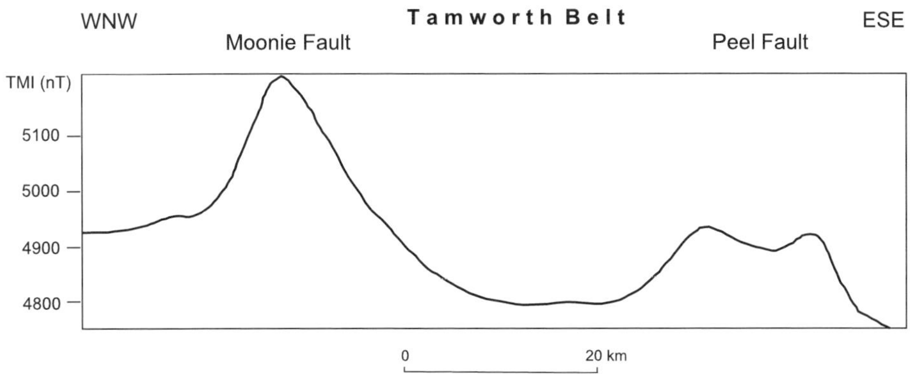

Fig. 9. Aeromagnetic anomaly traverse across the Tamworth Belt showing the total magnetic intensity in nT. See Figure 7 for location.

Mesozoic succession is at least up to 2000 m thick and mafic rocks do not occur within the succession. To explain this, it is assumed that the rock types of the Surat Basin are the same throughout the entire working area and are therefore of a constant magnetic intensity. The Surat Basin sits directly and disconformably on a basement of Bowen Basin and Tamworth Belt, and the geometry of this basement could well explain at least some of the magnetic responses. Although it is not possible to state categorically that the magnetic anomalies relate to the

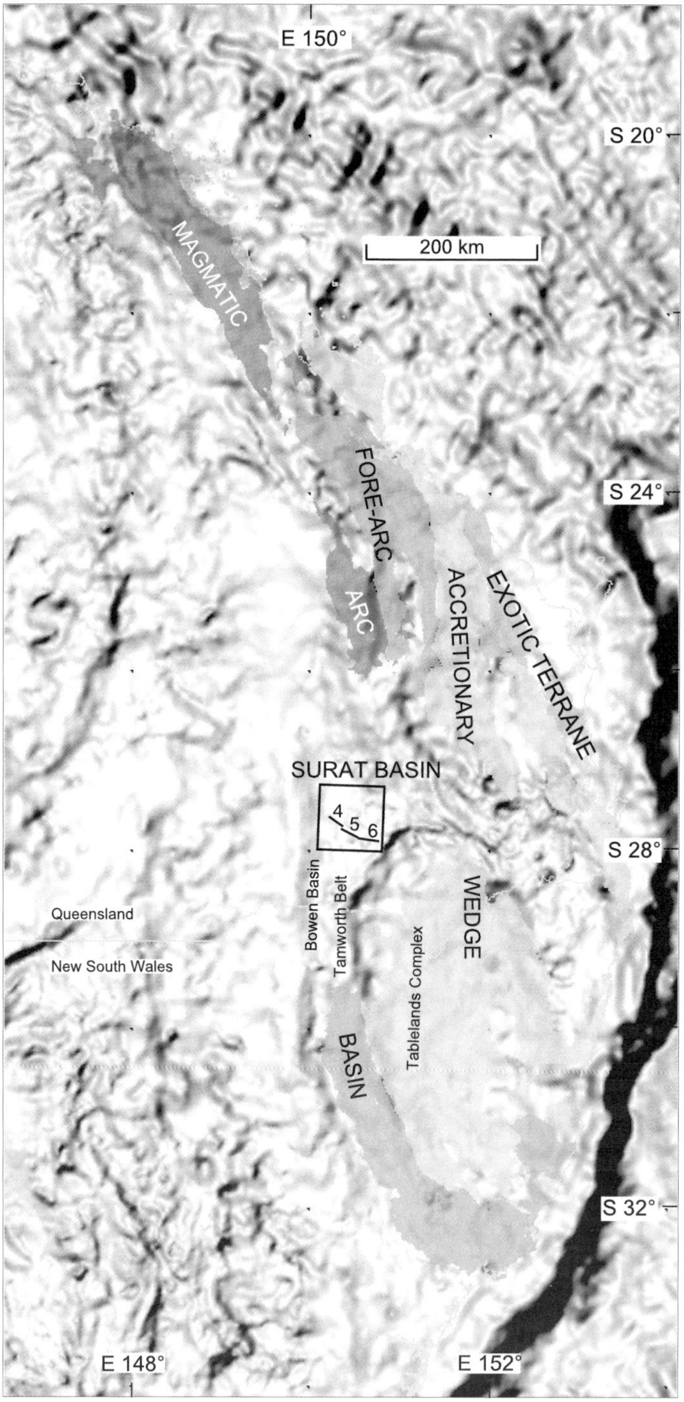

Fig. 10. Gravity anomaly map of the New England Orogen based on onshore residual Bouguer gravity anomalies and on offshore free air gravity anomalies (μm s^{-2}). The map is overlain by a simplified terrane map of the New England Orogen, identifying magmatic arc, fore-arc and accretionary complex rocks using different shades of grey. The box represents the study area, showing the location of the three seismic and magnetic profiles as to be seen in the Figures 4, 5 & 6 respectively.

basement's observed fault and fold geometry, it is possible that the stronger magnetic features in the images are due to Middle to Late Palaeozoic rocks. The relatively deep magnetic signals would penetrate, but be diluted by the Mesozoic sedimentary cover. Thus, it is probable that the Surat succession functions as a filter, which would explain the aeromagnetic lineaments along the western and the eastern border of the Tamworth Belt.

On the A82-LT-24 seismic profile, we interpret the base of the thrust sheet at the southeastern end and a ramp with NW-directed movement (Fig. 4). The asymmetrical high of the magnetic data confirms the structural geometry of the Moonie Fault ramping up west-directed beneath the Surat Basin succession and, due to the faulting, bringing up more magnetic rocks of the Tamworth Belt (i.e. volcanogenic compared with more quartz-rich in the Surat and Bowen basins) closer to the surface (see TMI path in Fig. 4).

Southeast of traverse A82-LT-24, in H82-T-109 (Fig. 5), the total magnetic intensity gently decreases southeastwards for another 130 nT, starting at 4975 nT in the northwest and reaching its minimum at the southeastern end of the profile at approximately 4845 nT. Although the fore-arc succession dips moderately to the northwest (Fig. 5), it does not coincide with the TMI direction of incidence. However, the decrease in magnetic intensity as one goes down the section implies that the upper sequences in the Tamworth Belt are more magnetic than the lower ones.

On the Millmerran profile (BMR86.M01, Fig. 6) located farther to the east, two small bulges are noted on the eastern end of the magnetic intensity section. These bulges may be correlated with two major west-dipping faults that are covered by the base of the Surat Basin on the corresponding seismic profile, identifying the northern extension to the Peel Fault (cf. Figs 1, 6, 7 & 8). The TMI graph on the western end of the traverse correlates directly to the values seen in traverse H82-T-109 (Figs 5 & 11). The magnetic low in the middle of the Millmerran traverse coincides with the end of continuous seismic reflectors of the Tamworth Belt sequences (Fig. 6).

Immediately south of the study area, we constructed an east–west-oriented aeromagnetic anomaly profile (see Fig. 7 for location, Fig. 9 for details) in the region where the Tamworth Belt narrows significantly and thus the structural geometry of the belt is similar to that observed in the exposed part of the fore-arc basin farther south. This profile confirms the observed magnetic anomalies as described above and supports the fault-related asymmetries (such as the asymmetry of the magnetic signal for the Moonie Fault in Fig. 4 and the westward-dipping faults in Fig. 6, respectively).

Gravity data – presentation and interpretation

On the Bouguer anomaly map (Fig. 10), a number of features can be picked out that substantiate the outcome of the gradient-enhanced residuals of the total magnetic intensity. However, because of the small-scale grid coverage (a maximum of 11 kilometres between the gravity stations), we did not construct any gravity traverses along the key seismic lines. In the region east of the Moonie Fault system, the most noticeable feature is that of a stronger gravity signal along the exposed part of the Peel Fault in northern New South Wales (Figs 1 & 10) which is due to the higher Bouguer gravity values of the serpentinites and related mafic–ultramafic rocks that occur along the fault. This gravity signal is part of a distinctive gravity ridge (approximately 10 to 50 $\mu m\ s^{-2}$) at the eastern end of the investigated area, identifying the boundary between Tamworth Belt and Tablelands Complex and, moreover, the Texas Orocline in southern Queensland (cf. Figs 1 & 10).

Discussion

The tectono-stratigraphic evolution of Eastern Australia in the Silurian–Carboniferous commenced with subduction of the Panthalassan Ocean plate beneath the eastern margin of Gondwana. This was accompanied by the development of a series of subduction-related basins, where individual basin stratigraphy was strongly influenced by the ongoing orogenic activity. Such changes included possible stop–start subduction activity (Korsch *et al.* 1997) or reversals in subduction polarity (Aitchison & Flood 1995) from Late Devonian to Late Carboniferous. In the Early Permian the region was subjected to oroclinal bending (i.e. Texas and Coffs Harbour oroclines), which resulted in the development of a complex structural pattern. As previously noted, the study area is located immediately to the west of the Texas Orocline. Thus, the development of the Tamworth Belt was influenced both by the initial period of subduction-related activity (which included basin formation and infill) and subsequent tectonic deformation. The region and its location adjacent to the orocline bend is, therefore, of key interest in elucidating both the pre-deformation history of the area, and the pathways of later tectonic activity.

Age control

Precise dating of the sequences in the study area is difficult since palynological control is sparse, and provided by only a few drill core samples across the entire region. The oldest palynoflora is dated in the Early Carboniferous as Viséan (De Jersey, in C. G. Murray 1994), confirming the correlation with the exposed Tamworth and Yarrol belts. Palynological samples from five wells to the west of the study area have been dated as Late Carboniferous to earliest Permian (C. G. Murray 1994). These are equivalent to the uppermost parts of the exposed belts. In three other wells (J. L. McKellar pers. comm. 1998), the palynoflora was dated as Early Permian (APP2.1 stage of Price 1997 see Fig. 2), which provides a correlation with the sequence at the base of the Bowen Basin (cf. Supersequence A in Fig. 2, defined by Brakel *et al.* in press; Korsch *et al.* 1998). In the Bowen Basin, Supersequence A occurs immediately below the B30 sequence boundary, implying that the seismic sequence in the subsurface Tamworth Belt between B30 and reflector HA is its equivalent (Fig. 2). However, the stratigraphic relationship between the Supersequence A section and the uppermost Tamworth Belt succession is unclear as no seismic lines tie the two successions. Unfortunately, interpretation of the remaining grid of seismic lines across the eastern Surat Basin did not provide further estimates of the ages of the sequences.

Structure of the concealed Tamworth Belt – initiation of the fore-arc succession

Initial formation of the Tamworth Belt fore-arc basin began in the Late Devonian (Fig. 2). However, the precise dimensions and extent of the original basin are uncertain due to subsequent tectonic deformation and sequence repetition (see below). Leitch (1975) suggested a possible length of approximately 400 km and a width of approximately 50 km, which is considerably larger than values obtained for current fore-arc basins (see Dickinson 1995). Fore-arc basin environments are extremely complex and variable both in terms of the basin architecture and the infill pattern (see Dickinson 1995). He recognizes a range of morphological types (Dickinson 1995, fig. 6.4), but our limited database precluded the recognition of any specific morphology.

Within the fore-arc basin, five reflectors, defining six seismic sequences, could be recognized. More precise interpretation of these sequences could not be attempted due to the poor quality of the seismic data. Variations in reflector amplitude and continuity within individual sequences, however, may be interpreted in terms of lateral changes in the depositional environment. The identified seismic sequences appear to thicken slightly towards the eastern part of the region (Fig. 5), reaching their maximum thickness close to the eastern faulted margin (adjacent to the Peel Fault). The succession between the B30 and HD reflectors on the A82-LT-24 seismic line (Fig. 4) is about 6 km thick (2 s TWT). Farther east, on the Millmerran profile (Fig. 6), the fore-arc succession below the HD reflector is approximately 2 s TWT thick. The entire thickness of the Tamworth Belt succession is, therefore, at least 12 km. This is similar to the values obtained by Liang (1991) and Woodward (1995) in the exposed part of the Tamworth Belt in New South Wales.

The top of the Tamworth Belt succession is defined by the B30 reflector (which marks the transition to the Bowen Basin). Seismic data shows that the folded Tamworth Belt succession appears to be structurally conformable with the overlying Bowen Basin units (Figs 4, 5 & 11), although there is evidence of some reflector truncation between the two.

Structure of the concealed Tamworth Belt – post-depositional deformation

Following initial basin formation and infill, the Tamworth Belt was subjected to compressional and oroclinal deformation. Evidence for this may be seen in the pattern of reflector offset within the basin fill sequence (Fig. 4). Two major fault sets developed within the Tamworth Belt – to the west the Moonie Fault and to the east the two westward-dipping faults which may be traced to corrrelate with the Peel Fault. Owing to the lack of cross-cutting relationships it is not possible to determine precisely which of the westward-dipping faults developed first. However, the easternmost faults define the boundary between the fore-arc basin and the accretionary wedge sequences (Fig. 6), suggesting that they were part of the original fore-arc basin-bounding structures. In contrast, the Moonie Fault is a later structure, related to Middle–Late Triassic deformation, which offsets basin infill reflectors (see above).

As described above, two major westward-dipping faults image a fault zone defining the eastern limit of the Tamworth Belt in the seismic reflection profiles (Fig. 6). If the easternmost of these faults is projected upwards beyond the limits of the seismic line, it intersects the base of the Surat Basin at the position of the northern extension of the Peel Fault (as interpreted from the aeromagnetic data) (Fig. 8). To the south of the study area, the surface expression of the Peel Fault contains serpentinites which result in higher magnetic intensity along the fault trace. Thus, the aeromagnetic data suggests that the true limit of the subsurface Tamworth Belt lies along the easternmost of the Peel Fault extension structures (c_2 on Fig. 6). Deformation within this fault zone precludes precise definition on the seismic profiles of the margin of the Tamworth Belt, but the aeromagnetic data suggests that the boundary is coincident with the easternmost limit of the fault array (Figs 6 & 8) (see Wartenberg, in press, for more details).

Fault initiation and propagation as a result of externally controlled tectonic activity, led to further deformation of the basin fill of the Tamworth Belt succession. Although the precise base of the Tamworth Belt is difficult to determine, it can be locally constrained in the area by the two basin-bounding fault sets, namely, the Moonie Fault and the Peel Fault extension (Figs 3 & 11). In the southern part of the study area, the Moonie Fault is oriented approximately north–south, curving to the NE (from approximately 28°05'S), with the highest part of the fore-arc succession located along the concave part of the fault curve (Fig. 3). The internal geometry of the Tamworth Belt can only be determined by more detailed examination of the hanging-wall sequence which developed between the Moonie Fault and the Peel Fault extension. The base of the hanging-wall succession, as defined by the Moonie Fault, lies less than 1 second TWT beneath the surface (approximately 700 ms TWT), while the deepest part is to the east, at a depth of over 3.8 seconds TWT (Fig. 3). The schematic structure contour map of the Tamworth Belt succession (Fig. 3) shows an asymmetrical syncline trending NE–SW (which is broad and flat bottomed in the northern part of the study area, while steepening to the south). This feature, which forms part of a larger fault-bend-fold structure (as seen elsewhere in the region), is related to NW-directed thrust movement (Fig. 4).

The NE–SW-trending steep asymmetrical syncline that broadens to the north, together with the change in direction of the Moonie Fault from north–south (to the south) to NE–SW-trending (to the north) underlines the anomalous geometrical situation within the study area. In this region, the Tamworth Belt sediments are positioned at a point where oroclinal bending begins to influence the geometry of the basin (Texas Orocline). To the south, the fore-arc basin geometry may be best described by the faulted character of the Tamworth Belt succession with the eastward-dipping Moonie Fault to the west and the westward-dipping Peel Fault extension to the east. Farther north, however, the basin geometry widens along strike, as can be seen in the deep seismic line BMR84.14 farther north (Wake-Dyster *et al.* 1987).

The previously noted apparently stratigraphic relationship between the successions in the Tamworth Belt and Bowen Basin raises an interesting issue. The Tamworth Belt has been interpreted as a fore-arc basin sequence (e.g. Korsch & Totterdell 1996), whereas the Bowen Basin formed in a back-arc setting (e.g. McKelvey & McPhie 1995). This relationship would imply that there was little deformation of the Tamworth Belt succession during the period when subduc-

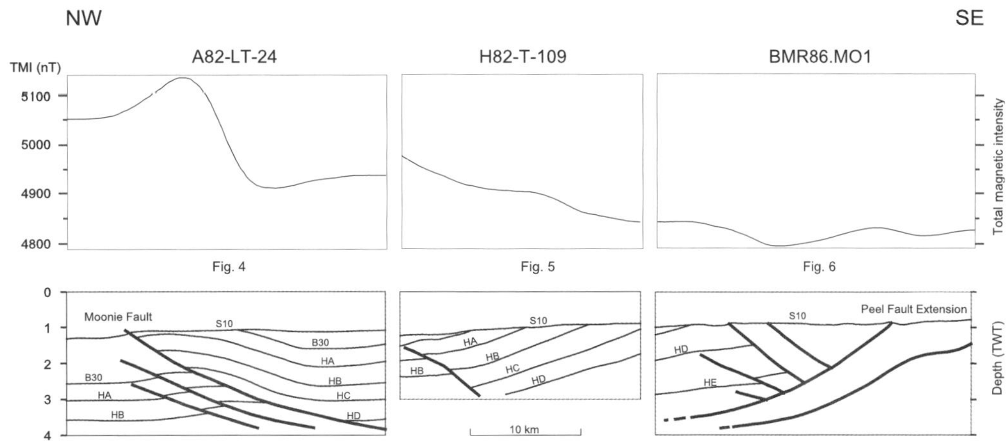

Fig. 11. Composite cross-section across the Tamworth Belt. Terminology as per Figures 4, 5 & 6.

tion migrated (Korsch *et al.* 1997) or jumped (Aitchison & Flood 1995) eastwards, and the magmatic arc was relocated to the east of the current Bowen Basin position. In the vicinity of the seismic lines A82-LT-24 (Fig. 4) and H82-T-109 (Fig. 5) a small remnant of the Bowen Basin is preserved to the east of the Moonie Fault. The presence of this remnant of the Bowen Basin in a thrust sheet to the east of the main basin suggests that the original area covered by the basin extended much farther to the east than its present limits. It also indicates that the original eastern margin of the Bowen Basin has been destroyed and has been incorporated into the Late Permian to Middle Triassic retro-foreland thrust belt of the New England Orogen. In the eastern part of the investigated region, the situation is less clear, partly due to the lack of sufficient seismic profile coverage (see Fig. 3), partly due to seismic record depths of only 3 s TWT.

Structure of the concealed Tamworth Belt – fault history

Two significant deformational events are recorded in the study area. Following initial formation and infill of the Tamworth Belt fore-arc basin, these tectonic events marked the various stages of subsequent thrusting. Seismic section A82-LT-24 shows evidence for the first of these deformational phases – the Goondiwindi Event of Korsch *et al.* (1998) – where strata of the Tamworth Belt are thrust over Middle Triassic rocks of the Bowen Basin along the Moonie Fault (Fig. 4). The fault itself provides convincing evidence for at least 3 km of vertical displacement. Erosion and peneplanation occurred in the Late Triassic before the Tamworth Belt and Bowen Basin were overlain by Early Jurassic rocks of the Surat Basin. Thus, the major period of fault movement was Middle–Late Triassic, cutting Middle Triassic rocks. Deposition could have been synchronous with thrusting but it is also possible that sediments were removed during the peneplanation and prior to the depositional phase in the Early Jurassic within the Surat Basin.

During the Late Cretaceous, there is evidence for the second deformational phase – the Moonie Event of Korsch *et al.* (1998) – in two of the seismic sections (Figs 4 & 6). Of note is the reactivation of the Moonie Fault, with the fault tip propagating into the Surat Basin succession and uplift of the Surat Basin on the hanging-wall side of the fault relative to the footwall (Fig. 4). Further evidence for the Moonie Event is provided by folding and uplift of the Surat Basin succession above an eastwards-dipping fault near station 2200 in seismic line BMR86.M01 (Fig. 6). Although the faults were reactivated in the basement, in most cases they have not propagated upwards into the Surat Basin succession, that is, usually there is no displacement of the Surat Basin succession above the faults.

Comparison with previous studies

The subsurface Tamworth Belt has been deformed by a series of thrust faults, most of which appear to be west-directed, and probably form part of a westward-propagating foreland thrust belt (Figs 4, 5 & 6). In the vicinity of seismic line A82-LT-24, in the hanging wall of the Moonie Fault, the sedimentary succession is folded into an anticline–syncline pair. Farther east, on seismic line H82-T-109, the Tamworth Belt succession dips moderately to the west, being interrupted by another west-directed thrust of minor vertical displacement (Fig. 5). On the eastern side of the belt, a series of minor eastward-dipping thrusts appear to be backthrusts which developed from a major westward-dipping fault, with the amount of displacement on the backthrusts decreasing to the west (Fig. 6). The overall geometry of the subsurface Tamworth Belt to the south of the study area is similar to that seen on the deep seismic line BMR91.G01 in the Boggabri–Manilla area (Glen *et al.* 1993; Korsch *et al.* 1993, 1997) (see Fig. 1 for location) and on geological cross-sections from the same region (Liang 1991; Glen & Brown 1993; Woodward 1995). The specific fault-bend fold in the hanging wall of the Moonie Fault (Fig. 4) is similar to the ramp anticline forming the Tulcumba Ridge, as described by Liang (1991), while the hanging-wall syncline (Figs 4 & 5) is equivalent to the Rocky Creek Syncline seen in the deep seismic profile BMR91.G01 (Korsch *et al.* 1997 fig. 4) and to the Belvue Syncline in Liang's (1991) cross-section. However, the syncline that has been identified within the study area tends to widen as we approach the northern part of the investigated area. Farther east, the eastward-dipping faults in the Millmerran profile appear to be in the same structural position and equivalent to the Baldwin Fault in BMR91.G01 (Korsch *et al.* 1997 fig. 3). The fact that two major westward-dipping faults can be identified within the study area, of which the westerly structure appears to die out just to the northeast of the investigated area at a latitude of approximately 27° 20'S (Figs 7 & 8), may indicate the stronger influence of the oroclinal bending to the east. The cross-sections constructed by Liang (1991) and Woodward (1995) invoke an eastward-dipping thrust system associated with the eastern limit of the Tamworth Belt, whereas the seismic sections of this study suggest that this boundary is defined by a major westward-dipping fault zone (c_1 and c_2 in Fig. 6) (see also Korsch *et al.* 1997). The precise character of this fault zone is uncertain due to the sparse seismic data within the eastern part of the investigated area (Fig. 3).

The eastern boundary of the Tamworth Belt, the Peel Fault in New South Wales, is known to dip steeply to the east at the surface. In the subsurface, it is interpreted on the deep seismic line BMR91.G01 (for location see Fig. 1) as an eastward-dipping splay off a major westward-dipping structure that corresponds with a mélange zone of serpentinite and related rocks and extends to the base of the crust (Korsch *et al.* 1993, 1997). This would suggest that the entire Tamworth Belt has been thrust westwards over basement rocks. In the vicinity of seismic line BMR91.G01, the intensity of deformation within the Tamworth Belt increases eastwards towards the Peel Fault, while surface studies have indicated that cleavage development also increases markedly (Durney & Kisch 1994). This eastward increase in deformational intensity, together with a concomitant steepening of the structures, is the probable explanation for the decrease in the quality of the seismic data seen on both lines BMR91.G01 and BMR86.M01 (Fig. 6).

From south to north, the Tamworth Belt shows great variability in its width, ranging from over 90 km in the very south, to only about 25 km near the border between Queensland and New South Wales (Fig. 1). In the subsurface, in the vicinity of our three seismic sections, the belt is just over 50 km wide, and we estimate (geometrical reconstruction of our cross-section based on the interpreted seismic fault structures) that there has been a minimum of at least 10 km (*c*.20%) of shortening across the belt. Here, the width of the belt is very similar to where seismic line BMR91.G01 crosses the belt (see Fig. 1 for location). In that area, Liang (1991) estimated that the belt had been shortened by 25% (*c*.12.5 km). Both of these results are significantly less than the 48% (75 km) shortening determined by Woodward (1995) in the vicinity of BMR91.G01. Woodward showed much greater complexity in his balanced cross-sections, including a series of footwall horses in his section closest to BMR91.G01.

Conclusions

(1) Interpretation of seismic and aeromagnetic profiles A82-LT-24, H82-T-109 and BMR86.M01 in the eastern Surat Basin of southern Queensland has led to an increase of our understanding of the geometry of the Tamworth Belt beneath the younger sedimentary cover.

(2) The Tamworth Belt succession appears to be at least 12 km thick and is moderately folded. Within it, six seismic sequences have been identified. Each sequence is separated by a significant sequence boundary. Due to poorer resolution of the seismic data in the eastern study area, it is likely that there are other sequence boundaries not imaged seismically.

(3) The western Tamworth Belt succession is truncated by the Moonie Fault, leaving the extent of the fore-arc basin in the footwall uncertain. The Moonie Fault has a classic fault–bend fold geometry, with a hanging-wall anticline–syncline pair immediately to the east of the thrust. Farther to the east, the sedimentary package dips moderately to the west.

(4) To the east – on the western side of the Texas Orocline – the boundary between the Tamworth Belt and the Tablelands Complex is identified by two westward-dipping faults, with a series of eastward-dipping backthrusts occurring farther to the west.

(5) In the vicinity of the seismic profiles, the subsurface Tamworth Belt is over 50 km wide and has been shortened by at least 10 km. Its overall geometry is similar to that observed farther south in the exposed part of the belt, with an increasingly anomalous widening across strike to the north. The thrust movement was NW-directed.

(6) The aeromagnetic traverses A82-LT-24, H82-T-109 and BMR86.M01 correlate, at least in part with the corresponding seismic profiles. Some of the magnetic responses, especially those of broad scale, can be explained by the geometry of the subsurface Bowen Basin and the underlying Tamworth Belt, although the Mesozoic sedimentary cover is up to 2000 m thick.

(7) The eastern boundary of the Tamworth Belt coincides with a gravity and magnetic ridge, mimicking the gravity and magnetic pattern of the serpentinites and iron-rich rocks, respectively, that occur along the Peel Fault.

Acknowledgement is given to T. McCann for stimulating discussions that initiated this paper and for critical comments on versions of this manuscript. We wish to thank the Deutsche Forschungsgemeinschaft and the Australian Geodynamics Cooperative Research Centre for financial assistance, and Geoscience Australia, the Geological Survey of Queensland, the Geological Survey of New South Wales and Werrie Gold Ltd. for providing access to the aeromagnetic data. Also, we thank C. Krawczyk and O. Clausen for reviewing this paper. R. J. Korsch publishes with permission of the CEO, Geoscience Australia.

References

AITCHISON, J. C. & FLOOD, P. G. 1995. Gamilaroi terrane: a Devonian rifted intra-oceanic island-arc assemblage, NSW, Australia. *In:* SMELLIE, J. L. (ed.) Volcanism associated with extension at consuming plate margins. Geological Society of London, Special Publications, **81**, 155–168.

BMR 1976. *Magnetic Map of Australia, residuals of total intensity, scale* 1:2 500 000, Bureau of Mineral Resources, Geology and Geophysics, Canberra.

BRAKEL, A. T., WELLS, A. T., TOTTERDELL, J. M., HOFFMANN, K. L., SIMPSON, G., DIXON, O. & NICOLL, M. G. in press. *Sequence Stratigraphy and Fill history of the Bowen Basin, Queensland,* Geoscience Australia, Canberra.

CATUNEANU, O., BEAUMONT, C. & WASCHBUSCH, P. 1997. Interplay of static loads and subduction dynamics in foreland basins: reciprocal stratigraphies and the 'missing' peripheral bulge. Geology, **25**, 1087–1090.

DAY, R. W., WHITAKER, W. G., MURRAY, C. G., WILSON, I. H. & GRIMES, K. G. 1983. *Queensland Geology,* Geological Survey of Queensland, Publication, **383**, 194 pp., Brisbane, Australia.

DICKINSON, W. R. 1995. Forearc basins. *In:* BUSBY, C. J. & INGERSOLL, R. V. (eds) *Tectonics of Sedimentary Basins.* Blackwell Science, Oxford, 221–261.

DURNEY, D. W. & KISCH, H. J. 1994. A field classification and intensity scale for first-generation cleavages. *AGSO Journal of Australian Geology and Geophysics*, **15**, 257–295, Canberra.

FIELDING, C. R., GRAY, A. R. G., HARRIS, G. I. & SALOMON, J. A. 1990. The Bowen Basin and overlying Surat Basin. *In*: FINLAYSON, D. M. (ed.) *The Eromanga–Brisbane Geoscience Transect: a Guide to Basin Development Across Phanerozoic Australia in Southern Queensland,* Bureau of Mineral Resources, Geology and Geophysics, Canberra, 105–116.

GLEN, R. A. & BROWN, R. E. 1993. A transect through a forearc basin: preliminary results from a transect across the Tamworth Trough at Manilla, N.S.W. *In*: FLOOD, P. G. & AITCHISON, J. C. (eds) *New England Orogen, Eastern Australia,* Department of Geology and Geophysics, University of New England, Armidale, 105–111.

GLEN, R. A., KORSCH, R. J. & WAKE-DYSTER, K. D. 1993. A deep seismic cross section through the Tamworth Belt: preliminary interpretation of 4 seconds two-way time data. *In*: FLOOD, P. G. & AITCHISON, J. C. (eds) *New England Orogen, Eastern Australia,* Department of Geology and Geophysics, University of New England, Armidale, 101–104.

HOLCOMBE, R. J., STEPHENS, C. J., FIELDING, C. R., GUST, D., LITTLE, T. A., SLIWA, R., KASSAN, J., MCPHIE, J. & EWART, A. 1997. Tectonic evolution of the northern New England Fold Belt: the Permian-Triassic Hunter-Bowen event. *In*: ASHLEY, P. M. & FLOOD, P. G. (eds) *Tectonics and Metallogenesis of the New England Orogen,* Geological Society of Australia, Special Publication, **19**, 52–65.

KIRKEGAARD, A. G. 1974. Structural elements of the northern part of the Tasman Geosyncline. *In*: DENMEAD, A. K., TWEEDALE, G. W. & WILSON, A. F. (eds) *The Tasman Geosyncline: a Symposium.* Geological Society of Australia, Queensland Div., 47–62.

KORSCH, R. J. & HARRINGTON, H. J. 1987. Oroclinal bending, fragmentation and deformation of terranes in the New England Orogen, eastern Australia. *In*: LEITCH, E. C. & SCHEIBNER, E. (eds) *Terrane Accretion and Orogenic Belts.* American Geophysical Union Geodynamic Series, **19**, 129-139.

KORSCH, R. J. & TOTTERDELL, J. M. 1995a. Eastern margin of the Bowen–Gunnedah–Sydney Basin: Geometry of the Burunga–Leichhardt–Moonie–Goondiwindi–Mooki–Hunter fault system. *In*: Geological Society of Australia, *Specialist Group in Tectonics and Structural Geology, Field Conference*, Clare Valley, Geological Society of Australia, Abstracts, **40**, 85–86.

KORSCH, R. J. & TOTTERDELL, J. M. 1995b. Structural events and deformational styles in the Bowen Basin. *In*: FOLLINGTON, I. L., BEESTON, J. W. & HAMILTON, L. H. (eds) *Bowen Basin Symposium 1995 . . . 150 years on . . .* Proceedings, Geological Society of Australia, Coal Geology Group, Brisbane, 27–35, Mackay, Queensland.

KORSCH, R. J. & TOTTERDELL, J. M. 1995c. Permian and Mesozoic tectonic and structural events in the Bowen and Surat basins and New England Orogen, Southwest Pacific Rim. *In*: MAUK, J. L. & ST GEORGE, J. D. (eds) *Proceedings of the 1995 PACRIM Congress,* Australasian Institute of Mining and Metallurgy, Publication **9/95**, 305–310.

KORSCH, R. J. & TOTTERDELL, J. M. 1996. *Mesozoic Deformational Events in Eastern Australia and Their Impact on Onshore Sedimentary Basins.* Proceedings of the Mesozoic Geology of the Eastern Australia Plate Conference, Geological Society of Australia, Extended Abstracts, **43**, 308–318.

KORSCH, R. J., HARRINGTON, H. J., MURRAY, C. G., FERGUSSON, C. L. & FLOOD, P. G. 1990. *Tectonics of the New England Orogen.* Bureau of Mineral Resources, Australia, Bulletin, **232**, 35-52.

KORSCH, R. J., WAKE-DYSTER, K. D. & JOHNSTONE, D. W. 1992a. Seismic imaging of Late Palaeozoic - Early Mesozoic extensional and contractional structures in the Bowen and Surat basins, Eastern Australia. *Tectonophysics*, **215**, 273–294.

KORSCH, R. J., WAKE-DYSTER, K. D., O'BRIAN, P. E., FINLAYSON, D. M. & JOHNSTONE, D. W. 1992b. Geometry of Permian to Mesozoic sedimentary basins in Eastern Australia and their relationship to the New England Orogen. *In*: RICKARD, M. J. *ET AL.* (eds) *Basement Tectonics,* **9**, Kluwer, Amsterdam 85–108.

KORSCH, R. J., WAKE-DYSTER, K. D. & JOHNSTONE, D. W. 1993. The Gunnedah Basin – New England Orogen deep seismic reflection profile: implications for New England tectonics. *In*: FLOOD, P. G. & AITCHISON, J. C. (eds) *New England Orogen, Eastern Australia,* Department of Geology and Geophysics, University of New England, NEO '93, Conference Proceedings, NSW, Armidale, 85–100.

KORSCH, R. J., JOHNSTONE, D. W. & WAKE-DYSTER, K. D. 1997. Crustal architecture of the New England Orogen based on deep seismic reflection profiling. *In*: ASHLEY, P. M. & FLOOD, P. G. (eds) *Tectonics and Metallogenesis of the New England Orogen*, Geological Society of Australia, Special Publications, **19**, 29–51.

KORSCH, R. J., BOREHAM, C. J., TOTTERDELL, J. M., SHAW, R. D. & NICOLL, M. G. 1998. Development and petroleum resource evaluation of the Bowen, Gunnedah and Surat Basins, Eastern Australia, *APPEA Journal*, **38**, 199–237.

LEITCH, E. C. 1975. Plate tectonic interpretation of the Palaeozoic history of the New England Fold Belt. *Geological Society of America Bulletin*, **86**, 141–144.

LIANG, T. C. K. 1991. Fault-related folding, Tulcumba Ridge, western New England, *Australian Journal of Earth Sciences*, **38**, 349–355.

MCKELVEY, B. C. & MCPHIE, J. 1995. Tamworth Belt. *In*: DIAZ, C. M. (ed.) *The Carboniferous of the World II: Australia, Indian Subcontinent, South Africa, South America and North Africa.* International Union of Geological Sciences Publication, **20**, 15–23.

MCPHIE, J. 1987. Andean analogue for Late Carboniferous volcanic arc and arc flank environments of the western New England Orogen, New South Wales, Australia. *Tectonophysics*, **138**, 269–288.

MURRAY, A. S., MORSE, M. P., MILLIGAN, P. R. & MACKAY, T. E. 1997. *Gravity Anomaly Map of the Australian Region (second edn), scale 1:5 000 000*, Australian Geological Survey Organisation, Canberra.

MURRAY, C. G. 1994. *Basement Cores from the Tasman Fold Belt System Beneath the Great Artesian Basin in Queensland.* Queensland Department of Minerals and Energy, Geological Record **1994/10**.

MURRAY, C. G. and the YARROL PROJECT TEAM 1997. The Yarrol Project – increasing the prospectivity of the New England Orogen in the Rockhampton–Monto region, central coastal Queensland. *In*: BEESTON, J. W. (compiler) *Proceedings of the Queensland Development 1997 Conference*, Queensland Department of Mines and Energy, Brisbane, 39–56.

MURRAY, C. G., FERGUSSON, C. L., FLOOD, P.G., WHITAKER, W. G. & KORSCH, R. J. 1987. Plate tectonic model for the Carboniferous evolution of the New England Fold Belt. *Australian Journal of Earth Sciences*, **34**, 213–236.

O'BRIEN, P. E., KORSCH, R. J., WELLS, A. T., SEXTON, M. J. & WAKE-DYSTER, K. D. 1990. Mesozoic basins at the eastern end of the Eromanga–Brisbane Geoscience Transect: strike-slip faulting and basin development. *In*: FINLAYSON, D. M. (ed.) *The Eromanga-Brisbane Geoscience Transect: a Guide to Basin Development Across Phanerozoic Australia in Southern Queensland*, Bureau of Mineral Resources, Geology and Geophysics, Canberra, 117–132.

PRICE, P. L. 1997. Permian to Jurassic palynostratigraphic nomenclature of the Bowen and Surat basins. *In*: GREEN, P. M. (ed.) *The Surat and Bowen Basins, Southeast Queensland.* Queensland Minerals and Energy Review Series, Queensland Department of Mines and Energy, 137–178.

SANDWELL, D. T. & SMITH, W. H. F. 1997. Marine gravity anomaly from Geosat and ERS-1 satellite altimetry. *Journal of Geophysical Research* **102**, 10.039–10.054.

SCHEIBNER, E. 1998. *Geology of New South Wales - Synthesis. Volume 2 Geological Evolution.* Geological Survey of New South Wales, Memoir Geology, **13(2)**, 666 pp.

SCHEIBNER, E. & BASDEN, H. (eds) 1996. *Geology of New South Wales – Synthesis. Volume 1, Structural Framework*, Department of Mineral Resources, Geological Survey of NSW, Memoir Geology **13(1)**, 1–295.

SUPPE, J. 1983. Geometry and kinematics of fault-bend folding. *American Journal of Science*, **283**, 648–721.

TADROS, N. Z. (ed.) 1993. *The Gunnedah Basin, New South Wales*. Department of Mineral Resources, Coal & Petroleum Geology Branch, Geological Survey of NSW, Memoir Geology **12**, 1–649.

TARLOWSKI, C., MILLIGAN, P. R. & MACKAY, T. E. 1996. *Magnetic Anomaly Map of Australia, scale 1:5 000 000*, Australian Geological Survey Organisation, Canberra.

TOTTERDELL, J. M., WELLS, A. T., BRAKEL, A. T., KORSCH, R. J. & NICOLL, M. G. 1992. Sequence stratigraphic interpretation of seismic data in the Taroom region, Bowen and Surat basins, Queensland. *Bureau of Mineral Resources*, **1991/102**, 61 pp., Canberra, Australia.

WAKE-DYSTER, K. D., SEXTON, M. J., JOHNSTONE, D. W., WRIGHT, C. & FINLAYSON, D. M. 1987. A deep seismic profile of 800 km length recorded in southern Queensland, Australia. *Geophysical Journal of the Royal Astronomical Society* **89**, 423–430.

WARTENBERG, W. in press. *Subsurface geometry of the Tamworth Belt (New England Orogen) with implications of petrographical observations from conglomerates in the Bowen–Gunnedah Basin, northern New South Wales and southern Queensland, Australia*. PhD Thesis, Rheinische Friedrich–Wilhelms-Universität Bonn.

WARTENBERG, W., KORSCH, R. J. & SCHÄFER, A. 1999. Geometry of the Tamworth Belt in the New England Orogen beneath the Surat Basin, southern Queensland. *In*: Flood, P. G. (ed.) *Regional Geology; Tectonics and Metallogenesis; New England Orogen Symposium.* Earth Sciences, UNE, Armidale, Australia, 211–220.

WOODWARD, N. B. 1995. Thrust systems in the Tamworth Zone, southern New England Orogen, New South Wales. *Australian Journal of Earth Sciences*, **42**, 107–117.

Subsidence, stress regime and rotation(s) of a tectonically active sedimentary basin within the western Alpine Orogen: the Tertiary Piedmont Basin (Alpine domain, NW Italy)

B. CARRAPA[1], G. BERTOTTI[1] & W. KRIJGSMAN[2]

[1] *Vrije Universiteit, Faculty of Earth and life Sciences, De Boelelaan 1085, 1081 HV Amsterdam, The Netherlands* (*e-mail: carb@geo.vu.nl*)
[2] *Utrecht University, Paleomagnetic Laboratory 'Fort Hoofddijk', Faculty of Earth Sciences, Budapestlaan 17, 3584 CD Utrecht, The Netherlands*

Abstract: The Oligocene to Miocene Tertiary Piedmont Basin (TPB) is located in the NW part of Italy at the junction between the Apennine and the Alpine thrust belts. The position of the TPB on top of the Alpine/Apennine Orogen poses fundamental questions as to the tectonics of the basin subsidence. Having undergone little deformation, the TPB sediments provide an insight into the stress regime and rotations in the kinematically very complex area surrounding the basin itself. In this study we integrate subsidence and structural analysis with measurements of magnetic susceptibility anisotropy (AMS) and natural remanent magnetization (NRM) in order to better constrain the tectonic kinematics of the basin evolution. A major important period of subsidence occurred in the Middle Miocene involving the whole basin. During this period the TPB experienced NE–SW-directed compression and limited shortening. Some NW–SE-directed compressional features have been identified and they were probably active during post Tortonian times. Structures associated with north–south tension are quite common, but the amount of strain that they accommodate is minor. In addition this research provides new preliminary data suggesting counterclockwise rotation in the TPB by *c.* 20° which has taken place during Middle Miocene time.

The presence of strongly subsiding areas inside and overlying orogenic arcs is at the same time a common observation and a typically unexplained phenomenon. In Europe, two apparent examples are the Transylvania Basin, lying within the East Carpathian–South Carpathian Arc (e.g. Ciulavu & Bertotti 1994; Huismans *et al.* 1997; Matenco & Bertotti 2000), and the western Po Plain situated between the Western Alps and the Ligurian Alps. Both basins formed during and/or following deformation in the surrounding orogenic belt and are characterized by a surprisingly low amount of internal deformation. A discrepancy between the amount of subsidence and the paucity of tectonic deformation is apparent, thereby leaving open the question as to the cause of the observed subsidence.

The Tertiary Piedmont Basin (TPB), a well-exposed part of the western Po Plain (Fig. 1) is well suited for the purpose of investigating the indicated topics. The basin, which has a >4 km thick Lower Oligocene to Upper Miocene fill is located on the strongly shortened area of the Alpine–Apennine junction (Fig. 1). The TPB is particularly interesting because its most important features, such as the kinematics and dynamics of subsidence and the structural setting under which the accommodation space was created, have never been addressed at a basin scale. The TPB also plays a very significant role in the regional geological picture, since it overlies and seals Alpine and Apennine structures. The stress/strain evolution of the basin contains important information on this key area. The TPB has been chosen because of the well-preserved Oligo-Miocene clastic infill and the relatively good stratigraphic control of the sedimentary record. Because of the Pliocene and younger uplift (e.g. Lorenz 1984), basin sediments are presently lying at elevations of several hundred metres and are dissected by fairly deep valleys providing comparatively good outcrops.

From: McCann, T. & Saintot, A. (eds) *Tracing Tectonic Deformation Using the Sedimentary Record.* Geological Society, London, Special Publications, **208**, 205–227. 0305–8719/03/$15.00

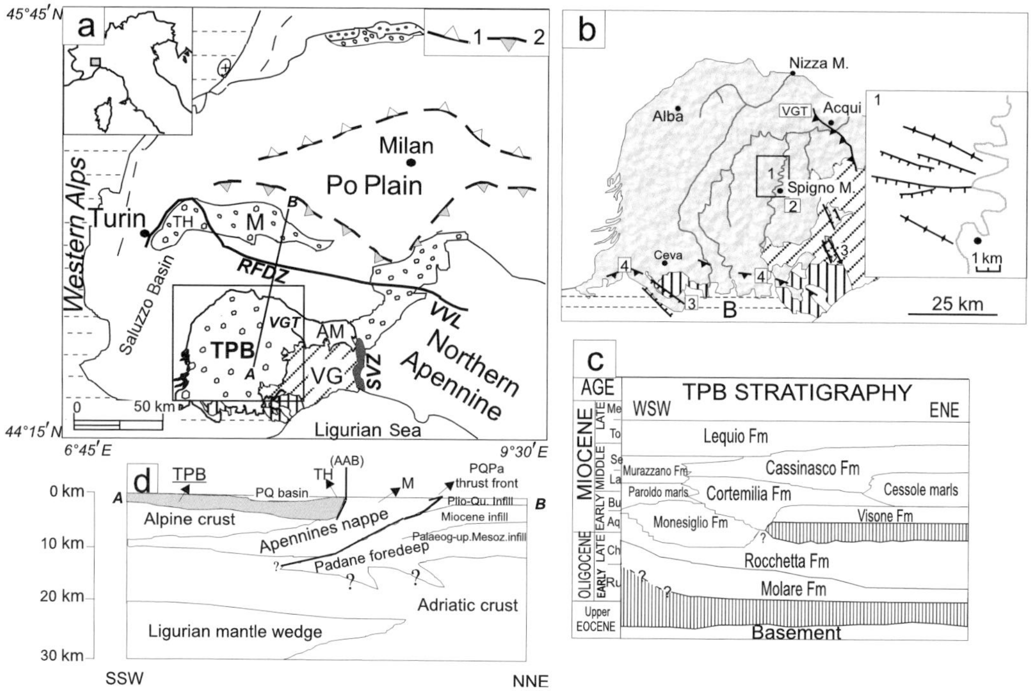

Fig. 1. (**a**) Tectonic map of NW Italy modified from Polino *et al.* (1995); 1, Apenninic thrusts, 2, South Alpine Thrusts. Square, study area; black vertical lines, Hercynian Crystalline Massifs; black close dots, Cretaceous to Eocene flysch; open dots, Oligo-Miocene sediments. TPB, Tertiary Piedmont Basin; AM, Alto Monferrato; TH, Torino Hill; M, Monferrato; VG, Voltri Group; RFDZ, Rio Freddo deformation zone; VVL, Villalvernia–Varzi Line; SVZ, Sestri Voltaggio Zone; VGT, Val Gorrini Thrust; (**b**) detailed enlargement of the study area with structural features mentioned in the text: (1) folds and faults described by Gelati & Gnaccolini (1998); (2) the area studied by Bernini & Zecca (1990) (Mioglia fold) and Mutti *et al.* (1995) (NE–SW trending normal faults). Note that the small scale (*c.*2 km) of the latter structures does not allow direct visualization in Fig. 1(b); B=Briançonnais domain; (3) Oligocene grabens by Lorenz (1979); (4) thrusts found inside the magnetic basement after Cassano *et al.* (1986). (**c**) stratigraphic scheme modified after Gelati (1968). (**d**) Structural profile after Clari *et al.* (1995) (M, Monferrato; PQ basin, Plio-Quaternary basin; PQPa thrust front, Plio-Quaternary Padan thrust front).

In this contribution, we report on three fundamental issues, the subsidence, the stress/strain history and tectonic rotation(s) of the basin. To reconstruct the subsidence history we build on the vast stratigraphic and sedimentological database from the literature (Gelati 1968; Gnaccolini *et al.* 1990; Gelati *et al.* 1993; Gnaccolini & Rossi 1994; Gelati & Gnaccolini 1996; Gelati & Gnaccolini 1998). We here derive the stress/strain regime during and following subsidence using structural analysis and, for the first time in the TPB, anisotropy of magnetic susceptibility (AMS) and natural remanent magnetization (NMR). Taking advantage of the low degree of internal deformation we furthermore derive information on possible rotations of the TPB. The final aim of this study is a better understanding of the TPB geodynamic evolution through a multidisciplinary approach.

Geological setting of the Tertiary Piedmont Basin

The Tertiary Piedmont Basin (TPB) is an episutural basin formed on top of the Mesoalpine edifice resulting from the collision between the Adriatic and the European plates (Rehault *et al.* 1985; Polino *et al.* 1990; Schumacher & Laubscher 1996; Biella *et al.* 1997; Schmid & Kissling 2000). The basin developed on a substratum consisting of allochthonous Alpine and Apennine units (Roure *et al.* 1990; Dela Pierre *et al.* 1995; Piana & Polino 1995; Piana 2000). The present day southern and southwestern limits of the TPB sediments are of an erosional nature and, consequently, it is unknown how far the basin extended above the Ligurian Alps. Towards the north, the Oligo-Miocene sediments of the TPB dip underneath the younger clastic sediments of the western Po Plain (Dalla *et al.* 1992; Schumacher & Laubscher 1996).

Sedimentological and structural evolution

During the Late Eocene–Early Oligocene, a prograding marine transgression, probably coming from the north–NE (Lorenz 1979; Lorenz 1984; Gelati & Gnaccolini 1988), followed by a progressive deepening of the basin floor, took place in the TPB. As a result, alluvial and nearshore sediments were deposited (i.e. the Molare Formation) (Lorenz 1979; Lorenz 1984; Gelati *et al.* 1993; Gelati & Gnaccolini 1996; Gelati & Gnaccolini 1998). Sediments of the Molare Formation show a clear source in the southern sector (present day Ligurian Alps and Voltri Group) (Gelati & Gnaccolini 1982). During the deposition of the Molare Formation, limited extension was taking place (Lorenz 1984; Hoogerduijn Strating *et al.* 1991; Hoogerduijn Strating 1994; Vanossi *et al.* 1994; Mutti *et al.* 1995).

At the end of the Early Oligocene, an increase in subsidence (Dela Pierre *et al.* 1995) was coeval with the deposition of a marly/sandy sequence known as the Rocchetta Formation (shallow-water sandstones and hemipelagic mudstones; Late Oligocene–Early Miocene). The transition between the alluvial and nearshore deposits of the Molare Formation and the overlying Rocchetta Formation marks a general deepening of the basin during the Oligocene and Early Miocene. Palaeocurrent indicators document sediment transport mainly from the SSW (present-day Ligurian Alps and Voltri Group) towards the NNE (Gelati *et al.* 1993). The contact between the Molare Formation and the Rocchetta Formation is progressively younger from NE to SW (Fig. 1c).

Evidence of shortening has been detected during Late Oligocene–Early Miocene times. A contractional structure (Mioglia fold) with NE–SW direction of shortening was active until Burdigalian times in the southern sector of the study area (Cazzola & Rigazio 1982; Bernini & Zecca 1990) (Fig. 1b). Some NW–SE to NNW–SSE-trending folds active in Late Oligocene times have been observed in the central-eastern part of the study area (Gelati & Gnaccolini 1998) (Fig. 1b). Northeast–southwest shortening during post-Oligocene times also formed the Val Gorrini Thrust (VGT) (Piana *et al.* 1997; Fig. 1a) across which the metamorphic rocks of the Voltri Group are thrust on to the Oligocene sediments of the TPB. The VGT deformation is sealed by the Lower Miocene Visone Formation (Capponi *et al.* 1999).

Some ENE–WSW normal faults, with displacements up to few hundred metres, have been described as being active during the deposition of the Rocchetta Formation, but with poor time constraints (Bernini & Zecca 1990; Mutti *et al.* 1995) (Fig. 1b).

During the Early Miocene, small platforms with both terrigenous and carbonate sedimentation developed in the eastern part of the basin while the central part was characterized by more widespread basinal conditions. Hemipelagic sediments characterize the NW area. The whole central area was characterized by silty/sandy sequences within which several different depositional bodies can be distinguished on the basis of the silt/sand ratio (Rocchetta–Monesiglio Group; see Gelati & Gnaccolini 1998). In the central-eastern part of the study area, normal faults directed west–east/WNW–ESE were active during Aquitanian–Burdigalian times (Gelati & Gnaccolini 1998) (Fig. 1b).

Since Late Burdigalian times, sedimentation became more homogeneous in the entire basin with the deposition of the Cortemilia Formation (classic turbidites with a flow direction from west to east) (Gelati *et al.* 1993). The homogeneity of Miocene palaeocurrent directions suggests a change/enlargement in the sediment source area (Gelati *et al.* 1993; Carrapa *et al.* 2000). This change is also supported by sandstone composition patterns which record an increase of rock fragments derived from quarzites, mica schists and gneisses, and a decrease in the percentage of rock fragments derived from ultramafic rocks (Gnaccolini & Rossi 1994; Gelati & Gnaccolini 1998).

During Langhian/Serravallian times the eastern part of the TPB was characterized by shallow-water shelf sedimentation (Caprara *et al.* 1985; Ghibaudo *et al.* 1985) while in the rest of the basin sedimentation was characterized by homogeneous deep-water tabular turbidites (Cassinasco Formation, Murazzano Formation, Lequio Formation) (Gelati *et al.* 1993).

Subsidence analysis

Method and input data

To derive the history of vertical movements in the TPB we have constructed subsidence curves from different localities within the basin (Figs 2 & 3). For its central and southern parts we have used a published stratigraphic reconstruction

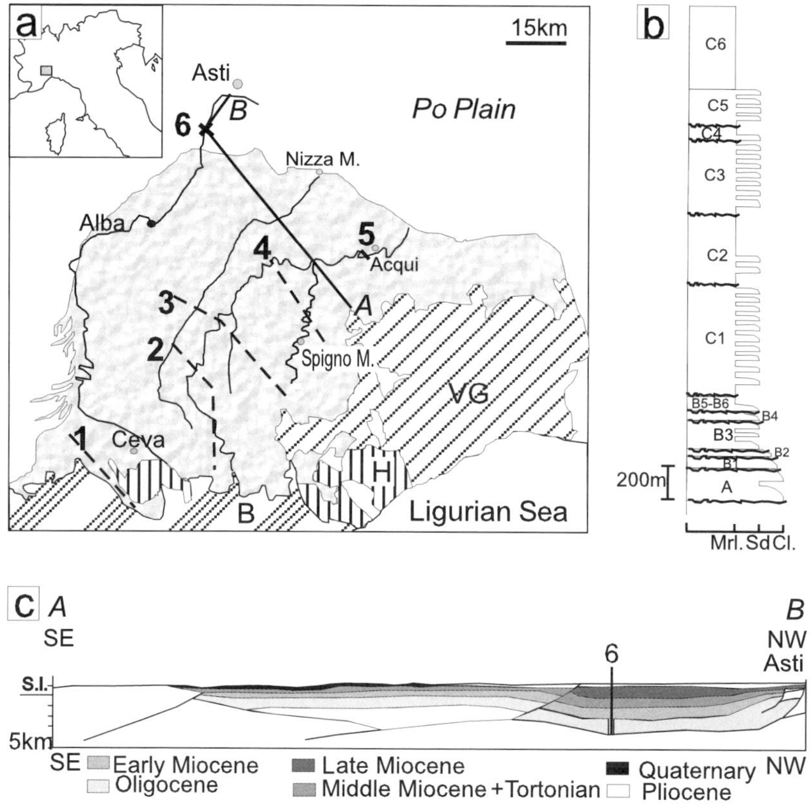

Fig. 2. (**a**) Map with location of the sections used for the stratigraphic reconstruction of Gelati *et al.* (1993) and of profile 2 from Pieri & Groppi (1981) used to construct the subsidence analysis in Fig. 3 (the cross corresponds with the synthetic well on which the subsidence curve 6 of Fig. 3 is based); B, Briançonnais domain; H, Hercynian Crystalline Massifs; VG, Voltri Group. (**b**) schematic lithostratigraphic column corresponding with section 4 in Figure 3 (Mrl, marls, Sd, sandstones, Cl, conglomerates); (**c**) Profile *A–B* after Cassano *et al.* (1986).

obtained by assembling stratigraphic sections along several km long transects (Gelati *et al.* 1993; Figs 2a & 2b). Given the gentle dip of the bedding, this was the only way to cover reasonably wide stratigraphic intervals. In the northern part of the basin we have extracted a synthetic stratigraphic column from an interpreted seismic line (plate II, line 2 from Pieri & Groppi (1981) interpreted by Cassano *et al.* (1996)) (Fig. 2a). In this region (section 6, Tanaro) the TPB kept on subsiding into post-Miocene times and was thus buried by Pliocene to Quaternary sediments (Figs 2a & 3). Lithologies and palaeobathymetries have been obtained from regional correlations (Gelati *et al.* 1993) (Table 1). Average palaeodepths between 25 and 50 m have been chosen for the Molare Formation due to its sedimentological features which indicate a shallow water and transitional environment. Palaeobathymetries for younger formations are based on the occurrence of particular benthonic foraminifer and on planktonic/benthonic species ratios indicating palaeodepths between 200 and 600 m for Upper Oligocene–Upper Miocene sediments (Gelati *et al.* 1993). We ended our analysis in Tortonian times (except for section 6) in order to avoid the uncertainties associated with the Messinian crisis. Stratigraphic columns have been backstripped adopting standard procedures (Sclater & Christie 1980; Bond & Kominz 1984; Bessis 1986). Subsidence curves describe the vertical movements of the basement and of the formational boundaries (chronostratigraphic horizons), taking into account compaction. To enable easier visualization of changes in vertical movements occurring during basin evolution, we have also constructed subsidence rate diagrams (Fig. 4).

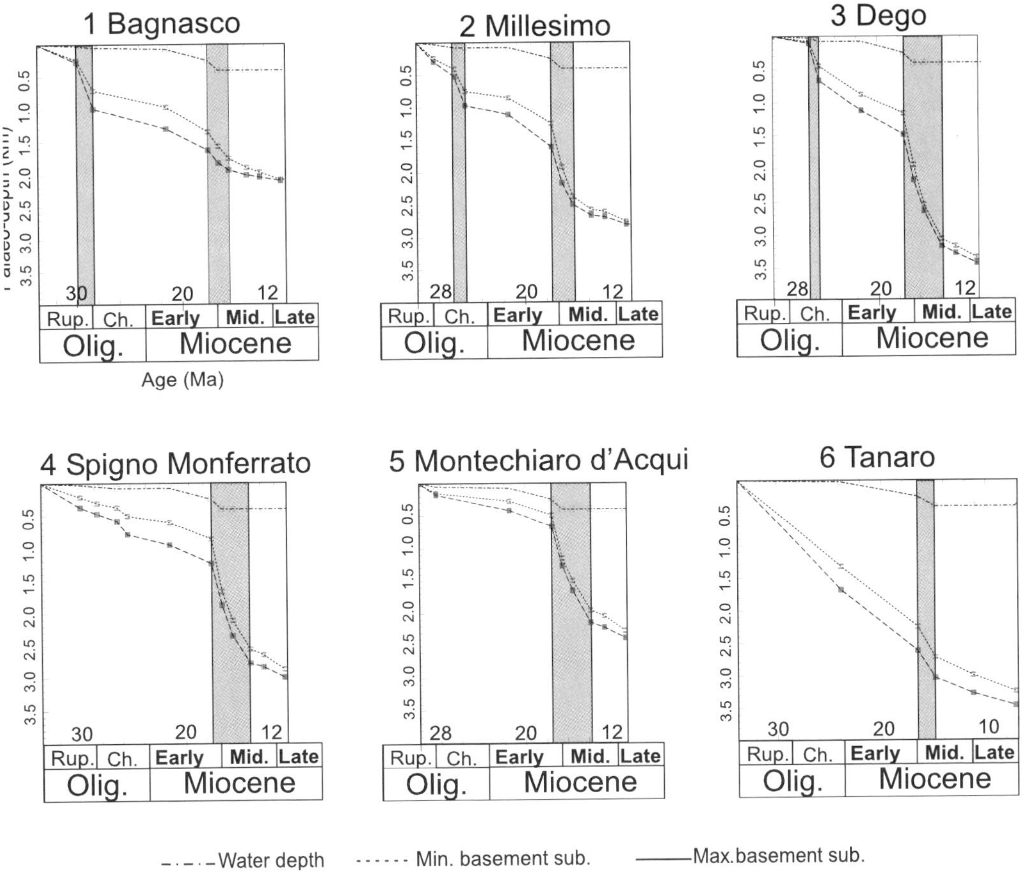

Fig. 3. Subsidence curves for the TPB. Note that section 6 shows a fairly constant subsidence through the whole Oligocene, probably due to poor stratigraphic control of the section, which does not allow the detection of accelerating subsidence during this time-span. Grey bars indicate times when the most subsidence occurred.

Results

Subsidence of the TPB began in the Early Oligocene and continued throughout the Miocene (Figs 3 & 4). During the Oligocene, subsidence was stronger in the SW part of the TPB (Bagnasco, Millesimo, Dego; Fig. 3) than in the NE (Montechiaro d'Acqui; Fig. 3), where little movement took place.

Towards the end of the Early Miocene, subsidence accelerated over most of the TPB, including the eastern sector which was a high structural domain during the Oligocene. Burdigalian subsidence affected the entire basin and, in particular, its central-eastern parts (Dego, Spigno, Montechiaro d'Acqui; Fig. 3). Magnitudes of vertical movements are higher in this period than during Oligocene time (<1 km during the Oligocene, >2 km during the Miocene).

A compilation of subsidence rates along the selected transects shows (Fig. 4) that the main periods of subsidence are restricted to a remarkably short time span (≥1mm/year between 17.5 and 15.5 Ma and <0.5 mm/year after 17.5 Ma). The central part of the basin (Dego, Spigno) remains the most subsiding area from Chattian until Burdigalian–Langhian times, with relatively high subsidence rates from 0.7 mm/year during the Chattian up to ≥1 mm/year during the Burdigalian–Langhian. The northeastern part of the basin was not strongly affected by the Rupelian subsidence, although it was strongly subsiding during the Langhian. This because of compressional tectonics active in the eastern sector until Late Aquitanian times and responsible for the uplift and erosion of the Rocchetta sediments (Piana *et al.* 1997).

Table 1. *Input data used to construct the subsidence curves of Figure 3*

	Deph	Age	Sand	Silt	Shale	Carbonate	Halite	Anhydr.	W.D. min	W.D. max	Aver. W.D.
Bagnasco											
C6	28	10.5	0	0	1	0	0	0	200	600	400
C5	143	12.5	0.8	0	0.2	0	0	0	200	600	400
C4	214	13.8	0.8	0	0.2	0	0	0	200	600	400
C3	357	15.5	0.5	0	0.5	0	0	0	200	600	400
C2	543	16.5	0.5	0	0.5	0	0	0	200	600	400
C1	614	17.5	0	0	1	0	0	0	200	600	400
B5+B6	828	21.5	0	0	1	0	0	0	100	400	250
B1−B4	1057	28.4	0	0	1	0	0	0	50	100	75
A	1500	30	0	0	1	0	0	0	0	100	50
A	1714	33.7	1	0	0	0	0	0	0	50	25
	1715	0	1	0	0	0	0	0	0	0	0
Millesimo											
C6	128	10.5	0	0	1	0	0	0	200	600	400
C5	280	12.5	0.7	0	0.3	0	0	0	200	600	400
C4	314	13.8	0.8	0	0.2	0	0	0	200	600	400
C3	514	15.5	0.6	0	0.4	0	0	0	200	600	400
C2	971	16.5	0.8	0	0.2	0	0	0	200	600	400
C1	1486	17.5	0.8	0	0.2	0	0	0	200	600	400
B5+B6	1714	21.5	0.2	0	0.8	0	0	0	100	400	250
B4	1814	25.5	0.2	0	0.8	0	0	0	50	100	75
B3	2157	26.5	0.5	0	0.5	0	0	0	50	100	75
B2	2271	28.4	0.5	0	0.5	0	0	0	50	100	75
B1	2485	30	1	0	0	0	0	0	0	50	25
	2486	0	1	0	0	0	0	0	0	0	0
Dego											
C6	314	10.5	0.5	0	0.5	0	0	0	200	600	400
C5	485	12.5	1	0	0	0	0	0	200	600	400
C4	600	13.8	0.8	0	0.2	0	0	0	200	600	400
C3	1114	15.5	0.8	0	0.2	0	0	0	200	600	400
C2	1714	16.5	1	0	0	0	0	0	200	600	400
C1	2342	17.5	1	0	0	0	0	0	200	600	400
B5+B6	2454	21.5	0	0	1	0	0	0	100	400	250
B4	2886	25.5	0.8	0	0.2	0	0	0	50	100	75
B3	3200	26.5	0.5	0	0.5	0	0	0	50	100	75
B1+B2	3257	30	1	0	0	0	0	0	0	50	25
	3257.1	01	0	0	0	0	0	0	0	0	0
Spigno M.											
C6	343	10.5	0.5	0	0.5	0	0	0	200	600	400
C5	571	12.5	1	0	0	0	0	0	200	600	400
C4	657	13.8	0.8	0	0.2	0	0	0	200	600	400
C3	1100	15.5	0.8	0	0.2	0	0	0	200	600	400
C2	1543	16.5	0.4	0	0.6	0	0	0	200	600	400
C1	2200	17.5	1	0	0	0	0	0	200	600	400
B5+B6	2286	21.5	0.2	0	0.8	0	0	0	100	400	250
B4	2385	25.5	0	0	1	0	0	0	50	100	75
B3	2514	26.5	0.2	0	0.8	0	0	0	50	100	75
B2	2557	28.4	0	0	1	0	0	0	50	100	75
B1	2628	30	1	0	0	0	0	0	0	100	50
A	2814	33.7	0.4	0	0.6	0	0	0	0	50	25
	2815	0	0.4	0	0.6	0	0	0	0	0	0

Table 1 *(continued)*

	Deph	Age	Sand	Silt	Shale	Carbonate	Halite	Anhydr.	W.D. min	W.D. max	Aver. W.D.
Montechiaro d'Acqui											
C6	343	10.5	0.5	0	0.5	0	0	0	200	600	400
C5	571	12.5	1	0	0	0	0	0	200	600	400
C4	657	13.8	0.8	0	0.2	0	0	0	200	600	400
C3	1114	15.5	0.8	0	0.2	0	0	0	200	600	400
C2	1457	16.5	0.4	0	0.6	0	0	0	200	600	400
C1	1971	17.5	1	0	0	0	0	0	200	600	400
B5+B6	2014	21.5	0.2	0	0.8	0	0	0	100	400	250
B2–B4	2114	28.4	0	0	1	0	0	0	50	100	75
A	2214	30	1	0	0	0	0	0	0	100	50
	2215	0	0.4	0	0.6	0	0	0	0	0	0
Tanaro											
Pliocene	410	5.3	0.2	0	0.8	0	0	0	50	100	75
Messinian	550	7.12	0.5	0	0.5	0	0	0	50	100	75
Tortonian	800	11.2	0.5	0	0.5	0	0	0	200	600	400
Serravallian	1080	14.8	0.5	0	0.5	0	0	0	200	600	400
Langhian	1400	16.5	0.5	0	0.5	0	0	0	200	600	400
Aquitanian	2100	23.8	0.5	0	0.5	0	0	0	50	400	225
Oligocene	3400	33.7	0.5	0	0.5	0	0	0	0	50	25
	3400.1	0	0.5	0	0.5	0	0	0	0	0	0

First column, sequences from Gelati *et al.* (1993) and Pieri & Groppi (1981). Age, chronological boundaries; w.d. max, maximum palaeodepth; w.d. min, minimum palaeodepth; aver. w.d., average between w.d. max and w.d. min.

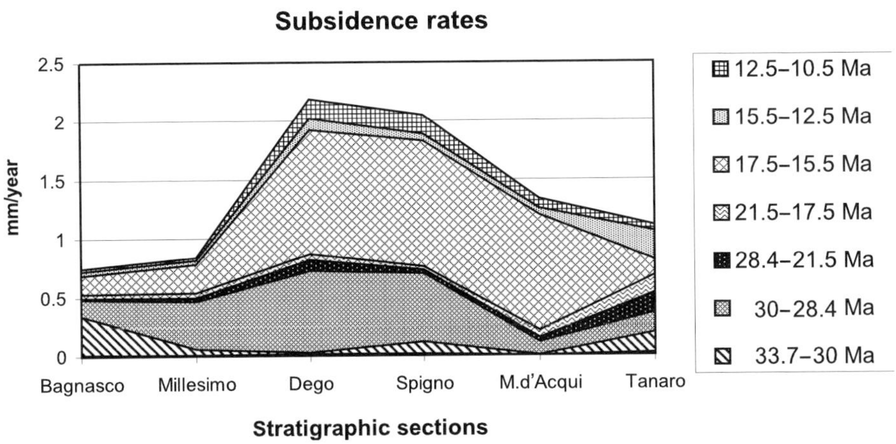

Fig. 4. Early Oligocene–Late Miocene subsidence rates based on data obtained from the subsidence analysis of Fig. 3.

Structural studies

The occurrence of a >4 km deep basin within an area of convergence and more interestingly overlapping an orogenic wedge, poses major questions as to the tectonic mechanism responsible for the observed subsidence. An apparent feature of the TPB is the lack of major tectonic structures. For instance, normal faults never show offsets larger than few hundred metres. Given the long history of investigations in the area, it is also unlikely that major structures have escaped field mapping. Smaller-scale structural features such as folds and faults ranging from several tens of meters to decimetres, on the contrary, are not uncommon in the TPB and have received surprisingly little attention. Despite the limited strain that they accommodate, these structures are of great importance because they constrain the stress regime during basin development. A determination of the stress regime during subsidence is obviously a necessary first step to understand the subsidence dynamics of the TPB.

Method and input data

Two approaches have been used to reconstruct the stress evolution of the TPB, classical structural analysis on folds and faults, and palaeostress analysis. In the first case, outcrop-scale structures have been described and measured to derive fold axes and transport directions in the case of asymmetrical folds and of thrusts. Contractional axes are considered to be perpendicular to the fold axes and parallel to the transport directions.

When enough faults were present in a single outcrop, we carried out a palaeostress analysis following well-established principles (Angelier, 1989 and references therein). To determine the position and shape of the stress tensor we have used the program TENSOR (Delvaux, 1993). Measurements were taken generally on sub-horizontal beds. The typical problem in reconstructing the stress evolution of a sedimentary basin is the dating of single tectonic structures. In some important cases we were able to date folds and faults on the basis of their synsedimentary character. These structures are described in detail in the following section. For other features, a minimum age has been derived from the age of the sediments affected by deformation. A similar approach has been used for the palaeostress measurements.

Results

Contractional structures

Our data seem to indicate the existence of two different families of folds: the first with axes NW–SE and the second with axes NE–SW (Fig. 5, Table 2).

NW–SE fold axes. NE–SW shortening has been observed in different localities of the TPB in sediments as young as Tortonian (Figs 5 & 6)

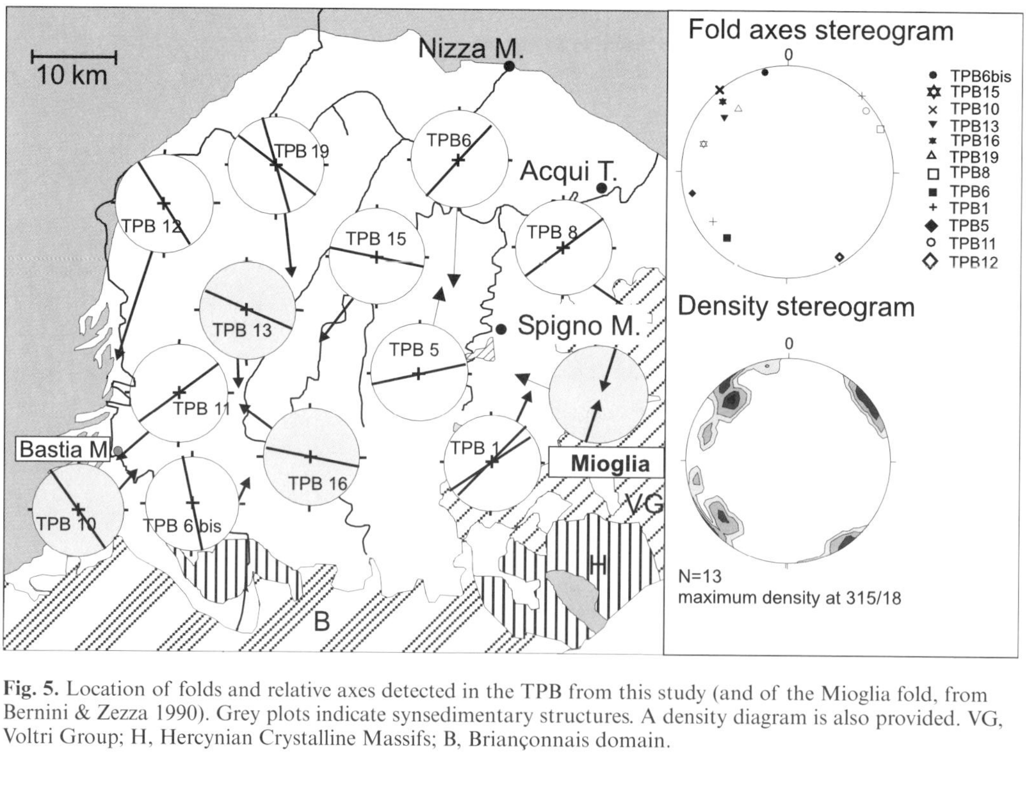

Fig. 5. Location of folds and relative axes detected in the TPB from this study (and of the Mioglia fold, from Bernini & Zezza 1990). Grey plots indicate synsedimentary structures. A density diagram is also provided. VG, Voltri Group; H, Hercynian Crystalline Massifs; B, Briançonnais domain.

Table 2. *Depositional time, formations, and coordinates (UTM, zone 32T) of sites used for structural studies mentioned in the text.*

Sites	Formations	Depositional time	UTM coordinates zone 32T
TPB1bis	Molare Fm	Rupelian	448390-493148
TPB7	Molare Fm	Rupelian	461000-493412
TPB8	Molare Fm	Rupelian	459700-493380
TPB4	Rocchetta Fm (muddy matrix, Gelati & Gnaccolini 1998)	Rupelian–Aquitanian	446720-493590
TPB2	Rocchetta Fm (Noceto system)	Rupelian–Aquitanian	444110-423250
TPB6	Rocchetta Fm (muddy matrix, Gelati & Gnaccolini 1998)	Rupelian–Aquitanian	445410-493563
TPB6bis	Paroldo Marls	Rupelian–Aquitanian	425530-491956
TPB9	Monesiglio Fm (system N3, Gelati & Gnaccolini 1996)	Chattian	415780-491670
TPB3	Monesiglio Fm (glauconite-rich hybrid arenites, Gelati & Gnaccolini 1998)	Aquitanian	446300-493691
Mioglia	Rocchetta Fm (see Bernini & Zecca, 1990)	Rupelian–Burdigalian	453000-492480
TPB1	Rocchetta Fm (Mioglia system, Cazzola & Sgavetti 1983)	Chattian–Burdigalian	452502-492717
TPB5	Rocchetta Fm (Piantivello body, Gelati & Gnaccolini 1998)	Burdigalian	443360-493583
TPB15	Cortemilia Fm	Aquitanian–Langhian	431500-493228
TPB11	–	Aquitanian–Serravallian?	412200-492207
TPB10	Murazzano Fm?	Langhian	414400-492041
TPB13	Cassinasco Fm	Langhian–Serravallian?	424110-492738
TPB16	Cassinasco Fm	Langhian–Serravallian	424110-492645
TPB17	Lequio Fm	Serravallian–Tortonian	493047-418980
TPB19	Lequio Fm	Serravallian–Tortonian	426920-493800
TPB 12	Lequio Fm?	Tortonian?	413400-492901

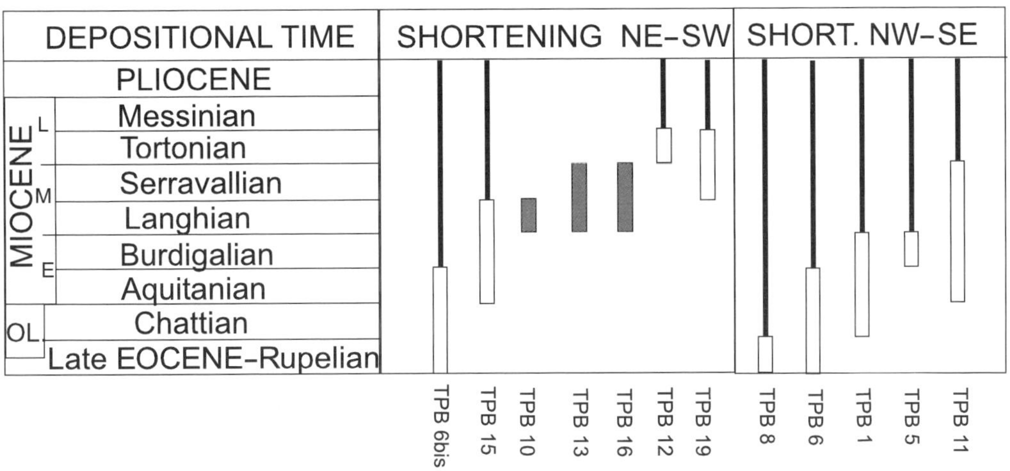

Fig. 6. Age constraints of shortening. White bars, age of sediments in which shortening has been found to correspond with the maximum deformation age. The possible age of shortening is younger than the white bars, and is indicated by thick black lines. Grey bars, synsedimentary structures.

represented mainly by open asymmetrical anticlines and a few closed folds from tens to hundreds of metres in dimensions.

A number of these structures have a synsedimentary character and developed during Langhian–Serravallian times (TPB 10, 13, 16; Figs 7–9). Possibly the most spectacular NW–SE synsedimentary-trending fold is the Ciglie anticline (TPB 10, Fig. 7) where the growing structure is unconformably onlapped by a subhorizontal turbidite sequence which has been dated as Middle Langhian (D'Atri, pers. comm.). Other structures with similar shortening geometries developed in soft sediments are presently well exposed along road cuttings near the Bossola Pass (TPB 13, TPB 16; Figs 8 & 9). Site TPB 13 corresponds with an asymmetrical closed syncline overturned toward the NE. This fold developed while the sediments were still soft and therefore it is representative of synsedimentary deformation (Fig. 10). A low-angle normal fault in the hinge of the anticline is associated with fold development and is sealed by an undeformed stratum (on the top of the SW flank, Fig. 8). Site TPB 13 is the only example we have of an overturned structure representing strong deformation. Because of the lack of subsurface data, we cannot say whether this deformation is associated with a major deep structure. Site TPB16 (Fig. 9) corresponds with a box fold with axis 317/11 and related reverse faults 220/18, 247/40 with movements towards the NE. This developed while the sediments were still soft and therefore it is representative of a synsedimentary deformation fold as site TPB 13.

TPB 12 and TPB 19 indicate that NE–SW shortening also affected Tortonian sediments. Therefore on the basis of the observed structures indicated above we conclude that NE–SW shortening was active during Langhian–Tortonian times although the lack of information on the Messinian does not allow a better upper age limit to be given.

NE–SW fold axes. NW–SE shortening has been observed in five different localities of the TPB (Fig. 5). A good example of a NE–SW fold axis is TPB 11 (near Bastia Mondovì) which has been detected in Lower–Middle Miocene sediments. The TPB 11 is a 200 m open asymmetrical anticline with subhorizontal strata in the SE and with dips up to 30° on the NW flank.

Fig. 7. Example of NE–SW-directed synsedimentary shortening: site TPB 10, Ciglie growing anticline. The sample indicated with the black dot has been dated as Middle Langhian (d'Atri, pers. comm.).

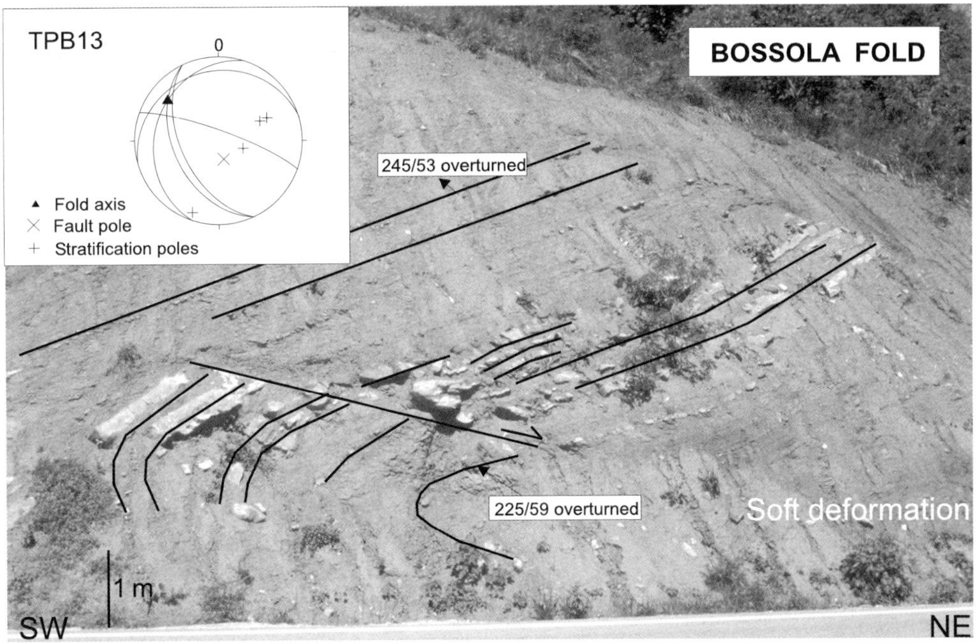

Fig. 8. Example of NE–SW directed synsedimentary shortening: site TPB 13, Bossola fold.

The NE–SW-trending folds never display a synsedimentary character (Fig. 6). For this reason we interpret NW–SE shortening to be younger than NE–SW shortening, namely post-Tortonian.

Extensional structures

Palaeostress analysis on small-scale structures suggests a fairly homogeneous north–south tension over the entire basin (Figs 10–12). All of these extensional structures have been detected in Rupelian to Tortonian sediments (Fig. 12). Synsedimentary extension has been detected only in one site (TPB 17) in the Lequio Formation (Serravallian–Tortonian), represented by a set of normal faults with a few centimetres offset and strata thickening toward the fault plane. Our data suggest a north–south tension active at least during Serravallian–Tortonian times.

Magnetic anisotropy

Method and input data

The anisotropy of magnetic susceptibility (AMS) of ferromagnetic minerals is widely used to provide information on the tectonic history of weakly deformed sediments (Scheepers & Langereis 1994; Duermeijer *et al.* 1998), such as those present in the TPB. The AMS is represented by a second-order tensor, which can be visualized using a three-axis ellipsoid (K_{max}, K_{int} and K_{min}). The total degree of anisotropy is defined by $P=K_{max}/K_{int}$; the magnetic foliation is defined by $F=K_{int}/K_{min}$ and the magnetic lineation, which is the degree of anisotropy in the magnetic foliation plane, is defined by $L=K_{max}/K_{int}$ (Tarling & Hrouda 1993). The orientation of the AMS ellipsoid is in most cases congruent with the strain ellipsoid. In fine-grained rocks, such as the studied sediments, the preferred orientation of phyllosilicates depends on the strain caused by compaction and tectonic processes (Clark 1970; Moore & Geigle 1974; Oertel 1983; Peterson *et al.* 1995), although depositional currents can also account for lineation. Several studies have shown the relationships between magnetic fabric and strain in compressional regimes (Kissel *et al.* 1986; Lee *et al.* 1990; Scheepers & Langereis 1994). In weakly deformed sediments, the orientation of K_{min} is perpendicular to the bedding plane, while the orientation of K_{max} is generally perpendicular to the direction of major shortening (Lee *et al.* 1990; Tarling & Hrouda 1993). An increase in strain causes the ellipsoid to have a more prolate structure (Lee *et al.* 1990; Tarling &

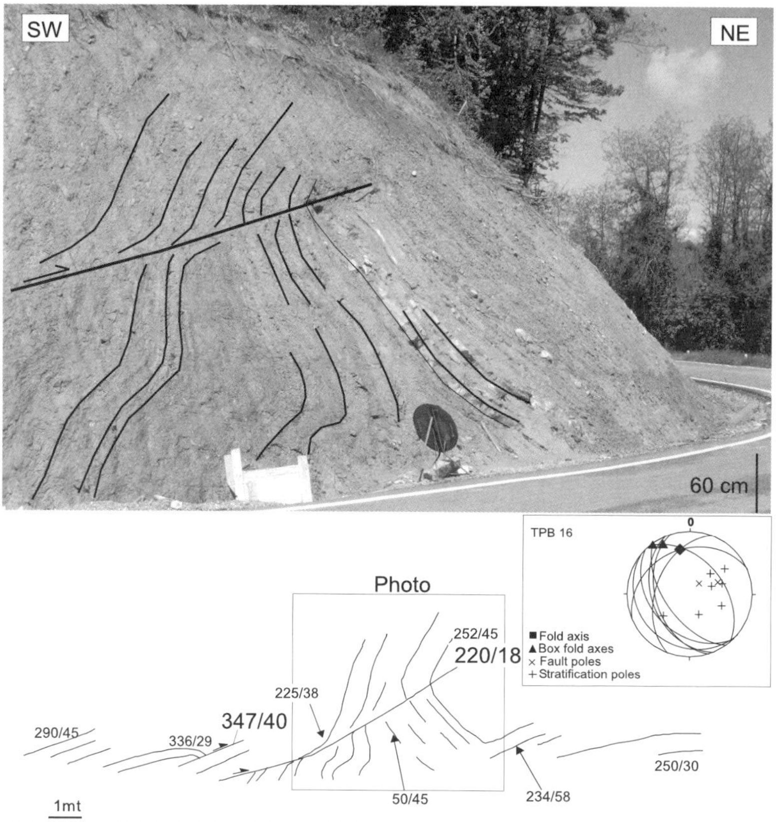

Fig. 9. Example of NE–SW-directed synsedimentary shortening: site TPB 16, fold box (100 m south of the Bossola fold); drag folds associated with the fault constrain the displacement direction.

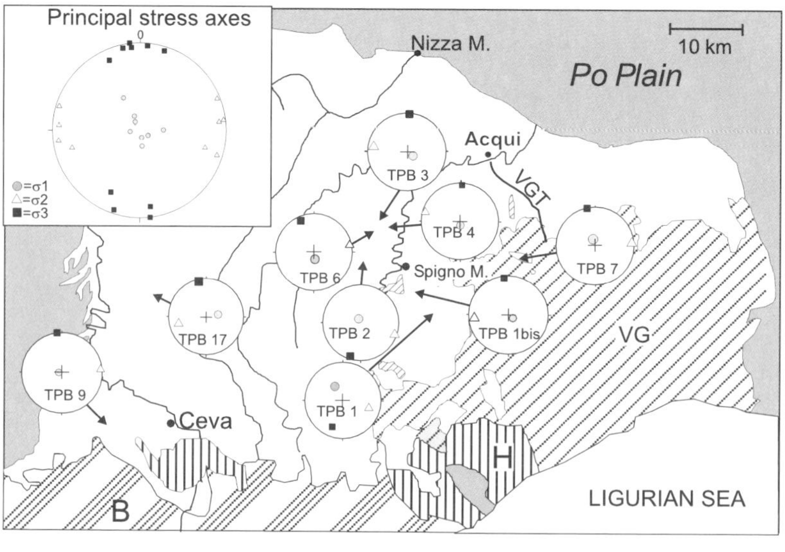

Fig. 10. Sites and results of palaeostress analysis. Summary of results in the upper left square. VG, Voltri Group; H, Hercynian Crystalline Massifs; B, Briançonnais domain. VGT, Val Gorrini Thrust.

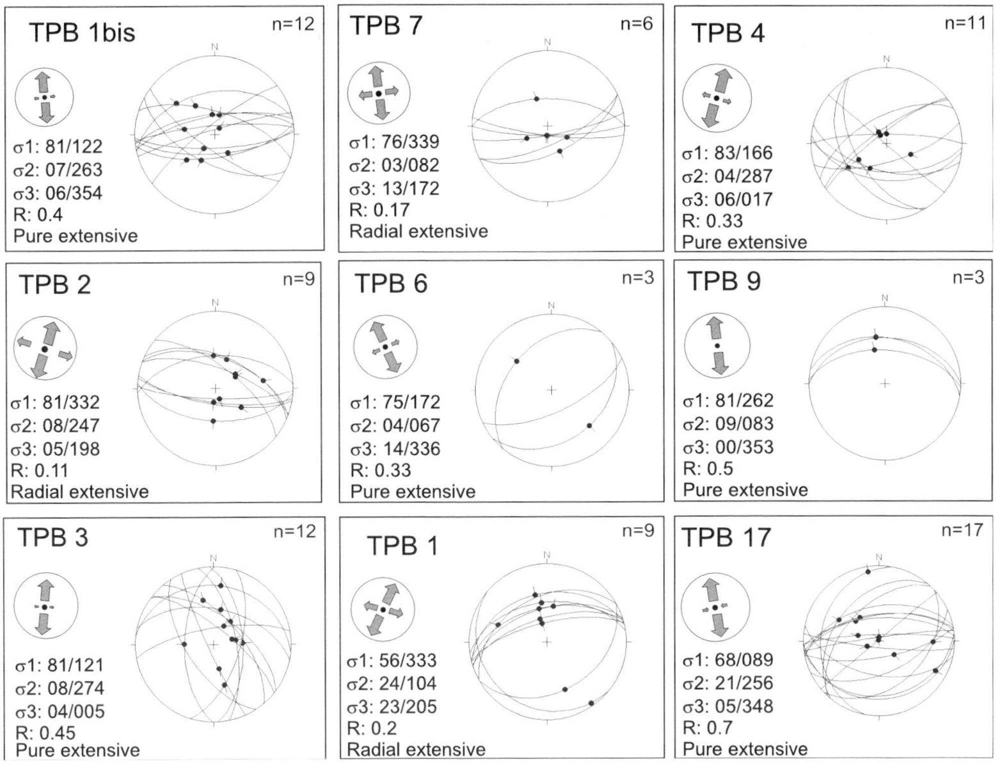

Fig. 11. Stereoplots of normal faults measured in the TPB; site locations are given in Figure 10. R, stress ellipsoid shape ratio.

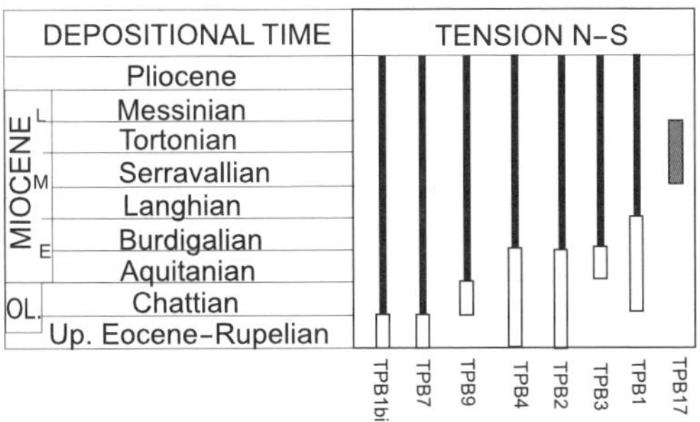

Fig. 12. Age determination for north–south tension. White bars, age of sediments in which tension has been found and therefore maximum deformation age. The possible age of tension is younger than the white bars and indicated by thick black lines. Grey bars: synsedimentary structures.

Hrouda 1993). Therefore the magnetic lineation (L), in weakly deformed sediment strongly depends on the stress field (Kissel *et al.* 1986) allowing comparison with structural data. A high value of L is most likely to be related to synsedimentary deformation, since the fabric of sediments is more easily affected by strain when they are relatively soft and unconsolidated (Borradaile 1988; Mattei *et al.* 1997).

Samples have been collected within shale/silt layers belonging to the entire stratigraphic sequence and covering the entire basin (Fig. 13; Table 3). AMS measurements were carried out using a high-sensitivity low-field susceptibility bridge (KLY-3) at Fort Hoofddijk, University of Utrecht. The mean ellipsoids have been calculated according to Jelinek (1978). Furthermore in this work only values of K_{max} with dD (errors) <25 have been considered as being representative of shortening directions. The locations of all sites are shown in Fig. 13, but with only the most representative results for the AMS axes while the complete AMS data are presented in Fig. 14.

Results

AMS data show the existence of two directions of shortening, respectively NE–SW and NW–SE, supporting the directions of deformation

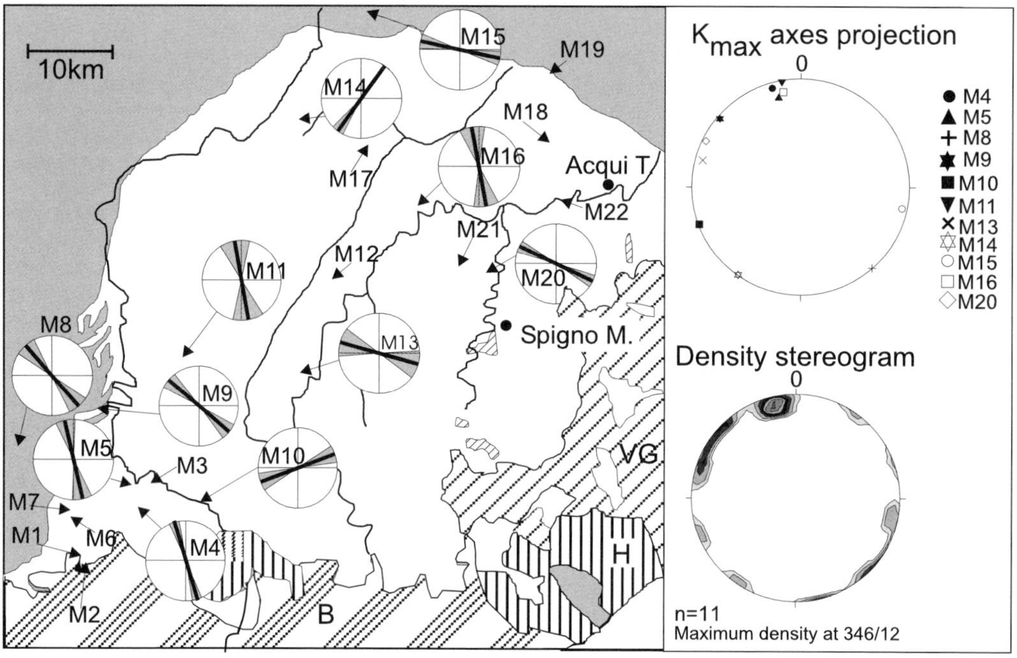

Fig. 13. Location of AMS and NRM analysis. Stereoplots indicate the results of AMS measurements; shaded segment represents error on mean K_{max} axes (d*Az*) with solid line as mean lineation (*L*) direction per section. Only sites with d*D* <25 have been plotted. Density diagram allows a comparison with the structural data in Figure 5. VG, Voltri Group; H, Hercynian Crystalline Massifs; B, Briançonnais domain.

detected with structural analysis (Figs 13 & 14). This confirms the statement that the AMS ellipsoid can be associated with regional deformations. This assumption is further supported by the trend of magnetic lineation, which differs from the palaeocurrent directions in the same area (Gelati *et al.* 1993).

In some cases, an age of deformation can be proposed. This is the case of sediments with high *L*. Very high values of *L* correspond in general with NW–SE AMS axes, which can be related to NE–SW shortening (Table 3; Fig. 13). In particular they occur in Lower Oligocene–Aquitanian sediments (site M20), in Upper Oligocene–Lower Miocene sediments (site M4), and in Middle Langhian sediments (site M5) (Table 3; Fig. 13). Furthermore, site M5 has been measured on the Ciglie Anticline (Fig. 7) supporting the relationship between shortening (in this case synsedimentary) and AMS data (especially high values of *L*).

NRM analysis for rotations

Palaeomagnetic results and palaeoreconstruction models from areas surrounding the TPB suggested that significant counterclockwise rotations took place during Oligo–Miocene times (Burrus 1984; Boccaletti *et al.* 1990; Vanossi *et al.* 1994; Vigliotti & Langenheim 1995; Bormioli & Lanza 1995; Muttoni *et al.* 2000). Since no palaeomagnetic data were available for the TPB itself, we have performed a demagnetization analysis on sediments from the entire basin in order to investigate the tectonic rotation history. NRM analysis was also performed on Pliocene sediments that seal the geometrical relationships between the TPB basin and the surrounding areas (e.g. the Po Plain), providing information on this particular time-span.

Method and input data

The study of the natural remanent magnetization (NRM) in rock samples was carried out in order to derive the characteristic remanent magnetization (ChRM), which can be used to estimate tectonic rotations. Samples were collected from the entire basin (Fig. 13) and covering the whole stratigraphic interval (Table 4). The ChRM component was obtained by means of pro-

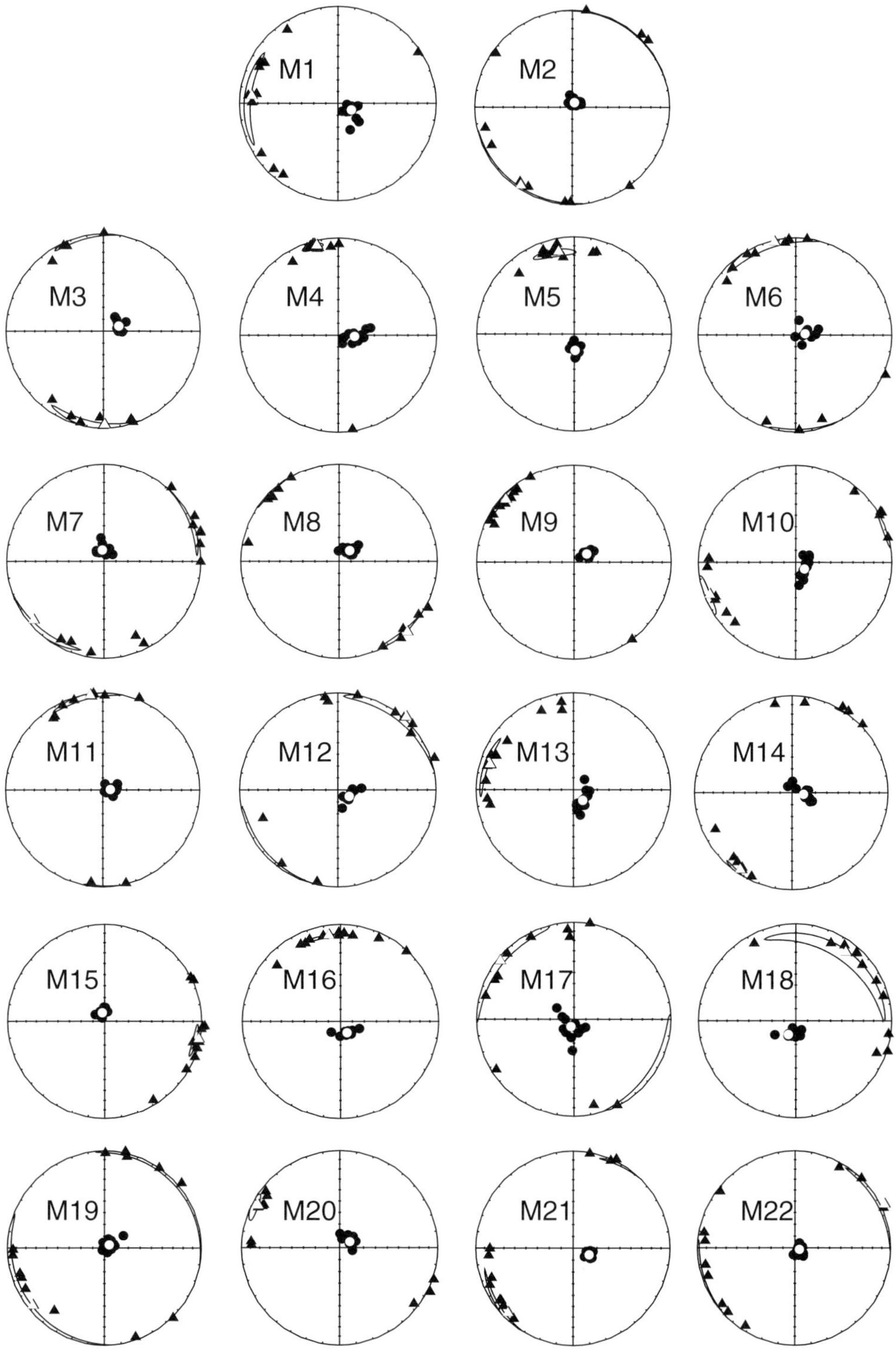

Fig. 14. Equal area projection of K_{max} (triangles) and K_{min} (circles) of the ellipsoid of the AMS for the individual samples, with the calculation of the mean ellipsoid according to Jelinek (1978).

Table 3. *AMS data. n, number of cores for each samples.*

Code	*n*	*D*	*I*	*D*(tc)	*I*(tc)	d*D*	d*I*	*L*	*F*	Formations	Depositional time	UTM coordinates Zone 32T
M 1	11	273.9	9.0	94.4	2.3	30.9	3.5	1.0019	1.0249	Rocchetta (N4-W2 by Gelati & Gnaccolini 1996)	Latest Olig–Early Miocene	415635-491086
M 2	11	214.8	0.5	214.8	0.5	41.5	2	1.0013	1.0551	Rocchetta (N4-W2 by Gelati & Gnaccolini 1996)	Latest Olig–Early Miocene	415885-491078
M 3	11	181.6	2.1	1.8	0.2	33.5	3.8	1.0058	1.1514	Rocchetta (N4-W2 of Gelati & Gnaccolini 1996)	Latest Olig–Early Miocene	413918-491887
M 4	11	344.0	6.3	343.5	0.3	6.6	3.2	**1.0277**	1.0685	Monesiglio (N4-W2, of Gelati & Gnaccolini 1996)	Latest Olig–Early Miocene	414788-491567
M 5	11	346.2	15.5	347	0	14.6	3.8	**1.0148**	1.183	–	Middle Langhian	415000-492038
M 6	11	344.4	2.8	164.4	2.7	31.8	4.8	1.0036	1.0429	–	Late Langhian (d'Atri, pers.comm)	408667-491503
M 7	11	230.8	1.1	230.8	0	36.2	3	1.0022	1.0395	–	Early Tortonian (d'Atri, pers.comm.)	408000-491670
M 8	11	138.8	0.9	138.9	0	15.2	2.8	1.0092	1.0517	–	Late Tortonian/Messinian (d'Atri, pers. comm.)	403065-492272
M 9	11	310.2	3.2	130.2	0.8	12.4	2	1.0072	1.0716	–	Late Miocene	411303-492559
M 10	11	249.4	1.4	249.5	0	12.6	5.4	**1.0116**	1.1408	Rocchetta	Late Oligocene–Aquitanian	424121-491714
M 11	11	349.7	2.8	169.8	4.2	22.4	2.8	1.0043	1.0649	Lequio	Tortonian–Serravalian	420284-493184
M 12	11	41.6	1.7	41.5	0.2	38.1	4.9	1.002	1.1723	Cassinasco	Serravallian–Langhian	429325-493630
M 13	11	285.2	7.8	285.7	1.8	19.7	2.3	1.0038	1.0668	Cortemilia	Aquitanian–Langhian	429767-492761
M 14	11	215.2	0.9	215.3	0.8	8.7	1.4	1.005	1.071	S. Agata fossiliferus Marls	Tortonian	429653-495209
M 15	11	102.0	5.7	102.2	4.6	10.3	2.2	1.0048	1.0826	Lugagnano Shales	Pliocene	435034-496416
M 16	11	349.6	12.3	169.6	0.7	17.9	1.5	1.0027	1.076	Cessole Marls	Langhian	440665-494478
M 17	11	306.1	0.7	126.1	0.2	39	7.2	1.0191	1.1645	Serravalle Sandstones	Serravallian	436940-494960
M 18	11	34.9	16.8	34.8	0	54.4	5.9	1.0007	1.0374	S. Agata fossiliferous Marls	Tortonian	450943-495264
M 19	11	232.2	0.9	232.7	3.4	57.8	4.1	1.0007	1.0371	Lugagnano Shales	Pliocene	454180-495717
M 20	11	295.8	3.9	295.7	0	8.7	3.8	**1.0178**	1.1642	Rocchetta–Monesiglio (muddy matrix of Gelati & Gnaccolini 1998)	Early Olig-Aquitanian	446547-493629
M 21	11	226.2	1.9	226.4	0	30.5	1.7	1.0037	1.1019	Rocchetta–Monesiglio (S.Sebastiano unit of Gelati & Gnaccolini 1998)	Burdigalian	443174-493557
M 22	11	64.0	0.8	63.9	2.2	31	2.7	1.0037	1.2362	Cortemilia	Aquitanian-Langhian	451486-494266

D, *I*, mean azimuth and dip of Kmax, (tc, after tilt correction), d*D*, d*I*, errors on mean Kmax; *L*, magnetic lineation (Kmax/Kint); *F*, magnetic foliation (Kint/Kmin). Bold characters refer to the high value of *L* with d*D* <25. Depositional time refers to the formation in which the AMS has been measured..

Table 4. *Results from NRM analysis from the different sites of the TPB (see Fig. 13 for location).*

Site	Pol.	N	Dec	Inc	k	a95	rot.	Dep. Age
M01	R	9	167	–46	147	4	–13	Latest Oligocene–Early Miocene
M01*	R	9	167	–47	199	4	–13	Latest Oligocene–Early Miocene
M03	R	11	157	–52	76	5	–23	Latest Oligocene–Early Miocene
M03*	R	11	141	–55	67	6	–39	Latest Oligocene–Early Miocene
M05	R	8	135	–45	45	8	–45	Middle Langhian
M05*	R	9	150	–51	84	6	–30	Middle Langhian
M06	N	4	340	53	167	7	–20	Late Langhian
M07	R	10	190	–36	3	35	10	Early Tortonian
M07*	R	–	–	–	–	–	–	Early Tortonian
M08	N	11	17	69	141	4	17	Late Tortonian/Messinian?
M09	R	6	193	–48	23	14	13	Late Miocene
M10	N	8	345	45	31	10	–15	Late Oligocene–Aquitanian
M13	N	11	355	52	80	5	–5	Aquitanian–Langhian
M14	N	11	3	62	15	12	3	Tortonian
M15	N	11	353	49	143	4	–7	Pliocene
M19	N	11	359	41	30	9	–1	Pliocene
M20	N	11	299	27	10	15	–61	Early Oligocene–Aquitanian

* AF demagnetization. N, number of specimens; Dec, Inc, site mean ChRM declination and inclination; k, Fisher's precision parameter; a95, 95% cone of confidence; rot, angle of rotations (counterclockwise). Grey areas corresponds with the age of sediments affected by counterclockwise rotation (>10°).

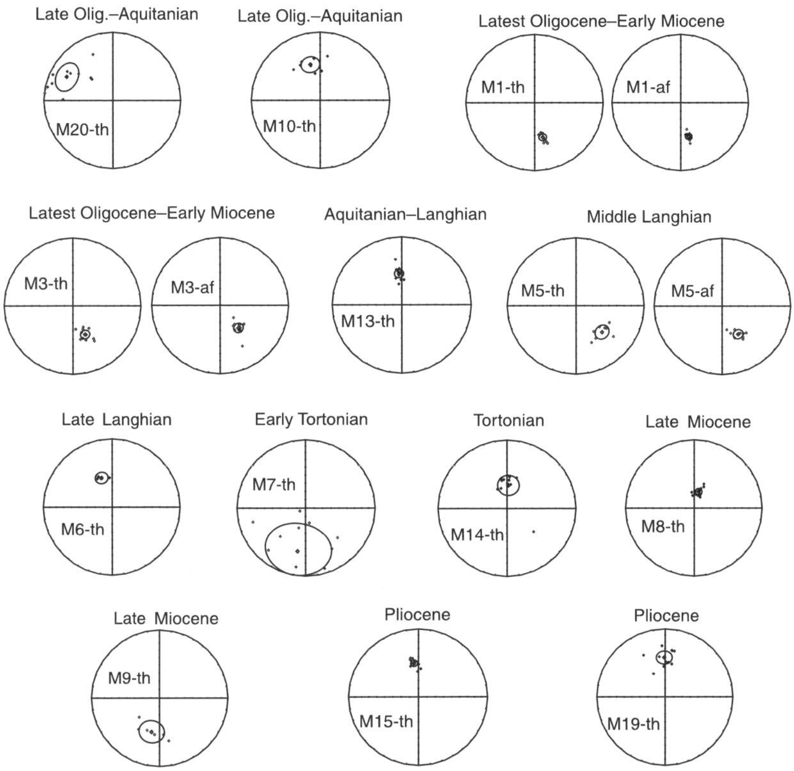

Fig. 15. Equal area projection of ChRM components of samples from the TPB. Dots represent the individual sample directions; full (open) dots represent downward (upward) projections. The circles give a_{95} (Fisher's 1953) for the different site means.

gressive stepwise thermal demagnetization (TH) using small temperature increments (30–50°C) (Fig. 15) and, for some samples, by alternating field demagnetization (AF) (Fig. 16). This latter method involves increasing at each step the alternating field strength instead of the temperature. Each mineral has its characteristic unblocking field in the same way that it has a typical unblocking temperature. In the perfect case the same rotation should be obtained by both techniques. The NRM was measured for all the samples with a 2G Enterprise DC SQUID cryogenic magnetometer. Demagnetization diagrams were used for the interpretation of representative samples (Fig. 16). The ChRM-components were determined by calculating best-fit lines through data-points belonging to specific temperature intervals. Demagnetization vectors were finally combined using Fisher statistics (Fisher, 1953) to calculate mean directions per site. Palaeomagnetic analyses were performed on 22 sites, each consisting of 11 samples, but only 13 sites gave reliable results (Figs 15 & 17; Table 4). In general all samples show an NRM which is largely removed at temperatures of 360–450°C or at fields of 80 mT. Demagnetization at higher temperatures commonly resulted in the generation of a randomly oriented viscous component.

Results

Thermal demagnetization analyses show that the Oligocene–Early Miocene sites of the TPB have been affected by rotation (Figs 15–17). Only sites with rotations >10° are here considered as representative.

Most sites, especially in the SW part, show a counterclockwise rotation of an average of few tens of degrees around 20° (sites: M1, M3, M5, M6, M10 and M20; Fig. 17; Table 4). The counterclockwise rotation from sites M1, M3, M5 is also confirmed by the AF demagnetization data (Figs 16 & 17; Table 4). Two sites (M20 and M5) show an anomalous high rotation between 45 and 61°, possibly due to local processes. Only two sites in the south-western area (M8 and M9) show a representative (>10°) clockwise rotation.

A tentative time constraint is here made by looking at the depositional time of sediments involved in counterclockwise rotations. In general, rotations greater than 10° are restricted to sediments as old as Late Langhian, while Tortonian–Pliocene sediments show no rotations (Fig. 15; Table 4). Therefore, a Middle Miocene age (Serravallian) is suggested for the tectonic phase responsible for the counterclockwise rotation of the TPB.

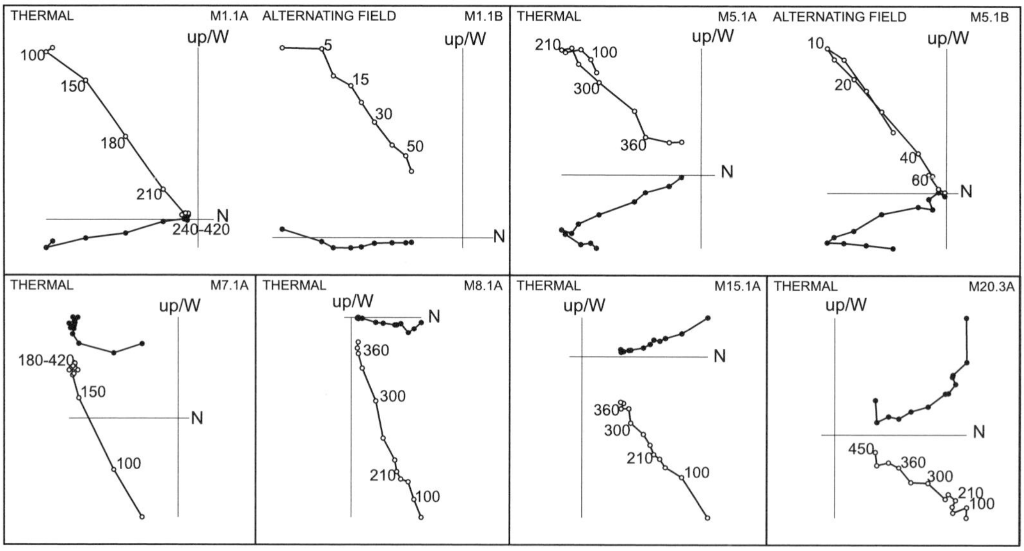

Fig. 16. Stepwise thermal demagnetisation diagrams of the individual specimens from the most representative site of the TPB, and comparison with the alternating field demagnetisation method. Dots are projections on the horizontal plane and circles are projected on the vertical north–south or east–west plane. Numbers denote demagnetisation steps in °C and mT, respectively.

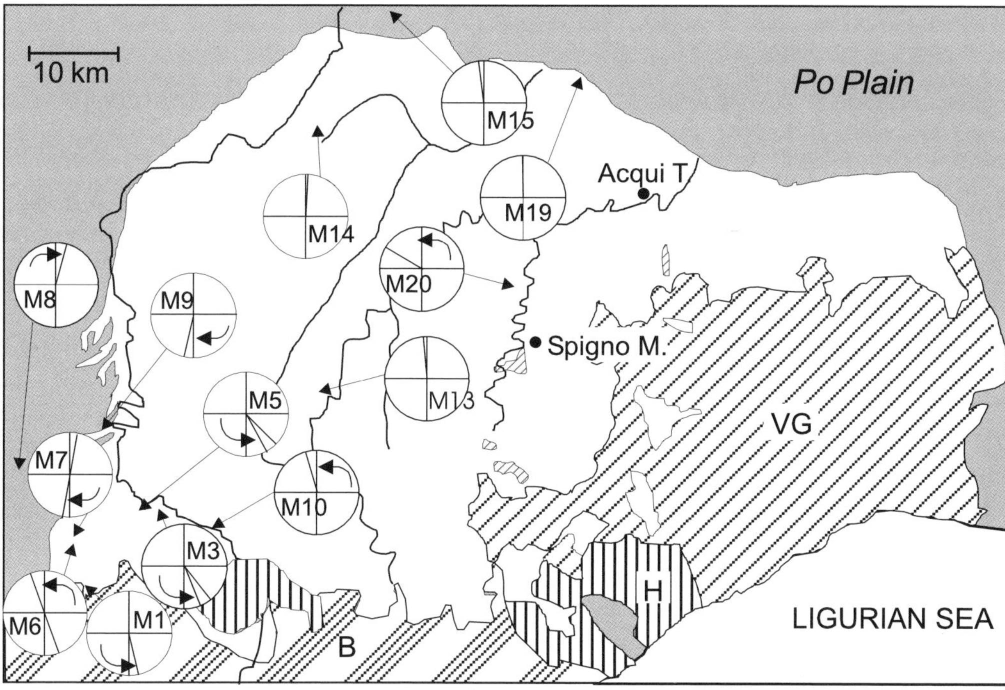

Fig. 17. Diagram showing rotations (from Fig. 15) related to different sites. Only angles >10° are considered as representative. VG, Voltri Group; H, Hercynian Crystalline Massifs; B, Briançonnais domain.

The evolution of the TPB

Subsidence affecting the westernmost segment of the Western Alpine orogen, that is, the Ligurian Alps, allowed the deposition of transitional sediments on top of the orogen, progressively younging from the NE (Late Eocene–Early Oligocene) to SW (Late Oligocene), thereby initiating the evolution of the TPB. Subsidence was fairly constant from the Oligocene until the Middle Miocene, with the exception of an Oligocene episode of subsidence acceleration in the SW part of the basin. The stress/strain regime during Oligocene to Early Miocene times is poorly resolved. Some north–south tensional stresses have been detected in several parts of the basin, especially in its SE portion. They are compatible with relatively small ENE–WSW- to WNW–ESE-trending normal faults mapped in the SE parts of the basin (Bernini & Zecca 1990; Mutti *et al.* 1995; Gelati & Gnaccolini 1998). The overall extension accommodated by these faults is of the order of several hundred metres (*c.*1700 m by measurement of the horizontal displacements) and, therefore, they are indicative of stress rather than substantial strain. During the same time interval, Late Oligocene to Early Miocene NE–SW compression caused the formation of few contractional structures such as the NNW–SSE-trending Mioglia fold in the south and open anticlines in the eastern parts of the basin (Bernini & Zecca 1990; Gelati & Gnaccolini 1998). This phase of shortening fits in well with our AMS data. Timing and thus the genetic relations between north–south tension and NE–SW shortening remain unclear.

From the Early to Middle Miocene, strong subsidence affected large parts of the TPB. Little is known from the southern part of the basin where Miocene sediments are lacking. Similarly to what was discussed for the previous time-span, the observed structures are quite widespread but systematically associated with small displacements. No significant extensional features have been detected. The NE–SW-directed compression and limited shortening remained active through Serravallian–Tortonian times, producing synsedimentary structures such as those observed at sites TPB 10, TPB 13 and TPB 16 (Figs 8 & 9).

Serravallian and older sediments also experienced NW–SE-directed compression associated with the formation of small- and some larger-scale folds. The age of this deformation stage is poorly constrained, but because of the consistent lack of synsedimentary activity we interpret the NW–SE directed compression to be younger than the NE–SW-directed one and therefore post-Tortonian. Despite these uncertainties, it is clear that the entire Miocene TPB evolution took place under a NE–SW- to NW–SE-trending prevailing compressional regime.

The TPB and its regional context

Our new palaeomagnetic data show that TPB sediments experienced a fairly small (*c*.20°) counterclockwise rotation in Middle Miocene times, following which the TPB has basically acquired its present-day position. The stress/strain geometries that we have obtained are thus not substantially different from their original position.

The TPB sediments transgress and only partly seal the pre-Oligocene structures developed in the Ligurian Alps contractional domain (Vanossi *et al.* 1984), since evidence of thrusts in the Ligurian Alps during Oligo-Miocene times over the TPB sediments (Hoogerduijn Strating *et al.* 1991; Piana *et al.* 1997) suggest that the compressional structures in the belts were still active during this time-span.

The Oligocene to Early Miocene basin was lying between the subsiding Po Plain `foredeep' in the north (Dela Pierre *et al.* 1995) and the extensional Liguro–Provençal Basin in the SW. During this time-span the TPB was undergoing general NE–SW shortening.

During Late Oligocene–Early Miocene times, the eastern margin of the TPB experienced NE–SW shortening responsible for the NE-verging Alto Monferrato thrusts (e.g. VGT in Fig. 1; Piana *et al.* 1997) and the overthrusting of the Ligurian units on to the Tuscan units along the Villalvernia–Varzi-Line (VVL in Fig. 1a; Miletto & Polino 1992). The same time-span corresponds with the activity period of the transpressive Rio Freddo Deformation Zone (RFDZ in Fig. 1a), which has been interpreted as the superficial expression of a palaeo-Apenninic thrust (Piana 2000). This phase of transpression and shortening falls within the palaeo-Apenninic phase of deformation, which in the Alpine domain coincides with the Insubric–Helvetic phase, caused mainly by the NW–SE Africa–Europe convergence (Laubscher 1991). The Po Plain was also undergoing subsidence, allowing the deposition of a thick clastic succession probably connected to the TPB.

To the SW of the TPB, a NE–SW continental rift developed in Oligocene times between France/Spain and the Corsica–Sardinia block, leading to Late Oligocene crustal separation and generation of oceanic crust (e.g. Burrus 1984; Jolivet *et al.* 1999). Spreading and drifting ended in the Early Miocene with the shift of the extension site to the east of the Corsica–Sardinia block.

Tectonic and dynamic relationships between the TPB on one side and the Po Plain and Liguro–Provençal Basin on the other are still unclear. Small normal faults found in the south of the basin possibly could be associated with the Liguro–Provençal rifting. However, the clear predominance of compressional stresses during the Oligocene to Early Miocene TPB evolution suggests that the basin was essentially a part of the western Po Plain compressional system.

During the Middle–Late Miocene, the western Po Plain was undergoing roughly *c*.N–S-directed shortening, with the development of the south-verging Milano belt (Jura–Lombardic deformation phase; Laubscher 1992; Schumacher & Laubscher 1996) and strong subsidence. Between the Langhian and the Serravallian the NW Apennine were undergoing roughly NE–SW shortening, responsible for the emplacement of the Ligurian units on to the Modino–Cervarola and Umbro-Marchean units (Pedeapennine Thrust Front, Boccaletti *et al.* 1985). On the western margin of the TPB in Middle Miocene to Pliocene times, active tectonic shortening was responsible (Saluzzo Basin; Fig. 1) for the formation of the Saluzzo fold (Pieri & Groppi 1981). During the same time-span the eastern margin of the TPB was still affected by shortening (*c*.NW–SE) responsible for the shifting towards the north of the Alto Monferrato thrust fronts (Falletti *et al.* 1995). By this time, deformation in the Ligurian Alps had ceased.

From the Pliocene to Quaternary, the Neo-Apenninic deformation phase is responsible for tectonic shortening leading to the formation of the Plio–Quaternary Padan thrust front (Fig. 1d) and for the translation towards the north of the Torino Hill (e.g. Piana & Polino 1995) and probably for the present-day TPB elevation.

Conclusion

The TPB, despite being located on top of an orogenic belt, subsided in Oligocene to Miocene times, allowing the deposition of up to 4 km of clastic sediments. Normal faults detected so far in the TPB are not enough to explain the total amount of subsidence (4–5 km) present in the TPB basin from Oligocene until Late Miocene times. Furthermore, no regional normal faults have been detected so far in the area. Small-scale normal faults detected in Oligocene–Miocene sediments could be related to the extensional phase responsible for the opening of the Liguro-Provençal Basin.

The Miocene is characterized by general strong subsidence and NE–SW to NW–SE shortening. In particular, the Middle Miocene constitutes an important period in the evolution of the TPB, with an acceleration of subsidence under a prevalent NE–SW tectonic shortening. General shortening was active in the same time-span on the western margin of the Po Plain and in the Northern Apennine. In particular, NE–SW to NW–SE directions of shortening detected in TPB sediments fit well with the post-Eocene pattern of NE–SW arc-normal transport direction of the Western Alpine arc proposed by Platt *et al.* (1989), and with the NE–SW and NW–SE directions of shortening detected in the Ligurian Alps (close to the boundary with the Northern Apennine) by Hoogerduijn Strating *et al.* (1991). The latter authors have suggested a possible superposition of two displacement directions of different ages, and in particular proposed a Late Miocene age for the NW–SE shortening related to the development of the Monferrato culmination (Hoogerduijn Strating *et al.* 1991), supporting our data. At this stage there is uncertainty exists regarding the mechanism responsible for the accommodation space in the TPB under prevalent tectonic shortening.

New preliminary NRM data show general 20° counterclockwise rotations of approximately Serravallian age. These new results show that the TPB was not affected by rotations during Tortonian to Quaternary times, despite ongoing shortening in the Apennine domain (e.g. Clari *et al.* 1995; Schumacher & Laubscher 1996). Our data seem to be fairly consistent with rotations detected in the eastern and northern parts of the TPB (Thio 1988; Bormioli & Lanza 1995) and in the Northern Apennine for the Oligocene–Miocene time span (Muttoni *et al.* 2000).

The authors gratefully acknowledge O. Lacombe and S. Sliaupa for their constructive criticism of the manuscript. R. Polino and F. Piana are also kindly acknowledged for very useful discussions, A. d'Atri for fundamental help in biostratigraphic dating, Jan Wijbrans for his interest and comments on the earlier version of the manuscript, and G. R. Murrell for the English review. Technical facilities for AMS and NRM analysis were provided by the Universiteit of Utrecht, Paleomagnetic Laboratory, 'Fort Hoofddijk'. The project was financed by grants from the NWO (the Netherland Organization for Scientific Research) and the NSG (Netherlands Research School of Sedimentary Geology; 20020401).

References

ANGELIER, J. 1989. From orientation to magnitudes in paleostress determinations using fault slip data. *Journal of Structural Geology*, **11**, 37–50.

BERNINI, M. & ZECCA, M. 1990. The deformations in the Oligocene–lower Miocene Molare and Rocchetta Formations. *Atti Ticinesi di Scienze della Terra*, **33**, 1–10.

BESSIS, F. 1986. Some remarks on the study of subsidence of sedimentary basins. Application to the Gulf of Lions margin (Western Mediterranean). *Marine and Petroleum Geology*, **3**, 37–63.

BIELLA, G., POLINO, R., DE FRANCO, R., ROSSI, P. M., CLARI, P., CORSI, A. & GELATI, R. 1997. The crustal structure of the western Po plain: reconstruction from integrated geological and seismic data. *Terra Nova*, **9**, 28–31.

BOCCALETTI, M., COLI, M., EVA, C., FERRARI, G., GIGLIA, G., LAZZAROTTO, A., MERLANTI, F., NICOLICH, R., PAPANI, G., POSTPISCHL, D. 1985. Considerations on the seismotectonics of the Northern Apennines. *Tectonophysics*, **117(1–2)**, 7–38.

BOCCALETTI, M., CIARANFI, N., COSENTINO, D., DEIANA, G., GELATI, R., LENTINI, F., MASSARI, F., MORATTI, G., PESCATORE, T., RICCI LUCCHI, F. & TORTORICI, L. 1990. Palinspastic restoration and paleogeographic reconstruction of the peri-Tyrrhenian area during the Neogene. *Palaeogeography, Palaeoclimatology, Palaeoecology*, **77**, 41–50.

BOND, G. C. & KOMINZ, M. A. 1984. Construction of tectonic subsidence curves for the early Paleozoic miogeocline, southern Canadian Rocky Mountains: implications for subsidence mechanisms. *Geological Society of America Bulletin*, **89,** 1389–1403.

BORMIOLI, D. & LANZA, R. 1995. Rotazioni antiorarie delle rocce terziarie delle Alpi Occidentali e dell'Appennino Settentrionale. *In:* POLINO, R. & SACCHI, R. (eds), A. N. *Scienze, Atti convegno rapporti Alpi-Appennino, Peveragno*, 277–289.

BORRADAILE, G. J. 1988. Magnetic fabrics, petrofabrics and strain. *Tectonophysics*, **156**, 1–20.

BURRUS, J. 1984. Contribution to a geodynamic synthesis of the Provençal basin (north-western Mediterranean). *Marine Geology*, **55**, 247–260.

CAPPONI, G., CRISPINI, L., SILVESTRI, R. & VIGO, E. 1999. The role of Early Miocene trhust tectonics in the structural arrangement of the Voltri Group (Ligurian Alps, Italy): evidence from the Bandita area. *Ofioliti*, **24(1)**, 13–19.

CAPRARA, L., GARZANTI, E., GNACCOLINI, M. & MUTTI, L. 1985. Shelf–basin transition: sedimentology and petrology of the Tertiary Piedmont Basin (Northern Italy). *Rivista Italiana di Paleontologia e Stratigrafia*, **90**, 545–564.

CARRAPA, B., BERTOTTI, G. & WIJBRANS, J. 2000. Mechanisms involved in the formation of the Tertiary Piemonte Basin in a collisional setting and relations between source area and basin infill from $^{40}Ar/^{39}Ar$ dating. *European Geophysical Society, (UN)COUPLED 2000, Continental Collision and the Tectonosedimentary Evolution of Forelands: Mechanisms of Coupling and Far-field Deformation*, Amsterdam, Czech Geological Society Publisher, **45(3–4)**, 216.

CASSANO, E., ANELLI, L. FICHERA, R. & CAPPELLI, V. 1986. Pianura Padana, interpretazione integrata di dati geofisici e geologici. *In:* PIERI, M. & GROPPI, G. (eds)*73° Congresso della Societá Geologica Italiana, Roma 1986*, 27.

CAZZOLA, C. & RIGAZIO, G. 1982. Sedimentology of Valla and Moglia turbidites, Rochetta Formation (Oligocene–Miocene) of Piedmont Tertiary basin. *Giornale di Geologia*, **45(1)**, 87–100.

CAZZOLA, C. & SGAVETTI, M. 1983. Geometria dei depositi torbiditici della Formazione di Rocchetta e Monesiglio (Oligocene superiore-Miocene inferiore) nell'area compresa tra Spigno e Ceva. *Giornale di Geologia*, **(2) XLV**(II), 227–240.

CIULAVU, D. & BERTOTTI, G. 1994. The Transylvania Basin and its Upper Cretaceous substratum. *Romanian Journal of Tectonics and Regional Geology*, **75(2)**, 59–64.

CLARI, P., DELA PIERRE, F., NOVARETTI, A. & TIMPANELLI, M. 1995. Late Oligocene–Miocene sedimentary evolution of the critical Alps/Apennines junction: the Monferrato area, Northwestern Italy. *Terra Research*, **7**, 144–152.

CLARK, B. L. 1970. Mechanical formation of preferred orientation in clays. *American Journal of Science*, **269**, 250–266.

DALLA, S., ROSSI, M., ORLANDO, M., VISENTIN, C., GELATI, R., GNACCOLINI, M., PAPANI, G., BELLI, A., BIFFI, U. & CATRULLO, D. 1992. Late Eocene–Tortonian tectono-sedimentary evolution in the western part of the Padan basin (northern Italy). *Paleontologia Evolucio*, **24–25**, 341–362.

DELA PIERRE, F., MIKHAILOV, V. & POLINO, R. 1995. The tectonosedimentary evolution of the Tertiary basins in the western Po plain: kinematics inferred from subsidence curves. *In:* POLINO, R. & SACCHI, R. (eds), *A. N. Scienze, Atti Convegno Rapporti Alpi–Appennino, Peveragno*, 129–146.

DELVAUX, D. 1993. The TENSOR program for paleostress reconstruction:: examples from the east A frican and the Baikal rift zones. EUG VII, **5**, abstract supplement 1 to TERRA Nova, Strasbourg, 216.

DUERMEIJER, C. E., VAN VUGT, N., LANGEREIS, C. G., MEULENKAMP, J. E. & ZACHARIASSE, W. J. 1998. A major late Tortonian rotation phase in the Crotone basin using AMS as tectonic tilt correction and timing of the opening of the Tyrrhenian basin. *Tectonophysics*, **287**, 233–249.

FALLETTI, P., GELATI, R. & ROGLEDI, S. 1995. Oligo-Miocene evolution of Monferrato and Langhe, related to deep structures. In: POLINO, R. & SACCHI, R., A. N. Scienze, *Atti Convegno Rapporti Alpi–Appennino, Peveragno*, 1–19.

FISHER, R. A. 1953. Dispersion on a sphere. *Proceedings of the Royal Society of London*, **A217**, 295–305.

GELATI, R. 1968. Stratigrafia dell'Oligo-Miocene delle Langhe tra le valli dei fiumi Tanaro e Bormida di Spigno. *Rivista Italiana di Paleontologia e Stratigrafia*, **74(3)**, 865–967.

GELATI, R. & GNACCOLINI, M. 1982. Evoluzione tettonico-sedimentaria della zona limite tra Alpi ed Appennini tra l'inizio dell'Oligocene ed il Miocene medio. *Memoria della Società Geologica Italiana*, **24**, 183–191.

GELATI, R. & GNACCOLINI, M. 1988. Sequenze deposizionali in un bacino episuturale, nella zona di raccordo tra Alpi ed Appennino settentrionale. *Atti Ticinesi di Scienze della Terra*, **31**, 340–350.

GELATI, R. & GNACCOLINI M. 1996. The stratigraphic record of the Oligocene–Early Miocene events at the south-western end of the Piedmont Tertiary Basin. *Rivista Italiana di Paleontologia e Stratigrafia*, **102(1)**, 65–76.

GELATI, R. & GNACCOLINI, M. 1998. Synsedimentary tectonics and sedimentation in the Tertiary Piedmont Basin, Northwestern Italy. *Rivista Italiana di Paleontologia e Stratigrafia*, **104(2)**, 193–214.

GELATI, R., GNACCOLINI, M., FALLETTI, P. & CATRULLO, D. 1993. Stratigrafia sequenziale della successione Oligo–Miocenica delle Langhe, Bacino Terziario Ligure–Piemontese. *Rivista Italiana di Paleontologia e Stratigrafia*, **98(4)**, 425–452.

GHIBAUDO, G., CLARI, P. & PERELLO, M. 1985. Litostratigrafia, sedimentologia ed evoluzione tettonico-sedimentaria dei depositi miocenici del margine sud-orientale del Bacino Terziario Ligure–Piemontese. *Rivista Italiana di Paleontologia e Stratigrafia*, **104**, 349–397.

GNACCOLINI, M. & ROSSI, P. M. 1994. Sequenze deposizionali e composizione delle arenarie nel Bacino Terziario Ligure–Piemontese: osservazioni preliminari. *Atti Ticinesi di Scienze della Terra*, **37**, 3–15.

GNACCOLINI, M., GELATI, R., CATRULLO, D. & FALLETTI, P. 1990. Sequenze deposizionali nella successione oligo-miocenica delle 'Langhe': un approccio alla stratigrafia sequenziale del Bacino Terziario Ligure–Piemontese. *Memoria della Societa Geologica Italiana*, **45**, 671–686.

HOOGERDUIJN STRATING, E. H. 1994. Extensional faulting in an intraoceanic subduction complex-working hypothesis for the Paleogene of the Alps–Appenine system. *Tectonophysics*, **238**, 255–273.

HOOGERDUIJN STRATING, E. H., VAN WAMEL, W. A. & VISSERS, R. L. M. 1991. Some constraints on the kinematics of the Tertiary Piemonte Basin (northwestern Italy). *Tectonophysics*, **198**, 47–51.

HUISMANS, R. S., BERTOTTI, G., CIULAVU, D., SANDERS, C. A. E., CLOETINGH, S. & DINU, C. 1997. Structural evolution of the Transylvanian Basin (Romania); a sedimentary basin in the bend zone of the Carpathians. *Tectonophysics*, **272**(2–4), 249–268.

JELINEK, V. 1978. Statistical processing of anisotropy of magnetic susceptibility on groups of specimens. *Studia Geophysics Geodynamics*, **22**, 50–62.

JOLIVET, L., FRIZON DE LAMOTTE, D., MASCALE, A. & SÉRANNE, M. 1999. The Mediterranean basins: Tertiary extension within the Alpine Orogeny–an Introduction. *In:* DURAND, B., JOLIVET, L., HORVÁTH, F. & SÉRANNE, M. (eds), *The Mediterranean Basins: Tertiary extension within the Alpine Orogeny*. Geological Society, London, Special Publications, **156**, 1–14.

KISSEL, C., BARRIER, E., LAJ, C. & LEE, T. Q. 1986. Magnetic fabric in 'undeformed' marine clays from compressional zones. *Tectonics*, **5**, 769–781.

LAUBSCHER, H. 1991. The arc of the western Alps today. *Eclogae Geologica Helvetica*, **84(3)**, 631–659.

LAUBSCHER, H. P. 1992. Jura kinematics and the Molasse Basin. *Eclogae Geologica Helvetica*, **85**, 653–675.

LEE, T., KISSEL, C., LAJ, C., HORNG & LUE, Y. 1990. Magnetic fabric analysis of the Plio-Pleistocene sedimentary formations of the Coastal Range of Taiwan. *Earth and Planetary Science Letters*, **98**, 23–32.

LORENZ, C. 1979. L' Oligo-Miocène ligure: un example de transgression. *Bulletin de Société geologique de la France*, **12(4)**, 375–378.

LORENZ, C. 1984. Evolution stratigraphique et structurale des Alpes Ligures depuis l'Eocene superieur. *Memoria della Società Geologica Italiana*, **28**, 211–228.

MATENCO, L. & BERTOTTI, G., 2000. Tertiary tectonic evolution of the external East Carpathians (Romania). *Tectonophysics*, **316(3–4)**, 255–286.

MATTEI, M., SAGNOTTI, L., FACENNA, C. & FUNICELLO, R. 1997. Magnetic fabric of weakly deformed clay-rich sediments in the Italian peninsula: relationship with compressional and extensional tectonics. *Tectonophysics*, **271**, 107–122.

MILETTO, M. & POLINO, R. 1992. A gravity model of the crust beneath the Tertiary Piemonte basin (northwest Italy). *Tectonophysics*, **212**, 243–256.

MOORE, J. C. & GEIGLE, J. E. 1974. Slaty cleavage: incipient occurrences in the deep sea. *Science*, **183**, 509–510.

MUTTI, E., PAPANI, L., DI BIASE, D., DAVOLI, G., MORA, S., SEGADELLI, S. & TINTERRI, R. 1995. Il Bacino Terziario Epimesoalpino e le sue implicazioni sui rapporti tra Alpi ed Appennino. *Memorie di Scienze Geologiche di Padova*, **47**, 217–244.

MUTTONI, G., LANCI, L., ARGNANI, A., HIRT, A., M., CIBIN, U., ABRAHAMSEN, N. & LOWRIE, W. 2000. Paleomagnetic evidence for a Neogene two-phase counterclockwise tectonic rotation in the Northern Apennines (Italy). *Tectonophysics*, **326**, 241–253.

OERTEL, G. 1983. The relationship of strain and preferred orientation of phyllosilicate grains in rocks–a review. *Tectonophysics*, **100**, 413–447.

PETERSON, S. R., HAO, Y. & OERTEL, G. 1995. Primary and tectonic intensities in mudrocks. *Tectonophysics*, **247**, 105–119.

PIANA, F. 2000. Structural setting of the Western Monferrato (Alps–Appennines Junction Zone, NW). *Tectonics*, **19(5)**, 943–960.

PIANA, F. & POLINO, R. 1995. Tertiary structural relationships between Alps and Apennines: the critical Torino Hill and Monferrato area, northwestern Italy. *Terra Research*, **7**, 138–143.

PIANA, F., D'ATRI, A. & ORIONE, P. 1997. The Visone Formation, a marker of the Early Miocene tectonics in the Alto Monferrato Domain (Tertiary Piemonte Basin, NW Italy). *Memoria della Società Geologica Italiana*, **49**, 145–162.

PIERI, M. & GROPPI, G. 1981. Subsurface Geological Structure of the Po Plain, Italy. *Progretto Finalizzato Geodinamica*, **414**, 1–13.

PLATT, J. P., BEHRMANN, J. H., CUNNINGHAM, P. C., DEWEY, J. F., HELMAN, M., PARISH, M., SHEPLEY, M. G., WALLIS, S. & WESTON, P. J. 1989. Kinematics of the Alpine arc and the motion of Adria. *Nature*, **337**, 158–161.

POLINO, R., DAL PIAZ, V. & GOSSO, G. 1990. Tectonic erosion at the Adria margin and accretionary processes for the Cretaceous orogeny of the Alps. *In:* ROURE, F., HEITZMANN, P. & POLINO, R. (eds) *Deep Structure of the Alps*. Société Geologique de France, Paris, Société Geologique Suisse, Zürich, Società Geologica Italiana, Roma, 345–367.

REHAULT, J. P., BOILLOT, G. & MAUFFRET, A. 1985. The Western Mediterranean Basin. *In:* STANLEY, D. J. & WEZEL-FORESE, C. (eds), *Geological Evolution of the Mediterranean Basin*. Springer-Verlag, New York, 101–129.

ROURE, F., POLINO, R. & NICOLICH, R. 1990. Early Neogene deformation beneath the Po plain: constraints on the post-collisional Alpine evolution. *Mémoire de la Société géologique de France*, **156**, 309–322.

SCHEEPERS, P. J. J. & LANGEREIS, C. G. 1994. Magnetic fabric of Pleistocene clays from the southern Appennines and its foreland: a magnetic lineation induced after tectonic rotations in the arc. *Tectonics*, **13(5)**, 1190–1200.

SCHMID, S. M. & KISSLING, E. 2000. The arc of the Western Alps in the light of geophysical data on deep crustal structure. *Tectonics*, **19**, 62–85.

SCHUMACHER, M. E., & LAUBSCHER, H. P. 1996. 3D crustal architecture of the Alps–Apennines join; a new view on seismic data. *Tectonophysics*, **260(4)**, 349–363.

SCLATER, J. & CHRISTIE, P. A. F. 1980. Continental stretching: an explanation of the post-mid-Cretaceous subsidence on the central North Sea basin. *Journal of Geophysical Research*, **85**, 3711–3739.

TARLING, D. H. & HROUDA, F. 1993. *The magnetic anisotropy of rocks*. Chapman and Hall, London.

THIO, H. K. 1988. Magnetotectonics in the Piemont Tertiary Basin. *Physics of the Earth and Planetary Interiors*, **52**, 308–319.

VANOSSI, M., CORTESOGNO, L., GALBIATI, G., MESSIGA, B., PICCARDO, G. & VANNUCCI, R. 1984. Geologia delle Alpi Liguri: dati, problemi, ipotesi. *Memoria della Società Geologica Italiana*, **28**, 5–57.

VANOSSI, M., PEROTTI, C. R. & SENO, S. 1994. The Maritime Alps arc in the Ligurian and Tyrrhenian systems. *Tectonophysics*, **230**, 75–89.

VIGLIOTTI, L. & LANGENHEIM, V. E. 1995. When did Sardinia stop rotating? New paleomagnetic results. *Terra Nova*, **7**, 424–435.

Time, place and mode of propagation of foreland basin systems as recorded by the sedimentary fill: examples of the Late Cretaceous and Eocene retro-foreland basins of the north-eastern Pyrenees

F. CHRISTOPHOUL, J.-C. SOULA, S. BRUSSET, B. ELIBANA, M. RODDAZ, G. BESSIERE & J. DERAMOND

Laboratoire des Mécanismes de Transferts en Géologie, UMR 5563, Université Paul Sabatier, 38 rue des 36 ponts, 31400 Toulouse, France (e-mail: christop@cict.fr)

Abstract: The relationship between tectonics and sedimentary fill has been studied in two syncontractional basins of the western Corbières (eastern North Pyrenean retro-foreland basin). The Late Cretaceous basin formed during *c.* 10–12 Ma as a result of left-lateral transpressional deformation, and is composed of forward-younging sub-basins characterized by reworking of the forelimbs of growing fold-propagation folds. Thrust-wedge advance and cratonward migration of the platform are recorded by a deepening-upward stacking pattern indicating increased regional subsidence with a limited contribution of the submarine orogen. Tectonic quiescence and erosional unloading lasting 29–30 Ma are recorded by a shallowing-upward stacking pattern, and fluvial sedimentation issued from widespread sources in the emerging inner orogen. The Early to Middle Eocene basin formed as a result of pure shortening normal to the range. The marine Early Eocene basin developed during *c.* 2 Ma by widening of a single basin provoked by the two-step propagation of a basement duplex. This is recorded by growth-stratal patterns and coarsening-upward depositional sequences indicating the increasing contribution of the emerged orogen. The Middle Eocene continental deposits infilled two sub-basins working synchronously and were transported by alluvial fans with a provenance in the inner orogen, during decreasing thrust-wedge advance and increasing erosional unloading.

Foreland basins have long been favoured as areas for tracing orogenic events using the sedimentary record. The propagation of a foreland basin is achieved by either the development of forward-younging distinct fault thrust-related sub-basins progressively integrated into the wedge (Déramond *et al.* 1993; DeCelles & Giles 1996; Horton & DeCelles 2001), or forward migration and/or deepening/widening of a single basin (Catuneanu *et al.* 1997, 2000; Sinclair 1997; Burkhard & Sommaruga 1998; Christophoul *et al.* 2002).

At the regional scale, deposition in foreland basins is widely controlled by flexural subsidence resulting from tectonic and sublithospheric static and dynamic loading (Beaumont 1981; Beaumont *et al.* 1993; Washbusch *et al.* 1996, Catuneanu *et al.* 1997, 2000). Tectonic loading is frequently considered to be followed by tectonic quiescence/unloading resulting from release of lithospheric forces and/or erosion (e.g. Blair and Bilodeau 1988; Heller *et al.* 1988; Heller & Paola 1992, Burbank 1992; Catuneanu *et al.* 1997, 2000), which implies that loading/unloading cycles are related to the major tectonic events. Catuneanu *et al.* (1997, 2000) have, however, described a succession of basin-scale loading/unloading cycles controlling second- and third-order sedimentary sequences with time-spans of *c.* 20–25 and 0.4 to 3 Ma.

The question of the transition from underfilled to overfilled is closely related to foreland basin propagation and loading/unloading cycles. Most frequently, this is intended as the transition from marine to non-marine deposition, when the volume of sediment delivered from the orogen becomes greater than accommodation space (e.g. Ricci Lucchi 1986; DeCelles and Giles 1996; Sinclair 1997). According to Jordan (1995), the transition from underfilled to overfilled marks the change from a subsiding foredeep parallel and close to the wedge front to a large-scale

From: McCann, T. & Saintot, A. (eds) *Tracing Tectonic Deformation Using the Sedimentary Record.* Geological Society, London, Special Publications, **208**, 229–252. 0305–8719/03/$15.00

transverse sediment transport widely outpacing the former foredeep. In spite of differing definitions, both Sinclair (1997) and Jordan (1995) invoke an increase in tectonic shortening. In contrast, other authors (Blair and Bilodeau 1988; Heller *et al.* 1988; Heller & Paola 1992; Burbank 1992) believe that the transition from longitudinal to transverse drainage is due to erosional unloading succeeding tectonic loading.

How are these events recorded in the sedimentary fill? Tectonism is often emphasized as a forcing mechanism of depositional sequences in active basins (e.g. Clifton *et al.* 1988; Carter *et al.* 1991; Vail *et al.* 1991; Sinclair 1993; Yoshida *et al.* 1996; Catuneanu *et al.* 1997, 2000; Serrano *et al.* 2001), even though attempts to relate sequence stratigraphy to thrust/fold sequence are rather few in number (Specht 1989; Wallez-Fondecave & Souquet 1991; Déramond *et al.* 1993; Brusset *et al.* 1997; Ito *et al.* 1999). Traces of tectonic events, however, are likely to have been recorded by other sedimentary features, including facies, provenance, and palaeocurrents (e.g. Sinclair 1992, 1997; Schlunegger *et al.* 1997*a*, *b*; DeCelles *et al.* 1998; Horton & DeCelles 2001) or growth stratal patterns (e.g. Suppe *et al.* 1992; Ford *et al.* 1997).

In the present paper, the mode of propagation of foreland basins, loading/unloading cycles and the transition from underfilled to overfilled are investigated using the methods listed above, coupled with the structural study in the western Corbières (eastern North Pyrenean retro-foreland basin). In this area, two syncontractional basins differing in age (late Cretaceous and Eocene) and in geodynamic context ('fully' underfilled in front of a submarine wedge and underfilled to overfilled in front of an emerged wedge) are separated in time by a rather long period of tectonic quiescence (latest Cretaceous and Paleocene) and in place by a large basement duplex.

Regional setting

The Pyrenees constitute a narrow asymmetrical double-wedged range, including a relatively wide southern pro-foreland basin and a narrower northern retro-foreland basin. The Pyrenean range (Fig. 1a) is traditionally divided into five longitudinal structural zones striking *c.*N110E, which are from south to north:

(1) the south Pyrenean zone consisting of Mesozoic and Tertiary terranes and comprising the southern foreland basin;
(2) the Axial Zone essentially made up of Palaeozoic and pre-Palaeozoic terranes affected by Hercynian deformation and metamorphism;
(3) the North Pyrenean Zone comprising Mesozoic strata overlying the Hercynian basement locally exhumed to form the so-called North Pyrenean massifs; and
(4) the sub-Pyrenean zone in which Upper Cretaceous and Tertiary deposits overlie the Palaeozoic basement.

The Axial Zone, the North Pyrenean and sub-Pyrenean Zones are separated from each other by two major faults, which are from south to north, the North Pyrenean Fault (NPF) and the North Pyrenean Frontal Thrust (NPFT). Eastwards, the general N110E trend of the structures changes to north–south in the Corbières bend (Virgation des Corbières).

The southern pro-foreland basin has long and intensely been studied from the double viewpoint of structure (ECORS Pyrenees Team 1988; Burbank *et al.* 1992; Muñoz 1992; Beaumont *et al.* 2000) and relationships between tectonics and sedimentation (Puigdefabregas & Souquet 1986; Puigdefabregas *et al.* 1986; Specht 1989; Déramond *et al.* 1993; Williams *et al.* 1998; Vergés *et al.* 1998; Nijman 1998, amongst others). Studies of the tectonics–sedimentation relationship are less numerous in the northern retro-foreland basin (Baby *et al.* 1988; Razin 1989; Déramond *et al.* 1993; Brusset *et al.* 1997; Serrano *et al.* 2001). From a tectonic viewpoint, the essential difference between the southern pro-foreland basin and northern retro-foreland basin is that in the former the basal décollement is constituted by the Upper Triassic evaporites, whereas, in the latter, Palaeozoic strata and even metamorphic/plutonic basement are largely involved in the Alpine structures.

Our study area is situated between the Aude valley and the Corbières bend (Fig. 1). A particularly interesting feature of this area is that Palaeozoic rocks are involved in Alpine structures as far as in the distal parts of the foreland basin (Fig. 1b).

Stratigraphy

The Hercynian series is composed of a basement including a group of medium-pressure granulite-facies rocks overlain by HT–LP migmatites and metasediments and unmetamorphosed strata of Early to Late Palaeozoic age. In the study area, metamorphic and anatectic Hercynian rocks are only present in the south of the North Pyrenean zone (Bessède, Salvezines and Agly massifs, Figs 1b & 2). Only unmetamorphosed Middle Ordovician to Late Carboniferous strata are

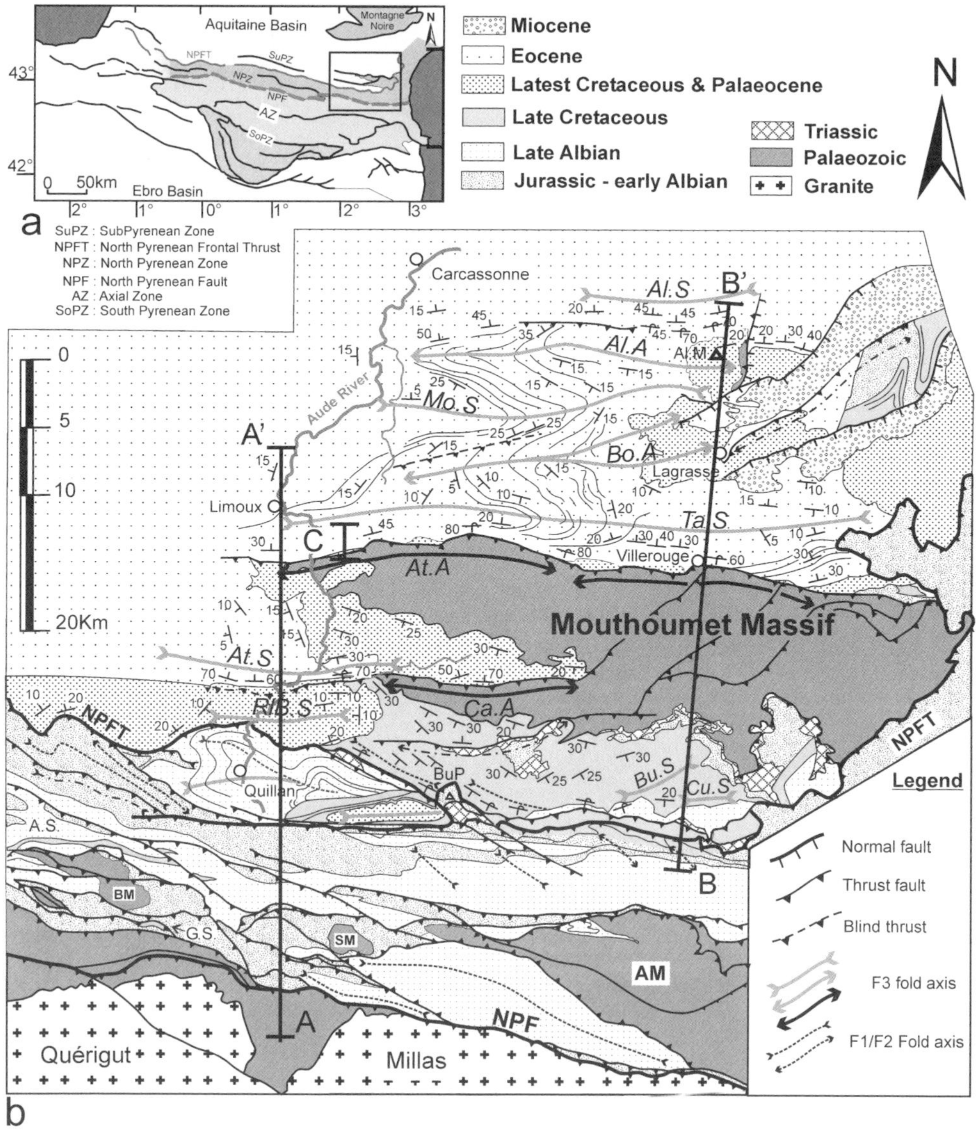

Fig. 1. (**a**) Location map. (**b**) Structural map of the studied area. AA′ and BB′ are the cross-sections shown in Fig. 2. C is the cross-section shown in Fig. 7. NPF, North Pyrenean Fault; NPFT, North Pyrenean Frontal Thrust; AM, Agly Massif; BM, Bessède Massif; SM, Salvezines Massif; A.S, Axat Syncline; Al.S, Alaric Syncline; At.S, Alet Syncline; BeS, Bezu Syncline; Bu.S, Bugarach Syncline; Cu.S, Cucugnan Syncline; G.S, Gesse Syncline; Mo.S, Montlaur Syncline; RlB.S, Rennes-les-Bains Syncline; Ta.S, Talairan Syncline; Al.A, Alaric Anticline; At.A, Alet Anticline; Bo.A, Boucher Anticline; Ca.A, Cardou Anticline; Al.M. Montagne d'Alaric; Bu.P, Bugarach peak.

observed in the sub-Pyrenean Zone (Mouthoumet and Alaric massifs). These were previously affected by flat-lying Hercynian thrusts (Arthaud *et al.* 1976; Bessière 1987; Bessière *et al.* 1989) reactivated during Alpine deformation (Bessière 1987; Bessière *et al.* 1989).

The Mesozoic series begins with a thin blanket of Lower Triassic sandstones and conglomerates often preserved in the Mouthoumet Massif. The Upper Triassic evaporite-bearing marls are widespread in the southern sub-Pyrenean Zone and northern North Pyrenean Zone, where they

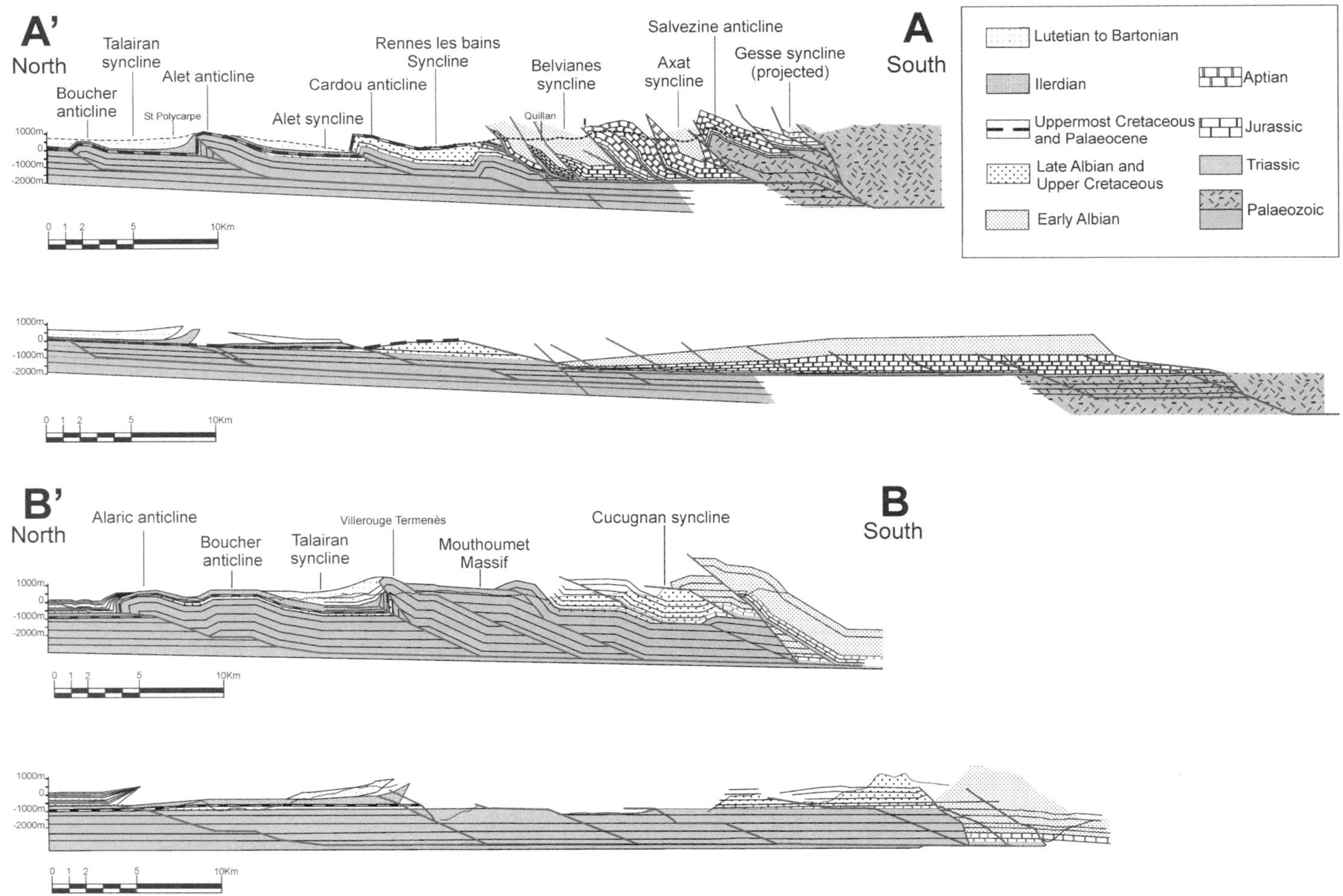

Fig. 2. Balanced cross-sections contructed using 2DMove software (Midland Valley Inc.). Flexural fold model. The Triassic is taken as the reference level. Locations are as indicated as in Fig. 1. A-A′ Aude valley, North Pyrenean Zone and sub-Pyrenean Zone. Total shortening of –30%. Late Cretaceous shortening of –42%. B-B′ Cucugnan-Alaric. Northern North Pyrenean Zone and sub-Pyrenean Zone. Total shortening of –24%.

constitute the upper décollement, but are frequently lacking in the southern North Pyrenean zone. These are succeeded by Jurassic and Lower Cretaceous shallow marine carbonates and marls, 1000 to 1100 m thick (Wallez, 1974) only present in the North Pyrenean Zone and northeastern sub-Pyrenean Zone (Fig. 1b).

Upper Cretaceous strata crop out in two small areas at the vicinity of the North Pyrenean Fault (Gesse syncline) and north of the Bessède North Pyrenean Massif (Axat Syncline) and in a much wider area in the northern North Pyrenean Zone and southern sub-Pyrenean Zone (Fig. 1b). The Paleocene is represented by shallow-marine to non-marine deposits (Plaziat 1984, 1987; Tambareau *et al.* 1995).

The Eocene is represented by Early Ilerdian platform limestones, succeeded by deep marine deposits grading northwards to a shallow-marine platform of Middle to Late Ilerdian age, and then by a thick series of fluvial deposits known as the Palassou conglomerates, of Late Ilerdian (Crochet 1991) through Early to Middle(?) Bartonian age (Berger & Rey 1990; Berger *et al.* 1993).

Structure

In the study area, the North Pyrenean Zone appears as a large-scale anticline with Hercynian rocks forming structural highs in its southern part (Figs 1b & 2). To the south, the North Pyrenean Zone is overthrust by the Axial Zone along the North Pyrenean Fault which is likely to represent a major strike-slip fault (Choukroune 1976; Choukroune and Mattauer 1978) reactivated as a thrust fault (Soula *et al.* 1986). To the north, the North Pyrenean Zone overrides the sub-Pyrenean Zone along the North Pyrenean Frontal Thrust (Figs 1b & 2). The internal structure is mainly determined by N120–100E-trending folds, F_1, with half-wavelengths ranging from 1–2 km to 200–300 m, subparallel to the overall direction of the range in the east (Leblanc and Vaudin 1984), but clearly oblique in the west (Soula *et al.* 1986). The F_1 folds are deformed by N135E to N110E-trending map-scale fold bands, F_2, and associated smaller (*c.*200–300 m half-wavelength) *en échelon* folds (Soula & Bessière 1980; Leblanc & Vaudin 1984; Soula *et al.* 1986) (Fig. 1b). F_1 and F_2 folds have been interpreted as a result of left-lateral strike-slip shearing accompanying regional shortening (so-called 'transpression') (Soula & Bessière 1980; Soula *et al.* 1986). F_1 and F_2 folds are north-verging fault-related folds, mostly fault-propagation folds, frequently related to blind thrusts passing sideways to merging thrust faults (Figs 1b & 2). The fault-propagation folds are responsible for the general strata overturning which characterizes this area. To the north, the main décollement is constituted by the Triassic evaporites and the basement is not involved. To the south, the Triassic evaporites are often lacking and the décollement lies within the Hercynian migmatites, as shown by the involvement of these migmatites in the thrust structures (Fig. 2). In the Albian marls and Upper Cretaceous flysch, F_1 and F_2 folds are accompanied by minor folds associated in the southern area with foliations S_1 and S_2 (Choukroune & Meurisse 1970; Wallez 1974; Choukroune 1976). The F_1/F_2 folds and thrusts are unconformably overlain by Campanian and younger strata (Mattauer & Proust 1965) and locally overriden by *c.*east–west to ENE–WSW-trending thrust faults and related folds, F_3 (Fig. 1b).

In the northern sub-Pyrenean Zone, only the F_3 longitudinal folds are present and deform the Palaeozoic as well as the Paleocene and Eocene strata (Figs 1b & 2). These folds are either fault-propagation folds associated with overturned through shallow-dipping Paleocene and Eocene growth strata (Mouthoumet–Talairan or Alaric), or open fault-bend folds with gently dipping limbs lacking growth stratal patterns (Boucher–Montlaur). In the eastern Mouthoumet Massif, the east–west-trending and moderately dipping longitudinal thrusts faults branch on to reactivated pre-existing Hercynian subhorizontal thrust faults similarly showing S-N tectonic transport (Fig. 1b).

Relationship between tectonics and deposition

Late Cretaceous and Paleocene

The Late Cretaceous deposits crop out discontinuously in the southern North Pyrenean Zone and continuously in the the northern sub-Pyrenean Zone, whereas the latest Cretaceous and Paleocene deposits are observed all over the sub-Pyrenean Zone. These deposits have a total thickness varying from *c.*800 m in the Quillan/Rennes-le-Château area to less than 80 m in the north, and represent the infill of the eastern termination of the Late Cretaceous 'flysch' basin which is much wider and deeper westwards (e.g. Déramond *et al.* 1993; Razin 1989; Brusset *et al.* 1997; Serrano *et al.* 2001).

The stratigraphy of the Late Cretaceous and Paleocene deposits has been intensely studied for the past 30 years (e.g. Gélard 1969; Wallez 1974; Bilotte 1985, 1992; Pélissier 1987; Bessière *et al.* 1989; Tambareau *et al.* 1995), which provides an excellent age control. Modern sequence stratigraphy and sedimentary facies studies are, however, rather recent and concern only the Coniacian, Santonian and Early Campanian depositional systems (Wallez-Fondecave and Souquet 1991). The following is a comprehensive description of the Late Cretaceous and Palaeocene cycle, including the middle Cenomanian to Turonian and the Latest Cretaceous and Paleocene depositional units (Fig. 3). The sedimentological and sequence analyses in the Coniacian to Early Campanian are derived from those of Wallez-Fondecave and Souquet (1991), including Bilotte's (1992) stratigraphic revision and also the personal observations of the present authors.

Middle Cenomanian and Turonian. In the south, Middle Cenomanian to Turonian deposits have been recognized in front of the Axial Zone Thrust (Magné and Mattauer 1968; Wallez 1974) and in front of the Bessède–Salvezines thrust fault (Bilotte *et al.* 1973; Wallez 1974; Bilotte 1985). In front of the Axial Zone, these deposits are unconformably resting over Jurassic, Lower Cretaceous and Upper Albian strata deformed by a fault-related syncline (Gesse Syncline; Wallez 1974), although being themselves deformed by the same syncline. The series commences by highly heterometric and poorly sorted breccias containing Jurassic through Aptian clasts up to *c.* 1 m^3 in size, derived from the southern limb of the Gesse Syncline, included in a calcareous–marly matrix (50 m). These breccias are succeeded by carbonate turbidites consisting of a repeated succession of calcareous microbreccias, laminated calcarenites and hemipelagic muds (60 m). This sequence overlying an erosional intra-foredeep unconformity denotes retrograding characteristics (a facies trend from debrite–turbidite couplet B to BC4, see Souquet *et al.* 1987) indicating slope and base-of-slope environments. This facies association indicates normal persistent turbiditic then hemipelagic deposition and then catastrophic processes such as episodic slope failure and large-scale debris-to-turbidite flows related to tectonic slumping and disruption (Souquet *et al.* 1987; Brusset *et al.* 1997). In front of the

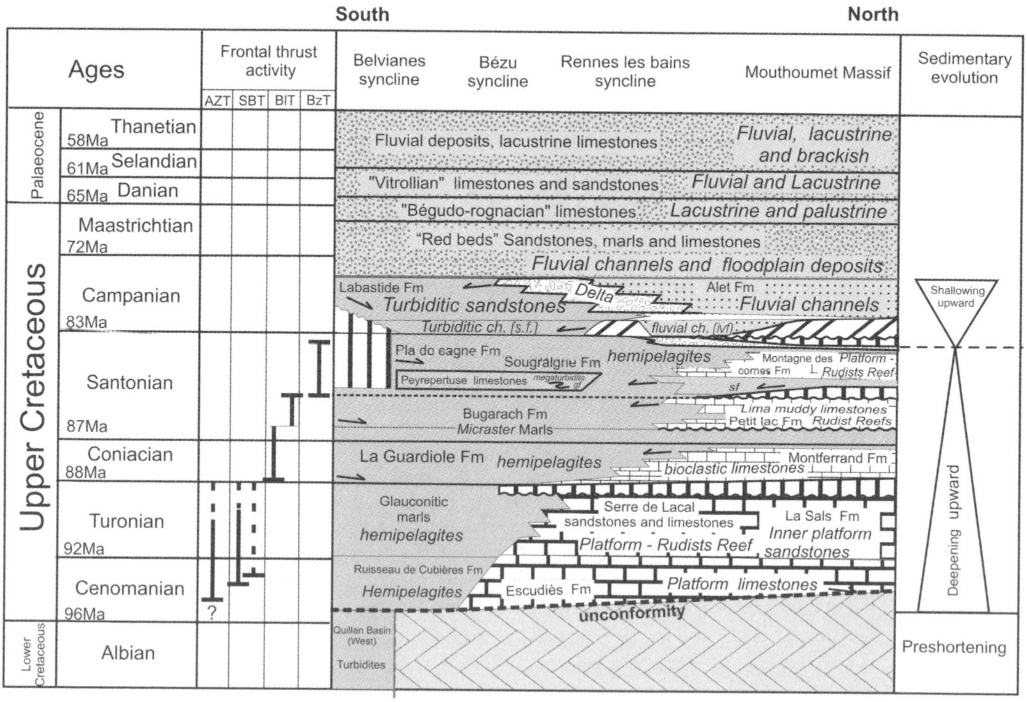

Fig. 3. Chronostratigraphic diagram of the late Cretaceous and Paleocene of the northern North Pyrenean Zone and the sub-Pyrenean Zone. Age scale after Odin (1994). No scale for Maastrichtian to Thanetian. AZT, Axial Zone Thrust (North Pyrenean Fault); SBT, Salvezines–Bessède Thrust; BIT, Belvianes Thrust; BzT, Bezu Thrust; gf, glide features; ivf, incised valley fill; s.f., slope fan. Arrows indicate sense of sediment transport. Data after Wallez-Fondecave & Souquet (1991); Bilotte (1992); Freytet (1970), and the personal observations of the authors.

Bessède–Salvezines–Agly thrust fault propagation fold (Axat Syncline), the same turbiditic system is observed (Bilotte *et al.* 1973; Wallez 1974). The basal erosional/unconformable contact here cuts across the Albian, but not the older strata of the overturned limb, and only Albian clasts are found in the basal breccias. This indicates that the deposits here were less mature than in the south. Olistoliths of Neocomian age observed within the upper part of the breccias (Bilotte *et al.* 1973; Wallez 1974) indicate, however, that erosion then penetrated deeper in the folded strata, but without reaching the Jurassic.

In the northern North Pyrenean/southern sub-Pyrenean Zones, the Middle to Late Cenomanian is represented in the Quillan–Cucugnan area by lower platform micritic limestones and white marls (Bilotte 1985) unconformably overlying the Late Albian strata (Gélard 1969), and still more to the north, by platform limestones deposited over an erosion surface cutting the Palaeozoic strata of the Mouthoumet Massif. The Turonian is represented by glauconitic marls (hemipelagites) indicating deep-water environments, passing northwards to mixed-platform deposits including reefal carbonate intervals interbedded with channelled sandstones representing shoreface facies (Serre de Lacal Formation; Bilotte 1985), and then to inner platform sandstones (La Sals Formation) onlapping the Mouthoumet High. This indicates a retrogradational evolution and thus a northwards migration of the platform, from the Middle Cenomanian to the Turonian (Fig. 3). The mixed and inner platforms were supplied with fine- and medium-grained clastics by the emerged Mouthoumet High, as shown by the nature of these clasts (quartz, Ordovician quartzitic pelites, Carboniferous black cherts). On the northern side of the Mouthoumet High, Turonian lignite-bearing dark clays representing lagoonal–brackish deposits have been observed unconformably resting on the Palaeozoic (Freytet 1970; Bilotte 1985).

Coniacian and early to middle Santonian. These deposits are observed between the Belvianes–Cucugnan Syncline and the Mouthoumet High (Fig. 3). The basal erosional unconformity cuts across the Turonian, Cenomanian and Upper Albian strata of the southern flank of the WNW–ESE-trending ($F_1/_2$) Belvianes Syncline (Gélard 1969; Wallez-Fondecave & Souquet 1991) in the south, and is marked in the north by Palaeozoic clast-bearing conglomerates infilling erosional scours. In the south, the deposits are fine-grained clastics and marls, in which three sequences characterizing slope and basin deposition have been recognized, each of them showing the succession of clay-rich bioclastic/sandy turbidites and homogeneous marls (Wallez-Fondecave & Souquet 1991). Palaeocurrents in the turbidites have approximately east-to-west directions indicating drainage parallel to the axis of the syncline (longitudinal flow: Burbank 1992; Jordan 1995). These turbidites include detritus of marls and planktonic microfaunas of Upper Albian age originating from the reworked southern limb of the fold (Wallez-Fondecave & Souquet 1991). In the north, the Coniacian and Early/Middle Santonian is represented by two retrograding depositional sequences, each including platform carbonates succeeded by a condensed section and offshore marls (Wallez-Fondecave & Souquet 1991). According to Wallez-Fondecave & Souquet (1991), these two sequences are correlated with the two upper southern sequences – the lowermost southern sequence corresponding with the basal erosional surface.

Middle to Late Santonian. These deposits are absent to the south of the North Pyrenean Frontal Thrust where the Latest Santonian/Early Campanian strata directly overlie the Early to Middle Santonian deposits. In front of the North Pyrenean Frontal Thrust, the deposits are homogeneous blue marls (Pla de Sagnes and Sougraigne Marls), *c.*300 m thick, infilling the WNW–ESE-trending Rennes-les-Bains Syncline and conformably resting on the Early to Middle Santonian marls (Fig. 3). At the eastern termination of the syncline near Cucugnan, the marls include a 50 m thick unit of calcarenites and bioclastic limestones containing rudist reefs, of Middle/Late Santonian age (Bilotte 1992). These limestones are affected by mesocopic slump folds, the asymmetry of which indicates top-to-the-north sliding. We interpret this unit as an earthquake-induced megaturbidite (Souquet *et al.* 1987), issuing from an unstable southern platform. In the north, the depositional sequence commences with an erosional unconformity overlain by offshore marls and channelled sandstones showing east to west palaeo-flows (Wallez-Fondecave & Souquet 1991). These are succeeded by a group of three parasequences (Montagne des Cornes; Bilotte 1985, 1992) showing an overall transgressive stacking pattern. The sequence ends with transitional shoreface argilites (Wallez-Fondecave & Souquet 1991) (Fig. 3).

Latest Santonian to Early Campanian. These deposits are observed all over the northern

North Pyrenean Zone (Quillan area) and in the sub-Pyrenean Zone to the south of the Mouthoumet Front, widely onlapping the older Late Cretaceous strata. In the south (Quillan and Cucugnan areas), the basal erosional unconformity cuts across the Santonian, Coniacian and Turonian strata (Fig. 3). The early deposits are parts of a slope fan turbiditic system (Labastide Sandstones; Bilotte 1985, 1992) constituted by marls and channelled sandstones showing longitudinal palaeo-flows toward the west (Wallez-Fondecave & Souquet 1991). Clasts originating from the northern Mouthoumet High, and clasts, olistoliths or olistostroms originating from the south (e.g. Middle Santonian limestones) are embedded in these deposits. The upper part of the formation is constituted by turbiditic marls interbedded with thin sandstones beds, which can be related to pro-delta facies (Pélissier 1987). In the north, the erosional basal unconformity cuts across the Middle to Late Santonian series (Bilotte, 1985, 1992; Wallez-Fondecave & Souquet 1991) and the Latest Santonian/Early Campanian strata directly overlie the Palaeozoic of the Mouthoumet Massif. The deposits are coarse-grained clastics infilling incised fluvial valleys, succeeded by fining-upward fluvial deposits (Alet Sandstones). Therefore, in contrast with the preceding systems, the Latest Santonian-Early Campanian system displays a shallowing upward stacking pattern (Fig. 3).

Late Campanian, Maastrichtian and Paleocene. The Late Campanian to Maastrichtian deposits rest conformably over the Early Campanian continental (Alet Sandstones) and marine (Labastide Sandstones) deposits of the southern sub-Pyrenean Zone and unconformably over the folded Albian and Aptian strata (Cucugnan Anticline). In the north, these deposits were deposited over the Palaeozoic substrate, widely onlapping the Alet sandstones (Mouthoumet and Alaric massifs). In the studied area, the sedimentation remained continental to lagoonal. The major part of the sediment corresponds with 'red beds' (known as 'Garumnian facies') comprising fine-grained fluvial floodplain deposits changing laterally to lacustrine/palustrine marls and limestones. The preserved channels are rather thin and most often fine grained. Conglomerates are observed at the base of three fining- and thinning-upward and downstream sequences, the two lower ones corresponding with the Begudian/Rognacian facies and the upper one to the Vitrollian facies (Freytet 1970). In the north (Alaric region), clastic material was derived from the north (Montagne Noire), the south (Mouthoumet High) and local sources. In the south (Rennes-les-Bains–Cucugnan) the clasts are derived from multiple sources in the whole North Pyrenean Zone (Freytet 1970). Provenance and palaeocurrent studies indicate south to north transport, that is, transverse drainage. In the lower part of the series ('Begudo-Rognacian' facies) only clastic material derived from the northern North Pyrenean Zone has been observed (Freytet 1970). In the upper part of the series ('Vitrollian' facies), clasts of metamorphic Mesozoic carbonates issued from the southern North Pyrenean Zone are common in the west of the studied area, but no Palaeozoic rocks originating from the North Pyrenean massifs have been observed (Freytet 1970; Plaziat 1984). The Thanetian is essentially represented by lacustrine/palustrine/lagoonal limestones, where three marine transgressions have been identified, succeeded by floodplain deposits including fine-grained thin and narrow fluvial channels (Tambareau *et al.* 1995).

Eocene cycle

Eocene strata crop out from the Mouthoumet Massif to the Montagne Noire (Fig. 1b). The depocentre of the Early Eocene marine strata (Early to Late Ilerdian) is situated in front of the Mouthoumet Duplex (Talairan Syncline), whereas the depocentres of the Middle to Late Eocene continental deposits are situated in front of the Mouthoumet Duplex and Alaric thrust-related fold (Talairan and Alaric Synclines) (Figs 2, 4 & 5).

The Ilerdian marine series. The lower unit of the Ilerdian marine series is constituted by foraminiferal and bivalve-bearing shelf carbonates ('basal marine limestones'; Doncieux 1912; Massieux 1973; Pautal 1985), 20–25 m thick, of earliest Ilerdian age (Shallow Benthic Zone 5; Serra Kiel *et al.* 1998). The middle unit is represented by silty to sandy marls and medium-grained sandstones (Rey & Bousquet 1981; Plaziat 1984; Pautal 1985; Tambareau *et al.* 1995) termed hereafter Blue Marls. The upper unit is represented in the north by tidal/lagoonal/continental sandstones usually termed Oyster Sandstones, laterally changing southwards to coarser-grained sandstones and conglomerates – first marine and then continental. The average deposition rate was 0.16 mm/year in the depocentre, and only 0.08 mm/year in the north of the basin (Montlaur) (Figs 4 & 5).

Two depositional sequences have been distinguished: The lower Blue Marls, of Early to Middle Ilerdian age (SBZ 6 and 7 and early SBZ

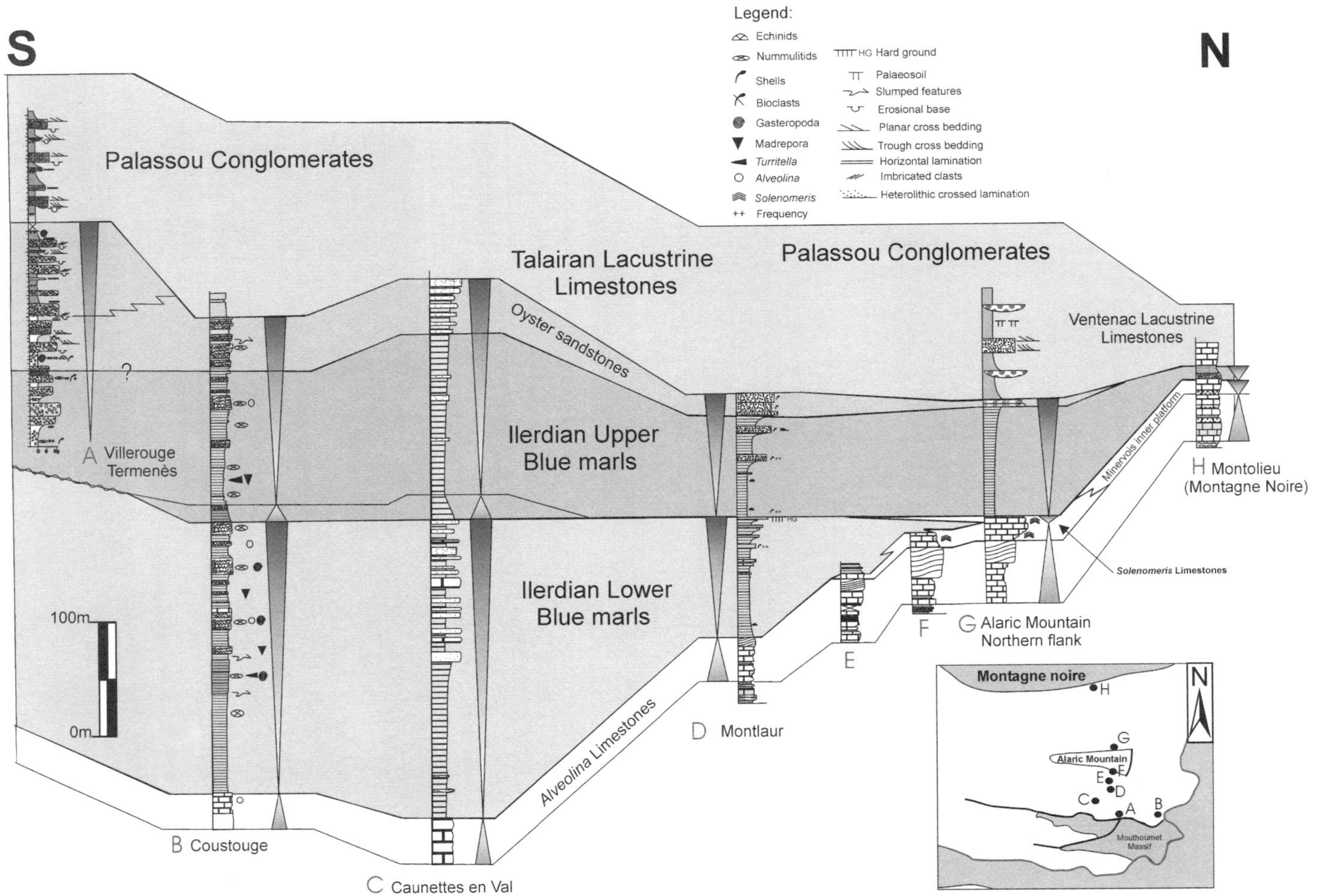

Fig. 4. Stratigraphic sections through the Eocene basin (location in bottom-right inset). Data after Massieux (1973), Rey & Bousquet (1981), Plaziat (1984, 1987), Pautal (1985), Tambareau (unpublished) and Christophoul (unpublished).

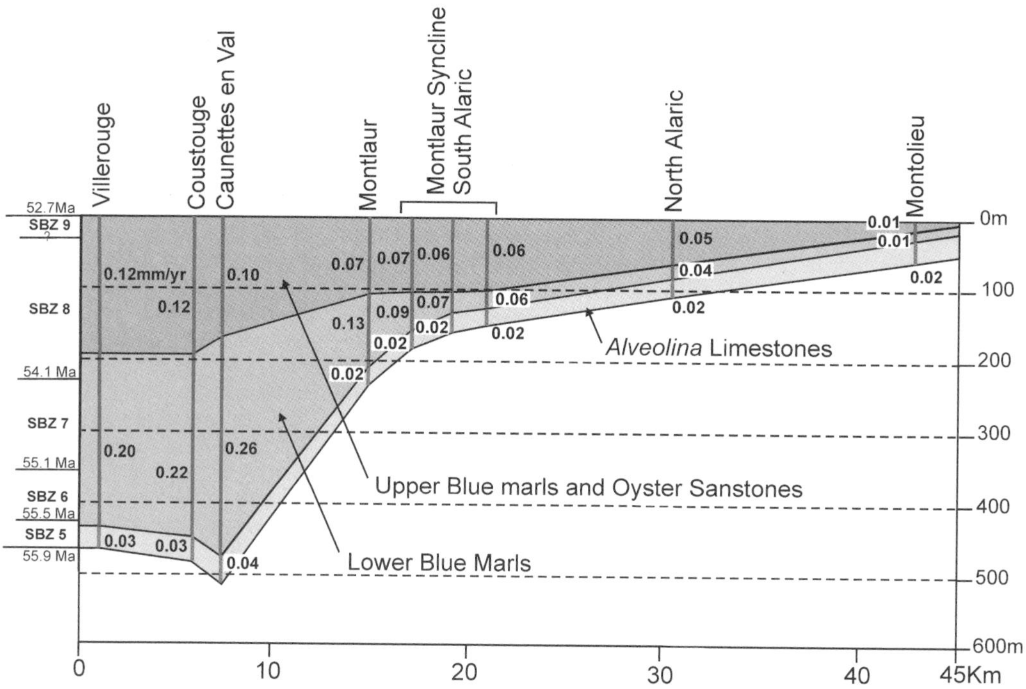

Fig. 5. Thicknesses and subsidence rates (mm/year) of the Ilerdian formations computed using Einsele's (1992) backstripping treatment. Water depths estimated after Pautal (1985). Ages of the boundaries between the shallow benthic zones (SBZ) after Serra Kiel *et al.* (1998).

8) and the Upper Blue Marls and Oyster Sandstones of Middle Ilerdian age (middle SBZ 8). Palaeoenvironmental analyses (Rey & Bousquet 1981; Plaziat 1984; Pautal 1985) have shown that the upper part of either sequence formed a prodelta slope or an inner lagoon within a deltaic environment fed from the south. The transport of shallow-water fossils (Alveolinae and corals) into deeper-water environments (Plaziat 1984) reflects the poor stability of the slope in the vicinity of the orogenic wedge (Fig. 8b). In the centre of the basin, the marls are enriched in planktonic foraminifera and impoverished in siliciclastics, and the grain size is markedly finer at any given level, which indicates a reduced siliciclastic supply from the orogen, coupled with increased water depth (Pautal 1985). Stratal growth patterns, including wedging, intra-formational unconformities, growth onlaps or, more rarely, growth offlaps (see Ford *et al.* 1997) are observed at the contact with the Mouthoumet front. The bases of both sequences are marked by major growth onlaps (Figs 4, 6 & 7).

The Lower Blue Marls overlie the basal limestones unconformably in front and on top of the Mouthoumet front (Figs 6 & 7) and conformably in the southern depocentre and the north (Fig. 4). In the southern depocentre, these are represented by a thick coarsening/shallowing-upward sequence, 170 m thick, ended by a conformable transgressive surface. In the centre (Montlaur; Fig. 4), the coarsening upward sequence is only *c.*85 m thick and ends with a hardground. In the north (Alaric), the Lower Blue Marls change laterally to platform limestones (Pautal 1985). These limestones, 40–80 m thick, of early Middle Ilerdian age (SBZ 6 and 7) become richer in siliciclastics towards the top and represent the outer platform (outer part of the 'Minervois platform'). In the inner platform, marls and sandy limestones containing quartzofeldspathic detritus derived from the north (Montagne Noire) are intercalated in the platform limestones. Two deltas have been identified (Issel and Caunes Minervois; Seguier 1972; Plaziat 1984). The scarcity of sand grains within the Lower Blue Marls south of the Alaric Mountain (Plaziat 1987) shows that the sandy sedimentation derived from the Montagne Noire did not overlap the edge of the platform, which indicates longitudinal drainage in the inner platform. The

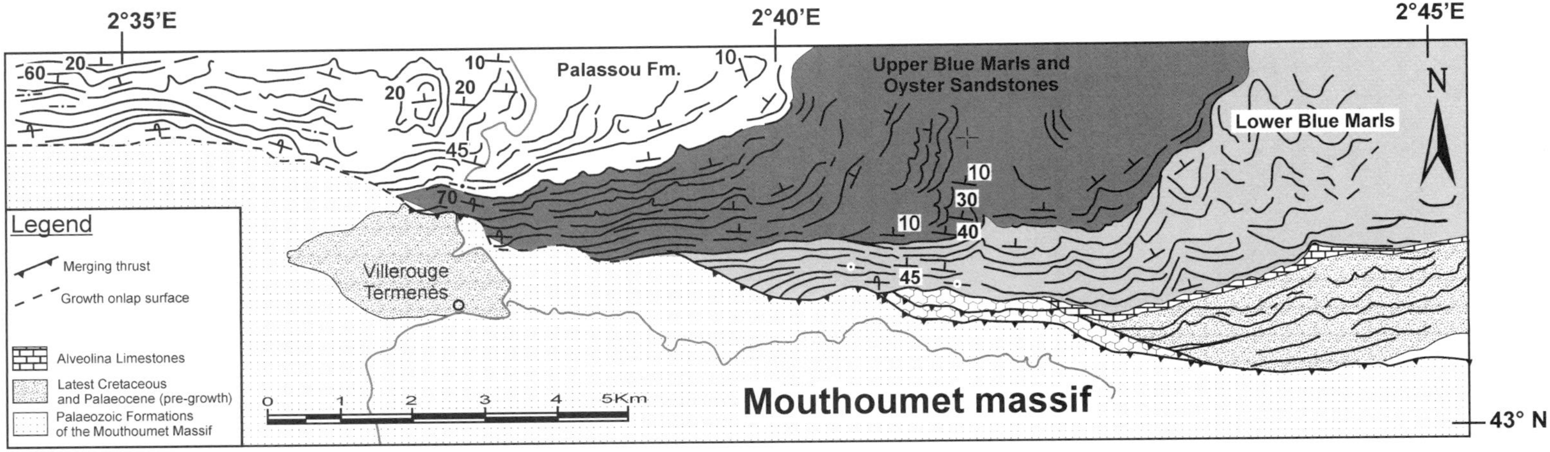

Fig. 6. Structural map of the north-eastern Mouthoumet Front and Eocene basin showing progressive unconformities and growth onlaps at the base of the syntectonic formations. The Lower Blue Marls unconformably overlie the latest Cretaceous, the Paleocene and the Early Ilerdian *Aveolina* limestones, and overlap the northernmost (lower) branch of the Mouthoumet frontal thrust system, although being overthrust by the median branch of this thrust system. The Upper Blue Marls overlap the Lower Blue Marls and the median branch of the thrust system although they are being overthrust by the southern (upper) branch. The Palassou Formation overlaps the Upper Blue Marls and the upper thrust branch.

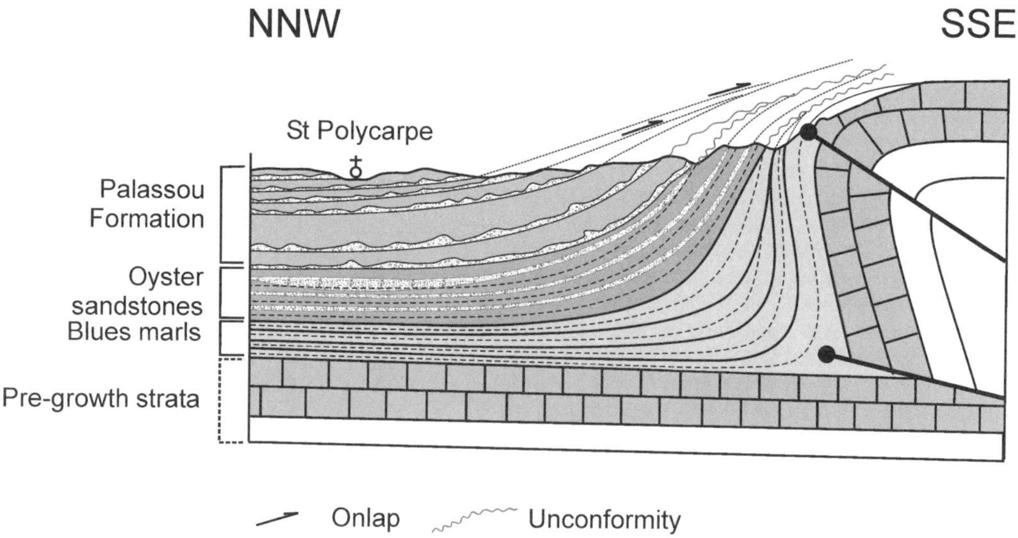

Fig. 7. Cross-section through the north western Mouthoumet Front showing growth structures involving the Blue Marls and Palassou Fotmsyion (see location in Fig. 1).

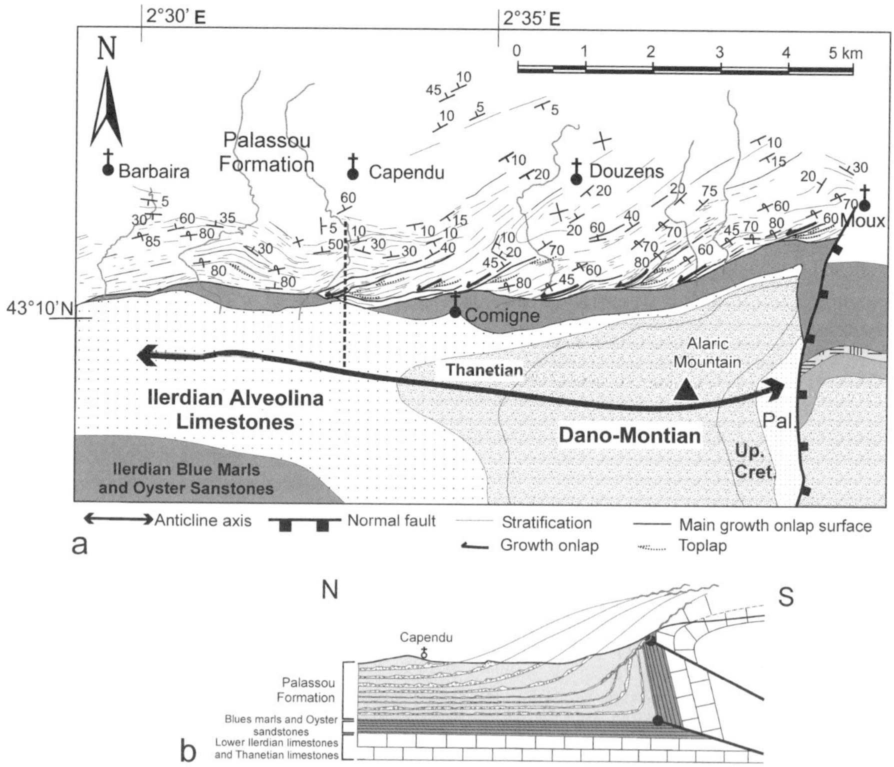

Fig. 8. (**a**) Structural map of the northern flank of the Alaric Anticline. The basal contact of the Palassou Formation is constituted by six major growth onlaps. Bed dip changes from overturned to shallow in *c.*1 km. (**b**) Cross-section showing the growth structures involving the Palassou Formation but not the Upper Blue Marls and Oyster Sandstones (interrupted line in (a)).

shelf-break is likely to have been situated near the present hinge of the Alaric Anticline – to the north of which *Solenomeris* reefs rimmed the platform, and to the south of which carbonate clasts and slipped blocks, including *Solenomeris* originating from this platform, are embedded in the Lower Blue Marls (Plaziat, 1987). Subsidence rates varied from 0.20–0.26 mm/year in the foredeep to 0.06 mm/year near the shelf break and 0.04 to 0.01 in the inner platform (Fig. 5).

The Upper Blue Marls and Oyster Sandstones sequence unconformably overlies the Lower Blue Marls and the Palaeozoic strata on top of the Mouthoumet Front (Figs 4, 6 & 7). In the southern depocentre, the basal contact is conformable and the deposition commences with a fining/deepening-upward parasequence, 15 m thick, succeeded by a thicker (*c*.95 m) coarsening/shallowing upward parasequence set (Fig. 9). The Oyster Sandstones which form the upper part of this parasequence set are made up of thicker (up to 1 m) and coarser-grained sandstone banks, and occasional conglomeratic lenses (Plaziat 1984) recording the northwards progradation of the deltaic system. Recognition of a Gilbert delta (Gilbert 1885; Nemec & Steel 1988; Reading & Collinson 1996) with palaeocurrents showing northwesterly, northerly and northeasterly sediment dispersal (Figs 4 & 9) indicates increased progradation. In the centre (Montlaur), the southern basal fining/deepening-upward parasequence set is absent. The coarsening/shallowing upward sequence is only 100 m thick. The Oyster Sandstones are predominantly medium-grained deltaic sandstones. In the north, the Upper Blue Marls overlap the former shelf-break and rest on the *Solenomeris* limestones. The deposits are richer in siliciclastic grains originating from the northern craton. On the northernmost margin of the basin, the Upper Blue Marls change laterally to more or less sandy lagoonal marls, several metres thick, containing beds of lacustrine limestones and sandstones (Valeron Marls; Seguier 1972; Plaziat 1987). The Oyster Sandstones are represented by tidal/lagoonal/continental fine- to medium-grained sandstones northwards onlapping the pre-Ilerdian strata. Subsidence rates vary from 0.10–0.12 mm/year in the foredeep to 0.05–0.01 mm/year near the craton (Fig. 5).

The Palassou formation. In the studied area, the continental Palassou Formation commences with the Late Ilerdian (SBZ 8–9) (Crochet 1991;

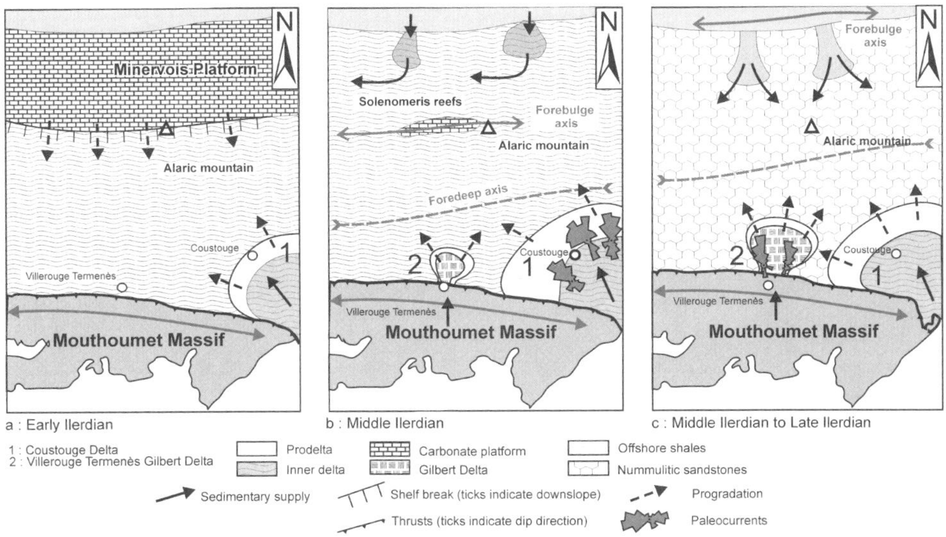

Fig. 9. Schematic evolution of the Ilerdian basin. (**a**) Early Ilerdian. Early Lower Blue Marls. A flexural basin initiates as a result of first step wedge advance. Clastic deposits are delivered by clastic deltas at outlets of major rivers coming from the wedge. (**b**) Late Lower Blue Marls. Foreland basin system is developed as a result of ongoing wedge advance. The forebulge separates the foredeep, fed by clastic deltas derived from the wedge, from the backbulge depozone draining deltas derived from the craton (Montagne Noire). (**c**) Middle to Late Ilerdian. Upper Blue Marls and Oyster Sandstones. Following tectonic quiescence, renewed wedge advance is responsible for basin widening and forebulge migration. The foredeep is now fed by deltas derived from both the craton and the orogen. A Gilbert delta forms at the outlet of a river that is incising the fault escarpment.

Berger *et al.* 1997). Only the lower part of the Palassou Formation is exposed here. The youngest strata dated from fossil evidence are exposed in the north and give early middle Bartonian ages (Plaziat 1987; Berger and Rey 1990). The total thickness can be evaluated there to more than 1000 m. Because no facies and architectural analyses have been yet published, this formation will be described here more thoroughly.

The Formation consists predominantly of marls and siltstones representing floodplain deposits. These are often pedogenetically modified, frequently showing rooted palaeosols, and contain thin, poorly incised and sheet-like (20-50 m wide, 1–10 m thick) sandstone-filled channels representing 'elements SB' of Miall (1996). The lithofacies consist mainly of St, Sp and less frequently Sh and Sl. Lateral accretion sets (LA elements) with cross-bed foresets up to 1–2 m high are common. Occasionally, the sand bodies are incised by narrower channels with a concave-up erosional base filled with sands and gravels rarely larger than 5 cm grain size, containing lithofacies Ss/Se, St, (Gt) and Sp. Very locally, coarser-grained conglomerates (maximum grain size >50 cm) filling deeper channels have been observed (Plaziat, 1987). On the scale of the entire formation, the deposits are coarsening upwards and westwards and fining northwards. The spacing of the channels is in general rather wide (some hundreds of metres). In the south, the channels are oriented roughly SSE–NNW (Plaziat 1984). On the northern margin, the directions of the palaeocurrents are SSE–NNW, south–north and, more rarely, east–west and south–north (Plaziat 1987; Berger and Rey 1990). The volumetric importance of the floodplain deposits and the lateral accretion patterns have led us to interpret the fluvial system as sandy meandering rivers. At any level and from south to north, the clasts are predominantly Mesozoic rocks (Berger and Rey 1990) originating from the Late Cretaceous basin and the North Pyrenean thrust sheets, and unmetamorphosed Palaeozoic rocks originating from the Mouthoumet massif. No southerly-derived granitoid or metamorphic Palaeozoic clasts are observed. Even in the northern margin, the northerly-derived materials are rare.

Small bodies of palustrine limestones are scattered throughout the whole formation (Figs 4 & 10). Much larger palustrine/lacustrine limestone and marl complexes are seen in the central part of the Talairan Syncline, where they are more than 20–30 m thick and crop out over areas of some square kilometres, and all over the northern margin where they have been mapped as two distinct units, 20 to 150 m thick (Ventenac and Agel limestones; Plaziat 1987; Berger and Rey 1990). There limestones pass laterally to, either lagoonal marls and deltaic sandstones/sandy marls, or fluvial sandstones (Plaziat 1987). Although the small bodies are likely to represent ephemeral overbank ponds or backswamps, the larger complexes can be considered as perennial. The Talairan Complex was the lateral equivalent of the lower part of the fluvial formation, and probably represented a swampy topographic depression between the Mouthoumet and Alaric highs. The northern complex persisted later and may be interpreted as a large longitudinal depression open to the west (Aquitaine marine basin), more or less swampy, which constituted the terminating environment of the fluvial system, as shown by the direction and sense of the palaeocurrents. The intercalations of lagoonal marls indicate that this depression remained with an altitude slightly above or below sea-level. The whole system may be interpreted as a shallow-slope meandering fan (similar to the middle–upper part of the 'losimean' model of Stanistreet and McCarthy 1993). The ubiquity of detritus derived from the orogen and the SSE–NNW direction of the channels, suggest multiple coalescing fans distributed by numerous outlets beyond the Mouthoumet frontal thrust.

The growth strata exposed at the contact of the active Mouthoumet frontal thrust are in continuity with those in the Ilerdian. Erosional unconformities associated with growth onlaps (Ford *et al.* 1997) and strata wedging are observed (Figs 6 & 7). To the north, growth strata are fairly well exposed at the contact of the Alaric Anticline. These developed there entirely in the Palassou Formation, where six intraformational unconformites and growth onlaps have been recognized (Fig. 8). As in the southern Pyrenean foreland basin (e.g. Ford *et al.* 1997), an upward change in dip from overturned to 15–10° and the progressive disappearance of strata wedging indicate that the growth folds decrease in amplitude upwards, so as to die out with the deposition of the younger exposed strata. Fold decay is also shown by the disappearance of the Talairan swamp.

Interpretation

Late Cretaceous and Palaeocene

Middle Cenomanian to Late Santonian. During Middle Cenomanian to Turonian times, the catastrophic sedimentation resulting from tectonic sloping and disruption of the forelimbs of the Gesse and Bessède–Salvezines anticlines together with the deformation of the basal erosional unconformity cutting across these

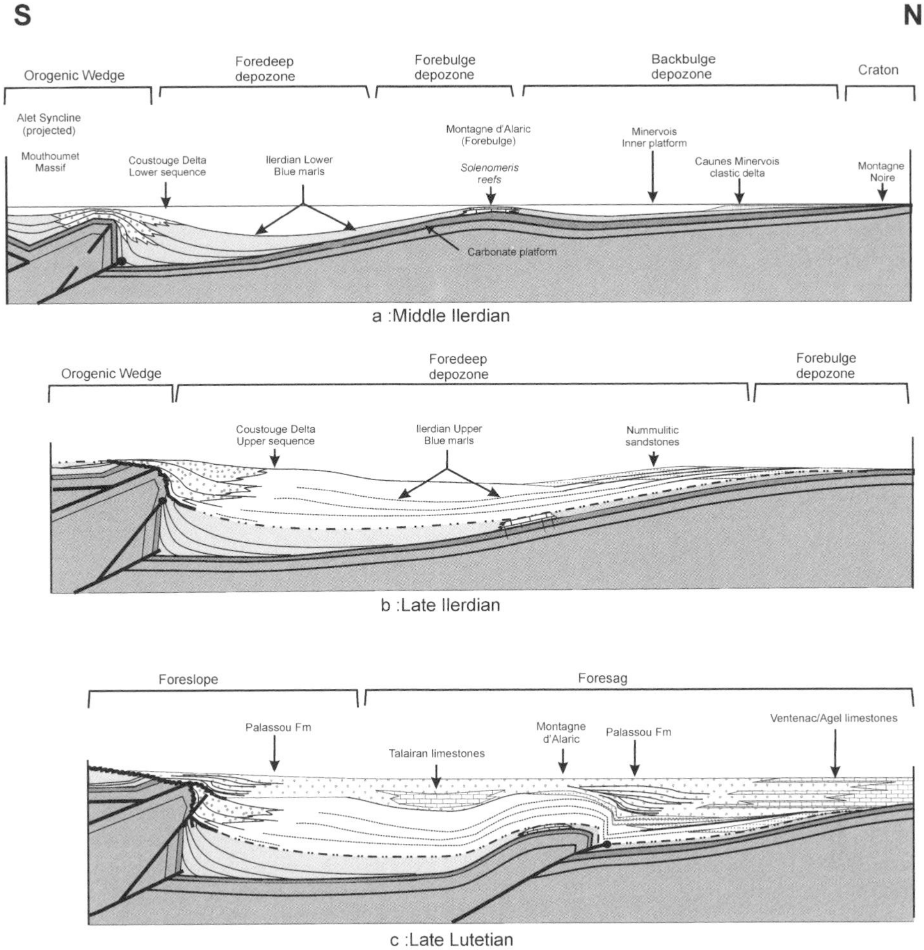

Fig. 10. Schematic evolution of the Eocene basin. No vertical scale. (**a**) Early to Middle Ilerdian. Lower Blue Marls deposited as a result of the propagation of the lower (northernmost) branch of the Mouthoumet frontal fault-propagation fold. The forebulge is located near the future Alaric Anticline. The backbulge depozone corresponds with the Minervois inner platform and drains northerly sourced clastic deltas. (**b**) Middle to Late Ilerdian. Upper Blue Marls deposited as a result of the propagation of an upper branch of the Mouthoumet frontal fault-propagation fold. The forebulge migrates on to the Montagne Noire. Northerly sourced clastic deltas feed the foredeep depozone (Oyster and Nummulitic sandstones. (**c**) Latest Ilerdian to Bartonian. Palassou Formation deposited during ongoing shortening and coeval erosional uplift. Growth structures form as a result of propagation of the Mouthoumet and Alaric frontal thrusts. Lacustrine limestones fill growth synclines. Upwards shallowing of bed dip and dying out of growth structures are attributed to decreasing shortening during ongoing erosional uplift (see text).

forelimbs, demonstrate syndepositional growth of the folds. Deeper erosion of the growth folds towards the south and the northwards-decreasing maturity of the deposits indicate basinward propagation of the thrusts. Syntectonic deposition continuing in the south shows that the rear thrust and related sub-basin incorporated into the wedge remained active, which characterizes the wedge-top depozone (DeCelles & Giles 1996). This and the coeval northwards migration of the platform in the northern area (Fig. 3) demonstrates craton-ward propagation of the entire basin during Middle Cenomanian to Turonian times. The marine deposits onlapping the southern side of the emerged Mouthoumet high and the thin lagoonal–brackish deposits onlapping its northern side enable us to interpret this high as the forebulge. The Cenomanian–

Turonian depositional system thus appears as an underfilled foreland basin system (DeCelles and Giles 1996) including:

(1) slope and base-of-slope facies in a turbiditic environment related to a northwards propagating thrust-and-fold wedge (wedge-top and proximal foredeep depozones);
(2) hemipelagites (distal foredeep) and platfom retrogradational deposits onlapping an emerged Palaeozoic forebulge (forebulge depozone); and
(3) to lagoonal–brackish sediment representing the backbulge depozone.

During Coniacian and Early to Middle Santonian times, the 'catastrophic' deposits interfering with the normal sedimentation were derived from the southern limbs of the more northerly Belvianes and Cucugnan Synclines, as shown by detritus of Late Albian age originating from the reworked southern limb of the fold (Fig. 3). The ongoing retrogradational evolution of the northern deposits bears evidence of coeval northwards migration of the platform. The two northern retrogradational sequences separated by an erosional contact are likely to be a result of the propagation of two branches of the Belvianes fault-propagation fold (see Déramond *et al.* 1993).

Northwards migration of the platform and wedge-front continued during Middle to Late Santonian times. The depocentre filled by the Pla de Sagne and Sougraigne marls must be related to the development of a more northerly fault-propagation fold (locally known as the Bezu Anticline) which was then transported over the catastrophic and other wedge-front deposits to form the North Pyrenean Frontal Thrust (Figs 1b & 2). Moreover, the northwards-transported Peyrepertuse megaturbidite indicates that a tectonically unstable carbonate platform was forming at this time on the wedge front.

Overall, the Middle Cenomanian to Late Santonian retrogradational evolution appears to have been a result of wedge-front advance with forward (*in-sequence*) propagation of thrusts and related sub-basins during increased regional subsidence (Fig. 3). This propagation may have lasted at most *c.* 10±3 or 12±1 Ma, depending on whether Odin's (1994) or Gradstein *et al.*'s (1994) time scale is used. The average deposition rate as inferred from the Coniacian to late Santonian deposits (800 m total thickness) was thus *c.*0.13 to 0.16 mm/year, which is close to that observed in most of the underfilled foreland basins (0.1 to 0.2 mm/year; Cross 1986; Homewood *et al.* 1986; Sinclair 1997). Three tectonically controlled sequences were deposited during that time (5 to 6 Ma), and the propagation of individual thrusts may be inferred to have lasted 1.5 to 2 Ma. The Middle Cenomanian to Turonian sequence was more long-lived (*c.*5–7 Ma); however, the outcrops still preserved show that at least two sub-basins formed successively during this time and that two others, now removed by subsequent tectonics and erosion, may have been associated with the two fault-propagation folds found between the Axat and Belvianes Synclines (Fig. 2). The duration of the development of each new sub-basin would be thus *c.*1.2 to 2.3 Ma, that is of the same order of magnitude as during the Coniacian and Santonian.

Average wedge advance rate and average strain rate estimated using these values and the balanced cross-section, are thus 1.9 to 2.25 mm/year and *c.* 1.3 to 1.6 × 10^{-15} s^{-1}, respectively, which is consistent with the values usually found in active orogens.

Since the propagating thrust-related fold were slightly oblique *en échelon* folds (WNW–ESE-trending F_1/F_2 folds), it should be inferred that basin propagation and thrusting occurred as a result of normal shortening combined with a minor component of left-lateral strike-slip shearing ('transpression').

Latest Santonian to Paleocene. Since the latest Santonian, the sedimentation displayed a shallowing-upwards evolution which went on until the Thanetian, that is, during a time-span of *c.*29–30 Ma (Gradstein *et al.* 1994; Odin 1994; Serra-Kiel *et al.* 1998), and includes the transition from underfilled to overfilled. The persistence of the same evolutionary trend during all this time precludes eustasy as the controlling process. The basal erosional unconformity and the depositional evolution may be interpreted as a result of either out-of-sequence thrusting (e.g. Jordan 1995; Schlunegger *et al.* 1997*a*, *b*), or unloading during tectonic quiescence, causing headward erosion of previous submarine highs (Catuneanu *et al.* 1997, 2000) and/or more internal emergent relief (Blair and Bilodeau 1988; Heller *et al.* 1988; Burbank 1992; Heller and Paola 1992; Burbank *et al.* 1996). Out-of-sequence thrusting might have accommodated the internal deformation required for the advancing tapered wedge (Davis *et al.* 1983) to be maintained or restored in a critical state as erosion proceeded (Boyer 1995; Horton 1999; Schlunegger 1999). This interpretation might be supported by:

(1) the presence of variously sized olistoliths and longitudinal palaeocurrents in the turbiditic

deposits (Labastide sandstones) in front of the Bugarach Thrust; and

(2) the occurrence in the 'Vitrollian' conglomerates of Mesozoic metamorphic carbonate clasts related to the reactivation of the Axial Zone and/or Bessède-Salvezines frontal thrusts.

Indeed, out-of-sequence thrusting is opposed by several lines of evidence:

(1) Out-of-sequence thrusting would have been responsible for additional loading and increased regional subsidence. If so, the shallowing-upwards evolution observed during the Latest Santonian and Early Campanian would have required the sedimentation rate to be higher than the subsidence rate. This could have occurred only if the basin were mainly fed by the orogen (see, for example, Jordan 1995; Sinclair 1997), which is not the case for the studied basin mainly fed from the northern craton at this time.
(2) Shallowing upwards, together with out-of-sequence thrusting, would have implied backward migration of the entire foredeep and of the loading point, which is inconsistent with the increase in regional shortening that out-of-sequence thrusting should have caused.
(3) The clasts in the conglomerates were issued from multiple point sources in the entire North Pyrenean Zone, and not only from the Mesozoic metamorphic zone, which indicates that no particular relief fed the dispersal system.
(4) Erosion was insufficient to reach the Palaeozoic core of the southern fold-propagation folds (Bessède, Salvezines and Agly massifs) which however, had previously been eroded during Cenomanian and Turonian times, which is inconsistent with the reactivation of the thrusts.

Headward erosion of pre-existing tectonic relief as a result of unloading is therefore more plausible. In the south of the studied area, the sequential evolution and the sources of olistoliths and clasts indicate that headward erosion attained, first submarine reliefs with 'passive infilling of the sub-basin (latest Santonian to Early Campanian Labastide sandstones), then the emergent northern North Pyrenean Zone (Late Campanian to Earliest Paleocene 'Begudo-Rognacian' facies), and finally the southernmost Mesozoic metamorphic zone (early Paleocene 'Vitrollian' facies). Southwards transport of clastics from the Mouthoumet High which continued during Latest Santonian and Early Campanian times ceased after that, and the Palaeozoic material was entirely transported northwards down to the Alaric Depression. The transition from underfilled to overfilled and from longitudinal to transverse drainage thus appear to have been related to a progressive uplift of the whole foreland basin system, and was not provoked by tectonic shortening. Lithospheric mechanisms capable of causing upper plate uplift, such as delamination (Bird 1979; Channel & Mareschal 1986) or slab-break-off (Davies & Von Blanckenburg 1995) are rather unlikely here, because of the absence of metamorphism and magmatism in the period and area considered. Traces of these events also cannot be found in the ECORS profiles (ECORS Pyrenees Team 1988). Erosional unloading accompanying tectonic quiescence (Blair and Bilodeau 1988; Heller *et al.* 1988; Heller and Paola 1992; Burbank 1992; Washbusch *et al.* 1996; Catuneanu *et al.* 1997, 1999, 2000) is therefore more likely. Erosional unloading may explain the wide but shallow erosion of the whole North Pyrenean Zone, as evidenced by the absence of Palaeozoic clasts. The northern depression, which is situated rather far beyond the previous underfilled foredeep, could be considered as the 'foresag' (Catuneanu *et al.* 2000), even though the total thickness of sediments accumulated here (*c*.200 to 300 m) indicates a moderate subsidence rate of 0.015 to 0.02 mm/year. Uplift is likely to have ceased or considerably decreased at the beginning of Thanetian times as suggested by the predominance of lacustrine/palustrine or lagoonal through marine deposits and of fine- to very fine-grained fluvial deposits. Deposition was probably mainly controlled at this time by eustatic sea-level fluctuations, even though incipient thrusting may have increased the accommodation space locally (Tambareau *et al.* 1995).

Eocene

Ilerdian Blue Marls and Oyster Sandstones. Their involvement in growth structures indicates that deposition of both the Lower and Upper Blue Marls was controlled by the development of the Mouthoumet frontal thrust-propagation fold during thrust-wedge advance (Figs 9 & 10). Ongoing fold growth is revealed by: firstly, the location of the depocentre in the frontal syncline where accomodation space created by regional subsidence was or less reduced or not reduced at all, and secondly, strata wedging which indicates a reduced deposition and deposition rate on to the anticline because of reduced accommodation space (Ito *et al.* 1999). In contrast with the Late Cretaceous sub-basins, the depositional sequences were predominantly coarsening/shallowing upwards, which indicates the increasing contribution of the advancing emerged thrust-wedge (see also DeCelles *et al.* 1998). However, the fining/deepening-upwards basal parasequence set of the Upper Blue Marls can be interpreted as a result of regional subsidence with no or reduced contribution from the wedge-front during initial fold growth. The duration of the thrust-fault controlled depositional sequences is here c.1–1.2 Ma, which is of the same order of magnitude as the higher order sequences identified in the Late Cretaceous.

The Lower Blue Marls were deposited during the propagation of a lower branch of the Mouthoumet frontal thrust, as shown by the basal growth onlap sealing a thrust fault which affects the pre-growth strata (Figs 6, 9a & 10a). The Talairan sub-basin of that period is likely to represent the foredeep (*sensu* DeCelles & Giles 1996). It is thus tempting to interpret the inner platform as the backbulge depozone. In this case, the outer boundary of the platform, which acted as a barrier for the northerly detrital material and on top of which are observed *Solenomeris* reefs, could be interpreted as the forebulge (Figs 9b & 10a). The presence of clasts and slipped blocks derived from this northern platform south of the shelf-break bears evidence of the instability of the slope, which may characterize the forebulge depozone. This interpretation may be opposed by the growth of the reefs (Plaziat & Perrin 1991) and the absence of a basal unconformity (Crampton & Allen 1995). However, since a long-term eustatic sea-level rise is signalled from 59 to 52 Ma, (Haq *et al.* 1987; Hardenbol *et al.* 1998) and forebulge uprise is rather slow (Beaumont *et al.* 1993), it should reasonably be envisaged that relative sea-level increased as a result of eustasy, while the forebulge rose – thus explaining reef growth and lack of subaerial erosion.

The Upper Blue Marls and overlying Oyster Sandstones are interpreted to be deposited in association with the propagation of an upper/younger thrust branch of the Mouthoumet frontal thrust locally cutting the Lower Blue Marls and sealed by the basal growth onlap (Figs 6 & 10b). A renewed and upwards-increasing contribution from the orogen was responsible for younger deltas prograding craton-ward such that the deltaic system finally covered the entire basin, as shown by the presence of the Oyster Sandstones all above the Upper Blue Marls, and of a Gilbert delta in the south (Figs 9c & 10b). The Upper Blue Marls overlapped the former shelf-break and invaded most of the former platform at the same time, indicating coeval northwards migration of the forebulge (Fig. 10b). Overall, the Blue Marls can be interpreted as having been deposited in a widening flexural basin. Wedge advance being less than basin widening, orogenic loading necessarily resulted from thrust stacking to the rear of the basin, which probably corresponds with the development of the Mouthoumet Duplex as also shown by the absence of detritus originating from the North Pyrenean or the southern sub-Pyrenean zones at this time.

The Palassou Formation. The growth stratal depositional patterns at the contact of the Mouthoumet and Alaric anticlines (Figs 6, 7 & 8) as well as the presence of perennial lacustrine/palustrine depressions in the growth synclines (Talairan and northern Alaric) indicate that deposition of the Palassou Formation was controlled by the propagation of the Mouthoumet and Alaric frontal thrusts (Fig. 10c). The absence of higher sedimentary discharges and debris flows and the presence of multiple channels crossing the growth fold front reveal, however, that these folds played a limited role in the production of sediment and were bypassed by alluvial fans derived from the orogen beyond the Mouthoumet Front. Moreover, the shallow slope depositional environment and the ubiquity of transverse north–south palaeocurrents show that the growing folds in the foreland basin have not significantly disturbed the drainage pattern, which indicates that erosion exceeded tectonic uplift (Burbank *et al.* 1996). According to Boyer (1995), erosion during wedge advance leads to isostatic adjustments that decrease the dip of the basement, and restoration of the critical taper (*sensu* Davis *et al.* 1983) requires thickening of the wedge. Therefore, the decrease in amplitude of the Mouthoumet and Alaric anticlines while the detrital deposits coarsened upwards indicates either internal deformation localizing in the inner orogen as a result of erosion during ongoing thrust-wedge advance, or progressive cessation of the wedge advance coupled with ongoing erosional unloading/uplift. Increasing internal deformation is consistent with decreasing accretion of frontal thrust sheets and limited thrust advance, because deformation absorbs most of the displacement (Boyer 1995). The presence of clasts of North Pyrenean Lower Cretaceous carbonates and sub-Pyrenean Upper Cretaceous sandstones at the basal part of this formation suggests that the Bugarach and other F_3 fold and thrusts of the southern sub-Pyrenean/northern North Pyrenean zones effectively propagated at this time. However, the absence of clasts derived from the southern basement (e.g. migmatites, granulites or even granites) in the Late Ilerdian to Early Bartonian conglomerates indicates no significant reactivation of the thrusts in the more internal orogen. Again, the accumulation of more than 1000 m of sediment in the terminal topographic depression between the latest Ilerdian and the Late Bartonian (subsidence rate of *c.*0.09 mm/year), suggests that this depression could be the 'foresag' resulting from the flexural uplift of the unloaded orogen (Beaumont *et al.* 1993; Washbusch *et al.* 1996; Catuneanu 1997, 2000). Although basement underthrusting in the inner pro-wedge (blind basement duplex?) cannot be ruled out, pure erosional unloading of the inner orogen twinned with sedimentary loading of the outer foreland basin may have been significant at the end of the deposition of the Palassou Formation studied here. The presence of deep-basement clasts in upper units of the Palassou Formation cropping out west of the study area (Crochet 1991) and the apatite fission-tracks ages of exhumation of the Agly Massif (*c.*47 to 40 Ma) and Axial Zone (*c.*35 to 26 Ma) (Morris *et al.* 1998) strongly suggest, however, that out-of-sequence thrusting and wedge advance occurred during late Eocene–early Oligocene times.

Discussion and conclusion: tracing tectonic events using sedimentary markers

The present sedimentological–tectonic study provides new insights into the modes of thrust-wedge advance and basin propagation, the transition from underfilled to overfilled during loading/unloading cycles, and the integration of the basins in the tectonic history of the range.

Thrust-wedge advance and foreland basin propagation

The Late Cretaceous basin system can be shown to have formed as a result of the progressive integration into the wedge-top depozone (DeCelles & Giles 1996) of individual sub-basins during forward propagation of thrust fault-propagation folds. Sedimentological evidence includes:

(1) reworking of the forelimbs of wedge-front fold-propagation folds, shown by 'catastrophic' deposits interdigitated with the 'normal' deposits derived from the craton;
(2) deepening-upwards stacking patterns in the fold-controlled sub-basins, resulting from the coeval migration of wedge-front and platform;
(3) overall deepening-upward stacking pattern indicating increased subsidence during thrust-wedge advance and craton-wards migration of the entire basin.

The depositional sequences controlled by the development of the Coniacian to Late Santonian sub-basins are well dated from fossil evidence (Bilotte 1992), and do not coincide with the 'global eustatic' sequences established by Hardenbol *et al.* (1998), even though using the same time-scale (Gradstein *et al.* 1994).

The Eocene marine basin propagated as a result of widening of a single foredeep with forebulge migration. Sedimentary evidence is borne out by:

(1) the presence of a well-characterized depocentre located in front of a basement duplex;
(2) the individualization of two deepening-upwards depositional sequences;
(3) the presence of growth stratal structures in front and on top of the wedge with a major growth onlap sealing a thrust fault branch at the base of either sequence;
(4) reworking of the forelimb of the frontal fold in both sequences and of the forebulge in the lower sequence.

The fact that craton-ward migration of the platform and forebulge largely exceeded the migration of the wedge-front indicates that there was limited wedge advance coupled with increasing regional flexural subsidence and orogenic loading, probably related to the development of the Mouthoumet basement duplex. Thus it can be stated that local depositional patterns were controlled by the propagation of the frontal fault-propagation fold of a developing basement duplex, which was responsible for an increase in regional subsidence.

The Eocene continental deposits infilled two synchronous fault-propagation-controlled sub-basins superimposed on the Eocene marine basin. The most conclusive sedimentological studies are here analyses of facies and architectural elements (according to Miall 1996), palaeocurrents and provenances. Coupled with the analysis of growth stratal patterns, these studies show that the folds, although synsedimentary, did not provide a large amount of sediment and were bypassed by shallow-slope alluvial fans derived from enhanced erosion of the inner orogen during ongoing and then decreasing thrust-wedge advance.

If we consider now the architecture of the depositional sequences with respect to wedge advance, differing indications have been given by the underfilled marine basins studied herein. In the Late Cretaceous basin, upwards-fining/deepening indicates that the subsidence rate was higher than the deposition rate (Catuneanu *et al.* 1997, 2000), which is likely to occur where the clastic supply is predominantly sourced on the craton ('normal sedimentation'). Apart from the basal coarse breccias of the Middle Cenomanian to Turonian basin, reworking of the forelimbs of the wedge-front folds, although qualitatively important, has provided relatively small amounts of material. This is believed to characterize a submarine wedge with little disruption of the fold limbs, i.e. prevailing ductile flexural folding, during *in-sequence* thrust propagation. In the Eocene basin, upwards shallowing of the depositional sequences is attributed to increasing erosion of the emerged wedge during overstep propagation of thrust fault branches, which reduced accommodation space despite increasing subsidence. Tectonic quiescence or initial fold growth (low-amplitude stage) during ongoing subsidence is, however, recorded in the foredeep by upwards fining/deepening at the base of the Upper Blue Marls. Fold growth-induced uplift is accommodated by strata wedging with no change in the sequence architecture, which differs from Ito *et al.*'s (1999) observations in the highest order sequences. This difference may be explained by the limited amplitude of eustatic fluctuations compared with tectonic subsidence and uplift for the time-scale and tectonic context considered here.

Loading/unloading cycles and the transition from underfilled to overfilled

In the Late Cretaceous–Paleocene basin system, the transition from underfilled to overfilled is interpreted as the change from regional contraction to tectonic quiescence and unloading uplift. This is recorded by the overall deepening-upwards stacking pattern indicating increasing subsidence during thrust-wedge advance, succeeded by a shallowing-upwards stacking pattern indicating progressive 'passive' filling of the basin with an increasing contribution of the emerging orogen and orogen-ward progradation of the platform, and then by fluvial sedimentation. A provenance study coupled with the tectonic study, appear, however, to be a necessary prerequisite for establishing that deposits derived that the deposits issued from the inner orogen were not produced by out-of-sequence thrusting.

During Eocene times, increased erosion of the inner orogen at the origin of the transition from underfilled to overfilled is shown by overall upwards coarsening, progressive decrease in amplitude of the synsedimentary folds in the outer basin, lack of control of drainage by these synsedimentary folds, and origin of the clasts. This can be interpreted to have been a result of erosional unloading coupled with either:

(1) internal deformation restoring the taper angle of the advancing wedge (Boyer 1995) as suggested by synsedimentary folding, or
(2) flexural uplift after cessation of the advance of the wedge, as suggested by the creation of a 'foresag' in the distal part of the basin and the absence of basement clasts which would be produced by out-of-sequence thrusting in the inner orogen, or, more probably,
(3) a combination of events (1) and (2).

Place of the foreland basin in the history of the range

The above results show that during Late Cretaceous to Paleocene times, the inferred loading and unloading events are clearly separated – the 'passive' phase of the cycle (erosion and isostatic rebound) being three times as long (*c.*29–30 Ma) as the 'active' phase (tectonic shortening, *c.*10–12 Ma). Although lasting a relatively short time in the history of the Pyrenean range, Late Cretaceous tectonic shortening thus appears to have been much longer-lived than previously thought on the basis of pure tectonic studies. The deformation rate inferred from balanced sections (*c.*10^{-15} s^{-1}) is however consistent with those usually observed during lithospheric deformation. On the other hand, the present study is in agreement with those of the previous tectonic studies: stating that the shortening at the origin of the Late Cretaceous folds and thrusts was a result of pure contraction normal to the range, combined with minor left-lateral strike-slip shearing parallel to the North Pyrenean Fault Zone.

The Eocene basins record a new tectonic pulse marked by the propagation of east–west folds and thrusts in a pure compressional context, which lasted at most *c.*15 Ma (late Early Ilerdian to Early Bartonian). The tectonic history inferred from the study of the sedimentary infill appears much more complex than previously envisaged, involving *in-sequence* (Early to Late Ilerdian), *out-of-sequence* (latest Ilerdian) and *synchronous* (Latest Ilerdian to Early Bartonian) thrusting.

We thank P. Baby, P. Souquet, Y. Tambareau, and J. Villatte for providing us with unpublished data and stimulating discussions during the completion of this work. Midland Valley Inc. is acknowledged for technical support. The manuscript also benefited from constructive reviews by F. Mouthereau and T. McCann.

References

ARTHAUD, F., BURG, J. P. & MATTE, P. 1976. L'évolution structurale hercynienne du massif de Mouthoumet (Sud de la France). *Bulletin de la Société Géologique de France*, 7, **XVIII(4)**, 967–972.

BABY, P., CROUZET, G., SPECHT, M., DERAMOND, J., BILOTTE, M. & DEBROAS, E. J. 1988. Rôle des paléostructures anté albo-cénomaniennes dans la géométrie des chevauchements frontaux nord-pyrénéens. *Comptes Rendus de l'Académie des Sciences, Paris*, **306(II)**, 307–313.

BEAUMONT, C. 1981. Foreland basins. *Geophysical Journal of the Royal Astronomical Society*, **137**, 291–329.

BEAUMONT, C., QUINLAN, G. M. & STOCKMAL, G. S. 1993. The evolution of the Western Interior basin: causes, consequences and unsolved problems. *In:* CALDWELL, W. G. E. & KAUFFMAN, E. G., (eds), *Evolution of the Western Interior Basin*. Geological Association of Canada, Special Paper, **39**, 97–117.

BEAUMONT, C., MUÑOZ, J. A., HAMILTON, J. & FULLSACK, P. 2000. Factors controlling the Alpine evolution of the Central Pyrenees inferred from a comparison of observations and geodynamical models. *Journal of Geophysical Research*, **105(B4)**, 8121–8145.

BERGER, G. M. & REY, J. 1990. Ilerdian. *In:* BERGER, G. M. & BOYER, REY, J., *Feuille de Lézignan-Corbières, Carte Géologique de la France à 1150000, no. 1038*, BRGM, Orleans, 70 pp.

BERGER, G. M., BOYER, F., DEBAT, P., DEMANGE, M., ISSARD, H., MARCHAL, J. P., FREYTET, P. & MAZEAS, H. 1993. *Feuille de Carcassonne, Carte Géologique de la France à 115000, n°1037*, BRGM, Orleans, 78 pp.

BERGER, G. M., ALABOUVETTE, B., BESSIERE, G., BILOTTE, M., CROCHET, B., DUBAR, M., MARCHAL, J. P., TAMBAREAU, Y., VILLATTE, J. & VIALLARD, P. 1997. *Feuille de Tuchan, Carte Ggéologique de la France à 1150 00, no. 1078*, BRGM, 113 pp.

BESSIERE, G. 1987. Modèle d'évolution polyorogénique d'un massif hercynien: le Massif de Mouthoumet (Pyrénées Orientales). Thesis, Université Toulouse III, 317 pp.

BESSIERE, G., BILOTTE, M., CROCHET, B., PEYBERNES, B., TAMBAREAU, Y. & VILATTE, J. 1989. *Feuille de Quillan, Carte géologique de la France à 1150 00, no. 1077*, BRGM, 98 pp.

BILOTTE, M. 1985. Le Crétacé supérieur des plateformes Est-pyrénéennes, *Strata*, **2(5)**, p. 438.

BILOTTE, M. 1992. Enregistrement sédimentaire et datation du passage de la marge stable à la marge convergente durant le Sénonien dans la zone sous-pyrénéenne orientale (Corbières, France). *Comptes Rendus de l'Académie des Sciences Paris*, **315**, 77–82.

BILOTTE, M., FONDECAVE, M.-J., PEYBERNES, B., SOUQUET, P. & WALLEZ J.-P. 1973. Distinction de l'Albien et du Crétacé supérieur dans le synclinorium d'Axat (Pyrénées). *Comptes Rendus Sommaires de la Société Géologique de France*, **5–6**, 119–121.

BIRD, P. 1979. Continental delamination and the Colorado Plateau. *Journal of Geophysical Research*, **84**, 7561–7571.

BLAIR, T. C. & BILODEAU, W. L. 1988. Development of tectonic cyclothems in rift, pull-apart and foreland basins: sedimentary response to episodic tectonism. *Geology*, **16**, 517–520.

BOYER, S. E. 1995. Sedimentary basin taper as a factor controlling the geometry and advance of thrust belts. *American Journal of Science*, **295**, 1220–1254.

BRUSSET, S., DERAMOND, J. & SOUQUET, P. 1997. Evolution tectono-sédimentaire des bassins flexuraux à taux de sédimentation réduite: exemple du basin flysch de St Jean-de-Luz (Pyrénées Atlantiques, France) au Crétacé supérieur. *Comptes Rendus de l'Académie des Sciences, Paris, IIa*, **325**, 265–171.

BURBANK, D. W. 1992. Causes of the recent Himalayan uplift deduced from deposited patterns in the Ganges basin. *Nature*, **357**, 680–682.

BURBANK, D. W., MEIGS, A. & BROZOVICS, N. 1996. Interactions of growing folds and coeval depositional systems. *Basin Research*, **8**, 199–223.

BURBANK, D. W., PUIGDEFABREGAS, C. & MUÑOZ, J. A. 1992. The chronology of the Eocene tectonic and stratigraphic development of the eastern Pyrenean foreland basin, Northeast Spain. *Geological Society of America Bulletin*, **104**, 1101–1124.

BURKHARD, M. & SOMMARUGA, M. 1998. Evolution of the western Swiss Molasse basin: structural relations with the Alps and the Jura Belt. *In:* MASCLE, A., PUIGDEFABREGAS, C., LUTERBACHER, H. P. & FERNANDEZ, M. (eds) *Cenozoic Foreland Basins of Western Europe*, Special Publication of the Geological Society, **134**, 279–298.

CARTER, R. M., ABBOT, A. T., FULTHORPE, C. S., HAYWICK, D. W. & HENDERSON R. A. 1991. Application of global sea-level and sequence-stratigraphic models in Southern Hemisphere Neogene strata from New Zealand. *In:* MACDONALD, D. I. M., (ed.) *Sedimentation, Tectonics and Eustasy*. International Association of Sedimentologists Special Publication, **12**, 41–65.

CATUNEANU, O., MIALL, A. D. & SWEET, A. R. 1997. Reciprocal architecture of Bearpaw T–R sequences, Uppermost Cretaceous, Western Canada. *Sedimentary Basin*, **45(1)**, 75–94.

CATUNEANU, O., SWEET, A. R. & MIALL, A. D. 2000. Reciprocal stratigraphy of the Campanian–Paleocene, Western Interior of North America. *Sedimentary Geology*, **134**, 235–255.

CHANNEL, J. E. T. & MARESCHAL, J. C. 1986. Delamination and asymmetric lithospheric thickening in the development of the Tyrrhenian Rift. *In:* COWARD, M. (ed.) *Alpine tectonics*, Special Publication of the Geological Society, **45**, 285–302.

CHOUKROUNE, P. 1976. *Structure et évolution tectonique de la zone Nord-Pyrénéenne. Analyse de la déformation dans une portion de chaîne à schistosité sub-verticale.* Mémoire de la Société Géologique de France, **55(217)**.

CHOUKROUNE, P. & MATTAUER, M. 1978. Tectonique des plaques et Pyrénées: sur le fonctionnement de la faille transformante nord-pyrénéenne: comparaison avec des modèles actuels. *Bulletin de la Société Géologique de France*, **20**, 689–700.

CHOUKROUNE, P. & MEURISSE, M. 1970. Phases de déformations superposées dans le Mésozoïque de la zone nord-pyrénéenne sur la transversale du massif de Salvezines (Aude). *Comptes Rendus de l'Académie des Sciences, Paris, D*, **270**, 14–17.

CHRISTOPHOUL, F., BABY, P. & DAVILA, C. 2002. Stratigraphic responses to a major tectonic event in a foreland basin: the Ecuadorian Oriente Basin from Eocene to Oligocene times. *Tectonophysics*, **345**, 281–298.

CLIFTON, H. E., HUNTER, R. E. & GARDNER, J. V. 1988. Analysis of eustatic, tectonic and sedimentologic influences on transgressive and regressive cycles in the upper Cenozoic Merced Formation, California. *In:* KLEINSPEHN, K. L. & PAOLA, C. (eds) *New Perspectives in Basin Analysis*, Springer-Verlag, New York, 109–128.

CRAMPTON, S. L. & ALLEN, P. A. 1995. Recognition of forebulge unconformities associated with early stage foreland basin development: example from the North Alpine Foreland Basin. *AAPG Bulletin*, **79**, 1495–1514.

CROCHET, B. 1991. Molasses syntectoniques du versant nord des Pyrénées: la série de Palassou. *Documents du BRGM*, **199**, 387 pp.

CROSS, T. A. 1986. Tectonic control of foreland basin subsidence and Laramide style deformation, western United States. *In:* ALLEN, P. A. & HOMEWOOD, P. (eds.) *Foreland Basins*. International Association of Sedimentologists Special Publication, **8**, 15–39.

DAVIES, J. H. & VON BLANCKENBURG, F. 1995. Slab Breakoff: a model of lithosphere detachment and its test in magmatism and deformation of collisional orogens. *Earth and Planetary Science Letters*, **129**, 85–102.

DAVIS, D. M., SUPPE, J. & DAHLEN, F. A. 1983. Mechanics of fold-and-thrust belts and accretionary wedges. *Journal of Geophysical Research*, **88**, 1153–1172.

DECELLES, P. G. & GILES, K. A. 1996. Foreland basin systems, *Basin Research*, **8**, 105–123.

DECELLES, P. G., GEHRELS, G. E., QUADE J. & OJHA, T. P. 1998. Eocene-Early Miocene basin development and the history of Himalayan thrusting, western and central Nepal. *Tectonics*, **17(5)**, 741–765.

DÉRAMOND, J., SOUQUET, P., FONDECAVE-WALLEZ, M. J. & SPECHT, M. 1993. Relationships between thrust tectonics and sequence stratigraphy surfaces in foredeeps: model and examples from the Pyrenees (Cretaceous–Eocene, France, Spain). *Geological Society of London Special Publication*, **71**, 193–219.

DONCIEUX, L. 1912. Révision de la faune lacustre de l'Eocène moyen des Corbières Septentrionales. *Bulletin de la Société des Etudes Scientifiques de l'Aude*, **13**, 25–50.

ECORS PYRENEES TEAM 1988. The ECORS deep reflection survey across the Pyrenees. *Nature*, **331**, 508–511.

EINSELE, G. 1992. Subsidence, *In:* EINSELE, G. (ed.), *Sedimentary Basins: Evolution, Facies and Sediments Budget*, Springer-Verlag, pp. 313–344.

FORD, M., WILLIAMS, E. A., ARTONI, A., VERGÈS, J. & HARDY, S. 1997. Progressive evolution of a fault-related fold pair from growth strata geometries, Sant Llorenç de Morunys, SE Pyrenees. *Journal of Structural Geology*, **19(3–4)**, 413–441.

FREYTET, P. 1970. Les dépôts continentaux du Crétacé supérieur et des couches de passage à l'Eocène en Languedoc. Thesis, Paris-Sud, 530 pp.

GÉLARD, J. P. 1969. Structure de la région située entre Quillan et le Pech de Bugarach. *Bulletin de la Société Géologique de France*, **7(11)**, 345–353.

GILBERT, G. K. 1885. Topographic features of lakes shores. *Annual Report of the US Geological Survey*, **5**, 75–123.

GRADSTEIN, F. M., AGTERBERG, F. P., OGG, J. G., HARDENBOL, J. G., VAN VEEN, P., THIERRY, J. & HUANG, Z. 1994. A Mesozoic time scale. *Journal of Geophysical Research*, **99(B12)**, 24051–24074.

HAQ, B. U., HARDENBOL, J. & VAIL, P. R. 1987. Chronology of fluctuating sea levels since the Triassic (250 million years ago to present). *Science*, **235**, 1156–1167.

HARDENBOL, J., THIERRY, J., FARLEY, M. B., JACQUIN, T., DE GRACIANSKY, P. C. & VAIL, P. R. 1998. Mesozoic and Cenozoic sequence stratigraphic framework of European basins. *In:* DE GRACIANSKY, P. C., JACQUIN, T. & VAIL, P. R. (eds) *Mesozoic and Cenozoic Sequence Stratigraphy of European Basins*. SEPM Special Publication, **60**.

HELLER, P. L. & PAOLA, C. 1992. The large scale dynamics of grain-size variation in alluvial basins, 2: application to syntectonic conglomerates. *Basin Research*, **4**, 91–102.

HELLER, P. L., ANGEVINE, C. L. T., WINSLOW, N. S. & PAOLA, C. 1988. Two-phase stratigraphic model of foreland-basin sequences. *Geology*, **16**, 501–504.

HOMEWOOD, P., ALLEN, P. A. & WILLIAMS, G. D. 1986. Dynamics of the Molasse basin of western Switzerland. *In:* ALLEN, P. A. & HOMEWOOD, P. (eds) *Foreland Basins*. International Association of Sedimentologists Special Publication, **8**, 199–217.

HORTON, B. K. 1999. Erosional control on the geometry and kinematics of thrust belt development in the central Andes. *Tectonics*, **18**, 1292–1304.

HORTON, B. & DECELLES, P. G. 2001. Modern and ancient fluvial megafans in the foreland basin system of the central Andes, southern Bolivia: implications for drainage network evolution in fold–thrust belts. *Basin Research*, **13**, 43–46.

ITO, M., NISHIKAWA, T. & SUGIMOTO, H. 1999. Tectonic control of high-frequency depositional sequences with durations shorter than Milankovitch cyclicity: an example from the Pleistocene paleo-Tokyo Bay, Japan. *Geology*, **27(8)**, 763–766.

JORDAN, T. E. 1995. Retroarc foreland and related basins. *In:* BUSBY, C. J. & INGERSOLL, R. V. (eds) *Tectonics of Sedimentary Basins*. Blackwell Science, Oxford, 331–362.

JORDAN, T. E., FLEMING, P. B. & BEER, J. A. 1988. Dating thrust-fault activity by use of foreland-basin strata. *In:* KLEINSPEHN, K. L. & PAOLA, C. (eds)

New Perspective in Basin Analysis. Springer-Verlag, Berlin, 307–330.

LEBLANC, D. & VAUDIN, J. L. 1984. Les déformations du Mésozoïque de la zone nord-pyrénéenne à l'est de l'Aude. *Bulletin du Bureau de Recherches Géologiques et Minières, Géologie de la France*, **4**, 57–68.

MAGNÉ, J. & MATTAUER, M. 1968. Sur la présence du Cénomanien–Turonien dans le série schisteuse de la couverture nord de la zone axiale des Pyrénées au Sud de Quillan (Aude). *Bulletin du Bureau de Recherches Géologiques et Minières, Géologie de la France*, **21(3)**, 39–43.

MASSIEUX, M. 1973. Micropaléontologie stratigraphique de l'Eocène des Corbières Orientales (Aude). *Cahiers de Paléontologie*, 146 pp.

MATTAUER, M. & PROUST, F. 1965. Sur la présence et la nature de deux importantes phases tectoniques dans les terrains secondaires des Pyrénées orientales. *Comptes Rendus Sommaires de la Société Géologique de France*, **5**, 132–133.

MIALL, A. D. 1996. *The Geology of Fluvial Deposits, Sedimentary Facies, Basin Analysis and Petroleum Geology*. Springer, Berlin, Heidelberg and New York, 582 pp.

MORRIS, R. G., SINCLAIR, H. D. & YELLAND, A. J. 1998. Exhumation of the Pyrenean orogen: implications for sediment discharge. *Basin Research*, **10**, 69–85.

MUÑOZ, J. A. 1992. Evolution of a continental collision belt: Ecors Pyrenees crustal balanced cross section. *In:* MCCLAY, K. R. (ed.) *Thrust Tectonics*. Chapman & Hall, London, 235–246.

NEMEC, W. & STEEL, R. J. 1988. *Fan deltas: Sedimentology and Tectonic Settings*. Blackie, London, 464 pp.

NIJMAN, W. 1998. Cyclicity and basin axis shift in a piggyback basin: toward modelling of the Eocene Tremp–Ager Basin, South Pyrenees, Spain. *In:* MASCLE, A., PUIGDEFABREGAS, C., LUTERBACHER, H. P. & FERNANDEZ, M. (eds) *Cenozoic Foreland Basins of Western Europe*, Special Publication of the Geological Society, **134**, 135–162.

ODIN, G. S. 1994. Geological time scale. *Comptes Rendus de l'Académie des Sciences, Paris*, **318**, 59–71.

PAUTAL, L. 1985. *Populations fossiles, associations micropaléontologiques et paléoenvironnements des séries deltaïques ilerdiennes des Corbières (Aude, France)*. Thesis, Université Toulouse III, 288 pp.

PELISSIER, P. 1987. *Etude sédimentologique des Grès de la Bastide (Santonien Supérieur–Campanien) dans le synclinal sous-pyrénéen St Louis (Hautes Corbières)*. Thesis, Université Toulouse III, 111 pp.

PLAZIAT, J. C. 1984. *Stratigraphie et évolution paléogéographique du domaine pyrénéen de la fin du Crétacé (phase maastrichtienne) à la fin de l'Eocène (phase pyrénéenne)*. Thesis, Université Paris VI, 1200 pp.

PLAZIAT, J. C. 1987. Ilerdian.. *In:* ELLENBERGER, F., FREYTET, P., PLAZIAT, J. C., BESSIERE, G., VIALLARD, P., BERGER, G. M. & MARCHAL, J. P. (eds) *Feuille de Capendu, Carte géologique de la France à 1150 00, n°1090*, BRGM, Paris, 88.

PLAZIAT, J. C. & PERRIN, C. 1991. Multikilometer-sized reefs built by foraminifera (*Solenomeris*) from the early Eocene of the Pyrenean domain (S France, N Spain)–Paleoecological relations with coral reefs. *Palaeogeography Palaeoclimatology Palaeoecology*, **96(3–4)**, 195–231.

PUIGDEFRABREGAS, C. & SOUQUET, P. 1986. Tecto-sedimentary cycles and depositional sequences of the Mesozoic and Tertiary from the Pyrenees. *Tectonophysics*, **129(1–4)**, 173–203.

PUIGDEFRABREGAS, C., MUÑOZ, J. A. & MARZO, M. 1986. Thrust belt development in the eastern Pyrenees and related depositional sequences in the southern foreland basin. *In:* ALLEN, P. A. & HOMEWOOD, P. (eds) *Foreland Basins*, Special Publication of the International Association of Sedimentologists, **8**, 229–246.

RAZIN, P. 1989. *Evolution tectono-sédimentaire alpine des Pyrénées Basques à l'Ouest de la transformante de Pamplona (Province du Labourd)*. Thesis, Université Bordeaux 3, 464 pp.

READING, H. G. & COLLINSON, J. D. 1996. Clastic coasts. *In:* READING, H. G. (ed.) *Sedimentary Environments: Processes, Facies and Stratigraphy*. Blackwell Science, 3rd edn. 154–230.

REY, J. & BOUSQUET, J. P. 1981. Observations préliminaires sur les paléoenvironnements de l'Ilerdien de Coustouge (Aude, France). *Geobios*, **14(5)**, 655–659.

RICCI LUCCHI, F. 1986. The Oligocene to Recent foreland basins of the northern Apennines. *In:* ALLEN, P. A. & HOMEWOOD, P. (eds) *Foreland Basins*. Special Publication of the International Association of Sedimentologists, **8**, 105–139.

SCHLUNEGGER, F. 1999. Controls of surface erosion on the evolution of the Alps: constraints from the stratigraphies of the adjacent foreland basins. *International Journal of Earth Sciences*, **88**, 285–304.

SCHLUNEGGER, F., MATTER, A., BURBANK, D. W. & KLAPER, E. M. 1997*a*. Magnetostratigraphic constraints on relationships between evolution of the central Swiss Molasse basin and Alpine orogenic events. *Geological Society of America Bulletin*, **109**, 225–241.

SCHLUNEGGER, F., MATTER, A., BURBANK D. W., LEU, W., MANGE, M. & MÀTYAS, J. 1997*b*. Sedimentary sequences, seismofacies, and evolution of depositional systems of the Oligo/Miocene Lower Freshwater Molasse Group, Switzerland. *Basin Research*, **9**, 1–26.

SEGUIER, J. 1972. *Etude stratigraphique du Paléozoïque du Cabardès (versant sud de la Montagne Noire) et de sa couverture Eocène*. Thesis, Université Toulouse III, 104 pp.

SERRA KIEL, J. *et al.* 1998. Larger foraminiferal zonation, Tethys ocean, IGCP 286 (Early Paleogene Benthos), IGCP 393 (neritic events at the Middle–Upper Eocene boundary). *Bulletin de la Société Géologique de France*, **169(2)**, 281–299.

SERRANO, O., GUILLOCHEAU, F. & LEROY, E. 2001. Évolution du bassin compressif Nord-Pyrénéen au Paléogène (bassin de l'Adour): contraintes stratigraphiques. *Comptes Rendus de l'Académie des Sciences, Paris*, **332**, 37–44.

SINCLAIR, H. D. 1992. Turbidite sedimentation during Alpine thrusting: the Taveyannaz Sandstones of eastern Switzerland. *Sedimentology*, **39**, 837–856.

SINCLAIR, H. D. 1993. High resolution stratigraphy and facies differentiation of the shallow marine Annot Sandstones, south-east France. *Sedimentology*, **40**, 955–978.

SINCLAIR, H. D. 1997. Tectonostratigraphic model for underfilled peripheral foreland basins: An alpine perspective. *Geological Society of America Bulletin*, **109**(3), 324–346.

SOULA, J.-C. & BESSIERE, G. 1980. Sinistral horizontal shearing as a dominant process of deformation in the Alpine Pyrenees. *Journal of Structural Geology*, **2**, 69–74.

SOULA, J.-C., LAMOUROUX, C., VIALLARD, P., BESSIERE, G., DEBAT, P. & FERRET, B. 1986. The mylonite zones in the Pyrenees and their place in the Alpine tectonic evolution. *Tectonophysics*, **129**, 115–147.

SOUQUET, P., ESCHARD, R. & LODS, H. 1987. Facies sequences in large-volume debris-and-turbidity flows deposits from the Pyrenees (Cretaceous; France, Spain). *Geomarine Letter*, **7**, 83–90.

SPECHT, M. 1989. *Tectonique de chevauchement le long du profil ECORS Pyrénées: un modèle d'évolution de prisme d'accrétion continental*. Thesis, Université de Bretagne Occidentale, Brest, 353 pp.

STANISTREET, I. G. & MCCARTHY, T. S. 1993. The Okavongo Fan and the classification of subaerial fan systems. *Sedimentary Geology*, **58**, 115–133.

SUPPE, J., CHOU, G. T. & HOOK, S. C. 1992. Rates of folding and faulting determined from growth strata. *In:* MCCLAY, K. R. (ed.) *Thrust Tectonics*. Chapman and Hall, London, 105–121.

TAMBAREAU, Y., CROCHET, B., VILLATTE, J. & DERAMOND, J. 1995. Evolution tectono-sédimentaire du versant nord des Pyrénées centre-orientales au Paléocène et à l'Eocène inférieur. *Bulletin de la Société Géologique de France*, **166(4)**, 375–387.

VAIL, P. R., AUDEMARD, F., BOWMAN, S. A., EISNER, P. N. & PEREZ-CRUZ, C. 1991. The stratigraphic signatures of tectonics, eustasy and sedimentology – an overview. *In:* EINSELE, G. & SEILACHER, A. (eds) *Cycles and Events in Stratigraphy.* Springer-Verlag, Berlin, 617–659.

VERGES, J., MARZO, M., SANTAEULARIA, T., SERRA-KIEL, J., BURBANK, D. W., MUÑOZ, J. A. & GIMENEZ-MONTSANT. J. 1998. Quantified vertical evolution of the SE Pyrenean foreland Basin. *In:* MASCLE, A., PUIGDEFABREGAS, C., LUTERBACHER, H. P. & FERNANDEZ, M. (eds) *Cenozoic Foreland Basins of Western Europe.* Special Publication of the Geological Society, **134**, 107–134.

WALLEZ, J. P. 1974. *Stratigraphie et structure de la partie méridionale du pays de Sault (Aude)*. Thesis, Université Toulouse III, 143 pp.

WALLEZ-FONDECAVE, M.-J. & SOUQUET, P. 1991. Signatures stratigraphiques de l'eustatisme et de la tectonique de chevauchement dans le Crétacé supérieur du versant nord des Pyrénées; exemple de la zone sous-pyrénéenne orientale (Corbières, France). *Comptes Rendus de l'Académie des Sciences Paris*, **312**, 631–637.

WASHBUSCH, P., CATUNEANU, O. & BEAUMONT, C. 1996. A combined tectonic/surface process model for the formation of reciprocal stratigraphies. *EOS – Transactions of the American Geophysical Union*, 701.

WILLIAMS, E. A., FORD, M., VERGÉS, J. & ARTONI, A. 1998. Alluvial gravel sedimentation in a contractional growth fold setting, Sant Llorenç de morunys, southeastern Pyrenees. *In:* MASCLE, A., PUIGDEFABREGAS, C., LUTERBACHER, H. P. & FERNANDEZ, M. (eds) *Cenozoic foreland basins of western Europe.* Special Publication of the Geological Society, **134**, 69–106.

YOSHIDA, S., WILLIS, A. & MIALL, A. D. 1996. Tectonic control of nested sequence architecture in the Castelgate Sandstone (Upper Cretaceous), Book Cliffs, Utah. *Journal of Sedimentary Research*, **66**, 737–748.

Active or passive continental margin? Geochemical and Nd isotope constraints of metasediments in the backstop of a pre-Andean accretionary wedge in southernmost Chile (46°30′–48°30′S)

C. AUGUSTSSON[1, 2] & H. BAHLBURG[1]

[1]*Geologisch-Paläontologisches Institut, Westfälische Wilhelms-Universität, Corrensstraße 24, 481 49 Münster, Germany (e-mail: augustss@uni-muenster.de)*
[2]*Zentrallaboratorium für Geochronologie, Institut für Mineralogie, Westfälische Wilhelms-Universität, Corrensstraße 24, 481 49 Münster, Germany*

Abstract: Provenance analysis of siliciclastic sedimentary rocks gives indications of the tectonic evolution and setting of source regions and the rocks contained in them. The composition of sedimentary rocks ideally reflects the nature of these regions, and only indirectly the tectonic setting of the basin where the erosional debris is deposited. This makes it possible to interpret Late Devonian to Early Carboniferous metasedimentary basement rocks of the Andes in southernmost Chile as having been deposited at a passive margin, despite geochemical indications of an active margin setting for the source rocks, and the position of the metasediments in the backstop of an accretionary wedge. Major and trace elements point to felsic source rocks from an active margin environment. The Nd model ages of 1170–1490 Ma indicate that the source rocks were part of an old continental crust in the Late Palaeozoic. The $\varepsilon_{Nd}(T)$ values range between –7 and –2. These characteristics, in combination with the regional geology, suggest that the geochemical signal is dominated by rocks formed at an active margin, which later acted as feeders for the sediments deposited in a passive-margin environment. If corroborated by research in progress this emphasizes the problem of deducing the tectonic setting of a depositional basin from provenance data.

Geochemistry and radiogenic isotopes are useful tools for characterizing the provenance of sediments when combined with petrographical methods (Bhatia & Crook 1986; Roser & Korsch 1988; McLennan *et al.* 1989, 1990). Processes such as weathering, sorting, diagenesis and metamorphism all affect and potentially modify the chemical and isotopic signals of sediments (McLennan *et al.* 1993; Mildowski & Zalasiewicz 1991; Condie *et al.* 1995; Roser & Nathan 1997). The Nd isotope system, which is in common use in provenance studies since it is generally unaffected by metamorphism, can be disturbed during weathering, diagenesis and metamorphism under certain circumstances (Zhao *et al.* 1992; McDaniel *et al.* 1994). By combining geochemistry and Nd isotopes, we track the origin of the source rocks of Late Palaeozoic metasedimentary basement rocks of the Andes in Andean Patagonia in southern Chile (46°30′–48°30′S), and the plate tectonic setting at the time of sedimentation.

Geological setting

In Andean Patagonia in southernmost Argentina and Chile, the oldest exposed rocks constitute the so-called basement of the southern Andes. The basement is mainly composed of siliciclastic rocks dominated by turbidites, and minor limestone intercalations. The rocks have ages, determined by fossils, from Devonian in the north to Permian and Triassic further south (Fig. 1a; Riccardi 1971; Ling *et al.* 1985; Ling & Forsythe 1987; Fortey *et al.* 1992; Fang *et al.* 1998). Estimations of maximum depositional time-spans, by a combination of zircon U–Pb and fission track dating, have confirmed some of these ages (Fig. 1a; Thomson *et al.* 2000). The basement sediments are interpreted as subduc-

From: McCann, T. & Saintot, A. (eds) *Tracing Tectonic Deformation Using the Sedimentary Record.* Geological Society, London, Special Publications, **208**, 253–268. 0305–8719/03/$15.00

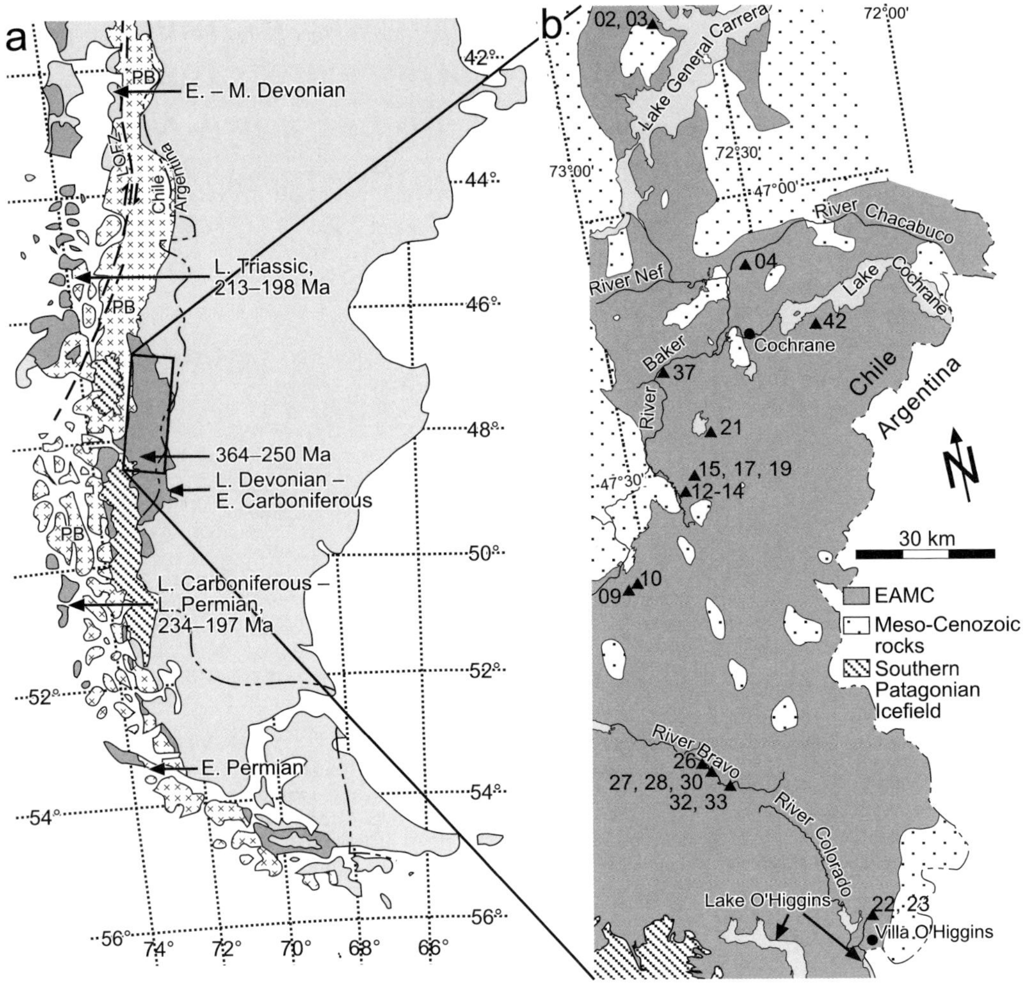

Fig. 1. (**a**) Southern Patagonia with age estimations of the Andean basement rocks (dark grey) determined from fossils, and the maximal depositional time-span determined by zircon U–Pb and fission track analysis. LOFZ, Liquiñe–Ofqui Fault Zone; PB, Patagonian Batholith (after Escobar 1980 and Caminos & González 1996). For further references, see the main text. (**b**) Sampling area in the Eastern Andean Metamorphic Complex. Numbered triangles are sampling points. After Lagally (1975) and Yoshida (1981).

tion complexes accreted to the margin of Gondwana in Late Palaeozoic to Early Mesozoic times (Forsythe 1982), as indicated by whole-rock Rb/Sr ages (*c.*280–140 Ma; Hervé 1988; Pankhurst *et al.* 1992; Hervé *et al.* 2000) and zircon fission track dating (264–209 Ma; Thomson *et al.* 2000, 2001). The incorporation of the western part of the deposits into an accretionary wedge, and the eastern part into its backstop, resulted in deformation and metamorphism of the basement rocks ranging from sub-greenschist to blueschist facies (Hervé *et al.* 1999; Willner *et al.* 2000). It has been unclear if the sediments were deposited at an active margin and directly incorporated into the accretionary wedge and its backstop, or if deposition took place at a passive margin with inclusion of the sediments into the accretionary wedge and its backstop at a later stage.

The study area is located in southern Chile at 46°30′–48°30′S (Fig. 1). It belongs to the Eastern Andean Metamorphic Complex (Hervé, 1993). This part is separated from the basement outcrops in the westward lying Chilean archipelago by the Mesozoic–Cenozoic Patagonian Batholith (Fig. 1a). In the north, the Liquiñe–Ofqui Fault Zone, a NNE–SSW-trending dextral shear zone which has been active at least since the Oligocene, may have displaced the western part northwards by 440–550 km from a former position closer to the Eastern Andean Metamorphic Complex (Fig. 1a; García *et al.* 1988).

Lagally (1975) described two distinct successions for the Eastern Andean Metamorphic Complex. The Lago General Carrera unit, dominated by mica schists, greenschists and marbles, is situated in the area of Lake General Carrera, and the Cochrane unit, mainly composed of greywackes and shales, is situated south of the rivers Chacabuco and Nef (*c.*47°S; Fig. 1b). If the units are stratigraphically equivalent or not has not been clarified (Bell & Suárez 2000, and references therein). The Cochrane unit extends down to Lake O'Higgins at 49°S, where the Bahía de la Lancha Formation crops out in the Argentinian sector of Lake San Martín (=Lake O'Higgins in Chile; Riccardi, 1971). This formation, and the Argentinian Río Lácteo Formation directly east of the study area, can be correlated with the Chilean Eastern Andean Metamorphic Complex (Leanza 1972). There is no fossil evidence of the depostitional age of the Chilean Eastern Andean Metamorphic Complex, but it is assumed to have the same age as the Argentinian Bahía de la Lancha Formation, from which Upper Devonian to Lower Carboniferous plant remains and tetrapod traces have been found (Fig. 1a; Riccardi 1971). From a combination of U–Pb and fission track data, a relatively broad depositional timespan of 364–250 Ma is indicated for the southern part of the Chilean Eastern Andean Metamorphic Complex (Fig. 1a; Thomson *et al.* 2000).

The present investigation concentrates on the Cochrane unit. The studied rocks are interpreted as turbidite deposits, which were deformed and low-grade metamorphosed in the Late Palaeozoic to Early Mesozoic. In the southern part of the area, single turbidite beds with typical sedimentary structures like grading and ripple cross laminations are relatively well preserved, whereas to the north the rocks are more deformed and metamorphosed. The deposits are mostly thin bedded, sand dominated $T_{a\text{–}c/d}$ or base absent $T_{b\text{–}c/d}$ turbidites with medium sand as the largest observed grain size. Bedding planes are usually parallel and major channelling is absent. The deposits were most likely formed in lobe environments (Mutti & Normark 1987). Four stages of post-depositional deformation, connected to the Late Palaeozoic to Early Mesozoic easterly directed subduction at the margin of Gondwana (present coordinates), have been determined by Bell & Suárez (2000).

Methods

17 greywacke samples and four pelites from the Cochrane unit, and one greywacke (CA-00-03-S) and one pelite sample (CA-00-02-M) from the Lago General Carrera unit, have been analysed for their geochemistry (Fig. 1b). Sm and Nd isotopes were measured for 10 of the greywacke samples from the Cochrane unit and the one from the General Carrera unit (CA-00-03-S).

The samples were powdered in an agate mill and chemical analyses were made by ACME lab in Vancouver, Canada. Major elements, Ba, Ni and Sc were measured by ICP-ES, and other trace elements by ICP/MS. The $C_{carbonate}$ was measured on a CS-MAT 5500 at the Geologisch–Paläontologisches Institut, University of Münster, Germany.

Isotopes were measured at the Zentrallaboratorium für Geochronologie, Münster, and the standard method for this laboratory was used. 100 mg of each sample was spiked with a mixed $^{150}Nd/^{149}Sm$ tracer. The dissolution procedure included a first step with HF in Teflon bombs at 175°C for 2 days, and a second step with $HClO_4$. The contribution of rare-earth elements from accessory zircon, which might not have completely dissolved, is negligible (Cherniak *et al.* 1997). The Sm and Nd were separated in cation exchange columns with HCl solutions. Measurements were performed with a VG Sector 54 mass spectrometer. The mean laboratory $^{143}Nd/^{144}Nd$ value for La Jolla standard solution is 0.511860±0.000011 (2σ, n=34). Laboratory Sm and Nd blanks of <200 pg are not considered to affect the calculations. The Nd_{sample}/Nd_{blank} ratios were always >10 000. Normalization was made to $^{146}Nd/^{144}Nd=0.7219$ (O'Nions *et al.*, 1977). The model of Goldstein *et al.* (1984), with $\lambda=6.54\times10^{-12}a^{-1}$, $^{143}Nd/^{144}Nd_{DM,\,today}=0.51315$, $^{147}Sm/^{144}Nd_{DM,\,today}=0.217$, $^{143}Nd/^{144}Nd_{CHUR,\,today}=0.512638$ and $^{147}Sm/^{144}Nd_{CHUR,\,today}=0.1967$, was used.

Results and interpretation

For this study we have preferred to analyse the greywacke parts of the turbidite deposits, although pelites have been considered as well. The greywackes have matrix contents of 40–60%. The clasts are dominated by angular to subrounded quartz and, to a minor degree, plagioclase grains. There are signs of alteration of feldspars to clay minerals, and secondary deposition of calcite as veinlets and in the matrix. Heavy minerals are sparce and dominated by zircon. The high quartz content indicates compositionally mature sediments. A former lower matrix content is to be expected (Galloway 1974). Due to the observed alterations, the petrography on its own is not a good enough provenance indicator for these sediments.

Geochemistry

Major and trace elements are presented in Table 1. The greywackes have SiO_2 contents of 71–81%, whereas the pelites range from 53 to 69%, with the exception of one calcite-rich sample. The wide CaO_{total} range between 0.1% and 15% is affected by the $CaO_{carbonate}$ content of the samples (Table 1).

The molecular proportions of Al_2O_3, $CaO_{silicate}$, Na_2O and K_2O have been used to calculate the chemical index of alteration (CIA; Nesbitt & Young 1982). This is a measure of the alteration of feldspars to clay minerals. For the nine carbonate-free samples from the Eastern Andean Metamorphic Complex, the values range from 60 to 72. Calculating with a carbonate phase content of 100% calcite gives maximum CIA values of 56 to 72 for all 23 samples (Table 1 & Fig. 2). These values indicate only a moderate degree of alteration. The K_2O content of up to 5% can be explained by K-metasomatism (Fedo *et al.* 1995). McLennan *et al.* (1990) used the K/Cs ratio as weathering indicator. Although both elements tend to be adsorbed on clay minerals during weathering, the K/Cs ratio decreases with increasing weathering. The narrow range for the ratios between upper continental crust (McLennan 2001) and post-Archean average Australian shale (Taylor & McLennan 1985) are in agreement with the medium high CIA values and a K-metasomatic effect (Fig. 2).

Provenance indicative elements such as Ti, Nb, Th, La, Co and Sc have values close to upper continental crust and post-Archaean average Australian Shale (Table 1 & Fig. 4). Trace-element concentrations of the pelites are higher, but with similar patterns. This points to a quartz dilution effect in sandstones. Element ratios of the incompatible elements La and Th v. the compatible elements Sc and Co have values slightly above those for the upper continental crust (Table 2 & Fig. 3).

Only Zr, Hf and Tl of the trace elements have higher concentrations in the greywackes than in the pelites (Table 1). The Zr and Hf concentrations show large scatter, but most greywacke samples have concentrations higher than the values for the upper continental crust at 190 and 5.8 respectively (Table 1 & Fig. 4). Zr and Hf concentrations are indicators of the maturity of the sediments, since zircons, the most important mineral for Zr and Hf, are sensitive neither to chemical nor to mechanical weathering and tend to concentrate in sandstones during sorting processes (e.g. Morton & Hallsworth 1999). The Zr/Sc and Hf/Sc ratios are also values of sorting and recycling (e.g. McLennan *et al.* 1993). All of the analysed Eastern Andean Metamorphic Complex greywackes have concentrations above the values for the upper continental crust at 14.0 and 0.43 (Table 2). The two Lago General Carrera samples differ from the others by having higher Zr and Hf concentrations in the pelite sample than in the greywacke sample (Table 1). The overall high values indicate that the studied sediments are relatively mature. Due to the immobility of Zr and Hf, it can be expected that the sediments were already mature at the time of deposition.

The rare earth elements (REEs) La to Lu have patterns similar to post-Archaean average Australian shale and upper continental crust (Fig. 5) and the La_N/Yb_N values are only slightly lower than those for the upper continental crust at 9.2 (Table 2). The REE Ce can occur as Ce^{4+}, and Eu as Eu^{2+}, as well as the normal REE 3+ state. Eu^{2+} readily replaces Ca^{2+} in plagioclase, thus, the Eu anomaly (Eu/Eu*) reflects the plagioclase fractionation. Felsic rocks and sediments usually have negative anomalies due to lithospheric or intracrustal feldspar fractionation or breakdown of feldspars during weathering processes (Condie *et al.* 1995). According to the study of McLennan *et al.* (1990), active-margin sediments, in contrast to passive margin sediments, often show lower Eu/Eu* values for shales than for greywackes. The studied rocks have negative anomalies with slightly lower values for pelites than for greywackes (Table 2 & Fig. 5). The Ce anomalies (Ce/Ce*) can indicate REE redistribution during weathering, resulting in possible fractionation also for Sm and Nd isotopes (McDaniel *et al.* 1994). The Ce/Ce* ratios are close to 1, and the small difference in Ce/Ce* for the studied rocks is within the uncertainties of the measurements. Thus no anomalous Ce/Ce* can be deduced (Table 2 &, Fig. 5).

Provenance. Several attempts have been made to use major elements as provenance indicators (e.g. Roser & Korsch 1986, 1988; Bhatia 1983), although the major elements are prone to changes in concentrations and ratios during weathering, sorting, diagenesis and metamorphism. However, the discrimination procedure of Roser & Korsch (1988), which takes into account

Table 1. *Major and trace elements.*

	CA-00 -03-S*	CA-00 -04-S	CA-00 -09-S	CA-00 -10-S	CA-00 -12-S	CA-00 -13-S	CA-00 -15-S	CA-00 -17-S	CA-00 -21-S	CA-00 -22-S	CA-00 -23-S	CA-00 -26-S
SiO_2	73.41	73.92	79.92	74.28	75.38	76.31	73.00	74.63	80.96	73.24	81.48	75.42
Al_2O_3	12.60	12.66	9.90	12.35	11.96	11.31	12.19	12.06	9.56	12.35	8.45	12.02
Fe_2O_3	3.36	3.71	2.56	4.08	3.28	2.90	3.49	3.19	2.22	4.28	2.59	3.62
MgO	1.26	1.28	0.94	1.53	1.10	0.89	1.52	1.14	0.64	1.44	0.90	1.28
CaO	0.90	0.17	0.46	0.91	0.72	0.59	1.53	0.75	0.16	0.38	0.54	0.29
Na_2O	3.78	3.29	1.72	0.77	2.49	2.91	3.35	3.49	2.85	1.97	1.84	2.57
K_2O	1.99	1.84	2.27	3.30	2.35	2.07	1.59	1.68	1.18	2.76	1.60	1.83
TiO_2	0.57	0.68	0.52	0.68	0.63	0.52	0.59	0.63	0.49	0.79	0.48	0.65
P_2O_5	0.07	0.12	0.06	0.12	0.12	0.10	0.06	0.08	0.09	0.11	0.09	0.10
MnO	0.04	0.05	0.03	0.04	0.05	0.05	0.06	0.05	0.03	0.04	0.04	0.04
Cr_2O_3	0.009	0.003	0.007	0.007	0.008	0.005	0.011	0.005	0.008	0.008	0.008	0.008
C	0.08	0.12	0.07	0.13	0.11	0.16	0.26	0.27	0.05	0.23	0.18	0.15
S	<0.01	<0.01	0.02	0.06	0.02	0.02	0.06	0.12	0.02	0.07	0.04	0.03
LOI	1.9	2.2	1.5	1.8	1.8	2.2	2.4	2.2	1.6	2.5	1.9	2.1
Sum	99.93	99.97	99.93	99.94	99.96	99.92	99.83	99.94	99.82	99.93	99.95	99.97
C†	0.03	<0.01	<0.01	<0.01	<0.01	0.05	0.20	0.08	<0.01	0.03	0.07	0.02
CaO‡	0.14	–	–	–	–	0.23	0.93	0.37	–	0.14	0.33	0.09
CIA§	56.4	62.1	61.8	65.5	60.1	59.6	59.4	59.4	60.4	65.0	62.1	64.7
Ba	344	369	411	590	601	548	350	329	286	548	316	336
Sc	10	10	7	11	8	7	11	10	6	11	6	10
Co	7.9	6.0	5.3	8.1	4.6	7.2	8.7	7.3	3.9	11.2	5.6	9.0
Cs	4.6	3.6	5.4	6.3	4.1	3.9	3.6	4.2	2.8	7.9	3.8	5.0
Ga	14.6	14.3	10.6	15.2	14.3	13.5	13.7	13.8	9.6	15.7	9.6	14.9
Hf	5.9	5.9	8.4	5.9	8.4	6.3	5.4	8.1	8.6	6.2	8.4	7.9
Nb	7.4	10.2	7.6	10.5	11.4	10.1	8.6	9.3	6.4	13.1	6.8	10.7
Rb	90.6	92.0	102.3	136.4	104.4	91.0	70.5	90.0	56.5	123.7	70.3	99.2
Sr	162.1	38.8	52.1	93.5	131.0	119.8	136.8	88.1	67.7	60.3	70.2	47.8
Ta	0.9	1.2	0.9	1.2	1.2	1.1	1.1	1.2	0.8	1.4	0.9	1.2
Th	7.8	9.9	7.6	9.2	12.1	10.9	6.4	8.7	8.8	11.3	7.3	9.8
Tl	1.4	1.2	1.1	1.2	1.1	0.8	0.9	0.7	1.1	0.9	0.7	0.7
U	2.2	2.8	2.5	2.4	3.3	3.0	1.9	2.3	2.4	3.0	2.0	2.8
V	60	74	50	74	47	39	59	61	35	74	42	75
Zr	187.1	188.5	278.1	191.2	272.6	211.8	178.7	264.7	278.3	197.9	275.1	249.4
Y	23.2	21.8	22.8	25.1	31.5	26.8	22.6	24.1	22.9	30.9	20.5	27.1
La	26.1	27.6	29.3	30.9	37.2	32.0	25.8	30.1	29.8	33.2	27.1	32.9
Ce	54.2	58.8	59.3	63.7	77.2	66.5	51.4	61.4	60.1	69.8	55.2	67.3
Pr	6.70	6.96	7.19	7.49	8.96	7.82	6.37	7.34	7.06	8.28	6.51	8.14
Nd	26.6	27.8	28.2	30.2	35.1	31.1	26.0	28.4	27.8	32.6	25.2	32.0
Sm	5.0	5.2	5.1	6.0	6.7	5.8	5.0	5.5	5.0	6.4	4.6	5.9
Eu	1.43	1.13	1.17	1.22	1.56	1.20	1.39	1.27	0.98	1.34	1.03	1.31
Gd	4.74	4.71	4.52	5.01	6.23	5.42	4.47	4.94	4.45	5.86	4.32	5.20
Tb	0.63	0.64	0.66	0.70	0.82	0.73	0.65	0.69	0.66	0.88	0.61	0.76
Dy	4.47	4.25	4.31	4.89	5.87	5.03	4.13	4.55	4.33	5.88	4.16	5.03
Ho	0.92	0.82	0.87	0.96	1.16	1.00	0.87	0.91	0.81	1.18	0.78	1.07
Er	2.50	2.26	2.44	2.69	3.40	2.83	2.33	2.74	2.46	3.33	2.33	2.90
Tm	0.37	0.33	0.35	0.39	0.47	0.41	0.37	0.39	0.35	0.47	0.33	0.45
Yb	2.63	2.19	2.56	2.74	3.13	2.78	2.37	2.73	2.40	3.28	2.36	2.94
Lu	0.38	0.29	0.36	0.36	0.43	0.36	0.35	0.37	0.32	0.48	0.33	0.40
Cu	5	9	6	<1	3	8	3	4	5	18	7	15
Pb	15	11	15	4	13	19	14	12	7	16	13	12
Zn	50	50	36	66	31	49	45	39	17	69	28	51
Ni	10	13	13	22	10	11	17	12	10	21	12	17
As	5	2	7	8	2	3	2	5	<2	7	4	3

Oxides, C and S in %, other elements in ppm. S and M denotes greywackes and pelites respectively.
*Lago General Carrera unit, other samples are from the Cochrane unit.
†C in carbonates.
‡Maximum CaO in carbonates calculated from the measured C in carbonates.
§CIA=$Al_2O_3/(Al_2O_3+Ca_{silicate}+Na_2O+K_2O)*100$ calculated from the molecular ratios.

Table 1. *Continued.*

	CA-00 -28-S	CA-00 -30-S	CA-00 -32-S	CA-00 -33-S	CA-00 -37-S	CA-00 -42-S	CA-00 -02-M*	CA-00 -14-M	CA-00 -19-M	CA-00 -25-M	CA-00 -27-M
SiO_2	80.36	79.68	79.83	74.94	74.28	71.61	67.89	64.96	68.73	53.61	43.33
Al_2O_3	9.48	9.60	7.86	11.00	12.46	9.43	15.13	16.38	14.40	22.28	12.51
Fe_2O_3	2.58	3.08	2.32	3.94	2.03	2.62	6.18	6.16	5.76	8.47	6.20
MgO	0.85	1.09	0.74	1.51	0.90	1.14	1.76	2.11	1.89	2.70	2.04
CaO	0.36	0.21	2.27	0.78	1.67	5.24	0.19	0.24	0.49	0.25	15.11
Na_2O	2.12	2.19	2.13	2.64	4.06	2.49	2.39	2.03	2.06	1.91	2.35
K_2O	1.58	1.38	1.24	1.66	1.35	1.35	2.44	4.02	2.67	4.84	1.79
TiO_2	0.57	0.63	0.50	0.65	0.36	0.41	0.77	0.82	0.81	1.02	0.64
P_2O_5	0.07	0.11	0.13	0.14	0.08	0.04	0.14	0.13	0.14	0.14	0.38
MnO	0.03	0.03	0.05	0.05	0.05	0.08	0.05	0.10	0.04	0.11	0.40
Cr_2O_3	0.009	0.010	0.009	0.009	0.002	0.006	0.008	0.010	0.010	0.019	0.007
C	0.14	0.08	0.53	0.21	0.39	1.17	0.24	0.29	0.36	0.44	3.74
S	0.06	0.08	<0.01	0.04	0.04	<0.01	<0.01	<0.01	0.10	0.55	0.03
LOI	1.9	1.9	2.7	2.6	2.7	5.5	2.8	2.9	2.9	4.3	14.6
Sum	99.94	99.94	99.81	99.96	99.97	99.95	99.81	99.98	99.96	99.75	99.40
C†	0.04	<0.01	0.43	0.11	0.28	1.05	<0.01	<0.01	0.05	<0.01	3.36
CaO‡	0.19	–	2.01	0.51	1.31	4.90	–	–	0.23	–	15.11
CIA§	63.2	63.7	59.6	62.4	58.6	60.4	68.6	66.8	68.1	71.6	68.3
Ba	284	254	215	334	244	285	502	1058	475	852	329
Sc	7	8	6	9	6	6	13	16	13	22	12
Co	5.4	7.1	4.8	8.0	3.7	4.9	11.3	14.2	17.3	17.7	12.9
Cs	3.1	3.4	2.7	4.4	2.5	3.2	6.9	8.8	7.2	14.6	4.3
Ga	10.6	11.0	7.6	12.8	14.4	10.3	18.7	25.3	18.4	29.9	16.0
Hf	8.9	8.9	10.3	6.0	5.0	5.0	6.9	4.8	6.2	4.3	3.4
Nb	8.0	9.3	6.2	9.4	7.0	6.2	12.2	15.1	13.6	17.7	11.1
Rb	75.4	72.3	48.8	74.3	66.9	66.3	121.6	186.0	135.9	218.0	89.3
Sr	73.5	55.4	122.7	93.6	142.4	272.4	73.3	72.1	70.1	37.7	575.2
Ta	0.9	1.1	0.7	1.1	0.9	0.7	1.1	1.4	1.3	1.7	1.1
Th	6.9	8.2	7.8	7.3	11.1	5.7	14.6	18.3	12.8	18.4	11.0
Tl	0.6	0.7	0.1	0.2	0.1	0.1	0.3	0.7	0.8	0.4	0.4
U	2.1	2.3	2.0	2.0	2.2	1.6	3.3	5.4	3.4	3.9	2.5
V	45	53	33	62	36	43	105	119	117	195	95
Zr	284.9	302.9	349.3	190.8	158.2	160.8	268.8	169.3	212.9	139.9	118.2
Y	22.4	22.6	19.7	26.4	19.9	18.6	28.4	35.9	31.9	34.8	29.4
La	27.6	29.9	27.9	29.1	34.2	22.5	26.8	44.3	37.2	50.3	33.3
Ce	55.3	59.4	56.2	59.7	66.9	43.8	62.4	96.1	80.1	104.2	68.5
Pr	6.67	7.01	6.28	7.26	7.82	5.13	6.89	11.30	9.23	12.15	7.77
Nd	25.7	27.3	25.3	29.1	28.8	19.9	26.7	44.5	37.5	47.0	31.3
Sm	5.1	5.1	4.5	5.6	4.9	4.0	5.5	8.8	7.4	9.8	6.2
Eu	1.15	1.18	1.02	1.42	1.02	1.02	1.12	1.53	1.47	1.69	1.25
Gd	4.71	4.34	3.77	5.43	4.00	3.45	5.11	7.16	6.38	6.65	5.61
Tb	0.59	0.65	0.54	0.80	0.56	0.48	0.78	1.05	0.93	1.08	0.75
Dy	4.19	4.14	3.68	4.88	3.58	3.18	5.14	6.62	6.01	6.74	4.66
Ho	0.82	0.84	0.73	1.01	0.73	0.66	1.17	1.39	1.33	1.35	1.08
Er	2.51	2.51	2.13	2.77	1.94	2.07	3.46	4.22	3.75	4.00	3.12
Tm	0.36	0.33	0.32	0.40	0.28	0.28	0.45	0.53	0.44	0.58	0.36
Yb	2.45	2.65	2.16	2.81	2.02	1.98	3.37	3.91	3.51	3.70	2.95
Lu	0.33	0.37	0.31	0.38	0.27	0.30	0.49	0.58	0.50	0.59	0.44
Cu	9	10	4	12	1	1	18	20	13	36	20
Pb	12	12	14	13	6	13	7	23	6	28	14
Zn	26	27	19	38	13	27	62	73	45	134	69
Ni	10	14	7	14	5	9	30	28	32	34	21
As	4	3	3	2	2	<2	6	5	11	19	< 2

Oxides, C and S in %, other elements in ppm. S and M denotes greywackes and pelites respectively.
*Lago General Carrera unit, other samples are from the Cochrane unit.
†C in carbonates.
‡Maximum CaO in carbonates calculated from the measured C in carbonates.
§CIA = $Al_2O_3/(Al_2O_3 + Ca_{silicate} + Na_2O + K_2O)*100$ calculated from the molecular ratios.

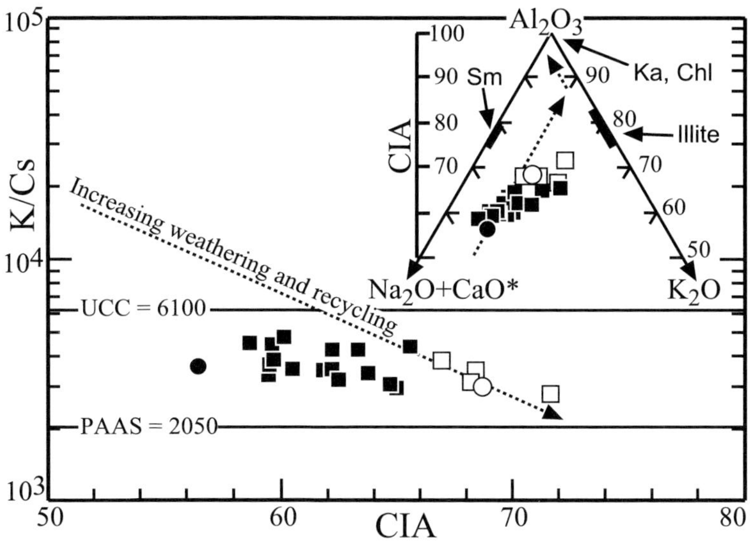

Fig. 2. Weathering condition of the Chilean metaturbidites, with oxides in molecular proportions. CaO* is CaO in silicates. The equation $CaO^* = CaO_{total} - CaO_{max.\ in\ carbonates}$ gives maximum CIA values. The arrow in the inset indicates ideal weathering from an approximate granodioritic composition. The observed trend towards illite composition is in concordance with a K-metasomatic effect. Squares, Cochrane unit; circles, Lago General Carrera unit; filled symbols, greywackes; unfilled symbols, pelites; Chl, chlorite; Ka, kaolinite; Sm, smectite. Main diagram after McLennan *et al.* (1990); inset after Fedo *et al.* (1995).

Table 2. *Selected element ratios.*

	K/Cs	Zr/Sc	Hf/Sc	Th/Sc	Th/Co	La/Sc	Ti/Nb	La_N/Yb_N	Eu/Eu*	Ce/Ce*
CA-00-03-S*	3591	15.8	0.59	0.78	0.99	2.61	462	6.71	0.90	0.96
CA-00-04-S	4243	16.1	0.59	0.99	1.65	2.76	400	8.52	0.70	0.99
CA-00-09-S	3490	25.5	1.20	1.09	1.43	4.19	410	7.73	0.74	0.96
CA-00-10-S	4348	17.0	0.54	0.84	1.14	2.81	388	7.62	0.68	0.98
CA-00-12-S	4758	23.6	1.05	1.51	2.63	4.65	331	8.03	0.74	0.99
CA-00-13-S	4406	27.3	0.90	1.56	1.51	4.57	309	7.78	0.65	0.98
CA-00-15-S	3666	17.4	0.49	0.58	0.74	2.35	411	7.36	0.90	0.94
CA-00-17-S	3321	19.8	0.81	0.87	1.19	3.01	406	7.45	0.74	0.97
CA-00-21-S	3498	35.3	1.43	1.47	2.26	4.97	459	8.39	0.63	0.97
CA-00-22-S	2900	22.7	0.56	1.03	1.01	3.02	362	6.84	0.67	0.99
CA-00-23-S	3495	44.1	1.40	1.22	1.30	4.52	423	7.76	0.71	0.97
CA-00-26-S	3038	27.3	0.79	0.98	1.09	3.29	364	7.56	0.72	0.96
CA-00-28-S	4231	39.3	1.27	0.99	1.28	3.94	427	7.61	0.72	0.95
CA-00-30-S	3369	34.8	1.11	1.03	1.15	3.74	406	7.62	0.77	0.96
CA-00-32-S	3813	46.4	1.72	1.30	1.63	4.65	483	8.73	0.76	0.99
CA-00-33-S	3132	31.7	0.67	0.81	0.91	3.23	415	7.00	0.79	0.96
CA-00-37-S	4483	50.5	0.83	1.85	3.00	5.70	308	11.44	0.70	0.96
CA-00-42-S	3502	58.2	0.83	0.95	1.16	3.75	396	7.68	0.84	0.96
CA-00-02-M*	2936	20.7	0.53	1.12	1.29	2.06	378	5.37	0.65	1.08
CA-00-14-M	3792	10.6	0.30	1.14	1.29	2.77	326	7.66	0.59	1.01
CA-00-19-M	3078	16.4	0.48	0.98	0.74	2.86	357	7.16	0.65	1.01
CA-00-25-M	2752	6.4	0.20	0.84	1.04	2.29	345	9.19	0.64	0.99
CA-00-27-M	3456	9.9	0.28	0.92	0.85	2.78	346	7.63	0.65	1.00

S and M denotes greywackes and pelites respectively.
$Eu/Eu^* = Eu_N/(Sm_N{*}Gd_N)^{1/2}$ and $Ce/Ce^* = Ce_N/(La_N{*}Pr_N)^{1/2}$.
*Lago General Carrera unit; other samples are from the Cochrane unit.

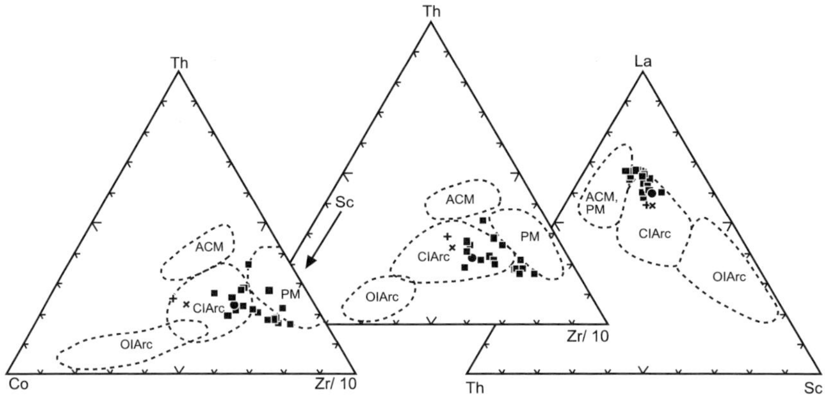

Fig. 3. Trace element diagrams with discrimination fields for greywackes after Bhatia & Crook (1986). ACM, active continental margin; CIArc, continental island arc; OIArc, oceanic island arc; PM, passive margin; squares, Cochrane unit; circles, Lago General Carrera unit; cross, upper continental crust (McLennan 2001); plus sign, post-Archaean average Australian shale (Taylor & McLennan 1985).

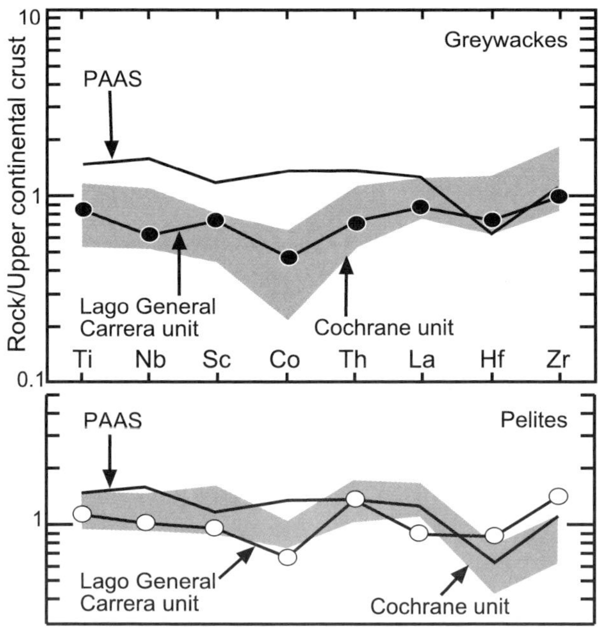

Fig. 4. Selected element ratios normalized to the upper continental crust (McLennan, 2001). PAAS=post-Archaean average Australian shale.

Al, Fe, Mg, Ti, K, Na and $Ca_{silicate}$, has been demonstrated to reliably indicate source-rock types for sediments, irrespective of grain size. The greywackes from the Cochrane unit are indicated to have felsic source rocks, concentrating across the dividing-line between the fields for recycled source rocks and primary felsic source rocks (Fig. 6). In the latter field, only the greywacke sample from the Lago General Carrera unit is positioned. The pelites have a larger scatter, due to some geochemical differences between the different grain sizes, and possibly as a result of a larger sampling area for finer grained sediments.

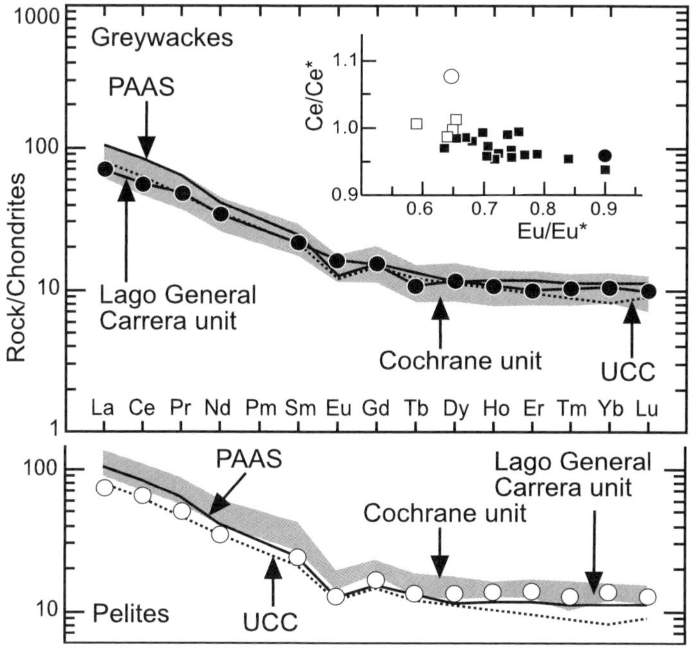

Fig. 5. REE pattern and Eu/Eu* v. Ce/Ce*. UCC, upper continental crust; PAAS, post-Archaean average Australian shale; squares, Cochrane unit; circles, Lago General Carrera unit; filled symbols, greywackes; unfilled symbols, pelites.

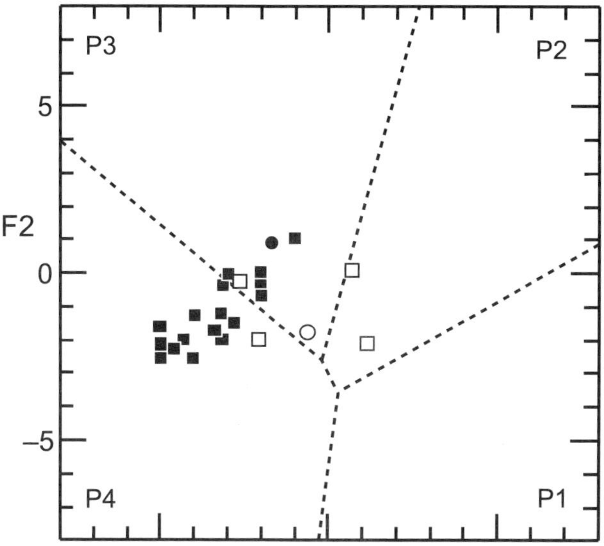

Fig. 6. Discrimination scheme for major elements. P1, primary mafic sources; P2, primary intermediate mafic–felsic sources; P3, primary felsic sources; P4, recycled sources; squares, Cochrane unit; circles, Lago General Carrera unit; filled symbols, greywackes; unfilled symbols, pelites. After Roser & Korsch (1988).

Bhatia & Crook (1986) showed that comparisons of the incompatible elements Th and La with the compatible elements Co and Sc can be powerful discriminators in provenance studies for greywackes, provided that mantle fractionation processes dominate the elemental signals. This assumes that all four elements are relatively immobile during weathering and diagenesis. The small range in La/Sc, Th/Sc and Th/Co values (Table 2) indicate that the samples have a similar history. The high values point to a continental, felsic origin for the source rocks. In the ternary diagrams of Bhatia & Crook (1986; Fig. 3 of this paper), the mentioned ratios, together with Zr, give indications for a continental island-arc setting for the source rocks of the greywacke samples, although the high Zr concentrations would make it possible to interpret them as passive-margin sediments. Continental island arc environments are classified by Bhatia & Crook (1986) as island arcs at convergent plate margins resting on a well developed, or thin, continental crust. Passive margins include extentional regimes.

In addition, the Ti/Zr ratio has been used by Bhatia & Crook (1986) as a provenance proxy for greywackes, although Zr is highly dependent on the zircon content. Titanium is usually moderately incompatible and concentrated in minerals typical of mafic rocks (Van Baalen 1993). It is immobile during weathering and diagenesis, but large concentrations of rutile, titanite and ilmenite in sediments will affect the concentration (Van Baalen 1993) and can disturb the provenance results. The analysed samples from the Eastern Andean Metamorphic Complex have no anomalous rutile, titanite or ilmenite concentrations. The Ti/Zr ratios are too high to indicate a passive margin setting for the sources, whereas La/Sc in general has too low values (Fig. 7). They rather indicate derivation from an active continental-margin setting, described by Bhatia & Crook (1986) as a continental margin of Andean type or a strike-slip continental margin.

Moreover, the Ti/Nb ratio can give some indications of the provenance for sedimentary rocks (e.g. Bonjour & Dabard 1991; Jenchen 2001; see also the oposite view of, for example, Condie *et al.* 1995). As well as Ti, highly incompatible Nb is considered immobile and often appears in the same heavy minerals as Ti (Bonjour & Dabard 1991). The Ti/Nb tends to decrease with increasing felsic components and with textural maturity, due to the higher incompatibility of Nb. The narrow Ti/Nb range of 300–500 for the Eastern Andean Metamorphic Complex metaturbidites is low enough to indicate a dominant origin from felsic source

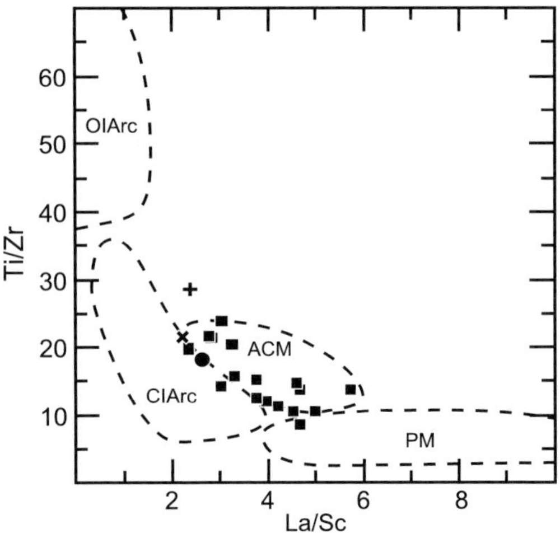

Fig. 7. La/Sc v. Ti/Zr, with discriminating fields for greywackes, after Bhatia & Crook (1986). Abbreviations and symbols as for Fig. 3.

rocks, and also emphasizes the chemical homogeneity across the sampling area (cf. Jenchen 2001).

The most commonly used provenance discrimination schemes for trace elements (Bhatia & Crook 1986; Floyd & Leveridge 1987; McLennan *et al.* 1993) discriminate felsic active margin source rocks from passive-margin rocks, based on the Zr or Hf concentrations in the sediments, since the zircon content increases with increasing maturity of the sediments and with recycling. Sand-sized sediments contain more zircons, i.e. more Zr and Hf, than finer grained sediments. They usually have higher incompatible v. compatible element ratios than accompanying mudstones, which often tend to reflect larger sampling areas. Hence, care must be employed when comparing sediment samples of different grain sizes. Furthermore, erosional debris can be transported across tectonic margins with their inherited petrographical and chemical signatures still preserved (McLennan *et al.* 1990; Potter 1994). Thus, provenance analysis based on the geochemistry of sediments often needs to be combined with other analytical techniques, e.g. petrographical and isotopic, in order to be convincing.

Nd-isotopes

To obtain age information for crust formation, Sm–Nd model age calculations assuming a depleted mantle source is preferred (DePaolo 1981). Taking into consideration the $^{147}Sm/^{144}Nd$

evolution of the crust, Frost & Winston (1987) and DePaolo *et al.* (1991) suggested a revision of the model age calculation for those cases where the isotopes were disturbed during a post-crustal formation stage. Model ages have for comparison been calculated both in the conventional way (T_{DM}), and with the considerations made by Frost & Winston (1987) and DePaolo *et al.* (1991; T_{DM}^*). The depositional age has been approximated to 350 Ma for the calculations, since this age overlaps both the Upper Devonian to Lower Carboniferous fossils of the Bahía de la Lancha Formation in Argentina (Riccardi 1971) and the maximum depositional time-span of 364–250 Ma for the southern Cochrane unit in Chile (Thomson *et al.* 2000). The T_{DM}^* ages for the Cochrane unit are in the range 1340–1490 Ma, which is in general 50–100 Ma younger than corresponding T_{DM} ages (Table 3). The only greywacke sample from the Lago General Carrera unit differs slightly with $T_{DM}^*=1177$ Ma and $T_{DM}=1320$ Ma. With an overall T_{DM}^* range of 1170–1490 Ma, the oldest crustal component in the sediments must be of an even older age, if the metasediments are not derived from one single source. Model ages for sediments are supposed to indicate the mean crustal formation age of the source components, but the difference in T_{DM} and T_{DM}^* ages indicates that minor post-crustal formation fractionation of Sm and Nd may have occurred, probably at a near syndepositional stage. Since the history of the source rocks is not known, pre- and post-depositional fractionations are also plausible.

The $\varepsilon_{Nd}(T)$, a value of the deviation of the $^{143}Nd/^{144}Nd$ ratio in a measured sample from CHUR calculated at the depositional age, can provide information about the nature of the source rocks. Sediments with their isotopic signature dominated by components from felsic, old crustal rocks tend to have negative values, while components from juvenile mafic rocks increase the values. For the analysed samples, $\varepsilon_{Nd}(T)$ is negative (–6.6 to –4.6 for the Cochrane unit and –2.3 for the Lago General Carrera unit; Table 3). These values are in agreement with the geochemical felsic signal of the samples. More negative $\varepsilon_{Nd}(T)$ values can be expected for sediments that derive from an old continental crust, but the obtained values are in a range possible for both passive and active margin turbidite greywackes (McLennan & Hemming 1992). This can be explained by sediments composed of a mixture of both mafic active-margin source rocks and older felsic continental source rocks. McLennan & Hemming (1992) noticed that old continental crust can be an important component even for sediments in active volcanic settings. The diagram of $\varepsilon_{Nd}(T)$ v. time in Fig. 8 shows the relationship of a possible influence of both old continental crust and more juvenile material in the isotopic signature of the Eastern Andean Metamorphic Complex greywackes.

The deviation of $^{147}Sm/^{144}Nd$ from CHUR is indicated by the $f^{Sm/Nd}$ values, which for the analysed metaturbidites range from –0.420 to –0.356 (Table 3). Ideally, no difference in $f^{Sm/Nd}$

Table 3. *Samarium-neodymium data*

	Sm/Nd	$^{147}Sm/^{144}Nd$	$f^{Sm/Nd}$	$^{143}Nd/^{144}Nd$	$\varepsilon_{Nd}(T)$	T_{DM} (Ma)	T_{DM}^* (Ma)
CA-00-03-S*	0.19	0.12583	–0.360	0.512360	–2.3	1320	1177
CA-00-10-S	0.20	0.11799	–0.400	0.512173	–5.6	1502	1416
CA-00-13-S	0.19	0.12055	–0.387	0.512215	–4.8	1475	1364
CA-00-15-S	0.19	0.12665	–0.356	0.512165	–6.1	1659	1456
CA-00-17-S	0.19	0.12004	–0.390	0.512225	–4.6	1451	1348
CA-00-21-S	0.18	0.11410	–0.420	0.512170	–5.4	1450	1408
CA-00-23-S	0.18	0.11595	–0.411	0.512143	–6.1	1517	1452
CA-00-28-S	0.20	0.11600	–0.410	0.512140	–6.1	1521	1456
CA-00-30-S	0.19	0.11514	–0.415	0.512115	–6.6	1545	1488
CA-00-32-S	0.18	0.11706	–0.405	0.512132	–6.3	1550	1471
CA-00-33-S	0.19	0.12382	–0.371	0.512205	–5.2	1543	1390

Maximum 2σ error for $^{143}Nd/^{144}Nd$ is 0.000017, which corresponds with $2\sigma\leq0.4$ for $\varepsilon_{Nd}(T)$.
$T_{DM}=\ln[(^{143}Nd/^{144}Nd_{sample,\ today}-{}^{143}Nd/^{144}Nd_{DM,\ today})/(^{147}Sm/^{144}Nd_{sample,\ today}-{}^{147}Sm/^{144}Nd_{DM,\ today})+1]/\lambda$ and $T_{DM}^*=\ln[(^{143}Nd/^{144}Nd_{sample,\ T_s}-{}^{143}Nd/^{144}Nd_{DM,\ T_s})/(^{147}Sm/^{144}Nd_{crust,\ T_s}-{}^{147}Sm/^{144}Nd_{DM,\ T_s})+1]/\lambda+T_s$, where T_s is the stratigraphic age, here 350 Ma. $^{147}Sm/^{144}Nd_{crust,\ T_s}$ was calculated assuming $^{147}Sm/^{144}Nd_{crust,\ today}=0.11$ (Albarède & Brouxel 1987).
*Lago General Carrera unit, other samples are from the Cochrane unit.

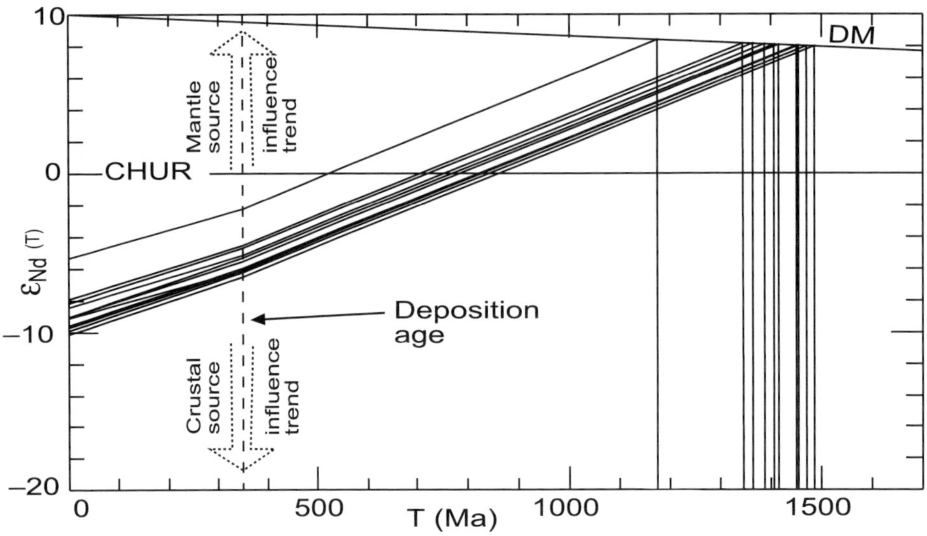

Fig. 8. $\varepsilon_{Nd}(T)$ as an indicator of source rock type v. time. The lines indicate developement for individual samples through time. Arrows show the effect of syndepositional mixing of old crustal and juvenile mantle rocks. A figure of 350 Ma has been used as the depositional age in the calculations. The small bend at 350 Ma for individual samples indicates the possibility of minor differentiation at a near syndepositional stage.

should occur between rocks of the same origin and history. A linear relationship of $f^{Sm/Nd}$ with corresponding model ages is an indication of REE fractionation after the crust formation age at a pre-, syn- or post-depositional stage (Zhao *et al.* 1992). No correlation between $f^{Sm/Nd}$ and T_{DM}^* (or T_{DM}) has been observed for the samples, thus a post-crustal formation fractionation is not indicated by this relationship.

McLennan *et al.* (1989) noticed that Nd isotope compositions can vary with grain size due to concentration of components from different sources during transport and sorting. This can result in differences in isotopic signatures for sediments of different grain size, reflecting differences in the concentration of mafic and felsic source components. The samples analysed for its Sm and Nd isotopes have generally REE patterns parallel to the pelites, and similar Sm/Nd ratios, which is interpreted to derive from the dilution of quartz in the greywackes. Furthermore, the Sm/Nd range is narrow (Table 3). Hence, differences also in Sm–Nd isotope systematics compared with pelites are assumed to be small.

Tectonic setting

The analysed sediments are petrographically mature, which is supported by the high Zr and Hf ratios. In concordance, major and minor elements indicate mainly felsic sources. The provenance-indicative element ratios used have not convincingly succeeded in discriminating an active margin from a passive margin origin for the Eastern Andean Metamorphic Complex. The ratios used give principal indications of an active margin origin, but ratios for individual samples are close to those typical of passive-margin settings. The isotopic signatures give indications of a dominance of old continental felsic source rocks, but with a possible minor contribution from younger mantle derived mafic rocks. An active-margin setting seems plausible.

The Lago General Carrera unit greywacke sample differs slightly from the samples of the Cochrane unit. The CIA value is the lowest, the Eu anomaly is among the highest, and the isotopic signature indicates slightly younger source rocks, than for the Cochrane unit, with a larger part of the signal coming from mantle-derived mafic sources. Due to the absence of further greywacke analyses from this unit, this difference may be purely accidental.

A plate-tectonic model with eastward subduction under Patagonia (present-day co-ordinates) was presented by Forsythe (1982) for the Late Palaeozoic to Early Mesozoic mainly on the basis of Rb–Sr isotope data from Northern Patagonia. Chilean Andean Patagonia and western Argentinian Patagonia was suggested to be part of a fore-arc province. Time constraints indicate that the subduction underneath southern Patagonia started at the latest in the Late Carboniferous (Hervé 1988; Pankhurst *et al.* 1992; Hervé *et al.* 2000). Before this time, there is no evidence of subduction in this region. The oldest basement rocks of the southernmost Andean basement, situated in the area of Chiloé (*c.*42°S, Fig. 1a), have been displaced northwards by the Liquiñe–Ofqui Fault Zone (García *et al.* 1988). The rocks have been dated by fossils as Lower to Middle Devonian (Fortey *et al.* 1992). Based on faunal similarities and provincialism, Fortey *et al.* (1992) suggested that a marine platform extended from Chiloé, over southern Patagonia, to the Malvinas, or Falkland, Islands east of the South American continent. Furthermore, Fortey *et al.* (1992) suggested, based on the stable platform affinity of the fossil assemblage, that at least part of the turbidite sequence at Chiloé was deposited at a passive margin. The Eastern Andean Metamorphic Complex of supposed Late Devonian to Early Carboniferous age is slightly younger than the basement rocks in the area of Chiloé. In the northern part, limestones of the Lago General Carrera unit are interpreted to be deposited in a platform environment (Hasegawa *et al.* 1971, Hervé *et al.* 2000). South of the study area evidence exists of continental pillow basalts interbedded in the sedimentary pile of the Chilean Eastern Andean Metamorphic Complex (Hervé *et al.* 1999). Hervé *et al.* (1999) suggested a transform fault zone as the setting of the subaqueous volcanism. Thus, at the time of deposition, at least a thin continental crust was present in the area. A recently dated Early Cambrian granodiorite from a drillhole on Tierra del Fuego (Söllner *et al.* 2000) probably formed part of this continental crust.

As stated above, sedimentary debris can be transported across tectonic margins with their inherited signal preserved (McLennan *et al.* 1990; Potter 1994). Thus, in a narrow sense, the presented results are not proof of the tectonic setting of the depositional basin, but rather an indication of the tectonic setting of the source rocks. Two different options in the interpretation of the Eastern Andean Metamorphic Complex are possible.

(1) The sediments were deposited in a basin with sedimentary detritus derived from rocks situated in a nearby active margin, but with its main contribution coming from more mature sources. This would explain the Grenvillian Nd model ages as well as the geochemical active margin indications.
(2) An alternative is that the sediments were deposited in a sedimentary basin at a passive margin with their geochemical and isotopic signals inherited from source rocks situated at an active margin, from which the erosional debris have been transported across a tectonic margin. This would not only explain the Nd model ages and the geochemical active-margin indications, but also the regional geological features, including the absence of coeval magmatism and that the first indications of subduction are of Late Carboniferous age.

In this context, it is important to remember that geochemical provenance analysis does not give any time constraints. Input of active-margin type magmatic source rocks does not necessarily indicate magmatism coeval with the sedimentation. The source rocks may have been produced during older active margin stages, and later included in a tectonically inactive area. The sediments produced can then give geochemical signals inherited from an active tectonic margin, although no such margin existed at the time for sedimentary deposition. The Nd isotopic signature gives indications of mixing between old crustal rocks and younger mantle-derived rocks. However, this does not mean that the mantle rocks need to be coeval with deposition, since only relative time is indicated.

With the regional geology in mind, including the Early to Middle Devonian stable platform fauna at Chiloé, as well as stable platform limestone and continental pillow basalts incorporated into the sedimentary pile of the Eastern Andean Metamorphic Complex, an interpretation of the plate tectonic setting for the metasediments can be made. They are suggested to be passive-margin deposits, with parts of its geochemical and isotopic signal inherited from source rocks of active margin origin. The formation of the active margin sources must not have been coeval with the deposition of the turbidites. If the deposition took place at a passive margin, the sediments were incorporated post-depositionally into the backstop of the Late Palaeozoic to Early Mesozoic accretionary wedge at the margin of Gondwana (present-day coordinates).

The evolution of the Palaeozoic southern Patagonian Andean basement coincides with the Devonian to Early Carboniferous evolution of northern Chile and northwestern Argentina farther north along this margin. Here, the absence of deformation, magmatism and prograde metamorphism defines a period of tectonic quiescence ranging from the Early Silurian to the end of the Early Carboniferous (Bahlburg & Hervé 1997; Lucassen *et al.* 2000). Facies patterns of thick and conformable Devonian and Early Carboniferous shallow-marine siliciclastic successions are interpreted as deposited at a passive margin, although the geochemistry indicates an active-margin origin for the source rocks (Bahlburg & Hervé 1997; Bock *et al.* 2000).

Concerning isotope data, only a few Nd model ages for the Andean basement rocks south of 39°S have earlier been published (Hervé *et al.* 1990; Pankhurst *et al.*, 1992, 1994). They are all in the range 1100–1500 Ma (both T_{DM} and T_{DM}^*) and with $\varepsilon_{Nd}(T)$ values of around –6. The Nd model ages of Precambrian gneisses and tonalites in northern Argentinian Patagonia (at San Martín de los Andes, *c.*40°S, 70°W) and pre-Devonian schists east of the study area (Río Deseado Massif, *c.*48°S, 68°W) indicate crust formation ages of *c.*1200–1300 Ma (Dalla Salda *et al.* 1991; Pankhurst *et al.* 1994). This is close to the T_{DM}^* ages of the rocks in this study. The extra-Andean ages are interpreted to indicate crustal genesis during Grenvillian time. Extra-Andean Patagonia is a possible source area for the turbidite detritus of the Eastern Andean Metamorphic Complex. Due to the similarity in model ages, the metasediments are suggested to mainly derive from the same continental mass as the pre-Devonian schists and the plutonic rocks of extra-Andean Patagonia. An extra-Andean origin was also suggested by Pankhurst *et al.* (1994) for the siliciclastic basement rocks of southern Chile at *c.*39°S and by Dalla Salda *et al.* (1999) for the basement rocks in the Chilean archipelago of southern Patagonia.

Conclusions

The Eastern Andean Metamorphic Complex metaturbidites were mainly felsic, as indicated by the geochemical signal. The Nd isotopes have a dominant signal from old continental crustal rocks, although $\varepsilon_{Nd}(T)$ values indicate possible mixing of both felsic and mafic source rock components. Model ages further indicate that a large part of the sedimentary material was being derived from an old continental crust. The tectonic setting of the siliciclastic metasediments of the Eastern Andean Metamorphic Complex was not necessarily the same as the one of its source rocks. The rocks are suggested to have been deposited on continental crust at a passive-continental margin with detritus derived from rocks generated at an older active margin. The principal source area for the sediments might be crustal rocks of Grenvillian age exposed in present-day extra-Andean Patagonia. If corroborated by further reseach in progess, the results presented emphasize the problem of deducing the tectonic setting of a depositional basin from provenance data.

K. Mezger and U. Jenchen (Münster), are thanked for commenting on an earlier draft of this chapter. This study was supported by grants Ba 1011/17-1 and 17-2 of the German Research Foundation. It is a contribution to IGCP projects 436 'Tectonic Evolution of the Pacific Gondwana Margin – Structure, Assembly and Break-up Events' and 453 'Uniformitarianism Revisited: a Comparison Between Modern and Ancient Orogens'. The suggestions of referees J. Wijbrans and I. Valladares improved the chapter and are gratefully acknowledged.

References

ALBARÈDE, F. & BROUXEL, M. 1987. The Sm/Nd secular evolution of the continental crust and the depleted mantle. *Earth and Planetary Science Letters*, **82**, 25–35.

BAHLBURG, H. & HERVÉ, F. 1997. Paleozoic geodynamic evolution and tectonostratigraphic terranes of northwestern Argentina and northern Chile. *Geological Society of America Bulletin*, **109**, 869–884.

BELL, C. M. & SUÁREZ, M. 2000. The Río Lácteo Formation of Southern Chile. Late Paleozoic orogeny in the Andes of southernmost South America. *Journal of South American Earth Sciences*, **13**, 133–145.

BHATIA, M. R. 1983. Plate tectonics and geochemical composition of sandstones. *Journal of Geology*, **91**, 611–627.

BHATIA, M. R. & CROOK, K. A. W. 1986. Trace element characteristics of graywackes and tectonic setting discrimination of sedimentary basins. *Contributions to Mineralogy and Petrology*, **92**, 181–193.

BOCK, B., BAHLBURG, H., WÖRNER, G. & ZIMMERMANN, U. 2000. Tracing crustal evolution in the southern central Andes from Late Precambrian to Permian with geochemical and Nd and Pb isotope data. *Journal of Geology*, **108**, 515–535.

BONJOUR, J.-L. & DABARD, M.-P. 1991. Ti/Nb ratios of clastic terrigenous sediments used as an indicator of provenance. *Chemical Geology*, **91**, 257–267.

CAMINOS, R. & GONZÁLEZ, P. D. 1996. *Mapa Geológico de la República Argentina*, Dirección Nacional del Servicio Geológico.

CHERNIAK, D. J., HANCHAR, J. M. & WATSON, E. B. 1997. Rare-earth diffusion in zircon. *Chemical Geology*, **134**, 289–301.

CONDIE, K. C., DENGATE, J. & CULLERS, R. L. 1995. Behavior of rare earth elements in a paleoweathering profile on granodiorite in the Front Range, Colorado, USA. *Geochimica et Cosmochimica Acta*, **59**, 279–294.

DALLA SALDA, L., CINGOLANI, C. A. & VARELA, R. 1991. El basamento pre-andino igneo metamórfico de San Martín de los Andes, Neuquén. *Asociación Geológica Argentina Revista*, **46**, 223–234.

DALLA SALDA, L., VARELA, R., & CINGOLANI, C. 1999. El basamento pre-gondwanico del centro-oeste del Macizo Nordpatagónico. *In*: CAMINOS, R. (ed.)

Geología Argentina. Instituto de Geología y Recursos Minerales Anales, **29**, 107–112.

DEPAOLO, D. J. 1981. Neodymium isotopes in the Colorado Front Range and crust–mantle evolution in the Proterozoic. *Nature*, **291**, 193–196.

DEPAOLO, D. J., LINN, A. M. & SCHUBERT, G. 1991. The continental age distribution: methods of determining mantle separation ages from Sm–Nd isotopic data and application to the southwestern United States. *Journal of Geophysical Research*, **96**, 2071–2088.

ESCOBAR, F. (ed.) 1980. *Mapa Geológico de Chile*. Servicio Nacional de Geología y Minería, Santiago de Chile.

FANG, Z., BOUCOT, A., COVACEVICH, V. & HERVÉ, F. 1998. Discovery of Late Triassic fossils in the Chonos Metamorphic Complex, Southern Chile. *Revista Geología de Chile*, **25**, 165–174.

FEDO, C. M., NESBITT, H. W. & YOUNG, G. M. 1995. Unraveling the effects of potassium metasomatism in sedimentary rocks and paleosols, with implications for paleoweathering conditions and provenance. *Geology*, **23**, 921–924.

FLOYD, P. A. & LEVERIDGE, B. E. 1987. Tectonic environment of the Devonian Gramscatho basin, south Cornwall: framework mode and geochemical evidence from turbiditic sandstones. *Journal of the Geological Society of London*, **144**, 531–542.

FORSYTHE, R. 1982. The late Palaeozoic to early Mesozoic evolution of southern south America: a plate tectonic interpretation. *Journal of the Geological Society of London*, **139**, 671–682.

FORTEY, R., PANKHURST, R. J. & HERVÉ, F. 1992. Devonian trilobites at Buill, Chiloé (42°S). *Revista Geológica de Chile*, **19**, 133–143.

FROST, C. D. & WINSTON, D. 1987. Nd isotope systematics of coarse- and fine-grained sediments: examples from the Middle Proterozoic Belt–Purcell Supergroup. *Journal of Geology*, **95**, 309–327.

GALLOWAY, W. E. 1974. Deposition and diagenetic alteration of sandstone in northeast Pacific arc-related basins: implications for graywacke genesis. *Geological Society of America Bulletin*, **85**, 379–390.

GARCÍA, A., BECK, M. E., BURMEISTER, R. F., MUNIZAGA, F. & HERVÉ, F. 1988. Paleomagnetic reconnaissance of the Región de los Lagos, southern Chile, and its tectonic implications. *Revista Geológica de Chile*, **15**, 13–30.

GOLDSTEIN, S. L., O'NIONS, R. K. & HAMILTON, P. J. 1984. A Sm–Nd study of atmospheric dusts and particulates from major river systems. *Earth and Planetary Science Letters*, **70**, 221–236.

HASEGAWA, K., MATSUSHITA, K., NISHIMURA, T., HERVÉ, F. & AKAGI, Y. 1971. Geological note at Las Chivas District, Aysén Province, Chile. *Bulletin of the Geological Survey of Hokkaido*, **44**, 25–29.

HERVÉ, F. 1988. Late Paleozoic subduction and accretion in southern Chile. *Episodes*, **11**, 97–113.

HERVÉ, F. 1993. Paleozoic metamorphic complexes in the Andes of Aysén, Southern Chile (West of Occidentalia). *Proceedings of the First Circum-Pacific and Circum-Atlantic Terrane Conference, Guanajuato, Mexico*, 64–65.

HERVÉ, F., PANKHURST, R. J., BROOK, M., ALFARO, G. FRUTOS, J., MILLER, H., SCHIRA, W. & AMSTUTZ, G. *c*. 1990. Rb–Sr and Sm–Nd data from some massive sulfide occurrences in the metamorphic basement of south central Chile. *In*: FONTBOTÉ, L., AMSTUTZ, G. C., CARDOZO, M., CEDILLO, E. & FRUTOS, J. (eds) *Stratabound Ore Deposits in the Andes*. Springer-Verlag, Berlin, pp. 221–228.

HERVÉ, F., AGUIRRE, L., SEPÚLVEDA, V. & MORATA, D. 1999. Contrasting geochemistry and metamorphism of pillow basalts in metamorphic complexes from Aysén, S. Chile. *Journal of South American Earth Sciences*, **12**, 379–388.

HERVÉ, F., DEMANT, A., RAMOS, V. A., PANKHURST, R. J. & SUÁREZ, M. 2000. The southern Andes. *In*: CORDANI, U. G., MILANI, E. J., FILHO, A. T. & CAMPOS, D. A. (eds) *Tectonic Evolution of South America*. 31st International Geological Congress, Rio de Janeiro, 605–634.

JENCHEN, U. 2001. Fazies und Geochemie in kontinentalen Trias-Becken im westlichen Argentinien und in Patagonien (30°–50°S). *Münstersche Forschungen zur Geologie und Paläontologie, Münster*, **91**, 1–441.

LAGALLY, U. 1975. *Geologische Untersuchungen im Gebiet Lago General Carrera–Lago Cochrane, Prov. Aisén/Chile unter besonderer Berücksichtigung des Grundgebirges und seiner Tektonik*. Dissertation, Ludwig–Maximilians-Universität, München.

LEANZA, A. F. 1972. Andes Patagónicos Australes. *In*: Leanza, A. F. (ed.) *Geología Regional Argentina*. Academia Nacional de Ciencias, Córdoba, pp. 689–706.

LING, H. Y. & FORSYTHE, R. D. 1987. Late Paleozoic pseudoalbaillellid radiolarians from southernmost Chile and their geological significance. *In*: MCKENZIE, D. G. (ed.) *Gondwana Six: Structure, Tectonics and Geophysics*. American Geophysical Union, Washington D.C., Geophysical Monograph, **40**, 253–260.

LING, H. Y., FORSYTHE, R. D. & DOUGLASS, R. C. 1985. Late Paleozoic microfaunas from southernmost Chile and their relation to Gondwanaland forearc development. *Geology*, **13**, 357–360.

LUCASSEN, F., BECCHIO, R., WILKE, H. G., FRANZ, G., THIRLWALL, M. F., VIRAMONTE, J. & WEMMER, K. P. 2000. Proterozoic–Paleozoic development of the basement of the Central Andes (18–26°S) – a mobile belt of the South American craton. *Journal of South American Earth Sciences*, **13**, 697–715.

MCDANIEL, D. K., HEMMING, S. R., MCLENNAN, S. M. & HANSON, G. N. 1994. Resetting of neodymium isotopes and redistribution of REEs during sedimentary processes: the Early Proterozoic Chelmsford Formation, Sudbury Basin, Ontario, Canada. *Geochimica et Cosmochimica Acta*, **58**, 931–941.

MCLENNAN, S. M. 2001. Relationships between the trace element composition of sedimentary rocks and upper continental crust. *Geochemistry, Geophysics, Geosystems (G^3)*, **2**, World Wide Web Address: *http://g-cubed.org*

MCLENNAN, S. M. & HEMMING, S. 1992. Samarium/neodymium elemental and isotopic systematics in sedimentary rocks. *Geochimica et Cosmochimica Acta*, **56**, 887–898.

McLennan, S. M., Mcculloch, M. T., Taylor, S. R. & Maynard, J. B. 1989. Effects of sedimentary sorting on neodymium isotopes in deep-sea turbidites. *Nature*, **337**, 547–549.

McLennan, S. M., Taylor, S. R., McCulloch, M. T. & Maynard, J. B. 1990. Geochemical and Nd–Sr isotopic composition of deep-sea turbidites: crustal evolution and plate tectonic associations. *Geochimica et Cosmochimica Acta*, **54**, 2015–2050.

McLennan, S. M., Hemming, S., McDaniel, D. K. & Hanson, G. N. 1993. Geochemical approaches to sedimentation, provenance and tectonics. *In*: Johnsson, M. J. & Basu, A. (eds) *Processes Controlling the Composition of Clastic Sediments*. Geological Society of America, Special Papers, **284**, 21–40.

Mildowski, A. E. & Zalasiewicz, J. A. 1991. Redistribution of rare earth elements during diagenesis of turbidite/hemipelagite mudrock sequences of Llandovery age from central Wales. *In*: Morton, A. C., Todd, S. P. & Haughton, P. D. W. (eds) *Developments in Sedimentary Provenance Studies*. Geological Society of London, Special Publications, **57**, 101–124.

Morton, A. C. & Hallsworth, C. R. 1999. Processes controlling the composition of heavy mineral assemblages in sandstones. *Sedimentary Geology*, **124**, 3–29.

Mutti, E. & Normark, W. R. 1987. Comparing examples of modern and ancient turbidite systems: problems and concepts. *In*: Leggett, J. K. & Zuffa, G. G. (eds) *Marine Clastic Sedimentology*, Graham & Trotman, London, 1–38.

Nesbitt, H. W. & Young, G. M. 1982. Early Proterozoic climates and plate motions inferred from major element chemistry of lutites. *Nature*, **299**, 715–717.

O'Nions, R. K., Hamilton, P. J. & Evensen, N. M. 1977. Variations in $^{143}Nd/^{144}Nd$ and $^{87}Sr/^{86}Sr$ ratios in oceanic basalts. *Earth and Planetary Science Letters*, **34**, 13–22.

Pankhurst, R. J., Hervé, F., Rojas, L. & Cembrano, J. 1992. Magmatism and tectonics in continental Chiloé, Chile (42°–42°30′S). *Tectonophysics*, **205**, 283–294.

Pankhurst, R. J., Hervé, F. & Rapela, C. W. 1994. Sm-Nd evidence for the Grenvillian provenance of the metasedimentary basement of southern Chile and Western Antarctica. *VII Congreso Geológico Chileno, Concepción, Actas*, **2**, 1414–1418.

Potter, P. E. 1994. Modern sands of south America: composition, provenance and global significance. *Geologische Rundschau*, **83**, 212–232.

Riccardi, A. C. 1971. Estratigrafía en el oriente de la Bahía de la Lancha, Lago San Marín, Santa Cruz, Argentina. *Revista del Museo de La Plata (nueva serie), Sección Geología*, **7**, 245–318.

Roser, B. P. & Korsch, R. J. 1986. Determination of tectonic setting of sandstone–mudstone suites using SiO_2 content and K_2O/Na_2O ratio. *Journal of Geology*, **94**, 635–650.

Roser, B. P. & Korsch, R. J. 1988. Provenance signatures of sandstone–mudstone suites determined using discriminant function analysis of major-element data. *Chemical Geology*, **67**, 119–139.

Roser, B. P. & Nathan, S. 1997. An evaluation of elemental mobility during metamorphism of a turbidite sequence (Greenland Group, New Zealand). *Geological Magazine*, **134**, 219–234.

Söllner, F., Miller, H. & Hervé, M. 2000. An Early Cambrian granodiorite age from the pre-Andean basement of Tierra del Fuego (Chile): the missing link between South America and Antarctica? *Journal of South American Earth Sciences*, **13**, 163–177.

Taylor, S. R. & McLennan, S. M. 1985. *The Continental Crust: its Composition and Evolution*. Blackwell Science, Oxford.

Thomson, S. N., Hervé, F. & Fanning, C. M. 2000. Combining fission-track and U–Pb SHRIMP zircon ages to establish stratigraphic and metamorphic ages in basement sedimentary rocks in southern Chile. *IX Congreso Geológico Chileno, Puerto Varas, Actas*, **2**, 769–779.

Thomson, S. N., Hervé, F. & Stöckhert, B. 2001. Mesozoic–Cenozoic denudation history of the Patagonian Andes (southern Chile) and its correlation to different subduction processes. *Tectonics*, **20**, 693–711.

Van Baalen, M. R. 1993. Titanium mobility in metamorphic systems: a review. *Chemical Geology*, **110**, 233–249.

Willner, A., Hervé, F. & Massonne, H.-J. 2000. Mineral chemistry and pressure–temperature evolution of two contrasting high-pressure–low-temperature belts in the Chonos Archipelago, Southern Chile. *Journal of Petrology*, **41**, 309–330.

Yoshida, K. 1981. *Estudio Geológico del curso superior del río Baker, Aysén, Chile (47°05′ a 47°42′ Lat. S., 72°28′ a 73°15′ Long. W.)*. Dissertation, Universidad de Chile, Santiago.

Zhao, J. X., McCulloch, M. T. & Bennett, V. C. 1992. Sm–Nd and U–Pb zircon isotopic constraints on the provenance of sediments from the Amadeus Basin, central Australia: evidence for REE fractionation. *Geochimica et Cosmochimica Acta*, **56**, 921–940.

Oligocene–Early Miocene tectonic evolution of the northern Apennines (northwestern Italy) traced through provenance of piggy-back basin fill successions

U. CIBIN[1], A. DI GIULIO[2] & L. MARTELLI[1]

[1]*Ufficio Geologico, Regione Emilia Romagna, Viale Silvani 4/3, Bologna, Italy (e-mail: ucibin@regione.emilia-romagna.it) (e-mail: lmartelli@regione.emilia-romagna.it)*
[2]*Dipartimento di Scienze della Terra, Università di Pavia, Via Ferrata 1, Pavia, Italy (e-mail: digiulio@unipv.it)*

Abstract: The provenance history of sediments deposited in the piggy-back basins of the Northern Apennines has been drawn by means of a petrographic study of nearly 200 sandstone samples collected over 250 km of the belt; it allows the evolution of the eroded part of the belt in Oligocene–Early Miocene times to be determined in detail, with special emphasis on the age of the exhumation and the onset of erosion of the high-pressure/low-temperature Pennine metamorphic units of the Ligurian Alps and Corsica that form the innermost part of the chain.

Five petrofacies were distinguished, representing three sources that were active separately (three 'pure' petrofacies) or together (two 'mixed' petrofacies). The resulting sandstone composition reflects the erosion of different source units, changing through time and space along the belt.

The stratigraphic distribution of petrofacies records a change in the main clastic source from Ligurian calcareous units to Penninic units. This change occurred over most of the study area, reflecting the complete exhumation of the Penninic metamorphic units within the innermost part of the belt. It occurred at different times along the chain, migrating from northwest to southeast from Late Rupelian to Aquitanian. This time shift is interpreted to be related to the obliquity of the Northern Apennines convergent system.

The study of clastic sediments infilling sedimentary basins related to thrust and fold belts represents a fundamental source of information about the evolution of the orogenic wedges that supplied sediment to the basins. Indeed, the basin-fill records the subsidence history within and around the wedge, and, at the same time, the petrology of the sediments records the rock types exposed within the developing orogen. Therefore, the time–space distribution of synorogenic basins, and the composition of their infill, provide fundamental insights into the dynamics of orogenic wedges and, most of all, into the composition of material eroded from the wedge, which is generally the most difficult to unravel (e.g. Evans & Mange-Rajetzky 1991; Critelli & Le Pera 1994; Garzanti *et al.* 1996).

In this respect, the study of orogen-related basins of the Northern Apennines has previously concentrated on the foredeep basin, which tells us very little about the composition and evolution of the inner part of the system, as it was mainly fed by the Alpine belt (e.g. Gandolfi *et al.* 1983; Ricci Lucchi 1986; Di Giulio 1999). In contrast, sediments deposited in piggy-back basins, whose distribution and composition are more directly influenced by the dynamic evolution of that part of the wedge, have received less attention. Only recently has a relatively large amount of data been collected from these piggy-back basins – mostly from the oldest, Middle Priabonian–Early Rupelian basin (e.g. Di Giulio 1990; Cibin 1993; Martelli *et al.* 1998) and the youngest Middle Miocene basin (Spadafora 1995); a first broad synthesis of these data has been published recently (Cibin *et al.* 2001). This paper develops that synthesis; it focuses on Upper Rupelian–Lower Burdigalian sediments, with the intention of discussing the meaning of provenance changes in clastic units within piggy-back basins, linking this to the geodynamic evolution of the innermost part of the Northern Apennines convergent system, and showing how the provenance study of sediments trapped in piggy-back basins can contribute to the knowledge of the dynamics of collisional settings.

From: McCann, T. & Saintot, A. (eds) *Tracing Tectonic Deformation Using the Sedimentary Record.* Geological Society, London, Special Publications, **208**, 269–287. 0305–8719/03/$15.00

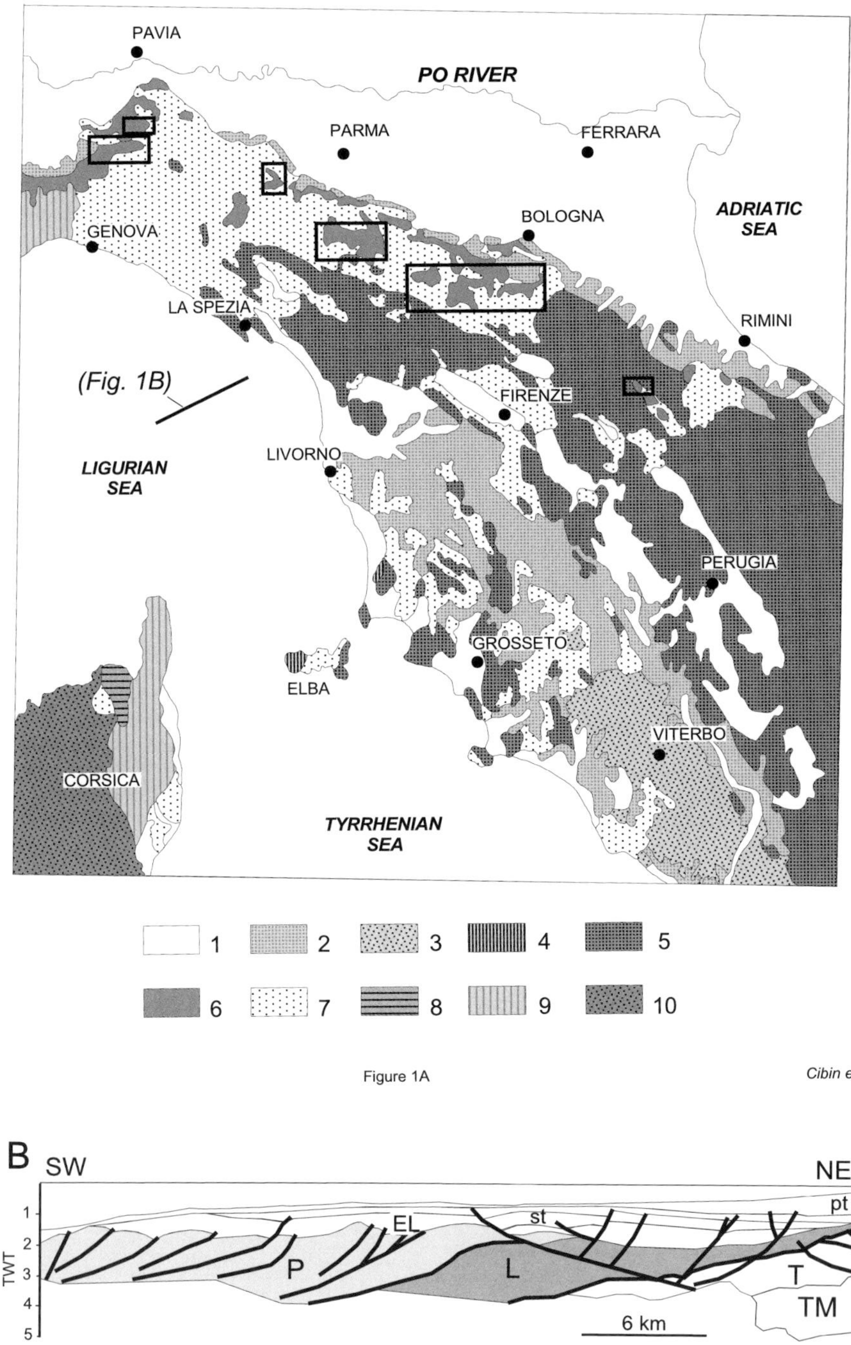

Fig. 1. (**a**) Tectonic sketch map of the Northern Apennines; the boxes report the location of sampling areas. Symbols: 1, Quaternary deposits; 2, Late Miocene–Pliocene deposits; 3, Pleistocene volcanics; 4, Neogenic plutons; 5, Tuscan and Umbro-Marchean units; 6, Epiligurian succession; 7, Ligurian units; 8, non-ophiolitic Penninic units; 9, meta-ophiolitic Penninic units; 10, foreland. (**b**) Interpretation of Northern Tyrrhenian Sea CROP seismic profile (after Bartole *et al.*, 1991). Symbols: pt, post-Tyrrhenian rift Neogene–Quaternary sediments; st, syn-Tyrrhenian rift Neogene sediments; EL, Epiligurian succession; L, Ligurian units; P, Penninic units; T, non-metamorphic Tuscan units; TM, metamorphic Tuscan units.

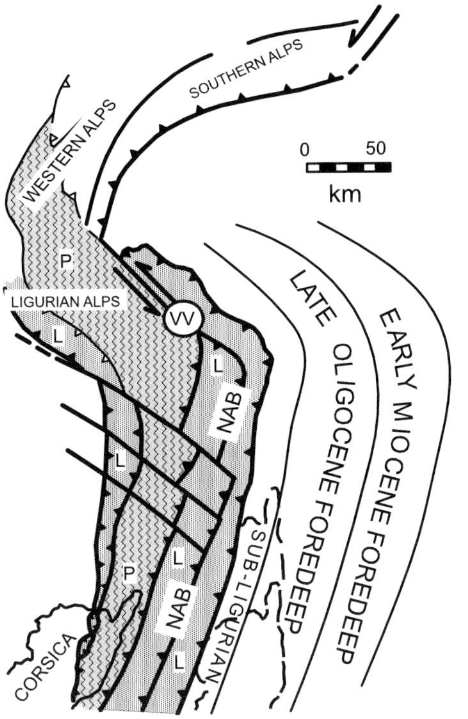

Fig. 2. Palaeotectonic scheme of the Northern Apennines during the Oligocene–Miocene (redrawn, simplified from Castellarin, 1994 and Cibin *et al.* 2001). NAB, Northern Apennines Belt; VV, Villavernia–Valzi Line; P, Penninic high-pressure/low-temperature metamorphic units; L, Ligurian units.

Geological setting

The Northern Apennines has experienced three quite different evolutionary stages during its 100 Ma history (e.g. Boccaletti *et al.* 1980; Finetti & Del Ben 2000); firstly, a Cretaceous–Early Palaeogene pre-collisional stage, related to the subduction of the Piedmont–Ligurian Basin (Alpine cycle); secondly, a collisional stage, related to the opening of the Balearic Sea with the anticlockwise rotation of the Corsica–Sardinia microplate from the Oligocene to the Burdigalian ('Balearic stage'; e.g. Castellarin 1992; Finetti & Del Ben 2000); and thirdly, an extensional stage developed in the innermost part of the belt since the Late Burdigalian, related to the opening of the Northern Tyrrhenian Sea, coeval with the eastwards shift of the thrust front in the outward part of the belt ('Tyrrhenian stage'; Finetti & Del Ben 2000).

The orogenic wedge that developed during the pre-collisional evolution of the convergent system, formed the westernmost ('innermost') part of the belt during the collisional deformation. It comprises an accretionary complex and a subduction-related high-pressure/low-temperature (HP/LT) metamorphic complex. The remnants of the former, mostly composed of accreted Cretaceous–Palaeocene oceanic turbidite units with minor ophiolites, constitutes the Ligurian complex of the Northern Apennines (Fig. 1a); the subduction complex is represented by the HP/LT ophiolite-rich metamorphic units exposed in the Ligurian Alps and northeastern Corsica, and detected by geophysics under the Ligurian Sea (Bartole *et al.* 1991; Fig. 1b).

During collision, the Ligurian complex was thrust eastwards on to the Adriatic continental margin, progressively incorporating allochthonous cover units (Tuscan and Umbro-Marchean units), which have at their top thick Upper Oligocene–Miocene turbidite successions accumulated in the Northern Apennines foredeep developed eastwards ('outwards'), in front of the accreting belt (Fig. 2). Contemporaneously, some basins developed on top of the overthrusting Ligurian units (Epiligurian piggy-back basins; EL; Fig. 1b), as well as on its suture zone with the HP/LT metamorphic complex in the Ligurian Alps (Piedmont Tertiary Basin; PTB; Fig. 1b).

The remnants of these basins are now scattered along the chain, from the southern Piedmont to the Montefeltro area (Rimini Province), mostly preserved in synclinal structures on top of the Ligurian units (Fig. 3). The stratigraphy and sandstone petrology of the Oligocene–Lower Miocene sediments, and particularly their use as tracers for the timing of exhumation of HP/LT metamorphic Alpine units during the Balearic stage of the belt evolution are the main topics of this paper. In this respect sandstone petrology is a particularly efficient tool, as sedimentary Ligurian units and HP/LT Pennine meta-ophiolite units delivered easily distinguishable detritus to piggy-back basins.

Stratigraphy of Middle Eocene–Lower Miocene piggy-back sediments

The EL and eastern PTB successions cover a time-span ranging from the Middle Eocene to the Late Miocene; Middle Eocene–Rupelian sediments are quite common all along the chain, Chattian–Burdigalian deposits occur in certain areas from the Piedmont to the Montefeltro region, whereas Middle–Upper Miocene deposits are basically restricted to the Vetto-Carpineti, Modena, Bologna and Montefeltro areas and to the Apennines margin, due to late Neogene uplift and erosion.

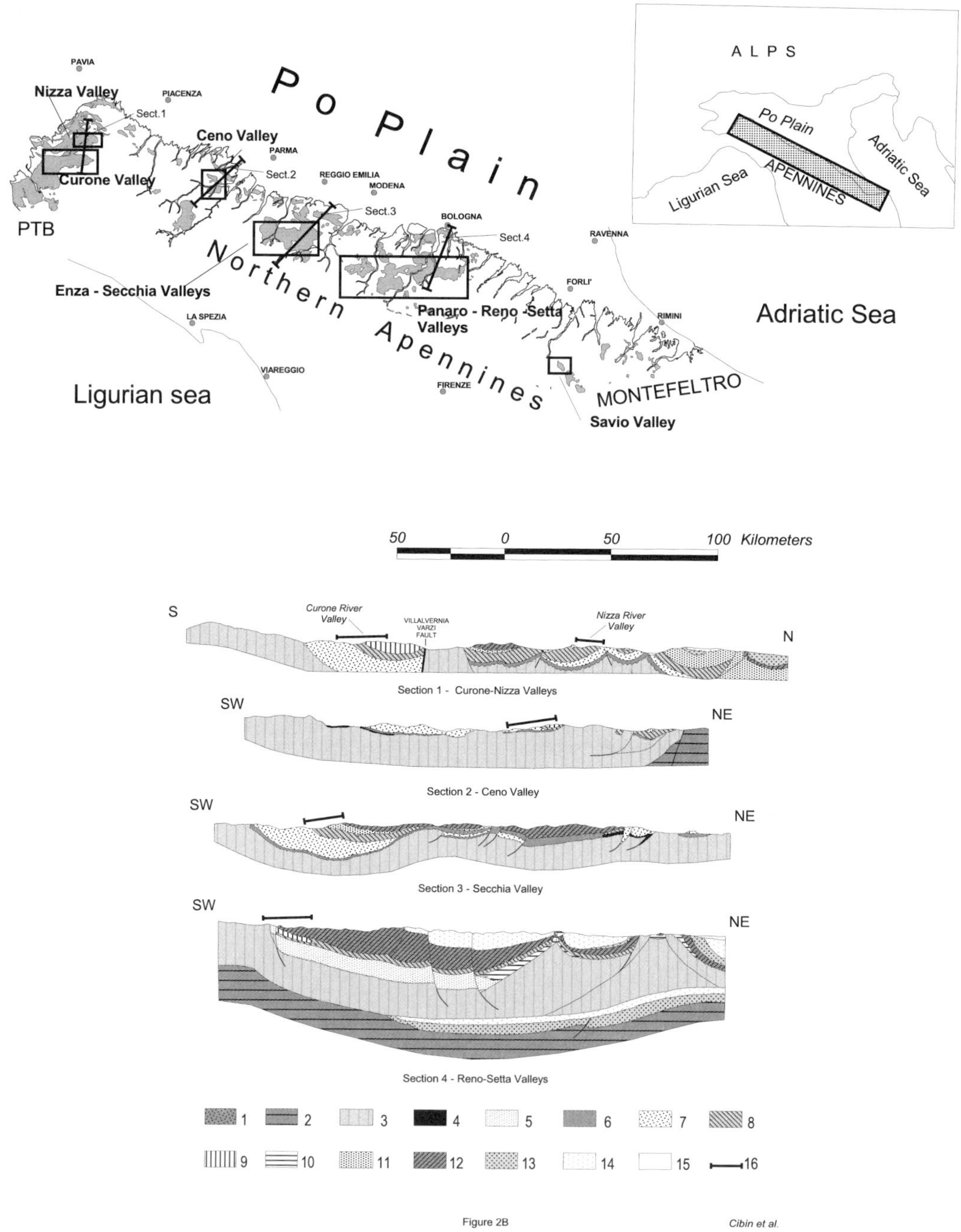

Fig. 3. (**a**) Distribution of main outcrop areas of Epiligurian sediments in the Northern Apennines belt and the location of the studied areas. (**b**) Cross-sections showing the main geological structures across the main study areas. Symbols: 1, Triassic evaporites; 2, Miocene foreland turbidites; 3, Ligurian units; (4–12), Epiligurian succession; 4, Basal olistostromes; 5, Loiano Formation; 6, Montepiano Formation; 7, Ranzano Formation; 8, Antognola/Rigoroso Formations; 9, Iatica/Anconella/Carnaio sandstone bodies and Castagnola Formation; 10, Eocene–Oligocene sedimentary mélanges; 11, Contignaco Formation and Mt Lumello Marl; 12, Upper Burdigalian–Serravallian shallow-water sediments; 13, evaporites and low-salinity Messinian deposits; 14, Pliocene marine deposits; 15, Quaternary alluvial deposits of the Po Plain; 16, location of the main sampled sections.

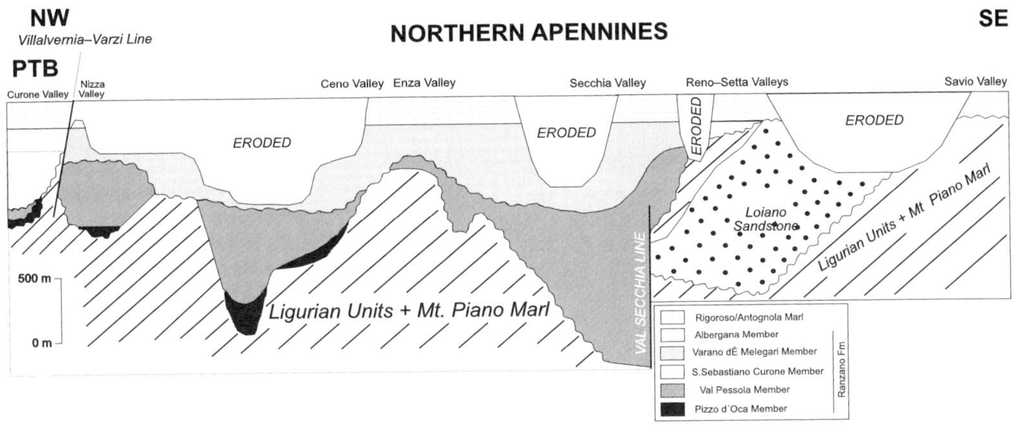

Fig. 4. Lithostratigraphic scheme of the Middle Eocene–Lower Oligocene part of the piggy-back Epiligurian and eastern Piedmont Tertiary Basin successions.

The EL–PTB successions include sediments ranging from turbidites to shelf facies, which form a depositional complex that includes two unconformity-bounded sedimentary cycles: a mostly deep-water, Middle Eocene–Early Miocene cycle, and a Middle–Late Miocene shallow- to deep-marine cycle.

Following Pieri (1961) and Sestini (1970), the Middle Eocene–Lower Miocene deposits have been subdivided into several lithostratigraphic units (Fig. 4). The Loiano Sandstone of Middle Eocene age is the oldest EL unit in the Modena and Bologna Apennines. It is a 1-km thick turbidite sandstone body with minor breccia and conglomerate beds (Dieci 1965; Cibin 1989), overlain by pelagic–hemipelagic marls (M. Piano Marl).

In the remaining part of the study area, the base of the EL succession comprises the Middle–Upper Eocene Monte Piano Marl, which is locally earliest Oligocene in age at the top (Bettelli *et al.* 1991; Catanzariti *et al.* 1997; Mancin & Pirini 2001).

From the latest Eocene to the Late Rupelian (Catanzariti *et al.* 1997), turbidite deposition prevailed across the study area, accumulating a thick clastic complex (Ranzano Formation) divided into five members. Each member has a different sandstone composition, which records changes in sediment sources during latest Priabonian–Rupelian times (Di Giulio 1991; Cibin 1993; Mutti *et al.* 1995; Martelli *et al.* 1998; Fig.4).

From the Late Rupelian onwards, turbidite deposition stopped abruptly and was replaced by monotonous hemipelagic mudstone deposition (Rigoroso and Antognola marls in the PTB and EL domains, respectively), which dominated throughout the Chattian–Aquitanian time-span (Fornaciari & Rio 1996; Catanzariti *et al.* 1997, and references therein; Andreoni *et al.* 1981; Mancin 1999). The mudstones contain scattered sandstone bodies and rare, relatively small lenses of turbiditic conglomerates, at several stratigraphic levels (Fig. 4). This coarse clastic input provides important evidence for the uplift and erosion of internal sectors of the orogenic belt, during a time-span characterized by reduced, mostly fine-grained terrigenous input from the belt. These rocks are described in detail in the next section.

The Middle Eocene–Lower Miocene sedimentary cycle is completed by Lower Miocene siliceous hemipelagic marls (Contignaco Formation and uppermost Castagnola Formation), that record a Mediterranean-scale biogenic–volcanogenic episode (Amorosi *et al.* 1995; Fornaciari and Rio 1996), but still includes some lens-shaped sandstone turbidite bodies. Late Burdigalian shallow-water sediments unconformably overlie the Contignaco and Antognola Formations and record the onset of the Middle–Upper Miocene sedimentary cycle (Amorosi 1992; Amorosi *et al.* 1993, 1996; Fornaciari *et al.* 1996).

Stratigraphy and petrography of Upper Rupelian–Burdigalian coarse-grained bodies

Coarse-grained turbidite lenticular bodies deposited into strongly confined depositional settings, interbedded at several levels in Upper

Rupelian–Burdigalian mudstones, have received local names in the geological literature, leading to a quite complex lithostratigraphic nomenclature summarized in Figure 5. They are tens to a few hundreds of metres thick and seldom exceed 10 kilometres in lateral extent.

The poorly organized coarse-grained facies of beds, the geometry of bodies, as well as their composition point to their deposition in very small turbidite systems directly linked to the sediment entry points (e.g. Mutti 1992; Stocchi *et al.* 1992). Thus, they provide limited but important information of the geology of the emergent part of the belt which fed clastic detritus to the piggy-back basins at different times during the collisional evolution of the chain.

In order to detail the geology of that part of the belt, the composition of 194 sandstone samples from all the main sandstone bodies known in the Oligocene-Early Miocene successions of the Epiligurian and eastern PTB basins was studied. Petrographic analyses were performed in thin section according to the Gazzi–Dickinson method, by means of a double point-counting for each sample, dealing with all the rock constituents (grains, matrix, cements; at least 250 essential grains counted) and fine-grained rock fragments (at least 100 grains counted). From this data set, a few key petrological parameters were calculated in order to highlight the most important differences occurring in the studied coarse-grained bodies (Table 1).

The stratigraphy and sandstone petrography of each studied body are described briefly in the next paragraphs, from northwest to southeast.

Eastern Piedmont Tertiary Basin (Curone River valley)

The stratigraphy of this area is summarized in Figure 6. Turbidite deposits of the Ranzano Formation are abruptly overlain by the hemipelagic Rigoroso Marl, which records a strong decrease in coarse-clastic terrigenous supply from the end of the Rupelian to the Oligocene–Miocene boundary (Andreoni *et al.* 1981; Cavanna *et al.* 1989). During this time interval, coarse-grained sediments occur only as confined, relatively thin bodies interbedded with marls (Fig. 6). They mostly comprise thick, amalgamated sandy beds of limited lateral continuity, deposited by structurally confined high-density turbidite currents flowing from west to east, into a channel (Cappella della Valle) and lobe (Nivione) upper bathyal environment (Cavanna *et al.* 1989; Stocchi *et al.* 1992; Di Giulio *et al.* 2002).

The depositional pattern changed at the very beginning of the Miocene, when a new increase

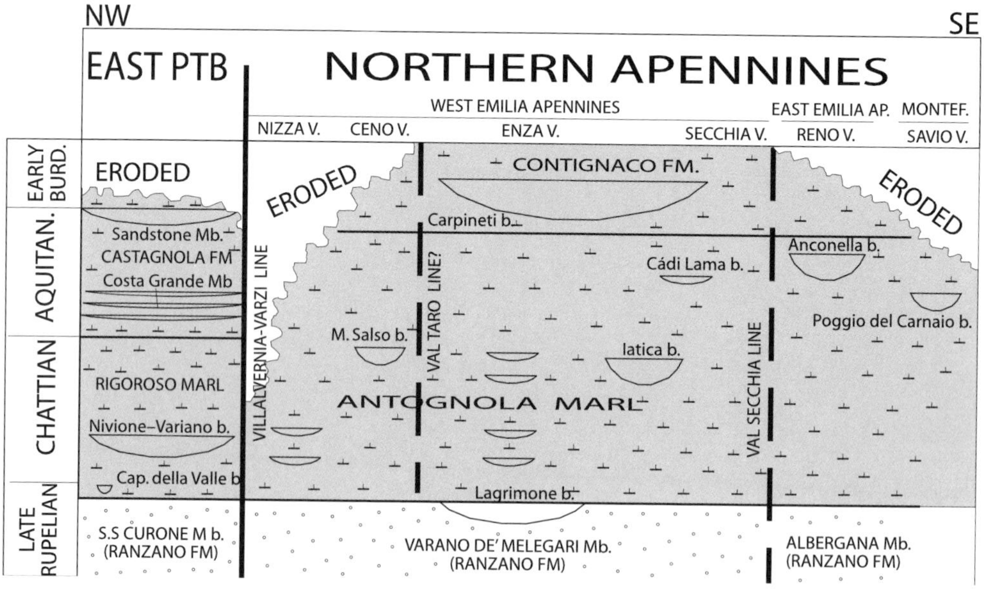

Fig. 5. Scheme of the lithostratigraphic nomenclature used in the geological literature for the Upper Rupelian–Lower Miocene sediments of the Epiligurian Northern Apennines and eastern Piedmont Tertiary Basin successions.

Table 1. *Summary of the main compositional characteristics of sandstone petrofacies recognized in the Oligocene–Lower Miocene Piedmont Tertiary Basin and Epiligurian sediments.*

Stratigraphic unit	Age	Area	Petrofacies	Q-F-L+C (average)	Lm-Lv-Ls+C (average)	Main rock fragment	Ssc/(L+C) (average)	Source-rock units	Sample
Castagnola Fm, Contignaco Fm, Rigoroso Marls, Enza Valley beds, M. Salso, S.S. Curone Mb.	Late Rupelian to Early Burdigalian	East PTB, West Emilia Apennine	A	31.0–8.9–60.2	74.8–13.3–11.9	Serpentine-schist, HP/LT and low-grade metamorphics	38.6	Penninic	142
Varano de Melegari Mb.	Late Rupelian	West Emilia Apennine	B	13.3–9.6–77.1	26.9–19.0–54.1	Limestone, siltstone, serpentinite	9.0	Ligurian (sedimentary)	11
Costa Grande Beds p.p. Ca' di Lama, Iatica, Anconella, Albergana Mb.	Late Rupelian to Aquitanian	East PTB, West and East Emilia Apennine	C	48.3–39.1–12.6	39.7–33.2–27.1	Granite, gneiss, medium and low-grade metamorphics, acidic volcanic	1.4	Continental basement (first cycle or recycled)	20
Enza Valley Beds, Lagrimone	Late Rupelian to Chattian	West Emilia Apennine	A+B	17.5–9.2–73.3	64.3–5.7–30.0	Serpentine–schist, HP/LT and low grade metamorphics, limestone, siltstone	12.3	Penninic and Ligurian (sedimentary)	15
Costa Grande Beds p.p P. del Carnaio	Aquitanian	East PTB, Montefeltro	A+C	41.3–29.6–29.1	74.4–12.2–13.4	Granite, gneiss, medium, low-grade and HP/LT metamorphic, serpentine–schist	15.0	Continental basement (first cycle or recycled) and Penninic	6

Q, quartz; F, feldspars; L, fine-grained lithics; C, carbonate rock fragments; Lm, metamorphic rock fragments; Lv, volcanic rock fragments; Ls, sedimentary rock fragments; Ssc, serpentine-schist. F and Lv do not include penecontemporaneous volcanic grains.

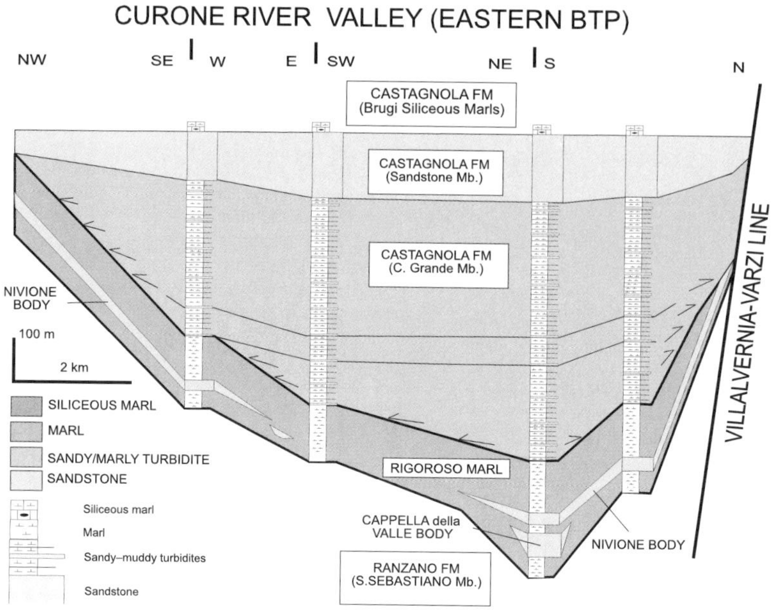

Fig. 6. Lithostratigraphic scheme of the Upper Rupelian–Lower Miocene succession of the eastern Piedmont Tertiary Basin.

in turbidite deposition occurred and thick sandstone/mudstone beds were deposited by means of more dilute turbidite currents in a structurally confined, oversupplied basin plain setting (Castagnola Formation: Andreoni *et al.* 1981; Cavanna *et al.* 1989; Stocchi *et al.* 1992).

Sandstones of Late Rupelian–Chattian and Aquitanian bodies show the same petrographic fingerprint of older Rupelian sediments (S. Sebastiano Curone member of Ranzano Formation; Martelli *et al.* 1998); they turn out to be mostly composed of metamorphic–lithic grains (Fig. 7), because of massive input from serpentinite–schist and HP/LT glaucophane-bearing blueschist rocks. In addition, arkosic sandstones containing granite fragments occur interbedded with lithic sandstones in the Aquitanian (Castagnola Formation).

Western Emilia Apennines (Nizza, Ceno, Enza and Secchia River valleys)

The stratigraphy of this area, mostly represented by the Vetto–Carpineti Syncline, is summarized in Figures 8–9. Above Rupelian sandstone–mudstone lobe turbidites of the Ranzano Formation (Varano de' Melegari Member), a very rapid transition to hemipelagic Antognola Marl generally occurs. However, in the Enza Valley there is an Upper Rupelian, 100–200-m thick arenitic–ruditic turbidite body (Lagrimone body) interposed between them (Fig. 8). Furthermore, the Antognola Marl includes isolated sandstone beds and bed-sets (Nizza River valley, Enza River valley and Ca' di Lama beds) or rare, tens to hundreds of metres thick, strongly confined, sandy turbidite bodies (M. Salso and Iatica bodies; Fig. 9).

Up-section, the occurrence of Late Aquitanian–Early Burdigalian silicified hemipelagic strata characterizes the Contignaco Formation, which includes another 15-km wide and 100–200-m thick sandstone turbidite lobe deposit (Carpineti body), one of the largest in the Late Oligocene–Early Miocene Northern Apennines piggy-back successions. Middle Miocene shallow-marine sandstones (Pantano Formation, Bismantova Group) unconformably cover the studied succession and mark the onset of the younger sedimentary cycle.

In this area, sandstone compositions show great variability in both time and space (Fig. 10), ranging from litharenites to litharenitic arkoses. Litharenites contain a great deal of sedimentary and low-grade metamorphic rock fragments, often showing HP/LT glaucophane-bearing blueschist paragenesis; litharenitic arkoses contain

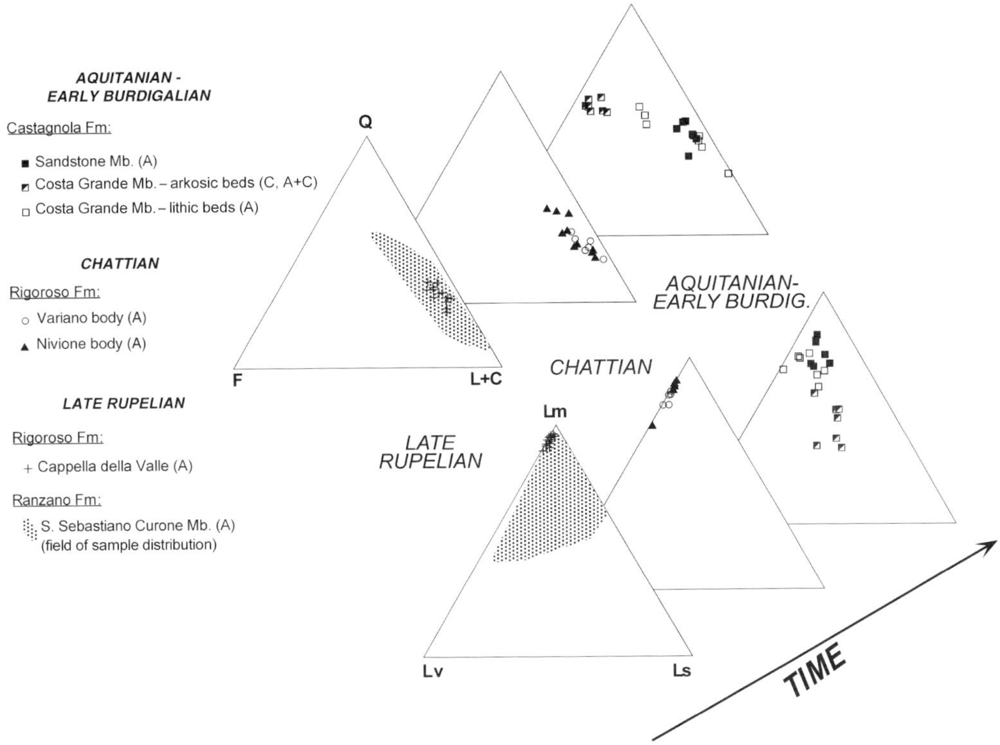

Fig. 7. Sandstone petrography of the Rupelian–Lower Miocene Piedmont Tertiary Basin; for lithostratigraphic nomenclature, see Figures 4–5. QFL+C and LmLvLs plots refer to matrix and fine-grained rock fragment compositions respectively.

mostly plutonic, medium- to low-grade metamorphic and minor volcanic rock fragments.

Eastern Emilia Apennines and Montefeltro area (Panaro, Reno and Savio river valleys)

For the sake of simplicity, the successions occurring in the Panaro–Reno and Savio River valleys are grouped together in this paragraph, even though their outcrops are several tens of kilometres apart (Fig. 3a).

In the Reno River valley area (Fig. 11), Upper Rupelian, fine-grained, thin-bedded mudstone–sandstone turbidites (Albergana Member of the Ranzano Formation) are overlain by the massive to finely stratified hemipelagic Antognola Marl. Above the marls, an Aquitanian coarse grained body occurs (Anconella body); it is tens of kilometres wide and hundreds of metres thick, and is made of sandstone and sandstone–mudstone beds deposited by high-density turbidity currents into a channel or sandy lobe environment (Cibin *et al.* 2001). It is in turn overlain by the hemipelagic chert-rich Contignaco Formation.

In the Savio River valley (Montefeltro area), only small outcrops of tectonically dismembered EL successions occur. They include an Aquitanian sandstone body, hundreds of metres wide and tens of metres thick, deposited by high-density turbidity currents in a channel or sandy lobe environment (Poggio del Carnaio body), interbedded with the Antognola Marl.

Petrographic data are relatively scarce in these areas; sandstone compositions range from arkoses to arkosic litharenites (Fig. 12). Arkoses of both late Rupelian and Aquitanian ages occur in the Reno Valley and contain mostly coarse-grained plutonic (granite) and gneissic rock fragments. Arkosic litharenites occur in the Savio Valley and include coarse-grained plutonic–gneissic rock fragments as well as low-grade fine-

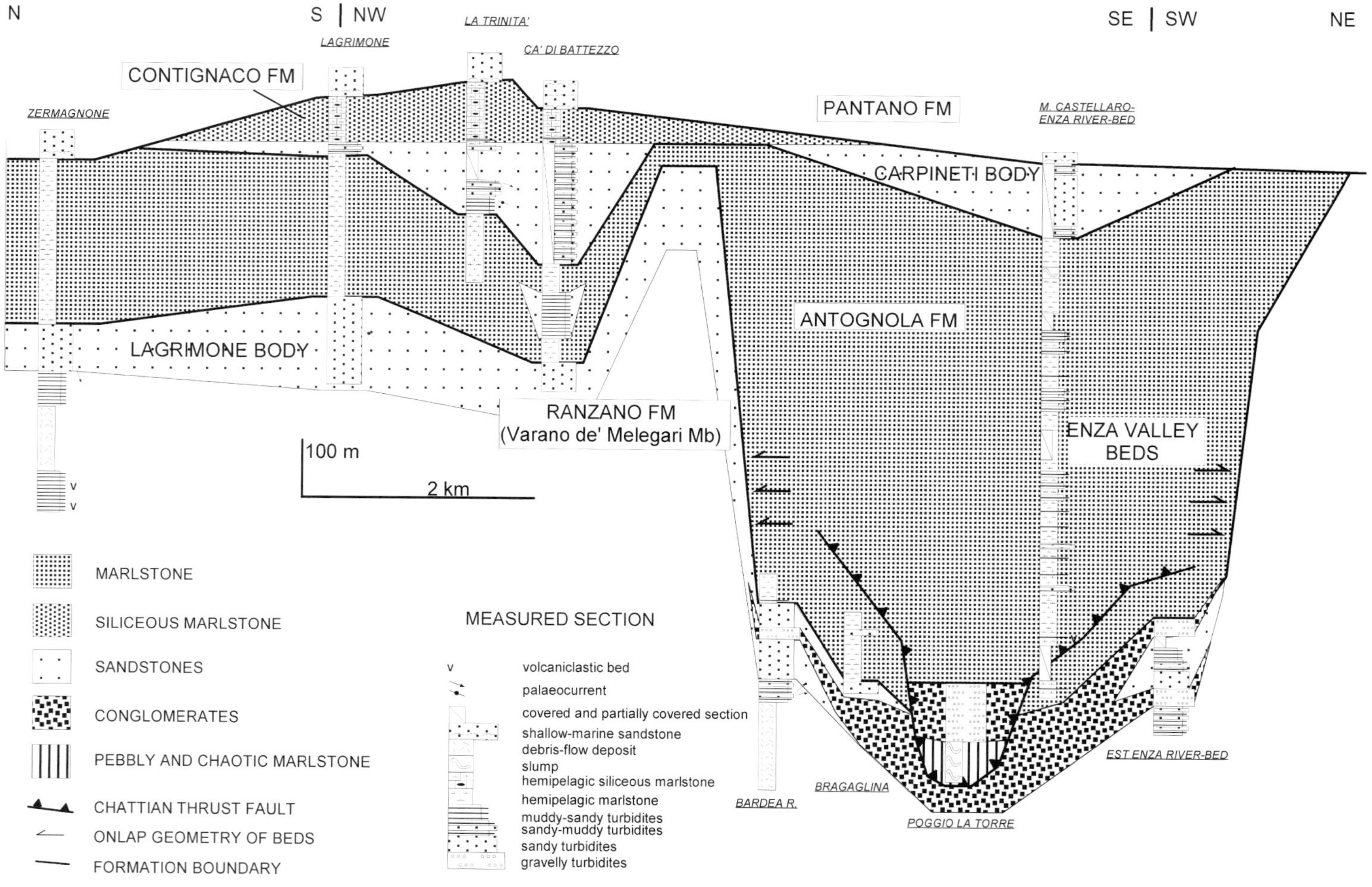

Fig. 8. Lithostratigraphic scheme of the Upper Rupelian–Lower Miocene Epiligurian succession of the Enza River valley. This new stratigraphic and structural scheme updates the interpretation of Cibin *et al.* (2001), where the Poggio La Torre conglomerate was erroneously attributed to the Antognola Formation.

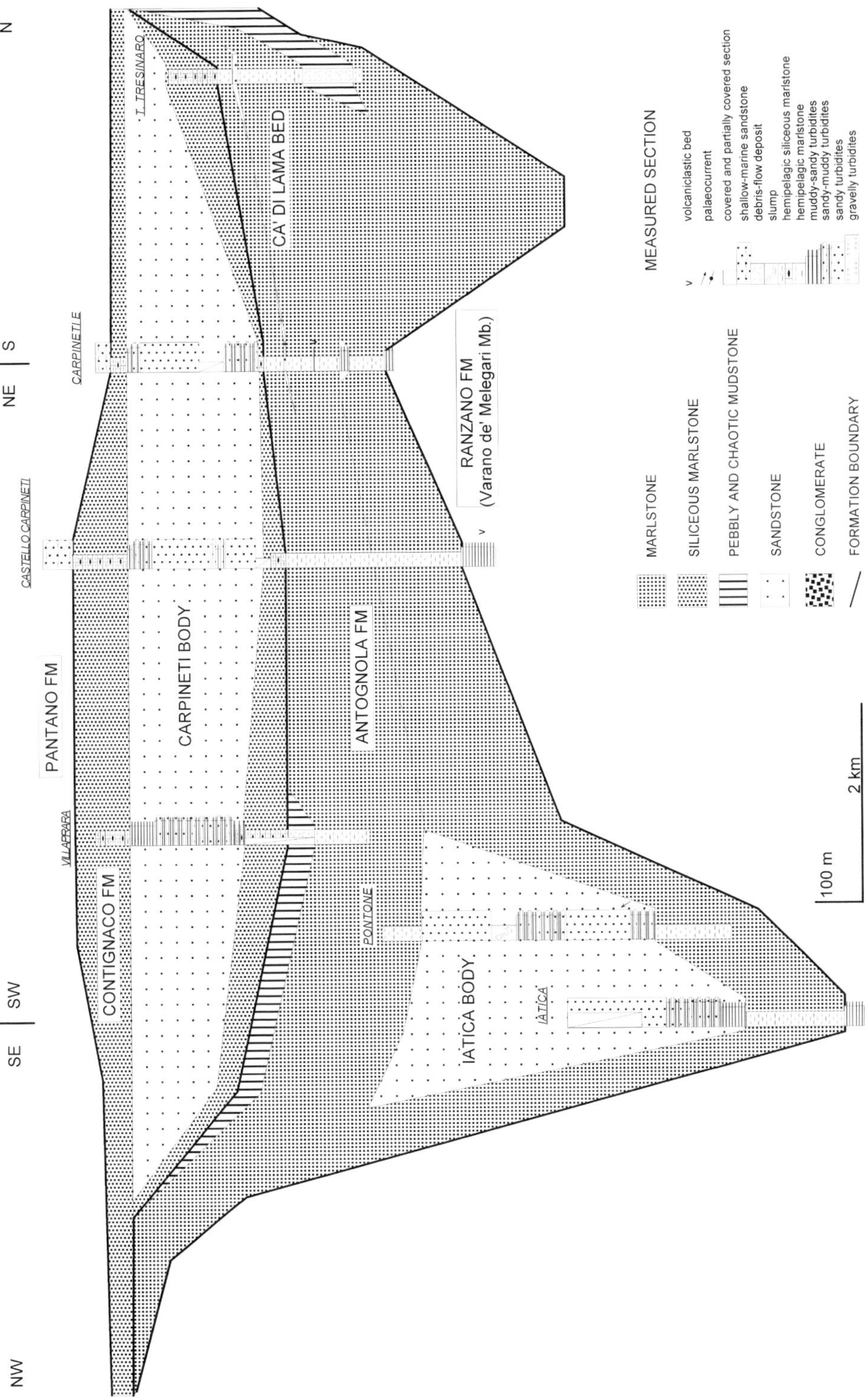

Fig. 9. Lithostratigraphic scheme of the Upper Rupelian–Lower Miocene Epiligurian succession of the Secchia River valley.

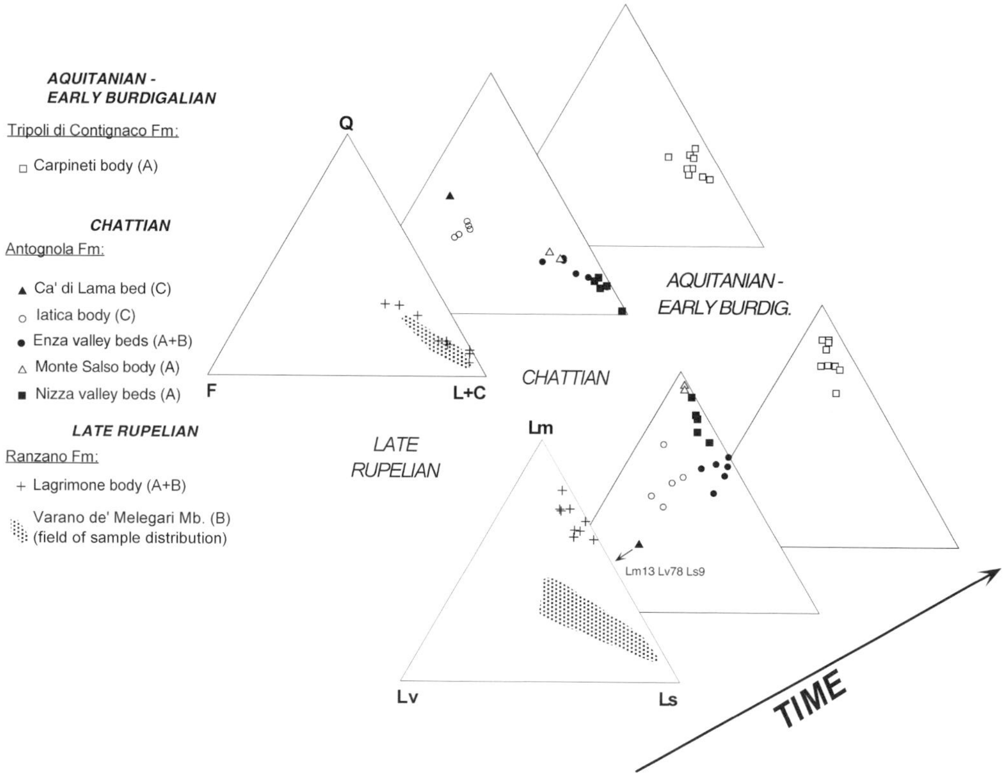

Fig. 10. Sandstone petrography of Rupelian–Lower Miocene Western Apennines Epiligurian sediments; for lithostratigraphic nomenclature, see Figures 4, 7 and 8. QFL+C and LmLvLs plots refer to framework and fine-grained rock fragment compositions respectively.

grained metamorphic lithics, sometimes showing HP/LT glaucophane-bearing blueschist paragenesis.

Petrofacies

Sandstone petrography highlights the great compositional variability of Upper Rupelian–Lower Burdigalian Northern Apennine sediments deposited in piggy-back basins, even if each sandstone body is usually homogeneous. In order to understand this complexity, samples have been grouped together as petrofacies, here described by the relative proportions of quartz (Q), feldspar (F) and fine-grained lithics (both carbonate C and non-carbonate L), and by the types and proportions of contained fine- and coarse-grained rock fragments, each interpreted in terms of source units (Table 1).

Five petrofacies can be distinguished. Three of them (petrofacies A, B and C) are characterized by relatively uniform rock-fragment associations and can be related to specific source-rock units; the remaining two contain heterogeneous rock-fragment associations due to a mix of the first petrofacies with the other two (petrofacies A+B and A+C).

In the following paragraphs each petrofacies is briefly described in terms of key petrological parameters, stratigraphic distribution and source units.

Petrofacies A

The majority of analysed samples belong to this petrofacies; they are all litharenites, although data are quite scattered around the average L+C values of 60%. Rock fragments are usually fine-grained and mostly comprise low-grade and HP/LT metamorphic rocks, including a great

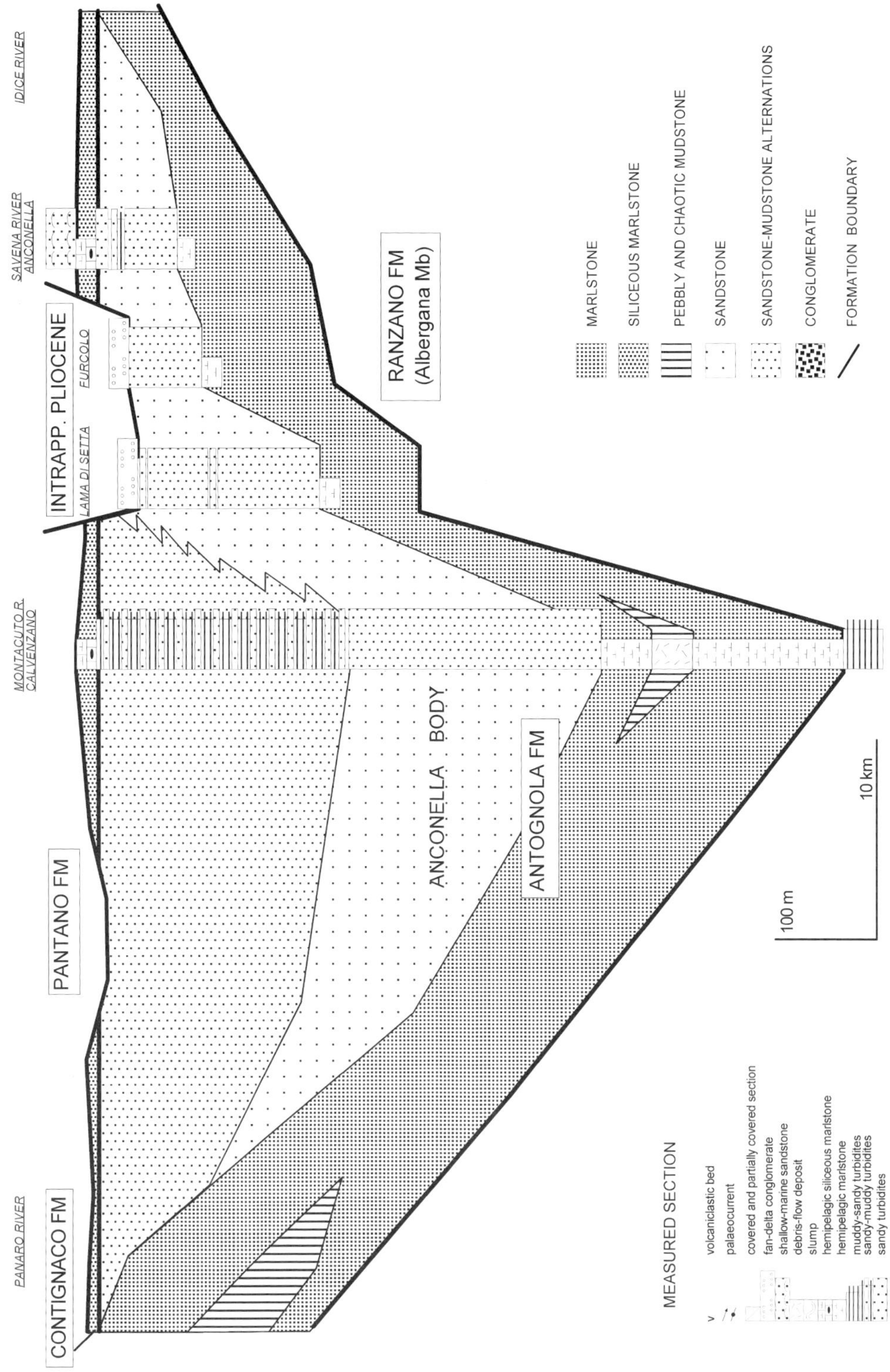

Fig. 11. Lithostratigraphic scheme of the Upper Rupelian–Lower Miocene Epiligurian succession of the Eastern Emilia Apennines.

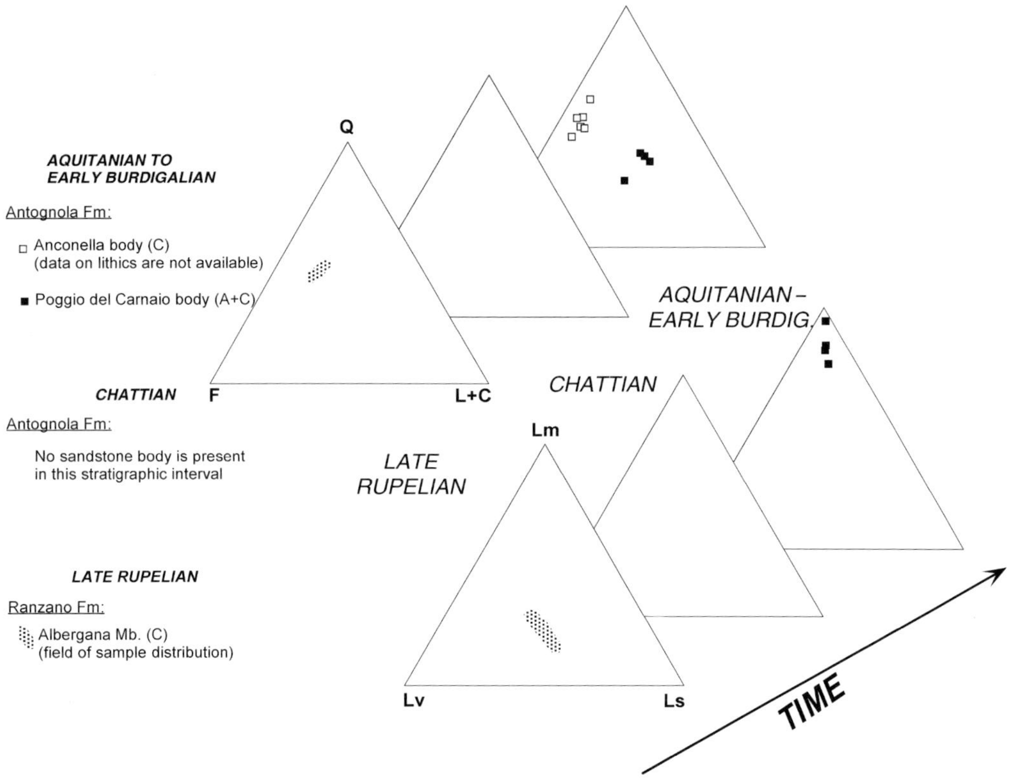

Fig. 12. Sandstone petrography of Rupelian–Lower Miocene Eastern Emilia Apennines and Montefeltro Epiligurian sediments; for lithostratigraphic nomenclature, see Figures 4–10 . QFL+C and LmLvLs plots refer to matrix and fine-grained rock fragment compositions respectively.

deal of serpentine schist (average 38.6% of total lithic grains), associated with many glaucophane–lawsonite-bearing HP/LT metamorphic rock fragments. Minor volcanic rocks and rare sedimentary rocks are present.

Sandstones of this petrofacies form almost all the Upper Rupelian–Lower Burdigalian bodies of the eastern PTB (with the exception of the Costa Grande arkosic beds), as well as part of the Chattian (Nizza valley beds, M. Salso body) and all the Aquitanian–Early Burdigalian (Carpineti body) units of the western Emilia Apennines.

In terms of source units, the abundance of blueschist, glaucophane–lawsonite-bearing metamorphic grains indicates a provenance mainly from HP/LT metamorphic units; in the eastern PTB, palaeocurrent patterns from the west-southwest in Rupelian, Chattian and Aquitanian sediments point to a provenance from the meta-ophiolite HP/LT massif of the Ligurian Alps (Voltri Massif) (Cavanna *et al.* 1989; Di Giulio 1991). More generally, HP/LT ophiolite Penninic units exposed in the innermost part of the developing belt provide an appropriate source for the detritus of this petrofacies (Di Giulio 1991, Martelli *et al.* 1998; Cibin *et al.* 2001).

Petrofacies B

Samples of petrofacies B are also litharenites, but contain a greater amount of lithic fragments than petrofacies A (average 77%); rock fragments are mostly of sedimentary rocks such as siltstones, shales and numerous deep-marine Cretaceous limestones. In addition, clasts of serpentinite, low-grade metamorphic rocks (including scarce HP/LT rocks) and minor serpentinite–schist clasts are present (average 9% of total lithics). This petrofacies characterizes the

Rupelian Varano de' Melegari Member of the Ranzano Formation in the western Emilia Appennines, but is not present in younger deposits (Antognola and Contignaco Formations).

The abundance of sedimentary rock fragments, mostly from Cretaceous deep-marine limestones, reflects a source from the Ligurian units (Cibin 1993; Martelli *et al.* 1998; Cibin *et al.* 2001), even if the small amount of HP/LT metamorphic rocks and serpentine–schist records minor contamination from Penninic sources.

Petrofacies C

Samples of petrofacies C show some variability (sub-petrofacies distinctions are possible; Cibin *et al.* 2001), as they range from arkoses to litharenitic arkoses. Rock fragments are not very abundant, particularly fine-grained lithics, thus, for highly arkosic samples (e.g. samples from the Anconella body) point-count data for lithics are not available. The rock fragments include mostly granite and gneiss, with some acidic volcanics and a variable amount of low- to medium-grade metamorphic rock with or without very little evidence of an HP/LT overprint (an average of 3.3% serpentine–schist out of the total lithics).

This petrofacies dominates the Rupelian–Aquitanian successions of the eastern Emilia Apennines (Albergana Member of Ranzano Formation and Anconalla body), as well as Chattian–Aquitanian sandstones of the easternmost sector of the western Emilia Apennines (Iatica, Ca' di Lama). In addition, some beds with this composition occur in the eastern PTB Aquitanian succession interbedded with beds of petrofacies A.

In terms of source units, the arkosic composition and the dominance of granitic–gneissic rock fragments should reflect a provenance from uplifted blocks of continental basement for this petrofacies, and the occurrence of minor amounts of acidic volcanic and metamorphic rocks does not contrast with this interpretation. On the other hand, the presence in the eastern Emilia Apennines of deeply eroded Late Cretaceous to Middle Eocene sandstones with a very similar composition (Ligurian sandstone units and EL Loiano Sandstone; Cibin 1989), strongly suggests that Rupelian–Aquitanian arkoses could have been fed by the recycling of such older units, rather than by a first-cycle erosion of a continental block. Because of the far western location, for arkosic beds of the eastern PTB, a first-cycle origin cannot, however, be ruled out.

Petrofacies A+B

Samples of this petrofacies are litharenites (L averages 73%) with a great amount and variety of rock fragments, the most common of which are HP/LT and low-grade metamorphic rock fragments, including serpentine–schist (average 12.3% of total lithics) and sedimentary rock fragments, such as deep-marine Cretaceous limestone. This petrofacies occurs in the Enza River valley in the western Emilia Apennines, from the top of the Rupelian to the Lower Aquitanian (Lagrimone and Enza River valley beds). It corresponds with a mixture of sediment derived from both Penninic units (petrofacies A) and Ligurian sedimentary units (petrofacies B).

Petrofacies A+C

Samples of this petrofacies are arkosic litharenites which represent an unusual combination of quartz, feldspars and coarse-grained rock fragments like granite and gneiss, with fine-grained rock fragments like HP/LT and low-grade metamorphic rocks, including serpentine–schist (average 11.1% of total lithics). It characterizes the small eroded remnants of Aquitanian sandstone bodies cropping out in the Montefeltro region (Poggio del Carnaio body), and is interpreted to result from a mixed source area fed both from the Penninic units (petrofacies A) and from the recycling of older arkosic sandstones (petrofacies C); this latter source is inferred because of the complete lack of known first-cycle granite source units in the Northern Apennines during theOligocene-Early Miocene, and, by contrast, because of the occurrence of several older arkosic clastic units (Late Cretaceous and Eocene) in the tectonic stack.

Space and time distribution of HP/LT clastics

The description of sandstone composition using petrofacies allows a better understanding of the complex evolution of the piggy-back basin sources, which mirrors the history of the emergent part of the developing orogenic belt.

The stratigraphic distribution of petrofacies in the studied piggy-back basins indicates that the main provenance changes occurred along the belt from the Late Rupelian to the Lower Burdigalian (Fig. 13). During the Late Rupelian the occurrence of three distinct petrofacies (end-members), virtually without sediment mixing, strongly suggests the existence of three basins fed by geologically different eroded areas, respectively made up of Penninic HP/LT metamorphic

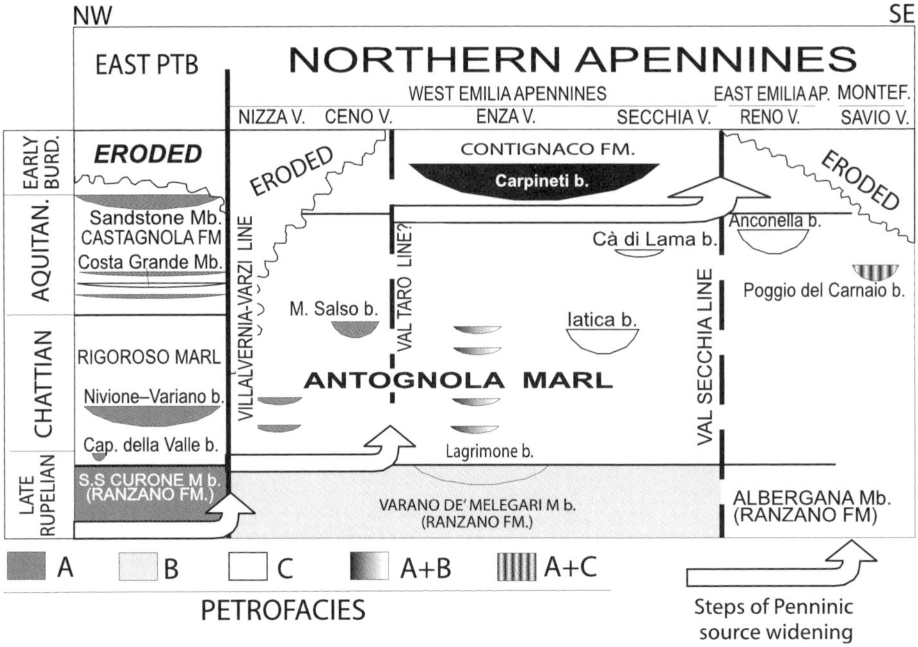

Fig. 13. Stratigraphic distribution of sandstone petrofacies in the Oligocene–Lower Miocene Piedmont Tertiary Basin and Epiligurian sediments.

units to the west, Ligurian calcareous units in the centre, and Cretaceous–Eocene siliciclastic units to the east. This picture strongly contrasts with the older homogeneous sandstone compositions (Martelli *et al.* 1998), and reflects the initiation of basin segmentation in the Late Rupelian.

Upwards, petrofacies A sandstones progressively appear from the NE to the SE, up to the Val Secchia area, which remained a strong provenance boundary for the entire Middle Eocene–Early Miocene time-span, likely due to the activity of the Val Secchia transverse line. This general trend is even more clearly distinguished by the time–space distribution of serpentine–schist grains (Fig. 14), which represent the most typical grain type of sandstones derived from the HP/LT Penninic units. Their distribution reflects an increasing influence of the Penninic source area which replaced the Ligurian units as the main source of detritus to the Epiligurian Basin, progressively from the eastern PTB to the western Emilia Apennines.

Through this source substitution, the complete exhumation and onset of erosion of Penninic units that form the southwesternmost part of the Northern Apennines occurred along approximately 150 km of the belt during a 10 Ma time-span.

This source evolution, which occurred in several stages, was the result of the increased uplift of the innermost part of the belt, possibly through out-of-sequence thrusting, as partly suggested by available northern Tyrrhenian crustal profiles (Bartole *et al.* 1991), or by means of enhanced erosive unroofing of the tectonic stack. In both cases, the time shift of that event, which becomes younger from northwest to southeast, strongly supports an oblique collisional geometry in the Northern Apennines.

Conclusions

The picture provided by the stratigraphy and sandstone composition of sediments deposited in piggy-back basins of the Northern Apennines is extremely complex. It reflects basin formation, segmentation and sediment source-area substitutions through time and space, highlighting the tectonic evolution of the innermost part of the orogenic wedge.

Basin segmentation first occurred during the Late Rupelian and seems to have been related to the activation of transverse lineaments crosscutting the submerged part of the belt (mostly the Villalvernia–Varzi and the Val Secchia Lines; Martelli *et al.* 1998).

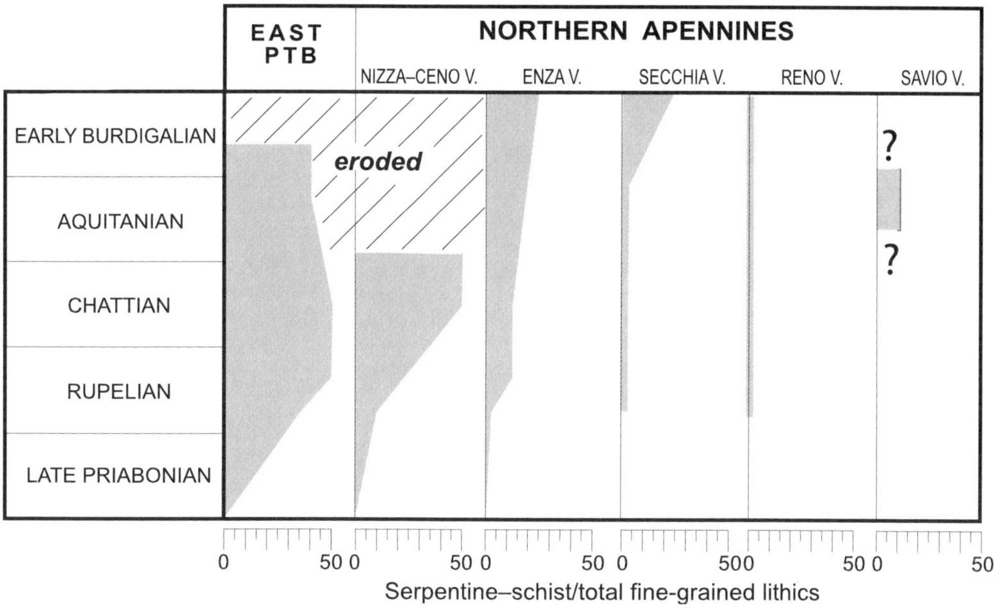

Fig. 14. Space–time distribution of HP/LT metamorphic rock fragments (expressed by the simple parameter serpentine–schist rock fragment v. total fine-grained rock fragments) in the Oligocene–Lower Miocene Piedmont Tertiary Basin and Epiligurian sediments.

Conversely, changes in the sediment supply areas mirrored the geological evolution of the emerged, innermost part of the belt. In this respect, even if at different times, the source-area evolution was quite similar along the belt; it experienced a substitution of a sedimentary Ligurian source by a HP/LT metamorphic Penninic one. This source-area substitution reflects the complete exhumation and onset of erosion of the Alpine subducted metamorphic units currently exposed in the Ligurian Alps and Alpine Corsica, and detected between them by geophysical surveys under the northern Tyrrhenian Sea (Bartole *et al.* 1991).

The time shift of that event, which becomes younger from northwest to southeast, migrating for 150 km along the belt during an approximately 10 Ma time-span, strongly supports an oblique collisional geometry in the Apennines. This result is quite intriguing, as a similar conclusion was drawn for the Late Miocene–Quaternary of the Southern Apennines from the study of the migration of foredeep depocentres (Casnedi 1991), suggesting an overall oblique collision for the Apennine convergent system in the last 30 Ma.

From a more general point of view, the Northern Apennines case shows how sediments accumulated in piggy-back basins provide fundamental tools for investigating the evolution of orogenic belts, and how they are also able to provide information on the overall geometrical aspects of collisional systems; among them obliquity seems to be of paramount importance, but has been widely ignored up to now.

R. Polino and W. Frisch are kindly acknowledged for their careful revision of a preliminary version of the paper. The authors are also grateful to A. Harley and an anonymous reviewer for their useful comments. Financial support was provided by Italian MURST, CNR and Emilia Romagna Region (CARG project) funds.

References

Amorosi, A. 1992. Correlazioni stratigrafiche e sequenze deposizionali nel Miocene epiligure delle Formazioni Bismantova, S. Marino e M. Fumaiolo (Appennino Settentrionale). *Giornale di Geologia*, **54**, 95–105.

Amorosi, A., Colalongo, M. L. & Vaiani, S. C. 1993. Le unità epiliguri mioceniche nel settore emiliano dell'Appennino Settentrionale. Biostratigrafia, stratigrafia sequenziale e implicazioni litostratigrafiche. *Paleopelagos*, **3**, 209–241.

Amorosi, A., Ricci Lucchi, F. & Tateo, F. 1995. The Lower Miocene siliceous lithozone: a marker in the paleogeographic evolution of the Northern Apennines. *Paleogeography Paleoclimatology Paleoecology*, **118**, 131–149.

AMOROSI, A., COLALONGO, M. & VAIANI, S. C. 1996. Revisione litostratigrafica dell'unità Bismantova (Miocene epiligure, Appennino Setettentrionale). *Bollettino Società Geologica Italiana*, **115**, 355–367.

ANDREONI, G., GALBIATI, B., MACCABRUNI, A. & VERCESI, P. L. 1981. Stratigrafia e paleogeografia dei depositi oligocenici sup. – miocenici inf nell'estremità orientale del Bacino Terziario Ligure Piemontese. *Rivista Italiana di Paleontologia e Straigrafia*, **87**, 245–282.

BARTOLE, R., TORELLI, L., MATTEI, G., PLEIS D. & BRANCOLINI, G. 1991. Assetto stratigrafico strutturale del Tirreno Settentrionale: stato dell'arte. *Studi Geologici Camerti*, Special Volume **1991/1**, 115–140.

BETTELLI, G., FIORONI, C., FREGNI, P. & PANINI, F. 1991. Nuovi dati stratigrafici sulla successione epiligure eo-oligocenica della Val Setta (Appennino Bolognese). *Memorie Descrittive Carta Geologica d'Italia*, **XLVI**, 221–227.

BOCCALETTI, M., COLI, M., DECANDIA, F., GIANNINI, E. & LAZZAROTTO, A. 1980. Evoluzione dell'Appennino Settentrionale secondo un nuovo modello strutturale. *Memorie della Società Geologica Italiana*, **21**, 359–373.

CASNEDI, R. 1991. Hydrocarbon accumulation in turbidites in migrating basins of the southern Adriatic foredeep (Italy). *In:* BOUMA, A. H. & CARTER, R. M. (eds) *Facies Models*, 219–233.

CASTELLARIN, A. 1992. Introduzione alla progettazione del profilo CROP. *Studi Geologici Camerti*, Special Volume **1992/2**, 9–15.

CASTELLARIN, A. 1994. Strutturazione Eo- e Mesoalpina dell'Appennino Settentrionale attorno al nodo ligure. *Studi Geologici Camerti*, Special Volume **1992/2**, 99–108.

CATANZARITI R., RIO, D. & MARTELLI, L. 1997. Late Eocene to Oligocene calcareous nannofossil biostratigraphy in Northern Apennines: the Ranzano Sandstone. *Memorie di Scienze Geologiche di Padova*, **49**, 207–253.

CAVANNA, F., DI GIULIO, A., GALBIATI, B., MOSNA, S., PEROTTI, C. R. & PIERI, M. 1989. Carta geologica dell'estremità orientale del Bacino Terziario ligure–piemontese. *Atti Ticinensi di Scienze della Terra*, **32**.

CIBIN, U. 1989. Petrografia e provenienza delle Arenarie di Loiano (Eocene sup.–Oligocene inf., Appennino modenese e bolognese). *Giornale di Geologia*, **51**, 81–92.

CIBIN, U. 1993. Evoluzione composizionale delle areniti nella successione epiligure eo-oligocenica (Appennino Settentrionale). *Giornale di Geologia*, **55**, 69–92.

CIBIN, U., SPADAFORA, E., ZUFFA, G. G. & CASTELLARIN, A. 2001. Continental collision history from arenites of episutural basins in the Northern Apennines, Italy. *Geological Society of America Bullettin*, **113**, 4–19.

CRITELLI, S. & LE PERA, E. 1994. Detrital modes and provenance of Miocene sandstones and modern sands of the Southern Apennines thrust-top basins (Italy). *Journal of Sedimentary Petrology*, **64/4**, 824–835.

DIECI, G. 1965. Età Luteziana delle 'Argille di Rio Giordano' (Appennino Settentrionale Bolognese). Documentazione micropaleontologica. *Bollettino Società Paleontologica Italiana*, **4**, 9–27.

DI GIULIO, A. 1990. Litostratigrafia e petrografia della successione eo-oligocenica del Bacino Terziario Ligure–Piemontese, nell'area compresa tra le valli Grue e Curone (provincia di Alessandria, Italia settentrionale). *Bollettino Società Geologica Italiana*, **109**, 279–298.

DI GIULIO, A. 1991. Detritismo nella parte orientale del Bacino Terziario Piemontese durante l'Eocene–Oligocene: composizione delle arenarie ed evoluzione tettono stratigrafica. *Atti Ticinensi di Scienze della Terra*, **34**, 21–41.

DI GIULIO, A. 1999. Mass transfer from the Alps to the Apennines: volumetric constraints in the provenance study of the Macigno–Modino source-basin system, Chattian–Aquitanian, north-western Italy. *Sedimentary Geology*, **124**, 69–80.

DI GIULIO, A., MANCIN, N. & MARTELLI, L. 2002. Geohistory of the Ligurian orogenic wedge: first inferences from Epiligurian sediments. *Bollettino Società Geologica Italiana*, Volume Speciale **1**, 375–384.

EVANS, M. J. & MANGE-RAJETZKY, M. A. 1991. The provenance of sediments in the Barreme thrust-top basin, Haut-Provence, France. *In:* MORTON, A. C., TODD, S. P. & HAUGHTON, P. D. W. (eds) *Developments in Sedimentary Provenance Studies*, Geological Society, London, Special Publications, **57**, 323–342.

FINETTI, I. & DEL BEN, A. 2000. Sismostratigrafia e tettono-dinamica crostale dell'Appennino Settentrionale da nuovi dati CROP. *80° Congresso Società Geologica Italiana,* Trieste, 252–253.

FORNACIARI, E. 1996. Biocronologia a nannofossili calcarei e stratigrafia ad eventi nel Miocene inferiore e medio italiano. *Ph.D. Thesis*, University of Padova, 320 pp.

FORNACIARI, E. & RIO, D. 1996. Latest Oligocene to early middle Miocene quantitative calcareous nannofossil biostratigraphy in the Mediterranean region. *Micropaleontology*, **42(1)**, 1–36.

FORNACIARI, E., DI STEFANO, A., RIO, D. & NEGRI, A. 1996. Middle Miocene quantitative calcareous nannofossil biostratigraphy in the Mediterranean region. *Micropaleontology*, **42(1)**, 37–63.

GANDOLFI, G., PAGANELLI, L. & ZUFFA, G. G. 1983. Petrology and dispersal pattern in the Marnoso Arenacea Formation (Miocene, Northern Apennines). *Journal of Sedimentary Petrology*, **53**, 493–507.

GARZANTI, E., CRITELLI, S. & INGERSOLL, R. V. 1996. Paleogeographic and paleotectonic evolution of the Himalayan Range as reflected by detrital modes of Tertiary sandstones and modern sands (Indus transect, India and Pakistan). *Geological Society of America Bulletin*, **108/6**, 631–642.

MANCIN, N. & PIRINI, C. 2001. Middle Eocene to Early Miocene planktonic foraminifer biostratigraphy in the Epiligurian succession (Northern Apennines, Italy). *Rivista Italiana di Paleotologia e Stratigrafia*, **107/3**, 371–393.

Martelli, L., Cibin, U., Di Giulio, A. & Catanzariti, R. 1998. Litostratigrafia della Formazione di Ranzano (Priaboniano–Rupeliano, Appennino Settentrionale e Bacino Terziario Piemontese). *Bollettino della Società Geologica Italiana*, **117**, 151–185.

Mutti, E. 1992. *Turbidite sandstones*. Edizioni Agip, S. Donato Milanese, 275 pp.

Mutti, E., Papani, L., Di Biase, D., Davoli, G., Mora, S., Segadelli, S. & Tinterri, R. 1995. Il Bacino Terziario Epimesoalpino e le sue implicazioni sui rapporti tra Alpi ed Appennino. *Memorie di Scienze Geologiche di Padova*, **47**, 217–244.

Pieri, M. 1961. Nota introduttiva al rilevamento del versante appenninico padano eseguito nel 1955–59 dai geologi dell'Agip Mineraria. *Bollettino della Società Geologica Italiana*, **80**, 3–34.

Ricci Lucchi, F. 1986. The foreland basin system of the Northern Apennines and related clastic wedges: a preliminary outline. *Giornale di Geologia*, **48**, 165–185.

Sestini, G. 1970. Sedimentation of the late geosynclinical stage. *Sedimentary Geology*, **4**, 445–479.

Spadafora, E. 1995. Compositional evolution of the Miocene Epiligurian succession of Emilia Apennines (Italy). *Giornale di Geologia*, **57**, 219–232.

Stocchi, S, Cavalli, C. & Baruffini, L. 1992. I depositi torbiditici di Guaso (Pirenei centro-meridionali), Gremiasco e Castagnola (settore orientale del BTLP). Geometria e correlazioni di dettaglio. *Atti Ticinensi di Scienze della Terra*, **35**, 153–177.

Precise tracing of exhumation and provenance using $^{40}Ar/^{39}Ar$ geochronology of detrital white mica: the example of the Central Alps

H. VON EYNATTEN[1] & J. R. WIJBRANS[2]

[1]*Institut für Geowissenschaften, FSU Jena, Burgweg 11, D-07749 Jena, Germany (e-mail: eynatten@geo.uni-jena.de)*
[2]*Faculty of Earth Sciences, De Boelelaan 1085, NL-1081 HV Amsterdam, The Netherlands*

Abstract: Single-grain $^{40}Ar/^{39}Ar$ dating of detrital white mica from Oligocene to Miocene (31–13 Ma) sediments of the North Alpine Foreland Basin in Switzerland reveals three prominent age clusters indicating cooling of the source rocks below 350–420°C in Carboniferous, Early Permian, and Tertiary times. Precise calibration of sedimentation age throughout the study area enables the thermal evolution of the hinterland in space and time to be precisely traced. Palaeozoic mica ages are documented in all samples and are used as additonal provenance indicators. Tertiary mica ages are restricted to sediments younger than 21 Ma, and are only found in central and western drainage systems. Tertiary micas document progressively increasing average cooling rates up to 34–41°C/Ma in the source area (Lepontine Dome), between 21 Ma and 14 Ma. The observed cooling rates and the time-span for rapid cooling in the source area (between 19 and 14 Ma) agree with thermal models derived from currently exposed rocks of the Lepontine metamorphic dome. This study proves that detrital mica geochronology is a robust tool for deciphering the thermal histories of ancient orogens which are no longer exposed today.

Spatial and temporal variations of cooling rates provide valuable insights into the thermal history of an orogen, especially in the late stages of its evolution. Because variations in the patterns of cooling and exhumation are strongly controlled by both endogenic and exogenic processes (e.g. Batt & Braun 1997), understanding of these parameters is crucial to the interpretation of orogenic processes. Because present-day exposures are limited in extent, it is becoming more and more necessary to use the detrital record of these processes found in synorogenic clastic sediments.

The last decade has shown that single-grain geochronology of detrital mineral grains is a potential tool for deciphering orogenic processes in both ancient and modern orogens (e.g. Cerveny *et al.* 1988; Copeland & Harrison 1990; von Eynatten *et al.* 1996, 1999; Najman *et al.* 1997; Sircombe 1999; Spiegel *et al.* 2000; Stuart *et al.* 2001). Out of the several geochronometers and suitable detrital minerals available, $^{40}Ar/^{39}Ar$ single-grain dating of detrital white mica has the following advantages:

(1) although not ultrastable like zircons, white micas are quite resistant to chemical breakdown during at least the first sedimentary cycle, implying that a robust age signal of the source rock is preserved in the sediment and may also withstand sedimentary recycling (Sherlock *et al.* 2001).
(2) The closure temperature of 350–420°C for Ar diffusion in white mica (von Blanckenburg *et al.* 1989; Hames & Bowring 1994; Kirschner *et al.* 1996) is especially suitable to date cooling after the last greenschist facies stages of metamorphism. Short-term thermal pulses below 250°C either in the source area or in the sedimentary basin generally do not affect the age signal. Even at higher temperatures ranging up to *c.* 500°C under favourable conditions white micas may at least in part retain their radiogenic argon (Wijbrans & McDougall 1986). Incremental heating techniques provide additional control of diffusion loss, but it was demonstrated that detrital white micas commonly have very uniform argon distributions (von Eynatten *et al.* 1996; Najman *et al.* 1997).
(3) Microprobe data on the chemical composition of the dated white mica (muscovite–phengite–Al-celadonite) make it possible to place additional constraints on source-rock petrology.

From: McCann, T. & Saintot, A. (eds) *Tracing Tectonic Deformation Using the Sedimentary Record.* Geological Society, London, Special Publications, **208**, 289–305. 0305–8719/03/$15.00

In this study detrital white micas from 31- to 13-Ma old clastic sedimentary deposits of the North Alpine Foreland Basin in Switzerland are dated using the $^{40}Ar/^{39}Ar$ method in order to constrain the time of unroofing of the Lepontine metamorphic rocks and the thermal history of the Oligocene to Miocene Central Alps. The results are compared with published cooling data derived from exposures in the Central and Western Alps, in order to test the significance of the detrital record where the hinterland is well known.

Case study

The Central Alps of Switzerland and the adjacent North Alpine Foreland Basin (NAFB) provide an excellent example for studying the sedimentary record of orogenic processes because, firstly, the Central Alps, which form the source area for NAFB deposits, are still exposed and among the most extensively studied orogens world-wide (e.g. Schmid *et al.* 1996), and, secondly, the sediments of the NAFB have been examined in detail with respect to facies and heavy-mineral analysis, and have been dated very precisely by magnetostratigraphy (e.g. Kempf *et al.* 1997).

The Central Alps

The Central Alps consitute a doubly-verging orogen with high-grade rocks of the Lepontine metamorphic dome in its core (Schmid *et al.* 1996; Fig. 1). The southern flank, comprising the unmetamorphosed south-verging South Alpine basement and sedimentary cover nappes, is separated from the Lepontine dome by the Insubric Line, which thus represents a major structural and thermal discontinuity. The northern flank comprises the north-verging nappe stack of low- to medium-grade meta-

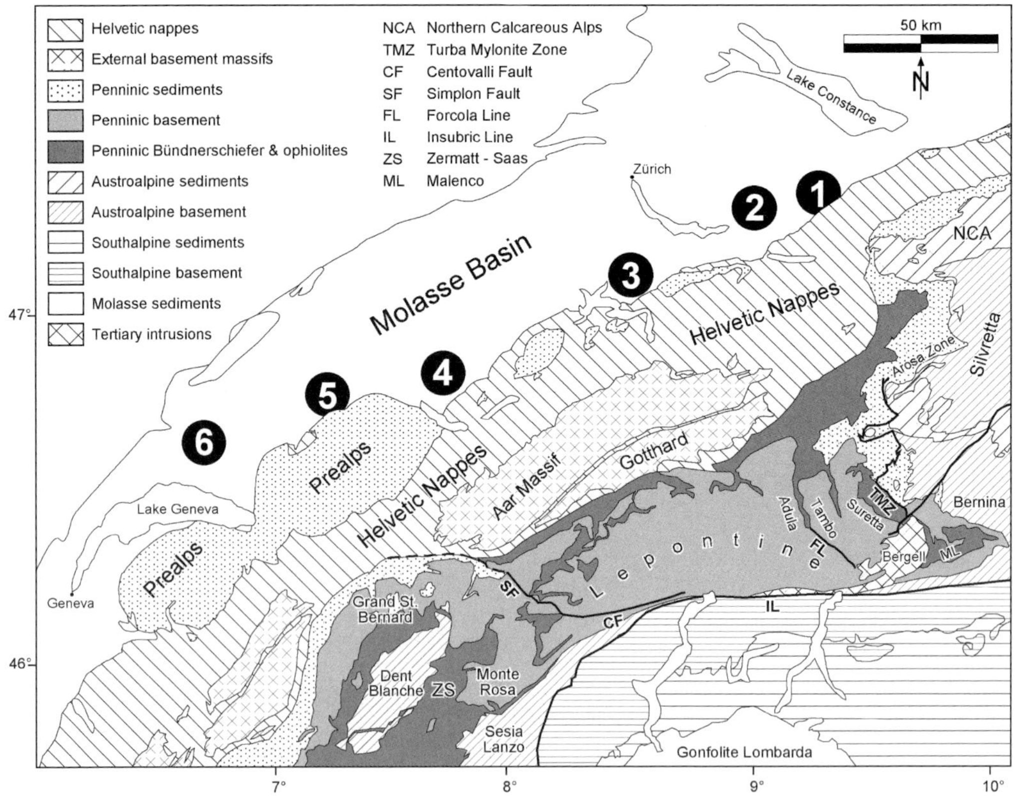

Fig. 1. Simplified map showing structural units (modified from Frey & Mählmann, 1999) and locations of sampled Oligocene to Miocene drainage systems: (1) Speer/Kronberg alluvial fans system; (2) Hörnli alluvial fan system; (3) Rigi/Höhrone alluvial fans system; (4) Honegg–Napf alluvial fans system; (5) Guggershörnli alluvial fan (and precursors); and (6) Lake Geneva axial dispersal system.

morphic to unmetamorphosed Helvetic, Penninic and Austroalpine basement and cover nappes (Fig. 1).

The peak of Alpine greenschist- to amphibolite-facies metamorphism in the Western Alps and Central Alps is dated at between 36 to 40 Ma (e.g. Steck & Hunziker 1994; Desmons *et al.* 1999) and 32 to 33 Ma (Gebauer 1999), respectively. Emplacement of the Penninic and Austroalpine nappes occurred prior to 35 Ma (Schmid *et al.* 1996; Markley *et al.* 1998). Subsequent crustal shortening and northward progradation of the thrust front led to accretion of Molasse strata to the orogenic wedge (e.g. Kempf *et al.* 1999).

The Lepontine metamorphic dome is separated from the hanging-wall units to the west (Austroalpine–Piedmont nappe stack; Dal Piaz 1999) by the Simplon Fault (SF; Fig. 1), which is a major detachment fault accommodating Late Oligocene to Recent lateral extension in the Central Alps. Exhumation of the Lepontine started at *c.*30 Ma (e.g. Hurford *et al.* 1989; Gebauer, 1999), and enhanced rates of displacement along the SF and exhumation of the Lepontine are suggested to have occurred from 18 to 15 Ma (Grasemann & Mancktelow 1993), or even prior to 20 Ma (Schlunegger & Willett 1999). Both studies are based on thermal modelling of cooling ages from high-grade metamorphic rocks presently exposed in the Lepontine Dome. To the east the Lepontine metamorphic dome is separated from higher Penninic nappes by steeply inclined eastward-dipping normal faults (e.g. Forcola Line, Fig. 1, Baudin *et al.* 1993).

The North Alpine Foreland basin (NAFB)

The NAFB formed on European upper crust in response to the tectonic load of the evolving Alpine orogen (e.g. Homewood *et al.* 1986). The sedimentary fill of the molasse stage of the basin is governed by two regressive megacycles, each starting with marine deposits and grading up-section into clastic fluvial deposits. The first cycle started in the Rupelian and lasted up to the Aquitanian and the second cycle started with the Burdigalian transgression (*c.*20 Ma) and ended with Serravalian fluvial clastics (Berger 1996; Kuhlemann & Kempf 2002). During the fluvial stages the basin was dominated by longitudinal meander belts with axial drainages flowing to the NE (first cycle) and later to the SW (second cycle). Both marine and fluvial deposits interfinger with transverse alluvial fan systems delivering coarse-grained detritus from the prograding orogenic wedge. The sediments of the Swiss part of the NAFB between Lake Constance and Lake Geneva are very precisely dated by mammal biostratigraphy as well as magnetostratigraphy (Schlunegger *et al.* 1996; Kempf *et al.* 1997; Strunck 2001).

In this study we focus on well dated sections from transverse drainage systems in order to minimize: firstly, uncertainties in the stratigraphic age of the radiometrically dated micas, and secondly, possible mixing with sediment derived from longitudinal drainage with a less well-constrained provenance. The sampled drainage systems are, from east to west (numbers refer to Fig. 1):

(1) Speer/Kronberg alluvial fan system (31 to 21 Ma sedimentation age; Kempf *et al.* 1999);
(2) Hörnli alluvial fan system (20 to 13 Ma, Kempf *et al.* 1999);
(3) Rigi–Höhrone alluvial fan system (30 to 22 Ma; Schlunegger *et al.* 1997);
(4) Honegg–Napf alluvial fan system (31 to 14 Ma; Schlunegger *et al.* 1996; Kempf *et al.* 1997);
(5) Guggershörnli alluvial fan and precursors (*c.*20 to 18 Ma; Strunck 2001), and
(6) the Lake Geneva drainage system (28 to 24 Ma; Strunck 2001).

The latter drainage system system is the only one with mostly axial transport from the SW into the study area. The entry point of the material into the foreland basin must have been located somewhere to the south or SW of the present-day Lake Geneva (Maurer 1983; Strunck 2001; Fig. 1).

Methodical approach

Cooling data of detrital minerals from sediments or sedimentary rocks deposited at some time in the geological past constrain the cooling of rocks which were exposed to the surface in the source area of these sediments at that time. Specifically, radiometric dating of a detrital mineral species dates the time when the host rock of that mineral cooled below a closure temperature specific to the geochronometer used. A major prerequisite for any interpretation in terms of provenance is that the post-sedimentation thermal overprint did not exceed that closure temperature.

Cooling rates for detrital mineral grains can be calculated using the closure temperature of the applied geochronometer for a certain mineral species and the difference between the cooling age of this mineral and its sedimentation age (Cerveny *et al.* 1988). The time-span between erosion and deposition is considered to be negligible in this case study because we are dealing with a mountainous setting with high topographic gradients and small distances between sediment source and site of deposition. The rate of diffusion of radiogenic argon in a crystal is a function of the temperature, grain size and the lattice geometry. As a consequence, isotopic closure in a cooling system is a function of the cooling rate. Mineral imperfections such as, for example, deformation-induced crystal defects, subgrains, and exsolution effects will tend to enhance the loss of argon, and hence lower the closure temperature. The erosion–sedimentation cycle acts as an effective filter for imperfect minerals, but because mechanical fracturing will occur, the measured grain size of sedimentary mineral can only be a lower estimate for the average grain size in the source rock. For these reasons it is prudent to assume that, for the mica grains studied here, the higher estimates for the closure temperature apply. In cases where temporary storage of the sediment during its transportation plays a significant role, it is not feasible to calculate meaningful cooling rates given the difficulty in distinguishing the time taken to reach the erosion surface from the time between erosion and final deposition.

Provided that the basin under investigation is proximal to the source region and contains first-cycle sediments, then dating detrital minerals from different stratigraphic levels and from different sites within the sedimentary basin allows the evolution of cooling rates from ancient orogenic source areas to be traced through space and time. Applying a range of geochronometers, each with different closure temperatures, enhances the resolution of differential cooling paths. To establish a strong relation between cooling data and possible source areas a good control on sediment provenance is required. This approach has a great potential for clarifying the geodynamics of modern and ancient mountain belts, because the differential pattern of cooling is strongly related to both the crustal and the surface processes within an orogen (e.g. Batt & Braun 1997).

To perform high-resolution $^{40}Ar/^{39}Ar$ laser incremental heating and single fusion dating of whole single detrital white mica grains, we applied the following procedure. Sandstones were disaggregated in 10% acetic acid to remove the carbonate cement. Mica crystals of 500–1000 μm diameter (sample EY 18-8) and 250–500 μm size-fractions (all other samples) were separated using a Faul vibration table. Small amounts of hand picked white mica, up to 50 grains per sample, were loaded in Al foil envelopes and stacked with Cu foil envelopes containing a mineral standard (TCR sanidine 85G003; Dalrymple & Duffield 1988) in a 6 mm internal diameter quartz tube. Several tubes containing different experiments were loaded in an Al irradiation-can and irradiated for 12 h in the CLICIT facility of the Oregon State University TRIGA reactor. The experimental techniques, including information on the standards and the correction factors for interfering nucleogenic isotopes as used at the VULKAAN $^{40}Ar/^{39}Ar$ laser-probe laboratory at VU Amsterdam, are described fully in Wijbrans *et al.* (1995): samples were loaded in a low-volume UHV system and heated using a defocused argon ion laser. Beam intensities of the argon spectrum were measured on a MAP 215-50 double focusing mass spectrometer using an SEM detector operated at a gain of 50 000. The CLICIT facility is particularly useful because its $(^{40}Ar/^{39}Ar)_K$ correction factor is one of the lowest reported (cf. McDougall & Harrison 1999; table 3–5). Most of the micas were dated by total fusion experiments, but some grains were measured by incremental heating experiments to test for the homogeneity of radiogenic argon. When measuring small argon ion beams in a mass spectrometer, the system blank values for ^{40}Ar and for ^{36}Ar limit the ultimate precision of the measurement (e.g. Sherlock & Kelley 2002). In our case, total system blanks were measured following every second unknown analysis, and blank values for ^{40}Ar were consistently lower than 0.01 times that of the unknown.

Results

$^{40}Ar/^{39}Ar$ geochronology

Some 351 single detrital white mica grains from 20 individual samples (Table 1) were dated. The overall age distribution shows three prominent age groups: Carboniferous (peak age 329.4±9.0 Ma), Early Permian (280.5±9.5 Ma), and Tertiary (34.8±5.3 Ma). These three age groups cover more than 90% of all dated micas. Almost no Mesozoic mica ages are recorded in Oligocene to Miocene sediments (Fig. 2).

To test for the distribution of radiogenic argon within individual mica grains we ran several incremental heating experiments. The resulting ^{39}Ar degassing spectra show rather flat and homogeneous plateau ages for mica from all of the three age groups (Fig. 2). We therefore interpret single-grain ages belonging to these age groups to represent geologically meaningful ages reflecting the cooling of host rocks below 350–420°C (von Blanckenburg *et al.* 1989; Hames & Bowring 1994; Kirschner *et al.* 1996). For some source rocks, especially low-grade

Table 1. *Sample description, sedimentation age with reference, and number of dated micas.*

Sample	Lithology	Locality†	Section	Age (Ma)	Reference	No. of micas
EY 19-3	Fine–medium litharenite	1	Necker	21.2±0.2	Kempf *et al.* (1999)	7
EY 19-4	Fine–medium litharenite	1	Necker	23.9±0.1	Kempf *et al.* (1999)	17
EY 19-2	Medium litharenite	1	Thur	28.9±0.2	Kempf *et al.* (1999)	19
EY 19-13	Medium litharenite	2	Hörnli	13.2±0.2	Kempf *et al.* (1999)	16
EY 19-14	Medium litharenite	2	Jona	15.1±0.1	Kempf *et al.* (1999)	15
EY 19-11	Fine–medium litharenite	2	Goldinger Tobel	20.0±0.2	Kempf *et al.* (1999)	16
EY 18-7	Medium–coarse litharenite	3	Höhronen/Nettenbach	22.1±0.3	Schlunegger *et al.* (1997)	19
EY 18-2	Medium litharenite	3	Rigi/Fischchrattenbach	29.8±0.2	Schlunegger *et al.* (1997)	10
EY 18-8*	Medium–coarse litharenite	4	Fontannen	13.6±0.1	Schlunegger *et al.* (1996) Kempf *et al.* (1997)	31
EY 18-13	Medium litharenite	4	Fontannen	15.4±0.6	Schlunegger *et al.* (1996) Kempf *et al.* (1997)	22
EY 18-12	Medium litharenite	4	Schwändigraben	18.0±0.3	Schlunegger *et al.* (1996) Kempf *et al.* (1997)	21
EY 18-11	Medium litharenite	4	Schwändigraben	18.9±0.2	Schlunegger *et al.* (1996) Kempf *et al.* (1997)	20
EY 18-1*	Medium litharenite	4	Fischenbach	20.9±0.1	Schlunegger *et al.* (1996)	22
EY 18-16*	Fine-medium litharenite	4	Prässerebach	24.9±0.2	Schlunegger *et al.* (1996)	17
EY 18-14*	Medium–coarse litharenite	4	Emme	31.0±0.2	Schlunegger *et al.* (1996)	17
EY 21-4	Medium litharenite	5	Sensegraben	17.9±0.2	Strunck (2001)	20
EY 21-1	Medium–coarse litharenite	5	Heitenried	20.4±0.2	Strunck (2001)	20
EY 21-10	Medium litharenite	6	Talent south	23.9±0.1	Strunck (2001)	17
EY 21-8	Medium litharenite	6	Talent south	27.2±0.8	Strunck (2001)	10
EY 21-7	Medium litharenite	6	Talent south	28.1±0.2	Strunck (2001)	15

*$^{40}Ar/^{39}Ar$ results for these samples are already reported in von Eynatten *et al.* (1999)
†Numbers refer to Fig. 1

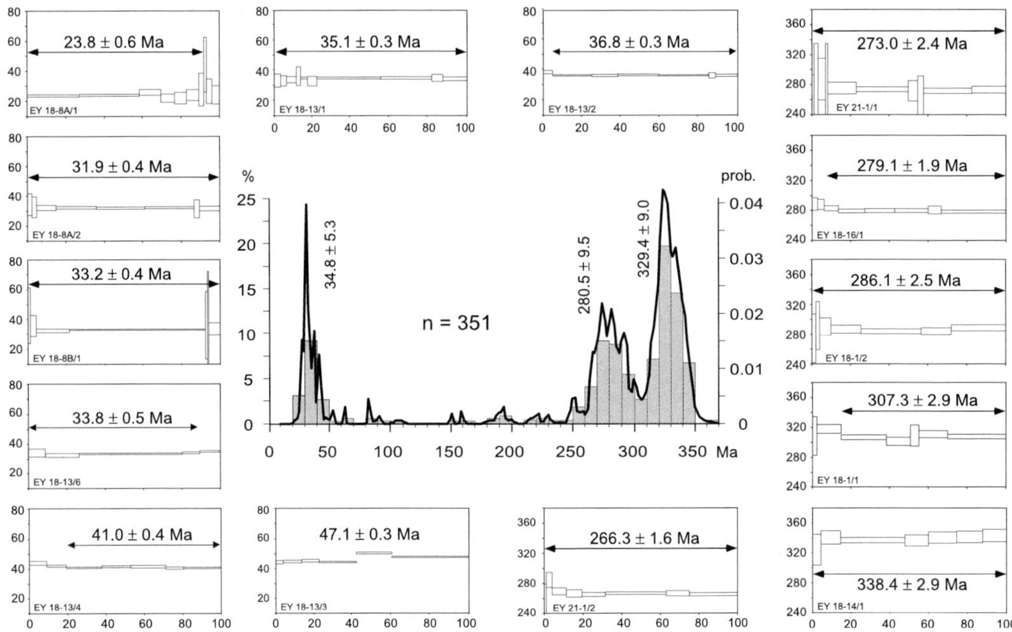

Fig. 2. Summary of $^{40}Ar/^{39}Ar$ data showing the age distribution of all analysed white micas (n=351). The data are presented in the form of a histogram, a probability distribution plot, and ^{39}Ar release spectra of incremental heating experiments. Average peak ages are calculated using the arithmetic mean and one sigma standard deviation of 20–50, 260–300, and 310–350 Ma age intervals. Probability distribution plots are calculated so that the area under the curve sums to 1 (see Sircombe, 1999 for details).

metasediments, the data may also represent mica crystallization ages under greenschist-grade metamorphic conditions (350–450°C, e.g. Markley *et al.* 1998). This alternative interpretation would not significantly affect cooling rates calculated from these micas.

The observed age data are not homogeneously distributed within the Swiss Molasse Basin, instead showing distinct changes with respect to both place and time of deposition. In the following paragraphs we will discuss the distribution of detrital white mica ages within the stratigraphic column from east to west.

In the eastern drainage systems (locations 1 and 2 in Fig. 1) the stratigraphic age of the samples ranges from *c.*31 to *c.*13 Ma. Mica cooling ages are mostly Carboniferous (83%) ranging from 300 to 340 Ma, with a peak age of 325 Ma (Fig. 3). The age data concentrate between 300 and 340 Ma throughout the stratigraphic column (Fig. 3). In addition, a few Cretaceous white mica ages are recorded from the oldest strata, and some scattered Permian ages are observed at 20 Ma stratigraphic age.

In the Rigi–Höhrone drainage system (location 3 in Fig. 1) the stratigraphic ages of the two samples are 29.8 and 22.1 Ma, covering the beginning and the end of sedimentation in this drainage system. At *c.*30 Ma, mica cooling ages show a very narrow distribution between 335 and 347 Ma, comparable with micas from the oldest deposits of the Honegg–Napf drainage system to the west (Fig. 3, loc. 4). At *c.*22 Ma, mica cooling ages show a wider distribution, ranging between 278 and 368 Ma, but concentrate around 320 to 335 Ma. The peak age of 324 Ma is very similar to the peak age of eastern sections (325 Ma). No Tertiary mica ages are observed in the Rigi–Höhrone drainage system.

In the Honegg–Napf drainage system further to the west (location 4 in Fig. 1) the stratigraphic ages of the samples range from *c.*31 to *c.*14 Ma. Mica cooling ages display two distinct Palaeozoic ('Variscan') groups and a very pronounced Tertiary group with a peak age of 32 Ma (Fig. 3). The older Variscan age group is Early Carboniferous (320–350 Ma, peak age 335 Ma) and thus slightly older than the amalgamated peak age in Fig. 2. The second Variscan age group is Early Permian (265–295 Ma, peak ages at 274 and 282 Ma; Fig. 3) in age, and thus fairly precisely resembles the Early Permian peak age

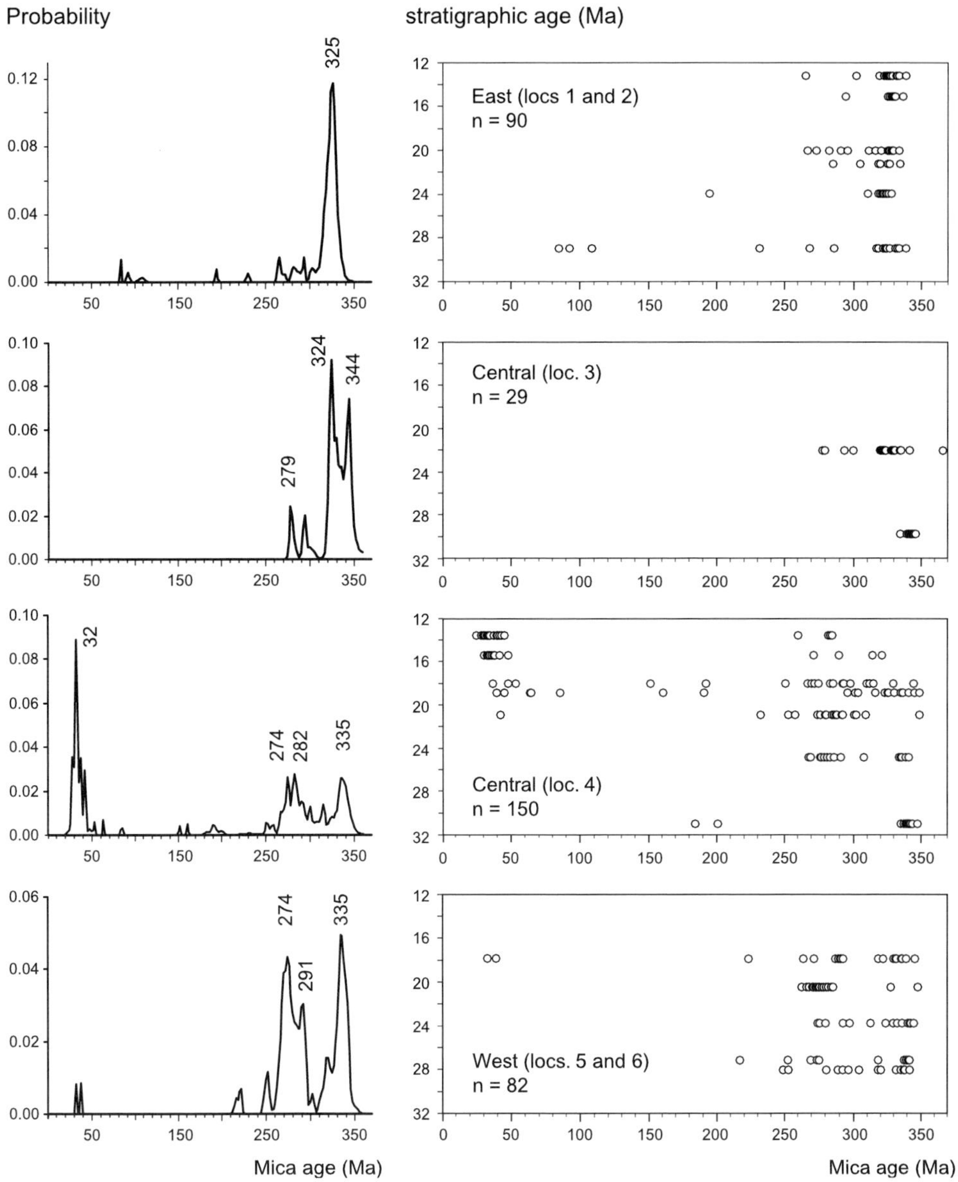

Fig. 3. Spatial differentiation of the age data in probability distribution plots (on the left) for the eastern (drainage systems 1 and 2, see Fig. 1), central (seperately for drainage systems 3 and 4) and western parts of the basin (drainage systems 5 and 6). Numbers give the age of the peak maximum. On the right, plots of mica age v. stratigraphic age for each area display the variation in mica ages within the stratigraphic column.

from the overall age distribution (Fig. 2). The occurrence of the Variscan age groups varies within the stratigraphic column. At the base of the system in the sediments deposited at around 31 Ma, only Early Carboniferous ages (330–350 Ma) are recorded, but, in the sediments deposited at around 25 Ma, Early Permian ages become frequent. From now on this age group is present in all sediments up-section, whereas Early Carboniferous ages disappear in the youngest sediments at 16 to 13 Ma sedimentation age. The first Tertiary ages occur in *c.*20-Ma old sediments, and their proportion strongly increases up section (Fig. 3). In the youngest sample, with

13.6 Ma sedimentation age, 87% of all dated grains are Tertiary with a peak age of 32 Ma. Almost 40% of them are <30 Ma, with the youngest grain dated at 23.8±0.6 Ma (Fig. 2).

In the western drainage systems (locations 5 and 6 in Fig. 1) the stratigraphic ages of the samples range from *c.*28 to *c.*18 Ma. Mica cooling ages display two distinct Variscan groups, and the first Tertiary mica grains, with cooling ages between 30 and 40 Ma, are observed in the youngest sediments (Fig. 3). The Carboniferous and Early Permian age peaks (335 Ma and 274/(291) Ma, respectively) precisely resemble those from the Honeg–/Napf system in locality 4 (Fig. 1). These two age groups are present throughout all stratigraphic levels, whereas the Tertiary grains are restricted to the youngest sample, which has a stratigraphic age of 17.9 Ma.

Mica chemistry

The pronounced change to Tertiary mica ages at around 20 Ma in drainage system 4 (Honegg–Napf) is not only related to a different cooling history of the source rocks but also to a different source rock lithology. This is reflected by a concomitant change in mica chemistry. Fig. 4 illustrates electron microprobe data from detrital white micas using Si–Mg plots for different stratigraphic levels. In 31 Ma and 25 Ma old sediments the detrital micas are exclusively muscovites with up to 3.20 Si p.f.u. The first phengitic mica, with up to 3.50 Si p.f.u., occurs in *c.*20-Ma old sediments. In the youngest sediments detrital micas are almost exclusively phengites with average values of 3.35–3.40 Si p.f.u. and maximum values of >3.50 Si p.f.u. The latter are comparable in composition with phengites from rocks exposed in the western Lepontine Dome (Hammerschmidt & Frank 1991), but also with several occurrences of phengitic mica from Penninic units and Austroalpine outliers from the hanging wall of the Simplon Fault zone (e.g. Dal Piaz *et al.* 2001).

We can conclude that the change to Tertiary ages in drainage system 4 is accompanied by a change in mica chemistry from muscovite to phengite. Although we do not know the details of the host-rock paragenesis of the detrital grains, the mica chemistry is in line with a predominance

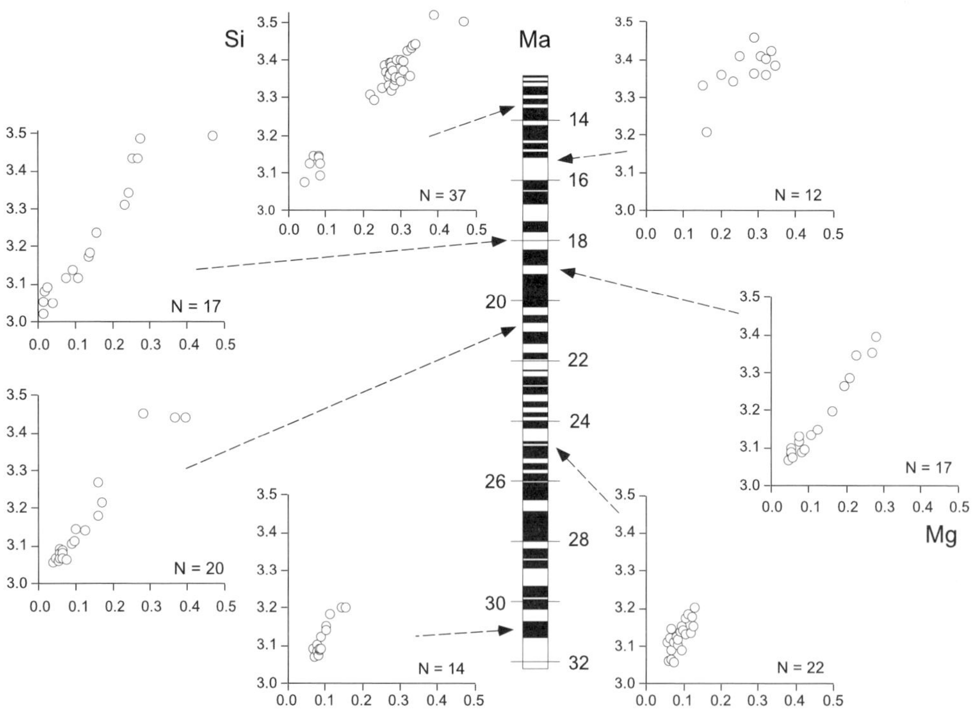

Fig. 4. Si v. Mg p.f.u. (11 cations) plots based on electron microprobe data of the dated white mica from central transverse drainage system 4 (see Fig. 1). The central column shows a precise magnetostratigraphic calibration of the sandstone samples, following Schlunegger *et al.* (1996) and Kempf *et al.* (1997).

of granitoid source rocks for sediments older than *c.*20 Ma, and with a significant contribution from metamorphic source rocks (low temperature, possibly high-pressure metasediments) for sediments younger than *c.*20 Ma (von Eynatten *et al.* 1999).

Discussion

In the following section we first discuss potential sources for detrital micas with Tertiary cooling ages and, subsequently, we use the Tertiary mica ages from this study to calculate cooling rates and to reconstruct palaeo-cooling paths for different time slices of the Early to Middle Miocene. Finally, we discuss potential sources for Variscan micas by relating the observed age cluster to geochronological data of the presently exposed hinterland.

Recent compilations of geochronological data from the present-day hinterland are given by Gebauer (1999), Schaltegger & Gebauer (1999), and Thöni (1999). Concerning Variscan micas, we argue that mica cooling ages of present-day exposed Variscan rocks closely resemble mica cooling ages of Variscan source rocks exposed in Oligocene to Miocene times. Both have experienced only a weak Alpine overprint, and the original Variscan age information should be the same regardless of whether the rocks were exposed in the Oligocene or today, under the provision that the source rocks have uniform cooling ages, i.e. they cooled quickly following Variscan magmatic emplacement, and remained in an upper crust environment. In contrast, thermochronological differences between Tertiary mica grains from Oligocene to Miocene source rocks and from present-day exposed rocks are obvious, because both were subjected to different stages of Tertiary Alpine metamorphism and were exhumed at different times within an active orogenic process.

For the following compilations it is important to note that published data from fine-grained (<100 μm) illite and/or white mica fractions are not considered because they are inappropriate sources for detrital grains of 250–500 μm and 500–1000 μm grain-size fractions analysed here. For the $^{40}Ar/^{39}Ar$ data we have cited the total fusion ages if not otherwise stated.

Sources for Tertiary micas

Tertiary micas from the Eastern Alps are restricted to Penninic windows which cannot be considered as sources for Swiss Molasse sediments because they were still covered by Austroalpine units at around 15 Ma (Frisch *et al.* 1998).

Presently exposed rocks in the Central and Western Alps show a wide range in Tertiary K–Ar white mica ages from Paleocene to Middle Miocene. The youngest ages, between 14 and *c.*25 Ma, are reported from the Lepontine metamorphic dome (e.g. Hurford *et al.* 1989; Steck & Hunziker 1994). The youngest known $^{40}Ar/^{39}Ar$ plateau age (14.1±0.3 Ma) on a coarse-grained fraction was measured on muscovite (180–250 μm) from basement rocks of the western part of the Lepontine (Hammerschmidt & Frank 1991). Towards the east, in the Tambo and Suretta nappes K–Ar white mica ages between 30 and 45 Ma become dominant (Hunziker *et al.* 1992; Steck & Hunziker 1994).

Towards the west, in the hanging wall of the Simplon detachment fault the Penninic–Austroalpine nappe stack yield a range of $^{40}Ar/^{39}Ar$ white mica ages from 30 to >65 Ma. Synkinematic phengitic white micas from cover units of the Siviez Mischabel Nappe (Grand St Bernard; see Fig. 1) are dated between 36 and 41 Ma and interpreted to reflect peak metamorphism (Markley *et al.* 1998). In the eastern Siviez–Mischabel the ages become younger (30–36 Ma) and probably reflect post-kinematic cooling due to extension along the Simplon detachment fault. In the internal Penninic Monte Rosa Massif, Tertiary $^{40}Ar/^{39}Ar$ white mica (mostly phengite) ages range between 30 and 45 Ma. Barnicoat *et al.* (1995) reported $^{40}Ar/^{39}Ar$ white mica ages of 40–44 Ma from the thrust contact zone between the Zermatt–Sass eclogites and gneisses of the Grand St Bernard. Dal Piaz *et al.* (2001) reported ages of 43 to 46 Ma from high-pressure phengites of the Zermatt–Saas nappe. Similar ages (42–45 Ma) are known from greenschist-facies gneisses in the contact zone between the Austroalpine Pillonet outlier and the underlying Penninic Combin Zone (Cortiana *et al.* 1998). The outlier itself yields uppermost Cretaceous ages (*c.*75 Ma) comparable with those reported from the Sesia Lanzo Zone (Hurford *et al.* 1989; Ruffet *et al.* 1995). In the Austroalpine Dent Blanche Nappe the K–Ar white mica ages are concentrated between 45 and 60 Ma (Hunziker *et al.* 1997). The latter were mostly obtained from fine-grained fractions. No $^{40}Ar/^{39}Ar$ ages on coarse-grained white mica from the Dent Blanche are known from the literature, but the Rb/Sr ages of phengitic micas are in the range of 40 to 49 Ma (Dal Piaz *et al.* 2001).

In a previous study (von Eynatten *et al.* 1999) we already concluded that detrital white mica in the Swiss Molasse basin, whose source rock had cooled below 350–420°C before 30 Ma must have been derived from the footwall of the Simplon detachment fault. This is because micas from currently exposed rocks from the hanging wall are generally older than 30 Ma and mostly range between 35 and 45 Ma (see above), whereas micas from currently exposed rocks of the footwall concentrate between 20 and 15 Ma. Therefore, white mica dated as <30 Ma in a 15 Ma sediment cannot have been derived from the hanging wall of the Simplon Fault and most probably originated from the eroded upper levels of the Lepontine Dome.

Ongoing exhumation and erosion of the Alps since at least the Miocene implies that areas containing micas with distinct cooling ages today, delivered older mica from higher levels in the section into the Swiss Molasse Basin during the Miocene. This effect is illustrated for Middle Miocene times by the mica age distributions from 13.6 and 15.4 Ma old sediments from drainage system 4: the age distributions are quite similar in shape but micas from the older sandstone sample are clearly shifted towards older cooling ages (Fig. 5). Assuming the same cooling rates for both samples, the shift towards older ages for the 15.4 Ma sandstone sample should be in the same range as the shift in stratigraphic age (*c.*2 Ma). The higher shift in mica ages of about 4 Ma results from higher cooling rates for the 13.6 Ma sandstone sample (see below). The very similar age distribution pattern for both samples together with an almost identical mica chemistry (Fig. 4) suggests a common source block of similar composition and metamorphic history, which exhumed with higher rates at 13.6 Ma sedimentation age compared with the 15.4 Ma sedimentation age.

To summarize, the Tertiary white mica in the Swiss Molasse basin may be derived from the Penninic units of the Central and Western Alps and from Austroalpine units ('outliers') of the Western Alps. Contrary to the case of the Variscan micas, which yielded ages in the same range as found in the source rocks today (see below), the observed age pattern found in the Alpine metamorphic rocks today cannot be directly related to that inferred for Miocene times from the detrital minerals.

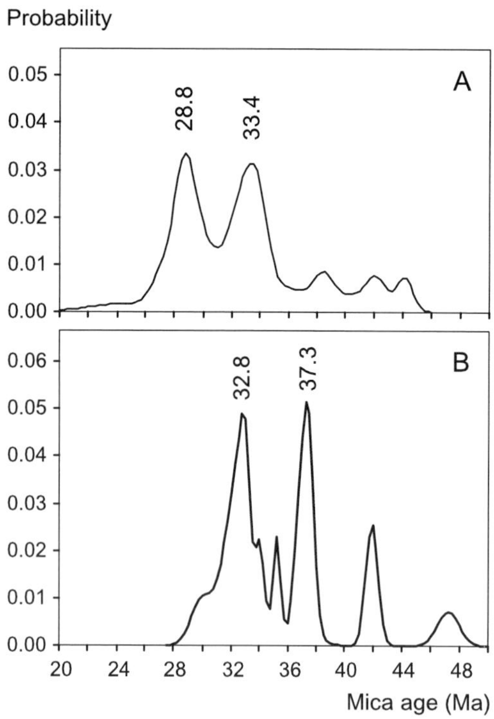

Fig. 5. Distribution of total fusion ages for 13.6 Ma (*A*, n=27) and 15.4 Ma (*B*, n=18) sandstone samples from central transverse drainage system 4 (see Fig. 1). The shape of the distributions is quite similar but shifted at about 4 Ma towards younger mica ages for the 13.6 Ma sandstone sample.

Paleo-cooling of the Central Alps

Figure 6 shows all observed Tertiary ages from drainage system 4 (Honegg–Napf; see Fig. 1), and while stratigraphic ages range from 20.9 to 13.6 Ma, the detrital white mica ages continuously decrease with decreasing sedimentation age. The difference in time Δt between the youngest mica age of each stratigraphic level and the age of deposition also decreases (italic numbers in Fig. 6). Taking this Δt and assuming a closure temperature for Ar diffusion in white mica of 350–420°C, a surface temperature comparable with that observed today, continuous erosion and immediate deposition, allows the calculation of average cooling rates. These cooling rates continuously increase with decreasing sample age (bold numbers in Fig. 6). The increase in cooling rates started at *c.*20 Ma stratigraphic age from rates of <20°C/Ma, and continues at least up to 13.6 Ma with rates of 34–41°C/Ma.

The $^{40}Ar/^{39}Ar$ data from detrital white mica presented here, and FT ages from detrital zircons (Spiegel *et al.* 2000) extracted from the same sandstone samples used in the present study, yield quite comparable cooling rates for each stratigraphic level (Fig. 7). Thus, zircon FT data support the continuous increase of cooling rates with decreasing stratigraphic age. Zircon data match almost perfectly the $^{40}Ar/^{39}Ar$ data when using the high end of the closure-temperature

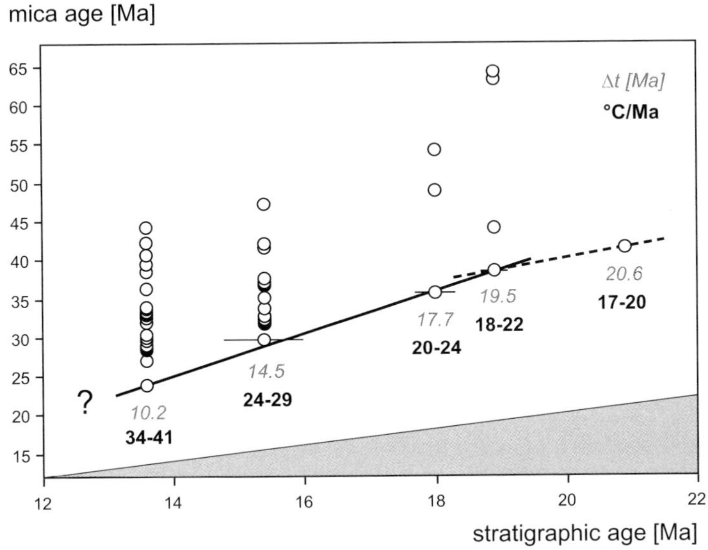

Fig. 6. Mica age v. stratigraphic age diagram for all Tertiary micas from the central transverse Honegg–Napf drainage system (no. 4 in Fig. 1). Both errors (2 sigma) are generally less than the size of the symbol, with the exception of some errors in the stratigraphic ages. The black line indicates the trend of decreasing mica ages with decreasing stratigraphic age. Italic numbers indicate the difference between the stratigraphic age and youngest mica age (Δt). Bold numbers give the corresponding range of cooling rates based on a closure temperature of 350–420°C. The grey area means that the mica ages would be lower than the stratigraphic ages.

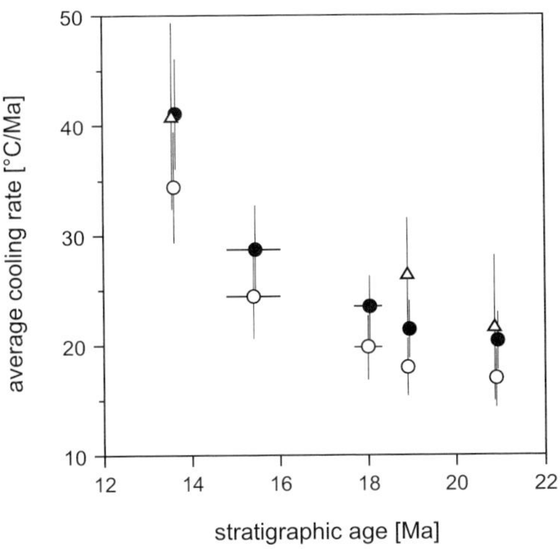

Fig. 7. Average cooling rates v. stratigraphic age calculated from $^{40}Ar/^{39}Ar$ white mica data (this study) and zircon FT data (Spiegel *et al.* 2000). Cooling rates are calculated, assuming closure temperatures of 350°C (open circles) and 420°C (filled circles) for Ar diffusion in white mica and a closure temperature of 240°C for fission tracks in zircon (open triangles). Calculated errors for the cooling rates include errors in the stratigraphic age (see Table 1), the mineral age (two sigma), and a ±50°C error for each closure temperature.

interval for Ar diffusion in white mica. The only exception, at *c.*19 Ma sedimentation age, suggests a significantly higher cooling rate of *c.*26°C/Ma based on a zircon FT model age of 28 Ma (Spiegel *et al.* 2000), but this age may be influenced by zircons derived from Bergell or related magmatic rocks which yield exactly the same ages (Giger & Hurford 1989).

Comparing the observed cooling rates for detrital mica from Miocene sediments with the spatial pattern of cooling rates calculated for the time interval between 15 and 20 Ma (Fig. 8), we have to conclude that rocks with cooling rates higher than 20°C/Ma derive from the area of the future Lepontine metamorphic dome. The spatial cooling pattern was calculated by Schlunegger & Willett (1999) using the cooling ages of presently exposed rocks.

The $^{40}Ar/^{39}Ar$ white mica and zircon FT data are used to construct palaeo-cooling paths of the Early to Middle Miocene Central Alps and to compare the result with the cooling paths of presently exposed rocks from the hanging wall and the footwall of the Simplon detachment fault (Fig. 9) To simplify this diagram, we used uniform closure temperatures of 350°C for Ar in white mica and of 240°C for zircon fission tracks (Hurford 1986). To construct such a palaeo-

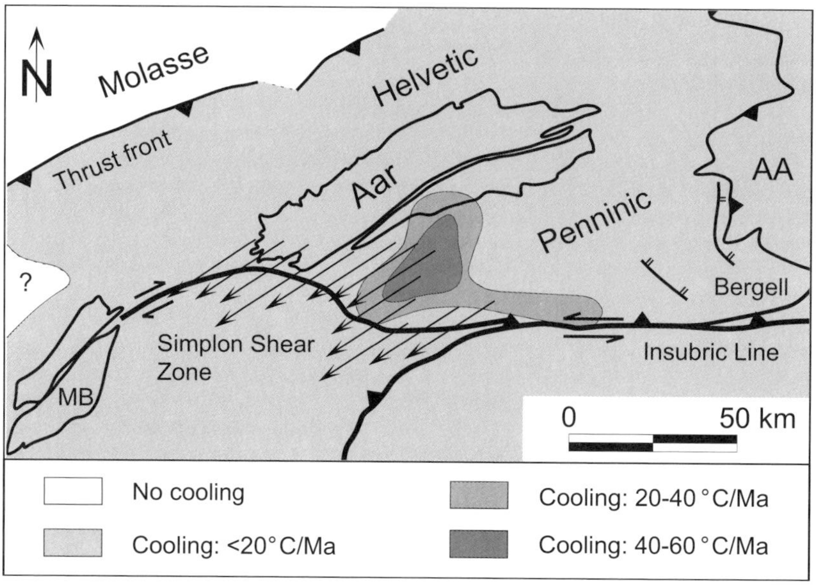

Fig. 8. Spatial pattern of cooling rates calculated from presently exposed rocks for the time interval 20–15 Ma (Schlunegger & Willett, 1999). AA, Austroalpine, MB, MontBlanc Massif. Arrows indicate the transport direction of hanging-wall units of the Simplon shear zone.

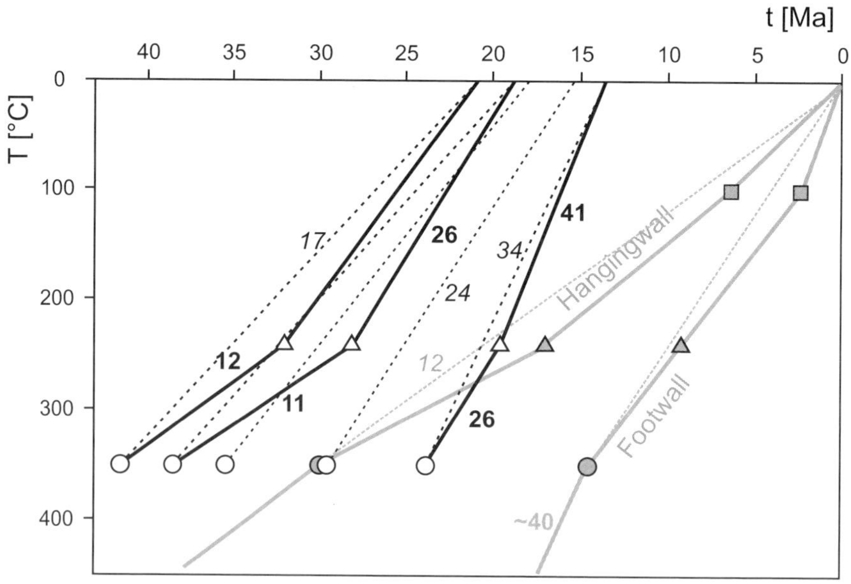

Fig. 9. Palaeo-cooling paths (solid black lines) for five time slices in the Miocene, ranging from 20.9 to 13.6 Ma (open circles, $^{40}Ar/^{39}Ar$ mica data (T_c=350°C; cf. Fig. 7), open triangles, zircon FT data). Stippled lines show palaeo-cooling paths based on $^{40}Ar/^{39}Ar$ white mica data alone. Grey lines and symbols indicate cooling paths of present-day exposed units from the hanging wall and the footwall of the Simplon detachment fault (including apatite FT data, grey squares). The latter were constructed using the youngest reported ages for each geochronometer (Seward & Mancktelow 1994; Steck & Hunziker 1994; Markley *et al.* 1998). Bold numbers indicate the cooling rates of solid lines, and italic numbers indicate the cooling rates of stippled lines (°C/Ma).

cooling path we have to assume that the detrital minerals originated not necessarily from the same rock but from the same source unit. Palaeo-cooling based on $^{40}Ar/^{39}Ar$ data alone (stippled lines) show a progressive steepening of gradients with decreasing stratigraphic age (Fig. 9). This steepening reflects the increase of average cooling rates from 17 to 34°C/Ma (T_c=350°C) or from 20 to 41°C/Ma (T_c=420°C, compare Figs 6 and 7).

Detrital minerals deposited around 20 Ma reflect a cooling path of the source rocks that displays moderate cooling rates of *c.*10°C/Ma from 350°C to 240°C, comparable with cooling rates of the presently exposed hanging wall rocks (Markley *et al.* 1998). Based on $^{40}Ar/^{39}Ar$ data alone, average cooling rates are slightly higher, but still <20°C/Ma. At a 19 Ma sedimentation age a zircon FT model age of 28 Ma already suggests an average cooling rate as high as 26°C/Ma for cooling from 240°C to around zero °C between 28 and 19 Ma (Fig. 9). This increase of cooling rates from *c.*10°C/Ma to 26 °C/Ma may be related to the beginning of the exhumation of the Lepontine, but the FT model age of 28 Ma may also be influenced by zircons derived from Bergell or related magmatic rocks (see above). However, a pronounced shift towards higher cooling rates between 24 and 41°C/Ma in the source area of *c.*16 to 13.6 Ma old sandstones is documented by both $^{40}Ar/^{39}Ar$ white mica and zircon FT data. The steep gradient for the youngest sample corresponds very well with the steepest part of the cooling path of the present-day footwall rocks exposed in the Lepontine (*c.*40°C/Ma; Fig. 9).

The $^{40}Ar/^{39}Ar$ data give quite similar cooling results compared to those derived from present-day crystalline rocks (e.g. Grasemann & Mancktelow 1993; Markley *et al.* 1998; Schlunegger & Willett 1999). This is valid for both rates of cooling in the source areas and the time of maximum cooling between *c.*19 and 13.6 Ma.

Sources for Variscan micas

Carboniferous mica ages are frequently reported from Austroalpine basement rocks and generally interpreted as recording cooling after the Variscan Orogeny. In the Austroalpine Ötztal–Stubai and Scarl units, the published K–Ar white mica data fall into a narrow age range between 297 and 318 Ma (Hoinkes & Thöni 1993). In a recent study on Ötztal basement rocks Hoinkes *et al.* (1997) reported $^{40}Ar/^{39}Ar$ plateau ages for different white mica grain-size fractions of a metagranitoid varying between 305 and 313 Ma. Total fusion ages fall into the same range, except for a slightly younger 295 Ma age for the finest fraction (100–160 μm). Frei *et al.* (1995) summarized published K–Ar white mica ages of the Austroalpine Silvretta Nappe, and calculated an average age of 314±17 Ma, representing about 70% of the data. The data are interpreted as cooling ages after late Variscan low-grade metamorphism (Magetti & Flisch 1993).

Brügel (1998) reported mica ages for gneissic pebbles from Miocene conglomerates of the Eastern Alpine Molasse Basin in Germany. These pebbles derive from Austroalpine basement units west and southwest of the Tauern window, and K–Ar white mica ages vary between 294 and 354 Ma (average 317±22 Ma).

From the Lower Austroalpine Err–Bernina Nappe no K–Ar ages of coarse-grained white micas are reported in the literature. Two phases of Variscan magmatic activity are documented by U–Pb zircon ages: a calc-alkaline granodiorite suite dated between 324 and 338 Ma and an alkali-granite suite around 295 Ma (von Quadt *et al.* 1994). These authors cite K–Ar biotite ages for these rocks scattered between 223 and 258 Ma (see also Hunziker *et al.* 1992). On these grounds we would expect that K–Ar ages from white micas derived from the Err–Bernina Nappe may range between >260 and *c.*330 Ma.

Recycling of Mesozoic clastic sediments may be another source for Variscan white mica grains. In the western part of the Austroalpine sedimentary cover nappes, synorogenic sediments of Cretaceous age contain exclusively Variscan white micas ranging from 301 to 378 Ma in age (von Eynatten *et al.* 1996). Quite similar $^{40}Ar/^{39}Ar$ plateau ages (303–360 Ma) are reported from detrital white micas from Upper Carboniferous and Permian sediments of the Austroalpine region (Handler *et al.* 1997). Variscan detrital white mica ages reported so far from the Rhenodanubian Flysch Zone of the Eastern Alps predominantly range from 299 to 330 Ma (Neubauer *et al.* 1999).

In the Central and Western Alps, Alpine greenschist to amphibolite metamorphism has mostly reset K–Ar and Rb–Sr isotope systems (see below). Variscan white micas are reported from Briançonnais basement units (orthogneisses and pegmatites) of the Siviez–Mischabel Nappe from the upper Penninic Grand St Bernard (see Fig. 1) nappe stack (Markley *et al.* 1998). The $^{40}Ar/^{39}Ar$ data range from 281 to 337 Ma for grain sizes >250 μm, from 200 to 250 Ma for grain size 180–250 μm, and go down to 60 Ma for grain sizes <63 μm. The data are thought to reflect an increasing Alpine imprint with decreasing grain size. These ages show good correspondence with the Early Permian ages observed for 250 μm to 500 μm detrial white mica in this study (Fig. 10). Metapelites from the Pontis Nappe (Grand St Bernard) lack the Alpine greenschist overprint and display $^{40}Ar/^{39}Ar$ total fusion ages between 288 and 307 Ma (Giorgis *et al.* 1999). Monié (1990) reported preserved Variscan $^{40}Ar/^{39}Ar$ ages from the Internal Penninic Ambin massif. Whereas phengites from the Triassic sedimentary cover rocks give Tertiary metamorphic ages (37±1 Ma), muscovites from basement rocks yield 340 and 346 Ma plateau ages (334 and 339 Ma total fusion ages, respectively).

Further possible relicts of Variscan mica ages are suggested by rare Variscan and more frequent Permo-Mesozoic K–Ar and Rb–Sr mica ages from the Dent-Blanche, Grand St Bernard, Tambo and Suretta nappes (Hunziker *et al.* 1992, 1997; Cortiana *et al.* 1998). These ages are often interpreted as Variscan–Alpine mixed ages, but may also be related to Mesozoic thermal events due to the opening of the Penninic Ocean (von Eynatten 1996; Bertotti *et al.* 1999) or to Permian magmatism documented by 270 to 280 Ma U–Pb ages of zircons from the Tambo, Suretta and Siviez–Mischabel nappes (Gebauer 1999).

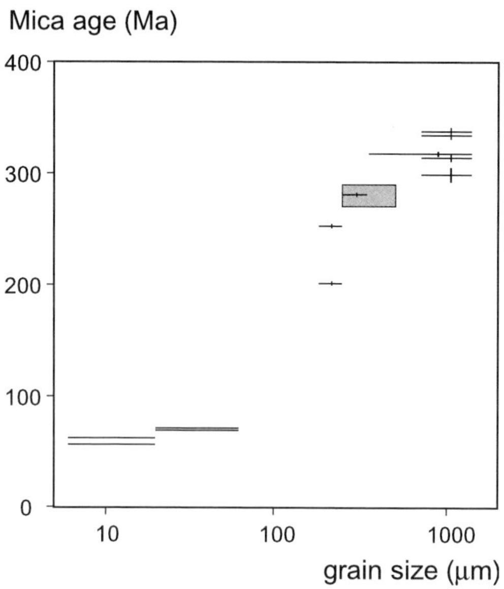

Fig. 10. White mica $^{40}Ar/^{39}Ar$ age v. grain-size fraction from the basement rocks of the Siviez–Mischabel Nappe (GSB; Markley *et al.* 1998). The shaded box indicate Early Permian age peaks (*c.*281±10 Ma, Fig. 2) observed for 250–500 μm detrital white mica from the central and western sections of the Swiss Molasse Basin.

To summarize, Austroalpine basement rocks from the western Eastern Alps are expected to provide Variscan detrital white micas with $^{40}Ar/^{39}Ar$ ages largely between 300 and 330 Ma. Recycled micas from the sedimentary cover may be slightly older – up to 350–360 Ma. Micas from the Lower Austroalpine basement (Err–Bernina) may be slightly younger. In the Central and Western Alps basement and sedimentary cover rocks lacking a pervasive Alpine greenschist-facies overprint are expected to yield detrital white micas with a larger K–Ar age range between 260–270 Ma and 340–350 Ma. Based on this compilation of data from present-day exposed rocks, we can propose that, firstly, Variscan detrital white micas from the eastern transverse drainage systems (numbers 1 and 2; see Fig. 1) of the Swiss Molasse Basin are largely derived from Austroalpine units, and secondly, Variscan white micas from central to western drainage systems (numbers 4, 5 and 6) show significant contributions from Penninic basement units. The latter is confirmed by frequent Early Permian ages of between 265 and 295 Ma, which are rare in Austroalpine units.

In contrast, Spiegel *et al.* (2001) proposed the erosion of Lower Austroalpine granitoids from a western prolongation of the Err–Bernina region in the hinterland of section 4 (Honegg–Napf) up to at least the Oligocene–Miocene boundary, and denied the possibility of Penninic sources. However, the increase of the Early Permian $^{40}Ar/^{39}Ar$ white mica ages towards the west and towards younger deposits (Fig. 3) allow us to argue in favour of Penninic source units from the Western Alps. If further geochronological and/or petrographic data support an Austroalpine origin for granitoid rocks exposed around the Oligocene to Miocene boundary in the hinterland of sections 4 to 6, the Dent Blanche region appears to be a more likely source area than the Err–Bernina region.

Conclusions

The results from $^{40}Ar/^{39}Ar$ dating of detrital white micas from precisely calibrated Oligocene to Miocene stratigraphic levels in the North Alpine foreland basin (NAFB) give valuable insights into the thermochronological evolution of the Oligocene to Miocene Central Alps. In detail, we draw can the following conclusions:

(1) The exhumation and cooling of the Lepontine metamorphic dome can be traced precisely by the $^{40}Ar/^{39}Ar$ geochronology of detrital white mica from the NAFB. This is valid for firstly, the evolution of cooling rates

up to 34–41°C/Ma, which correspond with cooling rates derived from detrital zircon FT thermochronometry and to cooling rates derived from presently exposed rocks, and secondly, the time of maximum cooling (19–14 Ma), which corresponds with results from thermal modelling of geochronological data from presently exposed rocks (Grasemann & Mancktelow 1993). To a certain extent even the place of maximum cooling is constrained because in the eastern drainage systems no Tertiary ages are observed in 21- to 13-Ma old sediments.

(2) White mica $^{40}Ar/^{39}Ar$ ages show two distinct Variscan age cluster that can be used as an additional provenance indicator for NAFB sediments. The frequent occurrence of Early Permian ages in the more westerly drainage systems 4, 5 and 6 most probably reflects erosion of Penninic basement rocks. Their first occurrence in the source area is documented at *c.*25 Ma in drainage system 4, and already at *c.*28 Ma in the westernmost drainage system 6.

(3) Detrital white micas from all sections deposited before 20 Ma appear to be not affected by Alpine greenschist-facies overprint, as documented by the lack of Cretaceous white mica ages. Fine-grained synkinematic phengitic micas of Cretaceous to Tertiary age are widespread within Upper Penninic and Austroalpine units of the Alps, but coarse-grained detrital micas in NAFB sandstones derived from these units only document Variscan to Early Permian ages (cf. Fig. 10). Whereas the FT-ages of detrital zircons are largely reset in the Cretaceous (Spiegel *et al.* 2000), Alpine greenschist facies metamorphism did not reset the Ar system of coarse-grained white mica in Oligocene–Miocene source rocks except for those of the Miocene Lepontine area.

(4) The results of this study clearly demonstrate that $^{40}Ar/^{39}Ar$ dating of detrital minerals is a precise recorder of tectonic processes in the source area, and thus a powerful tool for deciphering the geology of source areas that were largely eroded and possibly no longer exposed. Because white micas are very stable within the sedimentary cycle, they are among the most appropriate minerals for determining the geochronology of detrital material.

This study was supported by the Deutsche Forschungsgemeinschaft Grant EY 23/1. Field sampling strongly benefited from guidance and advice by O. Kempf, F. Schlunegger, and P. Strunk. A. Kronz gave valuable advice on the use of the electron microprobe. We appreciate the comments and suggestions of G. V. Dal Piaz, N. Froitzheim, R. Gaupp, N. Mancktelow, F. Schlunegger, C. Spiegel and B. van der Klauw. Thorough reviews by S. Sherlock and N. White have greatly helped to improve the manuscript. The contribution of JRW is covered under NSG No. 20020301.

References

BARNICOAT, A. C., REX, D. C., GUISE, P. G. & CLIFF, R. A. 1995. The timing of and nature of greenschist facies deformation and metamorphism in the upper Pennine Alps. *Tectonics*, **14**, 279–293.

BATT, G. E. & BRAUN, J. 1997. On the thermomechanical evolution of compressional orogens. *Geophysical Journal International*, **128**, 364–382.

BAUDIN, T., MARQUER, D. & PERSOZ, F. 1993. Basement–cover relationships in the Tambo nappe (Central Alps, Switzerland): geometry, structure, and kinematics. *Journal of Structural Geology*, **15**, 543–553.

BERGER, J.-P. 1996. Cartes paléogeographiques–palinspastiques du bassin molassique suisse (Oligocène inférieur–Miocéne moyen). *Neues Jahrbuch für Geologisch–Paläontologische Abhandlungen*, **202**, 1–44.

BERTOTTI, G., SEWARD, D., WIJBRANS, J., TER VOORDE, M. & HURFORD, A. J. 1999. Crustal thermal regime prior to, during, and after rifting: a geochronological and modeling study of the Mesozoic South Alpine rifted margin. *Tectonics*, **18**, 185–200.

BRÜGEL, A. 1998. Provenances of alluvial conglomerates from the eastalpine foreland: Oligo-/Miocene denudation history and drainage evolution of the Eastern Alps. *Tübinger Geowissenschaftliche Arbeiten, Reihe A*, **40**, 1–168.

CERVENY, P. F., NAESER, N. D., ZEITLER, P. K., NAESER, C. W. & JOHNSON, N. M. 1988. History of uplift and relief of the Himalaya during the past 18 million years: evidence from fission-track ages of detrital zircons from sandstones of the Siwalik Group. *In:* KLEINSPEHN, K. L. & PAOLA, C. (eds) *New Perspectives in Basin Analysis*. Springer, New York, 43–61.

COPELAND, P. & HARRISON, T. M. 1990. Episodic uplift in the Himalaya revealed by $^{40}Ar/^{39}Ar$ analysis of detrital K-feldspar and muscovite, Bengal fan. *Geology*, **18**, 354–357.

CORTIANA, G., DAL PIAZ, G. V., DEL MORO, A., HUNZIKER, J. C. & MARTIN, S. 1998. $^{40}Ar/^{39}Ar$ and Rb/Sr dating of the Pillonet klippe and Sesia–Lanzo basal slice in the Ayas valley and evolution of the Austroalpine–Piedmont nappe stack. *Memorie di Scienze Geologiche*, **50**, 177–194.

DAL PIAZ, G. V. 1999. The Austroalpine–Piedmont nappe stack and the puzzle of Alpine Tethys. *Memorie di Scienze Geologiche*, **51**, 155–176.

DAL PIAZ, G. V., CORTIANA, G., DEL MORO, A., MARTIN, S., PENNACCHIONI, G. & TARTAROTTI, P. 2001. Tertiary age and paleostructural inferences of the eclogitic imprint in the Austroalpine outliers and Zermatt–Saas ophiolite, western Alps. *International Journal of Earth Sciences*, **90**, 668–684.

DALRYMPLE, G. B. & DUFFIELD, W. A. 1988. High precision $^{40}Ar/^{39}Ar$ dating of Oligocene from Mogollon–Datil volcanic field using a continous laser system. *Geophysical Research Letters*, **15**, 463–466.

DESMONS, J., APRAHAMIAN, J., COMPAGNONI, R., CORTESOGNO, L. & FREY, M. 1999. Alpine metamorphism of the Western Alps: I. Middle to high T/P metamorphism. *Schweizerische Mineralogische und Petrographische Mitteilungen*, **79**, 89–110.

FREI, R., BIINO, G. G. & PROSPERT, C. 1995. Dating a Variscan pressure-temperature loop with staurolite. *Geology*, **23**, 1095–1098.

FREY, M. & FERREIRO MÄHLMANN, R. 1999. Alpine metamorphism of the Central Alps. *Schweizerische Mineralogische und Petrographische Mitteilungen*, **79**, 135–154.

FRISCH, W., KUHLEMANN, J., DUNKLE, I. & BRÜGEL, A. 1998. Palinspastic reconstruction and topographic evolution of the eastern Alps during late Tertiary tectonic extrusion. *Tectonophysics*, **297**, 1–15.

GEBAUER, D. 1999. Alpine geochronology of the Central and Western Alps: new constraints for a complex geodynamic evolution. *Schweizerische Mineralogische Petrographische Mitteilungen*, **79**, 191–208.

GIGER, M. & HURFORD, A. J. 1989. Tertiary intrusives of the Central Alps: their Tertiary uplift, erosion, redeposition and burial in the south-alpine foreland. *Eclogae Geologicae Helveticae*, **82**, 857–866.

GIORGIS, D., THÉLIN, P., STAMPFLI, G. & BUSSY, F. 1999. The Mont-Mort metapelites: Variscan metamorphism and geodynamic context (Briançonnais basement, Western Alps, Switzerland). *Schweizerische Mineralogische und Petrographische Mitteilungen*, **79**, 381–398.

GRASEMANN, B. & MANCKTELOW, N. S. 1993. Two-dimensional thermal modelling of normal faulting: the Simplon Fault Zone, Central Alps, Switzerland. *Tectonophysics*, **225**, 155–165.

HAMMERSCHMIDT, K. & FRANK, E. 1991. Relics of high pressure metamorphism in the Lepontine Alps (Switzerland) – $^{40}Ar/^{39}Ar$ and microprobe analyses on white K-micas. *Schweizerische Mineralogische und Petrographische Mitteilungen*, **71**, 261–274.

HAMES, W. E. & BOWRING, S. A. 1994. An empirical evaluation of the argon diffusion geometry in muscovite. *Earth and Planetary Science Letters*, **124**, 161–167.

HANDLER, R., DALLMEYER, R. D. & NEUBAUER, F. 1997. $^{40}Ar/^{39}Ar$ ages of detrital white mica from upper Austroalpine units in the Eastern Alps, Austria: evidence for Cadomian and contrasting Variscan sources. *International Journal of Earth Sciences*, **86**, 69–80.

HOINKES, G. & THÖNI, M. 1993. Evolution of the Ötztal–Stubai, Scarl–Campo, and Ulten basement units. *In:* VON RAUMER, J. F. & NEUBAUER, F. (eds) *Pre-Mesozoic Geology in the Alps*. Springer, Berlin, pp. 485–494.

HOINKES, G., THÖNI, M., LICHEM, C., BERNHARD, F., KAINDL, R., SCHWEIGL, J., TROPPER, P. & COSCA, M. 1997. Metagranitoids and associated metasediments as indicators for the pre-Alpine magmatic and metamorphic evolution of the western Ötztal basement (Kaunertal, Tirol). *Schweizerische Mineralogische und Petrographische Mitteilungen*, **77**, 299–314.

HOMEWOOD, P., ALLEN, P. A. & WILLIAMS, G. D. 1986. Dynamics of the Molasse basin of western Switzerland. *International Association of Sedimentology Special Publications*, **8**, 199–217.

HUNZIKER, J. C., DESMONS, J. & HURFORD, A. J. 1992. Thirty-two years of geochronological work in the Central and Western Alps: a review on seven maps. *Mémoires de Géologie (Lausanne)*, **13**.

HUNZIKER, J. C., HURFORD, A. J. & CALMBACH, L. 1997. Alpine cooling and uplift. *In:* PFIFFNER, O. A., LEHNER, P., HEITZMANN, P., MUELLER, ST & STECK, A. (eds) *Deep Structure of the Swiss Alps*. Birkhäuser, Basle, 260–263.

HURFORD, A. 1986. Cooling and uplift patterns in the Lepontine Alps, south-central Switzerland, and an age of vertical movement on the Insubric fault line. *Contributions to Mineralogy and Petrology*, **92**, 412–427.

HURFORD, A. J., FLISCH, M. & JÄGER, E. 1989. Unravelling the thermo-tectonic evolution of the Alps: a contribution from fission track analysis and mica dating. *Geological Society of London Special Publication*, **45**, 369–398.

KEMPF, O., BOLLIGER, T., KÄLIN, D., ENGESSER, B. & MATTER, A. 1997. New magnetostratigraphic calibration of Early to Middle Miocene mammal biozones of the North Alpine foreland basin. *In:* AGUILAR, J.-P., LEGENDRE, S. & MICHAUX, J. (eds) *Actes du Congrès BiochroM'97*. Mém. Trav. EPHE, Inst. Montpellier, **21**, 547–561.

KEMPF, O., MATTER, A., BURBANK, D. W. & MANGE, M. 1999. Depositional and structural evolution of a foreland basin margin in a magnetostratigraphic framework: the eastern Swiss Molasse Basin. *International Journal of Earth Sciences*, **88**, 253–275.

KIRSCHNER, D. L., COSCA, M. A., MASSON, H. & HUNZIKER, J. C. 1996. Staircase $^{40}Ar/^{39}Ar$ spectra of fine grained white mica: timing and duration of deformational events, and empirical constraints on argon diffusion. *Geology*, **24**, 747–750.

KUHLEMANN, J. & KEMPF, O. 2002. Post-Eocene evolution of the North Alpine Foreland Basin and its response to Alpine tectonics. *Sedimentary Geology*, **152**, 45–78

MAGETTI, M. & FLISCH, M. 1993. Evolution of the Silvretta nappe. *In:* VON RAUMER, J. F. & NEUBAUER, F. (eds) *Pre-Mesozoic Geology in the Alps*. Springer, Berlin, pp. 469–484.

MARKLEY, M. J., TEYSSIER, C., COSCA, M. A., CABY, R., HUNZIKER, J. C. & SARTORI, M. 1998. Alpine deformation and $^{40}Ar/^{39}Ar$ geochronology of synkinematic white mica in the Siviez–Mischabel nappe, western Pennine Alps, Switzerland. *Tectonics*, **17**, 407–425.

MAURER, H. 1983. Sedimentpetrographische Analysen an Molasseabfolgen der Westchweiz. *Jahrbuch der Geologischen Bundesanstalt Wien*, **126**, 23–69.

McDougall, I. & Harrison, T. M. 1999. *Geochronology and Thermochronology by the $^{40}Ar/^{39}Ar$ method.* 2nd edn, Oxford University Press, New York.

Monié, P. 1990. Preservation of Hercynian Ar/Ar ages through high-pressure low-temperature Alpine metamorphism in the Western Alps. *European Journal of Mineralogy*, **2**, 343–361.

Najman, Y. M. R., Pringle, M. S., Johnson, M. R. W., Robertson, A. H. F. & Wijbrans, J. R. 1997. Laser $^{40}Ar/^{39}Ar$ dating of single detrital muscovite grains from early foreland-basin sedimentary deposits in India: implications for early Himalayan evolution. *Geology*, **25**, 535–538.

Neubauer, F., Handler, R., Friedl, G., Genser, J. & Topa, D. 1999. Detrital mode and $^{40}Ar/^{39}Ar$ ages of detrital mica from the Rhenodanubian Flysch zone, Eastern Alps: significance for Alpine tectonics. *Tübinger Geowissenschaftliche Arbeiten, Reihe A*, **52**, 190.

Ruffet, G., Féraud, G., Balèvre, M. & Kiénast, J.-R. 1995. Plateau ages and excess argon in phengites: an $^{40}Ar/^{39}Ar$ laser probe study of Alpine micas (Sesia Zone, Western Alps, northern Italy). *Chemical Geology*, **121**, 327–343.

Schaltegger, U. & Gebauer, D. 1999. Pre-Alpine geochronology of the Central, Western and Southern Alps. *Schweizerische Mineralogische und Petrographische Mitteilungen*, **79**, 79–87.

Schlunegger, F. & Willett, S. 1999. Spatial and temporal variations in exhumation of the Swiss Alps and implications for exhumation mechanisms. *Geological Society of London Special Publication*, **154**, 157–179.

Schlunegger, F., Burbank, D. W., Matter, A., Engesser, B. & Mödden, C. 1996. Magnetostratigraphic calibration of the Oligocene to Middle Miocene (30–15 Ma) mammal biozones and depositional sequences of the Swiss Molasse Basin. *Eclogae Geologicae Helveticae*, **89**, 753–788.

Schlunegger, F., Matter, A., Burbank, D. W. & Klaper, E. M. 1997. Magnetostratigraphic constraints on relationships between evolution of the central Swiss Molasse basin and Alpine orogenic events. *Geological Society of America Bulletin*, **109**, 225–241.

Schmid, S. M., Pfiffner, O. A., Froitzheim, N., Schönborn, G. & Kissling, E. 1996. Geophysical–geological transect and tectonic evolution of the Swiss–Italian Alps. *Tectonics*, **15**, 1036–1064.

Seward, D. & Mancktelow, N. 1994. Neogene kinematics of the central and western Alps: evidence from fission-track dating. *Geology*, **22**, 803–806.

Sherlock, S. C. 2001. Two-stage erosion and deposition in a continental margin setting: an $^{40}Ar/^{39}Ar$ laserprobe study of offshore detrital white micas in the Norwegian Sea. *Journal of the Geological Society*, **158**, 793–799.

Sherlock, S. C. & Kelley, S. P., 2002. Excess argon evolution in HP–LT rocks: a UVLAMP study of phengite and K-free minerals, NW Turkey. *Chemical Geology*, **182**, 619–636.

Sircombe, K. N. 1999. Tracing provenance through the isotopic ages of littoral and sedimentary detrital zircon, eastern Australia. *Sedimentary Geology*, **124**, 47–67.

Spiegel, C., Kuhlemann, J., Dunkl, I., Frisch, W., von Eynatten, H. & Balogh, K. 2000. Erosion history of the Central Alps: evidence from zircon fission track data of the foreland basin sediments. *Terra Nova*, **12**, 163–170.

Spiegel, C., Kuhlemann, J., Dunkl, I. & Frisch, W. 2001. Paleogeography and catchment evolution in a mobile orogenic belt: the Central Alps in Oligo-Miocene times. *Tectonophysics*, **341**, 33–47.

Steck, A. & Hunziker, J. 1994. The Tertiary structural and thermal evolution of the Central Alps – compressional and extensional structures in an orogenic belt. *Tectonophysics*, **238**, 229–254.

Strunck, P. 2001. *The Molasse of western Switzerland.* Ph.D. Thesis, University of Berne.

Stuart, F. M., Bluck, B. J. & Pringle, M. S. 2001. Detrital muscovite $^{40}Ar/^{39}Ar$ ages from Carboniferous sandstones of the British Isles: provenance and implications for the uplift history of orogenic belts. *Tectonics*, **20**, 255–267.

Thöni, M. 1999. A review of geochronological data from the Eastern Alps. *Schweizerische Mineralogische und Petrographische Mitteilungen*, **79**, 209–230.

Von Blanckenburg, F., Villa, I. M., Baur, H., Morteani, G. & Steiger, R. H. 1989. Time calibration of a PT path from the Western Tauern Window, Eastern Alps: the problem of closure temperatures. *Contributions to Mineralogy and Petrology*, **101**, 1–11.

Von Eynatten, H. 1996. *Provenanzanalyse kretazischer Siliziklastika aus den Nördlichen Kalkalpen: Petrographie, Mineralchemie und Geochronologie des frühalpidisch umgelagerten Detritus.* Ph.D. Thesis, University of Mainz, 145 pp.

Von Eynatten, H., Gaupp, R. & Wijbrans, J. R. 1996. $^{40}Ar/^{39}Ar$ laser-probe dating of detrital white micas from Cretaceous sedimentary rocks of the Eastern Alps: evidence for Variscan high-pressure metamorphism and implications for Alpine orogeny. *Geology*, **24**, 691–694.

Von Eynatten, H., Schlunegger, F., Gaupp, R. & Wijbrans, J. R. 1999. Exhumation of the Central Alps: evidence from $^{40}Ar/^{39}Ar$ laserprobe dating of detrital white micas from the Swiss Molasse Basin. *Terra Nova*, **11**, 284–289.

Von Quadt, A., Grünenfelder, M. & Büchi, H. 1994. U–Pb zircon ages from igneous rocks of the Bernina nappe system (Grisons, Switzerland). *Schweizerische Mineralogische und Petrographische Mitteilungen*, **74**, 373–382.

Wijbrans, J. R. & McDougall, I. 1986. $^{40}Ar/^{39}Ar$ dating of white micas from an Alpine high-pressure metamorphic belt on Naxos (Greece): the resetting of the argon isotopic system. *Contributions to Mineralogy and Petrology*, **93**, 187–194.

Wijbrans, J. R., Pringle, M. S., Koppers, A. A. P. & Scheveers, R. 1995. Argon geochronology of small samples by laser heating. *Proc. Koninklijke Academie van Wetenschappen*, **98**, 185–218.

Effects of rate and nature of synkinematic sedimentation on the growth of compressive structures constrained by analogue models and field examples

T. NALPAS[1,3], D. GAPAIS[1], J. VERGES[2], L. BARRIER[1], V. GESTAIN[1], G. LEROUX[1], D. ROUBY[1] & J.-J. KERMARREC[1]

[1]*Géosciences Rennes, Université de Rennes 1, UMR 6118 du CNRS, Campus de Beaulieu, 35042 Rennes, France*
[2]*Institute of Earth Sciences 'Jaume Almera', CSIC, 08028 Barcelona, Spain*
[3]*IRD, Departamento de Geologia de la Universidad de Chile, Casilla 53390, Correo Central, Santiago, Chile*

Abstract: Compressive basins show various geometries of growth structures. In this paper, we examine the effects of the rate and nature of synkinematic sedimentation by comparing analogue models and field examples. We performed different types of experiments in order to simulate various natural conditions, the parameters tested including the rate of synkinematic sedimentation and the presence of a potential décollement layer deposited during deformation. Modelling techniques are similar to those usually used for experiments on brittle–ductile systems made of sand and silicone putty.

To study the influence of the synkinematic sedimentation rate, we used sets of experiments with different velocity ratios R between the rate of sedimentation (V_s) and the rate of uplift of the top of the structure (V_u). In standard experiments, we deposited only sand during deformation. In all experiments, growth folds and growth faults developed. The geometry of growth folds changes from a steep anticline for $R=1$, to a rounded broken anticline when R exceeds 2. The geometry of growth reverse faults changes: from segments which initiate with a dip angle of about 30°, but then flatten or steepen, for $R=1/2$ and for $R=1$, respectively. For some sets of experiments, a thin silicone layer deposited instead of sand at some stage of sedimentation simulated the introduction of potential décollement layers.

When a change in the nature of synkinematic sedimentation occurs between the two flanks of a fold, a fault develops in the flank where brittle material dominates. The direction of growth folds forming during compression in front of a ramp and above a décollement layer is parallel to the ramp and oblique to the direction of bulk shortening.

Good correlations are observed between experimental geometries and field examples from the southern Pyrenees (Eastern Ebro Basin and Jaca Basin, Spain) and the Apennines (Italy).

Introduction

The effect of synkinematic sedimentation in front of compressive systems has been studied recently using experimental or numerical modelling (Cobbold *et al.* 1993; Storti & McClay 1995; Hardy *et al.* 1998; Nalpas *et al.* 1999; Leturmy *et al.* 2000; Bonini 2001). A major outcome of these works is that high sedimentation rates in front of a compressive structure favour steepening of active thrusts (Tondji Biyo 1995; Mugnier *et al.* 1997; Nalpas *et al.* 1999; Nieuwland *et al.* 2000; Casas *et al.* 2001; Barrier *et al.* 2002).

On the other hand, it is widely accepted that the rheology of sediments can also influence the geometry of compressive structures, with the development of ramps and flats within layers showing brittle or ductile behaviour (Davis & Engelder 1985; Vergés *et al.* 1992; Sans & Vergés 1995). In general, only the prekinematic sedimentary pile is considered. However, facies variations can occur within synkinematic sediments deposited around compressive structures.

In this paper, we compare the relative effects of the synkinematic sedimentation of brittle layers and/or ductile layers on the geometry of thrust fronts. We present new experimental results and compare them with previously published experiments and with some natural examples.

From: McCANN, T. & SAINTOT, A. (eds) *Tracing Tectonic Deformation Using the Sedimentary Record.* Geological Society, London, Special Publications, **208**, 307–319. 0305–8719/03/$15.00

Geometries of natural compressive growth structures

In sedimentary basins, compressive growth structures vary from folds to faults that can show different geometries. For folds, a good example is the southern border of the Jaca Basin (SW Pyrenees, Spain) where we can clearly identify various types of folds (Fig. 1). Eocene folds within the southern Jaca basin developed during the late stages of sedimentation in the South Pyrenean foreland basin (Puigdefabregas 1975; Lafont 1994). From east to west, three folds (Gabardiella, Arguis and Bentué de Rasal) are observed. These have been mapped by field mapping and aerial photographs (Toledo 1990; Lafont 1994; Millán *et al.* 1994; Poblet & Hardy 1995). The three folds are contemporaneous with the E–W progradation of a fluvial-dominated delta (Puigdefabregas 1975; Toledo 1990; Lafont 1994; Millán *et al.* 1994). The fold shape changes from east to west, varying from a rounded broken fold for Gabardiella in the east to a sharp triangular, almost box-type fold, for Bentué de Rasal in the west (Nalpas *et al.* 1999). All of these structures are growth folds, but for each of them, the nature and rate of synkinematic sedimentation are different. Indeed, during the growth of these folds (in Eocene times), the rate of synkinematic sedimentation decreased from east to west, and sedimentation became less sandy toward the west, in relation to the progradation of the fluvial-dominated delta (Toledo 1990; Lafont 1994; Millán *et al.* 1994). These facies changes are especially important in the Gabardiella fold zone to the east.

For faults, a good example is the Apennines (Italy), where various geometries from flat reverse faults to steeply dipping faults can be observed on seismic profiles (Fig. 2). During fault growth, the rate of synkinematic sedimentation decreased from SW to NE (Storti & McClay 1995).

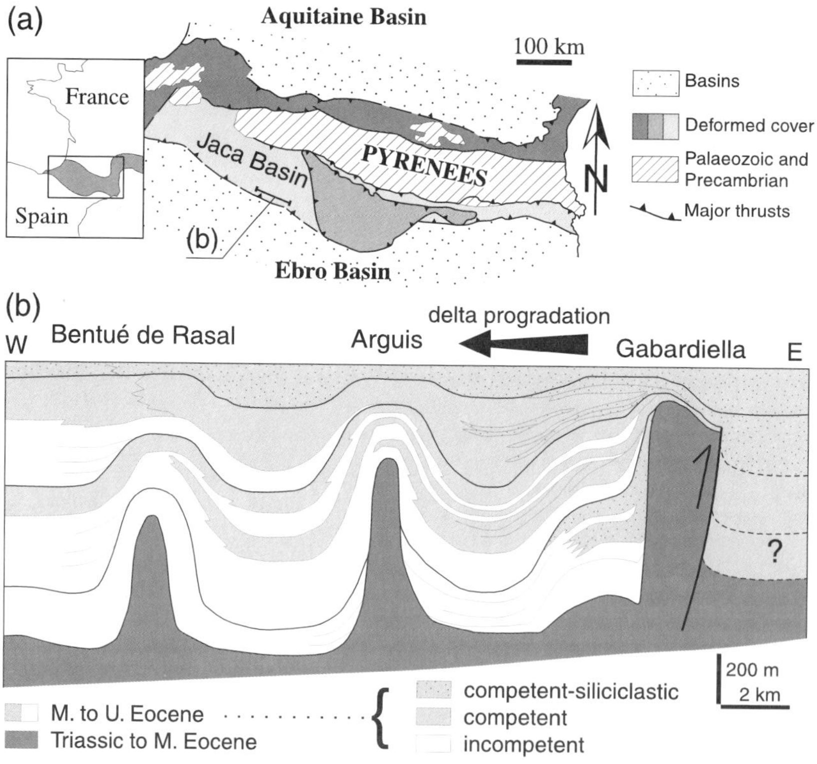

Fig. 1. Influence of the lithology on growth structures. (**a**) Location of the area; (**b**) cross-section of the north–south anticlines of the southern border of the Jaca Basin (modified from Millán et al., 1994). Synkinematic sedimentation rate decreases from east to west.

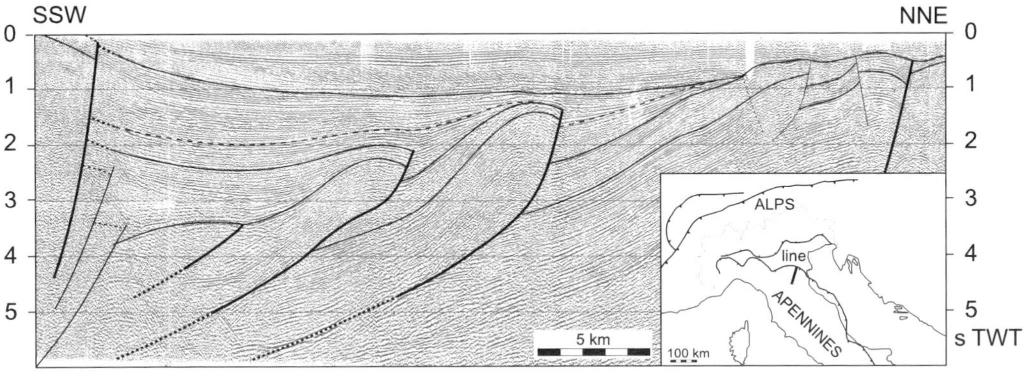

Fig. 2. Seismic profile of the thrust system of the northern part of the Apennines (modified from Pieri, 1989). Synkinematic sedimentation rate decreases from southwest to northeast.

Analogue models of compressive growth structures

Experimental procedure

The modelling techniques are similar to those usually used for experiments on brittle–ductile systems at the Laboratory of Experimental Tectonics of Géosciences Rennes (Rennes University, France) and which have been described in many previous studies (e.g. Faugère & Brun 1984; Vendeville *et al.* 1987; Davy & Cobbold 1991). Basement and brittle sediments (pre- and synkinematic) are represented by sand, with an angle of internal friction close to 30° (Krantz 1991) and a density (ρ) of 1400 kg m^{-3}. Weak ductile sediments such as shales, clay, marl or salt are represented by silicone putty with a viscosity (μ) around 10^4 Pa s^{-1} at 30°C and a density (ρ) around 1400 kg m^{-3}.

The experimental apparatus is composed of a fixed rigid basal plate over which a thin mobile plate fixed on a mobile wall is pushed at a constant rate (Fig. 3a, b and c). The boundary of the mobile plate induces an asymmetrical velocity discontinuity (VD) at the base of the model, which localizes the deformation (cf. Malavieille 1984; Balé 1986; Allemand *et al.* 1989; Ballard 1989). One series of experiments (Type IV, Fig. 3d) was performed without basal mobile plate. The model is contained within a 55×60 cm sand-box, wide enough to achieve a relatively large amount of shortening without edge effects.

For each type of experiment, the shortening rate is calibrated according to a series of test experiments, making it possible to determine the range of rates for which structure geometries comparable with those observed in nature are formed.

To simulate synkinematic sedimentation, fresh sand is continuously sprinkled manually on to the model during shortening. The sedimentation modes have been chosen in order to constrain possible sedimentation modes within continental compressive basins. Three modes are possible: firstly, one where the rate of sedimentation V_s is lower than the rate of uplift of the structure V_u, and the ratio R between V_s and V_u is lower than one; secondly, where V_s is equal to V_u ($R=1$); and thirdly, V_s is higher than V_u ($R>1$). To estimate V_s and V_u, we measured manually, at regular time intervals, the evolution of the topography of the experiment by reference to a fixed point. This evolution is verified on a cross-section at the end of the experiment.

Photographs of the model surface are taken at regular time intervals to observe structure development. After deformation, the internal structure is recorded from a series of cross-sections cut parallel to the compression direction (perpendicular to the VD). Brittle layers are built up with sand of various colours in order to reveal the structures and to observe them on photographs.

Four types of experiments are performed.

(1) Type I (Fig. 3a) – models are built with a two-layer brittle–ductile system. The initial pile (so-called precompression in Fig. 3a) is formed by a 1-cm thick layer of silicone overlain by 4 cm of alternating thin layers of sand and silicone. The basal silicone layer is designed to model a décollement level, and the overlying sandwich of sand and silicone represents a pre-tectonic sedimentary cover where flexural-slip-type folding is possible. Only syncompression fresh sand is continuously and manually sprinkled at different

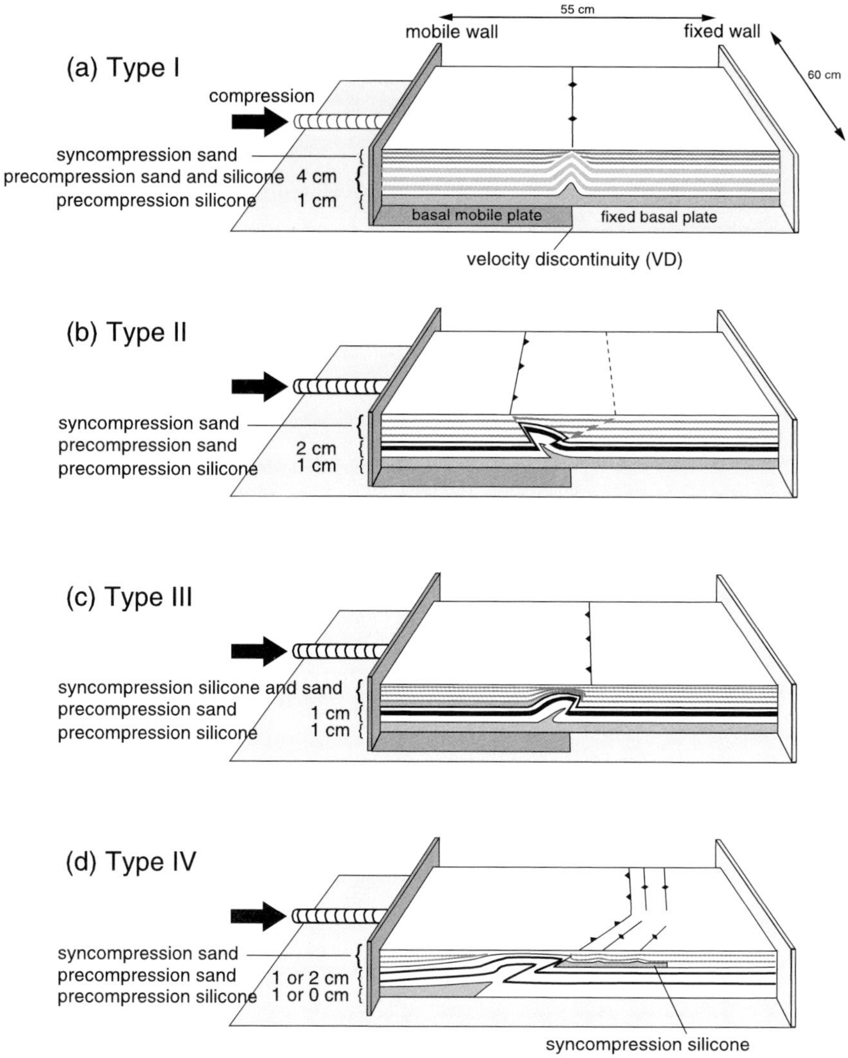

Fig. 3. Experimental setting for the four types of models.

velocities on to the model during shortening (Nalpas *et al.* 1999).

(2) Type II (Fig. 3b) – models are built with a two-layer brittle–ductile system, the initial pile being formed by a 1-cm thick layer of silicone overlain by 2 cm of black-and-white sand. The basal silicone layer is designed to model a décollement layer, and the overlying sand represents a brittle prekinematic sedimentary cover. Only fresh sand is continuously sprinkled manually at different velocities on to the model during shortening (Barrier *et al.* 2002).

(3) Type III (Fig. 3c) – models are built with two-layer brittle–ductile systems, with an initial pile formed by a 1-cm thick layer of silicone overlain by 1 cm of black-and-white sand. The basal silicone layer is designed to model a décollement level, and the overlying sand represents a brittle prekinematic sedimentary cover. Fresh sand is manually sprinkled at different velocities on to the model during shortening, and a thin silicone layer is deposited within the synkinematic pile at the beginning, during, or in the late stages of the experiments.

(4) Type IV (Fig. 3d) – models are deformed using a slightly modified experimental

setting. The basal mobile plate is absent and only the mobile wall induces shortening. Models are built with an initial pile made of a 1 cm thick layer of silicone of different geometries and a 1 cm layer of black-and-white sand. The basal silicone layer is designed to model a décollement level, the overlying sand representing a brittle pre-kinematic sedimentary cover. A thin silicone plate is deposited at the beginning of the shortening, in front of the emergence of the first ramp. Fresh sand is manually sprinkled at different velocities on to the model during subsequent shortening.

Experimental results

Influence of synkinematic sedimentation rate

Fold development, type I models (Fig. 4). For this series of experiments, the shortening velocity is 5 cm/h, and different velocity ratios (R) between the sedimentation rate (V_s) and the rate of structural uplift (V_u) are tested.

When the synkinematic sedimentation is slow ($R=1$), the fold shows a flat top and a limb dipping at 57°. With this sedimentation rate, there is no continuity of synkinematic beds above the anticline (Fig. 4a). When the sykine-

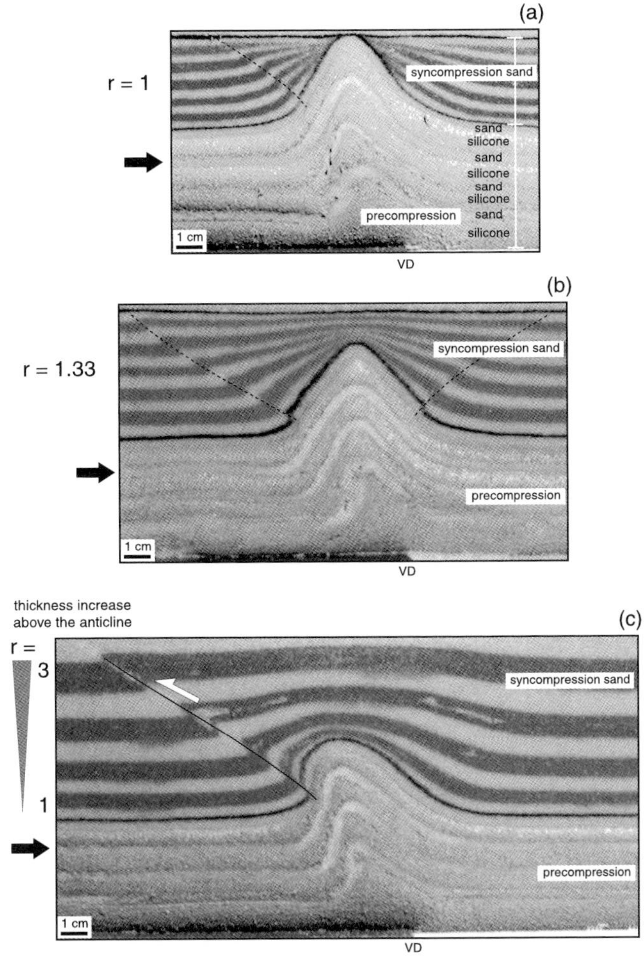

Fig. 4. Cross-sections from three models of type I, with different rates of synkinematic sedimentation. (**a**) $R=1$, no continuity of synkinematic beds on top of anticline; (**b**) $R=1.33$ continuity of synkinematic beds on top of anticline; (**c**) R increases from 1 to 3: less-developed folds and faults. The shortening rate is 5 cm/h (modified from Nalpas *et al.* 1999).

matic sedimentation rate is high (R=1.33), the fold is less developed, with a limb dip of 47°. Synkinematic beds are continuous above the anticline limbs. A fault developed when the ninth synkinematic layer was deposited (Fig. 4b). When the synkinematic sedimentation rate increases at each step (R varying from 1 to 3), the fold is less developed than previously. Concurrently, a reverse fault develops during the deposition of the seventh synkinematic layer. This fault controls the subsequent deformation (Fig. 4c), and the amplification rate of the anticline decreases progressively.

In these experiments, the rate of synkinematic sedimentation influences fold geometry, which changes from steep anticline for R=1, to broken anticline when R exceeds 2.

Fault development, Type II models (Fig. 5). For these experiments, the shortening velocity is 2.5 cm/h. The parameter tested here is the velocity ratio between synkinematic sedimentation and the structural uplift of the fault, for ratios R between the rate of sedimentation (V_s) and the rate of uplift (V_u) of 1/2 and 1. This second set of experiments was described in Barrier *et al.* (2002).

In all models, progressive shortening is accommodated by two conjugate reverse faults. The faults appear at the surface as straight scarps. The major fault roots on the VD. It is permanent and antithetic to the displacement of the mobile wall, with a hanging wall located on the mobile basal plate. The secondary fault set is synthetic with respect to the shortening direction, and consists of a series of short-lived faults (Ballard 1989; Tondji Biyo 1995; Merle Abidi 1995). The two conjugate thrust systems define a typical asymmetrical pop-up geometry. The antithetic motion is larger than the synthetic one. The fault pattern varies according to R: the dip of the main active fault, which is initially always about 30°, increases as synkinematic sediments accumulate in front of it.

In the experiment where R=1/2, the overall geometry of the model resembles a fault-bend fold nucleated at the VD (Fig. 5a). The bend of the anticline generates local extensional strains marked by a graben at the extrados of the fold. The main thrust band at the anticline front is thin and becomes convex upward. Along the main thrust, the fold overlaps the synkinematic sediments of the underthrusting part of the model without deforming them. The top of the anticline also backthrusts the synkinematic sediments through the conjugate thrust system.

In the experiment where R=1, the main thrust zone is broad and widens upward (Fig. 5b). The main thrust consists of a fan-shaped series of reverse faults whose dip increases during each depositional increment. The fault zone therefore

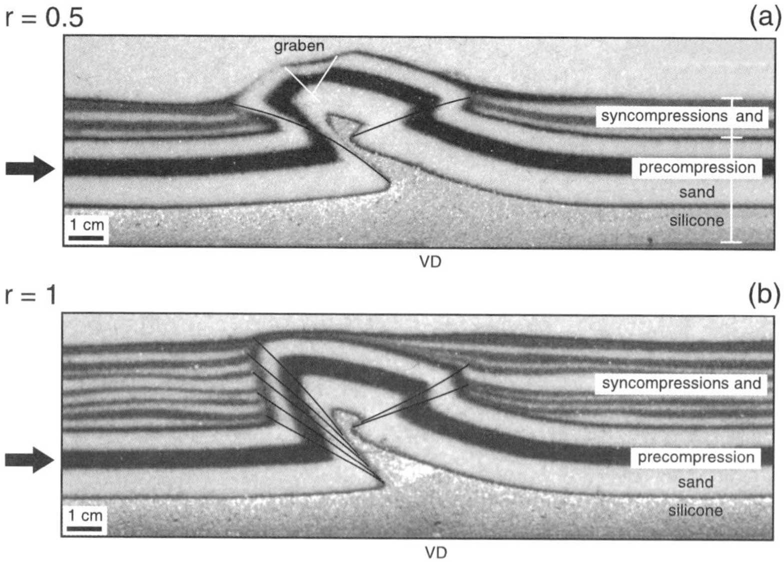

Fig. 5. Cross-sections from two models of type II, with different rates of synkinematic sedimentation. (**a**) R=0.5, fault dip decreases; (**b**) R=1, fault dip increases. The shortening velocity is 2.5 cm/h (modified from Barrier *et al.*, 2002).

develops more vertically than horizontally. Near the contact between the prekinematic and synkinematic sand, the growth strata are overturned. The divergent secondary thrust zone develops in a similar way, before it dies out – the last temporary fault being sealed by deposits which onlap the tilted roof of the pop-up.

In these experiments, synkinematic sedimentation influences thrust patterns in different ways, depending on the rate of deposition. The geometry of growth reverse faults changes from segments which initiate with a dip angle of about 30° and then flatten, when $R=1/2$, or steepen, when $R=1$. In any case, synsedimentary reverse faults always dip more steeply than those observed in models without synkinematic sedimentation for a similar bulk shortening (Ballard *et al.* 1987; Ballard 1989; Coletta *et al.* 1991).

Influence of the nature of synkinematic sedimentation

Fault development, Type III models (Fig. 6). For these experiments, the shortening velocity is 2.5 cm/h, and we tested the nature of the synkinematic sedimentation deposited during structural uplift. Only two of the experiments are shown and compared with the previous series. In all of the models, the progressive shortening is accommodated by two conjugate incipient reverse faults. These faults define an anticline pop-up geometry, which is symmetrical during the early stages of deformation.

In the first experiment (Fig. 6a), a thin silicone layer is deposited above pre-kinematic sand during the beginning of compression, just in front of the antithetic fault located above the mobile plate. At the same time, sand is deposited on the other side of the pop-up. Initially, a little wedge of sand penetrates the silicone layer, as expected for the onset of a flat development. However, this structure is rapidly abandoned. During wedge creation, synkinematic sedimentation of fresh sand above the previously deposited silicone continues with a ratio $R=1$. In the upper part of the wedge, the prekinematic sand is thus covered by synkinematic fresh sand. The antithetic fault is more developed than the synthetic one, as is usually observed in other experiments (Ballard 1989; Colletta *et al.* 1991; Tondji Biyo 1995; Barrier *et al.* 2002).

In the second experiment (Fig. 6b), following initial uplift, a thin silicone layer is deposited above prekinematic sand and above the antithetic fault at the top of the pop-up, whereas sand is deposited on the other side, where the mobile plate is. Initially, a very small wedge of sand penetrated the silicone layer, but this structure was rapidly abandoned. During wedge creation,

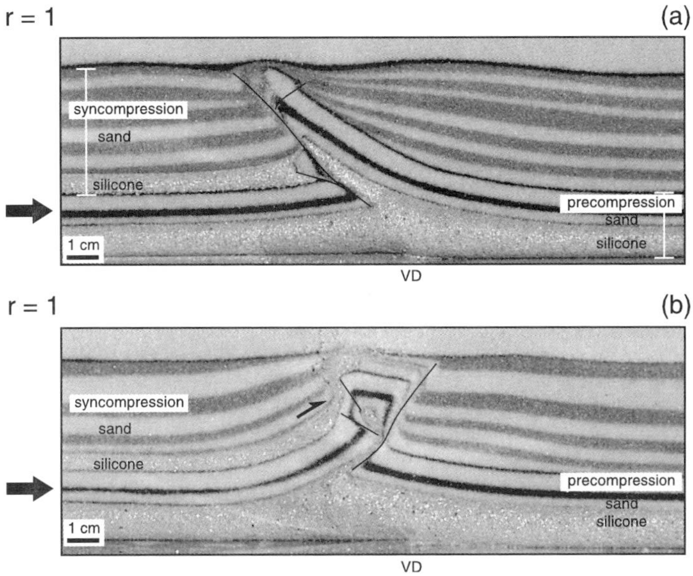

Fig. 6. Cross-sections from two models of type III, with the different locations of synkinematic ductile sedimentation. (**a**) $R=1$, synkinematic silicone layer above prekinematic sand in front of the antithetic fault, (**b**) $R=1$, synkinematic silicone layer above prekinematic sand and above the antithetic fault. The shortening velocity is 2.5 cm/h.

synkinematic sand is deposited above the silicone with a ratio $R=1$. In the upper part of the small wedge, the thin synkinematic silicone layer and the synkinematic sand thus cover the pre-kinematic sand. The pile is a brittle–ductile sandwich in the antithetic side of the structure and only brittle in the other side. In contrast with the usual observations in experiments where a velocity discontinuity is modelled (Ballard 1989; Coletta *et al.* 1991; Tondji Biyo 1995; Barrier *et al.* 2002, and the previous models), the major fault is synthetic and only a short-lived antithetic fault is developed. The latter fault disappears in the silicone layer, which facilitates the development of a flexural slip fold across the synthetic fault.

Fold development, Type IV models (Fig. 7). For these experiments the shortening velocity is 1 cm/h, and the parameters tested are the nature of the synkinematic sediments deposited in front of a basal décollement layer with various geometries. The aim of these experiments is to reproduce what could happen in front of a frontal and lateral ramp if a synkinematic ductile layer is deposited just in front of the ramp. We performed several experiments with different geometries of the pre-shortening basal silicone layer, in order to generate frontal and lateral ramps of different geometries. During compression, the basal silicone layer acts as a velocity discontinuity. A frontal ramp develops where the limit of the basal silicone plate is parallel to the mobile wall, and a lateral ramp develops where the limit of the basal silicone plate is oblique to the mobile wall. Three angles of obliquity, between the boundary of the basal silicone plate and the mobile wall, have been tested (30°, 45° and 60°). An experiment with 30° obliquity is shown on Figures 7 and 8. During the early stages of compression, a synkinematic silicone layer was deposited just in front of the frontal and lateral ramps after their initial uplift. The interbedded silicone layer located in front of the lateral ramp is a square, whereas in front of the frontal ramp it is a triangle. We chose a triangular shape for the synkinematic weak layer in order to examine the effects of the strikes of the ramp and of the boundary of the silicone plate on the direction of structures developed above the silicone. Sand was first deposited around the interbedded silicone layer, and subsequently above the silicone plate throughout the entire period of compression.

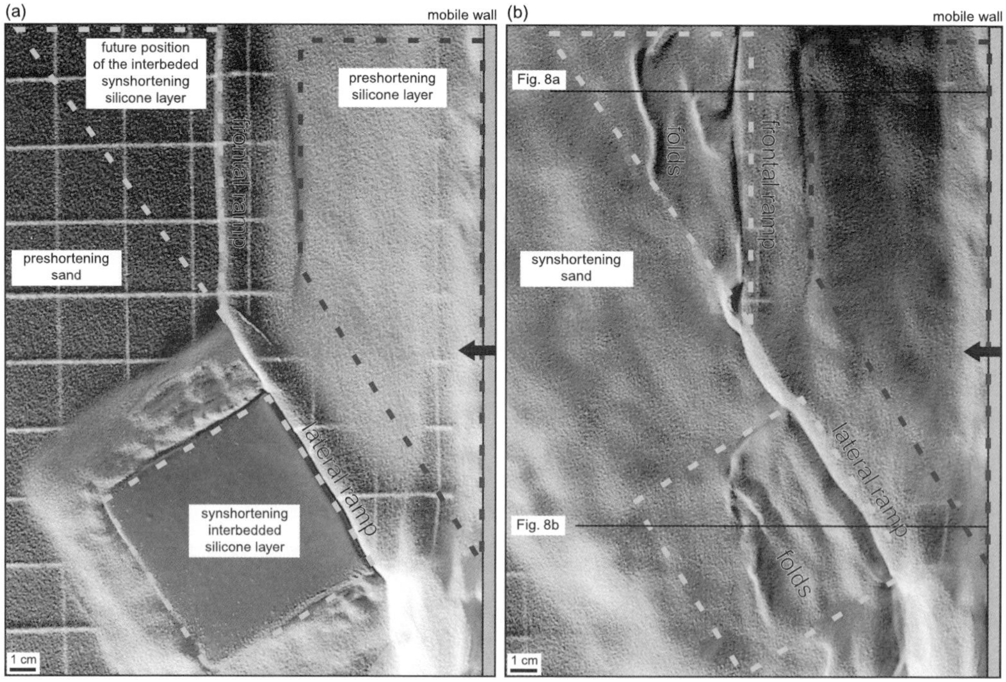

Fig. 7. Top views from a model of type IV. (**a**) Model surface just after the beginning of the shortening when the silicone layers are deposited; (**b**) model surface at the end of the experiment. The folds observed above the synkinematic silicone layers are parallel to the front of the ramps. The shortening rate is 1 cm/h.

Surface views of the model (Fig. 7b), show the frontal and lateral ramps related to the basal silicone plate at the end of the deformation. Above the interbedded silicone layer located in front of the frontal ramp, the folds that form are parallel to the frontal ramp. Thus, the triangular shape of the interbedded silicone layer has no influence on the direction of the folds. The frontal ramp is the only structure that influences their direction. Above the interbedded silicone layer situated in front of the lateral ramp, folds are parallel to the lateral ramp. Thus, the shape of the ramp influences drastically the orientation of folds. Indeed, they are perpendicular to the direction of the lateral ramp, and not to the general direction of compression.

At the end of the experiment, cross-sections (Fig. 8) show a ramp developed at the boundary between the basal silicone plate and the sand. It corresponds with a frontal ramp, in section (a) and to a lateral ramp in section (b). These two ramps are flat where they reach the interbedded silicone layer. As in previous experiments (Fig. 6a), a small wedge of sand penetrates slightly into the interlayered silicone plate. Above this wedge, sand layers are folded, which suggests that a décollement surface within the silicone translated the displacement from the ramp to the folds.

Discussion

Analogue models of growth folds and faults presented in this paper reproduce fundamental features of natural growth structures (Figs 1 & 2). In both nature and models the uplift of the structures is at a maximum when synkinematic sedimentation on the top of the structures is limited. This can be related to the load of the sedimentary overburden, i.e. to the sedimentation rate. In compression, the principal stress (σ_1) is horizontal, and the minimum stress (σ_3) is vertical and corresponds with the lithostatic pressure. The lower the σ_3, the better the compressive structures are developed.

Such a situation is illustrated by the geology along the southern border of the Jaca Basin (Fig. 1). Along the Sierras Exteriores that form the southern border of the Basin, there is a group of folds showing different geometries from east to west. These folds were formed in a more northerly trend, and were displaced to the south above a thrust ramp. As a consequence, these folds exhibit a significant northwards plunge as seen on the map, which is close to a cross-section. The final geometry of these growth folds changes from east to west, from the Gabardiella broken fold to the east, to the Bentué de Rasal sharp triangular fold to the west. During fold growth, a fluivial-dominated delta was prograding from east to west, resulting in a dcreasing sedimentation rate toward the west. The maximum synkinematic sedimentary load was first localized in the eastern part of the basin, and then migrated westwards during delta progradation. To the east, folding was associated with high sedimentation rates during the onset of fold growth. To the west, the sedimentation rate was low during the beginning of fold growth, and increased later. Consistently, the western most fold shows the maximum amplitude and the sharpest final geometry, as observed in

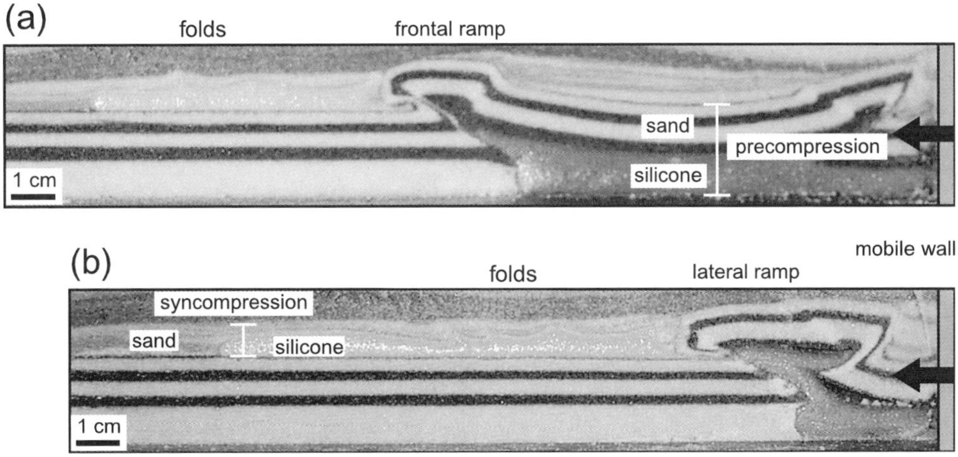

Fig. 8. Cross-sections from a model of type IV in front of the frontal ramp (**a**) and in front of the lateral ramp (**b**).

experiments with *R* equal to one (Fig. 4a). In the eastern part, the Gabardiella Anticline is a rounded broken fold because the sedimentary load was too high to allow strong fold amplification at the onset of fold growth, as observed in experiments with *R* higher than one (Fig. 4c).

The Apennines also show interesting features, well imaged with two-dimensional seismic profiles (Fig. 2). The dip of the thrusts increases from the southwest to the northeast. A fault in the centre of the profile dips at more than 60°, and the top of the structure is eroded or condensed. We know that in this basin the synkinematic sedimentary load is derived from the southwest, as illustrated on the seismic line by the decrease in sediment thickness toward the northeast. As in the example of the southern border of the Jaca Basin, the evolution of the fault geometry seems to be related to the sedimentation rate. The situation is the same as in the experiments where the modelled faults are steeply dipping when $R=1$ (Fig. 5b).

It seems clear in both nature and experiments that when the rate of synkinematic sedimentation (V_s), is equal to the rate of uplift of the fold (V_u), the development of steeply dipping structures if favoured.

Compressive structures are also influenced by the rheology of the synkinematic sedimentation. On the southern border of the Jaca Basin, the Gabardiella Anticline is a rounded broken fold antithetic to the general SE-directed displacement (Millán *et al.* 1994; Poblet & Hardy 1995). The vergence of the major faults oriented N160° (like Gabardiella) is toward the west-southwest. The lateral facies changes from sand-dominated to the east of the anticline, to clay-dominated to the west of the anticline, as a result of delta progradation. This facies change, which may be enhanced during fold amplification, seems to be important during deformation. The more brittle behaviour of the eastern fold limb can be related to the sandy deposits. From experiments presented in Figure 6, we suggest that the vergence of Gabardiella structure could be related to the facies changes on the top of the system during fold initiation. The deformation is mainly accommodated by flexural slip in the western part and by faulting in the eastern part. This could explain the opposite vergence direction of the Gabardiella broken fold with respect to other regional structures.

The influence of the lithology of synkinematic sedimentation during thrusting and folding is also observed along the southern front of the Pyrenean thrust sheets, in the Ebro foreland basin. There, the superposition of irregular geometries of former margins of sedimentary basins, reactivated as thrust sheets, and synkinematic evaporitic layers, produced a complex array of folds and thrusts showing different trends (Fig. 9). The most striking feature is the group of folds with an approximate N50° trend localized ahead of the major Segre oblique ramp in the Ebro foreland basin. These anticlines (Oliana, Vilanova, Cardona, Súria and Oló Callús) developed in Late Eocene–Oligocene times only above the salts of the Cardona Formation (Ramírez & Riba 1975; Vergés *et al.* 1992; Sans and Vergés 1995). Along the SW margin of the salt basin, a long N140°-striking anticline developed (Sanaúja). Where the major external thrust forms an east–west-striking frontal ramp, structures in the Ebro Basin also strike east–west (Bellmunt, Puig–Reig, Barbastro–Balaguer). Cross-sections show that the Eocene–Oligocene sedimentary cover is decoupled from other Cenozoic sediments along the Eocene salts of the Cardona Formation. The experiment presented in Figures 7 and 8 attempts to explain the relationships between the geometry of a major external thrust and the direction of associated folds developed during compression above a synkinematic décollement layer. To form oblique folds far from the major external thrust, a ductile level must be present in front of the major external lateral ramp. At ramp scale the regional deformation is partitioned into local strike-slip along the ramp and thrusting perpendicular to the ramp. The strike-slip motion is recorded along the ramp, and the compressive motion is translated above the décollement layer and recorded by structures parallel to the ramp; similar partitioning is shown by the fold geometry on Figure 7b.

Conclusions

The comparison between field examples and analogue modelling suggests the following conclusions:

(1) The geometry of compressive structures like growth folds and growth faults changes with the rate of synkinematic sedimentation. Growth folds or growth faults are vertically enhanced when the rate of synkinematic sedimentation around the structure equals the rate of structural uplift (i.e. when the synkinematic sedimentation on the top of the structure is nearly zero).

(2) The vergence of compressive structures, like broken folds, may change with the nature of the synkinematic sedimentation. Lateral changes of facies occurring across the grow-

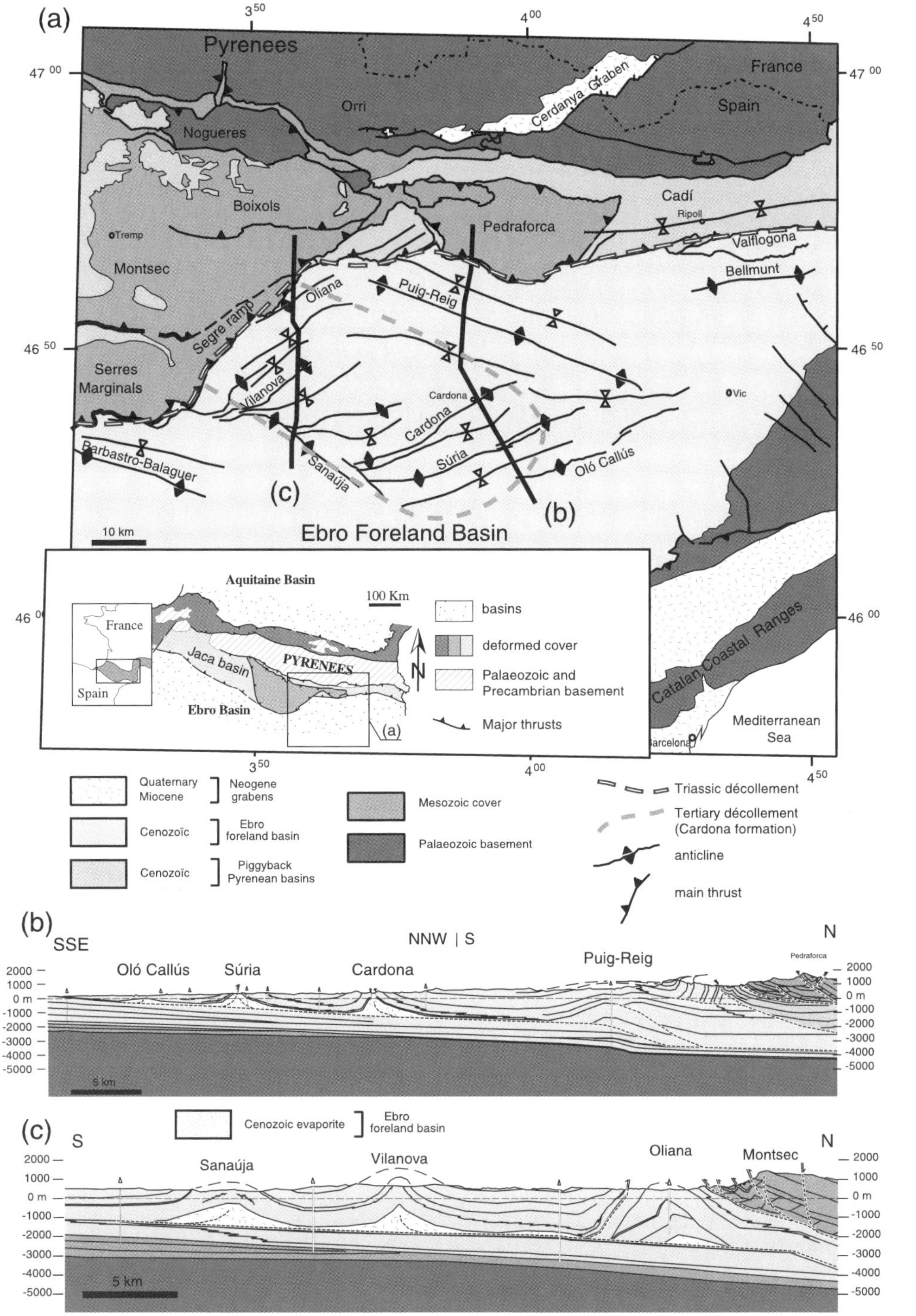

Fig. 9. Influence of the presence of evaporitic layer in the direction of growth structures. (**a**) Location of the area; (**b**) and (**c**) cross-sections in front of the external ramp. The folds situated above the Tertiary décollement (Cardona Formation) are parallel to the oblique Segre ramp (Vilanova, Cardona, Súria).

ing anticline may control the development of more complex structures, like a thrust through the limb constituted by more brittle material. In compressive basins, growth folds can evolve into broken folds and growth faults. Abnormal antithetic vergences of growth structures may be explained by spatial variations in the nature of the synkinematic sedimentation around structures.

(3) The direction of compressive structures like growth folds may change according to the lithology of synkinematic sedimentation and with the geometry of the major external thrust. Growth folds forming in front of a ramp and above a décollement layer are parallel to the ramp and oblique to the regional direction of shortening. In natural compressive basins, abnormal orientations of growth structures may be due to the presence of a décollement level in front of a lateral ramp.

Reviews and suggested improvements by A. Saintot and an anonymous reviewer were greatly appreciated.

References

ALLEMAND, P., BRUN, J.-P., DAVY, P. & VAN DEN DRIESSCHE, J. 1989. Symétrie et asymétrie des rifts et mécanismes d'amincissement de la lithosphere. *Bulletin de la Société Géologique de France*, **3(V)**, 445–451.

BALÉ, P. 1986. *Tectonique cadomienne en Bretagne nord. Interaction décrochement chevauchement: champs de déformation et modélisations expérimentales*. Thèse de 3ème cycle, Université de Rennes 1.

BALLARD, J. F. 1989. *Approche Géologique et Mécanique des Décollements dans la Croûte Supérieure*. Thèse de 3ème cycle, Université de Rennes 1, 299 pp.

BALLARD, J.-F., BRUN, J.-P., VAN DEN DRIESSCHE, J. & ALLEMAND, P. 1987. Propagation des chevauchements au-dessus des zones de décollement: modèles expérimentaux. *Comptes Rendus de l'Académie des Sciences de Paris*, **305(II)**, 1249–1253.

BARRIER, L., NALPAS, T., GAPAIS, D., PROUST, J.-N., CASAS, A. & BOURQUIN, S. 2002. Influence of syntectonic sedimentation on thrusts geometry. Field examples from the Iberian Chain (Spain) and analogue modelling. *Sedimentary Geology*, **146**, 91–104.

BONINI, M. 2001. Passive roof thrusting and forelandward fold propagation in scaled brittle–ductile physical models of thrust wedges. *Journal of Geophysical Research*, **106(B2)**, 2291–2311.

CASAS, A. M., GAPAIS, D., NALPAS, T., BESNARD, K. & ROMAN BERDIEL, T. 2001. Analogue models of transpressive systems. *Journal of Structural Geology*, **23**, 733–743.

COBBOLD, P. R., DAVY, P., *ET AL.* 1993. Sedimentary basins and crustal thickening. *Sedimentary Geology*, **86**, 77–89.

COLLETTA, B., LETOUZEY, J., PINEDO, R., BALLARD, J. F. & BALÉ, P. 1991. Computerized X-ray tomography analysis of sandbox models: examples of thin-skinned thrust systems. *Geology*, **19**, 1063–1067.

DAVIS, D. M. & ENGELDER, T. 1985. The role of salt in fold-and-thrust belts. *Tectonophysics*, **119**, 67–88.

DAVY, P. & COBBOLD, P. R. 1991. Experiments on shortening of a 4-layer model of continental lithosphere. *Tectonophysics*, **188**, 1–25.

FAUGÈRE, E. & BRUN, J.-P. 1984. Modélisation expérimentale de la distention continentale. *Comptes Rendus de l'Académie des Sciences de Paris*, **299(7)**, 365–370.

HARDY, S., DUNCAN, C., MASEK, J. & BROWN, D. 1998. Minimum work, fault activity and the growth of critical wedges in fold and thrust belts. *Basin Research*, **10**, 365–373.

KRANTZ, R. W. 1991. Measurements of friction coefficients and cohesion for faulting and fault reactivation in laboratory models using sand and sand mixtures. *Tectonophysics*, **188**, 203–207.

LAFONT, F. 1994. *Influences relatives de la subsidence et de l'eustatisme sur la localisation et la géométrie des réservoirs d'un système deltaïque*. Thèse d'Université, Université de Rennes 1.

LETURMY, P., MUGNIER, J. L., VINOUR, P., BABY, P., COLETTA, B. & CHABRON, E. 2000. Piggyback basin detachments levels as a function of interactions between tectonic and superficial mass transfer: the case of the Subandean Zone (Bolivia). *Tectonophysics*, **320**, 45–67.

MALAVIEILLE, J. 1984. Modélisation expérimentale des chevauchements imbriqués: application aux chaînes de montagnes. *Bulletin de la Société Géologique de France*, **XXVI(1)**, 129–138.

MERLE, O. & ABIDI, N. 1995. Approche expérimentale du fonctionnement des rampes émergentes. *Bulletin de la Société Géologique de France*, **166(5)**, 439–450.

MILLÁN, H., AURELL, M. & MELENDEZ, A. 1994. Synchronous detachment folds and coeval sedimentation in the Prepyrenean External Sierras (Spain): a case study for a tectonic origin of sequences and systems tracts. *Sedimentology*, **41**, 1001–1024.

MUGNIER, J. L., BABY, P., COLETTA, B., VINOUR, P., BALÉ, P. & LETURMY, P. 1997. Thrust geometry controlled by erosion and sedimentation: a view from analogue models. *Geology*, **25(5)**, 427–430.

NALPAS, T., GYÖRFI, I., GUILLOCHEAU, F., LAFONT, F. & HOMEWOOD, P. 1999. Influence de la charge sédimentaire sur le développement d'anticlinaux synsédimentaires. Modélisation analogique et exemple de terrain (bordure sud du basin de Jaca). *Bulletin de la Société Géologique de France*, **170(5)**, 733–740.

NIEUWLAND, D. A., LEUTSCHER, J. H. & GAST, J. 2000. Wedge equilibrium in fold-and-thrust belts: prediction of out-of-sequence thrusting based on sandbox experiments and natural examples. *Geologie en Mijnbouw*, **79(1)**, 81–91.

PIERI, M. 1989. Three seismic profiles through the Po plain. *In*: BALLY, A. W. (ed.) *Atlas of Seismic Stratigraphy*, American Association of Petroleum Geologists Studies in Geology, **27**, 90–110.

POBLET, J. & HARDY, S. 1995. Reverse modelling of detachment folds: application to the Pico del Aguila anticline in the South Central Pyrenees (Spain). *Journal of Structural Geology*, **17(12)**, 1707–1724.

PUIGDEFABREGAS, C. 1975. La sedimentación molásica en la cuenca de Jaca. Pirineos, **104**, 1–188.

RAMÍREZ, A. & RIBA, O. 1975. Bassin potassique catalan et mines de Cardona. In: *IX Congrés International de Sédimentologie, Nice*, **20**, 49–58.

SANS, M. & VERGÉS, J. 1995. Fold development related to contractional salt tectonics: Southeastern Pyrenean Thrust Front, Spain. *In*: JACKSON, P. A., ROBERTS, D. G. & SNELSON, S. (eds) *Salt Tectonics: a Global perspective*, American Association of Petroleum Geologists Memoir, **65**, 369–378.

STORTI, F. & MCCLAY, K. 1995. Influence of syntectonic sedimentation on thrust wedges in analogue models. *Geology*, **23(11)**, 999–1002.

TOLEDO, M. J. 1990. Séquences de dépôt et cyclicité tectonique dans l'Éocène du basin de Jaca (Espagne). Institut Français du Pétrole, Rueil-Malmaison, Internal Report No. **37983**.

TONDJI BIYO, J. J. 1995. *Chevauchements et Bassins Compressifs. Influence de l'Érosion et de la sédimentation.* Thèse d'Université Université de Rennes 1, 426 pp.

VENDEVILLE, B., COBBOLD, P. R., DAVY, P., BRUN, J. P. & CHOUKROUNE, P. 1987. Physical models of extensional tectonics at various scales. *In*: COWARD, M. P., DEWEY, J. F. & HANCOCK, P. L. (eds) *Continental Extension Tectonics*. Geological Society, London, Special Publications, **28**, 95–107.

VERGÉS, J., MUÑOZ, J. A. & MARTÍNEZ, A. 1992. South Pyrenean fold-and-thrust belt: the role of foreland evaporitic levels in thrust geometry. *In*: MCCLAY, K. R. (ed.) *Thrust Tectonics*. Chapman & Hall, 255–264.

Silurian sedimentation in East Siberia: evidence for variations in the rate of tectonic subsidence occurring without any significant sea-level changes

E. V. ARTYUSHKOV[1] & P. A. CHEKHOVICH[1,2]

[1]*Institute of Physics of the Earth, 10 B. Gruzinskaya, GSP-5, 123995, Moscow, Russia*
[2]*Institute of the Lithosphere of Marginal Seas, 22 Staromonetny, 119180, Moscow, Russia*

Abstract: It is widely accepted that major variations of sea-level have occurred in the Phanerozoic. Third-order cycles, 1–10 Ma long with amplitudes of 20–100 m, are of special interest for geochronology and petroleum geology. The amplitude of sea-level changes in the Silurian was estimated based on highly detailed data on the East Siberian Basin, which was 2×10^6 km^2 in size. Fischer plots were compiled based on the thickness of 54 chronostratigraphic units – chronozones, each corresponding to a time interval *c.*0.5 Ma long. The synchronicity of the chronozones ensures reliable comparison of the changes occurring with time in accommodation space in different regions. The subhorizontal Fischer plots derived for several regions indicate that sea-level changes were very small in the Silurian (≤5–10 m). A mathematical analysis of relative sea-level changes, which takes into account the finite rate of crustal subsidence and different possible forms of eustatic fluctuation, shows that from the observed structure of numerous Silurian successions in East Siberia, eustatic third-order sea-level changes could not have exceeded 6–20 m. In several regions of East Siberia, the rate of crustal subsidence varied as much as several hundred per cent at different times. These variations showed good similarity in form, but their amplitudes were different at different places in the basin. Most probably they were caused by variations in the rate of phase transformations in mafic rocks in the lower crust. Based on the example of the East Baltic, the absence of large-scale third-order cycles in eustatic changes of sea-level has been proven earlier for the Cambrian and earliest Ordovician. Probably, a similar situation was characteristic of many other epochs, when no large glaciations occurred, while many rapid changes of water depth in cratonic areas actually resulted from vertical crustal movements.

A lot of geological and seismic profiling data is available to prove that large-scale changes in water depth took place in Phanerozoic sedimentation basins. Such variations are commonly attributed to eustatic falls and rises in sea-level (Haq *et al.* 1987; Hallam 1992; Emery & Meyers 1996; de Graciansky *et al.* 1998). Eustatic curves include cycles of different lengths. Third-order cycles, 1–10 Ma long with amplitudes 20–100 m, which are often called 'eustatic events', are of special interest for hydrocarbon prospecting (Posamentier & Allen 2000) and geochronology (Cooper & Nowlan 1999). Many authors (e.g. Harris & Laws 1997; Maurer 2000; Cheng *et al.* 2001) have demonstrated the influence of crustal uplift and subsidence on water-depth variations of the third order. Moreover, the very existence of most eustatic events has been cast into doubt because the accuracy of biostratigraphic zonation is usually not enough to correlate events *c.*1 Ma long between different continents (Miall 1992, 1997; Miall & Miall 2001). Another serious problem is that no physical mechanisms have been proposed to explain the frequent sea-level changes with amplitudes of 20–100 m (Harrison 1990). Large-scale and rapid falls and rises of sea-level can be attributed to major glaciations. However, no large ice sheets existed during the many epochs when pronounced and short-term relative sea-level changes took place. For example, a number of sea-level peaks, 1–3 Ma long, have been proposed for the Late Cretaceous, Paleocene and Eocene (Haq *et al.* 1987) when no large-scale glaciations occurred. Moreover, most authors believe that numerous eustatic events occurred over most of the Phanerozoic (Haq *et al.* 1987; Hallam 1992; de Graciansky *et al.* 1998; Drzewiecki & Simo 2000; Schwarzacher 2000).

Based on the variations of water depth within six cratonic areas, eight eustatic events, from 1 Ma to several million years long, have been

From: McCann, T. & Saintot, A. (eds) *Tracing Tectonic Deformation Using the Sedimentary Record.* Geological Society, London, Special Publications, **208**, 321–350. 0305–8719/03/$15.00

proposed for the Silurian (Johnson 1996). Using conventional methods of water depth determination from geological data (Brett *et al.* 1993), the amplitude of these events has been estimated as *c.*30–130 m (Artyushkov & Chekhovich 2001). A lot of geological data has been obtained for the Silurian in East Siberia (Tesakov *et al.* 1986, 1998a, 1998b). On this basis, modelling of relative sea-level changes has been done for eustatic events 1–3 Ma long with an abrupt regressive phase, and for events with a harmonic form (*Ibid.*). The maximum possible amplitude of these events has been estimated as 20–30 m. Under such circumstances, large-scale variations of water depth in the Silurian (*c.*30–130 m) indicate rapid vertical crustal movements. Highly detailed data have been published recently on the East Siberian Basin in the Silurian (Tesakov *et al.* 2000). This allows us to make more accurate estimates of eustatic sea-level fluctuations in the Silurian, consider temporal and spatial variations in the rate of crustal subsidence in the basin, and to propose a possible physical mechanism for these variations.

The Silurian basin of East Siberia

Silurian deposits occur on the East Siberian Craton over an area of $c.2\times10^6$ km^2 (Fig. 1). Their thickness reaches *c.*600 m in the central part of the basin and increases to *c.*800 m towards the northwest. Practically unmeta-

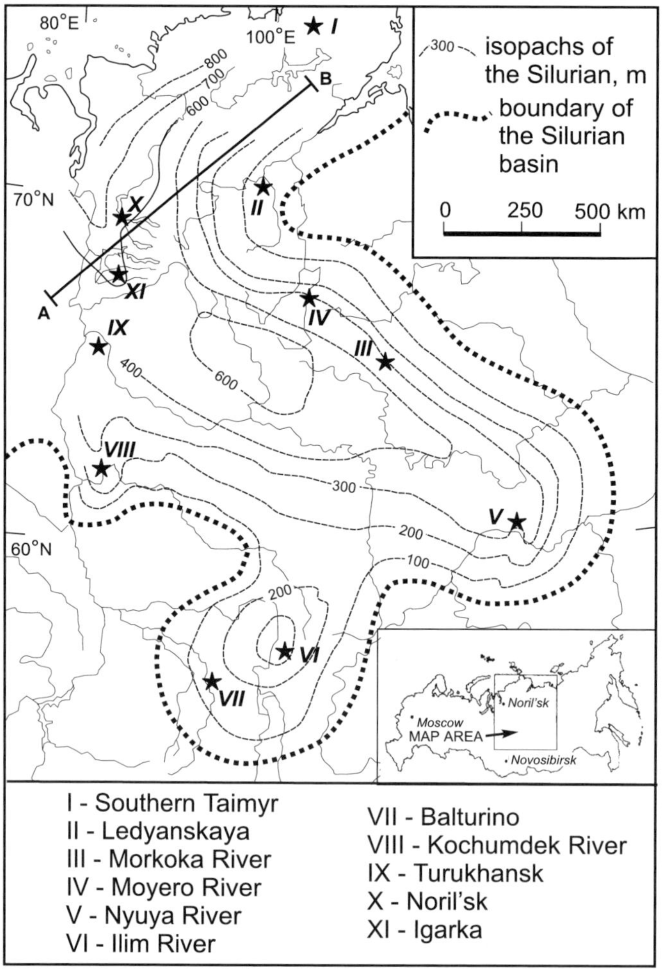

Fig. 1. Thickness of Silurian deposits in East Siberia (modified after Tesakov, 1981), and the regions of stratigraphic successions used for the analysis. AB – line of profile of Figure 10.

morphosed and gently folded Silurian deposits are found along numerous rivers. For example, in the middle part of the course of the Moyero river, they are found along a stretch 70 km long, in steep banks, up to 50–80 m high, which represent a continuous succession of the Silurian and the lowermost Devonian (Sokolov 1985, 1992; Tesakov *et al.* 2000). In many places, Silurian rocks have been also extensively drilled. Based on a detailed analysis of sedimentary facies and remnants of fauna, a number of continuous stratigraphic successions have been constructed for the Silurian in East Siberia (e.g. Figs A.1–A.3 in the Appendix) (Tesakov *et al.* 1986, 1998*a*, *b*, 2000; Sokolov 1985, 1992). The completeness of data is considerably better than for the other Silurian basins, where the successions are commonly constructed using fragments described from isolated regions.

Based on lithological and palaeontological data, sedimentary successions are subdivided into units of different type and rank (Murphy & Salvador 1999). The smallest standard (globally correlated) chronostratigraphic units are a 'stage' and, sometimes, a 'substage'. For the Phanerozoic, the duration of stages is usually several million years, and most substages are 1–2 Ma long. In some regions, for which highly detailed data are available, the smallest chronostratigraphic unit represents virtually a 'chronozone' corresponding to a time interval of 0.5–1 Ma. This is characteristic, for example, of the Middle and Upper Devonian of Euroamerica (Johnson *et al.* 1985) and the Mississippian (lower part of the Carboniferous) in the Midcontinent of the North American Craton (Ross & Ross 1987). For the Silurian, which is 26 Ma long (Gradstein & Ogg 1996), global correlation is based on graptolite and conodont zonations, which include 38 and 12 zones, respectively. Regional correlation is based on the coral, brachiopod, ostracode, and vertebrate zonations, which also include a number of zones. During the Silurian, the East Siberian Basin had a good connection to the ocean in the northwest. Subdivision of the Silurian in this area has been done using both global (graptolite and conodont) and regional scales based on hundreds of species (Tesakov *et al.* 1998*a*, *b*, 2000). The occurrence of numerous exposures along the rivers made it possible to establish reliable correlations between the successions all over the basin. As a result, the Silurian of East Siberia has been subdivided into 54 chronozones, which are equivalent to the time intervals (chrons), each corresponding with 0.48 Ma on average (Fig. 2).

The depth of water (h_w; see also Table 1) in palaeo-basins, in the Silurian in particular, is commonly estimated from benthic associations (fauna and flora) and sedimentary patterns (Brett *et al.* 1993; Scrutton 1998; Watkins *et al.* 2000). According to this approach, marginal parts of the basin (h_w from 0 to 150–200 m) are subdivided into a number of bathymetric zones with a certain h_w characteristic of each zone. This approach has been used for the East Siberian Basin in the Silurian (Tesakov *et al.* 1986, 1998a, 2000; Sokolov 1992; Johnson *et al.* 1997). Following Wilson (1975), for the shallowest part of the basin a more detailed subdivision has been used which includes a carbonate coastal shoal (h_w *c.*0–10 m), lagoon between the shoal and the coast (h_w *c.*0–10 m), and the uppermost shelf above the normal (fair-weather) wave basis (h_w *c.* 10–15 m). In the central part of the coastal shoal, composed of coral and algae build-ups and their debris, the depth of water is h_w *c.*0–5 m.

Based on a large amount of sedimentological and palaeontological data, the palaeo-depth has been determined for a number of places in East Siberia, for which the locations of bathymetric zones of the sea bottom were located at different time levels of the Silurian (Tesakov 1981; Sokolov 1992; Tesakov *et al.* 1986, 2000). Using these data, bathymetric curves have been plotted for several regions of East Siberia (Artyushkov & Chekhovich 2001). Let us briefly describe the evolution of the basin during the standard stages of the Silurian (Gradstein & Ogg 1996) as shown in Table 2. In the Late Ordovician, East Siberia was covered by a shallow epeiric sea. Around the transition from the Ordovician to the Silurian, subaerial exposure of short duration occurred over most of the basin – in its northern and central parts. Then a rapid transgression followed, and in 1–2 Ma a pelagic environment with water depths h_w *c.*100 m had established in these regions (e.g. I–IV, VIII–XI in Fig. 1). On the basin's margins, in the Ilim (VI) and Balturino (VII) regions in the south and in the Nyuya region (V) in the southeast, marine deposition was continuous across the Ordovician–Silurian boundary. At the start of the Silurian, the depth of water in these regions (V–VII) was $\leq$10 m, and slow deposition at these depths continued into the Silurian, but the upper part of the Silurian successions has since been eroded. In the Ilim region (VI), shallow-water ($h_w \leq 10$ m) Silurian deposits have been preserved from a time interval of *c.*12 Ma – from the earliest Silurian to the early Telychian. In the Balturino region (VII), the age of these sediments ranges from Early Rhuddanian to Early Wenlock (*c.*16 Ma). In the Nyuya region (V), Silurian sediments formed at $h_w \leq 10$ m have been preserved from the

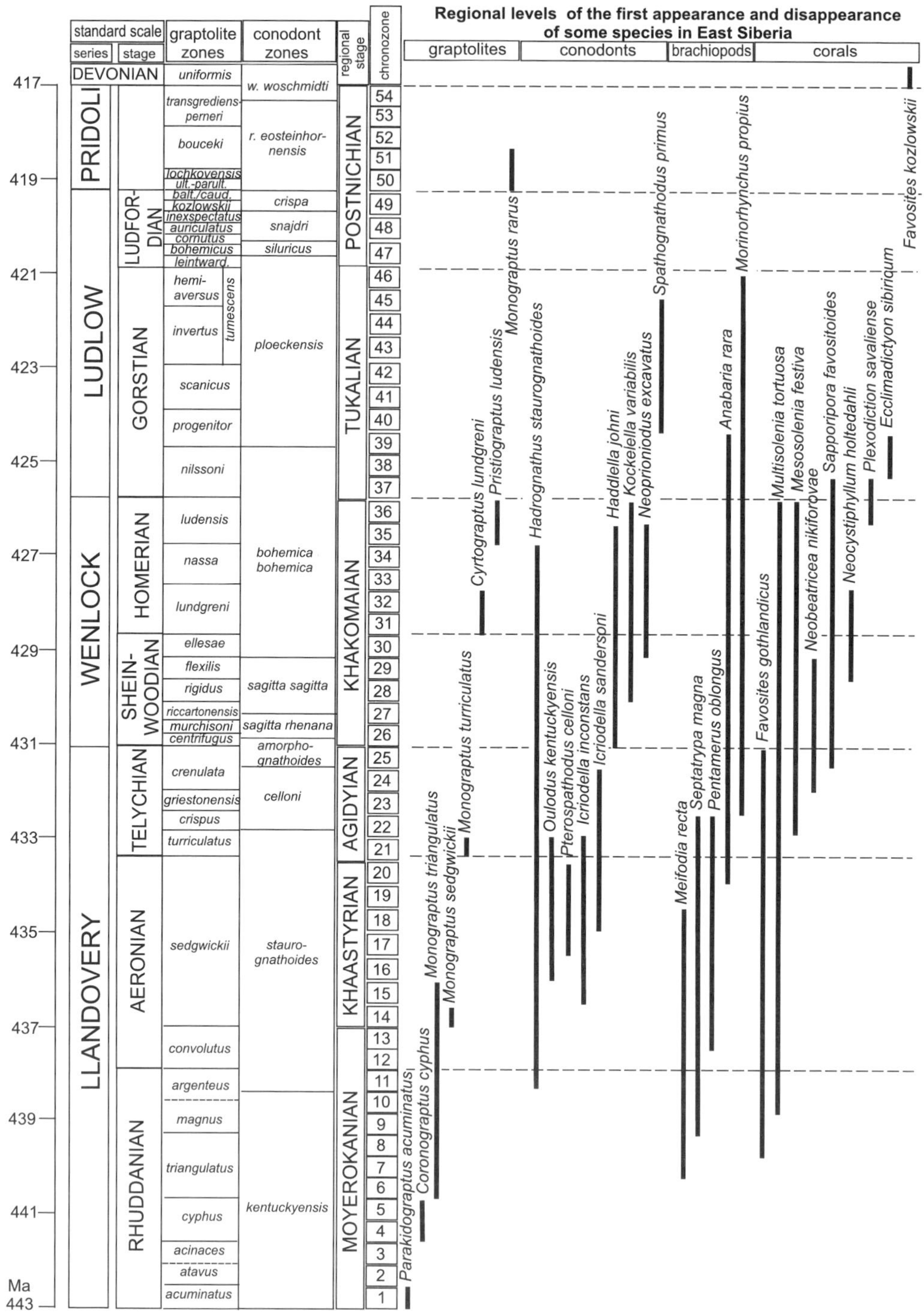

Fig. 2. Standard chronostratigraphic scale with graptolite and conodont zonations for the Silurian System (modified after Holland & Bassett, 1989), and regional scale for the Silurian of East Siberia (modified after Tesakov *et al.* 2000). Regional units and biochronology modified after Tesakov *et al.* (2000). Timing of the base and the top of the Silurian is after Gradstein & Ogg (1996).

Table 1. *List of symbols*

Notation	Description
h_w	Depth of water
h_w^0	Depth of water before eustatic event
d	Thickness of strong part of the lithosphere
d_0	Mean value of d
Δd	Lateral variations of d
L	Width of lateral variations of d
ΔF	Change in force acting along lithosphere
c	Vertical displacement of lithosphere
c_0	Amplitude of vertical displacement of lithosphere
ρ_m	Density of mantle — 3350 kg /m^3
ρ_s	Density of sediments
g	Acceleration due to gravity — 9.81 m/s^2
T	Duration of eustatic event
t	Time
τ	Dimensionless time (t/T)
Δt_{cz}	Duration of the interval corresponding to chronozone
Δt_{cz}^0	Initial duration of chronozone
$\delta(\Delta t_{cz})$	Change of chronozone's duration
n	Number of chronozones in certain time interval
Δt	Change of duration of time interval comprising n chronozones, due to change in their duration
Δh_{cz}	Thickness of chronozone
Δh_s	Change of thickness of sediments of time interval comprising n chronozones due to change in their duration
Δh_{cz}^0	Initial thickness of chronozone
Δt_{sa}	Duration of subaerial exposure and non-deposition
z_0	Minimum altitude at which subaerial exposure becomes noticeable
b	Amplitude of eustatic event
b_m	Minimum amplitude of resolvable events
a	Mean rate of crustal subsidence
a_0	Rate of crustal subsidence necessary for compensation of sea-level changes by change in chronozone duration
a'	Mean rate of crustal subsidence at certain time interval
ζ	Ordinate of crustal surface with respect to sea level
ζ_{eu}	Eustatic signal
$\Delta\zeta_{eu}$	Change of sea level during a chronozone
ζ_{eu}^0	Sea-level change required for subaerial exposure under local isostasy
ξ	Deviation from horizontal axis in Fischer plot
$\Delta\xi$	Change in deviation ξ during a certain time interval

Table 2. *Rates of crustal subsidence during the Silurian in the case when chronozone lengths are assumed to be variable*

Standard subdivision	Number of chronozones (n)	Length of standard subdivision, (Ma)*	Average chronozone length (Δt_{cz}), (Ma)	Minimum rate of crustal subsidence (a), (m/Ma)†	Region
Rhuddanian	11	4	0.36	11.2	Nyuya River
Aeronian	9	6.9	0.76	16.5	Nyuya River
Telychian	5	4.1	0.84	6.4	Balturino
Wenlock	11	5	0.45	7.1	Ledyanskaya
Ludlow	13	4	0.31	15	Kochumdek
Pridoli	5	2	0.4	10.6	Kochumdek

*The length of the Llandovery, which comprises the Rhuddanian, Aeronian and Telychian, Wenlock, Ludlow and Pridoli (Ma), was taken according to Gradstein & Ogg (1996). For this time scale, the relative lengths of the Rhuddanian, Aeronian and Telychian were taken according to M. E. Johnson (1996), and Tesakov *et al.* (1998*a*).

†Rock compaction is taken into account (Artyushkov & Chekhovich 2001).

base of the Silurian until the lower Ludlow (~20 Ma).

The fact that shallow-water conditions were preserved on the southern and southeastern basin margins in the Early Llandovery indicates that rapid deepening in the central and northern parts of the basin at the start of the Silurian was of a tectonic origin (Artyushkov & Chekhovich 2001). The water-filled depression, initially *c.* 100 m deep, was shoaling and filling with sediments in the Early Llandovery. In the middle of the Llandovery, large areas with water depths $h_w \leq 10$ m, probably up to 15–20 m in some places, existed in the Nyuya, Ilim and Balturino regions on the basin margins (V–VII in Fig. 3). The depth of water gradually increased to *c.* 100 m toward the northwest. In the Early Wenlock, water depths of ≤10 m were characteristic of most of the basin (Fig. 4). The depth of water reached 15–30 m in the basin's narrow axial part. This region was separated from a wide lagoon by a shoal which was several hundred kilometers wide and had water depth of h_w *c.* 0–5 m. By the end of the Silurian, the depth of water was $h_w \leq 10$ m all over the East Siberian basin (Tesakov *et al.* 2000).

Fischer plots based on elementary cycles

In this paper, we consider the temporal and spatial variations in the rate of crustal subsidence in East Siberia and sea-level changes during the Silurian. The study could have been done in different ways, but we chose the technique of Fischer plots (Fischer 1964), which is quite familiar to sedimentologists. Using this approach, many authors have evaluated relative sea-level changes of the third order (Osleger &

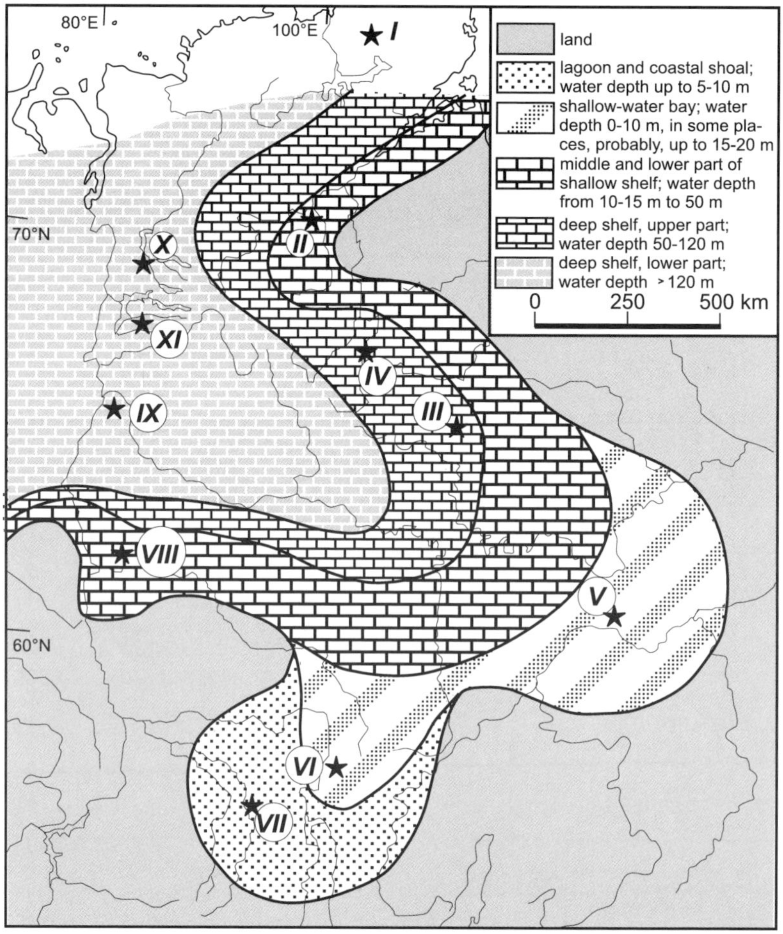

Fig. 3. The East Siberian sedimentary basin in the middle of the Llandovery (modified after Tesakov 1981).

Read 1991; Goldhammer *et al.* 1993; Elrick 1995; Bosence *et al.* 2000; Cheng *et al.* 2001). Since most scientists specializing in geodynamics are unfamiliar with these plots, we will now provide a brief description of how they are constructed.

Relative changes of sea-level produce changes in the accommodation space where the deposition of sediments takes place. Under conditions of continuous deposition in very shallow water, i.e. when the sea bottom is close to the surface, relative sea-level rises and falls cause increases and decreases in sediment thickness, respectively; large-scale falls can produce erosion and hiatuses. The sedimentary successions of shallow carbonate platforms include upward-shallowing elementary cycles of a metre scale with duration of 0.01-1 Ma. Fischer plots were constructed using large sets of these cycles. In each cycle, sediment thickness was compared with the thickness corresponding with crustal subsidence at a constant rate – line OO* in Fig. 5. Subsidence during one elementary cycle equals AB. If sediment thickness AC is greater than tectonic subsidence AB (Fig. 5a), line OC deflects upwards. If sediment thickness is less than crustal subsidence (Fig. 5b), line OC deflects downwards. In a sequence of elementary plots (Figs A.4 & 6), point C of plot *n* corresponds with point O of the next plot *n*+1.

It is often considered that elementary cycles are related to small-scale, quasi-periodic eustatic fluctuations most probably driven by Milankovich orbital forcing, thus the cycles are also quasi-periodic (Goldhammer *et al.* 1991; Read *et al.* 1991; Schwarzacher 2000). Then, under uniform tectonic subsidence, Fischer plots, which include a large number of elementary cycles, will

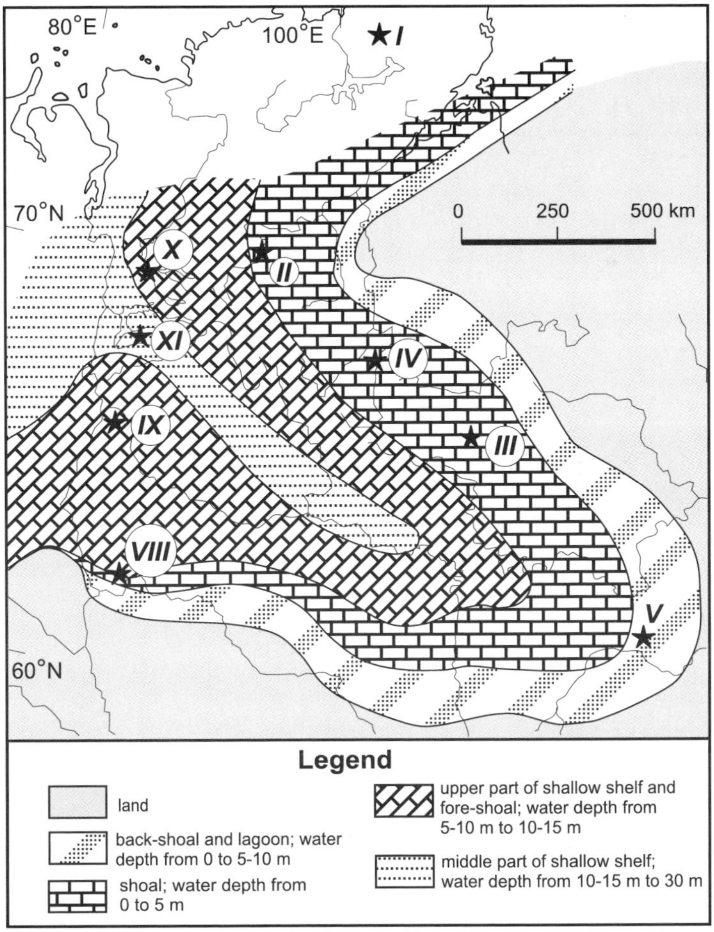

Fig. 4. The East Siberian sedimentary basin in the Early Wenlock (modified after Tesakov 1981).

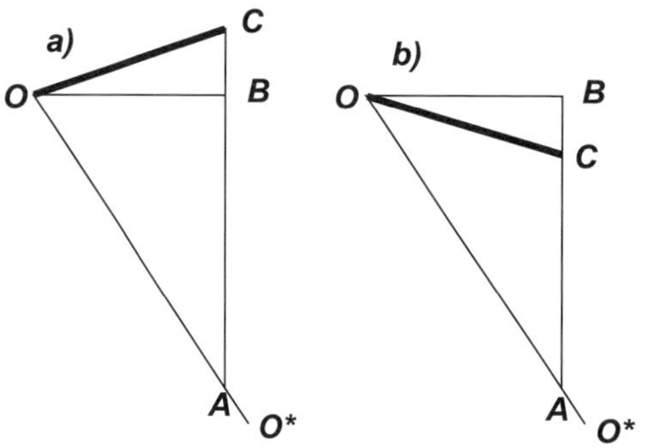

Fig. 5. Fischer plots for the elementary cycles in sedimentary successions. (**a**) for the case when sediment thickness AC exceeds tectonic subsidence AB during the cycle. (**b**) for the case when sediment thickness AC is smaller than tectonic subsidence AB. Under uniform crustal subsidence (line OO*), lines OC should approximate sea-level changes during the cycle.

describe eustatic sea-level changes of the third order, i.e. rises and falls in the plots will be equal to sea-level rises and falls, respectively (e.g. Osleger & Read 1991; Goldhammer *et al.* 1993; Elrick 1995; Bosence *et al.* 2000). As suggested by Ginsburg (1971), elementary cycles can be also produced by the well-known process of tidal-flat migration, which is due to the transport of sediments by currents in the shallow marginal parts of sedimentary basins. Then the lengths of the elementary cycles can change randomly over time, with corresponding random changes in the thicknesses of the elementary cycles, which are observed in many outcrops (Drummond & Wilkinson 1993 *a*, *b*; Wilkinson *et al.* 1997). An increase in cycle length will increase sediment thicknesses, thus producing rises in the Fischer plots, even under conditions of stable sea-level and uniform crustal subsidence. A decrease in cycle length will decrease sediment thicknesses, resulting in falls in the Fischer plots. Following on from the numerical modelling (Burgess *et al.* 2001), Fischer plots based on elementary cycles produced by random processes of near-shore deposition cannot be used for estimates of long-term sea-level fluctuations, even under a constant rate of crustal subsidence.

Fischer plots based on elementary cycles can describe sea-level changes of the third order, if two basic conditions are fulfilled: firstly elementary cycles are quasi-periodic, and secondly the rate of crustal subsidence is constant. The data on East Siberia can be used to check the validity of these assumption, as well as the applicability of Fischer plots based on elementary cycles for evaluation of sea-level changes. In Figure A.4, these plots are presented for five regions of East Siberia in the Silurian. They correspond with those epochs when deposition took place at very shallow depths. Eustatic sea-level fluctuations are global in character and should be similar for all the regions. The plots in Figure A.4 differ for the different regions. Hence they do not describe sea-level changes during the Silurian. As shown in the Appendix, at that epoch, the duration of elementary cycles in East Siberia changed in time, i.e. the first condition was not fulfilled. Moreover, elementary cycles were not synchronous in different regions, which caused large differences in the plots for different regions in Figure A.4. Based on the figure, it cannot be said that changes in the rate of crustal subsidence had an influence on the Fischer plots. However, it is well known that, in most sedimentary basins, the rate of crustal subsidence changed over time (e.g. Beloussov 1980; Sloss 1988). Hence the second assumption can also be unrealistic, at least for long periods of time. This will be considered below for the case of East Siberia in the Silurian.

Fischer plots based on sediment thicknesses of the Silurian chronozones

Large regions with deposition at very shallow depths $h_w \leq 10$ m existed in East Siberia throughout the Silurian. Deposition usually occurred in the peritidal zone or in the upper part of the subtidal zone (Tesakov *et al.* 2000). Due to the uncertainty of timing of elementary cycles in the Silurian successions, Fischer plots based on such cycles cannot be used for the reconstruction of sea-level changes. The Silurian successions of East Siberia are subdivided into chronozones, each lasting *c.*0.5 Ma. Their lengths fall within the range of the elementary cycles (0.01–1 Ma) used in Fischer plots. Therefore we can construct Fischer plots using the sediment thicknesses of the chronozones (Δh_{cz}) for those parts of sedimentary successions which correspond with the periods of deposition at $h_w \leq 10$ m. Plots constructed in this way for the 11 regions of East Siberia are presented in Figure 6. Chronozone thicknesses Δh_{cz} are taken from Tesakov *et al.* (1998*b*, 2000).

In Fischer plots based on elementary cycles, 'thin' cycles usually outnumber 'thick' cycles two to one (Sadler *et al.*, 1993). Then, taking into account the usual scatter of cycle thicknesses, it follows that more than 40 cycles, or, preferably, more than 50 cycles, are necessary for obtaining reliable Fischer plots (*Ibid.*). In the case of random distribution of chronozone lengths with a predominance of short chronozones, a large

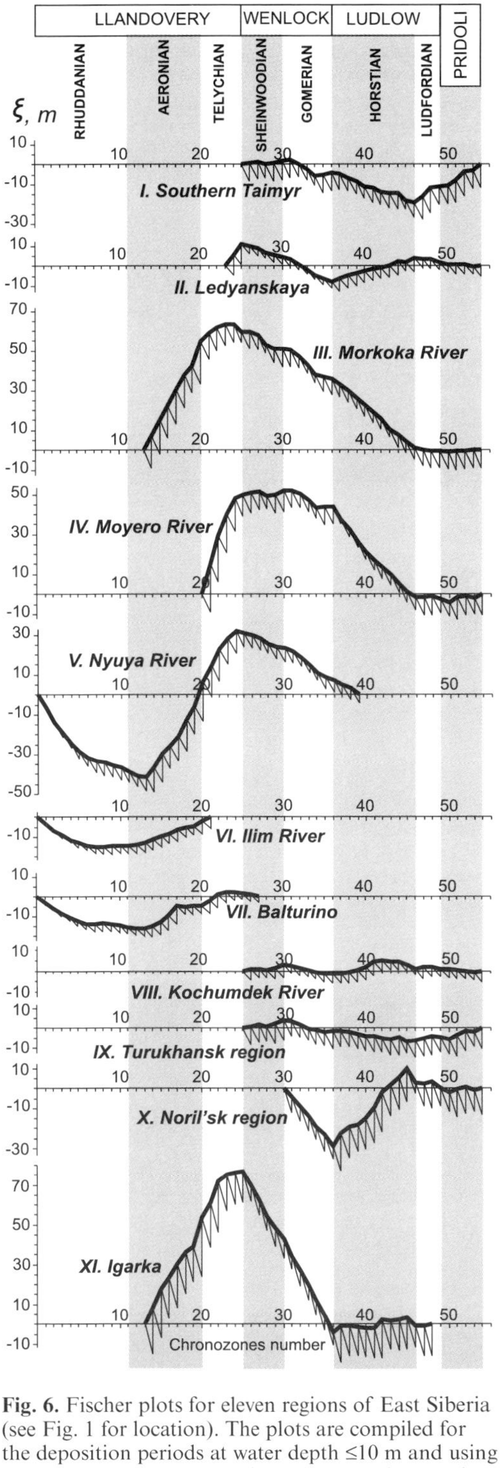

Fig. 6. Fischer plots for eleven regions of East Siberia (see Fig. 1 for location). The plots are compiled for the deposition periods at water depth ≤10 m and using regional chronozones as a time scale (Tesakov *et al.* 1998*a*, *b*). Compaction of sediments is taken into account (Artyushkov & Chekhovich 2001).

number of chronozones would be required for the construction of reliable plots. Fischer plots presented on Figure 6 were constructed on the basis of 21–41 chronozones, which is insufficient in view of what was said above. However, in Figure 6, 'thin' chronozones predominate only in three plots (III–V), while the other eight plots comprise comparable numbers of 'thin' and 'thick' chronozones. Also, the fact that the lengths of the Silurian chronozones (Δt_{cz}) were not as random as those of the elementary cycles cannot be ruled out. Moreover, in contrast to Fischer plots based on elementary cycles of uncertain timing, the plots in Figure 6 were based on synchronous units. This allows us to compare the plots for different regions, which leads to several important conclusions.

Plots on Figure 6 show cumulative sediment thickness deviations, corrected for compaction (Artyushkov & Chekhovich 2001), from the thickness corresponding to uniform crustal subsidence under the assumption of a constant duration of the chronozones Δt_{cz}. Deviation amplitudes reach 30–75 m, i.e. strongly exceed the deposition depths. Under such conditions, the main factors which can produce such deviations are:

(1) variations in the duration of the chronozones;
(2) eustatic sea-level variations; and
(3) variations in the rate of crustal subsidence.

Let us discuss the possible contributions of these factors.

Figure 6 includes two types of plots: with large and small deviations from the horizontal axis. In the plots of the first type (III–V and XI) deviations reach 50–75 m, i.e. are within the range commonly associated with third-order cycles (20–100 m). In plots I and X, deviations are 20–30 m, which is also within the range. However, the duration of the deviations equals *c.* 10–15 Ma which is typical of second-order rather than third-order cycles. In plots of the second type (II, VIII, and IX), deviations in the Late Silurian (Wenlock–Pridoli) do not exceed 5–10 m. In plots VI and VII, falls of 15 m occur in the first half of the early Silurian (Llandovery). Very small deviations of ≤5 m are also observed in some segments of the plots of the first type: plot I in the Sheinwoodian, plots III, IV, and X in the Ludfordian–Pridoli, and plot XI in the Ludlow. All these values are considerably smaller than the eustatic fluctuations of 20–100 m supposed for third-order cycles of sea-level changes.

Sub-horizontal plots. Indications for relative stability of the sea-level

Let us consider the conditions necessary for the formation of subhorizontal segments in plots I–IV and VIII–XI. Suppose that the duration of the chronozones Δt_{cz} varied with time, while sea-level remained stable and the rate of crustal subsidence was constant. Then, the thickness of the sediments corresponding with a chronozone will be as follows:

$$\Delta h_{cz} \approx a\Delta t_{cz}, \quad (1)$$

where a is the rate of crustal subsidence. Assume, for example, that chronozone duration Δt_{cz} increases by $\delta(\Delta t_{cz})$: $\Delta t_{cz}=\Delta t_{cz}^0+\delta(\Delta t_{cz})$. Then, according to equation (1), the chronozone thickness increases by $a\delta(\Delta t_{cz})$: $\Delta h_{cz}=\Delta h_{cz}^0+a\delta(\Delta t_{cz})$ (Fig. 7a). In an elementary Fischer plot, where the duration of the chronozones Δt_{cz} is supposed to be constant and equal to Δt_{cz}^0, this will result in a rise of the plot by $a\delta(\Delta t_{cz})$ (Fig. 7b). No rise will occur in the plot if there was a synchronous fall of sea-level ($\Delta\zeta_{eu}$) of the same amplitude (Fig. 7c):

$$\Delta\zeta_{eu}=-a\delta(\Delta t_{cz}). \quad (2)$$

To explain small deviations in plots I–IV and VIII–XI, it can be supposed that large-scale falls and rises of sea-level occurred in the Late Silurian; however, their influence on the plots was almost completely balanced by synchronous increases and decreases in Δt_{cz}. Sea-level changes $\Delta\zeta_{eu}$ are the same in all the regions. In different regions, crustal subsidence occurs at different rates. Equation (2) can be fulfilled only at a certain $a_0=|\Delta\zeta_{eu}|/|\delta(\Delta t_{cz})|$. In the regions where a is larger ($a>a_0$), sea-level fall $\Delta\zeta_{eu}$ will not completely balance the increase in Δh_{cz}, thus a rise will occur in the Fischer plot (Fig. 7d). In the region with lower a ($a<a_0$), there is a fall in the plot (Fig. 7e). Regions I–IV and VIII–XI of East Siberia are characterized by large variations of a: 17.4, 12.8, 20.6, 21.4, 11.7, 21.8, 25.6 and 33 m/Ma. With such variations of a from place to place (as large as 200%), plots will not show subhorizontal segments due to the compensation of sea-level changes by variations in chronozone lengths. This indicates the subperiodicity of the Late Silurian chronozones in East Siberia.

It should be also noted that an inverse correlation between the global sea-level changes and the lengths of regional chronozones Δt_{cz} is improbable for an area with a very small depth of water, especially taking into account that the Silurian chronozones are controlled by both regional and global zonations.

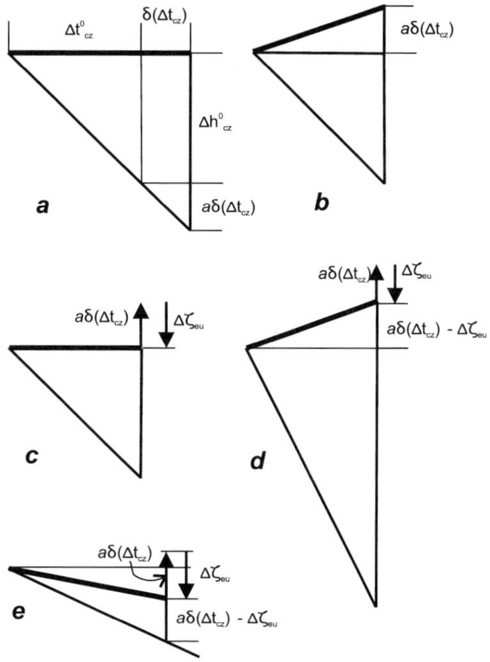

Fig. 7. Influence of increase $\delta(\Delta t_{cz})$ in chronozone length Δt_{cz} on Fischer plots for elementary cycles, assuming uniform crustal subsidence. (**a**) Increase $\delta(\Delta h_{cz})$ in the thickness Δh_{cz} of a chronozone caused by increase in Δt_{cz}. (**b**) A rise in the elementary Fischer plot resulting from the assumption of constant Δt_{cz}. (**c**) Compensation of this rise by a synchronous sea-level fall by $\Delta\zeta_{eu}=-\delta(\Delta h_{cz})$ at a certain rate of crustal subsidence $a=a_0=|\Delta\zeta_{eu}|/|\delta(\Delta t_{cz})|$. (d) A rise in elementary plot for a region where $a > a_0$. (e) A fall in the plot for a region with $a < a_0$.

Given that chronozone lengths Δt_{cz} are nearly constant, it can be presumed that in the sub-horizontal segments of plots I–IV, and VIII–XI of Figure 6, the influence of large-scale sea-level falls and rises was almost completely balanced by changes in the rate of crustal subsidence. To provide such a balance, similar changes in the rate of crustal subsidence must have occurred in all the regions mentioned above, despite the fact that the average rate of crustal subsidence had varied in them by up to 200%. Furthermore, global eustatic fluctuations and regional vertical crustal movements in cratonic areas are independent phenomena. Even in one region, the probability that they will balance each other is very low. Actually, no large deviations synchronously occurred in regions VIII and IX, which are *c.*400 km apart, as well as in region II which is located as far as 1000 km from region VIII. The Fischer plot is subhorizontal for region I in the Sheinwoodian, region XI (650 km from region VIII) in the Wenlock, and regions III and IV (which are 900–1000 km from region VIII) in the Ludfordian and Pridoli. The probability of balancing of large-scale sea-level changes by changes in the rate of crustal subsidence in all the above regions is negligible.

Thus, for the Late Silurian, the synchronous existence of three subhorizontal plots (II, VIII, and IX) and subhorizontal segments in plots I, III, IV, X, and XI is possible only if the following three conditions are simultaneously fulfilled:

(1) eustatic sea-level changes did not exceed 5–10 m;
(2) East Siberian chronozones were subperiodic; and
(3) crustal subsidence was nearly uniform in each region during those periods which correspond with subhorizontal segments of Fischer plots.

In the Early Silurian, falls of about 15 m followed by rises of similar amplitude are seen on plots VI and VII of Figure 6. The falls and rises on plot V are almost synchronous with those on plots VI and VII, but their amplitudes are about three times larger. The mean rate of crustal subsidence in region VII (15.8 m/Ma) is about two times larger than in regions VI and VII (7.8 and 8.5 m/Ma, respectively). As shown above, this precludes deviations due to changes in the lengths of the chronozones. A large difference in the amplitude of the deviations indicates a strong influence of the tectonic factor; however, the role of eustatic fluctuations still needs to be considered. Since global eustatic fluctuations and regional changes in the rate of crustal subsidence in cratonic areas are independent phenomena, they cannot be correlated in time and have different lengths and amplitudes. Superposition of such uncorrelated processes with a comparable intensity in plots VI and VII would have destroyed their synchronicity with plot V. Hence eustatic fluctuations could account for only a minor portion of the total deviations of *c.* 15 m in plots VI and VII, i.e. sea-level changes during the Early Silurian barely exceeded *c.* 5 m. In the Late Silurian these variations were ≤5–10 m. Thus, eustatic fluctuations were small throughout the Silurian Period, which lasted for 26 Ma. They were considerably smaller than sea-level changes of the third order (*c.* 20-100 m) supposed for the Phanerozoic in general (Haq *et al.* 1987; Ross & Ross 1987; Hallam 1992) and for the Silurian in particular (M. E. Johnson 1996).

Plots with large-scale falls and rises: indications of changes in the rate of crustal subsidence

Shown on Figure 6 are Fischer plots including both major (II–V, X, and XI) and minor (I, II, VI–IX) deviations. By comparing deviations (ξ) in different plots it is possible to determine the factors responsible for the major ones. First, let us once again consider the possible influence of changes in chronozone length. Suppose that within a time interval (Δt), which includes n chronozones, their average length changed by $\delta(\Delta t_{cz})$. Then, the interval's length would change by $\Delta t = n\Delta t_{cz}$ as compared with the case when chronozone length was constant. Given that the rate of crustal subsidence and the sea-level were constant, this would lead to a change (Δh_s) in the thickness of sediments formed during this time interval by:

$$\Delta h_s = a\Delta t. \tag{3}$$

Fischer plots were compiled using the assumption of constant chronozone (or elementary cycles) length. Therefore, a Δt change in deposition interval will lead to a Δh_s change in Fischer plot deviation, as determined by equation (3). Then, for synchronous time intervals, such deviation changes ($\Delta\xi$) will be proportional to the rates of crustal subsidence, a. Consider, for example, the interval of chronozones 21–24 in regions III–V, and VII. For this time interval (see plots III–V, VII on Fig. 6), the values of $\Delta\xi$ are 10, 46, 26, and 7 m, respectively, hence their ratio is 1:4.6:2.6:0.7. The average rates of crustal subsidence a in these regions were 18.8, 18.4, 15.8, and 8.5 m/Ma, respectively, and their ratios are quite different: 1:1.02:0.84:0.45. For the Wenlock (chronozones 26–36; see plots III–V, VIII, IX, and XI), the values of $\Delta\xi$ are 23, 5, 23, 0, 1.4, and 81 m, i.e. the ratios are 1:0.22:1:0:0.02:3.52. The average rates of crustal subsidence in these regions were 18.8, 18.4, 15.8, 9.2, 17.2, and 31.5 m/Ma and the ratios are 1:1.02:0.84:0.49:0.91:1.68, which is again quite different from the ratios of $\Delta\xi$. This confirms that no significant variations in chronozone length took place during the Silurian in East Siberia.

As follows from the above considerations, the existence of subhorizontal segments in the plots of Figure 6 indicates that sea-level remained nearly stable during the Silurian, and that the East Siberian chronozones were subperiodical. Then the horizontal scale in Figure 6 is sublinear, and the large-scale deviations in the plots approximately reflect changes in the rate of crustal subsidence: the rises correspond with epochs of accelerated subsidence, and the falls took place when subsidence slowed down. In plots III–V, X and XI, changes in the rate of crustal subsidence are quite large. For example, in plot III, the average rate of subsidence a' was 32 m/Ma during the rise in the Aeronian and 12.6 m/Ma during the fall in the Horstian. In plot IV, $a'=33.6$ m/Ma for the rise during chronozones 21–24 in the Telychian, and $a=10.9$ m/Ma during the fall in the Horstian. In plot V, the rate of subsidence was extremely low during the first five chronozones of the Silurian: $a'=4$ m/Ma, while $a'=29.7$ m/Ma for the rise during chronozones 14–24 in the Aeronian and Telychian. In plot XI, $a'=44.7$ m/Ma for the rise during chronozones 14–22, and $a'=16.2$ m/Ma for the fall in the Wenlock.

It has to be noted that the quasi-synchronous existence of large-scale deviations in plots III–V, X, XI and of subhorizontal plots II, VI–IX is already sufficient to prove a strong influence of tectonic movements on Fischer plots. These plots are based on synchronous units; hence it is possible to make a reliable comparison of the changes in accommodation space for a number of regions and time intervals. If such changes are of eustatic origin, they will be the same everywhere. Large differences in the plots show that in different regions of East Siberia accommodation space changed in time in different ways. This is a direct indication of the existence of considerable changes in the rate of crustal subsidence over time.

The major falls and rises seen on the plots of Figure 6 have different amplitudes, but in some cases they are nearly synchronous. This is characteristic of the falls seen in plots V, VI and VII in the Rhuddanian, and the rises in plots III–VII and XI in the Aeronian and Telychian. The fall in plot IV in the Horstian, and the fall in plot XI in the Wenlock are asynchronous. However, they both coincide in time with the long fall in plot III in the Wenlock and Horstian. Simultaneously with most of this fall, a fall is also observed on plot V in the Wenlock and earliest Ludlow; however, no younger sediments have been preserved in region V. Thus, accelerations and decelerations of crustal subsidence in East Siberia were, in most cases, roughly synchronous, but different in the amplitude. This suggests that they involved one and the same physical mechanism with a spatially variable intensity. Let us discuss the main possibilities.

Thermal relaxation and changes in the forces acting on the lithosphere

Cooling of the lithosphere causes thermoelastic contraction of rocks and hence crustal subsidence (Sleep 1971; McKenzie 1978). The characteristic time-scale for thermal relaxation is *c.* 10^2 Ma. In the Silurian, the Early Proterozoic lithosphere of East Siberia was $\geq 10^3$ Ma old (Rosen *et al.* 1994). Therefore, thermal relaxation of the crust and mantle could not have played any significant role in subsidence. Moreover, thermal relaxation is a gradual process, and it could not have produced strong accelerations and decelerations of subsidence during ≤ 10 Ma in regions II–V, X and XI.

Major forces are acting in the lithosphere (Zoback 1992), and considerable lateral variations of lithospheric thickness (d) are found in many areas. Under such conditions, to balance the momentum of the forces acting along the lithosphere, in the gravity field, vertical deflections of the lithosphere arise from its isostatical equilibrium position (Fig. 8) (Artyushkov 1974, 1983). It has been proposed that rapid, large-scale changes in the depth of water in sedimentary basins occurred due to lithospheric displacements caused by changes in the forces acting along the lithosphere (Cloetingh *et al.*

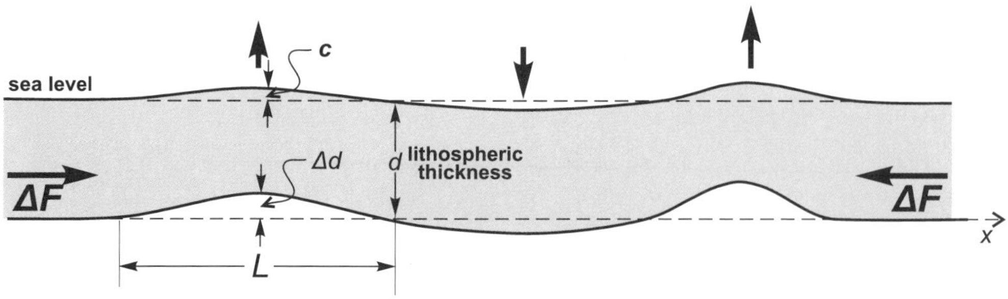

Fig. 8. Vertical deflections of the lithosphere with laterally variable thickness as a result of changes in the forces acting along this layer.

1985). Such displacements (c) are proportional to $1/L^2$, where L is the characteristic width of lateral variations in lithospheric thickness (Artyushkov 1974, 1983). Therefore, they are significant only in relatively narrow areas. Suppose, for example, that the thickness of the strong part of the lithosphere varies laterally as:

$$d = d_0 + \Delta d \sin(\pi x/L). \quad (4)$$

Assume that the force acting along the lithosphere changes by ΔF. Then, the vertical displacement of the lithosphere from its initial position (Artyushkov *et al.* 2000*a*) is:

$$c \approx c_0 \sin(\pi x/L) \quad (5)$$

where the amplitude of the displacement is:

$$c_0 = \pi^2 \Delta F \Delta d / [2(\rho_m - \rho_s) g L^2]. \quad (6)$$

Here $\rho_m = 3350$ kg/m^{-3}, ρ_s are the densities of the mantle and the sediments, respectively, and $g = 9.81$ m/s^{-2} is the gravitational acceleration. It is most probable that in the Silurian the characteristic horizontal dimension of the lithospheric thickness variations in East Siberia was approximately equal to that of the basin: $L \approx 1200$–2000 km. At that time, thickness variations of the strong part of the Early Precambrian lithosphere of East Siberia hardly exceeded $2\Delta d \approx 20$–30 km, which corresponds with $\Delta d \approx 10$–15 km. Suppose that force acting along the lithosphere changes by $\Delta F = 2 \times 10^{12}$ N m, which is roughly equivalent to the average present-day force caused by the spreading oceanic ridges (ridge push) (Artyushkov, 1973, 1983). Taking in equation (6) $\rho_s = 2500$ kg/m^{-3} with the above values of the other parameters, we obtain: $c_0 = 3$–12 m. This is too small to account for the falls and rises with amplitudes of up to 75 m which are seen in the plots of Figure 6.

Suppose now that regions with both increased and decreased lithospheric thicknesses existed in East Siberia in the Silurian, and their dimension L was by several times smaller than the basin width. Since $c_0 \approx 1/L^2$, this could increase lithospheric deflections by one order of magnitude and hence make them comparable with the observed ones. In regions with increased and reduced lithospheric thickness, $\Delta d > 0$ and $\Delta d < 0$, respectively. Then, as follows from equations (5) and (6), the vertical lithospheric displacements in these regions should have had opposite signs, and the basin should have remained close to regional isostasy. This was not the case for East Siberia in the Silurian where large-scale falls and rises (shown in plots of Fig. 6) were quasi-synchronous. Therefore, changes in forces acting along the lithosphere could have hardly been responsible for the observed changes in the rate of crustal subsidence.

This conclusion refers only to East Siberia in the Silurian. In many regions, the horizontal size L of lateral variations in d does not exceed several hundred kilometres. Changes in the forces acting along the lithosphere occurred from time to time in many regions. Then, as follows from the above estimates, vertical displacements in the lithosphere caused by these changes could have been between *c.* 20 m and *c.* 100 m, i.e. of the same order of magnitude as those characteristic of the third-order changes in the depth of water in sedimentary basins. Therefore, this mechanism could have been responsible for sea-level changes in some other cases.

Variations in the dynamic topography

Convective flows in the mantle cause lithospheric displacements from the equilibrium, and such displacements are called dynamic topography (Hager & Clayton 1989). In particular, dynamic topography can be generated by flows rising above subducting slabs of the oceanic lithosphere (Gurnis 1992; Burgess & Gurnis 1995; Coakley & Gurnis 1995; Burgess *et al.* 1997). Additional displacements of the lithosphere can be caused by its elastic bending under supracrustal loads (volcanoes, fold and thrust sheets and clastics). The dynamic topography above the subducting plates can change with time due to changes in the position of subduction zones at the surface, as well as due to changes in the rate and dip angle of subduction. Similarly, elastic displacements of the lithosphere will change with changing supracrustal loads.

This mechanism has been used to explain transgressive–regressive sequences in North American Phanerozoic cratonic strata, and vertical crustal movements in the Michigan Basin during the Middle Ordovician (*Ibid.*). In the Silurian, the East Siberian Basin was part of the Angara Craton (Fig. 9) (Şengör & Natal'in 1996). At that time, subduction took place under this craton along the line ABC, and under the adjacent Khanty–Mansi Ocean along the line DEF. These two subduction lines were separated by a transform fault CD. Therefore, the direction of subduction was approximately parallel to the fault CD, and the oceanic slab subducting along line DEF did not reach the mantle beneath the Angara Craton (Fig. 9). Only the plate subducting along line ABC, 1100–1200 km long, penetrated under this craton.

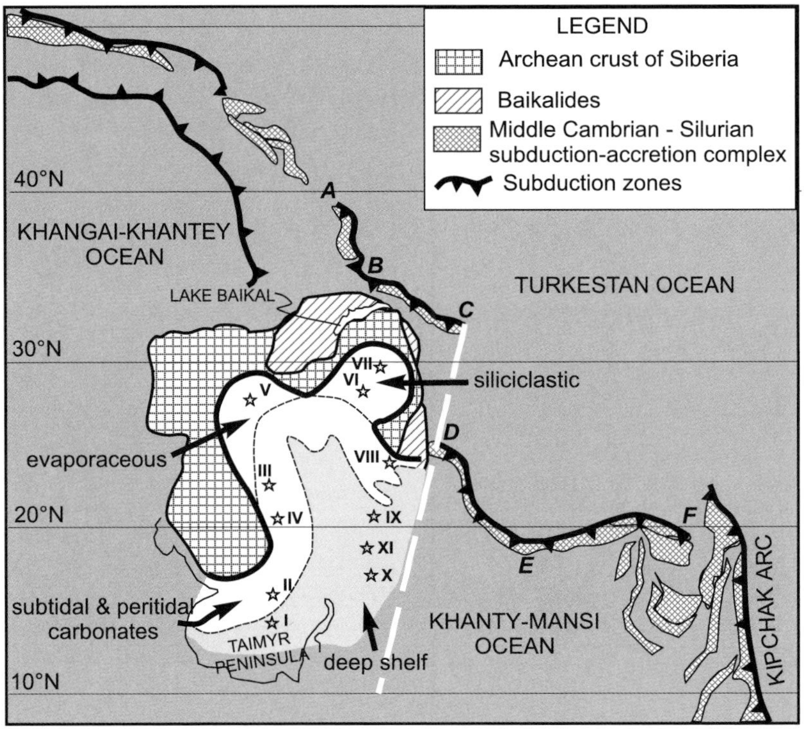

Fig. 9. East Siberia and adjacent areas in the Silurian (modified after Şengör & Natal'in, 1996). Positions of regions I-XI are shown so that they can be compared with their present locations in Figure 1.

Flows in the mantle have been considered for a particular value of upper mantle viscosity (Gurnis 1992; Coakley & Gurnis 1995). It is impossible to say for certain if the mantle viscosity beneath the Angara Craton in the Silurian was the same as in the model. It is unknown whether any changes in dip angle of subduction and its rate took place in the Silurian. Similarly, changes in supracrustal loads on the active margin of the Angara Craton are difficult to quantify. It is only possible to say that no drastic changes occurred in the position of subduction zones during the Silurian.

Under such circumstances, we only can compare changes in the rate of crustal subsidence in East Siberia with those expected from the mechanism under consideration. Lithospheric displacements generated in accordance with this mechanism decrease with distance from the subduction line (Coakley & Gurnis 1995; Burgess *et al.* 1997). In the Early Silurian, minor changes in the rate of crustal subsidence occurred in regions VI and VII, which are located 500 km and 700 km from the convergent boundary ABC. Regions III, IV and V were considerably farther from this line: 1600 km, 1700 km, and 1200 km, respectively. However, synchronous changes in the rate of crustal subsidence were several times larger in them (see Fig. 6). In the Late Silurian, almost no changes in the rate of crustal subsidence occurred in the Kochumdek and Turukhansk regions, 1100 and 1400 km from the convergent boundary, while major changes in the rate of subsidence took place in regions III, IV, V, X, and XI, which are 1600 km, 1700 km, 1200 km, 1800 km, and 1600 km from the line ABC, respectively. Furthermore, dynamic topography is reversible, which produces hiatuses in stratigraphic successions (Burgess *et al.*, 1997). No significant erosion occurred in the shallow-water Silurian successions of East Siberia. This is evidenced by the presence of all of the chronozones which, in many cases, are only several metres thick. Therefore, it is very improbable that large-scale variations in the rate of crustal subsidence in East Siberia were caused by changes in dynamic topography or by elastic lithospheric displacements generated by changes in the subduction regime and supracrustal loads.

Changes in the rate of metamorphism in the lower crust

The absence of any significant hiatuses in the Silurian shallow-water successions in East Siberia indicates that no crustal uplift occurred in the area; only subsidence took place at different rates which varied in time and space. An important feature of the plots (Fig. 6) is a correlation, in most cases, between the amplitude of the deviations and the average rate of crustal subsidence *a* during the epochs of shallow-water deposition. The largest falls and rises in plots III–V, X and XI are characterized by higher *a* (18.8, 18.4, 15.8, 25.6 and 31.5 m/Ma, respectively). Smaller deviations are observed in plots II, VI–VIII, with lower *a* (10.3, 7.8, 8.5, and 9.2 m/Ma, respectively). Among the subhorizontal plots, *a* is rather high only in plot IX (17.2 m/Ma). Therefore, it can be supposed that large-scale falls and rises seen in the plots resulted from changes in the intensity of the process, which caused crustal subsidence in the basin. The common origin of the variations in the rate of crustal subsidence in different regions is also indicated by the synchroneity of large-scale falls and rises seen in plots of Figure 6. This was typical for many cratonic basins, where a quasi-synchronous increase or decrease in the rate of subsidence often took place with intensities variable in space (Beloussov, 1980).

Up to 600–800 m of sediments were formed in East Siberia during the Silurian (Fig. 1). Silurian deposits constitute only a small portion of the Upper Proterozoic–Mesozoic 10–15-km thick sedimentary cover of East Siberia, which overlies the Early Proterozoic crystalline basement (Fig. 10) (Egorkin *et al.* 1987; Rosen *et al.* 1994; Pavlenkova 1996; Surkov 2000). Most of the deposits were formed on Early Proterozoic lithosphere prior to the onset of effusion of traps in the Triassic, which could be due to a thermal event. A subsidence of such amplitude in a cool lithosphere required a considerable increase of density in the layer. Contraction of the mafic rocks in the lower crust due to a transformation of gabbro into garnet granulites is the only known mechanism that could be responsible for subsidence of such a scale in a cratonic lithosphere at places located far away from plate-tectonic activity (Haxby *et al.* 1976; Artyushkov & Baer 1983; Baird *et al.* 1995). Since subsidence in East Siberia in the Silurian was only a short stage of the general and much more pronounced subsidence, it can be presumed that Silurian subsidence also resulted from phase transformations in mafic rocks of the lower crust.

Reaction rate strongly rises with temperature and increases in the presence of small amounts of water-containing fluid (Ahrens & Schubert 1975; Austrheim 1998). Temperature changes in

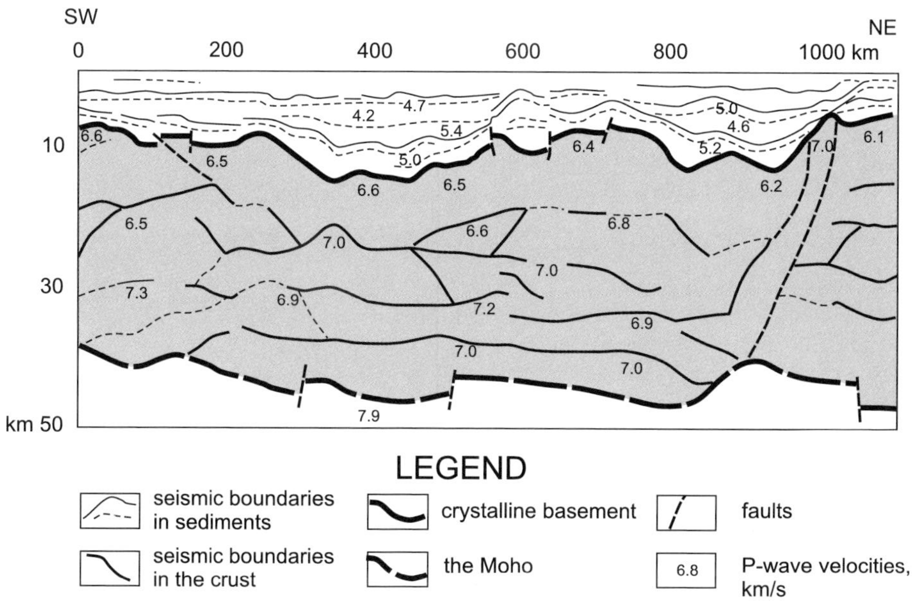

Fig. 10. Crustal structure in East Siberia along the profile AB in Figure 1 (modified after Pavlenkova 1996).

old and thick lithosphere could be caused by changes in heat flow from the asthenosphere. The durations of the falls and rises seen in the plots in Figure 6 (7–10 Ma) are much smaller than the time taken for thermal relaxation in cratonic lithosphere (*c.* 10^2 Ma). Therefore, acceleration in subsidence could not have resulted from temperature changes in the lower crust. Most probably, the increase in the rate of contraction in mafic rocks from phase transformations was caused by a temporary increase in the content of volatiles in the lower crust of East Siberia. Synchronous segregation of volatiles caused by metamorphism in the cool lower crust within the region, *c.* 1000 km wide, is highly improbable. We figure out that accelerations in the rate of crustal subsidence in East Siberia occurred during the epochs when infiltration of small volumes of water-containing fluids from the asthenosphere took place. This mechanism has been proposed earlier for many other sedimentary basins (Artyushkov *et al.* 1991, 2000*b*; Artyushkov 1993). Volatiles could have been generated by small mantle plumes which welled up to the base of the lithosphere and spread rapidly along the boundary. This process was considered in more detail in Artyushkov & Hofmann (1998).

Recently, the mechanism has received additional confirmation, provided by geochemical data (Artyushkov *et al.* 2000*b*). In regions where rapid crustal subsidence occurred during certain epochs without significant lithospheric stretching, evidence has been found for significant changes in the composition of the groundwaters percolating the sediments. Thus, over the period of 1–3 Ma at the beginning of the Late Devonian, a pelagic basin was formed on the shallow-water shelf of the Volga–Urals and Timan–Pechora hydrocarbon basins in the eastern part of the East European Craton (Artyushkov & Baer 1986). Sediments in these basins show extremely high contents of Se, As, Mo, Hg, U and Re (Pushkarev *et al.* 1994; Pisotsky 1999). These elements are found in the layers with both high and low contents of organic matter; therefore any significant influence of organic substances on either the origin or concentration of such elements can be ruled out. Such changes in the content of these elements suggest that they derive from infiltration of dry volatiles from the mantle (*Ibid.*). Oils and bitumens in the above basins include Nd and Sr with isotopic characteristics (ε_{Nd}=–(9–12), and $^{87}Sr/^{86}Sr$=0.708–0.719) (Gottikh *et al.* 2000) which are typical of Australian lamprophyres (Nelson 1992). Carbonates with low organic content were accumulated on the northern and western margins of the Peri-Caspian basin during rapid subsidence in the Viséan. These sediments have a high uranium content (Pisotsky 1999) – by an order of magnitude higher than the usual uranium content in marine carbonates (Taylor & McLennan 1985). This has been attributed to mantle fluids entering the basin. A rapid infiltration of fluids into the lithosphere is possible only if they are surface-active, i.e. a decrease occurs in free energy at the grain boundaries (Gibbs conditions). In this case, such fluids easily penetrate between the crystals in the form of thin films (10^{-5} cm).

Estimates of maximum possible sea-level changes from the structure of sedimentary successions

Maximum possible amplitudes of eustatic fluctuations (b_m) in the Silurian have been earlier estimated for two specific modes: events with an abrupt sea-level fall and subsequent linear rise, and events of a harmonic form (Artyushkov & Chekhovich 2001). The estimates were based on the following two assumptions: firstly, the absence of significant erosional features, and, secondly, the presence of all Silurian chronozones in stratigraphic successions. In this study, using the method described in the above paper, and based on detailed stratigraphic data on Silurian successions in East Siberia (Tesakov *et al.* 1998*b*, 2000), we have obtained more accurate estimates of b_m. This approach can be applied to other epochs and areas with detailed stratigraphic subdivision.

Eustatic fluctuations with an abrupt initial fall and subsequent linear rise (Fig. 11)

Let us denote the initial depth of water by h_w^0, and the fall's amplitude – by b. The water pressure drop caused by the fall results in an isostatic uplift of the lithosphere. Since isostatic recovery is very rapid (Artyushkov 1983), the uplift ought to have occurred almost synchronously with sea-level falls, $\geq 10^{-2}$ Ma long. For the lithosphere *c.* 10^3 Ma old, which includes the crust, 45 km thick, the effective elastic thickness of the lithosphere should be about 70 km (Burov & Diament 1995; Cloetingh & Burov 1996). In this case, the characteristic width of lithospheric flexure under surface load is of *c.* 200 km. The size of the East Siberian Basin was considerably larger: 1200×2000 km. Therefore, in the first approximation, vertical crustal displacements under changing water pressure corresponded with the case of local isostasy, and subaerial exposure began when the sea-level fall reached the amplitude as follows:

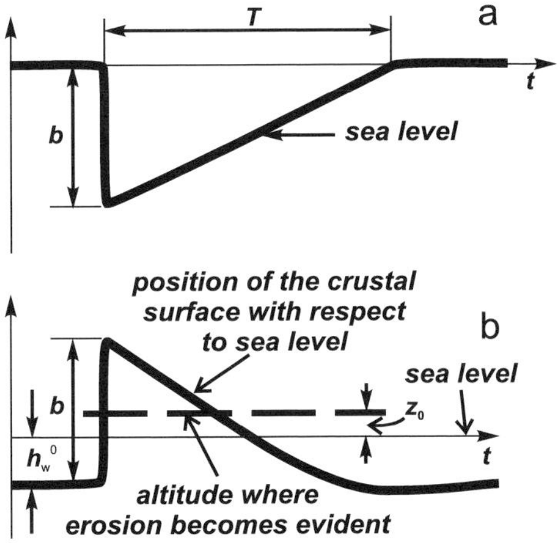

Fig. 11. (**a**) Eustatic event involving an abrupt sea-level fall of amplitude b followed by a uniform rise of the same amplitude b over the period of time T. (**b**) Change of position of crustal surface with respect to sea level, which corresponds with this event. Due to crustal subsidence at a constant rate a, the rate of decrease in the crustal surface's altitude is higher than that of the sea-level rise. h_w^0 – initial depth of water, z_0 – altitude above the sea-level where subaerial erosion becomes significant.

$$\zeta_{eu}^0=[(\rho_m-\rho_w)/\rho_m]h_w^0\approx 0.69h_w^0. \quad (7)$$

An abrupt fall can be detected when crustal surface reaches the altitude z_0 above sea-level where erosion becomes significant. During abrupt sea-level falls, deposition can be neglected. In this situation erosional features will be formed, if the fall's amplitude b reaches the minimum value b_m or exceeds it: $b \geq b_m$, where

$$b_m=0.69h_w^0+z_0. \quad (8)$$

In the Nyuya region (V in Fig. 1), deposition of shallow-water carbonates had been taking place since the start of the Silurian. Beginning from the Telychian, a carbonate platform had existed over most of the basin. The depth of water on the shoal was 0–5 m. Then the initial depth of water can be assumed to be equal $h_w^0=5$ m for the Rhuddanian and Aeronian, and $h_w^0=3$ m for the Telychian–Pridoli. Microkarst develops rapidly on carbonates if the altitude of the crustal surface exceeds the maximum amplitude of the tides (D'Argenio *et al.* 1999). Over most of East Siberia, far from the ocean, this amplitude could not exceed $z_0\approx 3$ m. Stratigraphic successions in East Siberia include numerous diastems – levels of weak subaerial erosion indicating short-term uplifts of sea bottom above the sea-level. However, no significant erosional features have been found in the successions. Then, according to (8), amplitudes of the abrupt sea-level falls could not have exceeded:

$b_m \approx 7$ m for the Rhuddanian and Aeronian, (9a)

$b_m \approx 5$ m for the Telychian–Pridoli. (9b)

Values of b_m can also be estimated from the variations of chronozone thickness in Silurian successions. Subaerial exposure of a long duration Δt_{sa} results in non-deposition and disappearance of sediments of a certain age from the succession. In East Siberia, sediments from all chronozones are present in shallow-water Silurian successions. This indicates that Δt_{sa} did not exceed the duration of two chronozones $2\Delta t_{cz}$: $\Delta t_{sa} \leq 2\Delta t_{cz}$ (Artyushkov & Chekhovich 2001). This constraint, which was used in that paper, is very strong. Not only all chronozones are present in the successions, but their thicknesses change gradually. Superposition of a period of subaerial exposure with two adjacent chronozones will result in a decrease in chronozone thicknesses Δh_{cz}. For example, given a constant deposition rate a and $\Delta t_{cz}=\Delta t_{sa}$, the sum of thicknesses of two chronozones overlapped by a period of non-deposition, equals the thickness of one chronozone (Fig. 12a). In Fischer plots

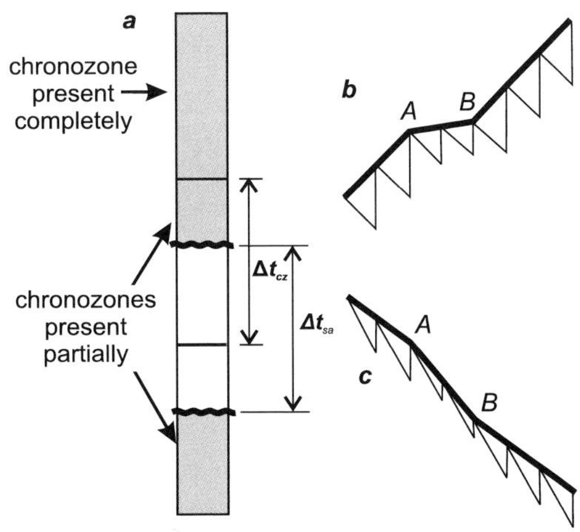

Fig. 12. (**a**) Superposition of the period of subaerial exposure, Δt_{sa} long, on to the adjacent chronozones, Δt_{cz} long, for the case when $\Delta t_{sa}=\Delta t_{cz}$. (**b**) Influence of this superposition on the rise in the Fischer plot, which reduces sediment thicknesses in two chronozones in segment AB. (**c**) Same for the fall in the Fischer plot.

this will be manifested in deflections in the graphs of the types shown in Figure 12b and 12c. Furthermore, such deflections should take place in all plots, or, at least, most of them. This is not seen in the plots of Figure 6. Therefore, if subaerial exposures occurred in regions I–XI, their duration Δt_{sa} did not exceed that of one chronozone Δt_{cz}:

$$\Delta t_{sa} \leq \Delta t_{cz}. \quad (10)$$

No significant subaerial erosion occurred in East Siberia in the Silurian. For the eustatic event of Figure 11a, at times, when crustal surface was above sea-level, its altitude (z) decreased with time as $\zeta = b - 0.69h_w^0 - b\tau - aT\tau$ where $\tau = t/T$ (Fig. 11b). The crustal surface subsided to sea-level ($\zeta=0$) at $\tau_0 = (b - 0.69h_w^0)/(b + aT)$. Then, according to equation (10), gradual changes in the chronozone thicknesses constrain the amplitudes of abrupt sea-level falls as:

$$b \leq b_m = (0.69h_w^0 + a\Delta t_{cz})/(1 - \Delta t_{cz}/T). \quad (11)$$

This parameter decreases with the period of eustatic event T, and increases as the average rate of crustal subsidence a rises. In this situation, the minimum values of b_m can be obtained for regions with the lowest rate of subsidence. Under a constant duration of the chronozones $\Delta t_{cz} = 0.48$ Ma, the minimum average values of a for the Telychian, Wenlock, Ludlow and Pridoli are shown in Table 3; regions with the slowest deposition are indicated. For the Rhuddanian and Aeronian, we take a for the Nyuya region, which at that time was the only one with shallow-water carbonate deposition. In Figure 13, values b_m determined by equation (11) are plotted as the functions of T for these values of a and $\Delta t_{cz} = 0.48$ Ma. The initial depth of water is taken as $h_w^0 = 5$ m for the Rhuddanian and Aeronian, and $h_w^0 = 3$ m for the Telychian–Pridoli.

The values of b_m in Figure 13, are generally larger than those determined by equations (9a) and (9b) under the constraint 1. However, the altitudes, where erosion takes place in sediments of various types, can be estimated only approximately. In the presence of reliable data, abrupt changes in the chronozone thicknesses can be established quite reliably, therefore constraint 2 appears to be more definite than constraint 1. As can be seen in Figure 13, the values of b_m are comparatively large only for the Aeronian ($b_m \geq 18$ m). For the rest of the Silurian, in the interval of $T = 1$–5 Ma, they are in the range of 6–14 m. This is smaller or even much smaller than the amplitudes of *c.*20–100 m commonly attributed to the third-order eustatic events.

Suppose now that the chronology of the Silurian was that to Gradstein & Ogg (1996), with the relative lengths of the Rhuddanian, Aeronian and Telychian according to Johnson *et al.* (1996) and Tesakov *et al.* (1998a) (Table 2). In Figure 14, $b_m(T)$ determined according to equation (11) are plotted for the main units of the Silurian under a and Δt_{cz} characteristic of this chronology. As in Figure 13, the values of b_m are comparatively large only for the Aeronian ($b_m \geq 19$ m). For the Telychian, $b_m = 9$–20 m in the range of $T = 1.34$–5 Ma. Low values of $b_m = 6$–12 m are characteristic of the Rhuddanian, Wenlock, Ludlow and Pridoli. For both chronologies, constraint 2 also leads to conclude that the amplitudes of eustatic events during most of the Silurian were low.

Eustatic events with a gradual sea-level fall

Let us consider a eustatic event of a harmonic form (Fig. 15):

Table 3. *Rates of crustal subsidence during the Silurian in the case when chronozone lengths are assumed to be constant and equal to 0.48 Ma*

Standard subdivision	Number of chronozones (n)	Length, Ma	Minimum rate of crustal subsidence (a) (m/Ma)*	Region
Rhuddanian	11	5.3	8.5	Nyuya River
Aeronian	9	4.3	26.3	Nyuya River
Telychian	5	2.4	11.2	Balturino
Wenlock	11	5.3	6.7	Ledyanskaya
Ludlow	13	6.2	9.7	Kochumdek River
Pridoli	5	2.4	8.8	Kochumdek River

*Rock compaction is taken into account (Artyushkov & Chekhovich, 2001).

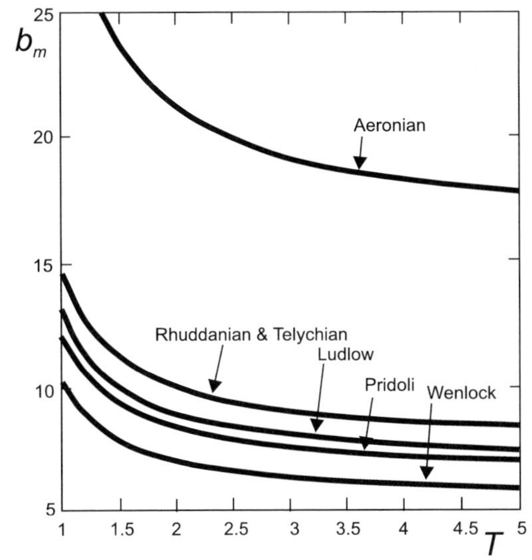

Fig. 13. Maximum amplitudes b_m (in metres) of eustatic events involving an abrupt sea-level fall and a linear sea-level rise (Fig. 11a) which could have occurred in the Silurian judging by the observed gradual changes in chronozone thicknesses (constraint 2, condition (10)). T, events' length (Ma). Chronozone length is $\Delta t_{cz}=0.48$ Ma. The minimum rates of crustal subsidence a were taken according to Table 3.

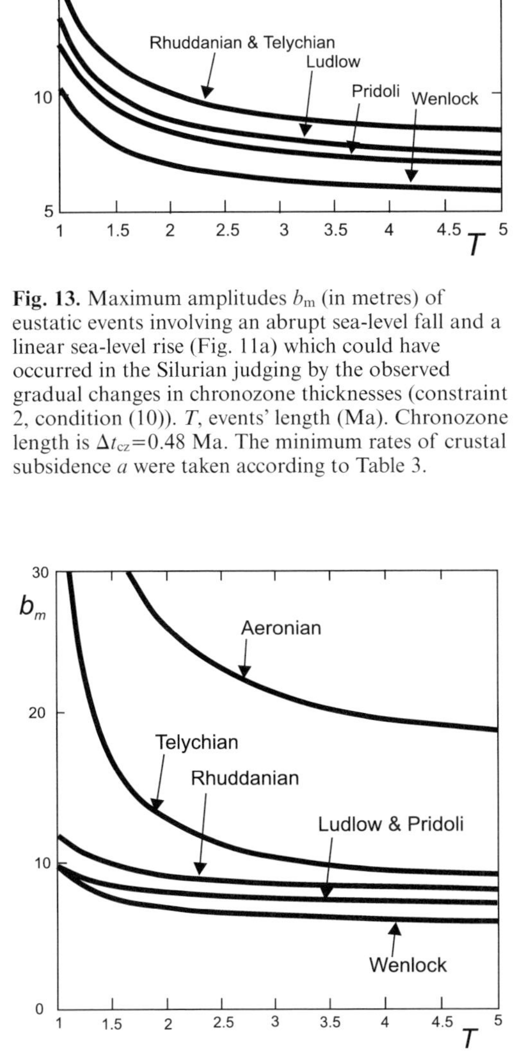

Fig. 14. Maximum amplitudes b_m (in metres) of eustatic events involving an abrupt sea-level fall and a linear sea-level rise (Fig. 11a), which could have occurred in the Silurian judging by the observed gradual changes in chronozone thicknesses (constraint 2, condition (10)). T, events' length (Ma). Chronozone lengths Δt_{cz}, and minimum rates of crustal subsidence a, were taken according to Table 2.

$$\zeta_{eu}=-\,b\sin^2(\pi\tau). \quad (12)$$

In the Rhuddanian and Aeronian, siliciclastic deposition took place in regions VI and VII (Ilim and Balturino) in the southwestern part of East Siberia. Its rate, most probably, was controlled by the rate of erosion within the vast adjacent landmasses, and had not changed during small-scale eustatic fluctuations. Let us assume that initial depth of water equalled $h_w{}^0=5$ m and subsidence rates were as shown in Table 3 at $\Delta t_{cz}=0.48$ Ma. Then, neglecting small displacements caused by isostatic recovery in response to sea-level falls, and using relations (5, 6) in Artyushkov & Chekhovich (2001), it follows that the maximum possible amplitudes b_m of eustatic events (equation (11)) which could have occurred under the conditions of equaton (10) (constraint 2) are equal to those shown in Fig. 16. Large $b_m \geq 17$ m are observed only at $T \leq 1.5$ Ma in the curve Aeronian-2. In the Early Silurian, shallow-water carbonate deposition took place in the Nyuya region (V). Using constraint 1, taking the altitude of erosion as being equal to $z_0=3$ m and using relation (6) (*Ibid.*), we obtain b_m as shown in Fig. 17. Here, at $T \leq 1.7$ Ma, $b_m \leq 13$ m. Thus, the combined application of constraints 1 and 2 leads us to produce values of b_m which are considerably smaller than those supposed for third-order eustatic events (20–100 m).

During the Telychian–Pridoli, deposition took place over most of the basin on the carbonate shoal (Tesakov *et al.* 2000). Usual water depths for these regions were *c.*0–5 m (Johnson *et al.* 1997). Under such conditions, deposition practically compensated for accommodation space variations. In this case, the average rate of formation of shallow-water carbonates ought to have increased with the increase in the depth of

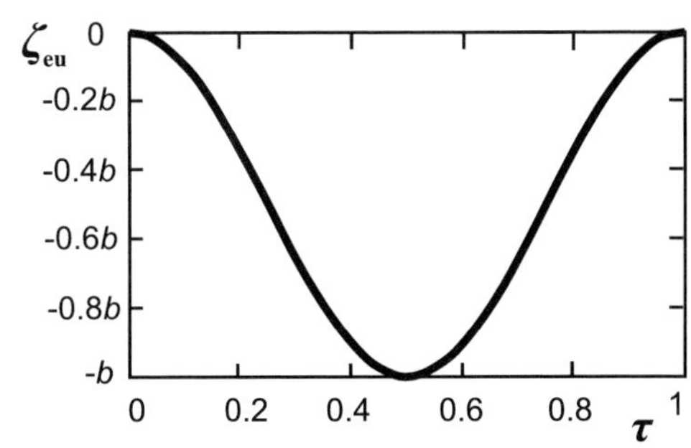

Fig. 15. Eustatic event of a harmonic form ζ_{eu} determined by equation (12), involving a gradual sea-level fall of the amplitude b. τ is the dimensionless time.

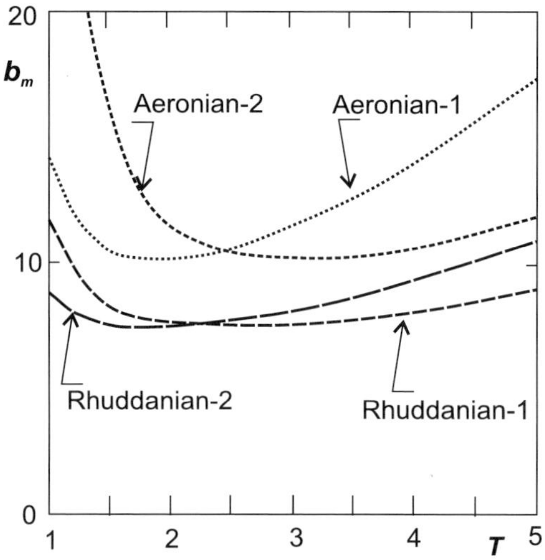

Fig. 16. Maximum amplitudes b_m (m) of eustatic events of a harmonic form (Fig. 15) and lengths T (Ma), which could have occurred in the Rhuddanian and Aeronian judging by the observed gradual changes in the chronozone thicknesses (constraint 2, condition equation (10)). The initial depth of water is $h_w{}^0=5$ m. Curves Rhuddanian-1 and Aeronian-1 were plotted for chronozone lengths $\Delta t_{cz}=0.48$ Ma (Table 3). The average rates of crustal subsidence are $a=5.1$ m/Ma in the Ilim region, and $a=10.7$ m/Ma in the Balturino region, respectively. Curves Rhuddanian-2 and Aeronian-2 were plotted for chronozone lengths of $\Delta t_{cz}=0.36$ Ma, and $\Delta t_{cz}=0.76$ Ma, respectively (Table 2). The average rates of crustal subsidence in both regions were $a=6.8$ m/Ma.

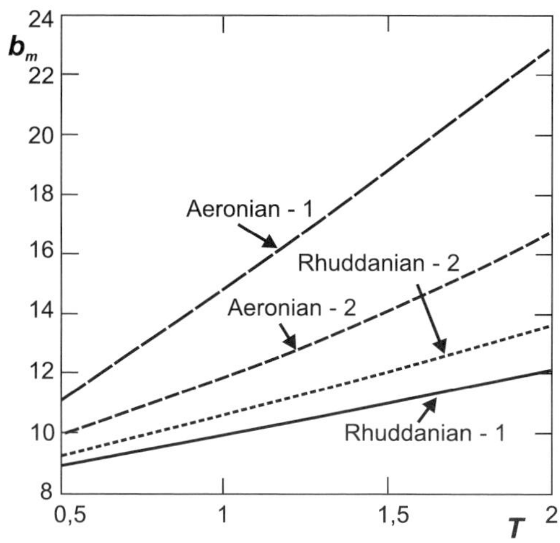

Fig. 17. Maximum amplitudes b_m (m) of eustatic events of a harmonic form (Fig. 15) and lengths T (Ma), which could have occurred in the Rhuddanian and Aeronian without producing any significant erosional features in the sedimentary successions of the Nyuya region. The initial depth of water is $h_w{}^0=5$ m, and the minimum altitude where erosion becomes significant is $z_0=3$ m. Curves Rhuddanian-1 and Aeronian-1 were plotted for constant chronozone lengths $\Delta t_{cz}=0.48$ Ma, and $a=8.5$ m/Ma, $a=26.3$ m/Ma, respectively. Curves Rhuddanian-2 and Aeronian-2 were plotted for chronozone lengths $\Delta t_{cz}=0.36$ Ma, $a=11.2$ m/Ma, and $\Delta t_{cz}=0.76$ Ma, $a=16.5$ m/Ma, respectively.

water. Let us assume that, at $h_w \leq 5$–10 m, the rate of shallow depth deposition was proportional to the depth of water h_w. From relations (9)–(11) in Artyushkov & Chekhovich (2001), it follows that the values of b_m were equalled to those shown in Figures 18–21. In the curves plotted under constraint 2 (Figs 18 & 20), large values of $b_m \geq 20$ m were found to be characteristic only of relatively short periods $T \leq 1.3$ Ma using the chronology of Table 2.

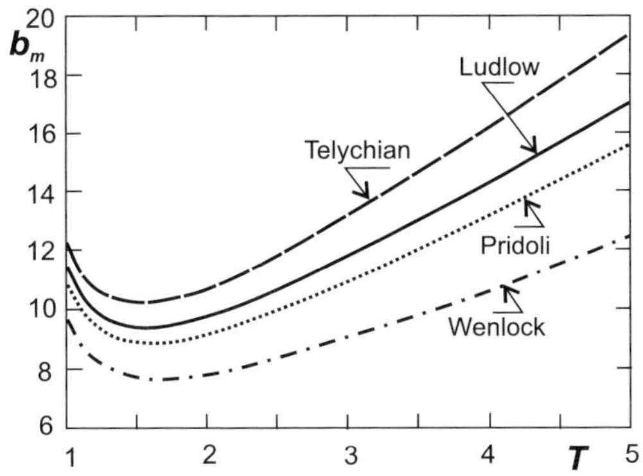

Fig. 18. Maximum amplitudes b_m (m) of eustatic events of a harmonic form (Fig. 15) and lengths T (Ma), which could have occurred in the Telychian–Pridoli according to the observed gradual changes in chronozone thicknesses (constraint 2, condition equation (10)). The initial depth of water is $h_w^0=3$ m, chronozone length is $\Delta t_{cz}=0.48$ Ma. The rates of crustal subsidence a were taken according to Table 3.

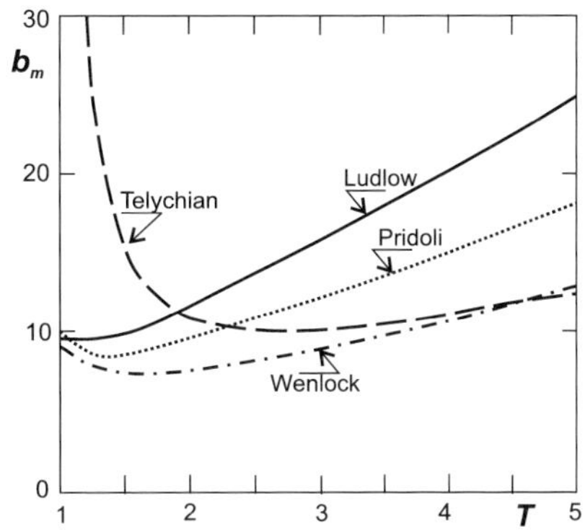

Fig. 20. Maximum amplitudes b_m (m) of eustatic events of a harmonic form (Fig. 15) and lengths T (Ma), which could have occurred in the Telychian–Pridoli according to the observed completeness of the stratigraphic successions and gradual changes in chronozone thicknesses (constraint 2, condition equation (10)) are shown. The initial depth of water is $h_w^0=3$ m. The rates of crustal subsidence a and chronozone lengths Δt_{cz} were taken according to Table 2.

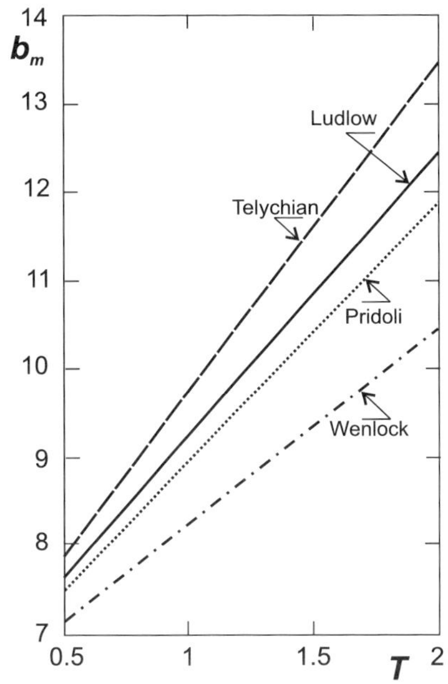

Fig. 19. Maximum amplitudes b_m (m) of eustatic events of a harmonic form (Fig. 15) and lengths T (Ma), which could have occurred in the Telychian–Pridoli without producing any significant erosional features in sedimentary successions. The initial depth of water is $h_w^0=3$ m, and the minimum altitude where erosion becomes significant is $z_0=3$ m. Chronozone lengths and rates of crustal subsidence were taken according to Table 3.

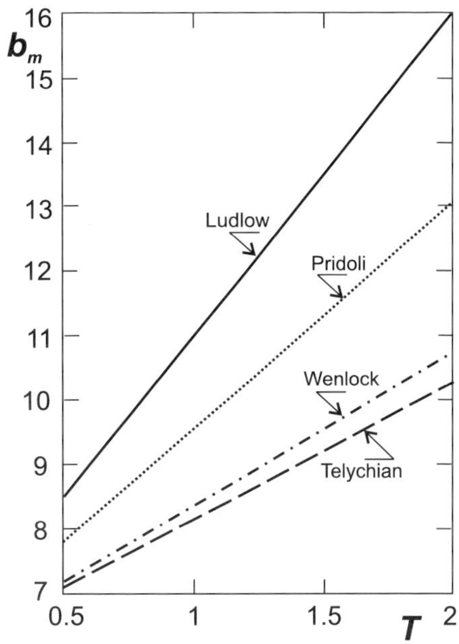

Fig. 21. Maximum amplitudes b_m (m) of eustatic events of a harmonic form (Fig. 15) and lengths T (Ma), which could have occurred in the Telychian–Pridoli without producing any significant erosional features in sedimentary successions. The initial depth of water is $h_w^0=3$ m; the minimum altitude, where erosion becomes significant, is $z_0=3$ m. The rates of crustal subsidence a were taken according to Table 2.

Conclusions and discussion

Sea-level changes of the third order with amplitudes of 20–100 m and lengths of *c.* 1–10 Ma (eustatic events) are believed to have occurred widely in the Phanerozoic (Haq *et al.* 1987; de Graciansky *et al.* 1998). Tectonic movements are commonly considered to be a complicating factor. Wide occurrence of third-order cycles in sea-level changes has been seldom doubted (Miall 1992; Miall & Miall 2001). Moreover, a number of eustatic events have been proposed for the Cambrian and earliest Ordovician (Webby & Laurie 1992; Cooper & Nowlan 1999). However, as follows from analysis of the available data on shallow-water deposition in East Baltic, eustatic fluctuations did not exceed 10–20 m at that time, while rapid crustal uplifts of *c.* 100 m occurred in southern Sweden and eastern Lithuania (Artyushkov *et al.* 2000a). Based on the changes of water depth in cratonic areas, eight large-scale eustatic events have been proposed for the Silurian (Johnson 1996). A large amount of highly detailed stratigraphic data was obtained for this epoch in the East Siberian Basin, (which was 2×10^6 km^2 in size) (Tesakov *et al.* 2000). Comparison of results obtained from the modelling of changes of water depth due to eustatic fluctuations, with the structure of shallow-water successions in this area leads us to conclude that, in the Silurian, eustatic events could not have exceeded 20–30 m (Artyushkov & Chekhovich 2001). During the same period of time, rapid, large-scale vertical crustal movements occurred in some regions of East Siberia and in other cratonic areas.

In this paper, we have analysed in more detail the role of eustatic and tectonic factors in the formation of Silurian sedimentary successions of East Siberia. Fischer plots have often been used for the identification of eustatic events from sedimentary records (e.g. Goldhammer *et al.* 1993). Such plots are commonly based on the thickness of elementary cycles of a metre scale in shallow-water successions. There are serious doubts about the applicability of this approach because elementary cycles can be random (Burgess *et al.* 2001), i.e. there is no certainty about their timing. The Silurian successions of East Siberia are subdivided into chronostratigraphic units – chronozones, each corresponding with a time interval *c.* 0.5 Ma long. On this basis, we have compiled Fischer plots in a different way – based on chronozone thickness. This has allowed us to make a reliable comparative analysis of the changes over time in the cumulative sediment thickness in a number of regions. Following this approach we have found that eustatic fluctuations in the Silurian were even lower (≤5–10 m) than those estimated earlier (≤20–30 m, according to Artyushkov & Chekhovich 2001). This is in agreement with the results of modelling of changes in the depth of water due to sea-level changes and their comparison with highly detailed stratigraphic successions which was done in this paper. This proves that eustatic events could not have exceeded *c.* 20 m; in certain periods of the Silurian they were only ≤5–7 m. The values obtained are considerably smaller than those commonly supposed for eustatic events (i.e. 20–100 m).

In Fischer plots based on elementary cycles it is assumed that crustal subsidence was a uniform process. However, it is known that in many sedimentary basins the rate of subsidence had changed in time (Beloussov 1980; Sloss 1988). An analysis of Fischer plots based on the Silurian chronozones indicates that in some regions of East Siberia the rate of subsidence remained almost constant over 10–15 Ma, while in the other regions it had changed by several hundred per cent. These changes were of the same sign, but their intensities varied strongly from place to place in the basin. The data can be used to verify the applicability of different mechanisms of crustal subsidence to East Siberia. Rapid crustal uplift and subsidence is sometimes explained by lithospheric deflections due to changes in the forces acting along the lithosphere with laterally variable thickness (Cloetingh *et al.* 1985). Such displacements tend to decrease with the width of the area (Artyushkov 1974), and are too small to explain the large-scale changes in the rate of subsidence in East Siberia, ≥1000 km wide. In the Silurian, subduction took place to the south of the area. Vertical crustal movements can be generated by changes in dynamic topography above subducting plates (Burgess *et al.* 1997; Burgess & Moresi 1999). The intensity of such movements decreases with distance from the convergent plate boundary. In the Silurian, changes in the rate of crustal subsidence were larger in the northern part of East Siberia. Therefore, it can be concluded that changes in dynamic topography could not have had any strong influence on crustal movements in the area. Contraction of mafic rocks in the lower crust caused by phase transformations, has also been suggested as a cause of crustal subsidence (Haxby *et al.* 1976; Artyushkov *et al.* 1991; Baird *et al.* 1995). In the absence of strong lithospheric stretching, this is the only mechanism that can explain crustal subsidence of the amplitude of up to 10–15 km in East Siberia during the Late Proterozoic and Phanerozoic.

The same mechanism most probably accounts for subsidence during the Silurian (up to 600–800 m) which was only a short episode of this major subsidence. As suggested for many other basins (Artyushkov *et al.* 1991, 2000b), the accelerations of subsidence in East Siberia could have occurred due to the infiltration of small volumes of surface-active fluids into the lower crust from the asthenosphere.

The epochs of relative sea-level stability in the Early Paleozoic and the Silurian lasted for about 70 Ma, or 13% of Phanerozoic time. These epochs were chosen for research just because original stratigraphic data was available, i.e. the choice was rather accidental. Therefore, it is quite probable that sea-level remained stable during most of the Phanerozoic, while the eustatic events proposed earlier, apart from the epochs of major glaciations, were connected with the rapid tectonic movements.

Rapid changes of water depth in cratonic areas were responsible for the formation of numerous stratigraphic traps for oil and gas. This phenomenon is commonly explained by eustatic fluctuations of sea-level (Posamentier & Allen 2000). If such changes in water depth were really caused by vertical crustal movements, the method of prospecting stratigraphic traps should be completely revised. Studies of rapid regional crustal uplift and subsidence in cratonic areas, and their basic regularities and driving mechanisms are essential for reliable prospecting. Rapid crustal movements can presently occur in some cratonic regions, which are supposed to be inactive. This is characteristic, for example, of the East European Platform (Kashin 1989). Such motions are sometimes accompanied by large and destructive earthquakes, as happened, for example, on the North American Craton and Indian Shield. Therefore, identification of such regions is of importance for earthquake prediction.

We thank Yu.I. Tesakov for valuable discussions and for providing a large amount of original data on the Silurian in East Siberia. The comments of the reviewers, E. B. Burov and P. M. Burgess, were helpful for clarifying presentation. The work was supported by the Russian Foundation for Basic Research (Grant 00-05-64095).

Appendix: Fischer plots based on elementary cycles for some regions of East Siberia in the Silurian

Silurian sedimentary successions of East Siberia include large stacks of elementary cycles on the scale of metres (Sokolov 1985; Tesakov *et al.* 2000). Typical examples are shown in Figures A.1–A.3. Using the shallow-water parts of the successions, Fischer plots were compiled (Fig. A.4) for regions III–VII. The plots include falls and rises, up to 40–60 m in the amplitude and from *c.*1 Ma to *c.*15 Ma in duration. Under a constant rate of crustal subsidence and quasi-periodic elementary cycles, the deviations in Fischer plots should reflect sea-level changes. If crustal subsidence is uniform in several regions, the Fischer plots will be similar, as when describing global eustatic fluctuations. In the plots on Fig. A.4, to a first approximation some falls and rises are synchronous, for example, the general long-term falls by 40–50 m in segment AB of plots III and V during most of the Llandovery. However, during the same period of time AB, asynchronous falls and rises took place in regions VI and VII, with amplitudes of only *c.*10–20 m. A rapid rise of *c.*30 m in segment BC of plot V in the Late Telychian was not accompanied by any significant changes in the other plots. A fall of 40 m in segment DF of plot IV in the Wenlock and Early Ludlow took place when no comparable changes occurred in segment DE of plots III and V, and a rapid rise of *c.*40 m took place in segment EF of plot III. No significant changes are seen on plot III within the epoch corresponding to the rise by *c.*30 m in segment FG of plot IV. Such differences in plots III–VII show that they do not describe large-scale eustatic sea-level changes.

In different regions, the numbers of elementary cycles in Figure A.4 differ considerably. For example, the Rhuddanian and Aeronian in regions V, VI, and VII include 95, 43, and 48 cycles, respectively. In the regions III, IV and V, the Wenlock comprises 16, 62 and 25 cycles, respectively. The Ludlow includes six cycles in region III and 32 cycles in region IV. Such differences in the numbers of high-frequency cycles lead us to conclude that they could not have had a eustatic origin. This could be the reason why the plots are so markedly different.

Considerable distortions of the plots can also occur if cycle lengths are not constant, but rather vary with time. The lengths of the Llandovery, Wenlock and Ludlow are estimated as 15 Ma, 5 Ma and 4 Ma, respectively (Gradstein & Ogg 1996). Taking the relative lengths of the Rhuddanian, Aeronian and Telychian as 1:1.725:1.025 according to Johnson (1996) and Tesakov *et al.* (1998*a*), the total duration of the Aeronian and Telychian can be estimated as 11 Ma. In plot III (Fig. A.4), the Aeronian with the Telychian, Wenlock and Ludlow include 59, 16, and 6 cycles, respectively. This corresponds with

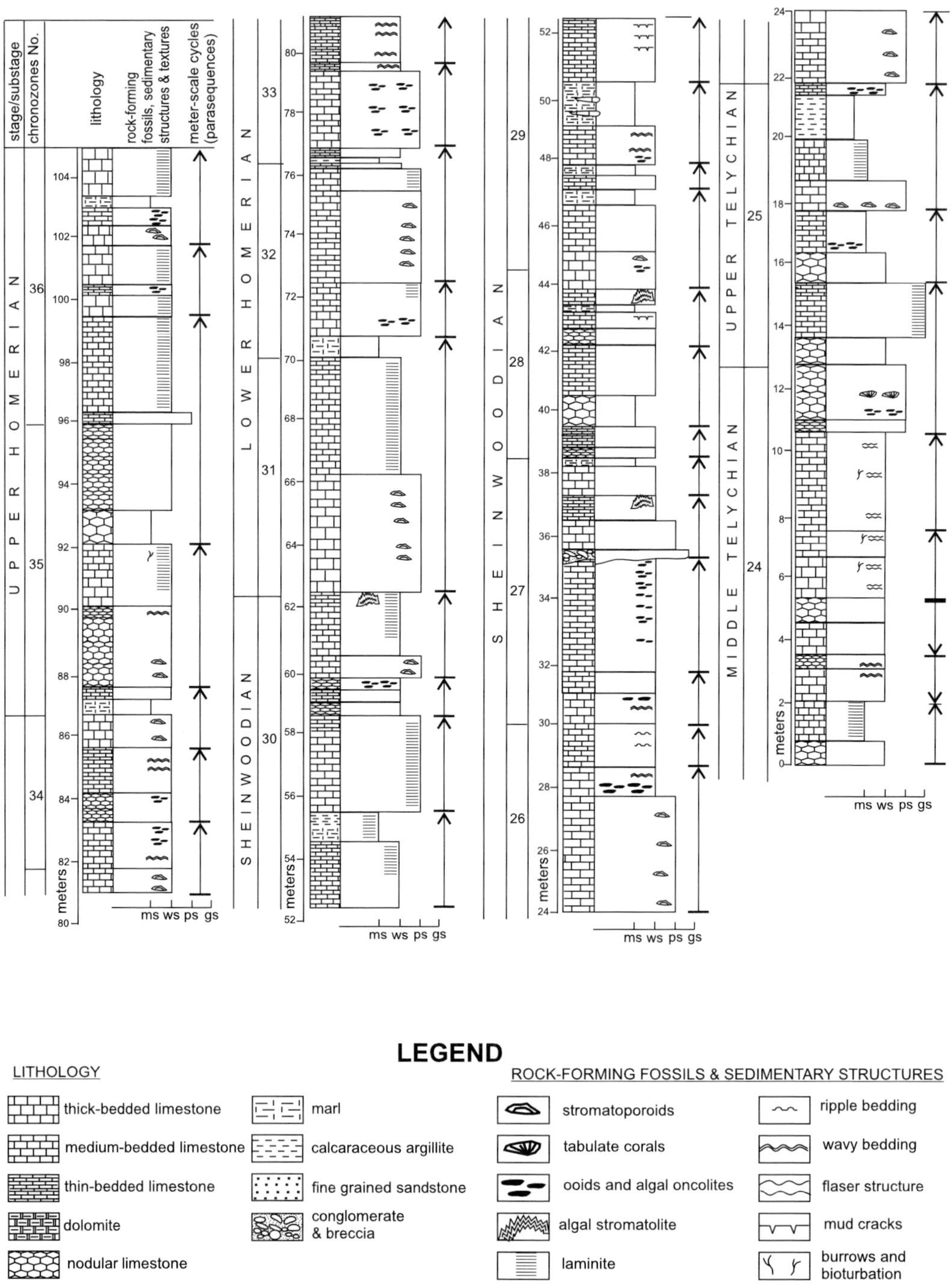

Fig. A.1. A fragment of a shallow-water succession (≤10 m) in the Moyero region (IV in Fig. 1) for the Telychian and Wenlock (modified after Sokolov 1985); ms, mudstone; ws, wackestone; ps, packstone; gs, grainstone

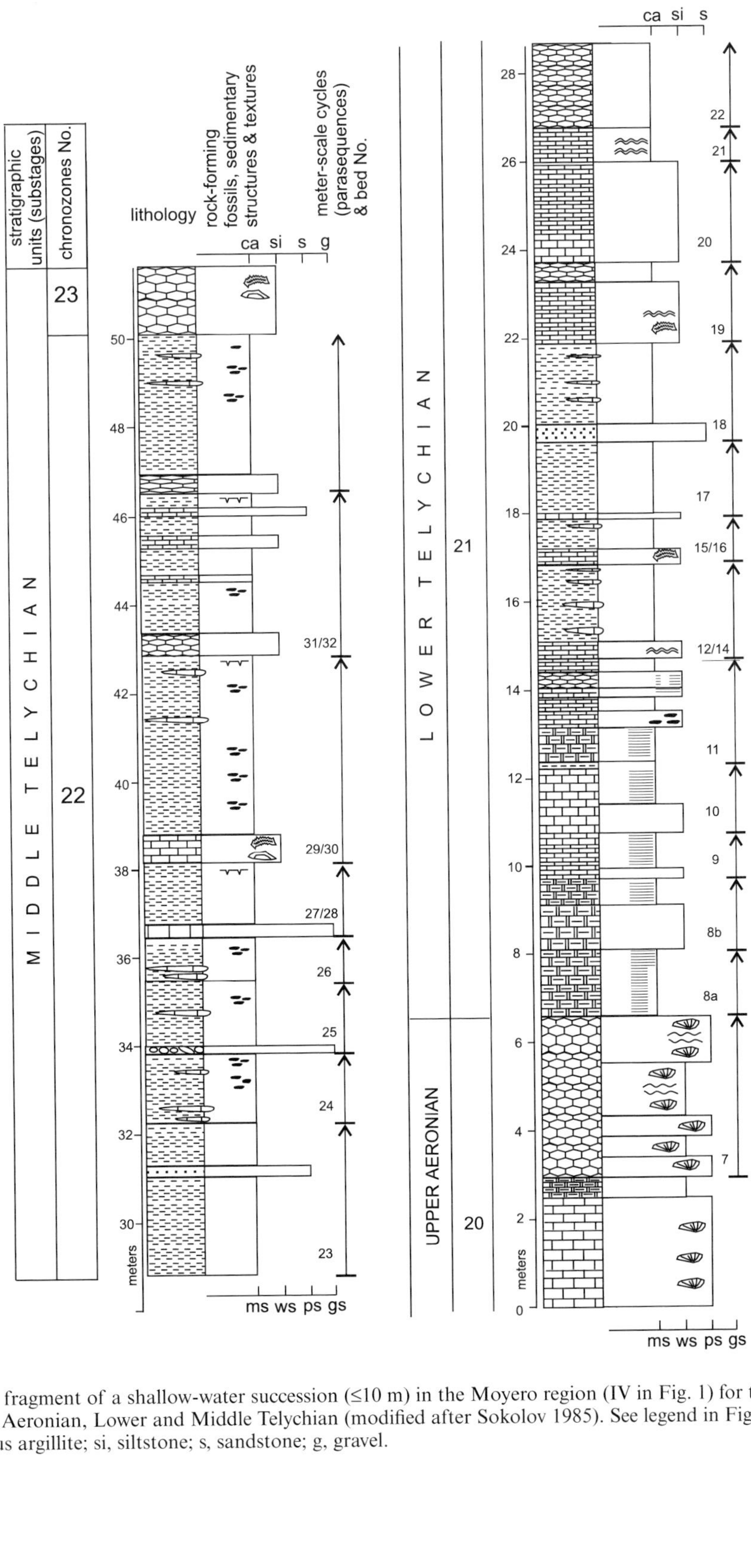

Fig. A.2. A fragment of a shallow-water succession (≤10 m) in the Moyero region (IV in Fig. 1) for the uppermost Aeronian, Lower and Middle Telychian (modified after Sokolov 1985). See legend in Figure A.1; ca, calcaraceous argillite; si, siltstone; s, sandstone; g, gravel.

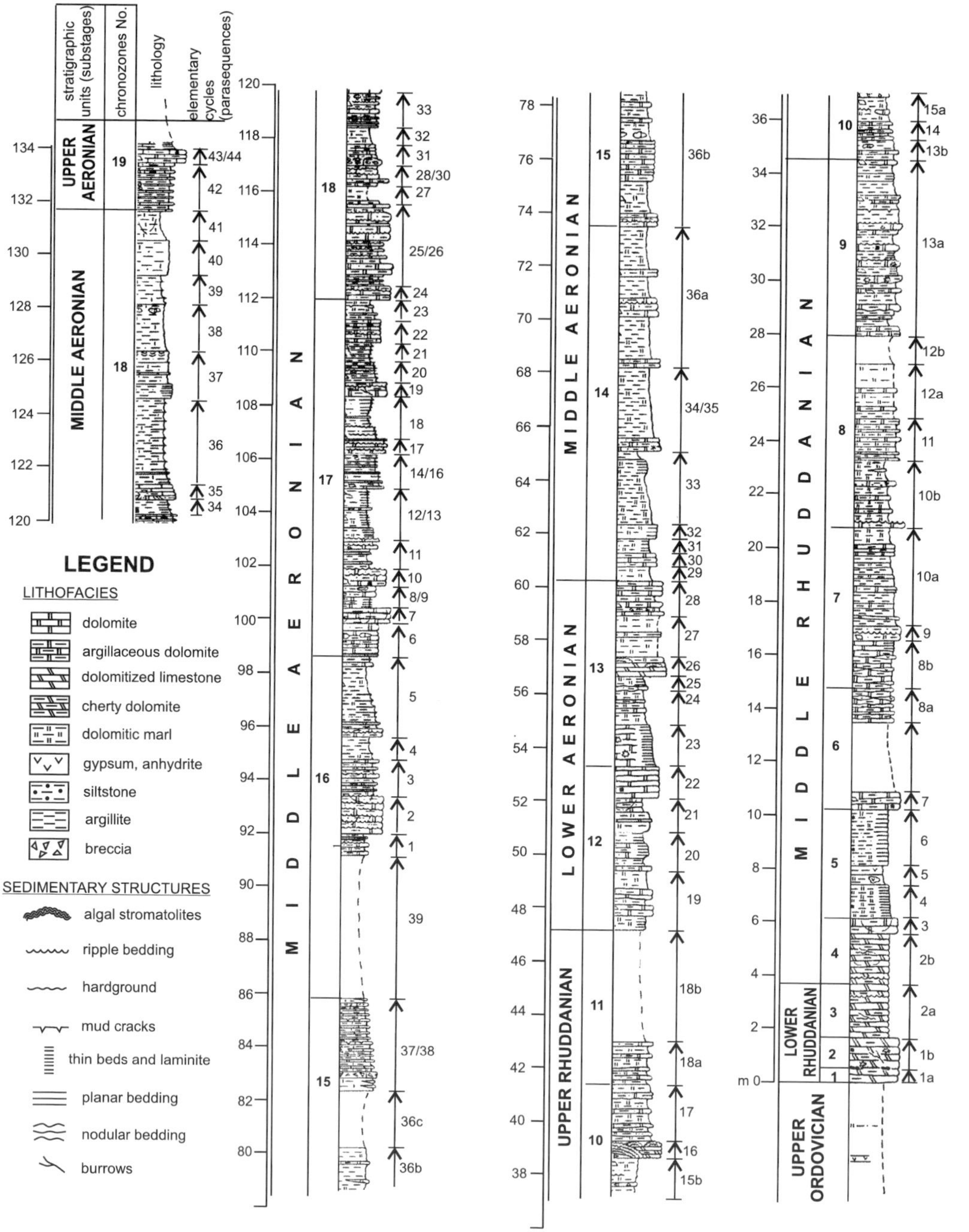

Fig. A.3. A fragment of a shallow-water succession (≤10 m) in the Nyuya region (V in Fig. 1) for the Rhuddanian to the middle Aeronian (modified after Tesakov *et al.* 2000).

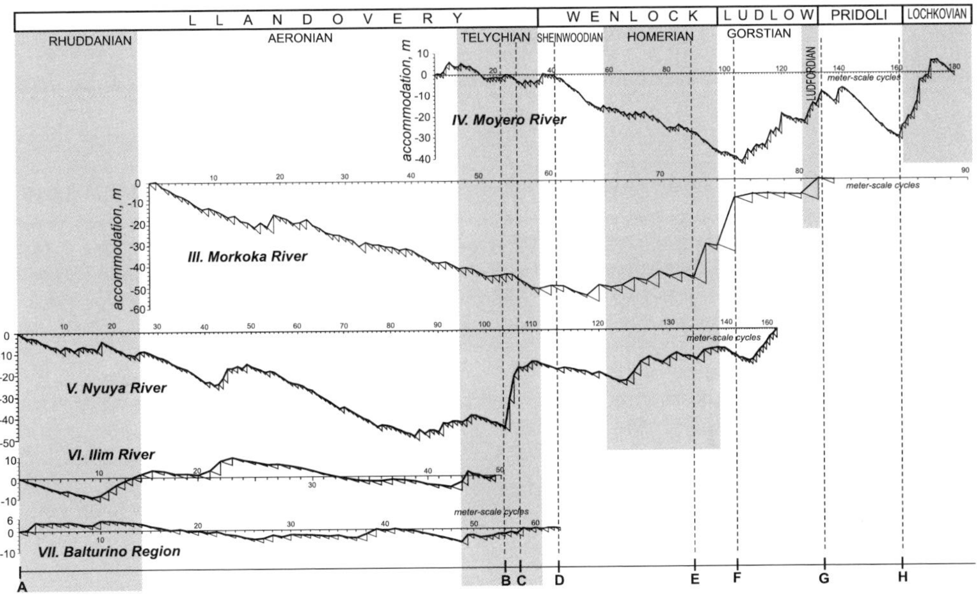

Fig. A.4. Fischer plots for five regions in East Siberia. The plots are based on elementary cycles identified for the periods of deposition at water depths ≤5–10 m using data from Sokolov (1985) and Tesakov *et al.* (1986, 2000).

the average lengths of the elementary cycles of 0.19, 0.31 and 0.67 Ma, respectively, with ratios of 1:1.6:3.5. Of course, an accurate determination of the lengths of the main units of the Silurian was not very high. However, it is very improbable that this inaccuracy could alter the relative lengths of these units by as much as three and a half times. Plot IV includes 62 cycles in the Wenlock and 36 cycles in the Ludlow. The average lengths of the cycles are 0.08 Ma and 0.11 Ma with a ratio of about 1:1.4. In plot IV, the ratio is 1.6:3.5=2.2, i.e. quite different from that for plot III. This also indicates that the lengths of metre-scale cycles changed considerably from one unit to another. Thus, in the Silurian, elementary cycles were not synchronous in different regions of East Siberia, and their lengths changed with time. Under such circumstances, Fischer plots based on elementary cycles cannot describe eustatic sea-level changes.

References

AHRENS, T. J. & SCHUBERT, G. 1975. Gabbro–eclogite reaction rate and its geophysical significance. *Review of Geophysics and Space Physics*, **13**, 383–400.

D'ARGENIO, B., FERRERI ,V., RASPINI, F. B., AMODIO, S. & BUONOCUNTO, F. B. 1999. Cyclostratigraphy of carbonate platform as a tool for high-precision correlation. *Tectonophysics*, **315**, 357–384.

ARTYUSHKOV, E. V. 1973. Stresses in the lithosphere caused by crustal thickness inhomogeneities. *Journal of Geophysical Research*, **78**, 7675–7708.

ARTYUSHKOV, E. V. 1974. Can the Earth's crust be in a state of isostasy. *Journal of Geophysical Research*, **79**, 741–752.

ARTYUSHKOV, E. V. 1983. *Geodynamics*. Elsevier, Amsterdam, 312 pp.

ARTYUSHKOV, E. V. 1993. *Physical Tectonics*. Nauka, Moscow, 456 pp. (in Russian).

ARTYUSHKOV, E. V. & BAER, M. A. 1983. Mechanism of continental crust subsidence in fold belts: the Urals, Appalachians and Scandinavian Caledonides. *Tectonophysics*, **100**, 5–42.

ARTYUSHKOV, E. V. & BAER, M. A. 1986. Mechanism of formation of hydrocarbon basins: the West Siberia, Volga–Urals, Timan–Pechora basins and Permian basin of Texas. *Tectonophysics*, **122**, 247–281.

ARTYUSHKOV, E. V. & CHEKHOVICH, P. A. 2000. The East Siberian sedimentary basin in the Silurian: evidence for a lack of rapid eustatic fluctuations. *Doklady Earth Sciences*, **373**, 793–797.

ARTYUSHKOV, E. V., CHEKHOVICH, P. A. 2001. The East Siberian basin in the Silurian: evidence for no large-scale sea-level changes. *Earth and Planetary Science Letters*, **193**, 183–196.

ARTYUSHKOV, E. V. & HOFMANN, A. W. 1998. The neotectonic crustal uplift on the continents and its possible mechanisms. The case of Southern Africa. *Surveys in Geophysics*, **19**, 369–415.

ARTYUSHKOV, E. V., BAER, M. A., LETNIKOV, F. A. & RUZHICH, V. V. 1991. On the mechanism of graben formation. *Tectonophysics*, **197**, 95–115.

ARTYUSHKOV, E. V., LINDSTRÖM, M. & POPOV, L. E. 2000*a*. Relative sea-level changes in Baltoscandia in the Cambrian and early Ordovician: the predominance of tectonic factor and the absence of large-scale eustatic fluctuations. *Tectonophysics*, **320**, 375–407.

ARTYUSHKOV, E. V., MÖRNER, N.-A., TARLING, D. 2000*b*. The cause of loss of lithospheric rigidity in areas far from plate tectonic activity. *Geophysical Journal International*, **143**, 752–776.

AUSTRHEIM, H. 1998. Influence of fluid and deformation on metamorphism of the deep crust and consequences for the geodynamics of collision zones. *In:* Hacker, B. R. & Liou, J. G. (eds) *When Continents Collide: Geodynamics and Geochemistry of Ultrahigh-Pressure Rocks.* Kluwer, Dordrecht, 297–323.

BAIRD, D. J., KNAPP, J. H., STEER, D. N., BROWN, L. D. & NELSON, K. D. 1995. Upper-mantle reflectivity beneath the Williston basin, phase-change Moho, and the origin of intracratonic basins. *Geology*, **23**, 431–434.

BELOUSSOV, V. V. 1980. *Geotectonics.* Springer-Verlag, Berlin. 330 pp.

BOSENCE, D. W. J., WOOD, J. L., ROSE, E. P. F. & QING, H. 2000. Low- and high-frequency sea-level changes control peritidal carbonate cycles, facies and dolomitization in the Rock of Gibraltar (Early Jurassic, Iberian Peninsula). *Journal of the Geological Society of London*, **157**, 61–74.

BRETT, C. E., BOUCOT, A. J. & JONES, B. 1993. Absolute depths of Silurian benthic assemblages. *Lethaia*, **26**, 25–40.

BURGESS, P. M. & GURNIS, M. 1995. Mechanisms for the formation of cratonic stratigraphic sequences. *Earth and Planetary Science Letters*, **136**, 647–663.

BURGESS, P. M. & MORESI, L. N. 1999. Modelling rates and distribution of subsidence due to dynamic topography over subducting slabs: is it possible to identify dynamic topography from ancient strata. *Basin Research,* **11**, 305–314.

BURGESS, P. M., GURNIS, M. & MORESI, L. N. 1997. Formation of sequences in the cratonic interior of North America by interaction between mantle, eustatic, and stratigraphic processes. *Geological Society of America Bulletin*, **109**, 1515–1535.

BURGESS, P. M., WRIGHT, V. P. & EMERY, D. 2001. Numerical forward modelling of peritidal carbonate parasequence development: implications for outcrop interpretation. *Basin Research*, **13(1)**, 1–25.

BUROV, E. B. & DIAMENT, M. 1995. The effective elastic thickness (T_e) of the continental lithosphere: What does it really mean? *Journal of Geophysical Research*, **100**, 3905–3927.

CHENG, D., TUCKER, M. E., JIANG, M. & ZHU, J. 2001. Long-distance correlation between tectonic-controlled, isolated carbonate platforms by cyclostratigraphy and sequence stratigraphy in the Devonian of South China. *Sedimentology*, **48**, 57–78.

CLOETINGH, S. & BUROV, E. B. 1996. Thermo-mechanical structure of European continental lithosphere: constraints from rheological profiles and EET estimates. *Geophysical Journal International*, **124**, 695–723.

CLOETINGH, S., MCQUEEN, H. & LAMBECK, K. 1985. On a tectonic mechanism for regional sea level variations. *Earth & Planetary Science Letters*, **51**, 139–162.

COAKLEY, B. & GURNIS, M. 1995. Far-field tilting of Laurentia during the Ordovician and constraints on the evolution of a slab under an ancient continent. *Journal of Geophysical Research*, **100**, 6313–6327.

COOPER, A. & NOWLAN, G. S. (eds.) 1999. *Proposed Global Stratigraphic Section and Point for Base of the Ordovician System.* International Working Group on the Cambrian–Ordovician Boundary, Circular March 1999, Calgary.

DRUMMOND, C. N. & WILKINSON, B. H. 1993a. Aperiodic accumulation of peritidal carbonates. *Geology*, **21**, 1023–1026.

DRUMMOND, C. N. & WILKINSON, B. H. 1993b. On the use of cycle thickness diagrams as recording long-term sea-level change during accumulation of carbonate sequences. *Journal of Geology*, **101**, 687–702.

DRZEWIECKI, P. A. & SIMO, J. A. 2000. Tectonic, eustatic and environmental controls on mid-Cretaceous carbonate platform deposition, south-central Pyrenees, Spain. *Sedimentology*, **47**, 471–475.

EGORKIN, A. V., ZUGANOV, S. K., PAVLENKOVA, N. A. & CHERNYSHEV, N. M. 1987. Results of lithosphere studies from long-range profiles in Siberia. *Tectonophysics*, **140**, 29–37.

ELRICK, M. 1995. Cyclostratigraphy of Middle Devonian carbonates of the eastern Great Basin. *Journal of Sedimentary Research*, **B65**, 61–79.

EMERY, D. & MYERS, K. J. 1996. *Sequence Stratigraphy*. Blackwell Science, Oxford.

FISCHER, A.G. 1964. The Lofer cyclothems of the Alpine Triassic. *Kansas Geological Survey Bulletin*, **169**, 107–149.

GINSBURG, R. N. 1971. Landward movement of carbonate mud: new model for regressive cycles in carbonates (abstract). *American Association of Petroleum Geologists Bulletin*, **55**, 340.

GOLDHAMMER, R. K., OSWALD, E. J. & DUNN, P. A. 1991. Hierarchy of stratigraphic forcing: example from Middle Pennsylvanian shelf of the Paradox basin. *In:* Franseen, E.K., Watney, L.W., Kendall, C.G. St. C. & Ross, W. (eds) *Sedimentary Modelling: Computer Simulation of Methods for Improved Parameter Definition. Kansas Geol. Surv. Bull.*, 361–413.

GOLDHAMMER, R. K., LEHMANN, P. J. & DUNN, P. A. 1993. The origin of high frequency platform carbonate cycles and third-order sequences (Lower Ordovician El Paso GP, West Texas): Constraints from outcrop data and stratigraphic modeling. *Journal of Sedimentary Petrology*, **63**, 318–359.

GOTTIKH, R. P., PISOTSKY, B. I. & ZHURAVLEV, D. Z. 2000. Geochemical pecularities of rocks of bitumens and oils of some hydrocarbon provinces: isotopic ratios of Neodim and Strontium. *Doklady Earth Science Section*, **375**, 83–88.

DE GRACIANSKY, P. C., HARDENBOL, J., JAQUIN, T. & VAIL, P. R. (eds.). 1998. *Mesozoic and Cenozoic Sequence Stratigraphy of European Basins*. SEPM Special Publication Series, **60**, Tulsa, 786 pp.

GRADSTEIN, F. M. & J. OGG, J. 1996. A Phanerozoic time scale. *Episodes*, **19(1–2)**, 3–5.

GURNIS, M. 1992. Rapid continental subsidence following the initiation and evolution of subduction. *Science*, **255**, 1556–1558.

HAGER, B. H. & CLAYTON, R. W. 1989. Constraints on the structure of mantle convection using seismic observations, flow models, and the geoid. *In*: Peltier, W.R. (ed) *Mantle Convection*. Gordon and Breach, New York, 657–766.

HALLAM, A. (ed.). 1992. *Phanerozoic Sea Level Changes*. Columbia Univ. Press, New York, 266 pp.

HAQ, B. U., HARDENBOL, J. & VAIL, P. R. 1987. Chronology of fluctuating sea levels since the Triassic. *Science*, **235**, 1156–1167.

HARRIS, W. B. & LAWS, R. A. 1997. Paleogene stratigraphy and sea-level history of the North Carolina Coastal Plain: Global coastal onlap and tectonics. *Sedimentary Geology*, **108**, 91–120.

HARRISON, G. G. A. 1990. Long-term eustasy and epeirogeny on the continents. *In*: Revell, R.R. (ed) *Sea-Level Change*. Natl. Acad. Press, Washington, DC, 141–158.

HAXBY, W. F., TURCOTTE, D. L. & BIRD, J. M. 1976. Thermal and mechanical evolution of the Michigan basin. *Tectonophysics*, **36**, 57–75.

HOLLAND, C. H. & BASSETT, M. G. (eds.). 1989. A global standard for the Silurian System. Cardiff. *Nat. Mus. of Wales. Geol. Ser.*, **9**.

JOHNSON, J. G., KLAPPER, G. & SANDBERG, CH. A. 1985. Devonian eustatic fluctuations in Euroamerica. *Geological Society of America Bulletin*, **96**, 567–587.

JOHNSON, M. E. 1996. Stable cratonic sequences and a standard for Silurian eustasy. *Geological Society of America Special Paper*, **306**, 202–211.

JOHNSON, M. E., TESAKOV, YU. I., PREDTECHENSKY, N. N. & BAARLI, B. G. 1997. Comparison of Lower Silurian shores and shelves in North America and Siberia. *Geological Society of America Special Paper* **321**, 23–46.

KASHIN, L. A. (compiler) 1989. *Map of Recent Crustal Movements According to Geodetic Data on the USSR Territory. Scale 1:5.000.000. GUGK*, Moscow.

MAURER, F. 2000. Growth mode of middle Triassic carbonate platforms in the western Dolomites (Southern Alps, Italy). *Sedimentary Geology*, **134**, 275–286.

MCKENZIE, D. 1978. Some remarks on the development of sedimentary basins. *Earth & Planetary Science Letters*, **40**, 25–32.

MIALL, A. D. 1992. The Exxon global cycle chart: an event for every occasion? *Geology*, **20**, 787–790.

MIALL, A. D. 1997. *The Geology of Stratigraphic Sequences*. Springer, Berlin, 433 pp.

MIALL, A. D. & MIALL, C. E. 2001. Sequence stratigraphy as a scientific enterprise: the evolution and persistence of conflicting paradigms. *Earth-Science Reviews*, **54**, 321–348.

MURPHY, M. A. & SALVADOR, A. (eds.). 1999. International Stratigraphic Guide: An abridged version. *Episodes*, **22(4)**, 255–271.

NELSON, D. R. 1992. Isotopic characteristics of potassic rocks: evidence for the involvement of subducted sediments in magma. *Lithos*, **28**, 403–420.

OSLEGER, D. A. & READ, J. F. 1991. Relation of eustasy to stacking patterns of meter-scale carbonate cycles, Late Cambrian, USA. *Journal of Sedimentary Petrology*, **61**, 1225–1252.

PAVLENKOVA, N. I. 1996. Crust and upper mantle structure in Northern Eurasia from seismic data. *Advances in Geophysics*, **37**, 1–133.

PISOTSKY, B. I. 1999. *Expression of the Processes of a Reducing Fluidization in Hydrocarbon Provinces and Role of Mantle Components in Genesis of Hydrocarbons*. Rehabilitation Thesis, VNIIGeosystems, Moscow, 357 pp. (in Russian).

POSAMENTIER, H. W. & ALLEN, G. P. 2000. *Siliciclastic Sequence Stratigraphy – Concepts and Applications*. SEPM Concepts in Sedimentology and Paleontology, **7**, SEPM.

PUSHKAREV, YU. D., GOTTIKH, R. P., PISOTSKY, B. I., ZHURAVLEV, D. Z. & AKHMEDOV, A. M. 1994. Isotope compositions of Sr, Pb, Nd, Os and REE distribution in oils, bitumens and black shales as indicators of matter sources, migration ways and genesis of hydrocarbon fluids. *Eighth Intern. Conference on Geochronology, Cosmochronology and Isotope Geology, Abstracts*, Berkley.

READ, J. F., KOERSCHNER, W. F. III, OSLEGER, D. A., BOLLINGER, G. A. & CORUH, C. 1991. Field and modelling studies of Cambrian carbonate cycles, Virginia Appalachians – reply. *Journal of Sedimentary Petrology*, **61**, 647–652.

ROSEN, O. M., CONDIE, K. C., NATAPOV, L. M. & NOZHKIN, A. D. 1994. Archean and Early Proterozoic evolution of the Siberia Craton: A preliminary assessment. *In*: Condie, K.C. (ed.) *Archean Crustal Evolution*. Elsevier, Amsterdam, pp. 411–459.

ROSS, C. A. & ROSS, J. R .P. 1987. Late Paleozoic sea-levels and depositional sequences. Cushman Foundation for Foraminiferal Research, Special Publication, **24**, 137–149.

SADLER, P. M., OSLEGER, D. A. & MONTANEZ, I. P. 1993. On the labeling, length, and objective basis of Fischer plots. *Journal of Sedimentary Petrology*, **63**, 360–368.

SCHWARZACHER, W. 2000. Repetitions and cycles in stratigraphy. *Earth-Science Reviews*, **50**, 51–75.

SCRUTTON, C. T. 1998. The Palaeozoic corals, II: structure, variation and palaeoecology. *Proceedings of Yorkshire Geological Society*, **52(1)**, 1–57.

ŞENGÖR, A. M. C. & NATAL'IN, B. A. 1996. Paleotectonics of Asia: fragments of a syntacsis in the tectonic evolution of Asia. *In*: JIN AN & HARRISON, M. (eds) *The Tectonic Evolution of Asia*, Cambridge, Cambridge University Press, 486–640.

SLEEP, N. H. 1971. Thermal effects of the formation of Atlantic continental margin by continental break up. *Geophysical Journal of Royal Astronomical Society*, **24**, 325–360.

SLOSS, L. L. 1988. Tectonic evolution of the craton in Phanerozoic time. *In*: SLOSS, L. L (ed.) *Sedimentary*

Cover – North American Craton, US. Geological Society of America Boulder, CO, pp. 25–51.

SOKOLOV, B. S. (ed.) 1985. *[Reference Section of the Moyero River of the Silurian of the Siberian Platform]* [in Russian], Nauka, Novosibirsk, 176 pp.

SOKOLOV, B. S. (ed.) 1992. *Sucessions and Fauna of the Silurian of the North of the Tunguska Syneclise* (in Russian). Nauka, Novosibirsk, 193 pp.

SURKOV, V. S. & KOROBEINIKOV, V. P. (eds) 2000. *Tectonic Map of Siberia. Scale 1:5 000 000*. Siberian Research Institute of Geology, Geophysics and Mineral Resources, Novosibirsk.

TAYLOR, S. R. & MCLENNAN, S. M. 1985. *Continental Crust: its Composition and Evolution*. Blackwell Scientific Publications, Oxford, 312 pp.

TESAKOV, YU. I. 1981. Evolution of ecosystems of ancient platform sedimentary basins. *In*: BOGOLEPOV, K. V. & ZHARKOV, M. A. (eds) *The Problems of Evolution of Geological Processes*. Nauka, Novosibirsk, 186–199.

TESAKOV, YU. I., PREDTECHENSKY, N. N., KHROMYKH, V. G., BERGER, A. YA. & BOGOLEPOVA, O. K. 1986. *Fauna and Flora of the Polar Areas of the Siberian Platform* (in Russian), Nauka, Novosibirsk, 216 pp.

TESAKOV, YU. I., JOHNSON, M. E., PREDTECHENSKI, N.N., KHROMYKH, V. G. & BERGER, A.YA. 1998*a*. Eustatic fluctuations in the East Siberian Basin (Siberian Platform and Taymyr Peninsula). In: LANDING, E. & JOHNSON, M. (eds) Silurian Cycles, Linkages of Dynamic Stratigraphy with Atmospheric, Oceanic, and Tectonic Changes. James Hall Centennial Volume, The New York State Museum Bulletin, **491**, 63–73.

TESAKOV, YU. I., PREDTECHENSKY, N. N. & KHROMYKH, V. G. 1998*b*. Stratigraphy of the Silurian of the Siberian Platform. *Russian Geology and Geophysics*, **39**, 1335–1356.

TESAKOV, YU. I., PREDTECHENSKY, N. N., LOPUSHINSKAYA, T. V., KHROMYKH, V. G., BAZAROVA, L. S., BERGER, A. YA. & KOVALEVSKAYA, E. O. 2000. [*Stratigraphy of the Hydrocarbon Basins of Siberia. Silurian of the Siberian Platform]* SO RAN, 'GEO, Novosibirsk, [in Russian].

WATKINS, R., COOROUGH, P. J. & MAYER, P. S. 2000. The Silurian Dicoelosia communities: temporal stability within an ecologic evolutionary unit. *Palaeogeography, Palaeoclimatology, Palaeoecology*, **162**, 225–237.

WEBBY, B. D. & LAURIE, J. R. (eds) 1992. *Global Perspectives on Ordovician Geology*. Balkema, Rotterdam, 524 pp.

WILKINSON, B. H., DRUMMOND, C. N., ROTHMAN, E. D. & DIEDRICH, N. W. 1997. Stratal order in peritidal carbonate sequences. *Journal of Sedimentary Research*, **B67**, 1068–1062.

WILSON, J. L. 1975. *Carbonate Facies in Geological History*. Springer-Verlag, Berlin.

ZOBACK, M. L. 1992. First- and second order patterns of stresses in the lithosphere: the world stress map project. *Journal of Geophysical Research*, **97**, 11703–11728.

Index

Note: Page references in *italics* refer to figures and tables